Advances in Intelligent Systems and Computing

Volume 1255

The series "Advances in Intelligent Systems and Computing" contains publications on theory, applications, and design methods of Intelligent Systems and Intelligent Computing. Virtually all disciplines such as engineering, natural sciences, computer and information science, ICT, economics, business, e-commerce, environment, healthcare, life science are covered. The list of topics spans all the areas of modern intelligent systems and computing such as: computational intelligence, soft computing including neural networks, fuzzy systems, evolutionary computing and the fusion of these paradigms, social intelligence, ambient intelligence, computational neuroscience, artificial life, virtual worlds and society, cognitive science and systems, Perception and Vision, DNA and immune based systems, self-organizing and adaptive systems, e-Learning and teaching, human-centered and human-centric computing, recommender systems, intelligent control, robotics and mechatronics including human-machine teaming, knowledge-based paradigms, learning paradigms, machine ethics, intelligent data analysis, knowledge management, intelligent agents, intelligent decision making and support, intelligent network security, trust management, interactive entertainment, Web intelligence and multimedia.

The publications within "Advances in Intelligent Systems and Computing" are primarily proceedings of important conferences, symposia and congresses. They cover significant recent developments in the field, both of a foundational and applicable character. An important characteristic feature of the series is the short publication time and world-wide distribution. This permits a rapid and broad dissemination of research results.

Indexed by SCOPUS, DBLP, EI Compendex, INSPEC, WTI Frankfurt eG, zbMATH, Japanese Science and Technology Agency (JST), SCImago.

All books published in the series are submitted for consideration in Web of Science.

More information about this series at http://www.springer.com/series/11156

Editors
Debotosh Bhattacharjee
Department of Computer Science and Engineering
Jadavpur University
Kolkata, West Bengal, India

Nilanjan Dey
Department of Information Technology
Techno India College of Technology
Kolkata, West Bengal, India

Dariusz Plewczynski
Center of New Technologies
University of Warsaw
Warsaw, Poland

Dipak Kumar Kole
Department of Computer Science and Engineering
Jalpaiguri Government Engineering College
Jalpaiguri, West Bengal, India

Subhadip Basu
Department of Computer Science and Engineering
Jadavpur University
Kolkata, West Bengal, India

ISSN 2194-5357 ISSN 2194-5365 (electronic)
Advances in Intelligent Systems and Computing
ISBN 978-981-15-7833-5 ISBN 978-981-15-7834-2 (eBook)
https://doi.org/10.1007/978-981-15-7834-2

This Springer imprint is published by the registered company Springer Nature Singapore Pte Ltd.
The registered company address is: 152 Beach Road, #21-01/04 Gateway East, Singapore 189721, Singapore

Debotosh Bhattacharjee · Dipak Kumar Kole ·
Nilanjan Dey · Subhadip Basu ·
Dariusz Plewczynski
Editors

Proceedings of International Conference on Frontiers in Computing and Systems

COMSYS 2020

Springer

Debotosh Bhattacharjee · Dipak Kumar Kole ·
Nilanjan Dey · Subhadip Basu ·
Dariusz Plewczynski
Editors

Proceedings of International Conference on Frontiers in Computing and Systems

COMSYS 2020

Springer

Editors
Debotosh Bhattacharjee
Department of Computer Science
and Engineering
Jadavpur University
Kolkata, West Bengal, India

Dipak Kumar Kole
Department of Computer Science
and Engineering
Jalpaiguri Government Engineering
College
Jalpaiguri, West Bengal, India

Nilanjan Dey
Department of Information Technology
Techno India College of Technology
Kolkata, West Bengal, India

Subhadip Basu
Department of Computer Science
and Engineering
Jadavpur University
Kolkata, West Bengal, India

Dariusz Plewczynski
Center of New Technologies
University of Warsaw
Warsaw, Poland

ISSN 2194-5357 ISSN 2194-5365 (electronic)
Advances in Intelligent Systems and Computing
ISBN 978-981-15-7833-5 ISBN 978-981-15-7834-2 (eBook)
https://doi.org/10.1007/978-981-15-7834-2

This Springer imprint is published by the registered company Springer Nature Singapore Pte Ltd.
The registered company address is: 152 Beach Road, #21-01/04 Gateway East, Singapore 189721, Singapore

Debotosh Bhattacharjee · Dipak Kumar Kole ·
Nilanjan Dey · Subhadip Basu ·
Dariusz Plewczynski
Editors

Proceedings of International Conference on Frontiers in Computing and Systems

COMSYS 2020

Springer

Editors
Debotosh Bhattacharjee
Department of Computer Science
and Engineering
Jadavpur University
Kolkata, West Bengal, India

Nilanjan Dey
Department of Information Technology
Techno India College of Technology
Kolkata, West Bengal, India

Dariusz Plewczynski
Center of New Technologies
University of Warsaw
Warsaw, Poland

Dipak Kumar Kole
Department of Computer Science
and Engineering
Jalpaiguri Government Engineering
College
Jalpaiguri, West Bengal, India

Subhadip Basu
Department of Computer Science
and Engineering
Jadavpur University
Kolkata, West Bengal, India

ISSN 2194-5357 ISSN 2194-5365 (electronic)
Advances in Intelligent Systems and Computing
ISBN 978-981-15-7833-5 ISBN 978-981-15-7834-2 (eBook)
https://doi.org/10.1007/978-981-15-7834-2

This Springer imprint is published by the registered company Springer Nature Singapore Pte Ltd.
The registered company address is: 152 Beach Road, #21-01/04 Gateway East, Singapore 189721, Singapore

Preface

COMSYS-2020, the First International Conference on Frontiers in Computing and Systems, was organized on January 13–15, 2020 to offer an intellectual platform for the scientists and researchers active in the domain of computing and systems. We sincerely hope that COMSYS-2020 helped the participating delegates to exchange new scientific ideas and to establish business or research collaborations.

COMSYS-2020 was hosted and organized by Jalpaiguri Government Engineering College (JGEC) in West Bengal, India. JGEC is a public-funded engineering institution in India, located in the picturesque city of Jalpaiguri, at the foothill of the Himalayas. Many thanks to the local organizing committee for managing every minute detail related to organizing the conference.

COMSYS-2020 conference proceedings constitute significant contributions to the knowledge in the scientific fields or machine learning, computational intelligence, VLSI, networks and systems, computational biology, and security. The conference was spanned over 3 days. The two half-day tutorials on the first day were delivered by Prof. Ananda Shankar Chowdhury of Jadavpur University, India, and Prof. Jacek Sroka of the University of Warsaw, Poland, giving researchers, practitioners, and students an excellent opportunity to learn about the latest trends in computing and systems. In addition to the technical sessions, COMSYS-2020 also included two keynote talks by Prof. Dariusz Plewczynski of the University of Warsaw, Poland, and Prof. Punam Kumar Saha of University of Iowa, USA. COMSYS-2020 also organized several technical competitions and provided a platform for start-up entrepreneurs.

COMSYS-2020 received an incredible response in terms of the submission of papers from across the globe. An eminent international program committee was constituted for a double-blind review process. Each submission was reviewed by at least two reviewers, and after rigorous evaluation, 86 papers were selected. We checked plagiarism using professional software, once at the time of submission, and once after acceptance, at the time of final preparation of the camera-ready copy. We convey our sincere gratitude to Springer for providing the opportunity to publish the proceedings of COMSYS-2020 in the prestigious series of Advances in

Intelligent Systems and Computing. We sincerely hope that the articles will be useful for researchers, pursuing in the field of computing and systems.

We are also indebted to the Patrons of the COMSYS-2020, Prof. Saikat Maitra, Vice-Chancellor, MAKAUT, and Prof. Amitava Ray, Principal, JGEC. Special thanks to our International Advisory Committee for their continued guidance and support, the learned reviewers for their voluntary assistance in the review process, different committee members who served to the best of their abilities, and the sponsoring organizations, especially the Department of Science and Technology, Government of India, for their generous financial support. Finally, we acknowledge the support received from the students, faculty members, officers, staff, and the authority of JGEC to make COMSYS-2020 a grand success.

In a word, it is always a team effort that defines a successful conference. We look forward to seeing all of you at the next edition of COMSYS.

Kolkata, India — Debotosh Bhattacharjee
Jalpaiguri, India — Dipak Kumar Kole
Kolkata, India — Nilanjan Dey
Kolkata, India — Subhadip Basu
Warsaw, Poland — Dariusz Plewczynski

Contents

Security Track

About the Editors

Debotosh Bhattacharjee is a Full Professor at the Department of Computer Science & Engineering, Jadavpur University. He has published more than 250 journal articles and holds two US patents. Prof. Bhattacharjee has been granted sponsored projects with total funding of ca. INR 2 Crores. During his postdoctoral research, Dr. Bhattacharjee visited various universities in Europe and the USA. He is a life member of the ISTE and IUPRAI, and a senior member of the IEEE (USA).

Dipak Kumar Kole received his Ph.D. in Engineering from Bengal Engineering & Science University (now the IIEST), Shibpur, India, in 2012. He is currently an Associate Professor at the Department of Computer Science & Engineering, Jalpaiguri Government Engineering College, India. His research interests include the synthesis & testing of reversible circuits, social network analysis, digital watermarking and agricultural engineering. He has published more than 50 articles in journals and conference proceedings.

Nilanjan Dey is an Assistant Professor at the Department of IT at Techno International New Town, India, and a Visiting Fellow of the University of Reading, UK. Holding a Ph.D. from Jadavpur University (2015), he is Editor-in-Chief of the International Journal of Ambient Computing and Intelligence, and Series Co-Editor of Springer Tracts in Nature-Inspired Computing. He is also the Indian Ambassador of the IFIP—the Young ICT Group.

Subhadip Basu received his Ph.D. from the Computer Science and Engineering Department of Jadavpur University in 2006, and has been working as a Full Professor at said department since 2017. He completed his postdoctoral research at the University of Iowa, USA, and University of Warsaw, Poland. He has also been an Honorary Research Scientist at the Department of Electrical and Computer Engineering, University of Iowa, since 2016. Dr. Basu has published over 200 research articles in the areas of pattern recognition and image processing and has received numerous fellowships and awards, including a DAAD Fellowship from Germany and a 'Research Award' from the UGC, Government of India. He is a senior member of the IEEE and life member of the IUPRAI (IAPR).

Dariusz Plewczynski interests are focused on functional and structural genomics. He is currently involved in several big data projects at three institutes: the Centre of New Technologies at the University of Warsaw (his main affiliation), Jackson Laboratory for Genomic Medicine (an international partner), and the Centre for Innovative Research at the Medical University of Bialystok (UMB). He is also participating in two large consortia projects, namely the 1000 Genomes Project (NIH, USA) and the biophysical modeling of chromatin three-dimensional conformation inside human cells using HiC and ChIA-PET techniques as part of the 4D Nucleome project (NIH, USA).

Computational Intelligence Track

HOG and LBP Based Writer Verification

Jaya Paul, Anasua Sarkar, Nibaran Das, and Kaushik Roy

Abstract We propose a writer-specific off-line writer verification procedure to improve the reliability of the verification system. It has to be a challenging task, particularly in the off-line scenario, which uses images of the scanned document, where the dynamic information is not available. In this paper, we propose a local textural based feature Histogram of Oriented Gradients (HOG) along with gradient-based feature like local binary pattern (LBP). KNN, SMO and MLP classifiers are experimented within this work. These different classifiers are then trained then individually for each writer and they learn to verify a writer from another randomly selected writer. Our method achieves 89.62% accuracy using KNN on 100 writers database, whose metric is better than both SMO and MLP classifiers. The experimental results show that the two chosen sets of features for the writer verification problem provides improved accuracy and reduces the error rate.

Keywords Histogram of oriented gradients · Local binary pattern · SMO · MLP

J. Paul (✉)
Government College of Engineering and Leather Technology, Kolkata, India
e-mail: jayapl2005@gmail.com

A. Sarkar · N. Das
Computer Science and Engineering Department, Jadavpur University, Kolkata, India
e-mail: anasua.sarkar@cse.jdvu.ac.in

N. Das
e-mail: nibaranju@gmail.com

K. Roy
West Bengal State University, Barasat, India
e-mail: kaushik.mrg@gmail.com

D. Bhattacharjee et al. (eds.), *Proceedings of International Conference on Frontiers in Computing and Systems*, Advances in Intelligent Systems and Computing 1255,
https://doi.org/10.1007/978-981-15-7834-2_1

1 Introduction

Handwriting biometric is a behavioral property that specifies the handwriting movements of the writers. The physical feature of a human such as a fingerprint, iris, or face, signature, handwriting document—all show intra-class(within the class) variability. The main aim of writer verification is to determine whether or not a handwritten document is written by the same writer. Signatures verification [14] is another similar verification technique that is applied on banking and official document, which are among important applications for off-line systems. Depending on the collecting procedures of the data, automatic writer verification can be classified into two segments, one is on-line (dynamic) [17] and other is off-line (static) [18].

In this paper, we present an off-line writer verification method based on the local textural feature that is produced from overall writers' writing pattern, which is established by writing strokes, shape variations, and ballistic movements of writer pen. These local features are based on pixel neighborhood patterns (local binary patterns—LBP) and on extracted gradient information (histogram of oriented gradients—HOG). We evaluate the dissimilarity of the image extracted features using MLP (Multilayer perceptron), SMO (Sequential minimal optimization), and KNN classifiers. We obtained state-of-the-art results using a combination of the above-mentioned two features.

1.1 Related Works

Off-line writer verification is an ongoing researched topic. We have studied a different number of features and chosen classifiers on this writer verification problem in this work. Srihari et al. [18] use two major categories of features, among them one is a macro and the other is micro. These macro (document level) and micro (character level)features are used for writer identification and verification methods. For dissimilarity measurement of writer verification, Euclidean distance is used and obtained 96% accuracy. Grayscale histogram features [5] and run-lengths of background intensity values are also used in [4]. A dataset of 3 samples of 13 writers, each one is used. In this experiment [4], the authors show the similarity between the histograms of two samples of intra-writer is greater than that of two samples of inter-writer. They have used the smoothing operation of each point on the histogram. For writer identification, texture-based features like Local Ternary Patterns (LTP), Local Binary Patterns (LBP), and Local Phase Quantization (LPQ) are enumerated from IFN/ENIT and IAM databases in [20]. IFN/ENIT and IAM databases are a combination of handwritten Arabic and English texts. In this method, each handwriting sample is divided into small parts and each part is considered as a texture. All these features are used on Arabic and English scripts. Bensefia et al. [3] are working on two common datasets, PSI DataBase and IAM DataBase with 150 writers, for writer identification and verification. They have used grapheme- based features.

Al-Maadeed [2] has proposed another document level work on writer identification using novel geometrical features including direction, curvature, chain code, and tortuosity. These proposed methods are applied to the IAM handwriting database for English and the QUWI database for Arabic script. The result of identification rates are 82 and 87 % in those two datasets, respectively.

Bulacu et al. [16] evaluate the performances of edge-based probability distributions on writer identification. They have compared to non-angular features, like a number of major style variations such as upper case to lower case. These forged styles are also used in forensic writer identification procedures.

In 2015, Halder et al. [6] present another Bangla writer verification work by textual features like FFT, GCT, and GLCM. These features are applied to superimposed isolated Bangla character images. They have considered a total number of 35500 Bangla characters from 100 writers as an input. Mahalanobis distances are used to calculate the distance of inter-writer and intra-writer. We have improved the verification rate by a combination of those features in this work. Kumar et al. [13] work on other Indic character recognition such as Gurumukhi characters. They have used lots of character-level features (namely directional, zoning, transition, intersection, open end points, diagonal, power curve fitting, and parabola curve fitting). Linear-SVM-, kNN-, RBF-SVM-, and Polynomial-SVM- based classifiers are used for this purpose. The proposed system achieves better performance using PCA (Principal Component Analysis). Halder et al. [9] have proposed a writer identification system of Devanagari characters. 64-dimensional gradient features are used for Devanagari characters (50 writers each having 5 copies). LIBLINEAR and LIBSVM classifiers of WEKA [10] data mining tool are used to obtain 99.12% accuracy. Karbhari et al. [12] have proposed a method for compound character recognition of the Devanagari script where moment-based techniques are applied. The overall compound character recognition rate of this proposed system using SVM and kNN classifiers is up to 98.37, and 95.82%, respectively. Another off-line handwritten digit recognition system based on HOG (Histogram of Oriented Gradients) is shown in [1] to capture the features of digit image. The achieved result using ANN in this experiment is 98.26%. The grids of Histograms of Oriented Gradient (HOG) descriptors [7] are a robust visual object recognition feature sets. They have used the original MIT pedestrian database to obtain very good results. The co-occurrence histograms of oriented gradients feature descriptors [19] are used to detect pedestrians from an image. This descriptor describes the complex shape of the object. Their proposed method performs well to reduce miss rates in comparison with HOG. In this paper, we have experimented with the HOG and LBP features for writer verification, and have proposed a very simple and powerful approach to build a user-dependent writer verification method.

The rest of the paper is organized as follows: in Sect. 2 we describe our proposed method using HOG and LBP features in detail. In Sect. 3, we describe the detail of the experimental results. Section 4 finalizes the conclusions of the paper with the future scopes.

2 Proposed Method

Writer verification is a two-class classification problem that evaluates whether two writers' handwriting is the same or different. To make this classification, dissimilarity measures between the queried handwriting and the reference handwriting is obtained. The combination of HOG and LBP features, which are significantly two types of features, is used in this work as the feature set on writer verification problem. The framework of this proposed method procedures is shown in Fig. 1. Detail of the model evaluation is shown in Fig. 2. In the first step, the input images are resized to 64 × 64 dimension. In the second step, for each image, we have calculated HOG feature values and LBP feature values, respectively. In the third step, a combination of HOG and LBP feature set is trained and tested by the user-specific model. In the following sections, we have given an overview of the two features extractions and some details of our proposed framework in this paper.

2.1 HOG Feature Extraction

Histogram of Oriented Gradients (HOG) [7] is a good feature descriptor, which is for the application of object recognition. It is similar to Scale-Invariant Feature Transform (SIFT)descriptors proposed by Lowe in [15]. HOG features are used for the purpose of edge information retrieval from the isolated Bangla numeral images. The histogram channels are calculated over Rectangle Histogram Oriented Gradient, which is called R-HOG [11]. The computation of unsigned gradients is done by R-HOG [11] cell. The gradient strengths are normalized over each cell to consider the changes in illumination and contrast. Each cell has a fixed number of gradient orientation bins. In this paper, we have computed HOG features for the isolated Bangla numeral image using 9 rectangular cells and 32 bin histograms per cell. The

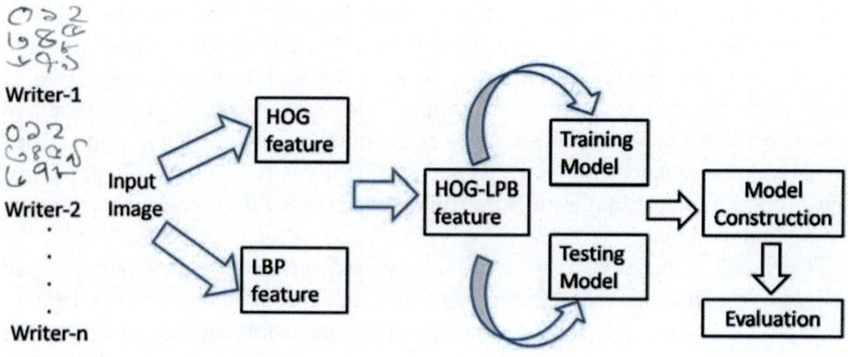

Fig. 1 Our proposed writer verification diagram

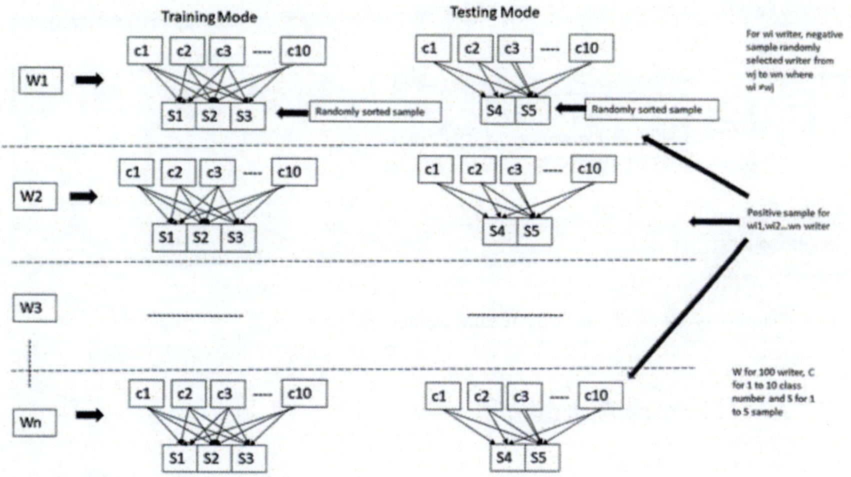

Fig. 2 Verification model construction

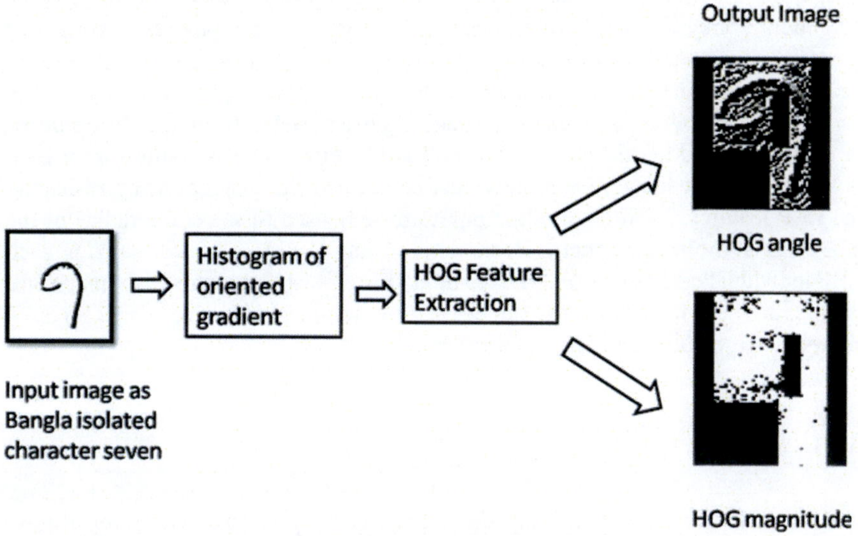

Fig. 3 Outputs as obtained after HOG feature extraction

9 histograms with 9 bins are concatenated and then produce 81-dimensional resulting feature for each image. Figure 3 shows the HOG output angle and magnitude of the input image. These angle and magnitude values are used directly during the training phase of the classifiers. The HOG features provide us with the edge information of the isolated Bangla numerals images. The HOG parameters are defined briefly in [11] (Fig. 3).

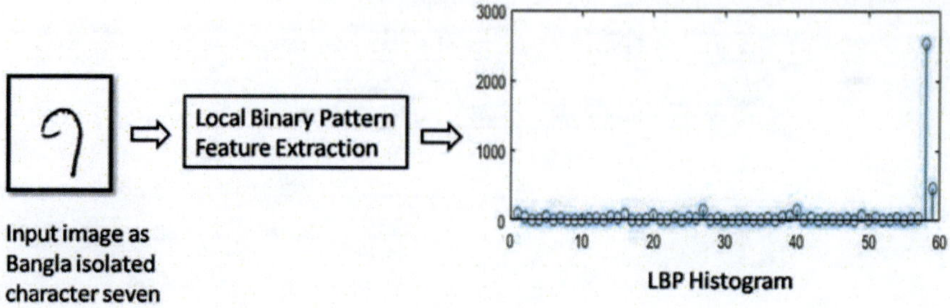

Fig. 4 Outputs as obtained after LBP feature extraction

2.2 *LBP Feature Extraction*

The second feature extraction model which is used in this work is the Local Binary Patterns (LBP) algorithm. For calculating the LBP code, for each pixel i of a gray image, the 8 neighbors of the center pixel are compared with the pixel i and the neighbors y are assigned a value which is 1 if $y \geq i$. This process is computed repeatedly across the whole gray image to generate 256×1 histogram. The feature vectors are obtained by this method on the input images. The input images are captured in a histogram through the normalization of the 64×64 images. The number of neighbors used in this work is 8. Then LBP feature vector is computed for each pixel in the input image. Each individual cell access the histograms by reshaping the LBP features. Each pixel in the input image is used to select the radius of the neighbors in a circular pattern. Here, we are using l1 norms for normalizing each LBP individual cell histogram. The resulting bins of histograms for each image will be of 238. These feature vectors can be used for training and testing mode. Figure 4 shows the procedures of LBP feature extraction.

3 Results and Analysis

In this section, we are describing the details of the proposed writer verification framework. At first, we describe the details of our database used in these experiments. Then, we discuss the details of the experimental results using different features along with their performances.

3.1 *Dataset Used*

We have used Halder et al. [8] database, which contains 9125 Bengali isolated numerical samples from 0 to 9 of 149 writers' Bangla numerical dataset. Here we have selected 100 writers' numerical Bangla characters from Halder et al. [8] database due to insufficient numerical characters in other writer sets. Each writer has five satraingmples of each character, in total for 50 isolated Bengali characters. We have divided the database for training and testing samples. We use 25 characters in the training phase and 25 characters in the testing phase.

3.2 Performances and Evaluations

Here, we have used Halder et al. [8] database to evaluate these experimental results. Halder et al. [8] have worked on 4500 numerals from 90 writers, in their work on the individuality of Bangla Numerals. For numeral characters four, five, and seven, individuality accuracies are very poor 25.17, 35.90, and 25.41%, respectively. Subsequently, we have used the same dataset with 100 writer Bangla numerals character set in our work in this writer verification model. All the images of the dataset are normalized to 64 × 64 pixels. According to the dataset, each writer has 5 different samples for each numerical character. Each writer has 10 numeral characters with 5 different sets. As in the machine learning algorithms, equal number of positive and negative samples are required for better performance. We first select the same number of positive and negative samples for the training and testing phase for each individual writer. We have considered 50 positive and negative (another writer) samples for each writer to build up the model. Therefore, this writer-specific model uses 25 numbers of positive samples which cover 0 to 9 characters, in both the training phase and the testing phase. For the same number of negative samples, we have selected writers randomly and cover 0–9 characters as used in the training and testing phases.

The writer verification performances are obtained on 100 writers with 5 samples' numerical data from the Halder et al. [8] database. As the original dataset consists a maximum of 5 different samples for each numerical class to maintain the proportion, we have to consider that. 25 characters are used as the training data for each writer, while 25 characters are used as the test data. In our verification framework, the writer's average top-10 verification accuracy rate is 97.79.20% and the worst-10

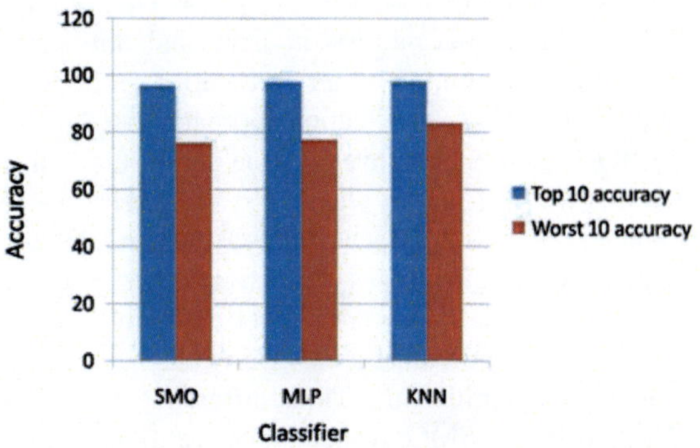

Fig. 5 Top-10 writers and Worst-10 writers verification accuracies for different classifiers

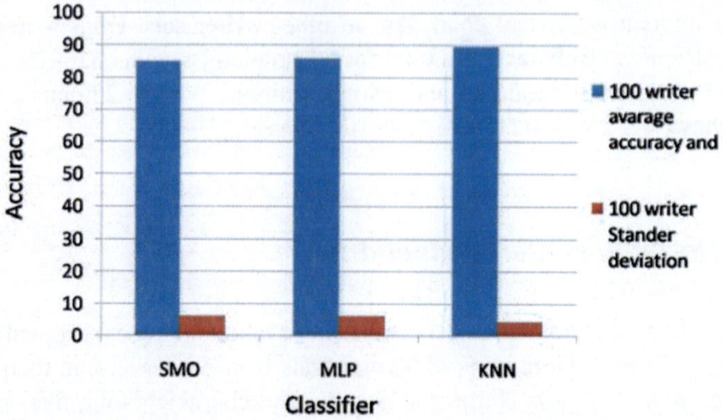

Fig. 6 100 writers' verification average and standard deviation accuracies for different classifiers

Table 1 Average accuracy and standard deviation results as obtained by different classifiers in our proposed framework

Classifier	Average accuracy	Standard deviation accuracy
SMO	97.64	6.33
MLP	98.74	6.35
KNN	99.20	4.53

verification accuracy rate is 83.18% using KNN classifier which is better than the accuracy values as obtained by MLP and SMO classifiers, respectively. 100 writers' average verification and standard deviation results as obtained by our verification framework are shown in Figs. 5 and 6 (Table 1).

4 Conclusions

Writer verification is a long-standing problem in computer vision. Writer verification methods can enhance multiple domains including forensics and finance. In this paper, we have focused on the writer verification model to improve verification accuracy. We propose a writer verification approach based on the combination of LBP feature and HOG feature. The investigation shows that our performance is comparable to the other state-of-the-art classifiers. The performances of Histograms of Oriented Gradients (HOGs) and Local Binary Patterns (LBP) features have proven that the two features are combinedly very effective for writer verification problem. In this paper, we have explored a very simple yet powerful writer-specific verification approach to consider the individual handwriting characteristics which affect the performance of the model significantly. Three different types of well-known classifiers, namely, KNN, MLP, and SMO, respectively, are trained in our framework to perform. We have also experimented with the combination of two features - HOG and LBP, which shows significant improvement in writer verification performances. Also, we have experimented with our model to calculate writer verification and writer identification accuracies using other Bangla datasets as well and those results are also encouraging using this proposed framework.

References

1. Lawgali, A.: Recognition of handwritten digits using histogram of oriented gradients. Int. J. Adv. Res. Sci. Eng. Technology **3**(7), 2359–2363 (2016)
2. Somaya, A.-M., Abdelaali, H., Ahmed, B., Muhammad, A.T.: Novel geometric features for off-line writer identification. Pattern Anal. Appl. **19**(3), 699–708 (2016)
3. Bensefia, A., Paquet, T., Heutte, L.: A writer identification and verification system. Pattern Recogn. Lett. **26**(13), 2080–2092 (2005)
4. Arazi, B.: Handwriting identification by means of run-length measurements. IEEE Trans. Syst. Man Cybern. **7**(12), 878–881 (1977)
5. Arazi, B.: Automatic handwriting identification based on the external properties of the samples. IEEE Transa. Syst. Man Cybern. **13**(4), 635–642 (1983)
6. Halder, C., Md Obaidullah, S.K., Paul, J., Roy, K.: Writer verification on bangla handwritten characters. In: Advanced Computing and Systems for Security, pp. 53–68. Springer (2015)
7. Dalal, N., Triggs, B.: Histograms of oriented gradients for human detection. Computer Society Conference on Computer Vision and Pattern Recognition **1**, 886–893 (2005)
8. Halder, C., Paul, J., Roy, K.: Individuality of bangla numerals. In: International Conference on Intelligent Systems Design and Applications (ISDA), pp. 264–268 (2012)
9. Halder, C., Thakur, S.P., Roy, K.: Writer identification from handwritten devanagari script. In: Information Systems Design and Intelligent Applications, pp. 497–505 (2015)
10. Hall, M., Frank, E., Holmes, G., Pfahringer, B., Reutemann, P., Witten, I.H.: The weka data mining software: an update. SIGKDD Explor. Newsl. **11**(1), 10–18 (2009)
11. Junior, O.L., Delgado, D., Goncalves, V., Nunes, U.: Trainable classifier-fusion schemes: an application to pedestrian detection. In: IEEE Conference on Intelligent Transportation Systems, pp. 1–6 (2009)
12. Kale, K.V., Deshmukh, P.D., Chavan, S.V., Kazi, M.M., Rode, Y.S.: Zernike moment feature extraction for handwritten devanagari (marathi) compound character recognition. Int. J. Adv. Res. Artif. Intell. **3**(1), 459–466 (2014)
13. Kumar, M., Sharma, R.K., Jindal, M.K.: Offline handwritten gurmukhi character recognition: study of different feature-classifier combinations. In: Workshop on Document Analysis and Recognition, number 6, pp. 94–99. ACM (2012)
14. Liang, W., Bin, W., Lin, Z.-C.: On-line signature verification with two-stage statistical models. In: Eighth International Conference on Document Analysis and Recognition (ICDAR'05), vol. 1, pp. 282–286 (2005)
15. Lowe, D.G.: Distinctive image features from scale-invariant keypoints. Int. J. Comput. Vis. **60**(2), 91–110 (2004)
16. Bulacu, M., Schomaker, L., Vuurpijl, L.: Writer identification using edge-based directional features. In: Seventh International Conference on Document Analysis and Recognition, p. 937. IEEE (2003)
17. Yoshikazu, N., Masatsugu, K.: Online writer verification using kanji handwriting. In: Bilge, G., Anil, K.J., Murat Tekalp, A., Bülent, S. (eds.) Multimedia Content Representation, Classification and Security, pp. 207–214. Springer (2006)
18. Srihari, S.N., Cha, S.H., Arora, H., Lee, S.: Individuality of handwriting. J. Forensic Sci. **47**, 856–72 (2002)

19. Tomoki, W., Satoshi, I., Kentaro, Y.: Co-occurrence histograms of oriented gradients for pedestrian detection. In: Advances in Image and Video Technology, pp. 37–47. Springer, Berlin, Heidelberg (2009)
20. Hannad, Y., Siddiqi, I., Kettani, M.E.Y.E.: Arabic writer identification using local binary patterns (lbp) of handwritten fragments. In: Pattern Recognition and Image Analysis, pp. 237–244. Springer (2015)

Implementation of Real-Time Virtual Dressing Room Using Microsoft Kinect SDK and Supervised Learning

Soma Bandyopadhyay, S. S. Thakur, and J. K. Mandal

Abstract Presently the usage of Electronic Commerce has been growing at a rapid space and many customers choose the online shopping option available, to avoid waiting in a long queue. Before buying an item specially clothing item, jewellery, or accessories the online shoppers have a desire to try them on. Female customers are worried about the security issue in the trial room of shopping complex. Using virtual dressing room people may virtually try and choose clothing, jewellery, accessories, etc. at their home without any security issues. In the proposed work augmented reality technology has been used to give an efficient trial experience to the customer by implementing virtual dressing room using Microsoft Kinect Software Development Kit (SDK), where Two-Dimensional (2D) model of an apparel is superimposed on to the user's image and it appears that the customers are wearing the actual dress and the results are satisfactory. In addition to this, work has been done using supervised learning algorithm and the same is in process. The proposed work eliminates the security issues as the privacy of woman customer is the major concern and they may feel like they are in trial room of shopping complex.

Keywords Virtual dressing room · Microsoft Kinect · Augmented reality · Web camera

S. Bandyopadhyay · S. S. Thakur (✉)
MCKV Institute of Engineering, Howrah 711204, West-Bengal, India
e-mail: subroto_thakur@yahoo.com

S. Bandyopadhyay
e-mail: somabanmuk@yahoo.co.in

J. K. Mandal
University of Kalyani, Nadia 741235, West-Bengal, India
e-mail: jkm.cse@gmail.com

D. Bhattacharjee et al. (eds.), *Proceedings of International Conference on Frontiers in Computing and Systems*, Advances in Intelligent Systems and Computing 1255,
https://doi.org/10.1007/978-981-15-7834-2_2

1 Introduction

Due to overwhelming growth of electronic commerce though it is easy to order any product online but in case of apparel industry customer never knows how the selected items would fit on them until they try them on. The customers are neither sure about the size of the garments nor they know how it looks on them without trying the clothes. So, customers are facing problem while purchasing dresses online from different shopping websites. Similarly, merchants are also worried about how much volume of apparel would be stored and what will be the possibility of returns. It has been observed that the cost of handling the garments return can cost up to four times than the cost of initial sale of garments. The analysists have estimated that 10% of basic clothing items and 35% to 40% of high-end clothing items returns to an electronic merchant. In addition to this, size of garments of one brand may differ with the size of garments of the other brand. Virtual dressing room helps the customer how each dress exactly looks on them before buying, and thus there will be fewer chances of returns. It can also save the time of retailers as they don't have to display many dresses to individual customers, and there will also be minimum chances of clothes getting torn due to repeated wear. This work proposes an implementation of virtual dressing room in shopping complex eliminating the above-stated problems and providing the users a better experience during shopping time. The rest of the paper has been organized as follows. The literature survey related to this work and existing system has been described in Sect. 2 and Sect. 3, respectively. We have discussed the overview of proposed methodology in Sect. 4. Next, in Sect. 5 the different design methodologies have been discussed. We have discussed about our experimental results in Sect. 6. Finally, conclusion and future work direction have been discussed in Sect. 7.

2 Literature Survey

In the mid 1980's hug magic mirrors were developed which is a collection of photographic slides of human dressed with clothing. But the major drawback of this system is that magic mirror does not establish any relationship between the clothing outfits and the human body measurement. A model using a single depth image was developed that predicts the 3D positions of the human body joints [1]. A web browser-based interactive 3D cloth wearing effect exhibition system was developed by Li et al. [2]. An approach was proposed transferring clothes from one user to another where a user can be displayed wearing previously recorded garments [3]. A virtual dressing room was developed by MVM Inc where the personalized virtual model seems like a user's body and it allows one to try on garments. In this case the hair style, skin color. etc. are stored in database as personal attributes [4]. A low-cost depth camera and commodity graphics hardware has been used to create accurate real-time depth mappings of complex and arbitrary indoor scenes in variable lighting conditions [5].

A methodology that can produce 3D clothes to provide users a feeling of realistic garments was developed [6]. Francisco Pereira et al. proposed an augmented reality solution for E-commerce which is based on computer vision [7]. It has been seen that the customer who purchased products online can get real shopping experience through augmented reality [8]. Augmented reality mirror was proposed which can be installed in a retail shop which provides motion-based interaction [9]. A methodology for retexturing garments has been proposed which uses a single static image [10]. In this case the depth information has been obtained using Microsoft Kinect 2 camera.

3 Existing System

Existing virtual dressing rooms are based on diverse approaches and are mostly virtual avatars or fiducial markers, they are mainly concerned with enhancing the experience of online shopping. The main drawbacks of virtual avatars are the capturing process accuracy and the possibility of missing of a virtual mirror, which limits the presentation to a virtual avatar. Some approaches which are already available in the market are Cisco Style Me and WatchBox. WatchBox is an online E-Commerce platform that allows for the buying, selling, and trading of preowned luxury watches. Customer can view how a watch would appear on their wrist by using this augmented reality system.

4 Proposed Work

In this work, the implementation of a real-time virtual dressing room (VDR) is proposed which utilizes depth cameras like the Microsoft Kinect and it is possible to track the movements of a body, extract the measurements of body and create a virtual mirror with the corresponding video stream. The video image can be combined with a piece of clothing and can be shown frame by frame. In this proposed system, the user can be tracked by using the user label data and depth data provided in the Kinect Sensor and the Graphical User Interface (GUI) is having the option to superimpose the virtual cloth on user's body, in front of the Kinect in real time. The proposed method deals with the methodology of how the customer can virtually try on different garments. In this case, the truth model of customer body and the constructed mathematical model have been taken into consideration. With the help of different fit factors analysis have been done. The best fit garment for the customer is chosen by comparing the ground truth model and constructed mathematical model of the customer.

In this work, Microsoft Kinect Sensor [11, 12] has been used to create a real-time virtual dressing room application which helps to develop augmented reality application. A plain wall behind the captured person is recommended as a background

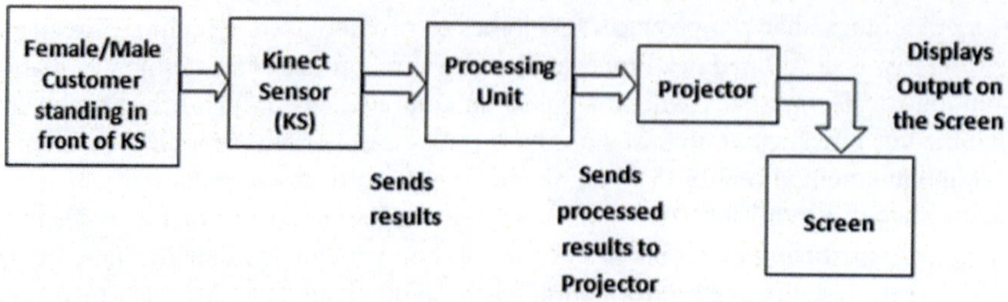

Fig. 1 Block diagram of the Proposed Virtual Dressing Room (VDR)

for the purpose of a stable skeleton tracking without interference. The consumers need to be in front of the sensor so that he/she can be recognized properly and the position of the sensor should be such that it can see the head and upper body portion of the customer. The user initially needs to stand in front of the Kinect Sensor which sends the depth image data to the computer. The Kinect Sensor consists of an infrared laser emitter, an infrared camera, and an RGB camera. The measurement of depth is a triangulation process [13]. The block diagram of the Proposed Virtual Dressing Room (VDR) is shown in Fig. 1.

The following steps are used while implementing the Virtual Dressing Room:

1. Reading the depth data and capturing the image by Microsoft Kinect
2. Displaying the user interface for cloth selection
3. Superimposing the selected garment on the recorded image
4. Executing algorithms for skeleton tracking
5. Superimposing dress on computed skeletal data
6. Exit

In Fig. 2 block diagram of the modified VDR is shown. The initial requirement is to attach the camera with the laptop and the customer needs to be standing in front of camera at a distance of approximately 1 m. The recording is started and the duration of recording is 30 s keeping in mind that the customer is wearing the dresses on trial. Once the recording has been completed it is saved in a file and the same is used for the processing. Later on the trimming is done on the video using available image processing software. Finally, we are getting the pictures in image form and the same has been stored in the databases.

In addition to this, images from databases are opened and shown to the customer whose recording has already been done earlier. The customer is requested to give feedback based on the existing images shown and a set of questionnaires has been asked. The questionnaires are

a. Do you like the fittings of the dress?
b. If the user gives the option yes then

    (i). The results are stored on a table else go to (iv).
    (ii). Do you go for online purchase? If the user gives the option yes then

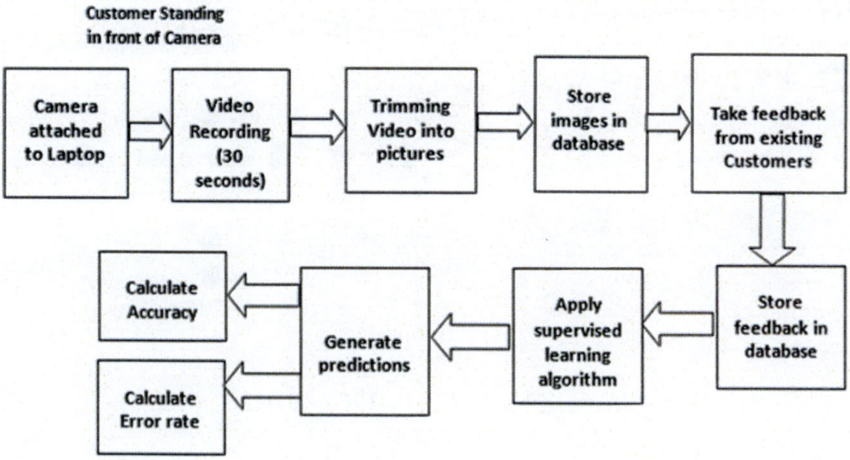

Fig. 2 Block diagram of the modified VDR

(iii). The results are stored on a table else
(iv). Exit

Based on user feedback, i.e., if the user like the fittings of the dresses, then there is a probability that the customer goes for online purchase. The results are stored in a table with attribute named as fitting (Y/N), with the value of 1/0, the value 1 indicates YES and 0 indicates NO. Again the customer is asked whether he/she is interested for online purchase, again the results of customer feedback are stored on the same table with attribute named as Online purchase(Y/N). The same can be used as a training set for classification problem using supervised learning algorithm.

5 Design Methodology

5.1 User Extraction

An augmented reality environment has been created by extracting and isolating the user. The depth images and user labels have been used to segment the foreground from the background. For segmentation the skin color has been used so as to allow the user to bring the hands in front of the cloth model. The image is in YC_bC_r color space [14].

$$Y = 0.299R + 0.587G + 0.114B < 70 \tag{1}$$

$$77 < C_b = 128 - 0.169R - 0.332G + 0.5B < 127 \tag{2}$$

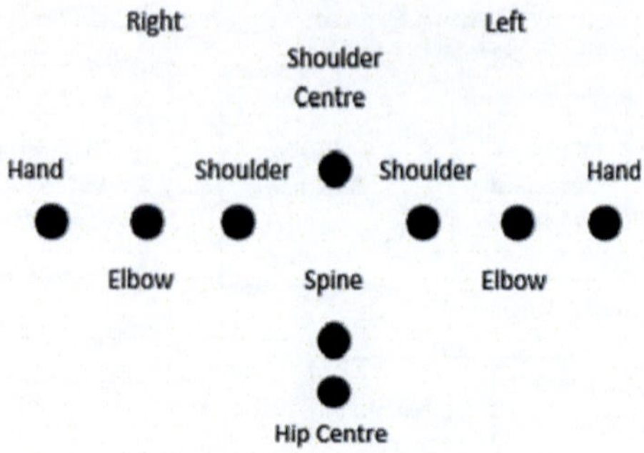

Fig. 3 Body joints required to superimpose clothes on user

$$133 < C_r = 128 + 0.5R - 0.419G - 0.081B < 173 \tag{3}$$

5.2 *Tracking*

The skeletal tracker estimates the depth of the body joints. Nine body joints have been used to superimpose the virtual cloth on the user which is shown in Fig. 3. The angle of rotation has been calculated by calculating the angle between the joints. For superimposing the virtual cloth on the user, the distance among the user and the Kinect Sensor and the distance between the joints of the user have been computed. In this case, the scaling factor is the ratio of the virtual cloth model when user is 1 m away from Kinect Sensor. We also define another shape-based scaling factor which depends on the Euclidian distance between the joints. Here, width to length ratio is denoted by α, and k represents either left or right [15–17].

$$H^k_{arm} = \sqrt{\left(x^k_{shoulder} - x^k_{elbow}\right)^2 + \left(y^k_{shoulder} - y^k_{elbow}\right)^2} \tag{4}$$

$$W^k_{arm} = H^k_{arm} \times \alpha \tag{5}$$

$$H_{body} = \left|x_{shouldercentre} - x_{hipcentre}\right| \tag{6}$$

$$W_{body} = \left|x^{left}_{shoulder} - x^{right}_{shoulder}\right| \tag{7}$$

6 Experimental Results

The time taken for the measurement of the body parameters and estimated clothes didn't take more than 1.10 s. We performed a study between the real measurements in the real feed and the estimated measurements in the virtual feed on 10 male and 10 female. The experiment tests performed on the consumers of different height and weight. Kinect Sensor was kept on a table which is 100 cm above the ground. Tests were carried out at a room with good lighting conditions. The experiment has been conducted by keeping customers in different distances from the Kinect Sensor and it has been observed that the body joints are satisfactorily detected when it has been kept at a distance of 1–2 m of the sensor. So, the system has been tested keeping the sensor at a distance of 1, 1.5, and 2 m from the user [18].

In Table 1 shoulder length of six different males has been shown. Here, shoulder length denotes the length from center back of the neck to shoulder tip.

In Table 2 arm length of six different females have been shown. The length from shoulder tip to the wrist is considered as the arm length.

In Table 3 body length of six different females has been shown where the body lengths are basically the fullest part of the chest.

Virtual dressing room implementation requires the image of customer/user in front view. Each virtual dress is superimposed to the user in 2D graphical representation. This approach is used to align the proposed model with the user/customer. And the same is repeated under different situations. The error percentage was calculated as

Table 1 Shoulder Length of Male

Real measurement in cm	Average estimated measurement in cm	Error in %
40.20	39.30	2.25
42.10	41.20	2.14
44.30	43.10	2.70
45.20	43.60	3.54
45.40	43.50	4.18
48.10	47.30	1.66

Table 2 Arm Length of Female

Real measurement in cm	Average estimated measurement in cm	Error in %
61.05	58.75	3.77
61.25	58.80	4.00
61.50	59.00	4.06
62.20	59.25	4.74
62.25	59.25	4.81
62.75	60.00	4.38

Table 3 Body Length of Female

Real measurement in cm	Average estimated measurement in cm	Error in %
66.40	60.80	8.40
66.80	61.50	7.93
67.50	62.25	7.78
68.00	63.00	7.35
68.50	63.25	7.66
69.00	64.00	7.24

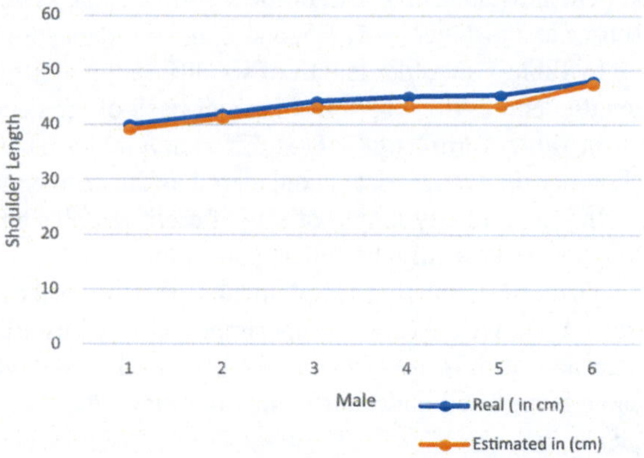

Fig. 4 Real and estimated shoulder length of six males

(Real Estimate/Real Value) * 100. The shoulder length of men is detected with an error of 2.745%, the arm length of women detected an error of 4.29%, and the body length of women detected an error of 7.72%. In each case, we estimated the size with less than 8% error. Figure 4 depicts the real and estimated shoulder length of male and it can be observed that both the plots, i.e., the estimated and real are overlapping each other, which shows the accuracy of the results.

Figure 5 depicts the real and estimated arm length of female. From this figure it is clear that real and estimated arm length are close to each other. As the error percentage is very less we can conclude that our proposed system works properly.

From Fig. 6 we can conclude that the estimated length has less than 8% error than that of real length. In this case the body length of the female has been taken, the error is more than that of previous two cases as the body curvature could not be addressed in estimated measurement.

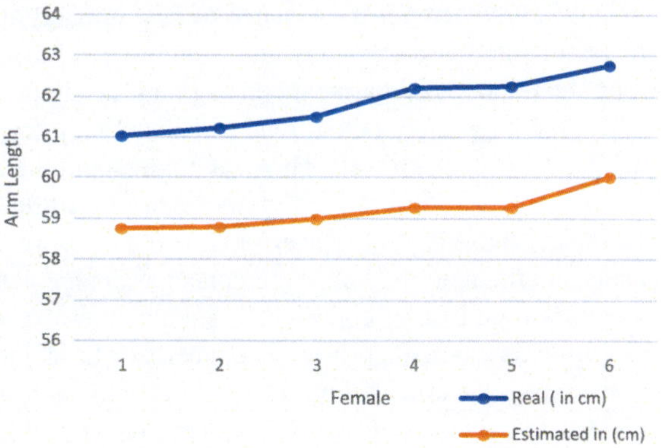

Fig. 5 Real and estimated arm length of six female

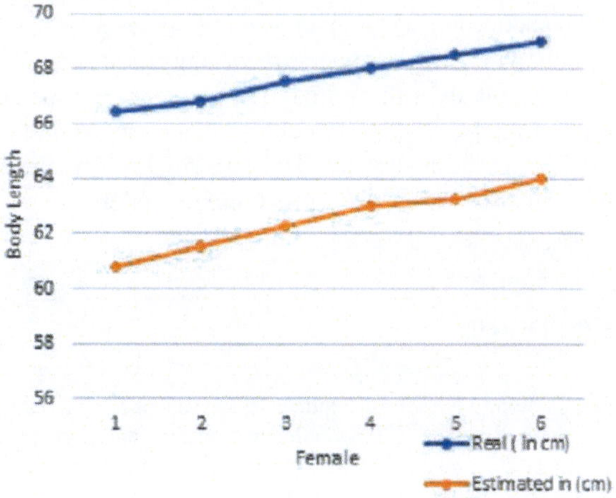

Fig. 6 Real and estimated body length of six females

7 Conclusion and Future Work

In this proposed work implementation of a complete system has been done, which can be used for fitting garments or different clothing items to an image of the body of user. In this case, the critical points of the garments are adjusted to match with the critical points of the image such that if the users are actually wearing the garments he/she can adopt such items. However as only joint positions are used for getting the body measurements of a person, it has limitations. Instead of wearing garments the user can examine the fitness of the garments virtually. Thus, this system has the ability to provide the experience of a trial room and the potential to minimize the number of articles returned or exchanged to the vendor. As the system is automatic the user can enjoy shopping experience independently. Virtual Dressing Room is one of the best solutions for the accurate trial of garments, before actual purchase. As it's a simple set up and can be assembled at home with a Kinect and computer with screen. People can get the essence of new apparel at his own house. Moreover, in the era of Internet technology and E-commerce business it is now a common trend of public to purchase garments through online shopping.

By creating augmented reality person can try on different clothes before purchasing it through web. They can understand the actual fittings of the garments and will able to decide which garments are best fitted for them and which garment may be purchased or not. The error percentage between the estimated body measurements and the real body measurements has been calculated and the accuracy of this system is found to be good which enables the use of Virtual Dressing Room and a solution as women's privacy is concerned. In our work, we were concerned only on the upper portion of the body and hence dealt with the nine joints. In future to implement the complete virtual dressing room, the lower portion of the body also should be taken care and we have to work with more joints. Similarly, in the second approach as the number of images stored in the databases are very limited, prediction may be done but accuracy may not be achieved. The same can be tested over a period of time when the customers' recordings of at least 1500 numbers are available with us. Efficient algorithms are required to enhance the performance of virtual dressing room. Finally, it can be concluded that a virtual dressing room application has been used where the apparel has been superimposed on the user and it can be more realistic if the dresses are superimposed on 3D model.

References

1. Shotton, J.: Real-time human pose recognition in parts from a single depth image. In: Proceedings of. IEEE Conference on Computer Vision and Pattern Recognition (CVPR), pp. 1297–1304(2011)
2. Li, R., Zou, K., Xu, X., Li, Y., Li, Z.: Research of interactive 3d virtual fitting room on web environment. In: Fourth International Symposium on Computational Intelligence and Design (ISCID), 2011, vol. 1, pp. 32–35. IEEE (2011)
3. Hauswiesner, S., Straka, M., Reitmayr, G.: Image-based clothes transfer. In: Proceedings of International Symposium on Mixed and Augmented Reality (ISMAR) (2011)
4. Brooks, A.L., Brooks, E.: Towards an inclusive virtual dressing room for wheelchair-bound customers. In: 2014 International Conference on Collaboration Technologies and Systems (CTS), 2014, pp. 582–589. Minneapolis, MN (2014)
5. Newcombe, R.A., Izadi, S., Hilliges, O., Molyneaux, D., Kim, D., Davison, A. J., Kohli, P., Shotton, J., Hodges, S., Fitzgibbon, A.: Kinect fusion: Real-time dense surface mapping and tracking. In: Proceedings of 10th IEEE International Symposium on Mixed and Augmented Reality (ISMAR'11), pp. 127–136. Basel, Switzerland (2011)
6. Protopsaltou, D.: A body and garment creation method for an internet based virtual fitting room. Advances in Modeling Animation and Rendering, pp. 105–122 (2002)
7. Pereira, F., Silva, C., Alves, M.: Virtual augmented reality techniques for e-commerce, ENTERprise information systems. Commun. Comput. Inf. Sci. **220**, 62–71 (2011)
8. Abed, S.S.: Opportunities and challenges of augmented reality shopping in emerging markets. In: Dwivedi, Y., et al. (eds.) Emerging Markets from a Multidisciplinary Perspective. Advances in Theory and Practice of Emerging Markets. Springer, Cham (2018)

9. Birliraki., C, Margetis,G., Patsiouras, N., Drossis G., Stephanidis, C.: Enhancing the customers' experience using an augmented reality mirror. In: Stephanidis C. (ed.) HCI International 2016–Posters' Extended Abstracts. HCI 2016. Communications in Computer and Information Science, vol. 618. Springer, Cham (2016)
10. Traumann, A., Anbarjafari G., Escalera, S.: A new retexturing method for virtual fitting room using Kinect 2 Camera. In: The IEEE Conference on Computer Vision and Pattern Recognition (CVPR) Workshops, pp. 75–79, (2015)
11. Microsoft Kinect for Windowsl Develop for the Kinectl Kinect for Windows. http://kinectforwindows.org/. Accessed 03 May 2012
12. Wai Mok, K., Wong C.T., Choi, S.K., Zhangand, L.M.: Development of virtual dressing room system based on Kinect, I.J. Inf. Technol. Comput. Sci. **9**, 39–46 (2018)
13. Salih, Y., Malik, A.S.: Depth and geometry from a single 2d image using triangulation. In: IEEE International Conference on Multimedia & Expo Workshops (ICMEW), pp. 511–515 (2012)
14. Chai, D., Ngan, K.N.: Face segmentation using skin-color map in videophone applications. IEEE Trans. Circuit Syst. Video Technol. **9**(4) (1999)
15. Kinect Quick Start Guide, http://support.xbox.com/en-GB/xbox-360/manuals-specs/manual-specs
16. Isıkdoga, F., Kara K.: A real time virtual dressing room application using Kinect. In: CMPE537 Computer Vision Course Project, January 2012, pp. 1–4(2012)
17. Gultepe, U., Gudukbay, U.: Real-time virtual fitting with body measurement and motion smoothing. Comput. Graph. **43**(1), 31–43 (2014) (Pergamon)
18. Kjærside, K., Kortbek, K.J., Hedegaard, H.: ARDressCode: augmented dressing room with tag-based motion tracking and real-time clothes simulation. In: Proceedings of the Central European Multimedia and Virtual Reality Conference (2005)

Multiple Radar Data Fusion to Improve the Accuracy in Position Measurement Based on K-Means Algorithm

Sourav Kaity, Biswapati Jana, P K Das Gupta, and Saikat Das

Abstract The position of any moving object can be easily determined with the help of radar. It can identify the object by using radio waves and determine the range, azimuth, and elevation. To achieve reliable and accurate position measurement instead of one more number of radar should be considered. At the same time if any of the radars have some wrong measurements then combined position measurement becomes erroneous. Data fusion techniques can be applied to integrate multiple radar measurements. Data fusion is a process to solve a problem based on the idea of integrating several pieces of information to obtain more consistent, accurate, and useful information. If the erroneous measurement is identified and eliminated from data fusion then final data fusion result becomes more accurate. Here we summarize how the k-means algorithm can be used to identify the position of any object by the process of combining data from various radars. Our main aim is to identify the erroneous radar measurements if any and establish a technique of combining the information from different radars to reach the best accurate solution.

Keywords Data fusion · Radar · Range · Azimuth · Elevation · K-means

S. Kaity (✉)
Integrated Test Range, DRDO Chandipur, Balasore 756025, India
e-mail: souravkaity@gmail.com

B. Jana
Dept of Computer Science, Vidyasagar University, Midnapore 721102, India
e-mail: biswapatijana@gmail.com

P. K. D. Gupta
Proof and Experimental Establishment, DRDO Chandipur, Balasore 756025, India
e-mail: pkdasgupta@gmail.com

S. Das
Dept of ECE, Jalpaiguri Government Engineering College, Jalpaiguri 735102, India
e-mail: saikatdas5032@gmail.com

D. Bhattacharjee et al. (eds.), *Proceedings of International Conference on Frontiers in Computing and Systems*, Advances in Intelligent Systems and Computing 1255,
https://doi.org/10.1007/978-981-15-7834-2_3

1 Introduction

Multiple radar data fusion is the process to combine all the data, which are given from several numbers of radar to produce the most specific information. Data fusion is an emerging technology applied to the defence areas such as battlefield surveillance and guidance, automated target recognition, and control of autonomous vehicles. It is also used in non-defence applications such as monitoring of medical diagnosis, complex machinery, and smart building [1]. We are using the data fusion technique to improve the accuracy in position measurement. Radar is a detection system that can easily track a moving object and determine the range, azimuth, and elevation. All the radar measurement data are not equally accurate. Successful application of data fusion technique makes the result acceptable. We apply a K-means clustering algorithm to get the final measurement over all the measurements. There are several types of clustering algorithm available but here we have applied K-means algorithm because it is very time efficient. It works in an iterative manner and leads to the final local centroid in each iteration. In this way, we fuse all the data and get final data from several numbers of data and we get the more accurate and acceptable result.

2 Literature Study

Radar is an object detection and tracking system that is used to identify objects by using radio waves and it determines what is the range of unknown objects and what an angle it is there and what is the velocity of that object. Radar systems come in a variety of sizes and have different performance specifications. It is widely used in the area of air traffic control, navigation, defence sectors, etc. [2]. Radar was developed before and during World War II by several nations secretly. The term RADAR was coined by the United States Navy as an acronym for Radio Detection and Ranging in 1940 [3].

Data collected from multiple sensors cannot be used directly because of the dissimilarity of data values. Data fusion techniques are used to combine those data to get the most accurate and complete unified data. Data fusion is used in military applications that include guidance for autonomous vehicles, remote sensing, automated target recognition, battlefield surveillance, and automated threat recognition systems. It is also in the nonmilitary application such as condition-based maintenance of complex machinery, monitoring of manufacturing process, medical pattern recognition, and robotics [4]. A prime example of data fusion is humans. We rely on a fusion of smelling, touching, and tasting food to ensure it is edible or not. Similarly, all senses are used to collect and brain fuses to perform most of the tasks in our daily lives. Our brain always fuses the data gathered from all senses of humans [5].

Clustering is the technique of classification of objects or entities into different groups. Clustering is used in the following fields widely such as statistics, pattern recognition, and machine learning. The main goal of clustering is to determine the intrinsic grouping in a set of unlabeled data [6]. There are several types of the clustering algorithm, but in this paper, we are using the K-means clustering algorithm [7]. K-mean is the most popular partitioning method of clustering. It was firstly proposed by MacQueen in 1967 [8]. The less variation we have within clusters, the more similar the data points are within the same cluster [9].

3 Procedure

In this paper, we are using radar as a sensor. Radar can track an object and determine the range, azimuth, and elevation of the object. So we have to convert the range, azimuth, and elevation into the Cartesian Coordinate (X, Y, Z). So first we have to convert the value of azimuth and elevation from degree to radian.

$$Range = r, \qquad Elevation = \alpha, \qquad Azimuth = \beta \tag{1}$$

In Cartesian coordinate,

$$X = r * \cos(\alpha) * \sin(\beta), \qquad Y = r * \cos(\alpha) * \cos(\beta), \qquad Z = r * \sin(\alpha) \tag{2}$$

So, from these formulae, we can easily convert the given Radar's data to the Cartesian Coordinate(X, Y, Z) [10].

Step 1: In this paper, We are using 15 radar measurements for experimentation. Firstly we have taken the radar measurement (range, azimuth, and elevation), actual location of the object, coordinate of radar and reference point as the inputs. Now we have to convert the radar measurement range, azimuth, and elevation to the Cartesian coordinate (X, Y, Z). To convert all the data we are using the Eq. 1. Now all the data are in Cartesian form.

Step 2: In this step, we are measuring the position of the object with respect to the common reference point for all radars. We have 15 different measurements from 15 radars for one real object due to inaccurate radar measurement. Then we find the difference between the exact object location and the radar measured object location for all radars in x, y, and z components. And also find the range difference for all radars. So, we are finding the error of the radar measurement. Let, Actual position of the object is (x, y, z). And after measuring the object's position with respect to the radar 1, we get the measured position of the object (x1, y1, z1).

So, error with respect to the X coordinate = (x − x1) error with respect to the Y coordinate = (y − y1) error with respect to the Z coordinate = (z − z1)

$$rangeerror = \sqrt{(x^2 + y^2 + z^2)} - \sqrt{(x1^2 + y1^2 + z1^2)} \tag{3}$$

Step 3: We have 15 locations (points) of the objects as per radar measurements. Some measurements are very near to the exact location of the object and some locations are far away from the exact location of the object. So, some position measurements are within acceptable limit and some are not. Now if we integrate all the nearest points ignoring all the points beyond the acceptable limit then the combined position measurement becomes acceptable. To combine all the points, we are using a K-means clustering [10] algorithm. We are considering two initial clusters. Then we calculate the initial centroid. The first centroid is chosen randomly and then calculate the Euclidean distance between the initial centroid and every point. Based on that distance we are choosing the second centroid, the furthest point from the first centroid is considered as the second centroid. Now we have to calculate the Euclidean distance between two centroids to all remaining points. Let position of a point = (x, y, z), and the position of the centroid = (cx, cy, cz), then the Euclidean distance is

$$d = \sqrt{(x - cx)^2 + (y - cy)^2 + (z - cz)^2} \tag{4}$$

Now we have to compare Euclidean distance between centroid two and every point and the Euclidean distance between centroid one and every point. And the centroid is getting updated based on the minimum distance. And the position of the updated centroid is the mean of the position of the centroid and the nearest point. Let D1 = distance from first centroid to first point, D2 = distance from second centroid to first point, position of the first point = (x1, y1, z1), position of the first centroid = (cx1, cy1, cz1), position of the second centroid = (cx2, cy2, cz2). Now, if D1 is less than the D2 then the position of the updated first centroid is updated else position of the second centroid is updated.

In this way, the centroid is updating in case of 15 points and finally, we get two centroids. Then we have to count the number of surrounding points of that centroids. Largest cluster centroid is considered as the final centroid. To specify the centroid more accurately, we are using a boundary condition. If the location of any point with respect to the centroid of a cluster is within this boundary then this point is entered into that cluster if not then we ignore that point.

4 Experimental Results and Analysis

For our experimentation, we have taken trajectory data of a projectile up to 80 km of range and 35 km of altitude. After completing all the steps successfully, we are given some plots by which we can verify and analyze our experiments and got some results. We considered many cases to analyze our results.

In ideal case when all the radars are free from noises and bias errors, in this time the total error in radar 1 is varying in the maximum limit of +0.06 m. This is very small. After applying the K-means clustering algorithm, we are getting the final total error as varying in the maximum limit slightly less than +0.06 m. So in the ideal case, there is also no effect of clustering.

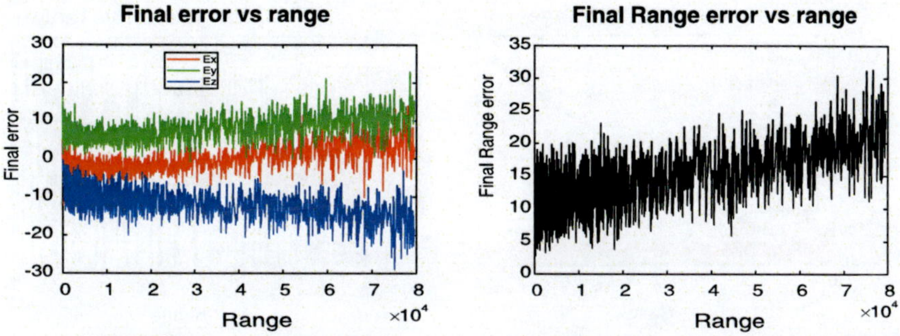

(a) Final error vs range graph (b) Final range error vs range graph

Fig. 1 Component wise error and range error graphs after clustering

4.1 Case 1: No Bias, Having Noise in Azimuth, Elevation, and Range

In this case, we are adding some noises on every radar. In the range, we are adding 25 m noise, in azimuth and elevation, we are adding 1-minute noise. Now for the effect of noise, the range error in radar 1 is rising up to the maximum range of +30 m approximately. After applying the K-means clustering algorithm, two clusters are produced, we are getting the final centroid, which is the final position of the object. From Fig. 1, we can see that the final range error is varying and maximum limit is slightly less than +30 m. So there is not much improvement occurred. In this way, we analyzed the results for every radar and we are getting the same results as radar1.

4.2 Case 2 : Bias in Azimuth but No Bias in Elevation and Range

In this case, we are adding some noises same in case 1. And we are adding some azimuth bias error in radar 1, radar 2, and radar 3. The value of the applied azimuth bias error is 0.5 degrees. Now after adding the azimuth bias error in the three radars, we can see from the Fig. 2a that the range error rose up to the maximum limit +650 m approximately. After applying the K-means clustering algorithm, we can see from Fig. 2b that the range error is varying up to the maximum limit +25 m approximately. In this way, we analyzed the results in case of every combination of radars and we are getting the same result like that.

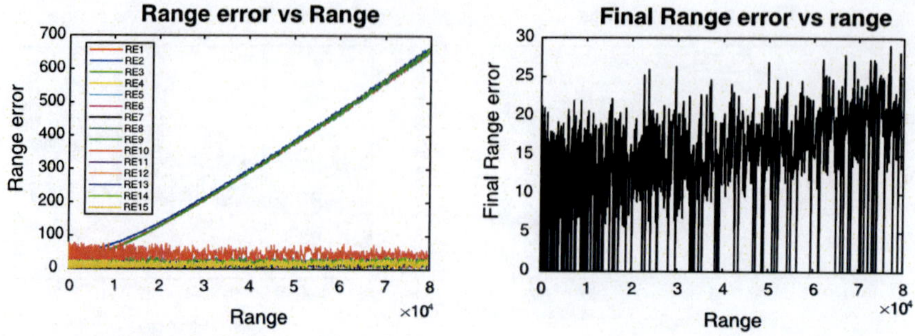

(a) Range error vs range graph for all radars

(b) Final range error vs range graph

Fig. 2 Range error graphs before and after clustering

4.3 *Case 3 : Bias in Elevation but No Bias in Azimuth and Range*

In this case, we are adding some noises same as case1. And we are adding some elevation bias error in radar 4, radar 5, and radar 7. The value of the applied elevation bias error is 0.5 degrees. Now after adding the elevation bias error in the three radars, we can see from Fig. 3a that the range error is varying maximum up to the +700 m approximately. After applying the K-means clustering algorithm, we can see the results from Fig. 3b that the final range error is varying maximum up to +25 m approximately. In this way, we analyzed the results in case every combination of radars and we are getting the same result like that.

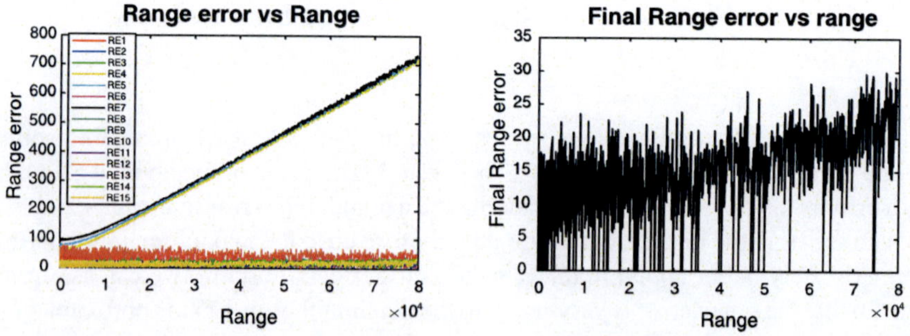

(a) Range error vs range graph for all radars

(b) Final range error vs range graph

Fig. 3 Range error graphs before and after clustering

4.4 Case 4 : Bias in Azimuth and Elevation but No Bias in Range

In this case, we are adding some noises the same as case 1. And we are adding some azimuth bias error in radar 7 and radar 12 and also adding some elevation error in radar 11. The value of applied azimuth and elevation bias error is 0.5 degrees. After adding the noise and bias errors in the three combinations of radar, we can see that the range error is varying maximum up to +700 m approximately. After applying the K-means clustering algorithm, final range error is varying maximum up to +25 m approximately. The figure is quite similar to case 3. In this way, we analyzed the results for every combination of radars and we are getting the same result like that.

4.5 Case 5 : Bias in Azimuth and Elevation but No Bias in Range in Other Combination

In this case, we are adding some noises same as case1. And we are adding some azimuth bias error in radar 7, radar 11, and radar 12 and also adding some elevation error in radar 7, radar 11 and radar 12. The value of applied azimuth and elevation bias error is 0.5 degrees. After adding the noise, azimuth bias error, and elevation bias error, we can see that the range error is varying maximum up to +1000 m approximately. After applying the K-means clustering algorithm, final range error is varying maximum up to +25 m approximately. So here the error is reduced. The figure is quite similar to case 3. In this way, we analyzed the results for every combination of radars and we are getting the same result like that.

4.6 Case 6 : Bias in Azimuth, Elevation, and Range

In this case, we are adding some noises the same as case 1. And we are adding some azimuth bias error in radar 12, radar 13, and radar 14, adding some elevation error in radar 12, radar 13, and radar 14 and also adding some range bias error in radar 13. The value of applied azimuth and elevation bias error is 0.5 degree and the value of applied range bias error is 10 m. After adding the noises, azimuth, elevation, and range bias error, we can see that the range error is varying up to the maximum limit +1100 m. After applying the K-means clustering algorithm, final range error is varying maximum up to +25 m approximately. So here the error is reduced. The figure is quite similar as case 3. In this way, we analyzed the results in case of every combination of radars and we are getting the same result like that.

Experimental results of all cases are depicted in Table 1. The observation is similar in all cases that maximum value of error is very high before clustering and maximum value of error is approximately 25 m after clustering.

Table 1 Maximum error values of all cases before and after clustering

Case	Max. value of total error before clustering(m)	Max. value of Total error after clustering(m)
Ideal	0.06	Less than 0.06
1st	30	25
2nd	650	25
3rd	700	25
4th	700	25
5th	1000	25
6th	1100	25

5 Conclusion and Future Scope

Measurement of the position from a single radar suffers from an accuracy and reliability problem. The accuracy and reliability problems can be rectified by getting measurements from multiple numbers of radars. That's why in this work, We proposed a real time data fusion technique to get more accurate measurements. We are using the k-means algorithm because it is most time efficient than other clustering algorithms. Its computational complexity is O(k*n*t), where k is the number of clusters, n is the number of objects, t is the number of iterations. When we are adding some noises and bias errors, then measurements are spread out. To increase the accuracy of the cluster centroid, boundary limit is established. We ignored the measurements that are beyond the boundary and able to achieve more accurate and reliable results. We added the same noises in every radar and also add the same bias errors in every combination of radars. Further experimentation can be carried out with variable bias errors and variable noise.

References

1. Hall, D.L., Llinas, J.: An introduction to multisensor data fusion. Proc. IEEE **85**(1), 6–23 (1997)
2. Anitha, R., Renuka, S., Abudhahir, A.: Multisensor data fusion algorithms for target tracking using multiple measurements. IEEE International Conference on Computational Intelligence and Computing Research (2013)
3. Ghoghre, D.A., Dhanshri, A., Priyanka, A.: Radar system using Ar-duino. IOSR J. Electron. Commun. Eng. (IOSR-JECE), 53–56. e-ISSN: 2278–2834, p-ISSN: 2278–8735
4. Shon, S.Y., Lee, S.H.: Data fusion, ensemble and clustering to im-prove the classification accuracy for the severity of road traffic accidents in Korea. Saf. Sci. **41**, 1–14 (2003)
5. Crowley, J.L., Demazeau, Y.: Principles and techniques for sensor data fusion, LIFIA (IMAG) 46 avenue Félix Viallet F-38031 Grenoble Cédex, FRANCE
6. Nirmala, A.M., Saravanan, S.: A study on clustering techniques on matlab. Int. J. Sci. Res. (IJSR), 1497–1502. ISSN (Online): 2319–7064

7. Yadav, J., Sharma, M.: A review of K-mean algorithm. Int. J. Eng. Trends Technol. (IJETT) **4**(7) (2013)
8. Steorts, R.C.: K-means Clustering. Duke University STA 325, Chapter 10 ISL
9. Dale B.W., Richards, M.A., Long, D.A.: Radar Measurements. Text Book (2010)
10. Wanner, L.: Introduction to Clustering. IULA (2004)

A Brief Survey of Steganographic Methods for ECG Signal

Pushan Bhattacherjee, Debayan Ganguly, and Kingshuk Chatterjee

Abstract The following paper represents a survey of steganography techniques suitable for usage in ECG. A few steganographic methods have been discussed below, it has been believed that these methods can be exploited to bring out more techniques suitable for ECG. The objective of this paper is to provide a comprehensive survey of existing steganographic techniques for ECG.

Keywords Image steganography · ECG based steganography · LSB based methods · Discrete wavelet transform · Singular value decomposition

1 Introduction

ECG provides the data in a graph, meanwhile, it also becomes a challenge for ensuring privacy and passing the information (maybe patient details or disease details) through a secured channel. The branch of security by obscurity has two branches, steganography and watermarking [1]. Both of them camouflage the existence of the secret information [2]. The motto of this paper is to aid researchers to design advanced steganographic techniques for other ECG signals.

P. Bhattacherjee
Heritage Institute of Technology Chowbaga Road, Kolkata P.O.-Anandapur, 700107, India
e-mail: pushan98@gmail.com

D. Ganguly (✉)
Government College of Engineering and Leather Technology Sector-III, Kolkata 700106, India
e-mail: debayan3737@gmail.com

K. Chatterjee
Government College of Engineering and Ceramic Technology, 73, Abinash Chandra Banerjee Lane, Kolkata 700010, India
e-mail: kingshukchaterjee@gmail.com

D. Bhattacharjee et al. (eds.), *Proceedings of International Conference on Frontiers in Computing and Systems*, Advances in Intelligent Systems and Computing 1255,
https://doi.org/10.1007/978-981-15-7834-2_4

2 Methods Specifically Made for ECG Signal

2.1 Frequency Domain Techniques for ECG Signals

Wavelet Domain techniques for ECG Signals Ibaida and Khalil [3] proposed an approach to perform a wavelet-based ECG steganography. The proposed method combines encryption and scrambling. It allows the ECG signal to hide and integrate the patient information.

There are five parts of the algorithm—encryption, wavelet decomposition, embedding operation using a scrambling matrix (128×32), inverse wavelet recomposition, and watermark extraction process. Encryption is provided by an XOR ciphering technique with an ASCII coded key that will be used as security. The technique is quite simple. In wavelet decomposition, there are five levels of decomposition which results in 32 sub-bands. Most of the important parts of the signals of ECG lie in the low frequency signals. So, they are further divided into two sub-bands approximate signal—low frequency—A and detail signal—D (mostly noise part). The embedding operation gives a scrambling matrix which is made by two rules—(a) Same row must not repeat elements (b) No two row can be identical.

Jero et al. [4] second way to use steganography on ECG signal somewhat similar. The process includes the conversion of 1D ECG to a 2D image. MIT-BIH arrhythmia database has been sampled at a frequency of 128 Hz.. The signal will be decomposed by discrete wavelet transform. Singular value Decomposition is then applied in both the secret information and the image (2D DWT transform). Then the image and the information are sent for the watermark embedding process. For the receiving end, there is an algorithm for watermark extraction. There is a key in this case which is composed of orthogonal matrices.

Jero et al. paper [5] constructs a 2D ECG matrix with a sampling rate of 360 samples/sec and a gain value of 100. The ECG is then segmented with respect to the R peaks. QRS detection algorithm has been applied to find the fiducial points or R peaks. DWT-SVD watermark embedding—DWT has been applied to decompose into three parts—low (LL), intermediate (LH and HL), and high part (HH). The third step is extraction of patient data by DWT and SVD.

Liji et al. [6] advise us to use XOR ciphering encryption followed by an integer-to-integer wavelet transform. After that comes an embedding operation that results in a scrambling matrix of 128×16. Each element must be an integer between 1 and 16. Inverse wavelet transform will be used followed by a watermark extraction process. The shared key and the scrambling matrix should be given to the receiver.

Dilip and Raskar [7] have done quite a frugal literature survey in their paper and their tools are similar to that of a few techniques mentioned above. XOR ciphering with ASCII coded key, Wavelet Decomposition, and Embedding with a scrambling matrix.

Tseng et al. [8] has proposed the following method. The first step is data preparation (ECG Signal is sampled at a rate of 360 Hz and 12-bit binary representation, the signal is eliminated from having any DC offset) followed by watermark insertion and extraction (basic tools used are DWT and an embedding algorithm). Jero et al. [9]—another technique has proposed that the 1D ECG Signal and patient data has been used as inputs. Binary conversion of patient data and 2D conversion of ECG Image are the next steps. DWT and SVD are two algorithms that are applied. After that watermark embedding is applied by using inverse SVD. The extraction process includes the conversion of the 1D watermarked signal to 2D ECG Image and DWT and then applies the key. This procedure uses Continuous Ant Colony Optimization (CACO). The related works here are—Mishra et al. [10]—using quantization and scaling factors in decreasing the deterioration of the cover signal. Ali and Ahn [11] worked on evaluating the Mishra et al. [10] process, by using a lesser scaling factor. Run et al. [12], has a similar algorithm. Loukhaoukha et al. [13] used Multi-objective Ant Colony Optimization. The last two of them are not scaled for ECG though.

Sankari and Nandhini [14]—Public key cryptosystem has been used to encrypt patient data. DWT based steganography method has been proposed. The related works are—Kaur et al. [15], Zheng and Quian [16], and Danyali and Golpira [17]. Although the last one is useful for MRI images—not ECG signal.

Jero and Ramu [18] in another one of his papers converts the ECG signal into 2D and uses SVD and DWT on them. Except, in this case, he uses BCH coded watermark.

Raeiatibanadkooki et al. [19] used Wavelet compression and Huffman coding. Mahmoud [20] proposed to use the ECG anonymisation method using wavelet packet decomposition. The signal shall again be reconstructed. Devi et al. [21] proposed a method using RSA encryption, wavelet decomposition, scrambling matrix, and extraction. Engin et al. [22] had used a wavelet-based watermarking process for ECG signal where the average power was used for each sub-band. The power is used as threshold parameters to select which sub-band to embed the data into. Asha [23], Meghani and Geetha [24], Sheeba et al. [25], Sivaranjani and Radha [26], Marakarkandy and Tiwari [27], Awasarmol et al. [28], and Tabash [29], proposed a method using wavelet transform—but it is similar to many methods mentioned above. PremChandran et al. [30] used integer-to-integer wavelet transform during the extraction. Sahu et al. [31] proposed an innovative and energy efficient method based on DWT. Mathivanan et al. [32–34] used Quick Response code and DWT.

Curvelet based ECG steganography. Jero et al. [35] used curvelet transform to perform steganography. The MIT-BIH normal sinus rhythm database of ECG signals has been used by the authors. The algorithm has two parts watermark embedding and watermark extraction. A key has been provided. Fast Discrete Curvelet Transform has been used to convert the 1D signal to 2D signal and then the curvelet coefficients are calculated. In order to construct a 2D ECG image from 1D ECG data, bandpass filtering is done followed by differentiation. Degadwala et al. [36] and Patil and Patil [37] have used similar methods that are based on curvelet transform.

Jero and Ramu [38] proposed that 1D ECG can be converted to 2D ECG and be subjected to a curvelet transform while the patient data is being transformed by a binary conversion algorithm. Both are then sent through a threshold selection algorithm. Then a quantization method has been proposed which involves both the original curvelet transform result, original binary conversion result, and the result after a threshold selection algorithm. The output of this threshold selection algorithm is subjected to an inverse curvelet transform. Thus, the watermarked ECG has developed with a key. The extraction procedure uses the key to extract the watermark.

Other frequency domain techniques. Ibaida et al. [39] enlightened us with a new technique. The proposed method asks us to shift up and scale the ECG signal to avoid negative values. It further advises us to convert the floating point to the integers. The magnitudes of shifting and scaling must be hidden. All the procedure up to this step is called a preprocessing. The techniques that are going to follow will be a shift special range transform, Data hiding, and ECG scaling.

$\mathrm{X} = \mathrm{s} + \mathrm{X}$ (used for shifting the ECG signal up)
$\hat{\mathrm{X}} = \mathrm{p} * \mathrm{X}$ (used for the scaling of the ECG)

Vu Mai et al. in his paper [40] used the same technique as proposed by Ibaida et al. [39]. Vu Mai's main work was in the process to protect the privacy of users and make it more efficient to upload and download a large amount of data.

Abuadbba and Khalil in their paper [41], claim that his technique has been able to balance both the security and the efficiency of the system. Their procedure of providing a distant point-of-care system has three steps. A Walsh-Hadamard Transform will be applied followed by a Hiding operation and inverse Walsh-Hadamard re-transform, respectively. In the end, a Key is used for retrieval. The related works are—Zheng and Qian [16], applied B-Spline wavelet into ECG samples and applied Arnold transform for scrambling. Golpira and Danyali [17] applied 2D wavelet on the host data (not very good with ECG). Kaur et al. [15] segmented the ECG sample using 10 bits and cut the signal into two parts. They used Window-Dependent Factor to modulate the signal. Patient ID is the key. Pandey et al. [42] proposed a technique of discrete cosine transform and ASCII encoding. Wu et al. [43] used integer-to-integer Harr Wavelet Transforms followed by histogram shifting and thresholding scheme. It has been proved in this paper that it has a high embedding capacity and low distortion. Vallathan et al. [44] used contourlet transform—though there was a shortcoming of this process. Mohsin et al. [45] did an impressive survey of not only ECG but also other healthcare images like MRI, Brain, and Iris. Here he mentioned the work of Premarathne et al. [46], and his methods to hide electronic health records data inside the ECG. Sivaranjani [47] proposed an efficient method that used Haar wavelet decomposition, Rivest-Shamir-Adleman based encryption technique, and SVD.

Liu et al. [48] proposed a technique using SVD. The proposed approach satisfies the need for the processing of biomedical signals directly in encrypted.

Pandey et al. [49] proposed to use chaotic map and sample value difference approach to hide patient data in ECG samples and also transmitting it wirelessly. Bethzia et al. [50] have also followed similar ventures.

2.2 *Spatial Domain Techniques for ECG Signals*

LSB based methods. Karakış et al. [51], made a technique with MRI as cover image and EEG along with patient information as payload. Karakış et al. [51] proposed three steps for embedding. The first step is message preprocessing step compression and encryption followed by an embedding stage. The embedding method is a similarity based LSB bit-based method or a fuzzy logic-based LSB bit method. This results in Stego-MRIs. Extracting the message requires the stego-image and a stego-key. Then it undergoes decompression and decryption.

Ibaida et al. [52] proposed another technique where he uses a slight scaling/preprocessing and watermarking using LSB. Shekhawat et al. [53] proposed the same technique to that of Ibaida—except for the fact that he introduced AES ciphering. Duy et al. [54], Neela et al. [55], Kumar and Raj [56], used LSB based methods and the data is hidden in the T-P waves. Acharya et al. [57, 58] is one of the oldest works done in this domain. The proposed measures are DCT is the main tool used here.

Other techniques in spatial domain. Yang and Wang [59] derived two methods to hide the ECG data—(a) lossy and (b) reversible. Lossy is subdivided into two more parts—a high quality and a high capacity. The reversible method was employed to preserve the originality of the initial ECG signal. The couple has another paper [60], on how to provide a high hiding-capacity.

Chen et al. [61] proposed a watermarking scheme that is built on a quantization auto marking technique. To apply this we have three steps, namely—preprocessing and digital watermark embedding and extraction, embedding and detecting algorithms, Embedding quantization and detection quantization. The patient's confidential data has been treated as a watermark for medical data.

Shiu et al. [62] proposed a reversible error-correcting-coding strategy that will be able to transform ECG signals into a bitstream. Then they use a matrix and Hamming code. The other end has the matrix and Hamming code as keys.

Rekha et al. [63] proposed a technique not only applicable to ECG but also applicable to other medical images. They had used it on ECG Signals and the patient's face is used as a cover image. The related works of this paper are Cherukari et al. [64] who used BSN for the security purpose. The key is transmitted through a Fuzzy Commitment scheme for security purposes. The RC5 algorithm has been used. Poon et al. proved that features derived from the ECG signal can be used for securing the key distribution. Raazi et al. [65], proposed a scheme called BAR IN. There are three types of secret keys—(a) communication key, (b) administrative key, and (c) basic key are used. Miao et al. [66], proposed a method that uses AES, RC4 and fuzzy vault schemes. Sufi et al. [67] developed a chaos-based encryption. Wu and Shimamoto [68] proposed an energy efficient procedure for communication.

Soni et al. [69] embedding the data in the non-QRS part of the ECG. Wang et al. [70] proposed a reversible data hiding technique for ECG. Khandare et al. [71] declared a hybrid method in her paper- using DWT and her measures were highly similar to many of the above-mentioned methods. Usman and Usman [72] used edge detection and swapped Huffman tree coding. Augustyniak [73] proposed an impressive list of suitable measures for ECG steganography. Sathya and Kumar [74] used blowfish algorithm. Mathivanan [75] had another research paper that explained an XOR based steganography method.

3 Conclusion

As we have mentioned in the past that an ideal image steganography method should have three things—(a) imperceptibility should be high, (b) it must be resistant against statistical, and (c) non-structural steganalysis detection attacks. We strongly believe that the number of methods for ECG steganography in the spatial domain shall increase with time.

References

1. Petitcolas, F.A., et al.: Information hiding-a survey. Proc. IEEE **87**, 1062–1078 (1999)
2. Johnson, N.F., et al.: Exploring steganography: seeing the unseen. Computer **31**, 26–34 (1998)
3. Ibaida, A., Khalil, I.: Wavelet-based ECG steganography for protecting patient confidential information in point-of-care systems. IEEE Trans. Biomed. Eng. **60**(12), 3322–3330 (2013)
4. Jero, S.E., Ramu, P., Ramakrishnan, S.: Discrete wavelet transform and singular value decomposition based ECG steganography for secured patient information transmission. J. Med. Syst. **38**(10) (2014)
5. Jero, S.E., Ramu, P., Ramakrishnan, S.: Steganography in arrhythmic electrocardiogram signal. In Proceedings of the 37th Annual International Conference of the IEEE Engineering in Medicine and Biology Society (EMBC) (2015)
6. Liji, C.A., Indiradevi, K.P., Babu, K.K.A.: Integer-to-integer wavelet transform based ECG steganography for securing patient confidential information. Proc. Technol. **24**, 1039–1047 (2016)
7. Dilip, P.K., Raskar, V.B.: Hiding patient confidential information in ECG signal using DWT technique. Int. J. Adv. Res. Comput. Eng. Technol. (IJARCET) **4**(2), 533–538 (2015)
8. Tseng, K.-K., He, X., Kung, W.-M., Chen, S.-T., Liao, M., Huang, H.-N.: Wavelet-based watermarking and compression for ECG signals with verification evaluation. Sensors **14**(2), 3721–3736 (2014)
9. Jero, S.E., Ramu, P., Swaminathan, R.: Imperceptibility—robustness tradeoff studies for ECG steganography using continuous ant colony optimization. Expert Syst. Appl. **49**, 123–135 (2016)
10. Mishra, A., Agarwal, C., Sharma, A., Bedi, P.: Optimized gray-scale image watermarking using DWT–SVD and Firefly algorithm. Expert Syst. Appl. **41**(17), 7858–7867 (2014)
11. Ali, M., Ahn, C.W.: Comments on optimized gray-scale image watermarking using DWT–SVD and Firefly Algorithm. Expert Syst. Appl. **42**(5), 2392–2394 (2015)
12. Run, R.S., et al.: An improved SVD based watermarking technique for copyright protection. Expert Syst. Appl. **39**(1), 673–689 (2012)
13. Loukhaoukha, K., Chouinard, J.-Y., Taieb, M.H.: Optimal image watermarking algorithm based on LWT-SVD via multi-objective ant colony optimization. J. Inf. Hiding Multimedia Sign. Proces. **2**(4), 303–319 (2011)

14. Sankari, V., Nandhini, K.: Steganography technique to secure patient confidential information using ECG signal. In: Proceedings of the International Conference on Information Communication and Embedded Systems (ICICES) (2014)
15. Kaur, S., Singhal, R., Farooq, O., Ahuja, B.: Digital watermarking of ECG data for secure wireless communication. In: Proceedings of the 2010 International Conference on Recent Trends in Information, Telecommunication and Computing. IEEE, pp. 140–144 (2010)
16. Zheng, K., Qian, X.: Reversible data hiding for electrocardiogram signal based on wavelet transforms. In: Proceedings of the International Conference on Computational Intelligence and Security, CIS'08, vol. 1, pp. 295–299. IEEE (2008)
17. Golpira, H., Danyali, H.: Reversible blind watermarking for medical images based on wavelet histogram shifting. In: 2009 IEEE International Symposium on Signal Processing and Information Technology (ISSPIT), pp. 31–36. IEEE (2009)
18. Jero, S.E., Ramu, P.: A robust ECG steganography method. In: 2016 10th International Symposium on Medical Information and Communication Technology (ISMICT) (2016)
19. Raeiatibanadkooki, M., Quchani, S.R., KhalilZade, M., Bahaadinbeigy, K.: Compression and encryption of ECG signal using wavelet and chaotically Huffman codein telemedicine application. J. Med. Syst. **40**(3) (2016)
20. Mahmoud, S.S.: A generalised wavelet packet-based anonymisation approach for ECG security application. Secur. Commun. Netw. **9**(18), 6137–6147 (2016)
21. Devi, A., Shiva Kumar, K.B.: Novel audio steganography technique for ECG signals in point of care systems (NASTPOCS). In: Proceedings of the 2016 IEEE International Conference on Cloud Computing in Emerging Markets (CCEM) (2016)
22. Engin, M., Çıdam, O., Engin, E.Z.: Wavelet transformation based watermarking technique for human electrocardiogram (ECG). J. Med. Syst. **29**(6), 589–594 (2005)
23. Asha, N.S., Anithadevi, M.D., Shivakumar, K.B., Kurian, M.Z.: ECG signal steganography using wavelet transforms. Int. J. Adv. Netw. Appl. (IJANA) In: Proceedings of the 1st International Conference on Innovations in Computing & Networking (ICICN16), CSE, RRCE, pp. 355–359 (2016)
24. Meghani, D., Geetha, S.: ECG Steganography to secure patient data in an E-Healthcare System. In: Proceedings of the ACM Symposium on Women in Research 2016—WIR'16 (2016)
25. Sheeba, G., et al.: Secure crypto and ECG steganography based data communication for wireless body sensor network. Int. J. Innovative Res. Comput. Commun. Eng. **3**(3) (2015) (An ISO 3297:2007 Certified Organization)
26. Sivaranjani, B., Radha, N.: Securing patient's confidiential information using ECG steganography. In: Proceedings of the 2017 2nd International Conference on Communication and Electronics Systems (ICCES) (2017)
27. Marakarkandy, B., Tiwari, M.R.: Secure steganography, compression and transmission of ECG signal for protecting patient confidential information in point-of care systems. Int. J. Appl. Innovation Eng. Manag. (IJAIEM) **4**(7), 94–99 (2015)
28. Awasarmol, S. P., Ashtekar, S., Chintawar, A.: Securely data hiding and transmission in an ECG signal using DWT. In: Proceedings of the International Conference on Energy, Communication, Data Analytics and Soft Computing (ICECDS) (2017)
29. Tabash, F.K.: A proposed technique for hiding the information of patients on ECG signals. In: IMPACT-2013 (2013)
30. PremChandran, K., et al.: ECG steganography using integer wavelet transform. In: Proceedings of the 2015 International Conference on Computer Communication and Informatics (ICCCI) (2015)
31. Sahu, N., Peng, D., Sharif, H.: Unequal steganography with unequal error protection for wireless physiological signal transmission. In: IEEE International Conference Communication (2017)
32. Mathivanan, P., et al.: QR code-based highly secure ECG steganography. In: Proceedings of the International Conference on Intelligent Computing and Applications, pp. 171–178 (2018)
33. Mathivanan, P., et al.: QR code-based patient data protection in ECG steganography. Australa. Phys. Eng. Sci. Med. (2018)

34. Mathivanan, P., Balaji Ganesh, A.: QR code-based color image cryptography for the secured transmission of ECG signal. Multimedia Tools Appl. (2018)
35. Edward Jero, S., Ramu, P., Ramakrishnan, S.: ECG steganography using curvelet transform. Biomed. Signal Process. Control **22**, 161–169 (2015)
36. Degadwala, S., et al.: High capacity image steganography using curvelet transform and bit plane slicing. Int. J. Adv. Res. Comp. Sci. **4** (2013) ISSN No. 0976–5697
37. Patil, V., Patil, M.: Curvelet based ECG steganography for protection of data. Lecture Notes in Computational Vision and Biomechanics, pp. 238–248 (2018)
38. Jero, S.E., Ramu, P.: Curvelets-based ECG steganography for data security. Electron. Lett. **52**(4), 283–285 (2016)
39. Ibaida, A., Khalil, I., & Al-Shammary, D.: Embedding patient's confidential data in ECG signal for healthcare information systems. In: Proceedings of the 2010 Annual International Conference of the IEEE Engineering in Medicine and Biology (2010)
40. Vu Mai, Khalil, I., & Ibaida, A. (2013). Steganography-based access control to medical data hidden in electrocardiogram. In: Proceedings of the 2013 35th Annual International Conference of the IEEE Engineering in Medicine and Biology Society (EMBC)
41. Abuadbba, A., Khalil, I.: Walsh–hadamard-based 3-D steganography for protecting sensitive information in point-of-care. IEEE Trans. Biomed. Eng. **64**(9), 2186–2195 (2017)
42. Pandey, A., Singh, B., Saini, B.S., Sood, N.: A joint application of optimal threshold based discrete cosine transform and ASCII encoding for ECG data compression with its inherent encryption. Australas. Phys. Eng. Sci. Med. **39**(4), 833–855 (2016)
43. Wu, W., Liu, B., Zhang, W., & Chen, C. (2015). Reversible data hiding in ecg signals based on histogram shifting and thresholding. In: Proceedings of the 2015 2nd International Symposium on Future Information and Communication Technologies for Ubiquitous HealthCare (Ubi-HealthTech)
44. Vallathan, G., Devi, G. G., & Kannan, A. V. (2016). Enhanced data concealing technique to secure medical image in telemedicine applications. In: Proceedings of the 2016 International Conference on Wireless Communications, Signal Processing and Networking (WiSPNET)
45. Mohsin, A.H., et al.: Real-time medical systems based on human biometric steganography: a systematic review. J. Med. Syst. **42**(12) (2018)
46. Premarathne, U., et al.: Hybrid cryptographic access control for cloud-based EHR systems. IEEE Cloud Comput. **3**(4), 58–64 (2016)
47. Sivaranjani, D.N.R.B.: Securing patient's confidential information using ECG steganography, 540–544 (2017)
48. Liu, T.Y., Lin, K.J., Wu, H.C., et al.: ECG data encryption then compression using singular value decomposition. IEEE J. Biomed. Health Infor. **22**(3), 707–713 (2018)
49. Pandey, A., et al.: An integrated approach using chaotic map & sample value difference method for electrocardiogram steganography and OFDM based secured patient information transmission. J. Med. Syst. **41**(12) (2017)
50. Bethzia, S.H., et al.: An efficient steganography scheme based on chaos encryption. i-Manager's J. Digital Sign. Proces. **2**(2), 22–26 (2014)
51. Karakış, R., Güler, I., Çapraz, I., Bilir, E.: A novel fuzzy logic-based image steganography method to ensure medical data security. Comput. Biol. Med. **67**, 172–183 (2015)
52. Ibaida, A., Khalil, I., van Schyndel, R.: A low complexity high capacity ECG signal watermark for wearable sensor-net health monitoring system (2011)
53. Shekhawat, A.S., et al.: A study of ECG steganography for securing patient's confidential data based on wavelet transformation. Int. J. Comput. Appl. (0975–8887) **105**(12) (2014)
54. Duy, D., et al.: Adaptive steganography technique to secure patient confidential information using ECG signal. In: Proceedings of the 4th NAFOSTED Conference on Information and Computer Science (2017)
55. Neela, S., et al.: ECG steganography and hash function based privacy protection of patients medical information. Int. J. Trends Eng. Technol. **5**(2), 236–241 (2015)
56. Kumar, P.P., Raj, E.B.: An enhanced cryptography for ECG steganography to satisfy HIPAA privacy and security regulation for bio-medical datas

57. Acharya, R., Niranjan, U.C., et al.: Simultaneous storage of patient information with medical images in the frequency domain. Comput. Methods Programs Biomed. **76**(1), 13–19 (2004)
58. Acharya, R., et al.: Transmission and storage of medical images with patient information. Comput. Biol. Med. **33**(4), 303–310 (2003)
59. Yang, C.Y., Wang, W.F.: Effective electrocardiogram steganography based on coefficient alignment. J. Med. Syst. **40**(3) (2015)
60. Yang, C.Y., Wang, W.F.: High-capacity ECG steganography with smart offset coefficients. Smart Innovation, Systems and Technologies, pp. 129–136 (2017)
61. Chen, S.-T., et al.: Hiding patients confidential data in the ECG signal via a transform-domain quantization scheme. J. Med. Syst. **38**(6) (2014)
62. Shiu, H.-J., et al.: Preserving privacy of online digital physiological signals using blind and reversible steganography. Comput. Methods Programs Biomed. **151**, 159–170 (2017)
63. Rekha, R., et al.: Secure medical data transmission in body area sensor networks using dynamic biometrics and steganography. Bonfring Int. J. Softw. Eng. Soft Comput. **2**(1), 5 (2012)
64. Cherukuri, S., et al.: Biosec: A biometric based approach for securing communication in wireless networks of biosensors implanted in the human body. In: Proceedings of the IEEE International Conference Parallel Processing Workshops, pp. 432–439 (2003)
65. Raazi, S.M.K.U.R., et al.: Bar in a distributed key management approach for wireless body area networks. In: IEEE Conference (2009)
66. Miao, F., et al.: A novel biometrics based security solution for body sensor networks. In: IEEE Conference (2009)
67. Sufi, F., et al.: A chaos-based encryption technique to protect ECG packets for time critical telecardiology applications. Secur. Commun. Netw. (2010)
68. Wu, J., Shimamoto, S.: An energy efficient data secrecy scheme for wireless body area sensor networks. Comput. Sci. Eng. Int. J. (CSEIJ) **1**(2) (2011)
69. Soni, N., et al.: Robust steganography in Non-QRS regions of 2D ECG for securing patients' confidential information in E-healthcare paradigm. In: Medical Data Security for Bioengineers (2019)
70. Wang, H., Zhang, W., Yu, N.: Protecting patient confidential information based on ECG reversible data hiding. Multimed. Tools Appl. **75**, 13733–13747 (2016)
71. Khandare, M., et al.: An approach of ECG steganography to secure the patient's confidential information. Int. Res. J. Eng. Technol. (IRJET) **3**(03), 1867–1871 (2016)
72. Usman, M. A., Usman, M. R.: Using image steganography for providing enhanced medical data security. In: Proceedings of the 15th IEEE Annual Consumer Communications & Networking Conference (CCNC) (2018)
73. Augustyniak, P.: Analysis of ECG bandwidth gap as a possible carrier for supplementary digital data. In: 2012 Computing in Cardiology, Krakow, Poland, 9–12 Sept 2012
74. Sathya, D., Kumar, P.G.: Secured remote health monitoring system. Healthc. Technol. Lett. **4**(6), 228–232 (2017)
75. Mathivanan, P., et al.: Color image steganography using XOR multi-bit embedding process. In Proceedings of the International Conference on Energy, Communication, Data Analytics and Soft Computing (ICECDS) (2017). Author, F.: Article title. Journal **2**(5), 99–110 (2016)

Spectral–Spatial Active Learning in Hyperspectral Image Classification Using Threshold-Free Attribute Profile

Kaushal Bhardwaj, Arundhati Das, and Swarnajyoti Patra

Abstract Limited availability of training samples makes the classification of hyperspectral images a challenging task. A small number of informative training samples may provide more accurate results than randomly selected samples. For detecting informative training samples, active learning methods are suggested in the literature. These methods when applied on spectral values alone, are less effective than in cases when both spectral and spatial information are considered. Integration of spectral and spatial information in hyperspectral images is recently accomplished by constructing a threshold-free attribute profile which is state-of-the-art. In this paper, we present an overview of the state-of-the-art active learning techniques and propose a spectral–spatial active learning model based on threshold-free attribute profiles. To this end, first a threshold-free extended attribute profile is constructed on reduced dimension of the hyperspectral image. Then, the informative training samples are iteratively selected based on uncertainty, diversity, cluster assumption or combination of these criteria. Experiments are conducted on two benchmark real hyperspectral data sets where the active learning methods based on spectral values alone are compared to the proposed active learning model. The results reveal that the proposed active learning model can identify more informative training samples.

K. Bhardwaj
Department of CSE, Indian Institute of Information Technology Senapati, Manipur Imphal 795002, India
e-mail: kaushal@iiitmanipur.ac.in

A. Das · S. Patra (✉)
Department of CSE, Tezpur University, Tezpur 784028, Assam, India
e-mail: swpatra@tezu.ernet.in

A. Das
e-mail: arundha@tezu.ernet.in

D. Bhattacharjee et al. (eds.), *Proceedings of International Conference on Frontiers in Computing and Systems*, Advances in Intelligent Systems and Computing 1255,
https://doi.org/10.1007/978-981-15-7834-2_5

1 Introduction

Hyperspectral images (HSIs) are acquired in a large number of contiguous spectral bands with a small spectral interval [4]. The rich spectral information of HSI enables it to distinguish minute variations in the materials on the earth's surface. Because of this ability, it is widely used for several remote sensing applications including mineralogy, agriculture, forestry and land-use/land-cover mapping. The pixel-wise classification of HSI has to deal with limited availability of training samples since labeling a pixel is costly in terms of time and effort. In the literature, two different approaches exist to deal with this challenge. One is semi-supervised learning which employs both labeled and unlabeled samples for improving its decision boundaries [4]. Another approach is active learning (AL) which begins with a few labeled samples and iteratively selects the most informative samples to include in the training set [10, 11].

The part of AL procedure that is of utmost importance is the definition of a query function [9, 15]. The query function uses one or more criteria to judge how informative each sample is to discriminate among classes. In the literature, several AL methods exist with different query functions based on uncertainty, diversity, cluster assumption, query-by-bagging, *etc.* [11, 13, 15]. The query function based on uncertainty criteria aims at discovering the samples whose class assignment is most ambiguous [9]. While the query function based on diversity criteria aims at avoiding the selection of redundant training samples by analyzing the dissimilarity among the unlabeled samples and selecting the set of most dissimilar samples [3, 15], the cluster assumption criterion aims at selecting the samples from low density regions in the feature space assuming that the low density regions are near the decision boundary [11]. In the literature, the combination of more than one criteria is suggested for better results [9, 15]. However, considering only the spectral values of HSI may not lead to the best results. To achieve better such results, spatial information should also be considered along with the spectral information [1, 7, 14]. This is because the neighboring pixels provide correlated information. In the literature, some methods are proposed to integrate spectral and spatial information among which the morphological attribute profiles are a popular choice for this purpose [1, 7, 8]. The construction of attribute profiles normally requires some threshold values, however, recently a new approach is presented in the literature to construct attribute profiles that does not require any threshold values [2]. This method automatically integrates significant amount of spectral and spatial information in HSI.

In this paper, we recall the state-of-the-art AL methods and propose a spectral–spatial AL model where the threshold-free extended attribute profiles (TEAPs) are used for integrating spectral and spatial information of the HSI. In the proposed model, first the spectral and spatial information of HSI are integrated by constructing threshold-free extended attribute profiles. Then a combination of query functions based on uncertainty, diversity and cluster assumption are used to identify informative training samples. Experiments are conducted on two benchmark real hyperspectral data sets. In the experiments, the spectral-value-based AL model is compared to the extended attribute profile (EAP) (created using manual thresholds)-based spectral–

spatial AL model and the proposed TEAP-based spectral–spatial AL model. In the experimental results, the proposed model is found to be more effective than the rest of the models.

2 Proposed TEAP-Based AL Model

Figure 1 presents the proposed TEAP-based AL model. It has two phases. In the first phase, the dimension of HSI is reduced using principal component analysis (PCA) and a TEAP is constructed in its reduced dimension to integrate the spectral and spatial information. In the next phase, the informative training samples (pixels of HSI) are selected based on multiple criteria, for example, uncertainty, diversity and cluster assumption. In the following subsections, the two phases of the proposed model are described.

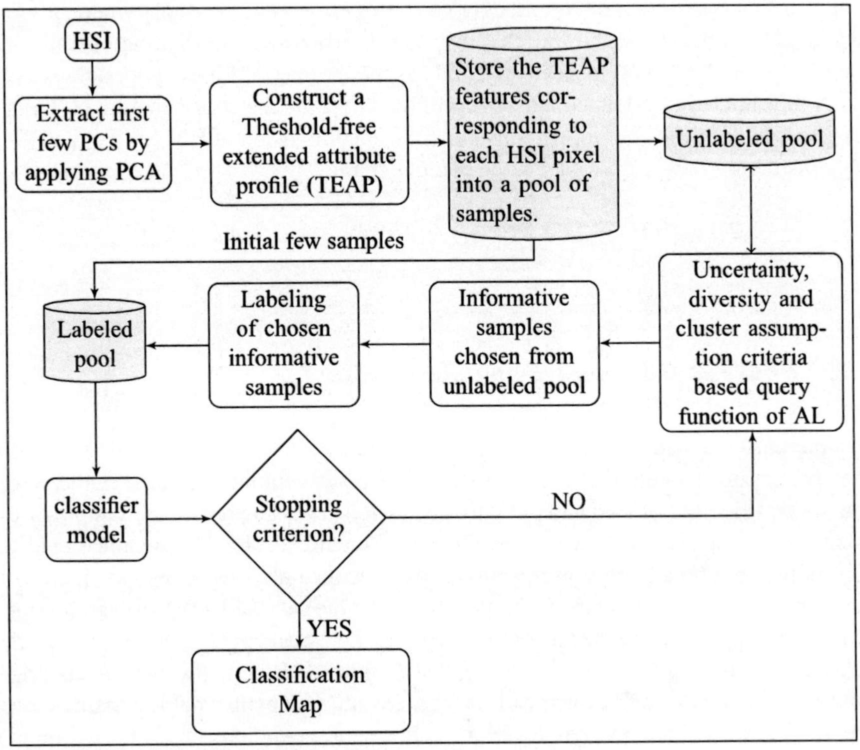

Fig. 1 Proposed TEAP-based spectral–spatial active learning framework

2.1 Phase 1: Construction of Threshold-Free Extended Attribute Profile

In order to integrate spectral and spatial information of an HSI, first its dimension is reduced using principal component analysis (PCA). Then for each component image in the reduced dimension, a threshold-free attribute profile (TAP) is constructed and concatenated to form a threshold-free extended attribute profile (TEAP). A TAP is the concatenation of original component image with its threshold-free attribute filtering results [2]. These filters create a component tree for a given image that represents nested components in a tree hierarchy and prune all the branches of the created tree from a position where sudden and significant differences in attribute values are observed. Such filtering operation using max-tree filters bright objects and is called threshold-free attribute thinning operation ($\gamma^i(I)$), whereas considering min-tree it is called threshold-free attribute thickening operation ($\phi^i(I)$). Multiple filtering operations can be performed on its previous results to obtain multi-scale filtering results, which can be concatenated to form a threshold-free attribute profile [2]. $TAP(I) = \{\phi^1(I), \phi^2(I), ..., \phi^t(I),\ I,\ \gamma^1(I), \gamma^2(I), ..., \gamma^t(I)\}$ where $\gamma^i(I)$ and $\phi^i(I)$ are the threshold-free thinning and threshold-free thickening operations on $\gamma^{i-1}(I)$ and $\phi^{i-1}(I)$, respectively, and t is the number of filtering operations.

A threshold-free extended attribute profile (TEAP) is the concatenation of TAPs constructed for each component image in reduced dimension of HSI [2]. For an HSI H considering ℓ PCs, a TEAP can be computed as

$$TEAP(H) = \{TAP(PC_1), TAP(PC_2), ..., TAP(PC_\ell)\}$$

.

2.2 Phase 2: Informative Sample Selection

In this phase, a batch of most informative training samples is selected from the unlabeled pool having TEAP features for HSI pixels. Initially, a few samples are randomly selected, labeled and put into the labeled pool. The rest of the samples are kept in the unlabeled pool. After that, the query function designed using uncertainty, diversity or cluster assumption criteria is applied on the unlabeled samples to identify a batch of the most informative samples. Let U be the set of all available samples in the unlabeled pool, and L be the set of samples in the labeled pool. In each iteration of the AL procedure, say, h unlabeled samples are selected from U for manual labeling and are appended to L. This process is continued until we obtain a stable classification result. The query function employed in each iteration exploits one or more criteria based on *uncertainty*, *diversity*, *cluster assumption* or their combinations. Next, we discuss these criteria and their possible combinations.

Uncertainty criteria: Several uncertainty criteria exist in the literature [5, 9]. These criteria aim at identifying the samples whose class assignment is most uncertain. Basically in binary classification, the sample nearest to the decision boundary is considered as most uncertain. However, in a multi-class scenario, approaches are not straightforward. In this work, we discuss state-of-the-art uncertainty criteria based on support vector machine classifier in a multi-class scenario. Entropy-based query bagging (EQB) is one such uncertainty criterion that decides based on maximum disagreement in the decision of a group of classifiers for the sample [3]. In case of margin sampling (MS), the AL procedure tries to identify the samples nearest to the separating hyperplane. The samples having the lowest classification certainty (CC) are selected for labeling. In a multi-class scenario, for c different classes, the distances from c hyperplanes are recorded and the minimum distance is considered for CC. This can be formulated for a sample s as $CC(s) = \min_{i=1,2...c} \{|f_i(s)|\}$. The multi-class level uncertainty (MCLU) criterion aims at identifying the sample that has the maximum difference among the distances from the farthest separating hyperplanes. For this, the distance from each separating hyperplane is recorded, the two largest distances are noted and the difference of the distances is recorded as CC. For a sample s, MCLU can be formulated as $r_{mx1} = \arg\max_{i=1,2...c} f_i(s)$, $CC(s) = r_{mx1} - \arg\max_{\substack{j=1,2...c \\ j \neq r_{mx1}}} f_j(s)$

Diversity: In the literature, several diversity criteria exist based on angle, closest support vector and clustering [3, 15]. Here, we present the state-of-the-art diversity criteria which are cluster-based diversity (CBD) and angle-based diversity (ABD). In CBD, the unlabeled samples are clustered into h clusters and one sample is selected from each group. In case of enhanced CBD (ECBD), the clustering is done in *kernel* space. In [10], the distance between their nearest samples is maximized to select diverse samples. In case of ABD, cosine angles between unlabeled samples are computed and the h samples with the maximum angle are selected for labeling.

Cluster assumption: In the literature, some cluster assumption-based criteria exist which try to select the samples from the low density region in feature space [10–12, 15]. In [11], an AL method based on cluster assumption with histogram thresholding (CAHT) is presented. In [10], the density of the sample is computed by calculating the average distance of their K-nearest neighbors.

For the combination of these criteria, v ($v > h$) samples are selected using uncertainty criteria, and h samples are selected out of the v samples based on diversity or cluster assumption or both. This approach is widely accepted [9, 15]. In this work, we present two single-criterion, two double-criteria methods based on spectral values alone and with spatial information. We also present a triple-criteria method as shown in [10] that is based on spectral–spatial information.

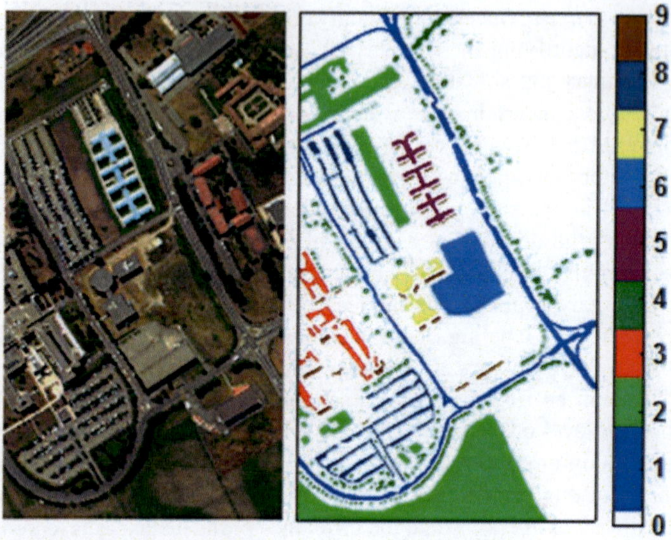

Fig. 2 Hyperspectral University of Pavia image and its reference map

3 Experimental Results

3.1 Data Sets and Experimental Setup

In order to assess the proposed TEAP-based spectral–spatial AL model, two benchmark real hyperspectral data sets[1] are considered in the experiments. The first data set is an urban scene of University of Pavia having 610 × 340 size, 103 bands available for processing with 1.3 m resolution. The data set has 42776 labeled samples and 9 thematic classes. A false color image and available reference samples for University of Pavia data set are shown in Fig. 2. The second data set is acquired at Kennedy Space Center (KSC), USA. It is of 512 × 614 size with 176 bands and 13 m resolution. It has 13 thematic classes and 5211 labeled samples. A false color image and available reference samples for KSC data set are shown in Fig. 3.

In the experimental analysis, the proposed TEAP-based spectral–spatial AL model is compared to the classic AL models that are based on spectral values alone as well as to the EAP-based spectral–spatial AL model. The methods based on spectral values alone are EQB, CAHT, MS-ABD and MCLU-ECBD whereas the EAP-based spectral–spatial AL methods are EAP-EQB, EAP-CAHT, EAP-MS-ABD, EAP-MCLU-ECBD and EAP-GA-Multic. The proposed TEAP-based methods are referred to as TEAP-EQB, TEAP-CAHT, TEAP-MS-ABD, TEAP-MCLU-ECBD and TEAP-GA-Multic. Initially, 3 labeled samples are randomly selected from each

[1] Available online: http://www.ehu.eus/ccwintco/index.php?title=Hyperspectral_Remote_Sensing_Scenes.

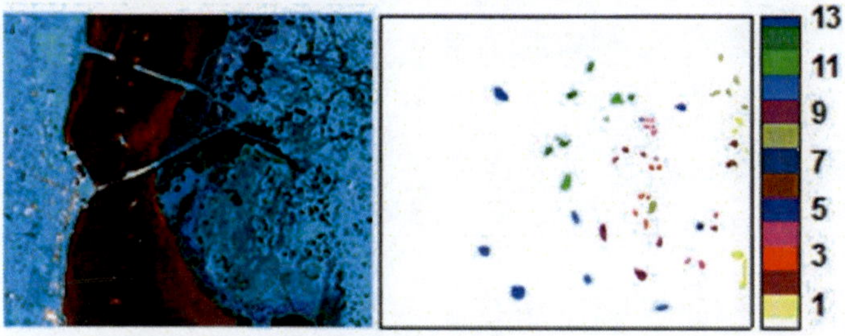

Fig. 3 Hyperspectral KSC image and its reference map

class and kept in a labeled pool. The rest of the samples are kept in the unlabeled pool. In MS-ABD, MCLU-ECBD, EAP-MS-ABD, EAP-MCLU-ECBD, TEAP-MS-ABD and TEAP-MCLU-ECBD, the first v samples are selected using the uncertainty criterion and the final h samples are selected out of v samples using the diversity criterion. The batch size h is kept as 20 in the experiments. The value of v is kept as $3 \times h$ in all the experiments. In case of EAP-GA-Multic and TEAP-GA-Multic, the number of clusters C is kept as 500. Here also, the first v samples are selected using the uncertainty criterion, and the final h samples are selected out of v samples using the diversity and cluster assumption criteria by exploiting genetic algorithms (GAs). The parameters of GAs are kept the same as presented in [10]. For the construction of EAP and TEAP, the dimension of HSI is reduced using PCA, and the first 5 PCs corresponding to maximum variance are considered. The EAP is constructed using the *area* attribute considering the threshold values $\{100, 500, 1000\}$ as used in [7]. The size of the EAP constructed for the HSI considering 5 PCs and 3 threshold values is 35 (7 for each PC). The TEAP is also constructed by considering 3 threshold-free filtering operations leading to the same size (i.e., 35) using *area* attribute.

All the methods used in experiments are implemented using MATLAB (R2015a). For classification purpose, a one-Vs-all SVM classifier with RBF kernel is used and is implemented with the help of LIBSVM library [6]. A fivefold cross-validation with grid search is carried out to obtain the parameters of SVM. The experimental results are reported in terms of the average of the class-wise accuracy, overall accuracy ($\overline{OA}$), kappa coefficient (kappa) and the standard deviation (std) for ten runs considering different randomly selected initial training samples.

3.2 Experimental Results

The first experiment is conducted on the University of Pavia data set. In the experimental results, as reported in Table 1, the AL methods based on EAP and TEAP outperform the methods based on spectral values alone. This shows the importance of integrating spectral and spatial information. One can also observe that the TEAP-based AL methods are performing better than the EAP-based AL methods. The meth-

Table 1 Classification accuracies obtained after ten runs of the experiment on University of Pavia data set. The best values are in boldface

Class	EQB	CAHT	MS-ABD	MCLU-ECBD	EAP EQB	EAP CAHT	EAP MS-ABD	EAP MCLU-ECBD	EAP GA-Multic	TEAP EQB	TEAP CAHT	TEAP MS-ABD	TEAP MCLU-ECBD	TEAP GA-Multic
1	91.436	89.036	89.469	91.081	92.752	94.132	95.728	96.329	95.802	96.203	98.631	98.043	99.526	**99.682**
2	71.397	96.936	97.240	97.162	74.793	97.783	98.508	98.358	98.020	90.123	99.533	**99.791**	99.703	99.755
3	39.967	70.276	67.132	69.238	82.925	80.300	84.912	90.176	86.289	95.112	99.028	99.538	**99.667**	99.552
4	86.172	92.249	92.288	92.428	85.940	97.161	97.226	97.657	**97.748**	96.191	94.856	97.092	97.461	97.438
5	98.684	97.755	98.401	98.379	99.204	99.628	99.294	99.093	99.613	99.859	99.911	99.874	99.933	**99.970**
6	82.913	86.874	86.107	86.244	84.337	87.634	90.616	89.960	90.537	99.871	99.771	99.928	99.954	**99.964**
7	45.256	71.774	66.376	73.677	83.429	79.977	82.368	86.714	83.759	99.895	99.677	99.827	**99.925**	99.895
8	94.864	85.970	86.896	86.279	93.110	92.357	94.495	93.558	94.006	99.443	98.800	99.033	99.424	**99.462**
9	97.276	99.155	98.828	99.250	96.262	99.820	96.917	99.440	99.736	**99.947**	99.916	99.884	99.937	**99.947**
$\overline{OA}$	78.012	91.233	91.116	91.644	82.985	94.204	95.533	95.876	95.500	94.521	99.023	99.271	99.538	**99.583**
kappa	0.7201	0.8834	0.8817	0.8888	0.7828	0.9230	0.9406	0.9452	0.9403	0.9291	0.9870	0.9903	0.9939	**0.9945**
std	3.3212	0.1760	0.4065	0.4138	3.9888	0.7577	0.4374	0.3910	0.3419	2.1624	0.1287	0.7945	0.1880	**0.0372**

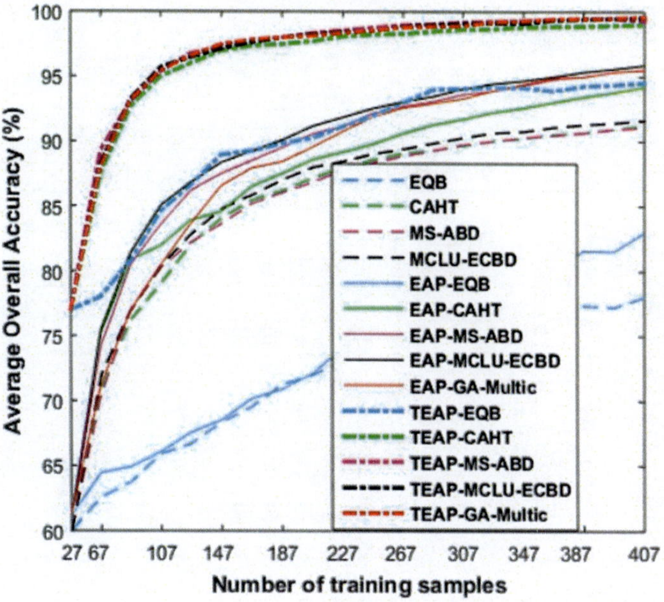

(a)

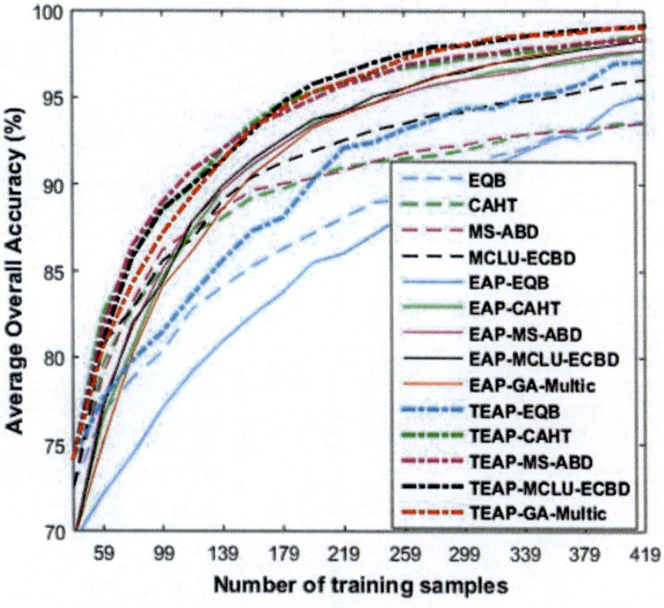

(b)

Fig. 4 $\overline{OA}$ against number of training samples obtained by AL methods based on spectral values alone (dashed line), based on EAP (solid line) and the proposed TEAP-based model (dash-dotted) on **a** University of Pavia and **b** KSC data sets

Table 2 Classification accuracies obtained after ten runs of the experiment on KSC data set. The best values are in boldface

Class	EQB	CAHT	MS-ABD	MCLU-ECBD	EAP EQB	EAP CAHT	EAP MS-ABD	EAP MCLU-ECBD	EAP GA-Multic	TEAP EQB	TEAP CAHT	TEAP MS-ABD	TEAP MCLU-ECBD	TEAP GA-Multic
1	94.599	98.108	97.845	97.937	98.962	99.185	99.606	99.685	99.816	99.198	99.671	99.829	**99.934**	99.829
2	87.942	89.136	93.498	89.835	86.296	96.502	95.679	98.025	98.025	93.251	96.379	97.778	**98.601**	98.354
3	58.516	95.234	95.039	95.195	95.156	97.773	98.320	98.672	98.633	98.125	98.398	98.477	98.672	**98.711**
4	91.706	55.595	56.865	85.595	89.167	94.802	96.151	96.944	**97.302**	89.643	95.119	96.071	96.349	96.429
5	78.634	73.851	72.050	86.894	92.981	91.739	92.733	94.410	93.913	92.671	93.975	95.093	**95.404**	94.907
6	84.367	71.354	75.852	81.354	91.528	92.576	93.231	95.284	95.153	99.520	99.170	99.694	**99.956**	99.476
7	90.476	90.857	81.619	91.524	97.143	94.190	95.619	97.143	**97.333**	90.571	89.333	88.476	90.476	89.810
8	94.548	94.849	95.545	97.494	92.297	95.638	96.404	96.984	97.146	98.794	99.118	99.698	**99.861**	99.791
9	98.962	99.173	98.962	99.404	92.788	98.865	98.846	95.712	99.154	92.577	99.038	99.615	99.654	**99.865**
10	98.292	97.995	98.614	98.317	95.446	96.089	95.644	99.109	**99.134**	94.851	95.866	93.812	98.317	97.946
11	99.403	99.212	99.189	98.807	97.566	99.308	99.260	99.642	99.618	99.451	99.356	99.833	**99.928**	99.857
12	99.304	96.859	95.408	97.972	94.076	98.767	97.137	99.105	98.767	98.867	99.662	97.455	99.801	**99.861**
13	98.857	99.914	99.310	99.126	98.964	99.946	99.773	99.806	99.924	99.730	99.924	99.806	**100.000**	99.978
$\overline{OA}$	93.665	93.539	93.546	96.053	95.139	97.697	97.736	98.315	98.678	97.097	98.444	98.373	**99.146**	99.058
kappa	0.9295	0.9280	0.9281	0.9560	0.9459	0.9743	0.9748	0.9812	0.9853	0.9677	0.9827	0.9819	**0.9905**	0.9895
std	0.8602	0.4038	0.5462	0.2784	1.1244	0.1200	0.3508	0.5818	0.2307	0.6677	0.1212	0.3212	**0.1094**	0.1677

ods in the proposed model namely, TEAP-CAHT, TEAP-MS-ABD, TEAP-MCLU-ECBD and TEAP-GA-Multic are able to achieve $\overline{OA}$ more than 99%, whereas none of the methods in classic and EAP-based models could achieve $\overline{OA}$ of 99%. This can also be observed from Fig. 4a where the $\overline{OA}$ for the proposed model is always above the methods in the classic and EAP-based models. This shows that the proposed AL model is robust in identifying the informative pixels. The second experiment is conducted on the KSC data set and the results are reported in Table 2. The results show that the AL methods working in the classic model depending on spectral values alone lead to poor results than the models based on spectral–spatial information. One can observe from the table that the EAP-GA-Multic has outperformed all the state-of-the-art methods in the classic and EAP-based methods. The performance of the proposed model is also visible from the plotted graph for all the methods in Fig. 4b. Therefore, the experiments confirm that the proposed model is robust for the classification of HSI with limited labeled samples.

4 Conclusion

This paper presents an overview of the state-of-the-art active learning techniques and proposes a spectral–spatial active learning model for the classification of HSI. In the proposed model, first the spectral and spatial contents of HSI are integrated by constructing a threshold-free extended attribute profile. Next, the AL methods that select informative pixels based on uncertainty, diversity and cluster assumption criteria are used for labeling. The proposed TEAP-based AL model is compared to the classic AL model that uses only spectral values, and the EAP-based AL model considering two different benchmark data sets. The experimental results reveal that the AL methods based on the spectral–spatial model are able to identify more informative training samples than those considering spectral values alone. The experiments also confirmed that the proposed threshold-free extended attribute profile-based model is more robust than the state-of-the-art ones.

References

1. Bhardwaj, K., Patra, S.: An unsupervised technique for optimal feature selection in attribute profiles for spectral-spatial classification of hyperspectral images. ISPRS J. Photogramm. Remote Sens. **138**, 139–150 (2018)
2. Bhardwaj, K., Patra, S., Bruzzone, L.: Threshold-free attribute profile for classification of hyperspectral images. IEEE Trans. Geosci. Remote Sens. **57**(10), 7731–7742 (2019)
3. Brinker, K.: Incorporating diversity in active learning with support vector machines. In: Proceedings of the 20th International Conference on Machine Learning (ICML-03), pp. 59–66 (2003)
4. Bruzzone, L., Chi, M., Marconcini, M.: A novel transductive SVM for semisupervised classification of remote-sensing images. IEEE Trans. Geosci. Remote Sens. **44**(11), 3363–3373 (2006)

5. Campbell, C., Cristianini, N., Smola, A., et al.: Query learning with large margin classifiers. In: ICML, pp. 111–118 (2000)
6. Chang, C.C., Lin, C.J.: LIBSVM: a library for support vector machines. ACM Trans. Intell. Syst. Technol. (TIST) **2**(3), 27 (2011)
7. Dalla Mura, M., Atli Benediktsson, J., Waske, B., Bruzzone, L.: Extended profiles with morphological attribute filters for the analysis of hyperspectral data. Int. J. Remote Sens. **31**(22), 5975–5991 (2010)
8. Das, A., Bhardwaj, K., Patra, S.: Morphological complexity profile for the analysis of hyperspectral images. In: 2018 4th International Conference on Recent Advances in Information Technology (RAIT), pp. 1–6. IEEE (2018)
9. Demir, B., Persello, C., Bruzzone, L.: Batch-mode active-learning methods for the interactive classification of remote sensing images. IEEE Trans. Geosci. Remote Sens. **49**(3), 1014–1031 (2011)
10. Patra, S., Bhardwaj, K., Bruzzone, L.: A spectral-spatial multicriteria active learning technique for hyperspectral image classification. IEEE J. Selected Topics Appl. Earth Observ. Remote Sens. **10**(12), 5213–5227 (2017)
11. Patra, S., Bruzzone, L.: A batch-mode active learning technique based on multiple uncertainty for SVM classifier. IEEE Geosci. Remote Sens. Lett. **9**(3), 497–501 (2012)
12. Patra, S., Bruzzone, L.: A cluster-assumption based batch mode active learning technique. Pattern Recognit. Lett. **33**(9), 1042–1048 (2012)
13. Patra, S., Bruzzone, L.: A novel SOM-SVM-based active learning technique for remote sensing image classification. IEEE Trans. Geosci. Remote Sens. **52**(11), 6899–6910 (2014)
14. Rajbanshi, S., Bhardwaj, K., Patra, S.: Spectral–spatial active learning techniques for hyperspectral image classification. In: Computational Intelligence in Data Mining, pp. 339–350. Springer (2020)
15. Tuia, D., Ratle, F., Pacifici, F., Kanevski, M.F., Emery, W.J.: Active learning methods for remote sensing image classification. IEEE Trans. Geosci. Remote Sens. **47**(7), 2218–2232 (2009)

A Fuzzy Logic-Based Crop Recommendation System

Gouravmoy Banerjee, Uditendu Sarkar, and Indrajit Ghosh

Abstract Soil, geographical and meteorological parameters have major impacts on sustained crop production. Most of the rural farmers have no adequate knowledge about the effects of these parameters on crop production. The rural farmers generally rely on their traditional knowledge to select a crop which often leads to huge economic loss. A scientific system considering these site-specific parameters along with the traditional knowledge of the farmers may be an effective solution. This paper suggests a fuzzy logic-based crop recommendation system to assist rural farmers. The proposed model has been designed to deal with eight major crops grown in the state of West Bengal. Separate fuzzy rule bases were created for each crop to achieve faster parallel processing. The performance of the model has been validated by a diverse dataset and achieved an accuracy of about 92%.

Keywords Crop recommendation system · Fuzzy logic · Fuzzy system in agriculture

G. Banerjee
Department of Computer Science, Ananda Chandra College, Jalpaiguri, West Bengal 735101, India
e-mail: gouravmoy@gmail.com

U. Sarkar
National Informatics Centre, Ministry of Electronics & Information Technology, Government of India, Jalpaiguri, West Bengal 735101, India
e-mail: uditsarkar@yahoo.com

I. Ghosh (✉)
Department of Computer Science, Ananda Chandra College, Jalpaiguri, West Bengal 735101, India
e-mail: ighosh2002@gmail.com

D. Bhattacharjee et al. (eds.), *Proceedings of International Conference on Frontiers in Computing and Systems*, Advances in Intelligent Systems and Computing 1255,
https://doi.org/10.1007/978-981-15-7834-2_6

1 Introduction

The selection of suitable crops for agriculture is fundamentally dependent on several factors. Some of them include the nutrients of the soil, average rainfall and terrain. A crop that is not compatible with these features will not give a profitable return. Traditionally, rural farmers select crops for cultivation based on their experiences without having sufficient knowledge about the soil and other influencing factors.

Use of modern scientific tools and technologies can enable a farmer to make more accurate and site-specific decisions for the selection of crops. Several kinds of researches have been carried out in this direction. In 2016, Arooj et al., compared four different data mining-models for classifying soils based on different parameters such as pH, texture, electrical conductivity [1]. In 2017, some researchers in Bangladesh used multiple linear regression and k-nearest neighbour models for prediction of yield [2]. An expert system based approach was proposed by Nevo et al., in determining the crop suitability based on different parameters [3]. Some researchers have used Geographic Information Systems (GIS) and other parameters for suggesting favourable crops for cultivation [4, 5]. Based on Agricultural Information Systems (AIS), Laliwala et al., has devised a rule-based recommendation system that covers multiple aspects of farming [6]. A web-based recommendation system for primary crop and secondary crop selection with fertilizer recommendation was developed by Shinde et al. [7]. Balakrishnan et al., employed ensemble machine learning model to recommend the most suitable crop based on crop production data and meteorological parameters [8]. The system exhibited an average accuracy of about 90%.

Fuzzy logic, conceptualized by Lofti Zadeh is a powerful tool in decision making, especially where the features are linguistic in nature. Fuzzy logic exploits the notion of vagueness for the qualitative representation of the variables. In fuzzy logic, inference and decisions are based on some rules containing linguistic variables or hedges [9]. This feature permits many real-world problems to be modelled using fuzzy logic. In this paper, a fuzzy logic based crop recommendation system has been proposed using the site-specific chemical parameters of the soil, rainfall and nature of the terrain. Hopefully, this would assist the farmers to select the crops which will be economically viable and profitable. Fuzzy logic has been applied in different sectors [10–12], but no such work has been reported on crop recommendation system for the state of West Bengal.

2 Materials and Methods

West Bengal is one of the largest crops producing states in India and six districts in the state have high agricultural productivity rate with high Simpson's Index (0.804) [13–15]. For such a high productivity rate and diverse cropping pattern, the state of West Bengal is one of the most promising area for application of such a system.

2.1 Data Collection

The soil health card scheme of the Government of India primarily aims to improve the quality of the soil by optimal fertilizer application, based on soil parameters [16, 17]. The soil health card for a sample contains information regarding 12 different parameters of soil like pH value, EC (Electrical Conductivity), OC (Organic Carbon content), Nitrogen (N), Phosphorus (P), Potassium (K), Sulphur (S), Zinc (Zn), Boron (B), Iron (Fe), Manganese (Mn) and Copper (Cu). The soil health card also depicts different crops cultivated by the farmers and latitude and longitude of the site.

Terrain classification has been done by the method suggested by Darajat et al. [18]. NASA's Shuttle Radar Topography Mission (SRTM3) WEBGIS database which contains elevation data with respect to latitude and longitude was used [19].

Chattopadhyay in her doctoral thesis divided West Bengal into four different regions based on Mean Annual Rainfall [20]. The data related to rainfall were obtained from the Indian meteorological department's Customized Rainfall Information System (CRIS) [21]. The average rainfalls of the districts for the last five years have been considered as the mean annual rainfall for that district. The different rainfall regions were grouped into four categories as suggested by Chattopadhyay.

Soil health cards of 370 randomly selected samples distributed over fourteen districts of West Bengal were considered. An output parameter Cultivation Index (CI) has been incorporated to capture the cropping trends of the local farming. CI is the percentage of selection of a particular crop for samples having similar linguistic ratings for every input parameter. The CI was calculated as follows:

$$CI = O/NP \times 100 \quad (1)$$

where NP is the number of cases having similar linguistic ratings for each input parameter. O represents the number of occurrences of individual crop selected out of NP. The data of all soil parameters, elevation, rainfall, CI and their ratings in terms of linguistic parameters are presented in Table 1.

Out of the twelve soil parameters, EC was found to be irrelevant for the present work as almost 99% of the samples had the same value of EC. It was further found that 370 samples covered 27 variety of crops. Out of these 27 crops, 8 major crops having more than 25 instances, were used for the preparation of the final dataset. After the reduction of the dataset to 8 different crop varieties (Paddy, Jute, Potato, Tobacco, Wheat, Sesamum, Mustard and Green gram) the total number of instances were 352. Finally, 11 soil parameters, elevation, rainfall along with their ratings for 8 crops were used as input for constructing the proposed fuzzy system.

Table 1 Parameters and their rating

Attribute	Rating	Code	Attribute	Rating	Code
Elevation	Flat	FLT	pH	Strongly Acidic	SACD
	Undulating	UND		Highly Acidic	HACD
	Flat to Undulating	FUN		Moderately Acidic	MACD
	Undulating to Hilly	UNH		Slightly Acidic	SACD
	Hilly	HIL		Neutral	NTRL
Mean annual rainfall	Sub Humid	SBH		Moderately Alkaline	MALK
	Semi Humid	SMH	OC, N, P, K, Cultivation Index*	Very Low	VRL
	Humid	HUM		Low*	LOW
	Super Humid	SUH		Medium*	MED
S, Zn, B, Fe, Mn, Cu	Deficient	DEF		High*	HGH
	Sufficient	SUF		Very High	VRH

*Applicable only for Cultivation Index which is rated as Low, Medium and High

2.2 *Architecture of the Proposed Fuzzy System*

A fuzzy-based system typically has three phases of operation; a fuzzification phase where the input crisp values are converted to fuzzified values, an inference phase where a fuzzy inference system determines which rules are to be fired and a defuzzification phase where the fuzzy output is reconverted into crisp output. The inference rules are represented in the form of "IF (I) Then (O)" where both I and O are linguistic values. Fuzzy logic also provides AND and OR logical operators to combine more than one variable for decision-making. Such a rule representation strategy helps to model the system with less complexity.

2.2.1 Fuzzification of Input Parameters

In fuzzy logic several membership functions have been proposed for fuzzification which includes Triangular, Trapezoidal, Gaussian, etc. and the selection of the membership function is the sole decision of the researcher [22]. On analysing the dataset, it was observed that each of the inputs is expressed in terms of a linguistic variable (Table 1). To remove disparity, the entire range of the crisp value of each input parameter was mapped into a range from 0 to 100 by normalization. A tolerance τ (=20% of the range) at the lower (L) and upper (U) boundary of each parameter was considered for constructing the membership function (see Fig. 1). The shape of the

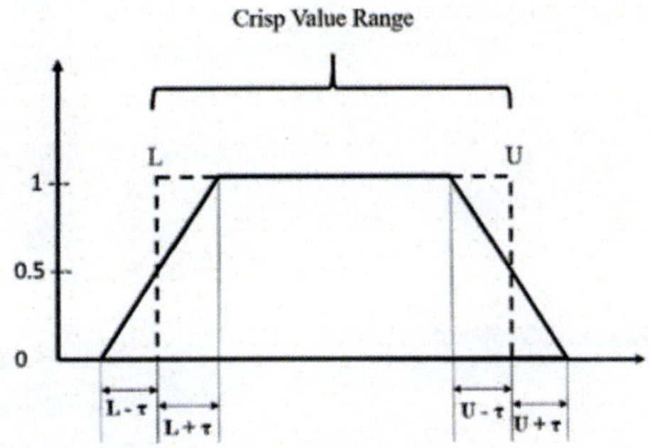

Fig. 1 Fuzzification of Input Parameters

membership function for fuzzy sets is problem specific. The trapezoidal membership function is chosen as it the best fit for the present problem.

The membership functions of the input are represented in Figs. 2, 3, 4, 5, 6, 7, 8, 9, 10, 11, 12, 13 and 14.

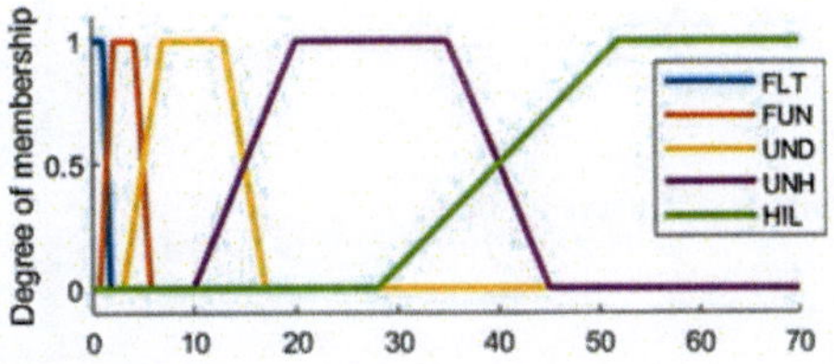

Fig. 2 Membership of Elevation

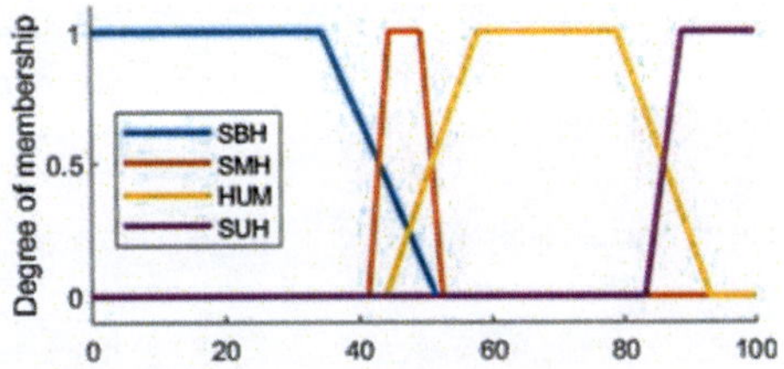

Fig. 3 Membership of Rainfall

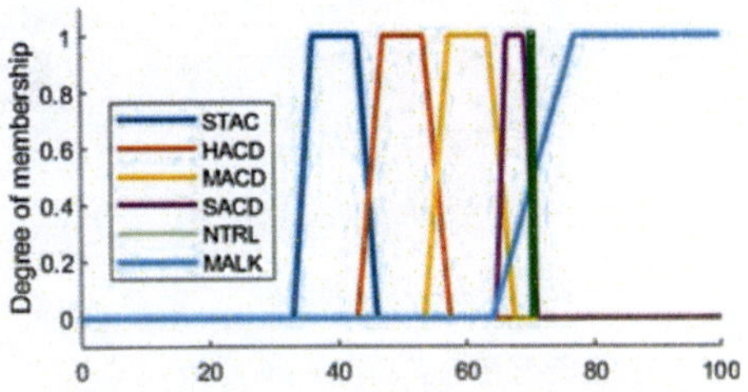

Fig. 4 Membership of pH

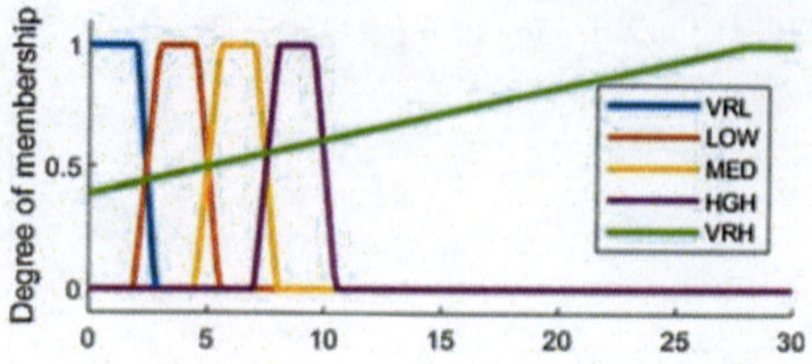

Fig. 5 Membership of OC

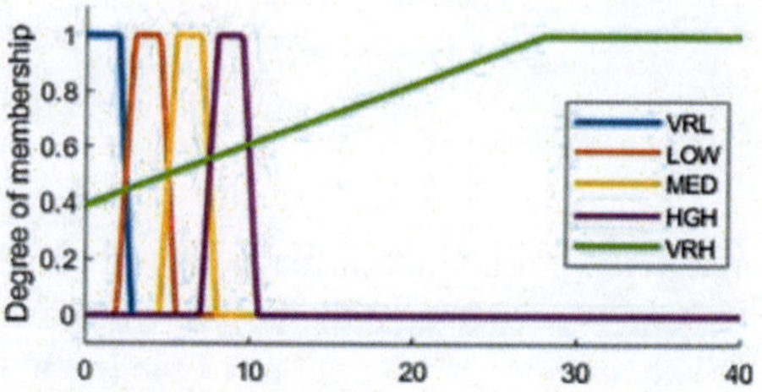

Fig. 6 Membership of N

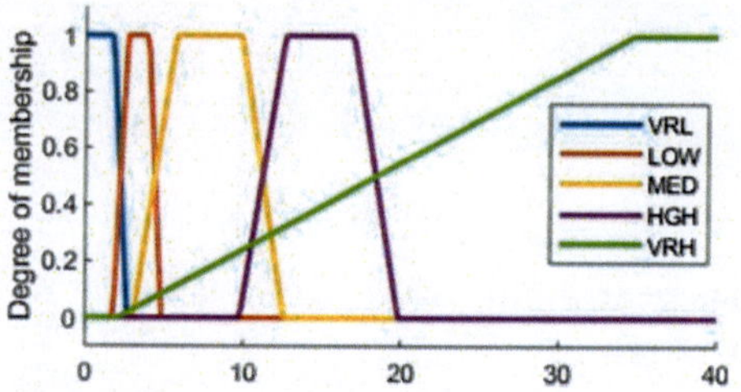

Fig. 7 Membership of P

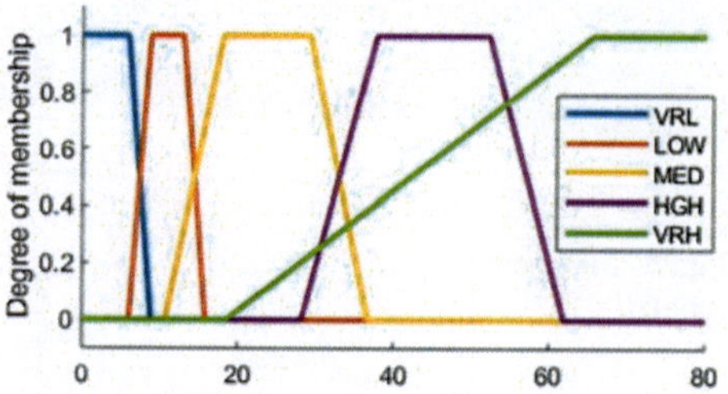

Fig. 8 Membership of K

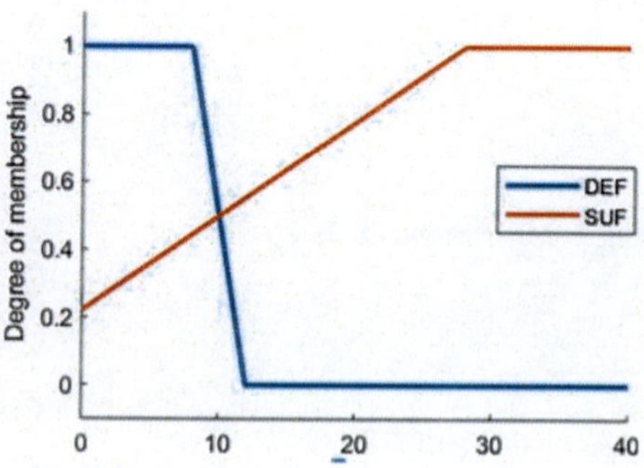

Fig. 9 Membership of S

Fig. 10 Membership of Zn

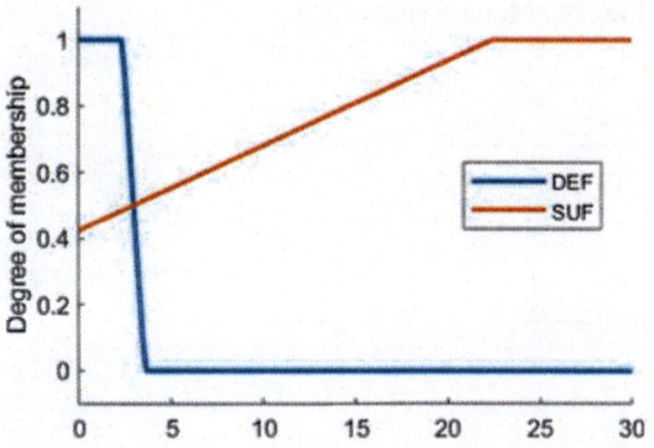

Fig. 11 Membership of B

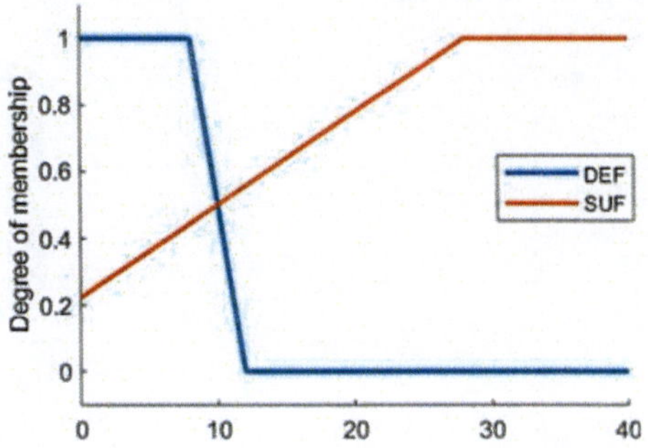

Fig. 12 Membership of Fe

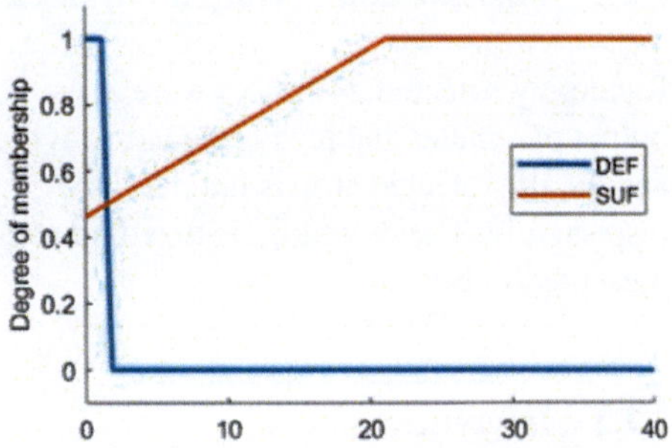

Fig. 13 Membership of Mn

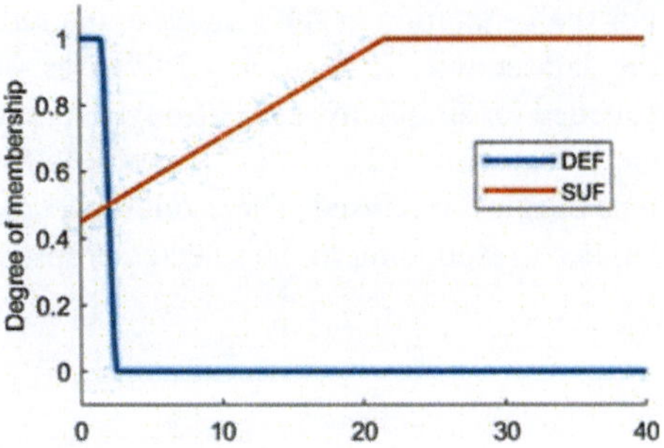

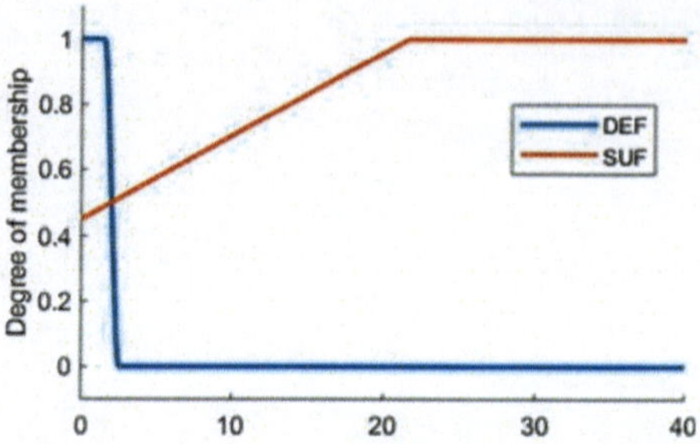

Fig. 14 Membership of Cu

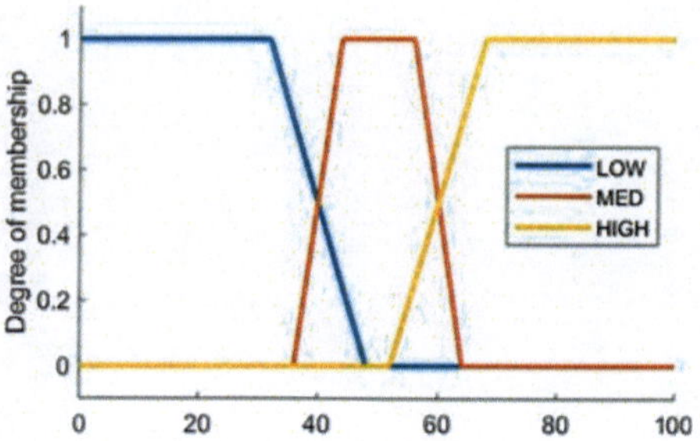

Fig. 15 Membership of Cultivation Index (CI)

2.2.2 Fuzzification of Output Parameter

Randomly selected 247 cases were considered for obtaining the membership of the output parameter and fuzzy rule bases. If the crisp output value of CI is in between 0 and 40, the selected crop is rated as LOW, for 40–60 MEDIUM and above 60 HIGH respectively. The output CI is fuzzified using a trapezoidal membership function as presented in Fig. 15.

2.2.3 Inference System

For the generation of the rule base, the redundant occurrences were removed from the dataset and the remaining 247 cases were considered. Each of the ratings was encoded for simplicity, e.g. "Strongly Acidic" was coded as STAC, Neutral as NTRL etc. as shown in Table 1, and a new field for CI was added. Finally, the "IF-THEN" type of rules were used. These rules were used to construct eight different fuzzy rule bases corresponding to the selected crops.

2.2.4 Defuzzification of Output Parameters

The output CI obtained from the fuzzy inference system was defuzzified by the centroid method to obtain the final cultivation index. A graphical representation of a case illustration is shown in Fig. 17.

3 Computation of Cultivation Index, a Case Illustration

For better understanding, a case study is being presented here. For a particular case, the inputs obtained were

$$\begin{aligned} X &= \{\text{Elevation, MAR, pH, OC, N, P, K, S, Zn, B, Fe, Mn, Cu}\} \\ &= \{48, 2782.025, 4.9, 0.22, 186.4, 22, 120, 0, 0, 0, 0, 0, 0\}. \end{aligned}$$

After normalization the values obtained were

$$X = \{9.6, 79.486, 49, 2.2, 18.64, 4.4, 12, 0, 0, 0, 0, 0, 0\}.$$

Now, the first input is Elevation = 9.6 which has membership values

$\mu(\text{Elevation})_{FLT} = 0$, $\mu(\text{Elevation})_{FUN} = 0$, $\mu(\text{Elevation})_{UND} = 1$, $\mu(\text{Elevation})_{UNH} = 0$, $\mu(\text{Elevation})_{HIL} = 0$.

Where μ is the membership function.

Similarly, the membership values for other twelve inputs are

MAR = 79.486 having $\mu(\text{MAR})_{SBH} = 0$, $\mu(\text{MAR})_{SMH} = 0$, $\mu(\text{MAR})_{HUM} = 0.9653$, $\mu(\text{MAR})_{SUH} = 0$.

pH = 49 having $\mu(\text{pH})_{STAC} = 0$, $\mu(\text{pH})_{HACD} = 1$, $\mu(\text{pH})_{MACD} = 0$, $\mu(\text{pH})_{SACD} = 0$, $\mu(\text{pH})_{NTRL} = 0$, $\mu(\text{pH})_{MALK} = 0$.

OC = 2.2 having $\mu(\text{OC})_{VRL} = 0.75$, $\mu(\text{OC})_{LOW} = 0.3077$, $\mu(\text{OC})_{MED} = 0$, $\mu(\text{OC})_{HGH} = 0$, $\mu(\text{OC})_{VRH} = 0$.

N = 18.64 having $\mu(\text{N})_{VRL} = 0$, $\mu(\text{N})_{LOW} = 1$, $\mu(\text{N})_{MED} = 0$, $\mu(\text{N})_{HGH} = 0$, $\mu(\text{N})_{VRH} = 0$.

P = 4.4 having $\mu(\text{P})_{VRL} = 0$, $\mu(\text{P})_{LOW} = 0.5$, $\mu(\text{P})_{MED} = 0.5$, $\mu(\text{P})_{HGH} = 0$, $\mu(\text{P})_{VRH} = 0.071$.

K = 12 having $\mu(\text{K})_{VRL} = 0$, $\mu(\text{K})_{LOW} = 1$, $\mu(\text{K})_{MED} = 0.1774$, $\mu(\text{K})_{HGH} = 0$, $\mu(\text{K})_{VRH} = 0$.

S = 0 having $\mu(\text{S})_{DEF} = 1$, $\mu(\text{S})_{SUF} = 0.222$

ZN = 0 having $\mu(\text{ZN})_{DEF} = 1$, $\mu(\text{ZN})_{SUF} = 0.4227$

B = 0 having $\mu(\text{B})_{DEF} = 1$, $\mu(\text{B})_{SUF} = 0.222$

FE = 0 having $\mu(\text{FE})_{DEF} = 1$, $\mu(\text{FE})_{SUF} = 0.4619$

MN = 0 having $\mu(\text{MN})_{DEF} = 1$, $\mu(\text{MN})_{SUF} = 0.449$

CU = 0 having $\mu(\text{CU})_{DEF} = 1$, $\mu(\text{CU})_{SUF} = 0.449$

These inputs fired two rules from the fuzzy rule bases

Rule 1:

IF Elevation is *UND* AND MAR is *HUM* AND pH is *HACD* AND OC is *LOW* AND N is *LOW* AND P is *VRH* AND K is *LOW* AND S is *DEF* AND Zn is *SUF* AND B is *DEF* AND Fe is *SUF* AND Mn is *SUF* AND Cu is *SUF*

THEN TOBACCO_CI is *HGH.*

The output CI for Tobacco is set to High (HGH) whose degree of membership is

$\mu(\text{TOBACCO\_CI})_{\text{HGH}} = \min\{\mu(\text{Elevation})_{\text{UND}}, \mu(\text{MAR})_{\text{HUM}}, \mu(\text{pH})_{\text{HACD}}, \mu(\text{OC})_{\text{LOW}}, \mu(\text{N})_{\text{LOW}}, \mu(\text{P})_{\text{VRH}}, \mu(\text{K})_{\text{LOW}}, \mu(\text{S})_{\text{DEF}}, \mu(\text{B})_{\text{DEF}}, \mu(\text{FE})_{\text{SUF}}, \mu(\text{MN})_{\text{SUF}}, \mu(\text{CU})_{\text{SUF}}\}$

$= \min\{1, 0.9653, 1, 0.3077, 1, 0.071, 1, 1, 0.4227, 1, 0.4619, 0.449, 0.449\}$

$= 0.071.$

Rule 2:

IF Elevation is *UND* AND MAR is *HUM* AND pH is *HACD* AND OC is *LOW* AND N is *LOW* AND P is *MED* AND K is *MED* AND S is *DEF* AND Zn is *SUF* AND B is *DEF* AND Fe is *SUF* AND Mn is *SUF* AND Cu is *SUF*

THEN TOBACCO_CI is *MED.*

The output CI for Tobacco is set to Medium (MED) whose degree of membership is

$\mu(\text{TOBACCO\_CI})_{\text{MED}} = \min\{\mu(\text{Elevation})_{\text{UND}}, \mu(\text{MAR})_{\text{HUM}}, \mu(\text{pH})_{\text{HACD}}, \mu(\text{OC})_{\text{LOW}}, \mu(\text{N})_{\text{LOW}}, \mu(\text{P})_{\text{MED}}, \mu(\text{K})_{\text{MED}}, \mu(\text{S})_{\text{DEF}}, \mu(\text{B})_{\text{DEF}}, \mu(\text{FE})_{\text{SUF}}, \mu(\text{MN})_{\text{SUF}}, \mu(\text{CU})_{\text{SUF}}\}$

$= \min\{1, 0.9653, 1, 0.3077, 1, 0.5, 0.1774, 1, 0.4227, 1, 0.4619, 0.449, 0.449\}$

$= 0.1774$

The numerical value of the output fuzzy values is computed by taking the maximum of the memberships of various rules fired. The membership value of CI for fuzzy label "HGH" is max {0.071} = 0.071 and that of "MED" is max {0.1774} = 0.1774 as shown in Fig. 16.

Based on the inputs, the final defuzzified value of CI is obtained by MATLAB R2018a using centroid method. The defuzzified value is 61.5041 as shown in Fig. 17.

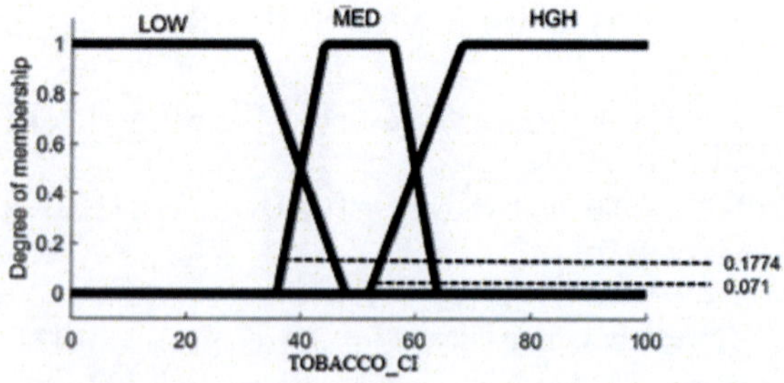

Fig. 16 Fuzzification of output parameter CI

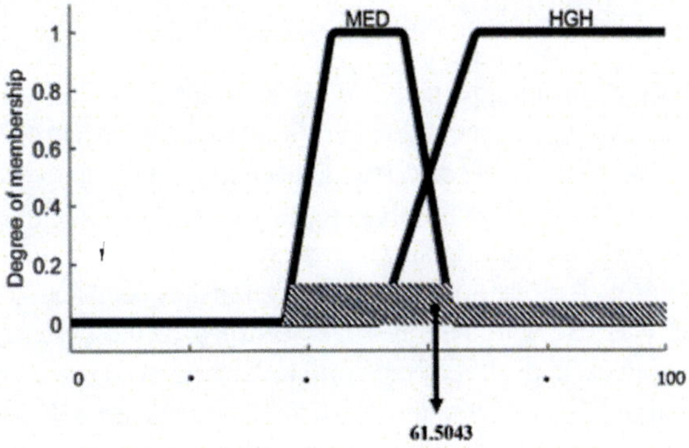

Fig. 17 Defuzzification of output parameter CI

4 Results and Discussion

The fuzzy rule base explained in the previous section was constructed using MATLAB R2018a fuzzy logic designer toolbox. The rule bases contained total 391 rules. The system was tested with 105 real field cases. It was observed that some improvised recommendations were made apart from the crops listed in the testing set. This feature exhibits the generality and completeness of the system. Detailed results for each of the cases are provided in Table 2.

The approach is novel in terms of procedure and parameters used to design the crop recommendation system as compared to existing works. Moreover, the accuracy achieved by this system exceeds that of existing crop recommendation systems [3–8].

Table 2 Summary of the results

Crop	No. of correct recommendations	No. of improvised recommendations	Percentage of accuracy
Paddy	75	4	75.23
Jute	91	9	95.23
Potato	68	23	86.66
Tobacco	92	10	97.14
Wheat	91	11	97.14
Seasumum	82	15	92.38
Mustard	84	19	98.09
Green gram	98	2	95.23
Average			**92.14**

5 Conclusion

This is an attempt to build an efficient and robust crop recommendation system considering soil parameters, rainfall and terrain pattern using fuzzy logic for the state of West Bengal, India. The dataset contained eleven soil parameters, land elevation and mean annual rainfall as input parameters and corresponding cultivation index as output. The membership functions of the inputs and output were derived from the dataset. The performance of the system has been tested for eight different major crops of West Bengal. The average accuracy of the system was measured to be 92.14% which exceeds that of the similar existing systems. Through this system, the farmers will be able to make more effective and accurate decision in selecting crops to have an enhanced productivity and better economy.

References

1. Arooj, A., Riaz, M., Akram, M.N.: Evaluation of predictive data mining algorithms in soil data classification for optimized crop recommendation. In: International Conference on Advancements in Computational Sciences, pp. 1–6. IEEE, Lahore, Pakistan (2018)
2. Siddique, T., Barua, D., Ferdous, Z., Chakrabarty, A.: Automated farming prediction. In: Intelligent Systems Conference (IntelliSys), pp. 757–763. IEEE, London, UK (2017)
3. Nevo, A., Amir, I.: CROPLOT-an expert system for determining the suitability of crops to plots. Agric. Syst. **37**(3), 225–241 (1991)
4. Kumar, V., Dave, V., Bhadauriya, R., Chaudhary, S.: Krishimantra: agricultural recommendation system. In: Proceedings of the 3rd ACM Symposium on Computing for Development, pp. 45. ACM, New York, USA (2013)
5. Zhang, H., Zhang, L., Ren, Y., Zhang. J., Xu, X., Ma, X., Lu, Z.: Design and implementation of crop recommendation fertilization decision system based on WEBGIS at village scale. In: International Conference on Computer and Computing Technologies in Agriculture, pp. 357–364. Springer, Berlin, Heidelberg (2010)
6. Laliwala, Z., Sorathia, V., Chaudhary, S.: Semantic and rule based event-driven services-oriented agricultural recommendation system. In: 26th IEEE International Conference on Distributed Computing Systems Workshops (ICDCSW'06), pp. 24–24. IEEE, Lisbon, Portugal (2006)
7. Shinde, K., Andrei, J., Oke, A.: Web based recommendation system for farmers. Int. J. Recent Innov. Trends Comput. Commun. **3**(3), 41–52 (2015)
8. Balakrishnan, N., Muthukumarasamy, G.: Crop production-ensemble machine learning model for prediction. Int. J. Comput. Sci. Softw. Eng. **5**(7), 148–153 (2016)
9. Binaghi, E.: A fuzzy logic inference model for a rule-based system in medical diagnosis. Expert Syst. **7**(3), 134–141 (1990)
10. Ghosh, I.: Measuring educational attainment: a proposed fuzzy methodology. Int. J. Comput. Sci. Softw. Eng. **5**(12), 651–657 (2015)
11. Banerjee, G., Sarkar, U., Das, S., Ghosh, I.: Artificial intelligence in agriculture: a literature survey. Int. J. Sci. Res. Comput. Sci. Appl. Manag. Stud. **7**(3), 1–6 (2018)
12. Ross, T.J.: Fuzzy logic with engineering applications, 3rd edn. Wiley, UK (2005)
13. Ghosh, B.K.: Essence of crop diversification: a study of West Bengal agriculture. Asian J. Agric. Res. **5**(1), 28–44 (2011)
14. Aktar, N.: Agricultural productivity and productivity regions in West Bengal. NEHU J. **13**(2), 49–61 (2015)

15. Majumder, K.: Nature and pattern of crop diversification in West Bengal. Int. J. Res. Manag. Pharm. **3**(2), 33–41 (2014)
16. Reddy, A.: Impact Study of Soil Health Card Scheme. National Institute of Agricultural Extension Management (MANAGE), Hyderabad, India (2017)
17. Soil Health Card. https://www.india.gov.in/spotlight/soil-health-card. Accessed 24 Jul 2019
18. Darajat, A.S., Susilowati, M. D.: Physical and facilities factors influencing tourist distribution in bantul regency, special region of Yogyakarta. In: E3S Web of Conferences (ICENIS 2017), pp. 1–5, EDP Sciences, Indonesia (2017)
19. GPS Visualizer Homepage. https://www.gpsvisualizer.com/elevation. Accessed 07 Aug 2019
20. Chattopadhyay, A.: Preservation system of college and university libraries of West Bengal in relation to climatic condition. Ph.D. Thesis, University of Calcutta (2007)
21. CRIS Homepage. http://hydro.imd.gov.in/hydrometweb/(S(o5bott45ve3bdm45jl1u0lm2))/DistrictRaifall.aspx. Accessed 24 Jul 2019
22. Klir, G.J., Yuan, B.: Fuzzy sets and fuzzy logic: theory and applications, 1st edn. Prentice Hall, New Jersey (1995)

Community Detection and Design of Recommendation System Based on Criminal Incidents

Sohom Roy, Sayan Kundu, Dhrubasish Sarkar, Chandan Giri, and Premananda Jana

Abstract In this cutting-edge, growth in computational areas have evolved rapidly. Nowadays people are communicating with each other using internet and more precisely it can be said that the online social network platforms are playing a major role to connect people around the globe. Now researchers are trying to dig up these online social networks and trying to implement in diverse fields, otherwise aim is to fetch valuable data from community sites. Here the intention is to use this ground to use victim's information to minimize the upcoming criminal activities through recommendation process. It will help the common people to know how and where antisocial activities are being occurred. The current paper aims to find out the relation between crime incidents and users profile using community detection algorithm where vector space model will play a key role and further to use recommendation algorithm on users to suggest the outcome of earlier analysis.

Keywords Vector space model · TF-IDF · Community detection · Recommendation in social media · Criminal incident · Social network analysis

S. Roy
IBM India Pvt. Ltd, Kolkata, India
e-mail: sohom.1988@gmail.com

S. Kundu
Silli Polytechnic, Jharkhand, Silli, India
e-mail: kundu.toni007@gmail.com

D. Sarkar (✉)
Amity University Kolkata, Kolkata, India
e-mail: dhrubasish@inbox.com

C. Giri
Indian Institute of Engineering Science & Technology, Shibpur, Howrah, India
e-mail: chandangiri@gmail.com

P. Jana
Netaji Subhas Open University, Kalyani, WB, India
e-mail: prema_jana@yahoo.com

D. Bhattacharjee et al. (eds.), *Proceedings of International Conference on Frontiers in Computing and Systems*, Advances in Intelligent Systems and Computing 1255,
https://doi.org/10.1007/978-981-15-7834-2_7

1 Introduction

In social network analysis, community analysis is one of the important features. Groups in social networks are important for several reasons such as; individuals most of the time create groups based on their common interests. Second, groups provide kind of high-level opinion of operator communication, but a local-interpretation of separate behavior is noisy and ad hoc in actual. Lastly, few actions be located only in a group and not observable on specific level, and for this reason, group collective behavior is more stable. In this method, different criminal incidents are considered as nodes, and victim's details related to those criminal incidents are defined as the node characteristics. Then the similar types of criminal activities form communities and according to the user profile and related crime incidents, recommendation will take place and people will be advised to take preventive actions.

This paper describes the model in following manner: Upcoming part is common terms and concepts accompanied by various modules of Statistics, Social Networks, and these are essential for the construction of the model. The related work area consults about the preceding researches in this domain. Remaining sections confers about the model to explore the communities, survey to get records and analysis of results. Final part covers conclusion and future scope.

2 Common Terms and Concepts

The common terminologies have been explained below, these are useful to create the model.

2.1 Member-Based Community Detection

One of the popular community detection algorithms is member-based community detection, where group members are being formed based on attributes or measures such as similarity, degree, or reachability [1].

2.2 Node Similarity

Let's assume there are two nodes v_i and v_j, then node similarity will try to find the similarity between two nodes. Same kind of nodes will create same groups in this case. Let $N(v_i)$ and $N(v_j)$, are the neighbors of vertices v_i and v_j, respectively. Here, calculation of vertex similarity (Eq. 1), can be defined as follows [1]:

$$\partial\left(v_i, v_j\right) = \left|N(v_i) \cap N(v_j)\right| \tag{1}$$

2.3 Recommendation in Social Media

For product recommendation, target is to recommend products to the users, based on previous details. Correctly, a recommendation algorithm considers a set of users U and a set of items I and studies a function f such that –f: U X I → IR [1]

2.4 Basic Concepts of Information Theory

Information theory is the process to study the coding of information where entropy is the key measure [2].

2.5 Vector Space Model AND TF-IDF

Gerard Salton developed VSM [3] and it convert texts into matrices and vectors, and then employ matrix analysis techniques to find the relations and key features in the document collection. Group D contains documents and document contains group of words. Document i with vector d_i is represented as (Eq. 2)

$$d_i = \left(w_{1,i}, w_{2,i}, \ldots\ldots\ldots, w_{N,i}\right) \tag{2}$$

where $w_{j,i}$ represents weight of the word j which occurs in document i & N represents word count which is used for the procedure. To calculate $w_{j,I}$, assign 1 when the word j present in document I, or 0. The count of the word j's presence in document i has been taken. Next to use term frequent—inverse document frequency (TF-IDF) where $w_{j,i}$ is calculated as (Eq. 3)

$$w_{j,i} = tf_{j,i} \times idf_i \tag{3}$$

where $tf_{j,i}$ is the frequency of word j in document i. idf_i is the inverse frequency of word j across all documents, (Eq. 4)

$$idf_i = \log_2 \frac{|D|}{|\{document \in D\}|\{j \in document\}} \tag{4}$$

this (tf-idf) allocates higher weights to words those are less frequent throw-out the documents and have higher frequencies within the document whatever has been used. Also, which are mutual in entire documents are assigned lesser weights [4].

3 Related Work

In last few years, the researchers are already working in the field of data analysis/mining, clustering techniques and they have shown the way to apply it in the field of criminology [4, 5]. Also, the process of analyzing social networks has been used in different areas such as alternate ranking findings in cricket or terror networks detection [6]. There are few published papers where [7] several data mining technics like classification, clustering, decision tree-based model [8] has been proposed for the analysis purpose of the crime scenes [9]. Community sites contain unidentified and various concealed activities and now researchers have already started to explore those areas [10]. NCRB has successfully configured the crime database to collect the crime records thoroughly in India. Crime Criminal Information system (CCIS)—The project has been carried out with a target to have regional level of database of crime incidents & GOI invested a lot for this purpose [11].

4 Proposed Model

Following example is a database table where relations among data has been explained. Initial three attributes are only for user details like name (Un), surname (Us) and email (Um). Next attributes have been defined as gender (Ug), location (Ul), social status (Uss), age category (Ua), and crime type (Uc) and these will participate to create the different classes. The model is shown in Table 1.

At the initial level, 3 different areas have been considered; Age can be taken as range of values (15–30, 30–45, etc.), thus 3 different age groups have been formed followed by 3 different social status and 4 different crime types. So, it will create different classes by depending on the selection of each attributes by the user. This process of creating different classes is not the prime target here, this is helpful to understand the depth of information received [12]. The model has been explained through a flow chart as shown in (Fig. 1). Initial target is to collect data of victims related to different types of criminal activities. Next part of the process is to create

Table 1 Table structure of victim's information

Un	Us	Um	Ug	Ul	Uss	Ua	Uc
K	Das	k@gmail.com	M	NK	A1	AG1	C1
N	Saha	n@gmail.com	F	CK	A2	AG2	C2

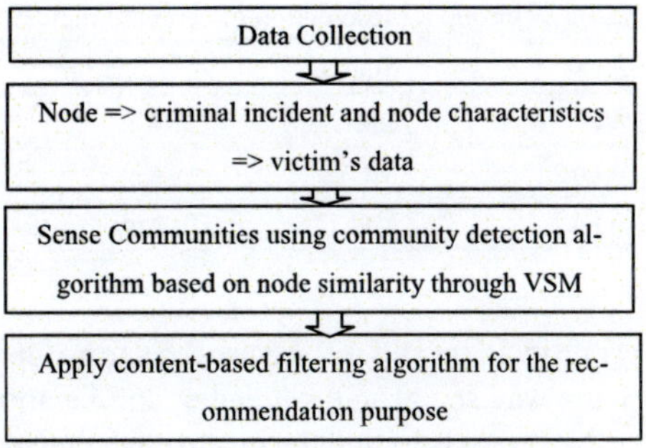

Fig. 1 Model representation of the process

groups of criminal Incidents (depending on the community detection algorithm and similarity between nodes, here node is crime incidence). Finally, the aim or target is to inform users and advise them to take preventive actions (Content-Based Method will be used).

5 Data Collection and Representation

In current analysis, 55 different types of criminal activities are chosen based on different crime reports. Then every criminal activity has been described by the victim's characteristics according to the age, social status, gender, and location. After collecting data from victims, each of the criminal activities are treated as nodes and victim's characteristics related to that node is being treated as the node's characteristics [13]. Please find below example of the google sheet as part of survey, which has been shared over the period of six months (Jan 30 2015 to June 30 2015) and more than 200 people had shared their reviews (Fig. 2):

Next similarities between the nodes are being calculated and depending on that, groups are being formed. Let's take the below example of Table 2, where d21 is a

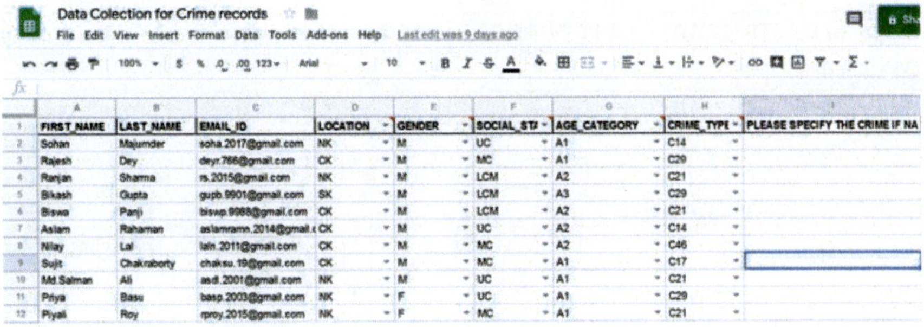

FIRST_NAME	LAST_NAME	EMAIL_ID	LOCATION	GENDER	SOCIAL_STA	AGE_CATEGORY	CRIME_TYPE	PLEASE SPECIFY THE CRIME IF NA
Sohan	Majumder	soha.2017@gmail.com	NK	M	UC	A1	C14	
Rajesh	Dey	deyr.786@gmail.com	CK	M	MC	A1	C29	
Ranjan	Sharma	rs.2015@gmail.com	NK	M	LCM	A2	C21	
Bikash	Gupta	gupb.9901@gmail.com	SK	M	LCM	A3	C29	
Biswa	Panji	biswp.9988@gmail.com	CK	M	LCM	A2	C21	
Aslam	Rahaman	aslamramn.2014@gmail.c	CK	M	UC	A2	C14	
Nilay	Lal	laln.2011@gmail.com	CK	M	MC	A2	C46	
Sujit	Chakraborty	chaksu.19@gmail.com	CK	M	MC	A1	C17	
Md.Salman	Ali	asdl.2001@gmail.com	NK	M	UC	A1	C21	
Priya	Basu	basp.2003@gmail.com	NK	F	UC	A1	C29	
Piyali	Roy	rproy.2015@gmail.com	NK	F	MC	A1	C21	

Fig. 2 Shared google sheet

Table 2 Group representation of criminal incidents

Age	Status	Location	Gender	Incident
A1	LMC	Any	NA	d20
A1A2	MC	Any	M	d21
A1A2	MC	Any	MF	d22

crime incident and A1, A2 are the age groups where that crime incident has happened. Same way social status, gender, and location has been considered.

Next part is to find the group characteristics, and for that purpose, need to sort out the collected information. Let's say, the three rows from the table is forming a group, then the following process will be applied to determine the group characteristics. Some threshold value should be determined for this process based on the situations. Let the threshold be 35%, then A1 has occurred 3 times in 3 rows, so A1's occurrence rate is 100%, whereas A2 occurred 2 times out of 3 rows. So, found that A2's occurrence rate is 66%, so both will represent the group. In the next column, occurrence rate of LMC is 1 out of 3 rows, and it is 33%. Let the threshold value be 35%, then LMC is removed and remaining items will be selected. So, the threshold value is not constant, and it must be selected depending on different situations. But whatever the threshold value will be set, should remain same for all the groups. After applying this process, finally the groups are populated. Next to proceed further, groups are being represented using vector space model and examples are shown in Tables 3, 4, and 5.

Now randomly select one user profile and represent it using vector space model and then apply the cosine similarity calculation with the groups to calculate the similarities. Finally, the details of similarities between user profile and groups have been generated in Table 6.

This information is required for the recommendation process.

Table 3 Vectorization process (1)

	m	F	a1	a2	a3	Lmc	Mc	Uc	Nk	Sk
G1	1	0	0	1	1	0	0	1	0	0
G2	1	0	0	1	1	0	0	0	0	0
G3	0	0	1	0	0	1	0	0	0	0
G4	1	1	0	1	1	0	0	0	1	0

Table 4 Frequency-idf values

m	F	a1	a2	a3	Lmc	Mc	Uc	Nk	Sk
10	5	4	10	6	5	4	2	3	1
0.263	1.263	1.584	0.263	1	1.263	1.584	2.584	2	3.584

Table 5 Vectorization process (2)

	m	F	a1	a2	a3	Lmc	Mc	Uc	Nk	Sk
G1	0.263	0	0	0.263	1	0	0	2.584	0	0
G2	0.263	0	0	0.263	1	0	0	0	0	0
G3	0	0	1.584	0	0	1.263	0	0	0	0
G4	0.263	1.263	0	0.263	1	0	0	0	2	0

Table 6 Similarity Calculation

G1	G2	G3	G4	G5	G6	G7	G8	G9	G10	G11	G12
0.797	0.021	0.204	0.348	0.192	0.012	0.011	0.209	0.646	0.191	0.144	0.379

6 Result Analysis and Discussion

In this process, each of the nodes are representing different types of criminal activities and their characteristics, which are basically victim's information. For 55 different criminal activities 12 groups will get selected and they will be treated as 12 nodes. At the time of selection, calculate the count of unique nodes available in the document & then from those unique rows, randomly 12 nodes are being selected [14] as 12 groups. The process has been implemented successfully using a piece of code written in java.

Next, 12 unique nodes have been represented using vector space model, where member-based community detection algorithm has been applied. Here, crime incidents with same functionalities are creating the groups among them using node similarities. Then a new node will be selected and converted using vector space model for further grouping purpose.

There are some attribute values like "NA", and "Any" can be considered as junk. Just an example of a group where a1, a2, a3 appears together then occurrences becomes 33.3% for each of them, then each of them has same probability as a1 = 1/3, a2 = 1/3, and a3 = 1/3.

Then $a1a2a3_{together} = 1/3 + 1/3 + 1/3 = 1$, $Information_{a1a2a3} \alpha$ (1/probability of $occurance_{a1a2a3}$), $I_{a1a2a3} = F(1/1)\ I_{a1a2a3} = F(1)\ I_{a1a2a3} = \log_b 1\ I_{a1a2a3} = 0$.

That's why for attribute value 'any' is taken as 0. And for attribute value NA which is not available, also be represented with 0, because, if it doesn't convey any information, then it can be counted as 0.

After forming 12 initial groups, next target is to create the grouping of other nodes, as they should be the part of a group. For that reason, randomly select 20 more nodes. Next, cosine similarity has been used to calculate the similarity for grouping process. In this process, the similarity between 12 groups and other 20 nodes have been calculated. Now represent the nodes using a graph where edges are representing the similarity between two nodes using weights. In Fig. 3, the thicker

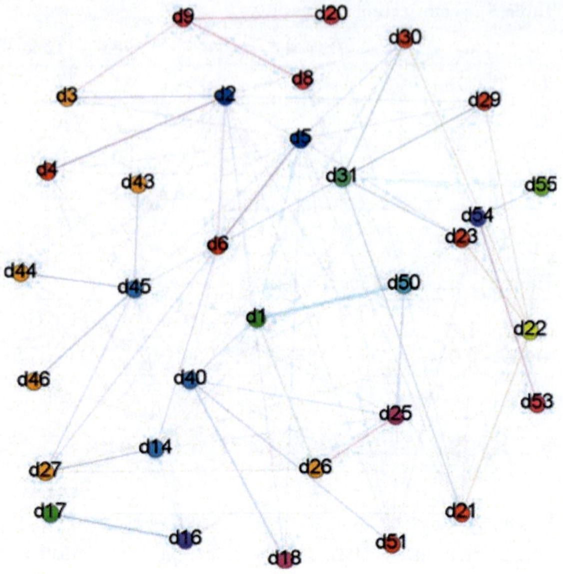

Fig. 3 Nodes representation of criminal incidents

Table 7 Edge details of criminal incidents

Source	Target	Type	ID	Label	Weight
13	5	Directed	82		0.332
13	7	Directed	83		0.283
13	10	Directed	84		0.1

edges between two nodes are having higher weightage than any other edges. Weight of the edges are calculated from the similarity. So thin edges are representing the lower similarity where as thick edges are representing the higher similarity [15].

Next, Table 7, will provide the example of details of the edges

At the initial level there were 12 groups. After applying vector space model and the similarity calculation, different communities have been formed as

{d2, d4}, {d5, d6}, {d9, d3, d8, d20}, {d14, d27}, {d22, d21, d23}, {d26, d25}, {d31, d29, d30}, {d40, d18, d51}, {d45, d43, d44, d46}, {d50, d1}, {d54, d53, d55}, {d17, d16}.

As a result, 12 groups and their members have been formed. Each group represent same types of criminal activities in terms of victim's records. Victims characteristics are same in each community in terms of age, gender, social status, location, etc. Different groups have been generated now and target is to suggest the topmost 3 communities to the user after relating user's details with the community characteristics. This procedure will be acting as knowledge transfer for the users to keep them safe from upcoming illegal happenings. Something like a person can be affected by one kind of illegal activity—cannot be said surely, so user will be provided few groups of criminal activities as recommendation. Now select top 3 similarities from Table 6, and they are G1, G9 and, G12 and these three will be recommended. Further communication to aware user could be done using mobile app or social networking sites.

7 Conclusion and Future Scope

The similarity between nodes has been calculated successfully. Some groups are detected using community detection algorithm. Then recommendation can be done by suggesting few top groups to the user by relating the profile details with the group behavior. For this kind of work, records are mostly available in database of crime records bureau, as crime department of the country has been modernized and they have started to build micro level database of crimes. The process of recommendation further can be expediate using mobile app or social networking sites as Facebook, WhatsApp, Twitter, etc. The process explained in this paper can be used further to find the gang activities from the crime incidents, or to analyze the herd behaviors from user's characteristics. At the initial level, expectation is to help the people to keep them safe from probable upcoming criminal activities.

References

1. Zafarani, R., Abbasi, M, A., Liu, H.: Social Media Mining an Introduction, Draft Version: April 20, 2014, Cambridge University press (2014)
2. MacKay, D.J.C.: Information Theory, Inference and Learning Algorithms, Version 6, (2003)
3. Wijewickrema P.K.C.M., Ratnayake, A.R.M.M.: Enhancing accuracy of a search output: a conceptual model for information retrieval. J. Univ. Librarians Assoc. Sri Lanka **17**(2), 119–135 (2013)
4. Xu, J.,Chen, H.: Criminal network analysis and visualization. Commun ACM **48**(6) (2005)
5. Papachristos, A, V.: The coming of a networked criminology. Adv. Criminol. Theory **17**, 101–140 (2011)
6. Roy. S., Dey. P., Kundu. D.: Social network analysis of cricket community using a composite distributed framework: from implementation viewpoint. IEEE Trans. Comput. Social Syst. **5**(1), 64–81 (2018). https://doi.org/10.1109/tcss.2017.2762430
7. Hassani. H., Huang. X., Silva. E.S., Ghodsi, M.: A review of data mining applications in crime. Stat. Anal. Data Min. **9**(3) (2016).https://doi.org/10.1002/sam.11312
8. Nasridinov, A., Ihm, S.Y., Park, Y.H.: A decision tree-based classification model for crime prediction. In: Proceedings of the 10th International Conference on Secure and Trust Computing (STA), Data Management, and Applications, pp. 531–538 (2013)
9. Hanson, R.F., Sawyer, G.K., Begle, A.M., Hubel, G.S.: The impact of crime victimization on quality of life.J Trauma Stress **23**(2), 189–197 (2010). doi:https://doi.org/10.1002/jts.20508
10. Prakash, D., Suren, S.: Detection and analysis of hidden activities in social networks. Int. J. Comput. Appl. **77**(16), 34–38 (2013)
11. National Crime Records Bureau, http://ncrb.gov.in/, last accessed 20/02/2018

12. Sarkar, D., Kole, D.K., Jana, P., Chakraborty, A.: Users activity measure in online social networks using association rule mining. In: Proceedings of the IEMCON 2014: 5th International 294 Conference on Electronics Engineering and Computer Science (Elsevier Science & Technology), Kolkata, India, pp. 172–178 (2014)
13. Sarkar, D., Jana, P.: Analyzing user activities using vector spacemodel in online social networks. In: Proceedings of the National Conference on Recent Trends in Information Technology & Management (RTITM 2017), India, pp. 155–158 (2017)
14. https://gephi.org/, last accessed 20/12/18
15. Sarkar, D., Roy, S., Giri, C., Kole, D.K.: A statistical model to determine the behavior adoption in different timestamps on online social network. Int. J. Knowl. Syst. Sci. (IJKSS) **10**(4), 1–17 (2019). https://doi.org/10.4018/IJKSS.2019100101

Personalized Word Recommendation System Using Sentiment Analysis

Subhra Samir Kundu, Krutika Desai, Soumyajit Ghosh, and Dhrubasish Sarkar

Abstract Online Social Networks is a place where a user is truly free to express himself or herself, and it is observed to be so. Social networks are used by users for not only socializing but to buy or sell products. This user behavior is quite adamant from the type of comments they post on different social networks. The word recommendation system is still not personalized rather generalized for all different websites. It can, however, be personalized by the use of sentiment analysis and the model here has done the same here. The model has used subjectivity and polarity for making a personalized recommender system by analyzing the behavior and classifying them. This provided the model with two greater ways to recommend, i.e., recommending on the basis of the user as prevalent from their comments and also by the topics been discussed by them.

Keywords Recommendation system · Sentiment analysis · Subjectivity · Polarity · Stopwords · TF-IDF

S. S. Kundu · K. Desai · S. Ghosh · D. Sarkar (✉)
Amity Institute of Information Technology, Amity University Kolkata, Major Arterial Road (South-East), Action Area II, Newtown, Kolkata, West Bengal 700135, India
e-mail: dhrubasish@inbox.com

S. S. Kundu
e-mail: subhrasamirk@gmail.com

K. Desai
e-mail: krutikadesai77@yahoo.com

S. Ghosh
e-mail: soumyajitghosh0094@gmail.com

D. Bhattacharjee et al. (eds.), *Proceedings of International Conference on Frontiers in Computing and Systems*, Advances in Intelligent Systems and Computing 1255,
https://doi.org/10.1007/978-981-15-7834-2_8

1 Introduction

Online Social Networks are observed to be places where a user is truly free to express him or herself. The social networks are used by users for not only socializing but to buy or sell products. Recommender systems, with an aim to provide personalized suggestions of products to the particular users, have been fiercely used by all these networks. Traditional recommendation methods usually focus on utilizing user profiles obtained from their past purchasing behaviors, from which the similarities [1] or relationships [2] between each user are generated. The work of Zhang, Y., Liu, R.D. and Li, A.Z., "A Novel Approach to Recommender System Based on Aspect-level Sentiment Analysis," Beijing University of Posts & Telecommunications, Beijing, 100876, China (2016) uses user review, as natural language text, in contrast to the traditionally used structured information(ratings) to obtain finer grained sentiment orientations toward different aspects of a single item. To obtain the details of author's opinion, and to discover the aspects of his comment [3] and determine whether the sentiment is positive or negative or the extent of positive/negative attitude on each aspect of a product. It helps in better filtering and recommendation of products as per the user's behavior [4]. This user behavior is quite adamant from the type of comments they post on different social networks. This is always the case be it FaceBook, Twitter, or LinkedIn where they provide contents to socialize or be it the product review sites where they give comments to make other users aware of the products. Here, in similar fashion, we employ Sentiment Analysis to design a personalized "word" recommendation system.

Often, their original behavior is observed in such situations due to the abstraction which they can expect from the same. But it creates a problem for the users whether they need a new vocabulary or not and want assistance in the same. This is a quite old area but is still not accurate. As a subclass of information filtering, recommendation system is used in areas based on user characteristics such as films, music, news, books, study papers, search queries, social tag [5], and products, in particular. A "Chatbot," short for Chat-Robot interacts by instant messaging, replicating human interaction patterns artificially by the use of this data. It simulates human conversations, formally permitting a form of human–machine interaction [6].

This model uses the dataset from the GitHub repository [7] which comprises the comment of each and every user from a social networking site and tries to predict their sentiment by classifying the same into its respective polarity and subjectivity. The polarity value ranges from positive (+1) to negative (−1) where the neutral is (0), meaning a score on the overall comment that either it is positive, negative, or neutral based upon the words used. Similarly, the subjectivity values range from (0) to (1). This is used in calculation and making of the recommender system, and this paper speaks of the same.

The organization of the remaining paper is as follows. Section 2 describes works previously done on sentiment analysis, recommendation systems, and how to connect both. Section 3 outlines the model in which the data is found in [7] and used. Section 4 discusses the proposed model and the algorithm used to perform the study on the data using the model. Section 5 outlines the results which we have received as a result of performing the algorithm on the data, and lastly to conclude, Sect. 6 gives the paper by outlining the model's ongoing and future work.

2 Related Work

Applying recommender systems on websites, many e-commerce and retail companies leverage the power of data to rocket-up their sales. Recommended systems work with two categories of information: i. Characteristic Information—The Information on items (keywords, categories, etc.) and users (preferences, profiles, etc.) [8]. ii. User Item Interactions—The information of ratings, purchase numbers, likes, etc.

Sentiment Analysis is the framework behind a sequence of phrases to determine the emotional tone used to understand the attitudes, beliefs, and thoughts articulated in an online reference. Analysis of sentiment is highly helpful in tracking social media as it allows us to obtain an overview of the broader public opinion behind some subjects. As shown by ViV.ai recently acquired by Samsung, there are also more sophisticated dynamic approaches. On Gmail, Google's software crawls billions of emails and then uses software to suggest/recommend phrases from that database that can be used as a response or confirmation message. For now, though, most general chatbots are considered still in their infancy. For simple "transactions" they are much more useful rather than enjoyable indulging conversations.

The survey [9] states that when queries are submitted to search engines, they are generally in natural languages and contain just one or two related words. In [10], they define an inferential strategy for integrating textual user reviews into collaborative filtering (CF) algorithms. The primary concept of their strategy is to obtain user preferences expressed in textual reviews, a problem known as sentiment analysis, and to map these preferences on certain rating scales that current CF algorithms can understand. Paper [11] proposes a user recommendation method based on a novel weighting function called sentiment-volume-objectivity (SVO) which takes into consideration not only the interests of the user, but also the feelings of the user. In paper [12], consideration is given to users' feelings about the services offered by e-shopping websites. People's opinions or feelings are indicated by reviews, ratings, and emoticons. Paper [13] uses content-based, memory-based, model-based, and hybrid models to recommend businesses to users. They also applied some natural language processing algorithms and sentiment analysis on reviews to find out what they think of the business in different aspects. Paper [14] proposes a multi-lingual recommendation system based on sentiment analysis to assist Algerian consumers decide on products, restaurants, films, and other services using online product reviews, combining suggestions and sentiment analysis to produce the most precise user suggestions. The study in paper [15] suggests a sentiment analysis scheme that works with user reviews as an extra source of information to address data sparsity issues. Paper [16] explores and analyzes the association between the objects (like photographs, posts, etc.) and its viewers (friends, acquaintances, etc.) for a given user and to find activity relationship among them by using the TF-IDF scheme of Vector Space Model. In paper [17], the authors use vector space models and term frequency—inverse document frequency techniques. The concepts of herd behavior and collective behavior have all been explained in the same manner with vividness.

In paper [18], the authors have used the process of behavior analysis to find out different influential nodes and also has applied sentiment analysis and hashtag analysis of the same for their model and give vivid idea of their usage.

3 Data Model

The data used for this model was collected from [7], and it had the following parts divided into columns as their attributes and rows as various users with their data(s):

- created_time: This field contains the exact date and time when the comment was created
- From_id: This field contains the id of the user who had created the comment.
- from_name: This field contains the names of the people who had created the comment.
- message: This field contains the original comment that was was created, i.e., the word(s), emojis, and other expressions used are stored.
- post_id: This field contains the identification number of the posts.

3.1 Preprocessing

1. The dataset had few rows with no values or with only emojis. Those rows with either N/A or only emojis are removed (Fig. 1).
2. There were few columns which were not required by the model but were there anyways, which were removed. Such as created_time and post_id . were removed.

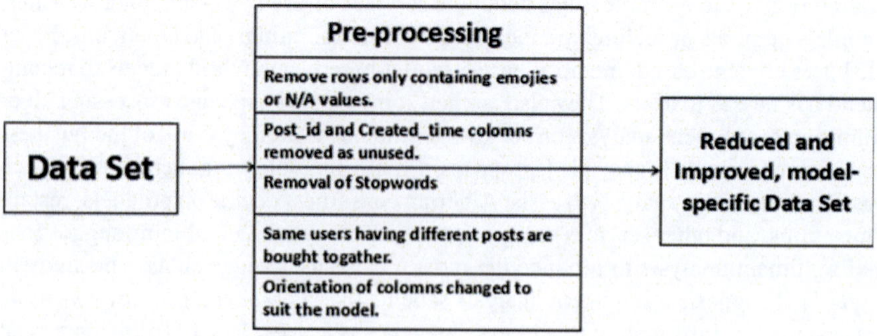

Fig. 1 The preprocessing of retrieved dataset

3. The comments had few stopwords, which were removed.
4. The dataset had same users in more than one row that was merged.
5. The orientation of the table from the original to the required form was changed [19].

4 Proposed Model

See (Fig. 2).

4.1 Retrieving Data and Making It Fit for Processing

Extracting data, preferably as Comment's Dataset from the source being facebook [20]. With the help of a web crawler, data can be extracted [21]. But in this case, the model used the dataset from [7].

Data cleaning is the process of ensuring that the data is accurate, consistent, and usable by identifying, correcting, deleting, or processing any errors or corruptions in the data as necessary to prevent the error from occurring again [22]. There are many things to go wrong with data, be it construction, arrangement, formatting, spelling, duplication, extra spaces, etc. Utility function uses simple regular expressions to clean text by removing links and special characters. Stopwords are also removed along with this. The commonly used word (such as "the," "a," "an," and "in") known as stopwords are ignored by a search engine during both times: when the search entries are indexed and when the search query results in them are being retrieved.

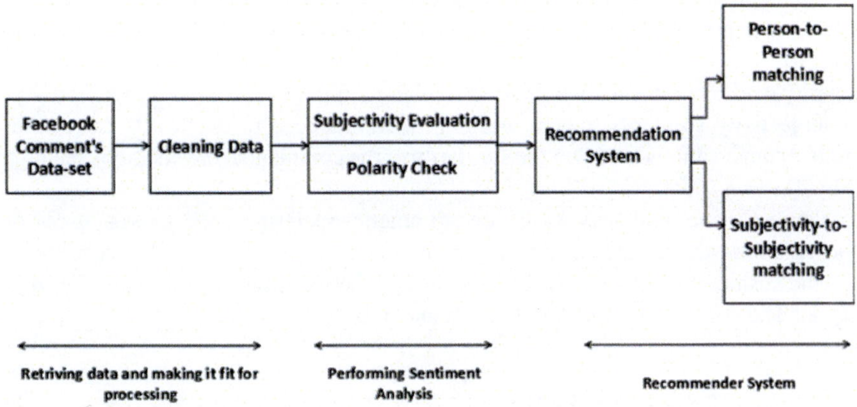

Fig. 2 The proposed work flow model for designing a personalized recommendation system

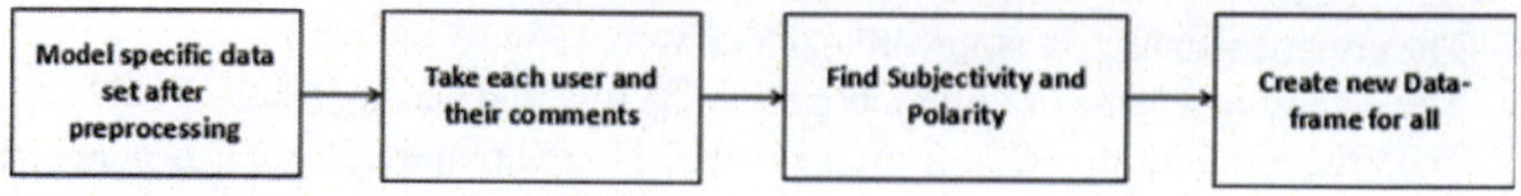

Fig. 3 Steps involved in sentiment analysis

4.2 Performing Sentiment Analysis

Analysis of sentiment is another case of primary use for the processing of natural language [23]. One can use it to evaluate comments that have been extracted from social media. TextBlob is a python library that provides simple functions for accessing its methods and carrying out basic NLP tasks. TextBlob's sentiment function returns many properties of which the model uses the most used two properties, polarity and subjectivity. Analysis of polarity takes into consideration the amount of positive or negative terms in a given sentence. It is a float within the range of $[-1, 1]$ where 1 is positive and -1 is negative. Subjective sentences usually refer to personal opinions, emotions, or judgments, while objective refers to factual information. Subjectivity is also a floating integer which is generally within the [0, 1] range [24].

The advantage of using TextBlob is that it is constructed on the shoulders of NLTK and Pattern, offering NLTK with an intuitive interface [25]. It also translates and detects languages that are powered by Google Translate (not provided with Spacy) (Fig. 3).

4.3 Recommender System

The model uses the memory-based approach of collaborative filtering. Two distinct approaches are used to filter here. Person-to-person approach takes a specific user, finds users similar to that user based on similarity in their subjectivity and polarity, and recommends words used by these similar users. Other is subjectivity-to-subjectivity similarity, where users are recommended to use words from other similar conversations based on the similarity of the comments/conversations feelings (Fig. 4).

Using Cosine similarity based on arithmetic operations, the closest users or conversations are calculated. By calculating the cosine of the angle between them, and the cosine similarity measures the similarity between two vectors (mentioned in Eq. 1). The cosine similarity for users' u and u′ is

$$sim(u, u') = \cos(\theta) = \frac{s_u . s_u^{'}}{s_u . s_u^{'}} = \sum_i \frac{s_u . s_{u'}}{\sqrt{\sum_i s_{ui}^2} . \sqrt{\sum_i s_{u'i}^2}} \tag{1}$$

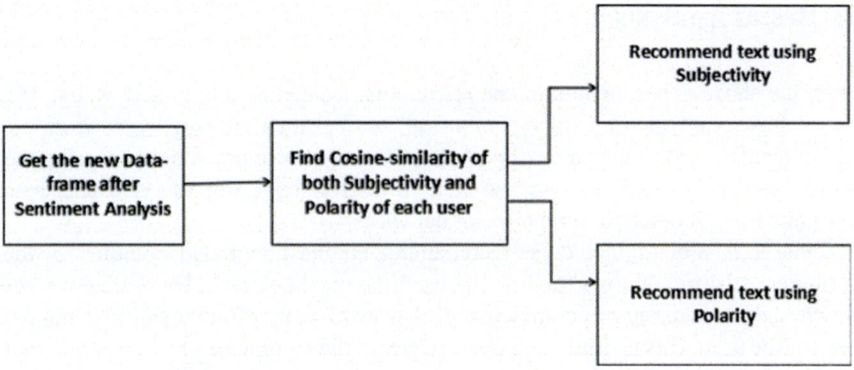

Fig. 4 Working of the recommendation system

Therefore, the cosine similarity between users and conversations along with the words commonly used between them is evaluated, one with the highest frequency [26] and recommend word used for similar conversations by the user or other similar users in the past. s_u' and s_u' are subjectivity and polarity score of users u and u' used variably.

4.4 Algorithm

Step 1: Get the Dataset [7].

Step 2: Check the Dataset for redundant and unimportant values.

Step 3: Clean the Dataset, i.e., remove all the unwanted rows and columns.

Step 4: Collect all the comments of the same person together.

Step 5: Perform sentiment analysis on each and every comment of each and every person and find out their polarity index and subjectivity index. Also find the most frequent words of each and every comment of the person.

Step 6: Generate the term frequency vector of the comments.

Step 7: Use the term frequency vector of the comments and generate the cosine similarity matrix. Similarly using the Subjectivity index (converted to a vector) to generate the cosine similarity matrix.

Step 8: Make the recommendation based upon the two matrices.

5 Result Analysis

From the above experimentation, the result obtained can be interpreted as first, if a person has commented on some similar topic earlier then the new words from the recommendation system can be provided to the new user or any other user using the same topic, and secondly, it can also be used to give suggestion to a new user who has same kind of behavior on the social networks.

Table 1 is the original dataset containing all the mentioned columns as the attributes and different rows for different users having their detail. There are few rows which had the same user's comments all that were merged into a single comment row for the user. This not only helped us to get all the comments of the user together but also helped the model in analyzing the overall behavior of the user by all the comments made by them at different instances of time.

Table 2 is the dataset which has the name of the person, their various comments, and the most used word in their comment(s), the polarity, its value, and the subjectivity value. This dataset is received after doing the sentiment analysis of the comments made by each and every user over the site at different instances of time.

Table 3 is the result of doing the sentiment analysis of the comments and finding their cosine similarity and using them recommending the user about the words they can use. This is done by passing the user name to the one that wants to recommend and the similarity matrix of that user to a function, which will return this table. Using this table one can easily use all the comments to recommend some new words to the user which will be either based upon the behavior of the user or based upon the topic the user is speaking upon or rather writing upon at the current time. From this, it is evident that the model's precision is 9/10 or 0.9, which corresponds to success of the same.

6 Conclusion and Future Scope

We were successful in doing what we intended, i.e., we made a recommender system which can recommend the user with a more personalized set of words as opposed to a general recommendation of words which is used everywhere. The model achieved both the methods by which it can be predicted, i.e., Topic-wise and Person-Wise.

In future, more work can be done in the area of "slang language" with words off the dictionary and with different figures of speech. We can also incorporate the feature of suggesting other languages and also the emojies based upon the emotions of the user presented by their language and behavioral patterns [27].

Table 1 The original dataset

Created_time	From_id	From_name	Message	Post_id
2017-07-14T14:43:54 + 0000	228735667216	BBC News	We are speaking to NRA supporters as well as W…	228735667216_10154890879532217
2017-07-14T14:41:59 + 0000	228735667216	BBC News	If you are just joining us we are outside of t…	228735667216_10154890879532217
2017-07-14T14:41:58 + 0000	10155675667923700	Dave Meredith	Do you know how backward America are in allowi…	228735667216_10154890879532217
2017-07-14T14:42:25 + 0000	10159128799640600	Ruth Wilson	People who legally own guns often seem all too…	228735667216_10154890879532217
2017-07-14T14:41:06 + 0000	1694899690817940	Dave Arnold	I don't understand why America wants to carry…	228735667216_10154890879532217
2017-07-14T14:41:02 + 0000	1617298378314690	Theresa Stevens	It is my right to legally protect my life as b…	228735667216_10154890879532217
2017-07-14T14:34:05 + 0000	10155480484879300	Ryan Churchill	I'm a gun owner, but the NRA are just terroris…	228735667216_10154890879532217
2017-07-14T14:42:47 + 0000	10155480484879300	Ryan Churchill	2nd amendment makes you think you are free fro…	228735667216_10154890879532217
2017-07-14T14:47:11 + 0000	1364364026946020	Billy Douglas	It should be a legal requirement that everyone…	228735667216_10154890879532217

Table 2 The dataset after doing the sentiment analysis

Names	Comments	Most_Used_words	ID	Polarity	Polarity_Value	Subjectivity
‘Audz Pretty	Looks like some of the people who commented he…	(someone)	10155473891197800	Positive	0.2	0.2
AJ Bullena	True!	(true)	239570526561594	Positive	0.4375	0.65
AJ Goodwin	[“Elouise Goodwin what’s the eggplant again?”,…	(elouise)	10213281175445500	Positive	0.02667	0.39167
AJ Santiago	Arman Sharif !!!!!!!!!	(arman)	10214009049406700	Neutral	0	0
Aafreen Karim	Over the top !!! First time my relative went a…	(flowers)	10210640460301700	Positive	0.08625	0.36667
Aaron Burdeau	Stephen I can always count on Aussies for havi…	(stephen)	1779580728725320	Neutral	0	0
Aaron Dixon	I highly doubt the “blue whale” was the reason…	(i)	1282580725202100	Positive	0.062	0.378
Aaron Doull	Peter Craighead toes for thumbs or thumbs for…	(toes)	1582124208498620	Neutral	0	0
Aaron Korozs	Marissa Toto first I thought it was u, then I…	(i)	1854924964534750	Positive	0.09167	0.46667

Table 3 Recommending after sentiment analysis

Names	Comments	Most_Used_words	ID	Polarity	Polarity_Value	Subjectivity
Joseph Quarshie	Wow	(wow)	496811410656693	Positive	0.1	1
Nancy Hope	Wow!	(wow)	10154683799876900	Positive	0.125	1
Nicodemus Singleton	Wow	(wow)	147890149099545	Positive	0.1	1
Caroline Hannah	Wow amazing ❤	(wow)	10209742058204400	Positive	0.35	0.95
Rumi VP	Wow! Thats amazing!! < 3	(wow)	10213315513271700	Positive	0.61667	0.96667
Tim Wheeler	Wow! Must come and see! Dippy I hope be put so…	(wow)	10213730466836700	Positive	0.4875	0.8
James C. Dalman	Wow, I hope everyone is alright. That came out…	(wow)	10211900870709900	Positive	0.125	1
Musengo Lihonde	[‘Can she play something please?	(can)	1345902378861820	Positive	0.1	0.6
Alan Shair	Wow, to be alive and see this, we are the firs…	(wow)	10211511806848200	Positive	0.2125	0.60833

References

1. Zhang, Y., Liu, R.F., Li, A.Z.: A novel approach to recommender system based on aspect-level sentiment analysis. In: 4th National Conference on Electrical, Electronics and Computer Engineering (NCEECE 2015)
2. Pazzani M.J., Billsus D.: Content-based recommendation systems. In: Brusilovsky P., Kobsa A., Nejdl W. (eds) The Adaptive Web. Lecture Notes in Computer Science, vol 4321. Springer, Berlin, Heidelberg (2007). https://doi.org/10.1007/978-3-540-72079-9_10
3. Du, Q., Zhu, D., Duan, W.: Recommendation system with aspect-based sentiment analysis (2018). http://ceur-ws.org/Vol-1520/paper29.pdf
4. Naw, N., Hlaing, E.E.: Relevant Words Extraction Method for Recommendation System. University of Technology (Yatanarpon Cyber City) and University of Computer Studies (Taung Ngu)
5. Halvey, M., Keane, M.T.: An assessment of tag presentation techniques. In: Archived 2017-05-14 at the Wayback Machine
6. Qiu, M., Li, F., Wang, S., Gao, X., Chen, Y., Zhao, W., Chen, H., Huang, J., Chu, W.: AliMe Chat: A Sequence to Sequence and Rerank based Chatbot Engine. Alibaba Group, Hangzhou, China
7. https://github.com/jbencina/facebook-news (March–May 2019)
8. Dong, Y., Tang, J., Wu, S., Tian, J., Chawla, N.V., Rao, J., Cao, H.: Link Prediction and Recommendation across Heterogeneous Social Networks. In: IEEE 12th International Conference on Data Mining, pp. 181–190 (2012)
9. Thies, I.M., Menon, N., Magapu, S., Subramony, M., O'Neill, J.: How do you want your Chatbot? An Exploratory Wizard-of-Oz Study with Young, Urban Indians. Conference paper-20 Sept 2017
10. Infanta, S.D., Chellammal, P.: A survey on sentiment analysis for product recommendation system using hybrid learning algorithm. Int. J. Res. Sci. Innov. (IJRSI) **VI**(I) (2019). ISSN 2321–2705
11. Leung, C.W., Chan, S.C., Chung, F.: Integrating collaborative filtering and sentiment analysis: A rating inference approach. In: ECAI 2006 Workshop on Recommender Systems, pp. 62–66
12. Gurini, D.F., Gasparetti, F, Micarelli, A., Sansonetti, G.: A sentiment-based approach to Twitter user recommendation. In: RSWeb@RecSys (2013)
13. Priyadharsini, R.L., Felciah, M.L.P.: Recommendation system in e-commerce using sentiment analysis. Int. J. Eng. Trends Technol. (IJETT) **49**(7) (2017)
14. Ziani, A., Azizi, N., Schwab, D., Aldwairi, M., Chekkai, N., Zenakhra, D., Cheriguene, S.: Recommender system through sentiment analysis. In: 2nd International Conference on Automatic Control, Telecommunications and Signals (Dec 2017), Annaba, Algeria. ffhal-01683511f, https://hal.archives-ouvertes.fr/hal-01683511. 13 Jan 2018
15. Hassan, A.K.A, Abdulwahhab, A.B.A.: Reviews Sentiment analysis for collaborative recommender system. Kurd. J. Appl. Res. (KJAR) **2**(3) (2017). https://doi.org/10.24017/science.2017.3.22. Print-ISSN: 2411-7684–Electronic-ISSN: 2411-7706, kjar.spu.edu.iq
16. Sarkar, D., Jana, P.: Analyzing user activities using vector space model in online social networks. In: National Conference on Recent Trends in Information Technology and Management (RTITM 2017)
17. Sarkar, D., Roy, S., Giri, C., Kole, D.K.: A statistical model to determine the behavior adoption in different timestamps on online social network. Int. J. Knowl. Syst. Sci. **10**(4) (2019)
18. Sarkar, D., Debnath, S., Kole, D.K., Jana, P.: Influential nodes identification based on activity behaviors and network structure with personality analysis in egocentric online social networks. Int. J. Knowl. Syst. Sci. **10**(4) (2019)
19. Pandas Documentation (March–June 2019). https://pandas.pydata.org
20. Manning, C.D., Raghavan, P., Schutze, H.: Scoring, term weighting, and the vector space model. In: Introduction to Information Retrieval. p. 100 (2008). https://doi.org/10.1017/cbo9780511809071.007. ISBN 978-0-511-80907-1

21. Rajaraman, A., Ullman, J.D.: T: Data Mining. Mining of Massive Datasets (2011)
22. Zafarani, R., Abbasi, M.A., Liu, H.: T: Social Media Mining, An Introduction. Cambridge University Press, 20 Apr 2014
23. Medhat, W., Hassan, A., Korashy, H.: Sentiment analysis algorithms and applications: a survey. In: School of Electronic Engineering, Canadian International College, Cairo Campus of CBU, Egypt. Ain Shams University, Faculty of Engineering, Computers & Systems Department, Egypt. Received 8 September 2013; Revised 8 April 2014. Accepted 19 April 2014
24. Pang, B., Lee, L.: Opinion Mining and Sentiment Analysis (2008). ISBN: 978-1-60198-150-9 c
25. NLTK 3.4 documentation (March–June 2019). http://www.nltk.org/
26. Koukourikos, A., Stoitsis, G., Karampiperis, P.: Sentiment Analysis: A tool for Rating Attribution to Content in Recommender Systems
27. Pandas Documentation (March–June 2019). http://danielhnyk.cz/limitations-of-pandas-0-18-1-hdfstore/

Simulative Performance Analysis of All Optical Universal Logic TAND Gate Using Reflective Semiconductor Optical Amplifier (RSOA)

Kajal Maji, Kousik Mukherjee, and Mrinal Kanti Mandal

Abstract Reflective Semiconductor Optical Amplifier (RSOA) is a suitable gain medium due to its double pass characteristics compared to ordinary semiconductor Optical Amplifier. RSOA-based gain dynamics is utilized to design and analyze the new RSOA-based all optical logic TAND gate using Soliton pulses for the first time. The proposed logic TAND is a universal gate and using this gate NOT, OR, and AND gates are also designed.

Keywords Universal logic · Soliton · RSOA · Q value · Gain saturation

1 Introduction

Reflective Semiconductor Optical Amplifiers (RSOAs) are important nonlinear devices for the design of all optical broadband communication applications as it can simultaneously accept and modulate signals in the downstream while demodulating them with end-user information in the upstream communication link [1]. Recent proposals of RSOA-based colorless transmitters have been demonstrated for Passive Optical Network (PON) applications [1–3]. RSOA is an effective gain media to design all optical logic processors [4, 5], and has higher gain at lower injection current compared to ordinary Semiconductor Optical Amplifiers (SOAs) due to its double pass gain and is shown in the Fig. 1. TAND gate is a universal gate with only one high output being proposed using Tera Hertz Optical Asymmetric Demultiplexer (TOAD) which is proposed in [6]. Therefore, any basic gates or complex circuits can

K. Maji · K. Mukherjee (✉)
Department of Physics, B.B. College, Asansol 713303, India
e-mail: klmukherjee003@gmail.com

K. Maji
e-mail: kajalmaji200@gmail.com

K. Maji · M. K. Mandal
Department of Physics, National Institute of Technology, Durgapur 713209, India
e-mail: nitmkm@yahoo.co.in

D. Bhattacharjee et al. (eds.), *Proceedings of International Conference on Frontiers in Computing and Systems*, Advances in Intelligent Systems and Computing 1255,
https://doi.org/10.1007/978-981-15-7834-2_9

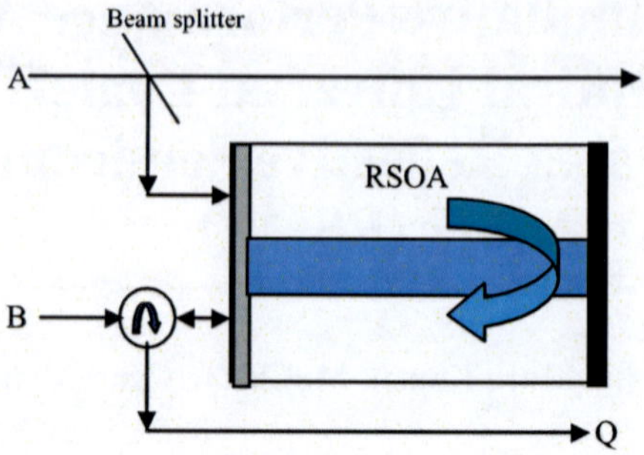

Fig. 1 RSOA-based TAND gate

be designed using this TAND gate. In this communication, a method of implementation of a new TAND gate using RSOA is proposed and analyzed for the first time as far as our knowledge goes. Using this TAND gate, three basic gates NOT, AND, and OR are designed which establish the fact that TAND is an universal gate. In Sect. 2.1, the basic principle of working of RSOA-based TAND gate is described. In Sects. 2.2–2.4, the designs of NOT, OR, and AND gates are described, which show that TAND gate is a universal gate. Section 3 shows the simulated results of operation of the proposed logic gates using MATLAB. Section 4 shows a comparison with other related work and gives a clear indication of improvement over those already established works. Section 5 concludes the paper.

2 Working Principle of the TAND Gate

The cross-gain modulation in RSOA is the basic mechanism behind the working of the proposed logic gate, which results in complementary or inverted output at probe wavelength due to pump modulation (Fig. 1). When pump or control signal is absent, the input probe experiences uncompressed gain resulting high output from RSOA. When pump or control is present, the input probe or data signal experiences compressed gain and output becomes low. Down conversion of wavelength in RSOA is better than up conversion in terms of BER [7], and pump wavelength and probe wavelength are selected 1550 nm and 1545 nm, respectively, for down conversion. The cross-gain modulation modeling can be found in many literatures [8, 9]. Now when both pump (or control) signals P_i and probe (or data) signal P_{in} passes through the RSOA, the data signal experiences lower gain due to gain compression by the high-intense pump. This results in low output. In the absence of pump signal, only probe signal is injected into the RSOA, and the data signal experiences unsaturated gain G_0 twice because of reflection at highly reflecting coating, and the output power from the RSOA is high at a probe signal wavelength.

When the control is injected into the RSOA, gain of the RSOA is [7–9]

$$G(t) = (\exp\{h(t)\})^2 \quad (1)$$

where h(t) is calculated by the following formula [7–9]:

$$h(t) = -\ln\left[1 - \left(1 - \frac{1}{G_0}\right)\exp\left(-\frac{E_c(t)}{E_s}\right)\right] \tag{2}$$

where E_s is the saturation energy, $E_c(t)$ is the control pulse energy of the RSOA. We consider the Soliton pulse train as control inputs or pump inputs described by [9]

$$P_i(t) = \sum_{n=1}^{n=N} a_{nA,B} P_{soli} \operatorname{sech}^2\left(1.763\frac{(t - n\chi)}{\tau_{fwhm}}\right) \tag{3}$$

where P_{soli} is the Soliton peak power [9].

2.1 Design and Principle of Universal TAND Gate

All optical universal logic TAND gates using Refractive Semiconductor Optical Amplifier is as shown in Fig. 1. It is made by a single RSOA. It has two inputs A and B and two outputs P = A and Q = ÃB as shown in the truth table (Table 1). The operation of the gate is as follows:

Case1: When both the inputs A and B are zero, there are no probe and pump signals present in the RSOA, so both the outputs (P and Q) of this RSOA becomes zero, i.e., P = 0 and Q = 0.

Case2: When the input A becomes '0' and B becomes '1', i.e., only the probe signal is present so that the output Q of the RSOA becomes high and output P becomes low, i.e., P = 0 and Q = 1.

Case3: When the input A becomes high, i.e., '1' and B becomes low, i.e., '0', and only the pump signal is present and the output Q of the RSOA becomes high and output P becomes low, i.e., P = 1 and Q = 0.

Case4: When both the inputs A and B become '1' both pump and probe signal are present so the output Q of the RSOA becomes low and output P is high, i.e., P = 1 and Q = 0.

Table 1 Truth table of TAND gate

Input		Output	
A	B	P	Q
0	0	0	0
0	1	0	1
1	0	1	0
1	1	1	0

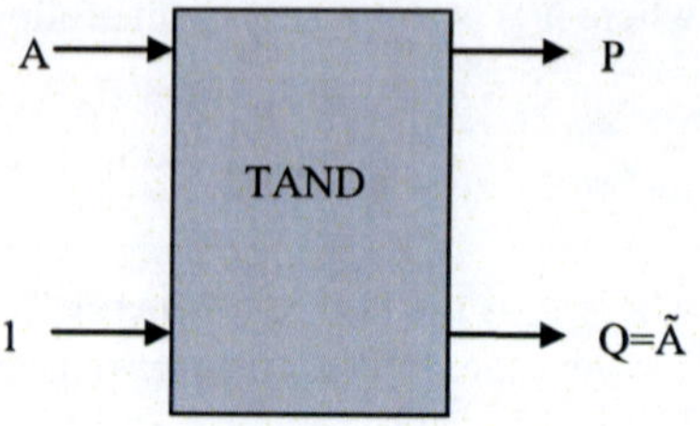

Fig. 2 RSOA-based NOT gate using TAND gate

2.2 NOT Gate Using TAND Gate

RSOA-based NOT gate using TAND gate is shown in Fig. 2. Here the input signal B = 1 is connected with RSOA as a data signal, and A is connected with RSOA as a control signal. Operation principle of this gate is given below.

Case1: When input A = 0, the output Q is high, i.e., in '1' state.

Case2: When input A = 1, the output Q is low, i.e., '0'. This gives the NOT operation of input control signal A. So the output of NOT gate is $Q = \overline{A}$.

2.3 AND Gate Using TAND Gate

RSOA-based AND gate using TAND gate is shown in Fig. 3. It consists of two TAND gates. The output Q of the TAND1 gate is connected to the control input of the TAND2 gate. Detail operational principle of AND gate is given below.

Case1: When A = 0 and B = 0, there is no control signal present of the RSOA-based TAND1 gate and data signal on the TAND2 gate so the output Q is low, i.e., '0'.

Case2: When A = 0 and B = 1, i.e., control signal is absent of the TAND1 gate and data signal is present on the TAND2 gate so the output Q is low, i.e., '0'.

Case3: When A = 1 and B = 0, i.e., control signal is present on the TAND1 gate and data signal is absent of the TAND2 gate so the output Q is low, i.e., '0'.

Case4: When both A = 1 and B = 1, output of the TAND 1 gate becomes low, i.e., control signal of the TAND2 is zero so the output Q is high, i.e., '1'.

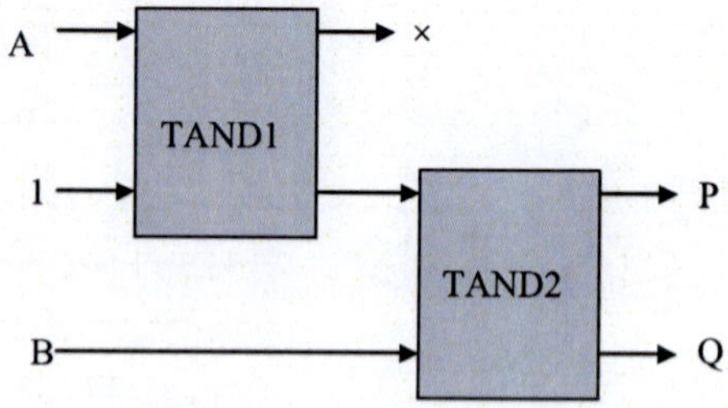

Fig. 3 RSOA-based AND gate using TAND gate

Fig. 4 RSOA-based OR gate using TAND gate

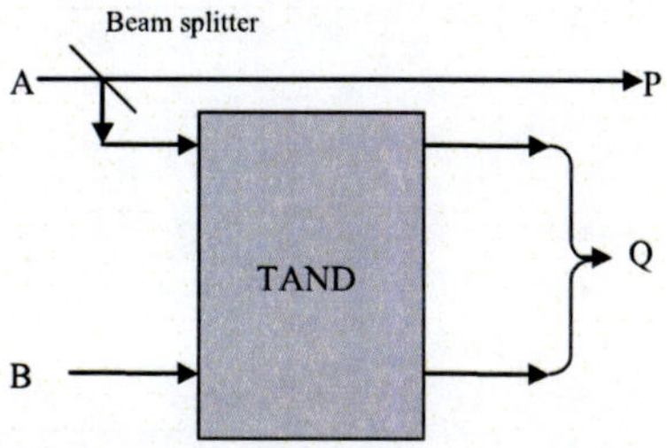

From the above discussion, we observed that the output of this gate is Q = AB which is the output of AND gate.

2.4 *OR Gate Using TAND Gate*

RSOA-based OR gate using TAND gate is shown in Fig. 4. It consists of only one TAND gate. Detail operational principle of OR gate is given below.

Case1: When both the inputs A and B are zero, there are no probe and pump signals present in the RSOA, so both outputs (P and Q) of this RSOA become zero, i.e., P = 0 and Q = 0.

Case2: When the input A becomes '0' and B becomes '1', i.e., only probe signal is present so the output Q of the RSOA becomes high and output P is low, i.e., P = 0 and Q = 1.

Case3: When the input A becomes '1' and B becomes '0', i.e., only pump signal is present so the output Q of the RSOA becomes high and output P is high, i.e., P = 1 and Q = 1.

Case4: When both the inputs A and B becomes '1' both pump and probe signal is present, so the output Q of the RSOA becomes high and output P is high, i.e., P = 1 and Q = 1.

From the above discussion, we observed that the output of this gate is $Q = A + \overline{A}B$ which is the output of OR gate.

3 Simulation Results

For the simulation of the performance of the TAND gate, the input control bits are considered as Soliton pulses given by Eq. (3). In Fig. 5a–c, the inputs A, B, and the TAND gate output bit patterns are shown. Inputs A and B are modulated at a rate of 200 Gbps. Figure 5a–c show a clear indication of A TAND B operation.

The outputs Extinction Ratio (E.R.) [7, 9], Contrast Ratio (C.R.) [7, 9], and Quality factor [9] of a TAND gate are also calculated. The variations of Extinction Ratio (ER) and Contrast Ratio (CR) and Q value with control pulse energy (E_c) are shown in

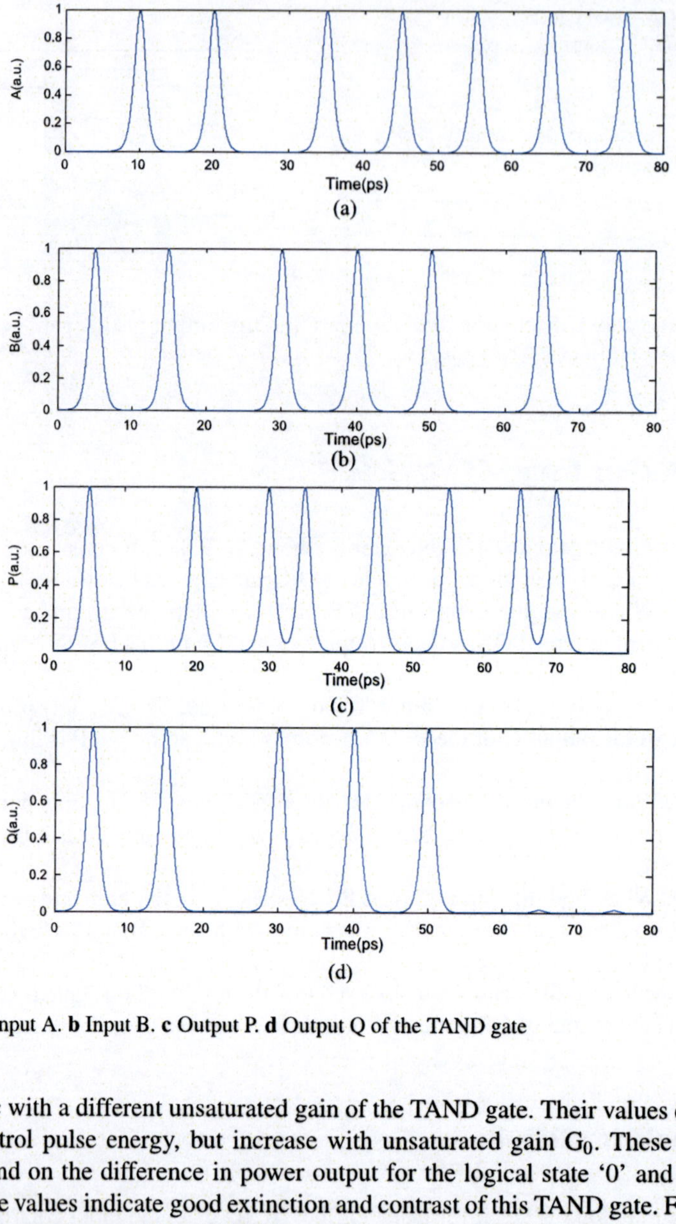

Fig. 5 **a** Input A. **b** Input B. **c** Output P. **d** Output Q of the TAND gate

Fig. 6a–c with a different unsaturated gain of the TAND gate. Their values decrease with control pulse energy, but increase with unsaturated gain G_0. These CR and ER depend on the difference in power output for the logical state '0' and '1', and their large values indicate good extinction and contrast of this TAND gate. Figure 6d shows the eye-diagram [7] of TAND gate. We used the RSOA parameters [9] Ec = 60fJ, $G_0 = 20$ dB, full-width half maximum 1 ps, the width of the RSOA 1.5um and depth of the RSOA 250 nm.

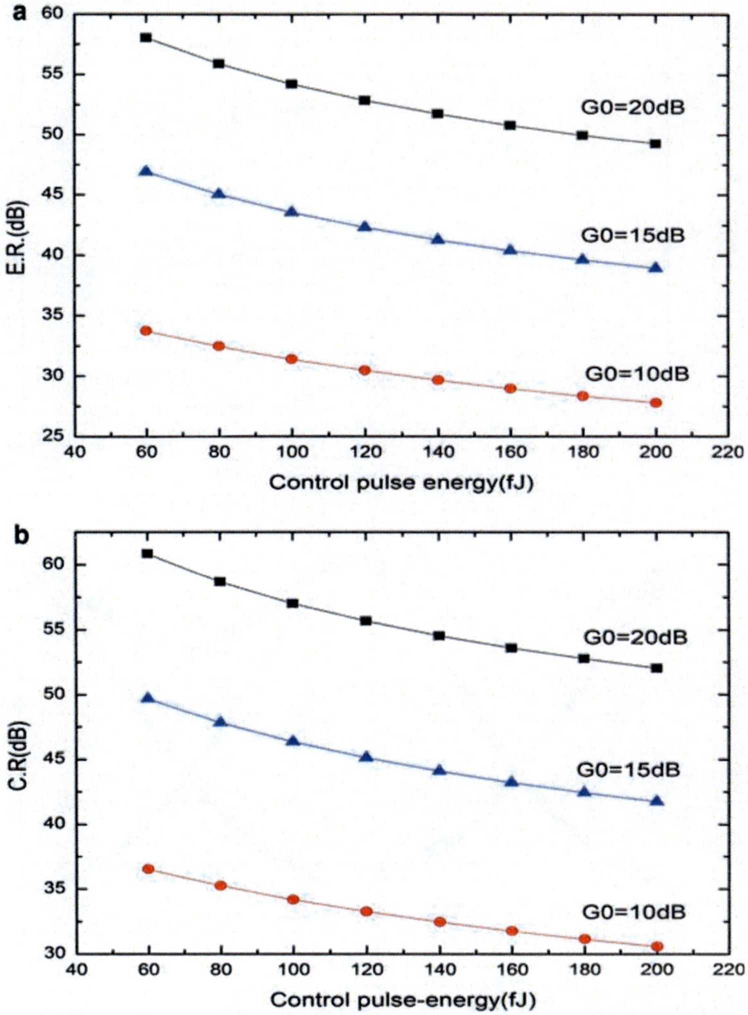

Fig. 6 **a** Variation of E.R. with E_c of the TAND gate. **b** Variation of C.R. with E_c of the TAND gate. **c** Variation of Q value with E_c of the TAND. **d** Eye-diagram of TAND gate

4 Comparison with Related Works

Table 2 shows comparison with similar types of gates. In works [6, 10–12], interferometric structures like TOAD and Mach Zehnder Interferometers(MZI) are used, but none have used RSOA. Most of the works have not calculated Q value, ER, CR except in [11, 12], where only ER and CR are calculated. But, the values of ER and CR is smaller compared to present work as clear from Table 2. None of the proposals have calculated Q value but present communication.

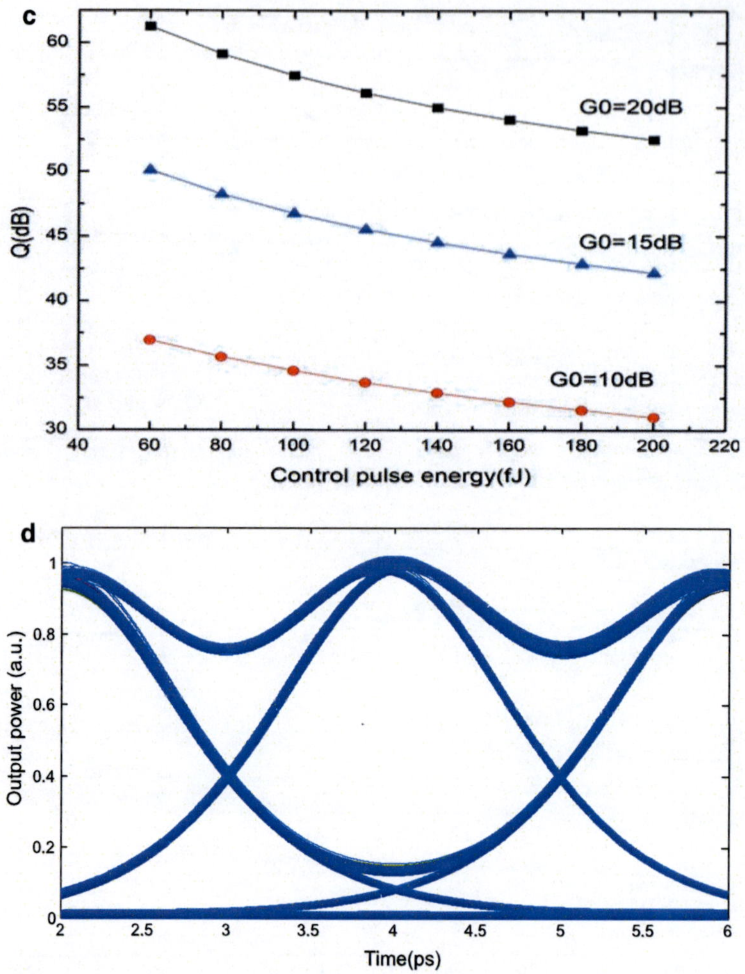

Fig. 6 (continued)

Table 2 Comparison table from previous work

Work	Bit pattern	E.R. (dB)	C.R. (dB)	Q value (dB)	Switching used
Ref. [6]	Yes	Not calculated	Not calculated	Not calculated	TOAD
Ref. [10]	No	Not calculated	Not calculated	Not calculated	TOAD
Ref. [11]	Yes	Not calculated	8.35	Not calculated	MZI
Ref. [12]	Yes	13.18	18.027	Not calculated	TOAD
Present work	Yes	58	60.83	61	RSOA

5 Conclusions

We have analyzed all optical universal logic TAND gate using RSOA. The maximum values of both ER and CR are found to be 58 and 60.83 dB for an unsaturated gain of 20 dB, and we also found higher Q value which is 61 dB. A higher Q value represents error-less performance. We also analyzed NOT, AND, and OR gates using TAND gate. Table 2 shows comparison with similar types of gates.

References

1. Rizou, Z.V., Zoiros, K.E.: Theoretical analysis of directly modulated reflective semiconductor optical amplifier performance enhancement by microring resonator-based notch filtering. Appl. Sci. **8**(2), 223 (2018). https://doi.org/10.3390/app8020223
2. Zhan, W., Zhou, P., Zeng, Y., Mukaikubo, M., Tanemura, T., Nakano, Y.: Optimization of modulation-canceling reflective semiconductor optical amplifier for colorless WDM transmitter applications. J. Lightwave Technol. **35**, 274–279 (2016)
3. Mandal, G.C., Mukherjee, R., Das, B., Patra, A.S.: Next-generation bidirectional triple-play services using RSOA based WDM radio on free-space optics PON. Opt. Commun. **411**, 138–142 (2018)
4. Kotb, A., Zoiros, K.E., Guo, C.: Performance investigation of 120 Gb/s all-optical logic XOR gate using dual-reflective semiconductor optical amplifier-based scheme. J. Comput. Electron. **17**(4), 1640–1649 (2018)
5. Mukherjee, K., Maji, K., Raja, A.: Design and performance analysis of all-optical soliton based 4-bit two's complement generator using Reflective Semiconductor Optical Amplifier (DevIC2019). https://doi.org/10.1109/devic.2019.8783531
6. Maity, G.K., Mandal, A.K., Samanta, S.: The all optical new universal gate using TOAD. IJACR, **4**(2), 15, 432–441 (2014)
7. Chattopadhyay, T.: All optical clocked delay flip flop using a single terahertz optical asymmetric demultiplexer based switch: a theoretical study. App. Opt. **49**(28), 5226 (2010). https://doi.org/10.1364/AO.49.005226
8. Mukherjee, K.: A novel frequency encoded all optical logic gates exploiting polarization insesensitive four wave mixing in semiconductor optical amplifier, filtering property of ADD/DROP multiplexer and non-linearity of reflective semiconductor optical amplifier. Optik **122**(10), 891–895 (2011). https://doi.org/10.1016/j.ijleo.2010.05.033
9. Maji, K., Mukherjee, K., Raja, A.: Performance of all optical logic soliton based AND gate using Reflective Semiconductor Optical Amplifier (RSOA) accepted for book chapter of Springer Lecture Notes in Electrical Engineering (LNEE), Book Series (Scopus Indexed). (ICCDC 2019). https://doi.org/10.1007/978-981-15-0829-5
10. Taraphdar, C.: Designing of all optical two bits full adder using TOAD, TMIN and feynman gate. Int. J. Comput. Intell. Res. **13**(5), 841–849 (2017)
11. Taraphdar, C., Chattopadhyay, T., Roy J.N.: Mach–Zehnder interferometer-based all-optical reversible logic gate. Opt. Laser Technol. **42**(2), 249–259 (2010)
12. Maity, G.K., Mandal, A.K., Samanta, S.: All-optical reversible hybrid new gate using TOAD. Int. J. Adv. Comput. Res. **4**(14), 2277–7970 (2014)

Development of a Publicly Available Terahertz Video Dataset and a Software Platform for Experimenting with the Intelligent Terahertz Visual Surveillance

Alexei A. Morozov and Olga S. Sushkova

Abstract A publicly available terahertz video dataset and a software platform for experimenting with the terahertz intelligent video surveillance are developed. The video dataset includes short videos of people with objects hidden under the clothing. The dataset is multimodal, that is, it contains synchronized videos of various kinds: terahertz, thermal, visible, near-infrared, and 3D. A special software platform is developed for the acquisition and preprocessing of the video data. The software platform includes a translator of the Actor Prolog language to Java and an open-source library of built-in classes for data acquisition and processing. In particular, the software enables one to project terahertz/thermal video data onto three-dimensional point clouds using 3D lookup tables. An experiment with the terahertz video data analysis based on various CNN architectures is described.

1 Problem Statement

Terahertz images/videos are wonderful test objects for development and validation of mathematical methods for image processing and visual pattern recognition because the images of this kind are inevitably fuzzy and one needs to tackle the problems linked with the low contrast, low signal-to-noise ratio, and low resolution of the images. The main reasons for these problems are a comparatively big wavelength of the terahertz radiation (usually from 0.1 up to 3 mm and more) and existent technical problems of terahertz video acquisition. In spite of these problems, the terahertz radiation remains to be an attractive object for researchers and engineers because of its unique useful properties [1, 4–6, 8, 11, 13, 15, 29–32, 36]. The terahertz radiation can penetrate textile, plastic, wood, ceramic, carton, and other dielectric materials;

A. A. Morozov (✉) · O. S. Sushkova
Kotel'nikov Institute of Radio Engineering and Electronics of RAS, Mokhovaya 11-7, Moscow 125009, Russia
e-mail: morozov@cplire.ru

O. S. Sushkova
e-mail: o.sushkova@mail.ru

D. Bhattacharjee et al. (eds.), *Proceedings of International Conference on Frontiers in Computing and Systems*, Advances in Intelligent Systems and Computing 1255,
https://doi.org/10.1007/978-981-15-7834-2_10

though this radiation is nonionizing, that is, it can be safely used in public places for detection of weapons and other dangerous objects hidden under the clothing. Moreover, it can be used for the detection of not only metallic but also ceramic and plastic weapons, for instance, the weapons created using a 3D printer.

The first generation of the industrial terahertz video surveillance systems was developed more than 10 years ago but, unfortunately, the terahertz video surveillance is still not used widely. The main application area of these systems is airports and other public places but real-life indicates that the manual examination of the passengers is a much more effective and cheap approach to reveal hidden objects. One can note that installed terahertz video surveillance systems are not applied in the airports and the airport personnel is not even interested in them. From the authors' point of view, the future application of the terahertz video surveillance is not the inspection of the passenger, but the fully automatic, wide, and hidden inspection of persons in public places. It is obvious that the modern generation of the terahertz video systems is not appropriate for this purpose; however, one can use them as labware for the investigation of the problem of automatic terahertz video surveillance and development of new methods for terahertz video analysis.

Even though there are a lot of papers that address the problem of terahertz video processing, there is still a lack of publicly available terahertz video datasets on the Web. We have created a publicly available terahertz video dataset [23] to make up for this deficiency. The dataset contains terahertz videos taken by the THERZ-7A industrial passive terahertz video surveillance system (Astrohn Technology Ltd.) [2] and is enhanced with additional information. The terahertz video is synchronized with several videos of other modalities: 3D cloud points, skeletons, and near-infrared video data taken by the Kinect 2 time-of-flight (ToF) camera (Microsoft Inc), thermal video data taken by the Thermal Expert V1 camera (i3system Inc), and RGB video data. The video dataset includes short videos of people with objects hidden under the clothing. Thus, the dataset is appropriate not only for experimenting with classical methods for improving the video data based on frame average, wavelet denoising, etc., [12, 14, 28, 33–35, 37, 38] but also for testing novel multimodal data fusion convolutional neural networks (CNN) architectures [9, 10]. To our knowledge, such publicly available multimodal data sets did not exist before.

A special software platform is developed for the acquisition and preprocessing of the multimodal video data. The software platform includes a translator of the Actor Prolog logic language [16–19, 21, 24, 25] to Java and open-source library [20] of built-in classes for data acquisition and processing. The publicly available installation pack of the software [22] contains ready-for-use logic scripts for terahertz/thermal video data acquisition, preprocessing, and logical analysis. In particular, the software enables one to project terahertz/thermal video data onto the three-dimensional point clouds using 3D lookup tables [26].

The paper is organized in the following way. In the first section, the structure of the terahertz video dataset is described. In the second section, the structure and main features of the terahertz video acquisition/processing software are discussed. In the third section, an experiment with the terahertz video data analysis is described.

2 The Structure of the Video Dataset

Each data file of the dataset contains one video of a person with an object hidden under the clothing. During the recording, the person moves and rotates a bit to provide a permanent change of the image (see Fig. 1).

The length of the video is 85 s and the frame rate is about 6 frames per second. The video record contains several video streams of different modalities. Each frame includes the following information:

1. A 2-byte integer matrix of terahertz image produced by THERZ-7A. The resolution is 110 × 180. The operating spectral range is 0.23–0.27 THz.
2. RGB image produced by the THERZ-7A system. The resolution is 576 × 704.
3. Point cloud produced by the ToF camera of the Kinect 2 device. The point cloud is a matrix of the resolution 512 × 424. Each element of the matrix contains the x, y, and z coordinates of the point. Note that Kinect 2 detects people in the video scene; thus, one can operate on the point cloud of the separate person in the 3D video.
4. RGB image produced by Kinect 2 and projected onto the 3D image. Thus, the 1920 × 1080 initial resolution of Kinect 2 RGB images is decreased up to 512 × 424 to save the memory.
5. Near-infrared image produced by Kinect 2. The resolution is also 512 × 424.
6. Images of skeletons of people detected by the Kinect 2 device.
7. 2D and 3D frames of person point clouds and skeletons detected in the video.
8. Thermal image 640 × 480 produced by the TE V1 thermal camera.

The data is published in Web Site [23]. Each data file is accompanied by an image of the object hidden under the clothing of the person and several examples of processed video images that give an idea of the location of the hidden object and the pose of the person in the video (Fig. 2). The video files are assembled in groups of four files. Each group is accompanied by two 3D lookup tables that allow you to project terahertz and thermal images onto 3D point clouds. Note that the data files

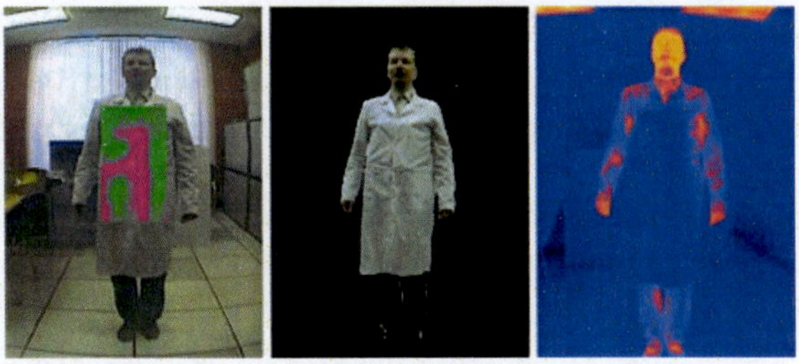

Fig. 1 An example of synchronized video streams recorded in a video file. From left to right: **a** terahertz and visual images produced by the THERZ-7A passive terahertz video system; **b** RGB image projected onto the 3D point cloud produced by Kinect 2; **c** thermal image produced by the Thermal Expert V1 camera

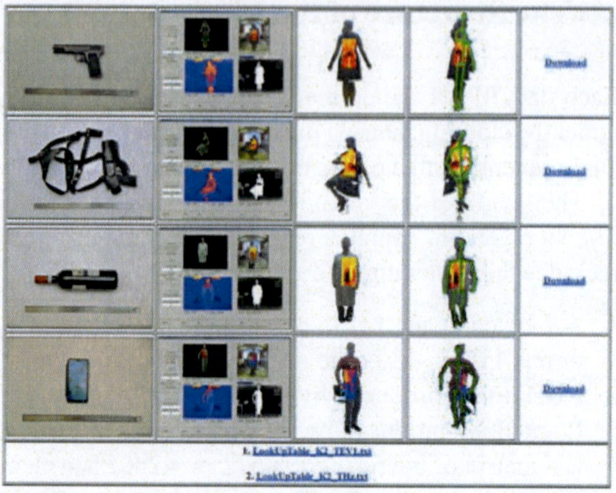

Fig. 2 The overview of the THz and thermal video data web site [23]

are recorded in the $AP2J$ multimodal video format that was specially developed in the Actor Prolog programming system. Thus, recording, reading, and preprocessing of these files require special software supplied in the Web Site.

3 The Software

The special software includes a translator of the Actor Prolog logic language to Java that can be freely downloaded from the Web Site [22] and open-source library of built-in classes for data acquisition and processing [20]. The built-in classes for the data acquisition from special hardware including the THERZ-7A terahertz camera and the TE V1 thermal camera were developed in the framework of the terahertz dataset development. One can find a set of ready-for-use programs based on these built-in classes in the installation pack of the Actor Prolog system. The main programs are the following ones:

1. The *Multimedia_01_3D_THz_film_recorder* program is used for the creation of the dataset. This program is useful also for viewing the video files and extracting video data frames of separate modality when necessary. The user interface of the program (see Fig. 3, left) includes four windows for the data view. The upper left window contains the image acquired by Kinect 2. The image of the skeleton is added to the image to indicate the location and pose of the person. The upper right window contains the image acquired by the THERZ-7A device. One can select the terahertz, RGB, or combined image. One can select also a color map for the visualization of the terahertz data. The lower left window contains the image produced by the thermal camera. A color map for the thermal data visualization can be also selected. The lower right window contains the near-infrared image produced by Kinect 2.

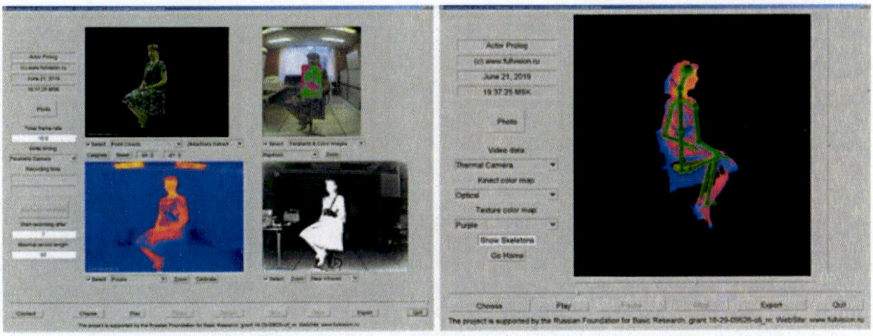

Fig. 3 The user interface of the multimodal video acquisition program (at the left) and the 3D viewer (at the right)

2. The *Multimedia_02_3D_viewer* program serves for the projection of the terahertz and thermal video data onto the point clouds. One can use it for 3D video playing and extracting fused 3D and terahertz/thermal frames (see Fig. 3, right). Note that the program automatically loads 3D lookup tables for data visualization; however, these lookup tables are to be placed manually to the directory where the corresponding video data files are stored.
3. The *Multimedia_04_3D_normalizer* program is an example of a program that performs a semantic fusion of multimodal video data and preparation of a training image set for experimenting with the machine learning methods. The idea of the semantic fusion is in that several images are combined into a united one in the following way: the semantic of one image is used to control the processing of another one [27]. In the program under consideration, the semantic fusion is used for the normalization of the video frames that is necessary for successful CNN training. The schema of the semantic video fusion consists in that the coordinates of the skeleton of the person recognized by the Kinect 2 device are used for correction of the point cloud position. The terahertz video data is projected onto the point cloud of the person; thus, all the terahertz images obtain a standard size and location in the frame (see Fig. 4).

4 An Example of Terahertz Video Analysis

Let us consider an example of terahertz video data analysis. The goal of the experiment is to check whether a neuronal network can automatically distinguish dangerous and safe objects placed under the clothing of people. Note that this problem statement differs from the standard one. We are going to check whether terahertz video surveillance can be used for automatic and hidden check of people in public places; this is not a case of checking passengers at the airport and people may have usual things in pockets. The neural network has to detect suspicious persons that hide

Fig. 4 The user interface of the program for semantic video data fusion and training dataset preparation

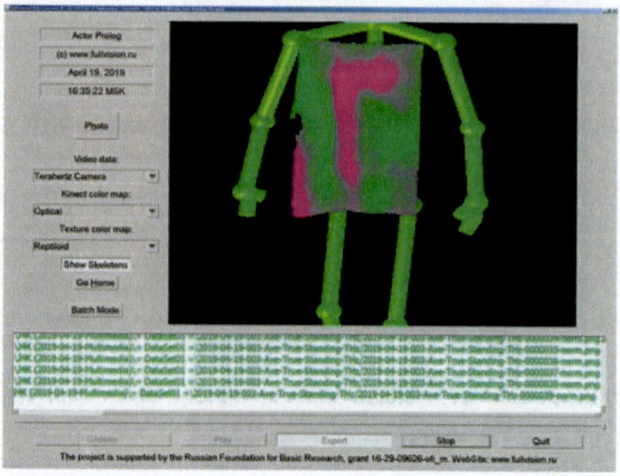

dangerous objects under the clothing. We will refer to the automatic terahertz video surveillance as intelligent terahertz video surveillance by analogy with the usual and infrared video surveillance.

We have prepared a dataset using the *Multimedia_04_3D_normalizer* program. The *Cool* color map was used and the skeletons were not included in the images (see Fig. 5). The size of the training dataset was 11520 frames. The size of the test dataset was 26020 frames. Note that the training and test datasets were formed using different video files because it is not correct to use the same videos for the estimation of the quality of object recognition; each video contains sequences of very similar frames and the usage of the same video sequences for the training and testing usually causes an overestimation of the recognition accuracy. The Darknet19, ResNet50,

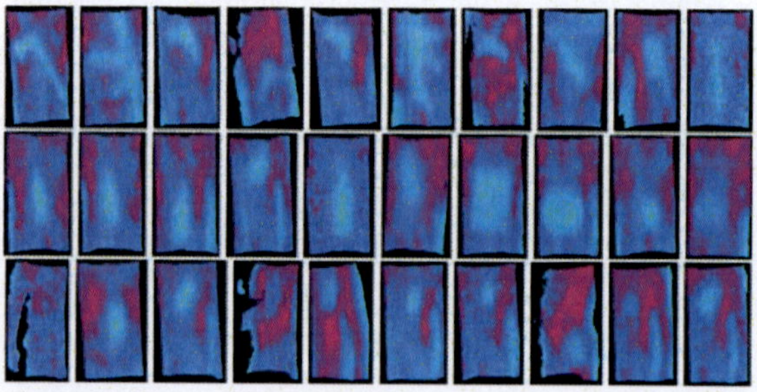

Fig. 5 Examples of terahertz images included in the training dataset. First row: dangerous objects like an axe, the Kalashnikov submachine-gun (AK), the Tokarev pistol, the Walther pistol, the M16 rifle, a shoulder holster with a gun inside, a knife, a meat knife, and a hummer. Second row: suspect and unusual objects like bottles of various shapes and sizes, a candy box lid, a saucepan lid, a tin, and a ceramic plate. Third row: safe objects like smartphones and mobile phones of various manufacturers and sizes, cigarettes, and USB disks; no hidden objects

Table 1 These are the results of the training of the convolutional networks of various architectures. The training process included two stages: 250 epochs without transformations and 250 epochs with the flip and warp transformations (115500 iterations in total). The image size was 224 × 224. The batch size was 50

Network	Accuracy	Precision	Recall	F1 score
Darknet19	0.8167	0.8136	0.7517	0.7686
ResNet50	0.8344	0.8195	0.8129	0.8122
AlexNet	0.6441	0.6400	0.5853	0.5966

and AlexNet CNN architectures implemented in the DeepLearning4J library [7] were tested (see Table 1).

The results demonstrate that modern CNN architectures like Darknet19 and ResNet50 can automatically distinguish safe and dangerous objects hidden under the clothes. Undoubtedly, the result can be substantially improved with the use of other modalities of the data and multimodal data fusion CNN architectures.

5 Conclusion

A publicly available terahertz video dataset [23] and special software were created to facilitate research in the area of intelligent terahertz video surveillance. Developed logic programming means include a set of built-in classes of the Actor Prolog language for multimodal (terahertz, 3D, infrared, and RGB) video data acquisition, writing, reading, and preprocessing [20]. A method of semantic fusion of multimodal video data is implemented for normalizing the terahertz video data. It is demonstrated that CNN can be successfully used for automatic terahertz video surveillance of people in public places. The intelligent terahertz video surveillance, of course, cannot guarantee the highest level of safety like a manual check of the passengers at the airport, but it can really detect a terrorist or just a crazy person who tries to carry a weapon in the public place.

The authors are grateful to Ivan A. Kershner and Renata A. Tolmacheva for the help in the preparation of terahertz/3D video samples, Margarita N. Khokhlova for the help in the experiments with CNN, and Angelos Barmpoutis for his J4K library [3] which was used for the data collection. We are grateful to the Astrohn Technology Ltd. and OOO ASoft, who provided us with the THERZ-7A terahertz scanning device. The work was carried out within the framework of the state task. This research was partially supported by the Russian Foundation for Basic Research (project number 16-29-09626-ofi-m).

References

1. Antsiperov, V.E.: Automatic target recognition algorithm for low-count terahertz images. Comput. Opt. **40**(5), 746–751 (2016)
2. ASTROHN: ASTROHN Technology Ltd. (2019). http://astrohn.com
3. Barmpoutis, A.: Tensor body: Real-time reconstruction of the human body and avatar synthesis from RGB-D. IEEE Trans Cybern **43**(5), 1347–1356 (2013)
4. Bhattacharyya K, Deka R, Baruah S (2017) Automatic RADAR target recognition system at THz frequency band. A review. ADBU J. Eng. Technol. **6**(3)
5. Chen, S., Luo, C., Wang, H., Deng, B., Cheng, Y., Zhuang, Z.: Three-dimensional terahertz coded-aperture imaging based on matched filtering and convolutional neural network. Sens (Basel, Switzerland) **18**(5), 1342 (2018). https://doi.org/10.3390/s18051342
6. CONSORTIS: CONSORTIS. Final Publishable Summary Report, Teknologian Tutkimuskeskus VTT (2018)
7. DeepLearning4J: Deep Learning for Java. Open-source, distributed, deep learning library for the JVM (2019). https://deeplearning4j.org
8. Dolganova, I.N., Zaytsev, K.I., Metelkina, A.A., Karasik, V.E., Yurchenko, S.O.: A hybrid continuous-wave terahertz imaging system. Rev Sci Instr **86**(113704) (2015). https://doi.org/10.1063/1.4935495
9. Gao, M., Jiang, J., Zou, G., John, V., Liu, Z.: RGB-D-based object recognition using multimodal convolutional neural networks: a survey. IEEE Access **7**, 43,110–43,136 (2019)
10. Guo, L., Qin, S.: High-performance detection of concealed forbidden objects on human body with deep neural networks based on passive millimeterwave and visible imagery hidden object detection and recognition in passive terahertz and mid-wavelength infrared. International Journal of Infrared and Millimeter Waves (2019). https://doi.org/10.1007/s10762-018-0558-3
11. Kowalski, M.: Hidden object detection and recognition in passive terahertz and mid-wavelength infrared. J. Infrared, Millimeter Terahertz Waves 1–18 (2019)
12. Kowalski, M., Kastek, M., Piszczek, M., Życzkowski, M., Szustakowski M.: Harmless screening of humans for the detection of concealed objects. WIT Trans. Built Environ. **151**, 215–223 (2015)
13. Liang, D., Pan, J., Yu, Y., Zhou, H.: Concealed object segmentation in terahertz imaging via adversarial learning. Optik—Int. J. Light Electron Optics **185**, 1104–1114 (2019). https://doi.org/10.1016/j.ijleo.2019.04.034
14. López-Tapia, S., Molina, R., de la Blanca, N.P.: Using machine learning to detect and localize concealed objects in passive millimeter-wave images. Eng. Appl. Artif. Intell. **67**, 81–90 (2018)
15. Mittleman, D.M.: Twenty years of terahertz imaging. Opt. Express **26**(8), 9417–9431 (2018)
16. Morozov, A.A.: The Prolog with actors. Programmirovanie **5**, 66–78 (1994). in Russian
17. Morozov, A.A.: Actor Prolog: an object-oriented language with the classical declarative semantics. In: Sagonas, K., Tarau, P. (eds.) IDL 1999, pp. 39–53. France, Paris (1999)
18. Morozov, A.A.: Logic object-oriented model of asynchronous concurrent computations. Pattern Recogn. Image Anal. **13**(4), 640–649 (2003)
19. Morozov, A.A.: Operational approach to the modified reasoning, based on the concept of repeated proving and logical actors. In: Salvador Abreu, V.S.C. (ed.) CICLOPS 2007, pp. 1–15. Porto, Portugal (2007)
20. Morozov, A.A.: A GitHub repository containing source codes of Actor Prolog built-in classes (2019). https://github.com/Morozov2012/actor-prolog-java-library
21. Morozov, A.A., Sushkova, O.S.: Real-time analysis of video by means of the Actor Prolog language. Comput. Opt. (Special issue 3), 97–105 (2017)
22. Morozov, A.A., Sushkova, O.S.: The intelligent visual surveillance logic programming web site (2019a). http://www.fullvision.ru
23. Morozov, A.A., Sushkova, O.S.: THz and thermal video data set (2019b). http://www.fullvision.ru/monitoring/description_eng.php

24. Morozov, A.A., Sushkova, O.S., Polupanov, A.F.: A translator of Actor Prolog to Java. In: Bassiliades, N., Fodor, P., Giurca, A., Gottlob, G., Kliegr, T., Nalepa, G., Palmirani, M., Paschke, A., Proctor, M., Roman, D., Sadri, F., Stojanovic, N. (eds.) RuleML 2015 DC and Challenge. CEUR, Berlin (2015)
25. Morozov, A.A., Sushkova, O.S., Polupanov, A.F.: Towards the distributed logic programming of intelligent visual surveillance applications, Part II. In: Pichardo-Lagunas, O., Miranda-Jimenez, S. (eds.) Advances in Soft Computing, pp. 42–53. Springer International Publishing, Cham (2017)
26. Morozov, A.A., Sushkova, O.S., Petrova, N.G., Khokhlova, M.N., Migniot, C.: Development of agent logic programming means for multichannel intelligent video surveillance. RENSIT **10**(1), 101–116 (2018). https://doi.org/10.17725/rensit.2018.10.101
27. Morozov, A.A., Sushkova, O.S., Kershner, I.A., Polupanov, A.F.: Development of a method of terahertz intelligent video surveillance based on the semantic fusion of terahertz and 3D video images. CEUR 2391 (2019). http://ceur-ws.org/Vol-2391/paper19.pdf
28. Murashov, D.M., Morozov, A.A., Murashov, F.D.: A technique for detecting concealed objects in terahertz images based on information measure. CEUR 2391 (2019). http://ceur-ws.org/Vol-2391/paper37.pdf
29. Ozhegov, R., Gorshkov, K., Vachtomin, Y.B., Smirnov, K., Finkel, M., Goltsman, G., Kiselev, O., Kinev, N., Filippenko, L., Koshelets, V.: Terahertz imaging system based on superconducting heterodyne integrated receiver. In: THz and Security Applications, pp 113–125. Springer (2014)
30. Semenov, A., Richter, H., Böttger, U., Hübers, H.W.: Imaging terahertz radar for security applications. In: Terahertz for Military and Security Applications VI, International Society for Optics and Photonics, vol. 6949, p. 694902 (2008)
31. Shen, X., Dietlein, C.R., Grossman, E., Popovic, Z., Meyer, F.G.: Detection and segmentation of concealed objects in terahertz images. IEEE Trans Image Process. **17**(12), 2465–2475 (2008)
32. Sizov, F.: Infrared and terahertz in biomedicine. Semicond. Phys. Quant. Electr. Optoelectr. **20**(3), 273–283 (2017)
33. Trofimov, V.A., Trofimov, V.V., Shestakov, I.L., Blednov, R.G., Kovalev, V.Y.: Effective algorithm based on Fourier transform for the passive THz image quality enhancement. In: Image Sensing Technologies IV, vol. 10209, p. 1020907 (2017)
34. Xu, L.M., Fan, W.H., Liu, J.: High-resolution reconstruction for terahertz imaging. Appl. Opt. **53**(33), 7891–7897 (2014). https://doi.org/10.1364/AO.53.007891
35. Yeom, S., Lee, D.S., Lee, H., Son, J.Y., Guschin, V.P.: Vector clustering of passive millimeter wave images with linear polarization for concealed object detection. Progr. Electromagn. Res. Lett. **39**, 169–180 (2013)
36. Zhang, J., Xing, W., Xing, M., Sun, G.: Terahertz image detection with the improved faster region-based convolutional neural network. Sensors **18**(7), 2327 (2018). https://doi.org/10.3390/s18072327
37. Zhao, Y., Qiao, Y., Zhang, C., Zhao, Y., Wu, H.: Terahertz/visible dual-band image fusion based on hybrid principal component analysis. J. Phys. Conf. Ser. **1187**(042), 096 (2019a). https://doi.org/10.1088/1742-6596/1187/4/042096
38. Zhao, Y., Sun, X., Zhang, C., Zhao, Y.: Using Markov constraint and constrained least square filter to develop a novel method of passive terahertz image restoration. J. Phys. Conf. Seri. 1187:042,094 (2019b). https://doi.org/10.1088/1742-6596/1187/4/042094

Identification of Plant Species Using Deep Learning

S. K. Mahmudul Hassan and Arnab Kumar Maji

Abstract Classification of plant species using machine learning is an automated task for recognizing the unknown plant species. Classification is very challenging due to the morphological similarity of different species of plants. In this paper, we have proposed a deep convolutional neural network (CNN) based model for the identification of plant species using plant leaf images. The main intuition of using CNN is learning the leaf features directly from the input images/data. Furthermore, it is observed that CNN-based techniques significantly increase the performances in case of different plant species having identical shape and sizes of leaves. The proposed model is compared with other existing techniques in the same domain. It is found that our model improves the recognition accuracy significantly.

Keywords Machine learning · Convolutional neural network(cnn) · Feature extraction

1 Introduction

Plants are the richest property of the earth and very essential for human life and the environment as well. Recognition of plant species provides collectable information in plant research and also can be useful for protecting useful plant species. To collect the information about plants, one needs to visit either to a botanist, or to a nursery, or needs to collect information from the Internet, which is very time consuming [19]. Therefore, automated identification of plant species is very effective and speed up the process. For classification/Identification of plants, extraction of feature is the essential task and leaves are the main visual organ that can be exploited by computer vision and pattern recognition [1]. Nowadays, convolutional neural network(CNN)

S. K. Mahmudul Hassan (✉) · A. Kumar Maji
Department of Information Technology, NEHU, Shillong 793022, India
e-mail: hassanmahmudul89@gmail.com

A. Kumar Maji
e-mail: arnab.maji@gmail.com

D. Bhattacharjee et al. (eds.), *Proceedings of International Conference on Frontiers in Computing and Systems*, Advances in Intelligent Systems and Computing 1255,
https://doi.org/10.1007/978-981-15-7834-2_11

Fig. 1 Some Sample leaf dataset [3] images

has achieved remarkable results in the field of image classification and pattern recognition. The plant leaf image feature representation is a crucial component of a plant leaf recognition algorithm. There are different methods to represent and describe the feature in machine learning for any classification problem. Among them, the most prominent methods are traditional hand-crafted features and deep learning (DL) based features. In hand-crafted features, we need to extract the different features like color, shape, and texture and then apply the classifier for identification. However, in DL-based methods, features are extracted and learned automatically as it is superior in providing deeper information about the image. For this reason, DL-based classification is so popular and widely used.

This paper begins with the introduction, which states about the processes for the identification of different plants and usefulness of Deep Learning. Next section describes a short review of different existing methodologies and their descriptions in context with identification of plants. Section 4 is about the implementation of DL in identification of plant species. Section 5 presents a discussion on the comparative results of the implemented methods. Next section is about conclusion along with the future aspects.

2 A Few Existing Plant Identification Techniques

Different approaches are there in the classification of plant species among them, shape and texture feature is the most popular feature in identification. Kumar et al. [9] use different shape features and edge-based features for classification of different plant species. Meyer et al. [18] introduced elliptic Fourier and discriminant analysis methods to distinguish different plant species based on their leaf shape. Zhao et al. [22] used Centroid Contour Distance Curve(CCDC) to describe the shape of leaf based upon their edge points. They converted the complex shape of a leaf into a graph structure that describes the topological skeleton of the leaves. Shape features, Hough histogram and Fourier histogram were used to describe leaf features by Mouine et al. [16]. Xiao et al. [20] used the Histogram of Oriented Gradient (HOG) to represent plant shape and used Maximum Margin Criterion(MMC) for dimension reduction.

The texture feature is also a widely adopted characteristic for leaf classification. Herdiyeni et al. [5] used a Local Binary Pattern(LBP) based texture feature, combined with shape and color feature (Morphological Feature). Then Probabilistic Neural Network (PNN) classifier was used for identification of Indonesian medical plants. Muthevi et al. [17] utilized the magnitude component of LBP combined with central pixel and sign component of LBP for classification and this method is rotation invariant. Shape, Color, Texture, and vein features were used by Kadir et al. [6] for identifying the plant leaf. Hu moments and uniform LBP histogram pattern feature were used by Lukic et al. [14] for feature extraction and Support Vector Machine (SVM) was used as the classifier.

DL has shown an eminent success in large-scale image recognition. Introduction of convolution neural network avoids the use of hand-crafted feature extraction. Grinblat et al. [4] use CNN in the plant leaf classification problem using the leaf vein pattern. They classify only the white bean, red bean, and soybean leaves. Liu et al. [13] used a hybrid DL network, namely, auto-encoder and convolution neural network. They applied CNN to extract the features and SVM is used for classification purposes. Deep CNN is used to recognize plant leaf by Lee et al. [11]. They applied the Deconvolution Network(DN) to gain intuition about the chosen features from different levels. It is found that venation is the best representation of features in comparison with the shape features of leaves. Further, Lee et al. [12] carried out their research and proposed one hybridized method for classification which includes both generic- and organ-based features. In this paper, at first, they extracted the organ information of leaves. Then the correlation between the organ information and generic-based features was established. Then, the plant leaves were classified based on those information.

3 Proposed Methodology for Identification of Plant Species Using Deep Learning

In this section, we are going to exploit the methodology that we have used in recognition of plant species from leaf images. In this work, the plant leaf images from different open data sets are used. The images are preprocessed and noises are removed using state-of-the-art methodologies. Then the leaf images are segmented and finally a CNN-based deep learning method is used for classification of plant leaf.

Prior to applying a CNN-based model, preprocessing steps are carried out to standardize the image orientation, scaling, and translation. In our experiment, we have used MATLAB 2016 for preprocessing and removal of noise. Different state-of-the-art filtering techniques like median filter and mean filter are used for noise removal. Median filter considers each pixel in the image and looks at its neighboring pixel to decide whether or not it is a representative of its surroundings. Median filter preserves the edges and lines in the image in the best possible ways. Median filter can be represented by the following equation:

$$y\left[m,n\right] = \text{median}\left\{x\left[i,j\right], (i,j) \in w\right\} \tag{1}$$

where w represents the neighbouring pixel value centered around the pixel (m,n). Mean filter is applied for smoothing the image by reducing the intensity variations among the neighboring pixels.

3.1 CNN Architecture

The CNN architecture of our system consists of different layers, such as Convolutional layer, Activation layer, Pooling layer, and Softmax classifier. Figure 2 represents the architecture of implemented CNN model. Table 1 summarizes the parameter that is used in the proposed CNN-based model.

3.1.1 Convolution Layer

This layer performs a convolution operation using a filter matrix of size $w \times h \times d$ (w: width, h: height, d: dimension) with the input image. The filter is passed over the input image to produce a filtered image by multiplying the filter with each patch of the image matrix. The output of the convolution layer is defined as

$$O = b + \sum_k W_k X_k \tag{2}$$

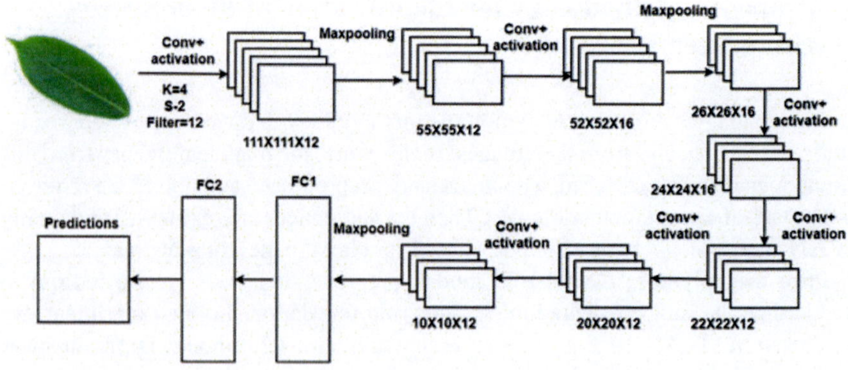

Fig. 2 Implemented CNN architecture

Table 1 A summary of different parameters used in the proposed CNN architecture

Layers	Output size ($w \times h \times c$)	Kernel window ($w \times h$/stride)	Pooling window ($w \times h$/stride)
Conv	$111 \times 111 \times 12$	$4 \times 4/2$	-
Pooling	$55 \times 55 \times 12$	-	$2 \times 2/2$
Conv	$52 \times 52 \times 16$	$4 \times 4/1$	-
Pooling	$26 \times 26 \times 16$	-	$2 \times 2/2$
Conv	$24 \times 24 \times 16$	$3 \times 3/1$	-
Conv	$22 \times 22 \times 12$	$3 \times 3/1$	-
Conv	$20 \times 20 \times 12$	$3 \times 3/1$	-
Pooling	$10 \times 10 \times 12$	-	$2 \times 2/2$
FC	4096	-	-
FC	4096	-	-
FC	32	-	-

where W_k is the weight in the particular location of the kernel and X_k is the pixel value of the image at the corresponding location. The output matrix size is given as

$$O_w = \frac{M - w}{S_w} + 1 \tag{3}$$

$$O_h = \frac{N - h}{S_h} + 1 \tag{4}$$

where M is the height and N is the width of the input image. S_w and S_h is the stride width and height of the convolutional window.

3.1.2 Activation Layer

Activation layer is used to monitor the firing rate of the neurons. The activation function is the nonlinear transformation that is applied over the input image. Activation function decides the final value of the neuron and whether the neuron needs to be fired to the next layer or not. Different activation functions are there such as Rectified Linear Unit (ReLU), Sigmoid, Tanh, etc. Among all these, ReLU is the most widely used activation function and we use ReLU as an activation function in our work.

3.1.3 Pooling Layer

The basic task of pooling layer is basically to reduce the weight in the image and cutting down the computation time. Pooling aggregates all of them into one single value. Different forms of pooling are there, namely, Max Pooling, Min Pooling, and

Average Pooling. Among these, we use Max Pooling where it takes the maximum value from the pooling window. Max pooling operation can be represented using the following equation:

$$\text{MaxPooling} = \max(w(i, j)) \tag{5}$$

where $i = 1, 2 \ldots\ldots \cdot w$ and $j = 1, 2 \ldots\ldots h$ $w \times h$ is the size of pooling window. Output matrix is of size is given as

$$O_w = \frac{M - w}{S_w} + 1 \tag{6}$$

$$O_h = \frac{N - h}{S_h} + 1 \tag{7}$$

3.1.4 Softmax Classifier

Softmax classifier gives us the probabilities for each class and used in the final layer. Softmax function can be defined as

$$f(x_i) = \frac{e^{x_i}}{\sum_j^k e^{x_j}}, i = 0, 1, 2 \ldots k \tag{8}$$

4 Dataset and Experiment Result

In this section, we are going to evaluate the performance such as accuracy, loss, and other parameters like precision, recall, F1-score of the proposed CNN model for identification of the plant. All the training and testing process of the CNN model implemented using Scikit-learn, OpenCV, and Keras libraries using Python programming language.

4.1 Dataset

In our experiment, we have used different leaf image dataset like Flavia [3], MalayaKew (MK), and LeafSnap dataset [10]. MalayaKew(MK) [15] consists of leaves from 44 different plant species. Two files associated with this dataset, namely, MK-D1 and MK-D2. MK-D1 consist of 2288 training and 528 testing images. The MK-D2 consists of 34672 and 8800 number of training and testing patches of leaf images, respectively, all images are of size 256 × 256. Flavia leaf image dataset consists of 1907 number of images of 32 different plant species and are of size 1600 × 1200. LeafSnap plant dataset consists of 7719 and 23147 number of field

Table 2 Performance comparison on different datasets

Dataset	Accuracy	Avg precision	Avg recall	F1 score
MK-D1	99.15	94.3	89.2	91.67
MK-D2	99.43	95.1	90.8	92.90
Flavia	99.75	96	92.5	94.21
LeafSnap	89.17	91.7	79.6	85.22

and lab images captured by the camera. Each image is of size 800 × 600. We consider 70% for training purpose and 30% for testing in our experiment. Figure 1 shows the sample leaf images of Flavia leaf image dataset.

4.2 *Experimental Result*

We evaluate the performance using various matrices as Accuracy rate, Precision, Recall, and F1-score. Matrices are evaluated as follows:

$$\text{Accuracy } rate = \frac{\mathrm{T}_c}{\mathrm{T}_n} \tag{9}$$

where T_c is the number of correct identification and T_n is the total number of images considered for evaluation.

$$\text{Precision}(P) = \frac{\mathrm{TP}}{\mathrm{TP} + \mathrm{FP}} \tag{10}$$

$$\text{Recall}(R) = \frac{\mathrm{TP}}{\mathrm{TP} + \mathrm{FN}} \tag{11}$$

$$\text{F1}-\text{score} = \frac{2 * \text{recall} * \text{precision}}{\text{recall} + \text{precision}} \tag{12}$$

where TP is the true positive values (i.e., correct predictions from the total number of images). FP is the false positive prediction to the class. FN is false negative which is equal to the number of wrongly predicted output from a given class [7].

Before training, all the images are standardized into the dimension 224 × 224 and filtering techniques are used to remove noise. Figure 3a, b represents the testing loss and training loss with respect to no. of epochs. We have calculated the cross-entropy loss to measure the performance of classification and it is found that it decreases with the increase of epoch. Figure 3c shows the accuracy of Flavia dataset and it increases monotonically. Table 2 and Fig. 3d show the maximum accuracy obtained on different datasets. It is found that our proposed CNN-based plant identification

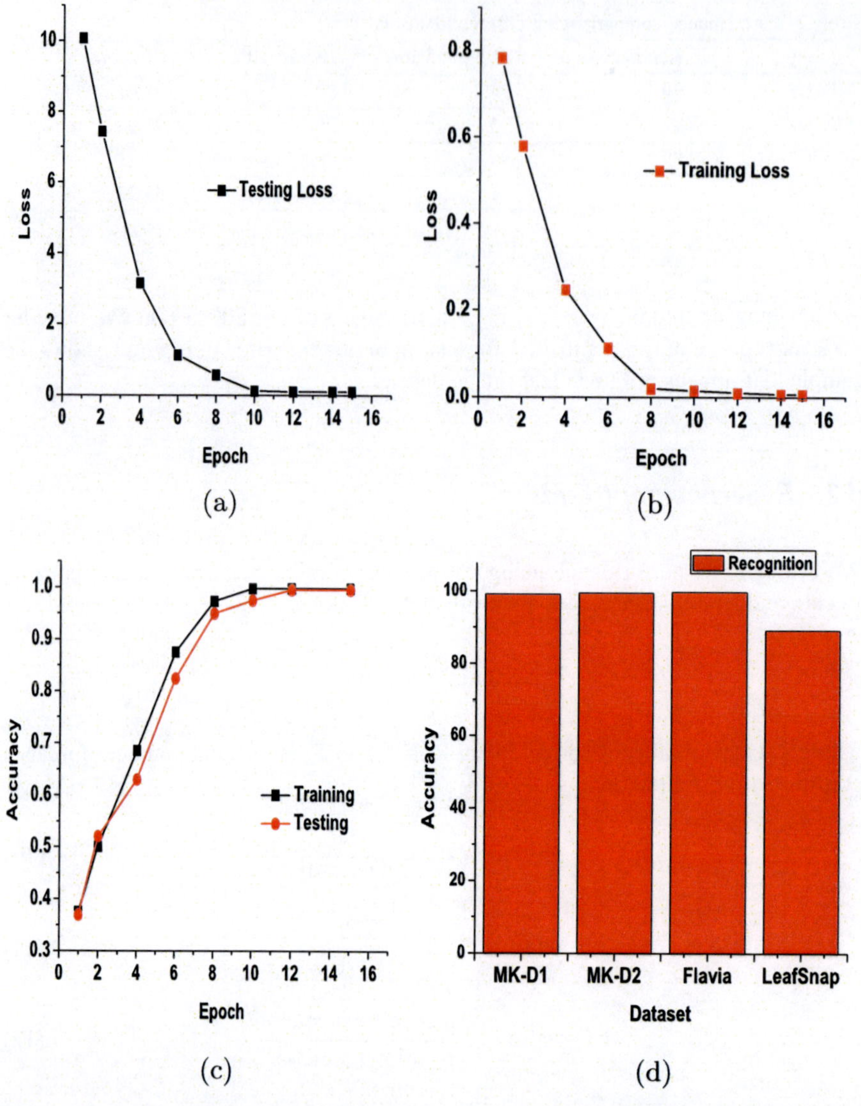

Fig. 3 **a** Testing loss versus Epoch **b** Training loss with respect to Epoch **c** Training and testing accuracy on Flavia dataset **d** Accuracy rate in compare with different datasets

Table 3 Performance comparison with some existing techniques

Author	Feature	Classifier	Accuracy(%)
Kadir et al. [6]	Texture feature, color moments, Fourier descriptor	PNN	93.4
Muthevi et al. [9]	LBP	-	87
Kazerouni et al. [8]	Scale invariant Feature transform (SIFT)	SVM	89.4
Oluleyeet et al. [2]	Fourier descriptor, Zernike moments	RBF	90.5
Lukic et al. [14]	HU moments, LBP	SVM	94.13
Yalein et al. [21]	CNN	Softmax	97
Proposed method	CNN	Softmax	99.4(MK) 99.7(Flavia)

system achieved an accuracy of 99.4 and 99.7% in the case of MK and Flavia datasets, respectively. It is having significantly better recognition accuracy in comparison with texture-, color-, and shape features-based methods. Kadir et al. [6] used texture, color, and Fourier descriptor feature and they got an accuracy rate of 93.4%. Oluleyeet et al. [2] used Fourier descriptor and Zernike moments to represent the leaf features and they got an accuracy rate of 90.5% using RBF classifier. Table 3 summarizes the accuracy of different existing techniques along with the features they have used for classification.

5 Conclusion

In our work, we apply the DL method for identification of plant species using plant leaf images. Identification using DL replaces the extraction of hand-crafted features as it automatically extracts and learns the features in every convolution. It is observed that the recognition rate monotonically increases and the loss decreases after each epoch. It is also figured out that CNN performs better in case of different plants having a similar leaf shape and size. Furthermore, in future, works can be carried to evaluate the performance of the proposed network based on real-time and drone-captured images. It will be interesting to develop a complete cloud-based architecture, where smart devices can be used to identify the plant species in a real-time environment.

References

1. Arrasco, C., Khlebnikov, S., Oncevay, A., Castañón, C.B.: Leaf venation enhancing for texture feature extraction in a plant classification task. In: 2018 IEEE Latin American Conference on Computational Intelligence (LA-CCI), pp. 1–4 (Nov 2018). https://doi.org/10.1109/LA-CCI.2018.8625221
2. Babatunde, O., Armstrong, L., Leng, J., Diepeveen, D.: A neuronal classification system for plant leaves using genetic image segmentation. Br. J. Math. Comput. Sci. **9**(3), 261–278 (2015)
3. Flavia Dataset: http://flavia.sourceforge.net/
4. Grinblat, G.L., Uzal, L.C., Larese, M.G., Granitto, P.M.: Deep learning for plant identification using vein morphological patterns. Comput. Electron. Agric. **127**, 418–424 (2016)
5. Herdiyeni, Y., Santoni, M.M.: Combination of morphological, local binary pattern variance and color moments features for Indonesian medicinal plants identification. In: 2012 International Conference on Advanced Computer Science and Information Systems (ICACSIS), pp. 255–259. IEEE (2012)
6. Kadir, A., Nugroho, L.E., Susanto, A., Santosa, P.I.: Leaf classification using shape, color, and texture features (2013). arXiv:1401.4447
7. Kamilaris, A., Prenafeta-Boldú, F.X.: Deep learning in agriculture: a survey. Comput. Electron. Agric. **147**, 70–90 (2018)
8. Kazerouni, M.F., Schlemper, J., Kuhnert, K.D.: Comparison of modern description methods for the recognition of 32 plant species. Signal & Image Process. **6**(2), 1 (2015)
9. Kumar, P.S., Rao, K.N.V., Raju, A.S.N., Kumar, D.N.: Leaf classification based on shape and edge feature with k-nn classifier. In: 2016 2nd International Conference on Contemporary Computing and Informatics (IC3I), pp. 548–552. IEEE (2016)
10. LeafSnap Image Dataset: http://leafsnap.com/dataset/
11. Lee, S.H., Chan, C.S., Mayo, S.J., Remagnino, P.: How deep learning extracts and learns leaf features for plant classification. Pattern Recogn. **71**, 1–13 (2017)
12. Lee, S.H., Chang, Y.L., Chan, C.S., Remagnino, P.: Hgo-cnn: Hybrid generic-organ convolutional neural network for multi-organ plant classification. In: 2017 IEEE International Conference on Image Processing (ICIP), pp. 4462–4466. IEEE (2017)
13. Liu, Z., Zhu, L., Zhang, X.P., Zhou, X., Shang, L., Huang, Z.K., Gan, Y.: Hybrid deep learning for plant leaves classification. In: International Conference on Intelligent Computing, pp. 115–123. Springer (2015)
14. Lukic, M., Tuba, E., Tuba, M.: Leaf recognition algorithm using support vector machine with hu moments and local binary patterns. In: 2017 IEEE 15th International Symposium on Applied Machine Intelligence and Informatics (SAMI), pp. 000485–000490. IEEE (2017)
15. MalayaKew(MK) Dataset: http://web.fsktm.um.edu.my/~cschan/downloads_MKLeaf_dataset.html
16. Mouine, S., Yahiaoui, I., Verroust-Blondet, A.: Advanced shape context for plant species identification using leaf image retrieval. In: Proceedings of the 2nd ACM International Conference on Multimedia Retrieval, p. 49. ACM (2012)
17. Muthevi, A., Uppu, R.B.: Leaf classification using completed local binary pattern of textures. In: 2017 IEEE 7th International Advance Computing Conference (IACC), pp. 870–874. IEEE (2017)
18. Neto, J.C., Meyer, G.E., Jones, D.D., Samal, A.K.: Plant species identification using elliptic fourier leaf shape analysis. Comput. Electron. Agric. **50**(2), 121–134 (2006)
19. Salman, A., Semwal, A., Bhatt, U., Thakkar, V.: Leaf classification and identification using canny edge detector and SVM classifier. In: 2017 International Conference on Inventive Systems and Control (ICISC), pp. 1–4. IEEE (2017)
20. Xiao, X.Y., Hu, R., Zhang, S.W., Wang, X.F.: Hog-based approach for leaf classification. In: Advanced Intelligent Computing Theories and Applications. With Aspects of Artificial Intelligence, pp. 149–155. Springer (2010)

21. Yalcin, H., Razavi, S.: Plant classification using convolutional neural networks. In: 2016 Fifth International Conference on Agro-Geoinformatics (Agro-Geoinformatics), pp. 1–5. IEEE (2016)
22. Zhao, A., Tsygankov, D., Qiu, P.: Graph-based extraction of shape features for leaf classification. In: 2017 IEEE Global Conference on Signal and Information Processing (GlobalSIP), pp. 663–666. IEEE (2017)

A Hybrid Approach for Segmenting Grey and White Matter from Brain Magnetic Resonance Imaging (MRI)

Ruhul Amin Hazarika, Khrawnam Kharkongor, Arnab Kumar Maji, Debdatta Kandar, and Sugata Sanyal

Abstract Magnetic Resonance Imaging (MRI) is a common medical imaging diagnostic tool for the identification of disease(s) during clinical investigation. Brain MRI is used for diagnosis of brain-related diseases such as brain tumours, Alzheimer's disease, etc. This has proven to be advantageous over other diagnostic techniques and also adds to the versatility and diagnostic utility for surgical treatment planning and clinical interventions. Brain tissues have grey matter and white matter whose intensity is almost similar, hence making the diagnosis of the brain-related disease difficult. Segmentation of grey matter and white matter is crucial to detect the various brain-related disease such as Alzheimer's, Migraine, Huntington, Multiple sclerosis and Dyslexia, which show significant volumetric changes in grey matter and white matter. Prior to the segmentation of brain regions, skull stripping is a necessity for accurate diagnosis of various brain-related diseases. In this paper, histogram- based skull stripping technique is applied to separate the skull and then a novel hybridised technique is proposed using Fuzzy Edge Detection and Region-Growing to Segment the Grey and White Matter from Brain MRI. The result of the proposed technique is compared with different existing techniques such as Region growing; Histogram-based method, fuzzy C- Means, K Means, etc. It is found that the proposed method produces convincing results.

R. A. Hazarika (✉) · K. Kharkongor · A. Kumar Maji · D. Kandar
Department of Information Technology, NEHU, Shillong 793022, India
e-mail: rahazarika@gmail.com

K. Kharkongor
e-mail: khraw2001@yahoo.co.in

A. Kumar Maji
e-mail: arnab.maji@gmail.com

D. Kandar
e-mail: kdebdatta@gmail.com

S. Sanyal
Tata Institute of Fundamental Research, Homi Bhabha Road, Mumbai 400005, India
e-mail: sanyals@gmail.com

D. Bhattacharjee et al. (eds.), *Proceedings of International Conference on Frontiers in Computing and Systems*, Advances in Intelligent Systems and Computing 1255,
https://doi.org/10.1007/978-981-15-7834-2_12

Keywords Magnetic resonance image · Segmentation · Skull stripping · Region growing · Fuzzy edge detection · Grey matter segmentation.

1 Introduction

Magnetic Resonance Imaging (MRI) is an efficient diagnostic tool used to study the internal structure of the body during medical diagnosis in clinical investigation and identification of diseases and disorders [11]. MRI provides chemical and physiological information of living tissues for which the investigation is undertaken.

Segmentation of the skull is also called skull stripping. Prior to segmentation of different brain regions for accurate diagnosis of various brain-related diseases, it is a prerequisite. Skull stripping can be defined as a morph metric study to remove the brain tissue from the non-brain tissues [25]. The Human Brain comprises of a complex structure of tissues. For better diagnosis of brain-related disease, researchers conducted several studies on the functions of the tissues/cells present in the brain. The cells of the human brain can be divided into two subcategories: (i) Neurons which help in the processing and communication within the brain (ii) Neuroglia, also called as the glial cells, which usually protects and supports the neurons [11].

The Central Nervous System (CNS) in the brain consists of two major categories of tissues, namely, Grey Matter and White Matter, which are an important subject of study in brain imaging. These tissues form the anatomical structure of the brain [18].

Grey Matter: It consists of mostly unmyelinated neurons. Grey matter is responsible for interconnections of nerves and processing [6]. Grey matter composed of neural and glial cells that control brain activity. These neuroglia or glia has grey matter nuclei, which are located deep within the white matter.

White matter: White matter consists of mostly medullary sheath (neurons), which connects the grey matter to each other and to the rest of the body parts. It is the information highway of the brain which speeds up the connections between distant parts of the brain and body [16]. White matter fibre consists of myelinated axons that connect the cerebral cortex with other brain regions.

Segmentation of Grey Matter and White Matter: Segmentation is the process of separating the image into distinct regions. Segmentation of medical images is a difficult task because unlike the other images, the medical images are complex in nature [21]. The anatomy of the human brain is complex and precise segmentation of brain tissues into White Matter, Grey Matter and Cerebrospinal Fluid is important in prescribing therapy for various brain-related diseases such as Epilepsy, Necrotic tissues, Multiple sclerosis, Tumours, Edema, Alzheimer's disease, Parkinson's disease, Cerebral atrophy, etc [14].

Segmentation of White Matter and Grey Matter is significant in various studies [21] like (i) Calculation of changes in grey and white matter volume in Migraine and Huntington's patients, (ii) Change in grey matter pathology in multiple sclerosis patients, (iii) A significant volumetric changes of grey matter is also observed during the medical diagnosis in Alzheimer's patients, etc.

There are several algorithms for segmentation of Brain Images. Region growing is one of the popular algorithms for the same. Edge detection plays a crucial role, as the absence of edge information may lead to over/under segmentation [22]. The introduction of an edge detection technique within region growing algorithm can help in accurate segmentation of the Region of Interest (RoI).

The use of fuzzy set theory based approach is really helpful in edge detection [4]. Hence, Fuzzy technique is used to detect the edges from the input brain MR Images. The proposed segmentation technique is a combination of two major steps that are: (i) Detection of edges in the input brain MR Images, (ii) Applying Region growing technique in the edge detected brain MR Images.

2 Various Existing MRI Brain Segmentation Methods

In order to segment the Grey matter and White matter from a brain MRI, different techniques have been implemented by the researchers. Amongst all the techniques, Fuzzy C-Means, Histogram-based, K-Means and Region-Growing are the most commonly used techniques.

Fuzzy C-Means(FCM) Technique: As fuzzy can preserve information more than a crisp approach, Fuzzy C-means(FCM) algorithm is one of the popularly used Image Segmentation technique [3]. FCM is a clustering-based approach, in which every pixel of an image is assigned with a degree of membership value. Based on the membership value, the pixel is included in a particular cluster. This method was first introduced by Joe Dunn in 1973 [5]. One of the major disadvantages of FCM is that FCM does not maintain any other information of the spatial domain and it is more sensitive towards noises. MR Images may be corrupted by noise for which FCM may not provide a good result.

Histogram-Based Algorithm: This is the simplest algorithm where the segmented parts of an input image can be obtained from the histogram of it. Histogram of the image represents the grey levels of background as well as the object [2]. The highest peak of the histogram represents the background, whereas the immediate highest peak represents the object. A threshold point is needed to be chosen from the histogram which should be in between the two peaks. The main objective of thresholding is to set a pixel value in such a way that it should clearly distinguish the object from its background. For the pixels whose value doesn't lie on the object side of the point is set to zero. One of the major challenges of histogram- based algorithm is to choose an appropriate threshold point. Inappropriate thresholding may lead to an improper result.

K-Means Algorithm: This algorithm is an unsupervised clustering-based technique, which is introduced in 1967 by MacQueen [15]. K-Means algorithm is an iterative method for dividing the input image into k number of clusters. For each cluster, the centre point is determined and all other pixels are compared with it and included in a cluster, based on the pixel colour, pixel intensity value, texture etc. The main challenge of this method is to choose the value of k. This technique works well only if the selection of the number of k is proper.

Region-Growing Algorithm: The idea behind this technique is to merge the similar pixels with a set of starting points (known as seed points). For each seed point, a number of neighbourhood pixels are considered (4 or 8). Based on similarity criteria, the seed points are grouped with their neighbourhood pixels. In the next step, all the newly added pixels are also considered as new seed points and the process repeats until all the pixels of the image are traversed. The major disadvantages of this technique are difficulty in selection of a seed point, since the process grows beyond the boundary as no edge information is present [24].

3 The Proposed Technique for Segmenting Grey and White Matter from Magnetic Resonance Imaging (MRI) of the Brain

The proposed hybridised technique is intended to overcome the drawback of region-growing based algorithm, which depends on the homogeneity of the criteria. The failure to provide their information on homogeneity can lead to flawed segmentation. The introduction of edge information to the region-growing algorithm can lead to a better segmentation result [13]. The block diagram of the proposed hybridised segmentation technique is provided in Fig. 1.

3.1 Input Image

The input image is a brain MRI. The input image can be a greyscaled image or a black and white image. However, for performance analysis, images are converted into black and white. All the input images are taken from the online dataset "MRBrainS13" [1]. No noise removal operation is performed in this work for further processing.

3.2 Skull Stripping

Five most commonly used image segmentation algorithms for skull stripping, namely, Region-Growing, Region Splitting–Merging, K-Means Clustering, Histogram-Based Thresholding and Fuzzy C-Means are implemented and tested for 50 MRI images and their results are evaluated using Performance Analysis Metrics. The result analysis shows that Histogram-Based Thresholding algorithm gives more accuracy in comparison with other algorithms [8]. Hence, for the proposed hybrid technique, the Histogram-Based Thresholding algorithm is used for skull stripping.

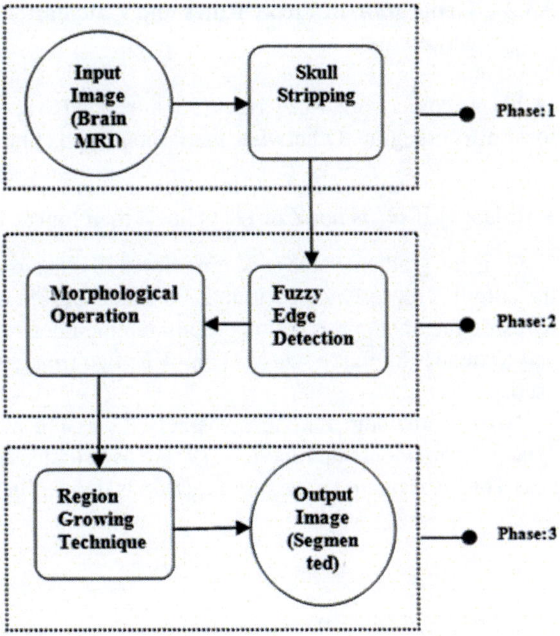

Fig. 1 Block diagram showing the proposed hybridised technique

3.3 *Fuzzy Edge Detection*

After removing the skulls from the brain MRI, the next step is to determine the edge points of the skull-free image. Some of the popular approaches for edge detection are described in the literature [17]. In the proposed technique, a fuzzy rule based approach is used to decide the edge pixels [4]. Steps for deciding edge pixels using a fuzzy logic based technique are described as follows:

3.3.1 Determination of Image Gradient

Image gradient means a directional change in intensity of the pixel of an image [10]. To obtain the gradient value for a pixel 'pi' in X-direction, with the help of a gradient filter mask [−1, 1], 'pi' is convoluted with its neighbourhood pixels. A similar operation is done for obtaining the gradient value in Y-direction with the help of a gradient filter mask [−1, 1] T. The process is applied for all pixels to obtain the image gradient.

3.3.2 Generation of Fuzzy Rules and Calculation of Membership Functions

In this proposed work, fuzzy rules [23] are added to mark a pixel as white if it belongs to a uniform region. Otherwise, mark the pixel as black, i.e.

- Rule 1 = If G_x is Z and G_y is Z then Iout is WH.
- Rule 2 = If G_x is not Z or G_y is not Z then Iout is BL.

G_x is the gradient value in X-direction, G_y is the gradient value in Y-direction and the output is represented by Iout. Z, WH and BL are the fuzzy sets, that are defined by using membership functions. Gaussian membership function is used to define Z and to define the fuzzy sets WH and BL; two triangular membership functions are used.

Two membership functions, namely, Gaussian membership function [20] and Triangular membership function [19] are used to determine the degree of belongingness.The Gaussian membership function is defined in Eq. 1.

$$\mu(x) = e^{\frac{-(x-x_e)^2}{2\sigma^2}} \tag{1}$$

In Eq. 1, x_e is the centre point and σ is the width. If the gradient value = $x_e = 0$, the belongingness of the pixel in fuzzy set Z is determined with a degree of 1. If the gradient value lies in between the transition zone of the Gaussian curve, then the belongingness of the pixel in fuzzy set Z is determined with a degree between 0 and 1. For any other gradient value, the belongingness of the pixel in fuzzy set Z is determined with a degree 0.

The Triangular membership function is defined in Eq. 2.

$$X = \begin{cases} 0, & \text{if } i \leq x \text{ or } z \leq i \\ \frac{i-x}{y-x}, & x \leq i \leq y \\ \frac{z-i}{z-y}, & y \leq i \leq z \end{cases} \tag{2}$$

Here, x, y, and z are the foot of the triangle. For the fuzzy set BL, the value of $x = y = 0$ and $z = 255$ whereas for the fuzzy set WH, the value of $x = 0$ and $y = z = 255$. When $x = y = 0$, the degree of belongingness of the pixel to the fuzzy set BL = 1. For $z = 255$, the degree of belongingness of the pixel to the fuzzy set BL = 0. For any value in between 0 and 255, the degree of belongingness of the pixel to the fuzzy set BL = 0 to 1. Similarly, when $x = 0$ the degree of belongingness of the pixel to the fuzzy set WH = 0. For $y = z = 255$, the degree of belongingness of the pixel to the fuzzy set BL = 1. For any value in between 0 and 255, the degree of belongingness of the pixel to the fuzzy set BL = 0 to 1.

3.4 *Morphological Operation*

After detecting the edges using a fuzzy approach, it is observed that there may be some broken points/edges. In order to get an efficient result, there is a necessity to fill those gaps. To connect the broken edges, the morphological dilation procedure is applied [7] which is a mathematical approach to enhance the size of an object, by filling the broken areas. For a greyscale image, dilation enhances the brightness of objects by taking the maximum value of neighbouring pixel. The structural element used in this operation is a 2×2 matrix.

After morphological dilation operation, the output image may include some unnecessary points which may make the edges bolder. To obtain the output accurately, it is necessary to remove unnecessary pixels and make the boundary/edges thinner/smoother. For edge thinning, the Morphological Erosion procedure is applied [9]. Erosion is a mathematical approach which decreases the sizes of objects and removes small anomalies. For a greyscale image, erosion reduces the brightness (and therefore the size) of bright objects on a dark background by taking the minimum value of neighbouring pixels. The Structural element used in this operation is a 2×2 matrix.

3.5 *Region-Growing Technique*

The thin edged image will be the input for the Region-Growing (RG) approach. The RG algorithm consists of three parts: (i) Selection of initial seed point, (ii) Determination of threshold value and (iii) Growing the region [12].

1. **Selection of initial seed point:** To choose the seed point, the centre pixel of the image is determined. The centre point is then considered as the initial seed point.
2. **Determination of threshold value:** The threshold value for the technique is determined based on the average pixel intensity values of the image as below: Threshold value = (Average intensity value)/2
3. **Growing the region:** The technique works as below.
   - Consider the eight neighbourhood pixels of the seed point.
   - Initial region size = 1.
   - Calculation of the pixel intensity difference between the seed pixel and its neighbourhood pixels.
   - If the intensity difference $\leq$ threshold value, add the pixel to the region, else move to the next pixel.
   - Consider the newly added pixel as a new seed point and repeat the earlier steps.
   - When all the pixels of the image are visited, stop the process.

4 Experimental Result

The implementation of the hybridised approach is done using MATLAB. For comparison of the result, ground truth images are taken from the online database 'MRbrains13' [1]. A total of 50 images are considered for the purpose. The output result of the hybridised approach is compared with some of the commonly used grey matter segmentation techniques, namely, Histogram- based approach, Region-growing technique, K-Means algorithm and FCM algorithm.

Accuracy, sensitivity, Specificity, Precision and Dice-coefficient of all the 50 output images are determined and compared with the existing techniques. Finally, the average performance metrics for each technique is as shown in Table 1.

The evaluation metrics are determined using Eqs. 3 to 7.

$$\text{Accuracy} = \frac{\text{TP} + \text{TN}}{\text{TP} + \text{TN} + \text{FP} + \text{FN}} \tag{3}$$

$$\text{Sensitivity} = \frac{\text{TP}}{\text{TP} + \text{FN}} \tag{4}$$

$$\text{Specificity} = \frac{\text{TN}}{\text{TN} + \text{FP}} \tag{5}$$

$$\text{Precision} = \frac{\text{TP}}{\text{TP} + \text{FP}} \tag{6}$$

$$\text{Dice} - \text{coefficient} = \frac{2\text{TP}}{2\text{TP} + \text{FP} + \text{FN}} \tag{7}$$

where True positive (TP) is defined as the number of correctly segmented pixels with the ground truth. False positive (FP) is defined as the number of mistakenly segmented pixels with the ground truth. True negative (TN) is defined as the number of correctly rejected pixels with the ground truth. False Negative (FN) is defined as the number of mistakenly rejected pixels with the ground truth.

The average performance analysis for 50 different MR Images is shown in Table 1. From the table as well as visually from the output image, it can be observed that the

Table 1 Performance comparison

Algorithm	Accuracy	Sensitivity	Specificity	Precision	Dice-coefficient
FCM	0.81	0.88	0.87	0.88	0.85
Histogram	0.85	0.91	0.88	0.89	0.87
K-Means	0.79	0.83	0.86	0.78	0.81
Region-Growing	0.83	0.84	0.87	0.88	0.85
Proposed Hybrid Technique	0.89	0.94	0.92	0.91	0.89

proposed hybrid approach is giving satisfactory accuracy and sensitivity than the other existing techniques. One of the visual results of the implementation is shown in Figs. 2 to 9 (Figs. 3, 4, 5, 6, 7, 8).

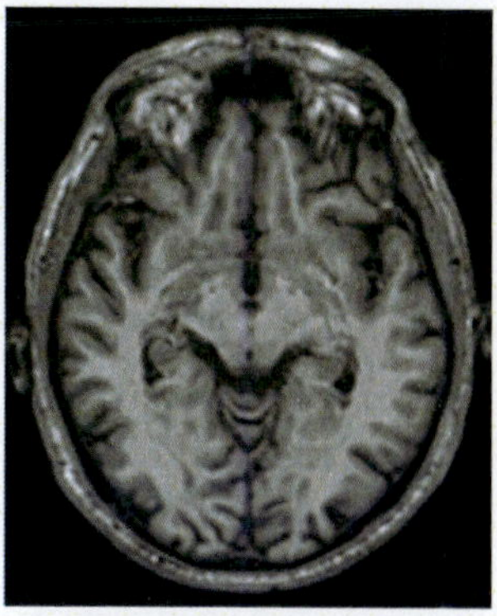

Fig. 2 Input Brain MR Image

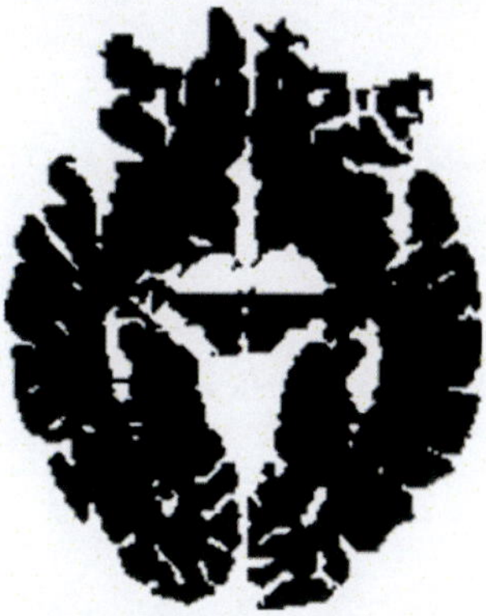

Fig. 3 Ground truth of the input image

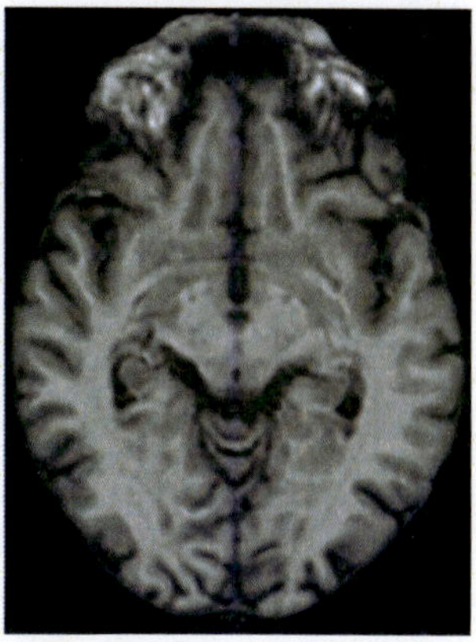

Fig. 4 Output image after removing skulls by Histogram-based thresholding technique

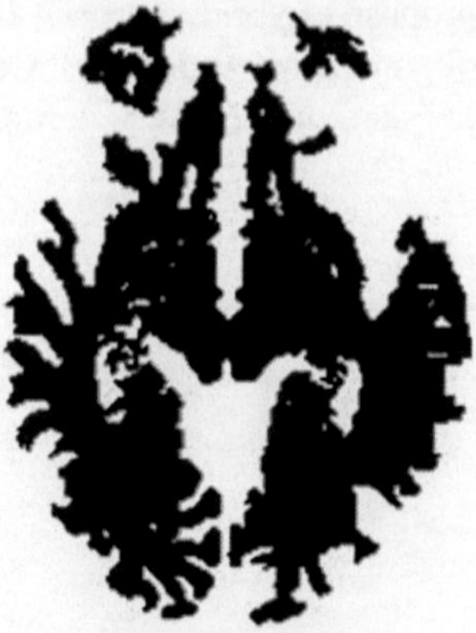

Fig. 5 Grey/white matter segmented image applying FCM algorithm

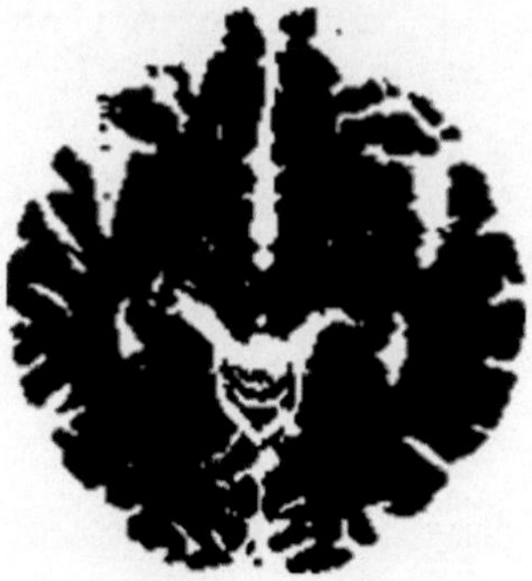

Fig. 6 Grey/white matter segmented image applying Histogram-based algorithm

Fig. 7 Grey/white matter segmented image applying K- Means algorithm

Fig. 8 Grey/white matter segmented image applying Region-Growing algorithm

Fig. 9 Grey/white matter segmented image applying the proposed hybrid algorithm

5 Conclusion

Some of the most commonly used brain MRI skull stripping techniques are implemented and their results are evaluated using Performance Analysis Metrics. From the performance analysis, it is found that Histogram-Based Thresholding algorithm has more accuracy compared to the other algorithms. Therefore, Histogram-based technique is used to remove skulls from the brain MR Image. Further, the proposed hybridised technique is applied to the skull-stripped MR Image to segment the Grey and White Matter. The output result of the proposed technique is compared with some of the existing segmentation technique. The input as well as the ground truth MR Images are obtained from the online dataset 'MRBrainS13' [1]. From the comparative analysis, it is observed that the proposed hybrid technique gives a more satisfactory result than the other popular existing techniques. Our proposed results will be highly beneficial for doing further research on diseases related to dementia.

References

1. Evaluation Framework for MR Brain Image Segmentation (2019). https://mrbrains13.isi.uu.nl/. Accessed 20 April 2019
2. Histogram-Based Image Segmentation, author=Segmentation, Histogram-Based Image
3. Agrawal, S., Panda, R., Dora, L.: A study on fuzzy clustering for magnetic resonance brain image segmentation using soft computing approaches. Appl. Soft Comput. **24**, 522–533 (2014)
4. Becerikli, Y., Karan, T.M.: A new fuzzy approach for edge detection. In: International Work-Conference on Artificial Neural Networks, pp. 943–951. Springer (2005)
5. Dunn, J.C.: A fuzzy relative of the ISODATA process and its use in detecting compact well-separated clusters (1973)
6. Giorgio, A., Santelli, L., Tomassini, V., Bosnell, R., Smith, S., De Stefano, N., Johansen-Berg, H.: Age-related changes in grey and white matter structure throughout adulthood. Neuroimage **51**(3), 943–951 (2010)
7. Haralick, R.M., Sternberg, S.R., Zhuang, X.: Image analysis using mathematical morphology. IEEE Trans. Pattern Anal. Mach. Intell. **4**, 532–550 (1987)
8. Hazarika, R.A., Kharkongr, K., Sanyal, S., Maji, A.K.: A comparative study on different skull stripping techniques from brain magnetic resonance imaging (2019) [Accepted for publication in International Conference on Innovative Computing and Communications, 2019]
9. Heijmans, H.J.: Morphological Image Operators, vol. 4. Academic Press Boston (1994)
10. Jacobs, D.: Image gradients. Class Notes CMSC **426** (2005)
11. Kalavathi, P., Prasath, V.S.: Methods on skull stripping of MRI head scan imagesa review. J. Digit. Imaging **29**(3), 365–379 (2016)
12. Kamdi, S., Krishna, R.: Image segmentation and region growing algorithm. Int. J. Comput. Technol. Electron. Eng. (IJCTEE) **2**(1) (2012)
13. Khwairakpam, A., Hazarika, R.A., Kandar, D.: Image segmentation by fuzzy edge detection and region growing technique. In: Proceedings of the Third International Conference on Microelectronics, Computing and Communication Systems, pp. 51–64. Springer (2019)
14. Lim, K.O., Pfefferbaum, A.: Segmentation of MR brain images into cerebrospinal fluid spaces, white and gray matter. J. Comput. Assisted Tomograp. **13**(4), 588–593 (1989)
15. MacQueen, J., et al.: Some methods for classification and analysis of multivariate observations. In: Proceedings of the fifth Berkeley Symposium on Mathematical Statistics and Probability, vol. 1, pp. 281–297. Oakland, CA, USA (1967)
16. Mori, S., Wakana, S., Van Zijl, P.C., Nagae-Poetscher, L.: MRI Atlas of Human White Matter. Elsevier (2005)
17. Muthukrishnan, R., Radha, M.: Edge detection techniques for image segmentation. Int. J. Comput. Sci. Inf. Technol. **3**(6), 259 (2011)
18. Navarro, A., Tolivia, J., Astudillo, A., del Valle, E.: Pattern of apolipoprotein D immunoreactivity in human brain. Neurosci. Lett. **254**(1), 17–20 (1998)
19. Pedrycz, W.: Why triangular membership functions? Fuzzy Sets Syst. **64**(1), 21–30 (1994)
20. Reddy, C.S., Raju, K.: An improved fuzzy approach for COCOMOs effort estimation using gaussian membership function. J. Softw. **4**(5), 452–459 (2009)
21. Sandhya, G., Kande, G.B., Satya, S.T.: An efficient MRI brain tumor segmentation by the fusion of active contour model and self-organizing-map. J. Biomimetics Biomater. Biomed. Eng. **40**, 79–91. Trans Tech Publ (2019)
22. Senthilkumaran, N., Rajesh, R.: Edge detection techniques for image segmentation-a survey of soft computing approaches. Int. J. Recent Trends Eng. **1**(2), 250 (2009)
23. Setnes, M., Babuska, R., Kaymak, U., van Nauta Lemke, H.R.: Similarity measures in fuzzy rule base simplification. IEEE Trans. Syst. Man Cybern. Part B (Cybernetics) **28**(3), 376–386 (1998)
24. Sreeji, C., Vineetha, G., Beevi, A.A., Nasseena, N.: Survey on different methods of image segmentation. Int. J. Sci. & Eng. Res. **4**(4) (2013)
25. Zhuang, A.H., Valentino, D.J., Toga, A.W.: Skull-stripping magnetic resonance brain images using a model-based level set. NeuroImage **32**(1), 79–92 (2006)

Retinal Vessel Segmentation Using Unsharp Masking and Otsu Thresholding

Sk Latib, Diksha Saha, and Chandan Giri

Abstract In the diagnosis of various optical and cardiovascular diseases, retinal blood vessel observation is an important task. So, it is required to accurately detect the retinal vasculature to assist in the diagnosis of complications in the eye. A method of segmentation of retinal blood vessel structure from fundus images is proposed in this paper. For preprocessing, the green channel of the image is extracted and subjected to an unsharp masking filter followed by contrast enhancement operations. Vessel extraction is done by using bit plane slicing and Otsu thresholding techniques and finally the morphological operation is used for restoration. The proposed algorithm comes out to 95.30% accuracy with 99.43% specificity.

Keywords Retinal vasculature · Unsharp masking · Contrast enhancement · Local thresholding · Morphological opening operation

1 Introduction

In the diagnosis of various optical and cardiovascular diseases, retinal blood vessels plays a vital role. The different abnormalities of the eyes as well as the body such as diabetic retinopathy are diagnosed by visual exploration of retinal blood vessels. They assist ophthalmologists in the diagnoses of [1], glaucoma [2], cardiovascular

Sk. Latib (✉)
Information Technology, St. Thomas College of Engineering & Technology, Kolkata 700023, India
e-mail: sklatib@gmail.com

D. Saha
Information Technology Department, St. Thomas College of Engineering & Technology, Kolkata 700023, India
e-mail: diksha7012@gmail.com

C. Giri
Information Technology, Indian Institute of Engineering Science & Technology, Shibpur 711103, India
e-mail: chandangiri@gmail.com

D. Bhattacharjee et al. (eds.), *Proceedings of International Conference on Frontiers in Computing and Systems*, Advances in Intelligent Systems and Computing 1255,
https://doi.org/10.1007/978-981-15-7834-2_13

ailment [3], high blood pressure [4], hypertension [5], arteriosclerosis [6], and age-related macular degeneration [7]. These can be detected by the state of many features that retinal blood vessel have such as length, tortuosity, diameter, width, branching factor, etc. Different eye diseases have different symptoms that aids in their detection. However, due to the complex structure of the eye, the manual inspection of retinal images by an ophthalmologist is expensive and tedious process. It requires a lot of skill and expertise. Various methods have been suggested to extract the retinal vessel structure. Fraz et al [8] suggested an automated method for segmentation of blood vessels in retinal images by means of multidirectional morphological top-hat operator and combination of first-order Gaussian differential filtering followed by bit plane slicing in their paper. In Ref. [9], the skeleton of main vessels is extracted by the application of directional differential operators, vessels are emphasized in a particular direction with a multidirectional top-hat operator having rotating structuring elements, and information is extracted using bit plane slicing, followed by an iterative region growing method to integrate intermediate results. De et al [10] proposed a two-step pipeline: first, the retinal vessels are segmented with the preference of preserving the skeleton network, i.e., retinal segmentation with a high recall, and second, a novel tracing algorithm is developed where the tracing problem is uniquely mapped to an inference problem in probabilistic graphical models. Aramesh et al [11] use mean filter in order to reduce the image noise, local thresholding to binarize the image, and finally line detector filters and mathematical morphology (top-hat filtering) is applied to the image. Mustafa et al [12] proposed to extract the blood vessel based on peak and valley detection in the green channel image and then the valleys and the inversion image peaks are imposed on one another.

In this paper, a new method has been proposed. A preprocessing is done using unsharp masking on the extracted green channel of fundus image and contrast enhancement operations. The vessels are then segmented by bit plane slicing and Otsu thresholding, followed by a restoration phase of morphological opening to obtain the final image.

2 Proposed Method

The proposed method uses preprocessing, segmentation, and restoration operations on the images taken from the drive dataset. These phases are explained below.

2.1 Preprocessing

2.1.1 Green Channel Extraction

For preprocessing, the green channel of the original RGB fundus image is obtained and stored. The contrast of an image is high in green channel so it is given preference over the red and blue channel.

2.1.2 Unsharp Masking

Unsharp masking is a process of sharpening an image, which is used to enhance the edges without increasing the noise. In this process, the smoothed image is subtracted from the original image to extract sharp edges. Here a nonlinear median filter M, of dimension $k \times k$, where k is a constant (here $k = 21$), is used for the smoothening step. The smoothening operation by a median filter can be given as

$$r_1(s, t) = \text{median}\{f(x + s, y + t), (x, y) \in W\} \tag{1}$$

where $g_1(s, t)$ is the smoothed image, $f(x, y)$ is the input image, and W defines the region covered by the median filter M.

The unsharp masking operation equation can be given as

$$r_2(x, y) = f(x, y) - r_1(x, y) \tag{2}$$

where $r_2(x, y)$ is the unsharp masked image, $f(x, y)$ is the original image and $r_1(x, y)$ is the smoothed image.

2.1.3 Contrast Enhancement

The brightness and contrast of the blood vessels are very low in the obtained image from the previous step, Hence to increase the contrast, the image is raised to the power of c, where c is a constant (experimentally $c = 1.5$ gives better performance).

Based on the properties of power transform, retinal blood vessels are enhanced and the background are suppressed, according to the following equation:

$$r_3(x, y) = r_2(x, y)^c \tag{3}$$

where $r_3(x, y)$ is contrast-enhanced image and $r_2(x, y)$ is the input image.

2.2 Segmentation

2.2.1 Bit Plane Slicing

The noise is also enhanced due to the effect of power transform. To remove the noise and extract the blood vessels, a bit plane slicing operation is used. The most significant bit planes contain the maximum information, so the logical AND of the 8th, 7th, and 6th bit planes are used to get vessel information.

$$h_1(x, y) = r_3(x, y).128 \tag{4}$$

$$h_2(x, y) = r_3(x, y).64 \tag{5}$$

$$h_3(x, y) = r_3(x, y).32 \tag{6}$$

where $h_1(x, y)$, $h_2(x, y)$, and $h_3(x, y)$ are the 8th, 7th, and 6th bit planes, respectively, of $r_3(x, y)$ obtained by AND-ing (.) the input image with the maximum corresponding bit plane value (here 128, 64, and 32, respectively, as it is a grayscale image).

The resultant image H is obtained AND-ing the three extracted bit planes as follows:

$$H = h_1.h_2.h_3 \tag{7}$$

2.2.2 Otsu Thresholding

Otsu thresholding is used to convert the resultant grayscale image to a binary image. It involves iteration through all the possible threshold values and calculating a measure of spread for the pixel levels on each side of the threshold, i.e., the pixels that either belongs to either in foreground or background. The aim is to find the threshold value for which the intra-class variance in the foreground and the background is minimum. But since that approach is computationally expensive, a threshold value can also be chosen in such a manner that the inter-class variance between the foreground and the background are maximized. The process is explained as follows:

Initially, the threshold t=0. *L-1* is the highest intensity level, i.e., 255, as it is grayscale image, and *n(i)* denotes pixels numbers in intensity level *i*. The background is represented by b and foreground by f. The background and foreground pixels weights are $\omega_b(t)$ and $\omega_f(t)$, respectively. The mean of background and foreground pixels are $\mu_b(t)$ and $\mu_f(t)$ and variance are $\sigma_b^2(t)$ and $\sigma_f^2(t)$ respectively, when threshold is set to t.

$$\omega_b(t) = \frac{\sum_{i=0}^{t-1} n(i)}{\sum_{i=0}^{L-1} n(i)}, \omega_f(t) = \frac{\sum_{i=t}^{L-1} n(i)}{\sum_{i=0}^{L-1} n(i)} \tag{8}$$

$$\mu_b(t) = \frac{\sum_{i=0}^{t-1} i.n(i)}{\sum_{i=0}^{t-1} n(i)}, \mu_f(t) = \frac{\sum_{i=t}^{L-1} i.n(i)}{\sum_{i=t}^{L-1} n(i)} \tag{9}$$

$$\sigma_b^2(t) = \frac{\sum_{i=0}^{t-1} (i - \mu_b(t))^2.n(i)}{\sum_{i=0}^{t-1} n(i)}, \sigma_f^2(t) = \frac{\sum_{i=t}^{L-1} \left(i - \mu_f(t)\right)^2}{\sum_{i=t}^{L-1} n(i)} \tag{10}$$

Using these parameters, the intra-class variance $\sigma_W^2(t)$ can be calculated as follows:

$$\sigma_W^2(t) = \omega_b(t).\sigma_b^2(t) + \omega_f(t)\sigma_f^2(t) \tag{11}$$

However, calculating the intra-class variance is computationally expensive, as it requires finding out σ_b^2(t) and $\sigma_f^2(t)$.Hence, inter-class variance $\sigma_B^2(t)$is calculated instead as follows:

$$\begin{aligned} \sigma_B^2(t) &= \sigma^2(t) - \sigma_W^2(t) \\ &= \omega_b(t).\{\mu_b(t) - \mu(t)\}^2 + \omega_f(t).\left\{\mu_f(t) - \mu(t)\right\}^2 \\ &\quad \left[\because \mu = \omega_b(t).\mu_b(t) + \omega_f(t).\mu_f(t)\right] \\ \sigma_B^2(t) &= \omega_b(t).\omega_f(t).\left\{\mu_b(t) - \mu_f(t)\right\}^2 \end{aligned} \tag{12}$$

In this way the inter-class variance for all t values ranging from 0 to $L-1$ is calculated. The t for which the inter-class variance is maximum that value is chosen as the threshold value.

The image obtained from the previous step is H, and then binarized to the resultant image I using the value t as follows:

$$I(x, y) = \begin{cases} 0, & \text{if } H(x, y) < t \\ 1, & \text{if } H(x, y) \geq t \end{cases}, \forall (x, y) \in H \tag{13}$$

2.3 Restoration

2.3.1 Morphological Operator

Morphological opening is applied on the segmented image to remove connected components having less than d pixels (here $d = 4$), which are usually observed to be noise. To morphologically open the image I, a structuring element S is used, which stores the 8-neighbour connectivity information. The resultant image J is thus obtained:

$$J = I \circ S = (I \ominus S) \oplus S \tag{14}$$

where, $I \ominus S = \{x|(S)_x \subseteq I\}$, is morphological erosion used for image thinning and $I \oplus S = \left\{x|\left(\hat{S}\right)_x \cap I \neq \phi\right\}$, $\hat{S} = \{x| - x \in S\}$, is morphological dilation used for image thickening.

A disk structuring element mask is multiplied with the image J to remove the outer optical disc.

$$J = \text{Final} \cdot \text{mask} \tag{15}$$

where Final is the resultant image thus obtained.

3 Summarization of Proposed Work

Step 1→Input RGB fundus image.
Step 2→Convert it to grayscale from RGB image.
Step 3→Follow it up with an Unsharp Median Mask (21 × 21).
Step 4→Use Contrast Enhancement operation to enhance foreground and suppress background with power function where c = 1.5.
Step 5→Extract 8th, 7th, and 6th bit plane and add the three bit planes followed by median filtering with a 2 × 2 median filter.
Step 6→Apply Otsu Thresholding to binarize image.
Step 7→Use Morphological Opening operation to remove connected components less than a certain value.
Step 8→Apply a mask of size < 4 to remove the optical disk outline and get the final output.

4 Result and Analysis

The proposed method's performance, i.e., specificity and accuracy calculated by comparing the black and white pixels of the final output with the ground truth image available in the DRIVE dataset. These indicate the number of properly classified pixels.

The number of white pixels correctly classified is called True Positives (TP). True Negatives (TN) are the correctly classified black pixels. Incorrectly classified white pixels are called False Positives (FP). Black pixels incorrectly classified are called False Negatives (FN).

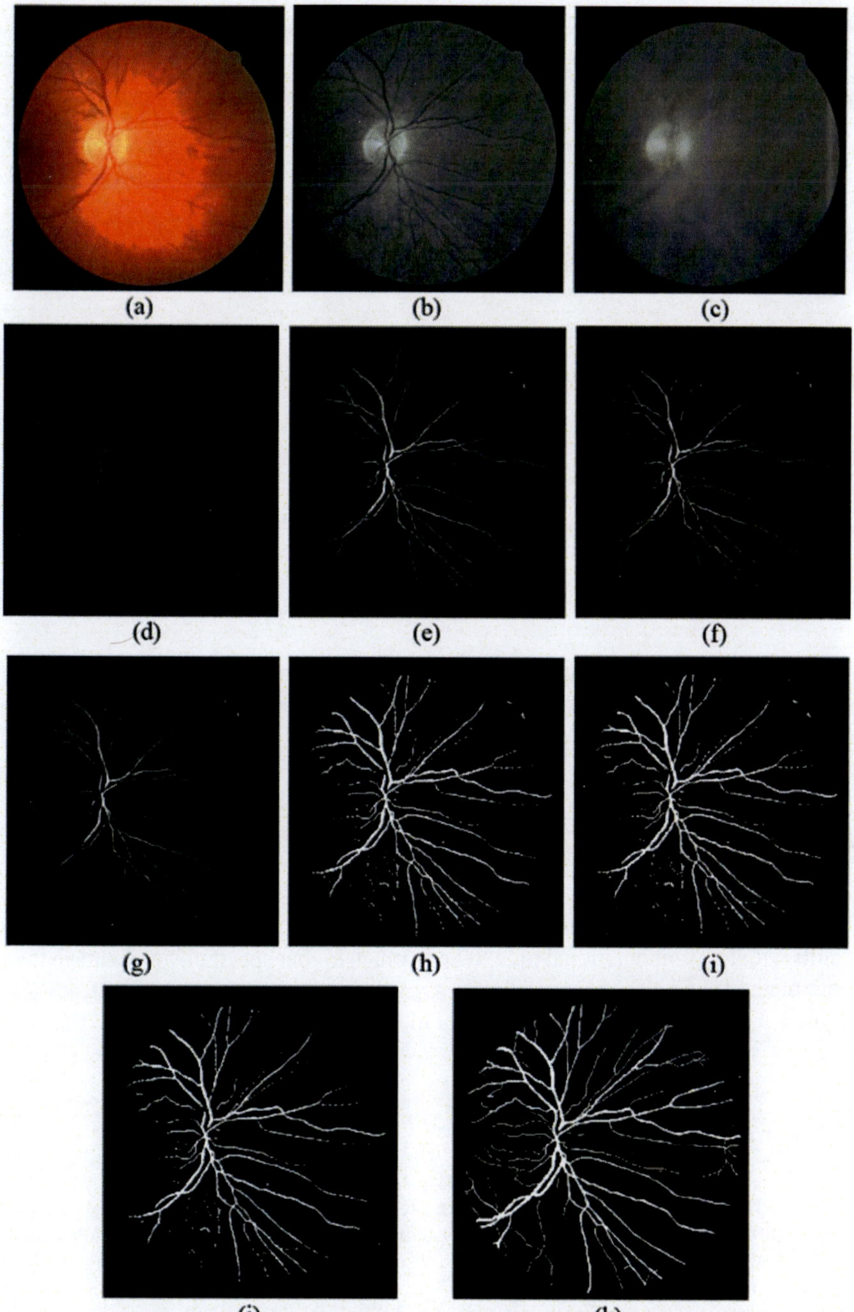

(a) (b) (c)
(d) (e) (f)
(g) (h) (i)
(j) (k)

Fig. 1 **a** Original RGB fundal image, **b** Green Channel Image, **c** Blurred image, **d** Unsharp masked image, **e** Contrast-enhanced image, **f** Bit-plane-sliced image, **g** Median-filtered image, **h** Otsu-thresholded image, **i** Morphologically opened image, **j** Optic disc fully removed image, **k** Ground truth image

Table 1 Comparison of the proposed method with the existing ones on DRIVE database

Methods	Specificity (%)	Accuracy (%)
Aramesh et al. [12]	98.26	94.80
Imani et al. [15]	97.53	95.23
Mann et al. [16]	98.32	94.71
Xue et al. [18]	98.61	94.65
Halder et al.[19]	–	94.31
Proposed method	99.43	95.30

Specificity (also called true negative rate, TNR) measures the fraction of black pixels that are correctly identified.

$$\mathrm{TNR} = \frac{\mathrm{TN}}{\mathrm{TN} + \mathrm{FP}} \tag{16}$$

Accuracy measures the fraction of pixels correctly identified.

$$\mathrm{Accuracy} = \frac{\mathrm{TP} + \mathrm{TN}}{\mathrm{TP} + \mathrm{FP} + \mathrm{TN} + \mathrm{FN}} \tag{17}$$

The proposed method produced the outputs as shown in Fig. 1.

The specificity and accuracy obtained from the proposed method are 99.43 and 95.30%, respectively (Table 1).

5 Conclusion

The proposed method is a novel approach to extract blood vessels and gives quite good results with less computational time. It has a much higher specificity percentage than other methods and also the overall accuracy value is good. Together the efficacy and easiness make this algorithm a convenient method for retinal blood vessel extraction.

References

1. Teng, T., Lefley, M., Claremont, D.: Progress towards automated diabetic ocular screening: a review of image analysis and intelligent systems for diabetic retinopathy. Med. Biolog. Eng. Comput. **40**(1), 2–13 (2002)
2. Jonas, J.B., Ngyuen, X.N., Naumann, G.O.H.: Parapapillary vessel diameter in normal and glaucoma eyes. Investig. Ophthalmol. Vis. Sci. **30**(7), 1604–1611 (1989)
3. Wong, T.Y., Klein, R., Sharrett, A.R., Duncan, B.B., Couper, D.J., Tielsch, J.M., Klein, B.E.K., Hubbard, L.D.: Retinal arteriolar narrowing and risk of coronary heart disease in men and women-the atherosclerosis risk in communities study. JAMA **287**(9), 1079–1212 (2002)

4. Leung, H., Wang, J.J., Rochtchina, E., Tan, A.G., Wong, T.Y., Klein, R., Hubbard, L.D., Mitchell, P.: Relationships between age, blood pressure, and retinal vessel diameters in an older population, investigative ophthalmology & visual science. Assoc. Res. Vis. Ophthalmol. **44**(7), 2900–2904 (2003)
5. Wong, T.Y., Klein, R., Klein, B.E.K., Tielsch, J.M., Hubbar, L., Nieto, F.J.: Relationship with hypertension. Cardiovasc. Dis. Mortal. Surv. Opthalmol. Elsevier **46**(1), 59–80 (2001)
6. Pedersen, L., Grunkin, M., Ersbøll, B., Madsen, K., Larsen, M., Christoffersen, N., Skands, U.: Quantitative measurement of changes in retinal vessel diameter in ocular fundus images. Pattern Recogn. Lett. **21**(13–14), 1215–1223 (2000)
7. Green, W.R.: Histopathology of age-related macular degeneration, Mol. Vis. 5–27 (1999)
8. Fraz, M.M., Barman, S.A., Remagnino, P., Hoppe, A., Basit, A., Uyyanonvara, B., Rudnicka, A.R., Owen, C.G.: An approach to localize the retinal blood vessels using bit planes and centerline detection. Comput. Methods Programs Biomed. Elsevier **108**(2), 600–616 (2012)
9. Fraz, M.M., Basit, A., Barman, S.A.: Application of morphological bit planes in retinal blood vessel extraction. J. Digit. Imaging **26**(2), 274–286 (2013)
10. De, J., Ma, T., Li, H., Dash, M., Li, C.: Automated tracing of retinal blood vessels using graphical models. In: Scandinavian Conference on Image Analysis, pp. 277–289. Springer-Verlag, Berlin, Heidelberg (2013)
11. Aramesh, R., Faez, K.: A new method for segmentation of retinal blood vessels using morphological image processing technique. Int. J. Adv. Stud. Comput. Sci. Eng. **3**(1), 1–6 (2014)
12. Mustafa, W.A.B.W., Yazid, H., Yaacob, S.B., Basah, S.N.B.: Blood vessel extraction using morphological operation for diabetic retinopathy. In: 2014 IEEE Region 10 Symposium, Kuala Lumpur, pp. 208–212 (2014)

Region Growing-Based Scheme for Extraction of Text from Scene Images

Ranjit Ghoshal and Ayan Banerjee

Abstract Extraction of text from images is a vital work for every OCR system. Binarization plays a key task for any text segmentation scheme form scene images. Therefore, an effective scheme for binarizatin as well as text extraction from digital images is required. This work presents two effective schemes. First one is binarization and second one is text separation from images. In binarization scheme, Canny's edge information is added in Savola's binarization scheme. This binarization scheme provides a number of connected components. Further, we proposed an effective region growing scheme for extraction of text components from the binary image. A number of text specific criteria are defined. Based on these criteria two seed points for non-text and text are generated. Connected component-based region growing scheme is applied based on these seed points. For selection of seed points, information from *ground-truth* (GT) images of text information and our hand made non-text components information are used. Our schemes are tested on the freely accessible Icdar 2011 Born Digital data set. The performances are quite satisfactory.

Keywords Region growing · Seed point · Text extraction · Binary image · Connected component

1 Introduction

Nowadays camera captured images are highly used due to increasing development of technology. Capturing of the images which contains text is a easy task, but proper binarization and extraction of text from images is a difficult work. Therefore, text portion from images has to be extracted properly prior to recognized through OCR. To

R. Ghoshal (✉)
St. Thomas' College of Engineering and Technology, Kolkata 700023, India
e-mail: ranjit.ghoshal@rediffmail.com

A. Banerjee
Lexmark Research and Development Corporation, Kolkata 700156, India
e-mail: ayanbanerje.stcet@gmail.com

D. Bhattacharjee et al. (eds.), *Proceedings of International Conference on Frontiers in Computing and Systems*, Advances in Intelligent Systems and Computing 1255,
https://doi.org/10.1007/978-981-15-7834-2_14

transfer gray scale image to black and white image is a challenging research, mainly where result of the binary image can improve directly the OCR rate. Binarization is a prime job of extracting text. Numerous techniques are available for document image binarization but these techniques are not appropriate for scene images. Images of natural scene are more critical than document images. These images are difficult due to complex background. Moreover, shadows, noise drastically enhance the problem extra critical. Therefore, effective novel schemes are needed. Usually, there are two types of binarization approaches: *Global* and [1] and *Local* ([2, 3]). In recent times, a number of enhancements over those schemes are also introduced in the analysis of document. Further, researchers are trying to enhance these schemes for binarization scene text. *Gatos* et al. [4] illustrated a recent technique for document image binarization by adding a few binarization techniques and edge features. Existing algorithms for text separation [5] can generally be classified into two groups: template and CC-based algorithms.

Connected component-based schemes separate text portions from images of natural scene through analysis of the CC. CC-based algorithms are normally considered because implementation is quite simple. In this work, we develop an effective binarization scheme and a region growing-based text extraction scheme from scene images.

2 Binarization Scheme

Preprocessing:

Input of this algorithm is a color image. So, at first the color image is converted into a gray scale image. Normally, edge searching algorithms are applied to select the text border. So, we aim to search edge information from the gray scale image by using Canny's algorithm.

Proposed Scheme:

Step 1: Apply Canny's algorithm to detect edges (ED) on the input image II_s. These edges are stored in EDGE1.

Step 2: Initialize matrx *img11* and *img22* (same size as II_s) by 0's.

Step 3: II_s is binarize using Sauvola's scheme and keep it in *img22*. Further, complement the image *img22* and keep it in *img11*.

Step 4: Move a $N \times N$ (here N=7)mask on II_s based on the identified edge points (pixels). This mask (BW) is binarized using Sauvola's scheme. Increase a pixel value by 1 in *img22* if its equivalent pixel value is 1 in the mask (BW). Here, the occurrence of 1's in each point (pixel) is stored in *img22*.

Step 5: In the same way, the occurrence of 0's in each point (pixel) is stored in *img11*. Reverse the local mask (BW) and increase a point (pixel) value of *img11* by 1 if its equivalent pixel value is 0, in the reverse binarize mask.

Step 6: Let $Bin(x, y)$ is the outcome of our binarization algorithm. Now, it can be expressed as

$$Bin(x, y) = \begin{cases} 1\,; \forall p_j, \text{ if } img22(p_j) > img11\,(p_j) \\ 0\,; \text{else} \end{cases}$$

3 Criteria Selection for Seed Points

Binarization creates connected components (CCs). To select the seed points, following criteria are considered.

Sr: *Size ratio*(Sr) [5].
Ar: *Aspect ratio* [5].
Er: *Elongatedness ratio* (Er) [6].
Or: *Ratio of the object pixels to background pixels* [6].
AXr: *Axial ratio* (AXr) [5].
Th: *Thickness*(*Th*) [6].
Lr: *Ratio of Lengths* [5].
Wv: *Variance of width* [5].

Considering the above criteria, a vector is generated for each sample, i.e., $\Re = \{Sr, Ar, Er, Or, AXr, Th, Lr, Wv\}$.

4 Region Growing Scheme for Text Segmentation

The Region Growing Scheme is a bottom up technique to image segmentation. This approach is a connected component aggregation scheme which initializes a region to a user-specified seed point(s) and spreads on acquiring neighboring connected components if they satisfy a similarity condition with the cluster. Our proposed scheme is as follows:

Proposed Region Growing Procedure:

Step 1: Selection of Seed points:
We have created two criteria files for text and non-text connected components. These files are used to find the mean of each criteria (i.e., Sr, Ar, Er, Or, AXr, Th, Lr ans Wv) for both text as well as non-text components. These are considered as seed points for the proposed region growing-based segmentation. Let the seed points for non-text and text components are $\boldsymbol{\beta}$ and $\boldsymbol{\alpha}$, respectively.

Step 2: Selection of similarity criteria:
Here, feature distance measurement is used as a similarity criteria. Let CC_i is a connected component. Euclidean distance from CC_i to seed points (i.e., $\boldsymbol{\alpha}$ and $\boldsymbol{\beta}$) are calculated. Based on these distances, the CC_i is decided, whether this component belongs to text region or non-text region.
Step 3: Region Growing Criteria
All the connected components from the input binary image are assessed for similarity by checking the feature distance. If the distance value is small it is deemed that the new connected component is similar to the cluster. In this way the cluster grows. Main objective is to group the characters.

- Connected Components growing Scheme.
- Predefined similarity condition is applied for growth.
- Group Connected Components into larger regions based on similarity condition.
- Region Growing starts from a set of seed points

Step 4: Stopping Condition:
The above procedure is repeated for all the connected components in the binary image. Let the binary image contains N number of connected components. So, the procedure is repeated N times.

Step 5: Outcome of the Procedure:
The connected components grouped with $\boldsymbol{\alpha}$ (i.e., seed point of text component) are considered as text components.

5 Experimental Results

The outcomes of our work are made from Born Digital Data set [7] of ICDAR 2011. Experimental results are divided into two parts which are described as follows based on our objective of this work.

5.1 Results of Image Binarization Scheme

Now, think about a number of image binarization results. A few example results are presented in Fig. 1. We evaluate our binarization scheme with the help of *Precision*, *Recall* and *F-measure* [8]. The value of parameters, generated from the data set are respectively 93.73%, 71.23% and 80.95%. Further, we judge the performance against Otsu's scheme for binarization. Outcomes of Table 1 says that our binarization technique performs better than others.

Table 1 Comparison with a number of known binarization techniques

	Recall	*Precision*	*FM*
Our scheme	**93.73**	**71.23**	**80.95**
Otsu [1]	88.98	65.36	65.05
Niblack [2]	87	36	38.17
Sauvola [3]	91	14	20.4
Bhattacharya et al. [9]	91.14	47.85	53.81
Kumar et al. [10]	85.56	47.09	46.81

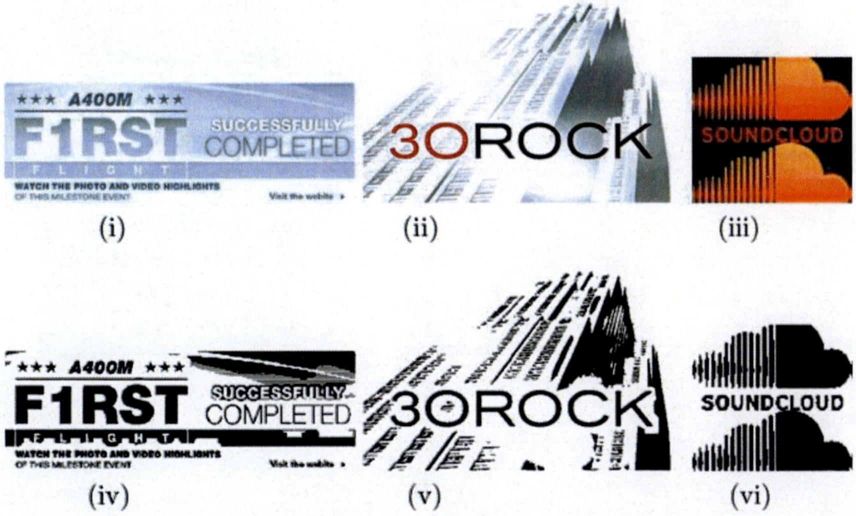

Fig. 1 (i), (ii), (iii) Sample images and (iv), (v), (vi) the corresponding binarized images

5.2 *Text Extraction Results*

Text extraction results obtained by our region growing-based scheme are presented in this subsection. Number of text and non-text components in our data set are 20723 and 10000. Table 2 represents a number of images and their subsequent extracted text. Our algorithm is performed well toward extraction of text. Further, we evaluate our scheme by evaluation standard. Clavelli et al. [11] described several features to estimate the text extraction excellence of each component of text presented in the GT image.

Further based on Clavelli et al. the text portions are separated as Well segmented, Merged and Lost. The Recall, Precision and F-Measure parameters of our region growing-based scheme generated on the basis of our data set images are 79.13, 80.72, and 79.91%. Lastly, we evaluate our method with other well-known algorithms. ICDAR 2011 Robust Reading Competition published results of numerous techniques

Table 2 A few images (1st, 3rd and 5th rows), the corresponding segmented text (2nd, 4th and 6th rows)

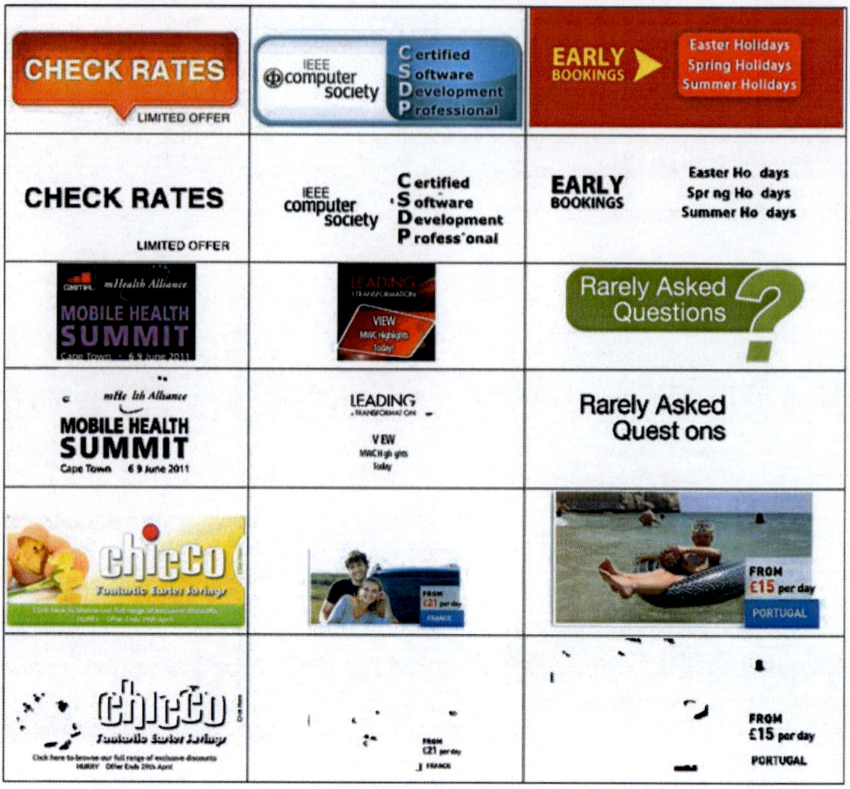

Table 3 Comparison of our region growing-based text segmentation scheme with other techniques

Scheme	Well segmented	Merged	Lost	Recall	Precision	F-Measure
Our scheme	**72.37**	11.47	**15.91**	**80.25**	**80.37**	**80.31**
Kumer et al. [10]	64.15	15.68	20.14	80.61	72.05	76.11
Adaptive edgedetection	66.55	9.23	24.20	78.23	70.97	74.42
Textorter	58.13	9.50	32.37	65.22	63.64	64.32
SASA	41.58	10.96	47.43	71.68	55.44	62.52

from a number of contributors. A number of such schemes are presented in Table 3. Our proposed technique is compared with these schemes. Our scheme has achieved precision and FM 80.72 and 79.91, respectively, which are superior to the other techniques. In terms of lost and well segment, our method performs well.

6 Conclusion and Scope of Future Work

This work offers an enhanced image binarization and region growing-based text separation scheme from images of natural scene. In the binarization, Canny's edge information is added in Savola's scheme. Further, a number of criteria are considered for selection of seed points which are required for region growing-based text extraction scheme. By adding the above schemes, a robust and effective text extraction scheme is developed that demonstrated better outcomes on Born Digital Data set of ICDAR 2011. In future, we shall explore advanced learning schemes to progress our text extraction performance.

References

1. Otsu, N.: A threshold selection method from gray-level histograms. IEEE Trans. Syst. Man Cybern. **9**(1), 377–393 (1979)
2. Niblack, W.: An Introduction to Digital Image Processing. Prentice Hall, Englewood Cliffs (1986)
3. Sauvola, J., Pietikinen, M.: Adaptive document image binarization. Pattern Recognit. **2**, 225–236 (2000)
4. Gatos, B., Pratikakis, I., Perantonis, S.J.: Document image binarization by using a combination of multiple binarization techniques and adapted edge information. In: Proceedings of the International Conference on Pattern Recognition (ICPR). (2008)
5. Ghoshal, R., Roy, A., Banerjee, A., Dhara, B., Parui, S.: A novel method for binarization of scene text images and its application in text identification. Pattern Anal. Appl. 1–15 (2018)
6. Ghoshal, R., Roy, A., Bhowmik, T.K., Parui, S.K.: Decision tree based recognition of bangla text from outdoor scene images. In: Eighteen International Conference on Neural Information Processing (ICONIP), pp. 538–546 (2011)
7. Karatzas, D., Robles Mestre, S., Mas, J., Nourbakhsh, F., Roy, P.P.: Icdar 2011 robust reading competition-challenge 1: Reading text in born-digital images (web and email). In: Proceedings of the 11th International Conference of Document Analysis and Recognition (ICDAR), pp. 1485 – 1490 (2011)
8. Dance, C.R., Seegar, M.: On the evaluation of document analysis components by recall, precision, and accuracy. In: Proceedings of the Fifth International Conference on Document Analysis and Recognition (ICDAR), pp. 713–716 (1999)
9. Bhattacharya, U., Parui, S.K., Mondal, S.: Devanagari and bangla text extraction from natural scene images. In: Proceedings of the International Conference on Document Analysis and Recognition (ICDAR), pp. 171–175 (2009)
10. Kumar, D., Ramakrishnan, A.G.: Octymist:otsu-canny minimal spanning tree for born-digital images. In: Proceedings of the 10th IAPR International Workshop on Document Analysis Systems. DAS '12, pp. 389–393 (2012)
11. Clavelli, A., Karatzas, D., Lladós, J.: A framework for the assessment of text extraction algorithms on complex colour images. In: Proceedings of the 9th IAPR International Workshop on Document Analysis Systems. DAS '10, pp. 19–26. ACM (2010)

An Automated Reflector Based Traffic Signal System

Somasree Bhadra, Sunirmal Khatua, and Anirban Kundu

Abstract Smart traffic automation is an integral part of smart cities in recent days. Huge amount of energy and power is consumed by the usage of high end computers and electronic devices. An energy optimization technique is proposed in this paper to minimize energy dissipation and power consumption by the existing traffic systems. Green computing concept is implemented in automated traffic system. It is observed that a remarkable amount of energy is dissipated and eventually wasted due to the use of regular electric lights at traffic points. 'Traffic Reflector' concept is introduced in place of regular traffic lights. LED lights or incandescent bulbs in the traffic signal posts would be replaced by the reflectors. Sunlight would be focused on these reflectors by properly placed mirrors. Mirrors would be adjusted automatically to change their angles according to sun's position. Light source would be replaced by solar panel driven light in the absence of sunlight at night or other weather conditions. Energy consumption by traffic signal lights at traffic signal points is minimized by our proposed approach.

Keywords Smart traffic · Green computing · Energy · Reflector · Traffic signal lights

S. Bhadra (✉) · S. Khatua
University of Calcutta, Kolkata 700106, West Bengal, India
e-mail: cse.soma@gmail.com

S. Khatua
e-mail: skhatuacomp@caluniv.ac.in

A. Kundu
Netaji Subhash Engineering College, Kolkata 700152, West Bengal, India
e-mail: anik76in@gmail.com

D. Bhattacharjee et al. (eds.), *Proceedings of International Conference on Frontiers in Computing and Systems*, Advances in Intelligent Systems and Computing 1255,
https://doi.org/10.1007/978-981-15-7834-2_15

1 Introduction

A vital role in urban sustainability is ensured by the transportation system of a city. With the ever growing population and urbanization the transportation infrastructure is growing too. High end computers and electronic devices are involved to implement smart traffic automation in smart cities. Increasing energy cost and growing environmental concerns would be handled by applying green computing concepts to traffic infrastructure and management system, especially in metropolitan areas. In this paper, in our effort to minimize overall energy consumption of a smart traffic automation system by reducing energy dispatched by LED traffic lights is made. Amount of energy consumption from traffic lights has been significantly minimized from recent past. Transformation of traffic lights from incandescent light to LEDs costs significant minimization of energy consumption and money. Advantages of LEDs in traffic signal points are shadowed by the different collateral problems. It is observed that visibility of LED signals is sometimes not adequate enough for the drivers due to solar glare, different reflections and refractions in real time. Usage of reflectors is proposed in this work to replace LEDs. Traffic lights are a means of communicating with the drivers, as well as pedestrians. It is observed that drivers literally spend no more than a moment to read and interpret the traffic signs, while driving at high speed or during night. Hence, the signs need to be brighter, legible, and conspicuous even from distance. Since reflectors are based on the principle of retro reflection, all these properties are satisfied. This would ensure lesser number of road accidents.

2 Literature Review

In [1], an analytical study has been done on the economic impacts when the regular incandescent traffic lights got replaced by Korean-type LEDs in various cities. It was predicted that there would be 85% savings in energy and 75% savings in maintenance charges. In 2001, majority of the regular incandescent traffic lights were replaced by LEDs. In Portland, USA, it led to an annual energy savings of 4.9 kWh and reduction of annual CO_2 emissions by about 2,880 tons [2]. In [3], traffic lights comprising a LED panel for directing complete road traffic, for on road displays notifying traffic related information and a personal device interface is implemented. In [4], high efficiency traffic signal light has been developed. A hybrid lighting technology is introduced in [5], to overcome the visibility problem of LED based traffic signals. Traffic signals are given after detection of traffic status on selective roads through usage of variety of sensor in [6]. In [7] and [8], sensor controlled traffic signal systems. A particular lane would be released dynamically and emergency vehicles would be given priority. Lane Status would be updated through GSM network [7, 8]. Communication between ambulance and traffic signal system is established using IoT for smooth and safe journey in [9]. Traffic system is automated using fuzzy logic in [10]. Automatic traffic road sign detection using segmentation and neural networks is proposed in [11]. In [12], a hybrid traffic lighting system is introduced to reduce road accidents caused by drivers. In [13], a fuzzy model is designed that measures the average delay times on a T-section road. An analytical study revealing 85% less energy consumption in Chicago, through usage of LED Traffic lights has been done in [14]. In [15], a secure intelligent traffic light control scheme has been developed using computing which is capable of avoiding the problem of single point failure. In [16], an easily replaceable LED traffic signal application with heat dissipation capabilities has been developed. In [17], a study has been done to calculate energy savings through usage of LED lights and part-night street lighting.

3 Proposed Work

In urban traffic signal system, LEDs are active throughout the whole day, resulting in a significant amount of energy consumption. A power saving approach is proposed by replacing LEDs with reflectors. Implementation of use of natural sunlight in place of traffic lights through a sun position sensing traffic signal controlling unit is proposed in this work.

The proposed Traffic signal processing unit (TPU) is categorized into Light mode Traffic signal processing Unit (TPU_L) and Dark mode Traffic signal processing Unit (TPU_D). Appropriate TPU would be selected based on the availability of natural light. A Light Dependent Resistor is used to detect the intensity of natural light. Figure 1 depicts the operational activity of TPUs. LDR detects the intensity of natural light.

In the proposed automated reflector based Traffic Signal System, Light Dependent Resistors (LDRs) are used to select TPUs based on the natural light's availability. LDRs comprise of photoconductive cells with spectral responses similar to human eyes. Hence, it is the best choice to differentiate between darkness and light, as well as sufficient and insufficient light. The cell resistance falls with increasing light intensity. A threshold value is considered to demarcate between sufficient and

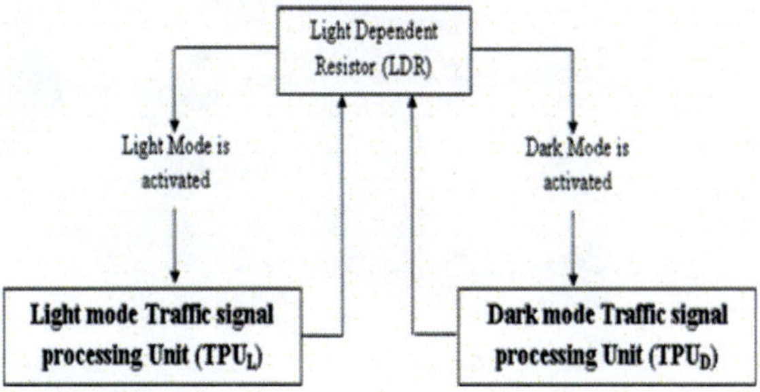

Fig. 1 Selection of TPU_L and TPU_D

insufficient light. When the intensity of light is less than the threshold value, Dark mode gets activated. Otherwise, Light mode is operational.

3.1 Light Mode Traffic Signal Processing Unit (TPU$_L$)

In TPU$_L$ maximum utilization of natural light is ensured. Sunlight is used as the light resource. Traffic lights in the signal post are replaced by different colored 'Reflectors'. A 'Mirror' is also placed at signal post to reflect the sunlight onto the reflectors.

TPU$_L$ comprises of 3 mirror setup, Red-Green-Yellow Reflector Disc, one microcontroller, one servo motor, one low speed dc motor, one LDR, and a 1 k resistor. LDR and 1 K resistor have been used for voltage divider circuit to ensure input of an analog signal to the microcontroller. A triangle shaped 3 mirror arrangement setup reflects the sunlight toward a window where a Red-Yellow-Green colored reflector disc is installed as shown in Fig. 2.

The angular rotation of the servo motor is controlled by a microcontroller. The angular rotation depends on the position of sun in the sky. A light dependent resistance LDR is used to implement the position dependent rotational movement of the servo motor. Analog readings of light sensitive LDR would be regulated by position of sun due to changes in sunlight intensity. Variation in servo motor rotation would be ensured by the varied analog reading of the sensor. The projection of light toward the reflector disc is ensured. The mirror arrangement and the Red-Green-Yellow disc are encased in a box. The reflector disc is coupled to a low speed dc motor behind the window. The dc motor would be controlled by an electronic and programmable switch. A long shaft coupled with the servo motor and the mirror arrangement is set on the shaft. The shaft is placed slightly slanting downwards to adjust deflection of sunlight. Light would enter through doors on the front and back walls of the box. The top section of the box is kept transparent for free flow of light. Sunlight would be available through the signal window, while the sun is in the eastern zone. Incoming

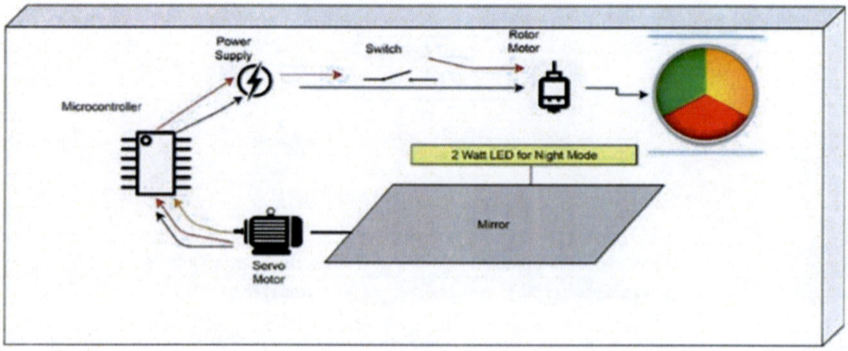

Fig. 2 Dynamic traffic signal controlling unit setup

sunlight would be reflected by the properly placed mirror and focused on the colored discs to glow. Sunlight would be entering through the top section of the box, while the sun is in the middle of the sky. Backside opening would be used as the sunlight access point, while the sun is in the western part of the sky. An additional light reflector would be placed at the backside of the box to increase the light intensity. Maximum amount of sunlight goes toward the signaling window. The case would be designed in a way that a specific color at a specific time instance as specified by traffic management software (i.e., Red or Green or Yellow) will be visible. This system is capable of supporting Full Circle, Pedestrian, and Arrow Signals without expending extra energy. Omission of electric light usage in daytime causes less power consumption.

3.2 Dark Mode Traffic Signal Processing Unit (TPU$_D$)

Sunlight is replaced by a combination of strategically placed focus lights at night. TPUD, however, also gets activated in case of insufficient natural light due to varied weather conditions. Operational activity of TPUD is same as TPUL.

Power consumption at night is comparatively low in comparison to the existing system. In our experimental setup, a combination of few electric lights placed strategically has been used at night to replace the natural light. It is observed that the total power consumption by the above mentioned combination of collection of lights does not exceed 2 W. Additional energy is required for the activation of dark mode automatically. The summation of all energy requirements is much lower than the existing system. The detailed calculation has been shown in the next Section.

4 Experimental Results and Analysis

In typical traffic system, Energy consumption by LED traffic lights comprising of Red-Yellow-Amber is 10–18 watts per day. The average consumption is about 10.01 Watts for 200 mm signal traffic point using only full circle and about 13.32 Watts for a similar 300 mm traffic signal (1).

In our proposed system, energy consumption by the proposed Traffic Signal Processing Unit (TPUL) is calculated as follows (Table 1).

In real time scenario a regular traffic signal comprises of regular Full Circle Red-Green-Amber, as well as pedestrian signals. It is also observed that on an average 60% of the total time Red Signal is active and the remaining 40% of time Green and Yellow is active. Taking these points into account, energy consumed at existing 200 mm traffic signal intersection,0.6 * (Average Energy Consumption by RED) + 0.4 * (Average Energy Consumption by YELLOW and GREEN) Watts = 10.066W atts (2).

Table 1 Energy consumption by proposed traffic signal processing unit (TPUL) (200 mm and 300 mm)

Energy usage for	Amount of energy expended (in watts) by traffic signal processing unit (TPUL) (200 mm signal)	Amount of energy expended (in watts) by traffic signal processing unit (TPUL) (300 mm signal)
Natural light (for light mode traffic signal processing unit) (TPUL)	0	0
Light source (dark mode traffic signal processing unit) (TPUL)	2	4
Rotor motor	0.42	0.42
Servo disk motor	2.75	2.75
Microcontroller	0.29	0.29
Total energy consumed	3.46	5.46

And at existing 300 mm traffic signal intersection,60% * (Average Energy Consumption by RED) + 40% * (Average Energy Consumption by YELLOW and GREEN) Watts = 15.042 Watts (3).

Based on calculations shown in Table 2, a comparative study of energy consumption between existing system and proposed system is shown using bar graph in Fig. 3a (only Full Circle Red-Green-Yellow)and Fig. 3b (Full Circle Red-Green-Yellow as well as Pedestrian symbol and Arrow signal).

Total energy consumption at a regular traffic signal is calculated based on the dataset provided in [18] and depicted here in Table 3.

Total annual energy consumption (KWh) by a regular 300 mm traffic signal comprising of Full Circle Red-Yellow-Green-Flash = $\sum$ (Annual power consumption by Full Circle Red + Annual power consumption by Full Circle Yellow + Annual power consumption by Full Circle Green + Flash Lamp) = 5574426.01 kWh (4)

In [18], 10000 KWh of power dissipation generates 5.65 short ton of CO2. Thus, TPU will reduce carbon footprint. It has been mathematically and experimentally shown that the TPU consumes lesser energy than existing system. However, if visibility of the TPU from a distance is insufficient, all efforts to implement this would fail. Thus, it is important to gauge the intensity at which the brightness appears to the human eyes. Thus, a graph depicting LUX value emanated from TPU versus time has been shown in Fig. 4, which shows that the TPU generates sufficient LUX.

This system would be implemented using a rigid structure to avoid any natural calamities. Spiked edges would be installed to avoid alignment disturbance due to bird sitting. Mirror would be placed inside a transparent box to avoid physical damage. The entire setup has been divided into a functional unit comprising of tri-colored rotating disc, rotor motor, LDR, and an intelligent unit made up of microcontroller and servo motor. The alignments of the various components have been done in such a way such that the intelligent unit does not get affected during regular, mundane maintenance work. The entire system would be placed in the most suitable position

Table 2 Energy consumption by existing regular traffic signal (200 mm & 300 mm)

Lamp type	Power consumption (Watt)	Number of modules in Istanbul	Average operating time per hour (hours)	Annual power consumption (kWh)	Annual cost (t) Turkish Lira
300 mm Red Lamp	100	5664	0.539	2674336.90	833039.90
300 mm Yellow Lamp	100	5664	0.022	109156.61	34001.63
300 mm Green Lamp	100	5664	0.439	2178170.50	678487.04
300 mm Flash Lamp	100	1399	0.50	612762.00	190871.69
200 mm Red Lamp	75	9789	0.539	3466510.05	1079797.08
200 mm Yellow Lamp	75	9789	0.022	141490.21	44073.35
200 mm Green Lamp	75	9789	0.439	2823372.75	879463.67
200 mm Flash Lamp	75	228	0.50	74898.00	23330.28
200 mm Red Pedestrian	75	11221	0.591	4356968.43	1357169.52
200 mm Green Pedestrian	75	11221	0.409	3015228.57	939225.61
300 mm Red Arrow	100	292	0.539	137871.89	42946.27
300 mm Yellow arrow	100	292	0.022	5627.42	1752.91
300 mm Green Arrow	100	292	0.439	112292.69	34978.50
200 mm Red Arrow	75	1855	0.539	656898.17	204619.84

(continued)

Table 2 (continued)

Lamp type	Power consumption (Watt)	Number of modules in Istanbul	Average operating time per hour (hours)	Annual power consumption (kWh)	Annual cost (t) Turkish Lira
200 mm Yellow Arrow	75	1855	0.022	26812.17	8351.83
200 mm Green Arrow	75	1855	0.439	535024.67	166656.97
Total		76869		20927421	65518766.08

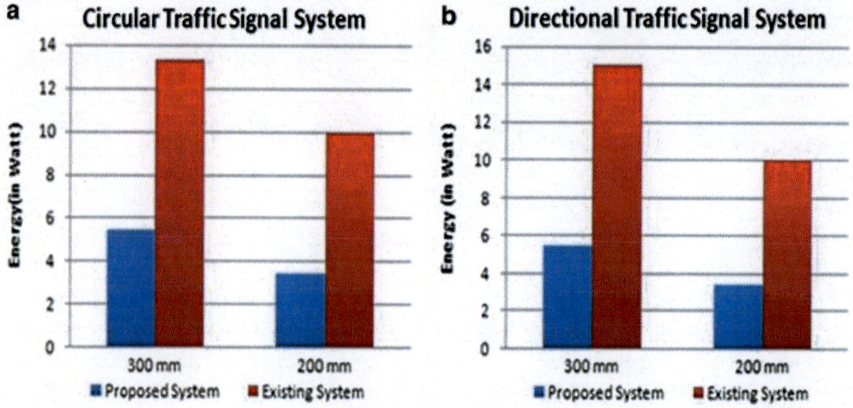

Fig. 3 Figure 3a, b Comparative study on energy consumption by proposed system and existing system

of the road so that it would be visible from every side of the road. A top view image of a miniature model of the proposed reflector-based traffic signal system is depicted below in Fig. 5.

Table 3 Dataset based on traffic signalization in Istanbul

Red signal	Green signal	Yellow signal						
Different instances	Energy consumption	Different instances	Energy consumption	Different instances	Energy consumption			
	In a 200 mm signal	In a 300 mm signal		In a 200 mm signal	In a 300 mm signal		In a 200 mm signal	In a 300 mm signal
	8 Watts	12 Watts		8 Watts	12 Watts		8 Watts	12 watts
	12.22 Watts	18 Watts	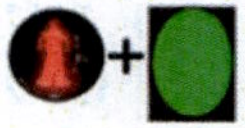	12.22Watts	18 Watts		NA	NA
	12 Watts	18.23 Watts		NA	NA		NA	NA
Average energy consumption	10.74Watts	16.07 Watts	Average energy consumption	10.11 Watts	15 Watts	Average energy consumption	8 Watts	12 Watts

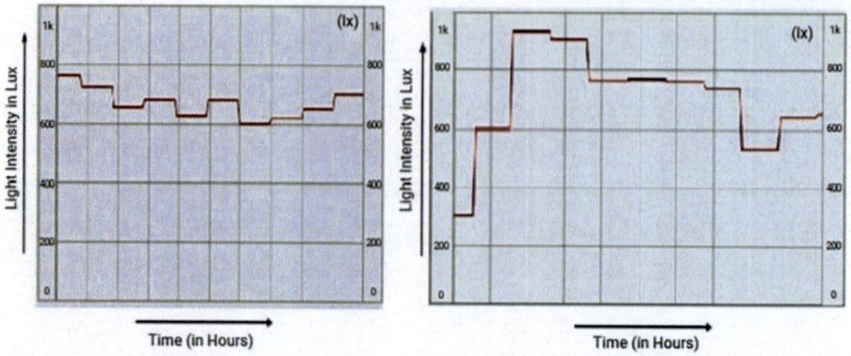

Fig. 4 Light intensity (LUX) of proposed traffic signal processing unit during 24 h slot

Fig. 5 Top view of miniature model of proposed reflector based traffic signal system

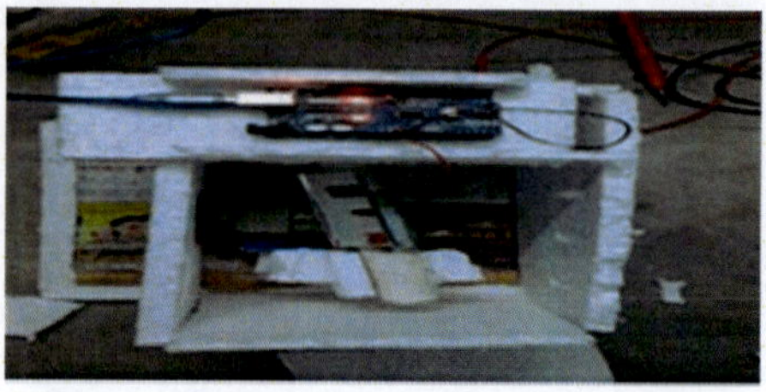

5 Conclusion

In this paper, a reflector-based traffic signal system is proposed to reduce the energy consumption due to traffic signal lights. Traffic signal lights are replaced by reflectors. Mirrors will be installed in front of the reflectors to gather and reflect the sunlight at day time to focus it onto the reflector. These mirrors will change their angles according to the sun position. Sensors are used to detect the sunlight path and the mirror would be adjusted by a control unit. In absence of sufficient natural light, sunlight will be replaced by a focus light. Natural lights are used in daytime and LED Focus light is used at night. A significant amount of the energy would be saved by implementing this approach in traffic signal system.

References

1. Jung, B.M., Jeong, H.G., Han, S.B.: A status of energy efficient led traffic signals in korea. In: Right Light 5, Proceedings of the 5th International Conference on Energy Efficient Lighting, Nice, France (2002)
2. Energy Sector Management Assistance Program (ESMAP): United States—LEDs for Traffic Signalss (2010). https://www.esmap.org/sites/esmap.org/files/CSEECIPortlandfinal.pdf

3. Teddy Yeung Man Lo: LED traffic light, US 7375650 B24. Jeffrey Kuo, Kwo-Sou Huang, Chan-Chuan Tai, US6019493 A
4. Osigbemeh, M., Onuu, M., Asaolu, O.: Design and development of an improvedtraffic light control system using hybrid lighting system. J. Traffic Trans. Eng. **4**(1), 88–95 (2017)
5. Yawle, R.U., Modak, K.K., Shivshette, P.S., Vhaval, S.S.: Smart traffic control system. SSRG Int. J. Electron. Commun. Eng. (SSRG-IJECE) **3**(3), 22–25 (2016)
6. Thakare, V.S., Jadhav, S.R., Sayyed, S.G, Pawar, P.V.: Design of smart traffic light controller using embedded system. IOSR J. Comput. Eng. **10**(1), 30–33 (2013)
7. Alzubaidi, A.J., Mohsen, A.A., Hassan, A.: Design of semi-automatic traffic light control system. Int J Sci Technol Res. **3**(10), 84–86 (2014)
8. Balamurugan, A., Kumar, G.N.S., Thilak, S.R., Selvakumar, P.: Automated emeregency system in ambulance to control traffic signals using Iot. Int. J. Eng. Comput. Sci. **4**(4), 11533–11539 (2015)
9. Jain, P.: Automatic traffic signal controller for roads by exploiting fuzzy logic. In: Computer Networks and Information Technologies, Communications in Computer and Information Science, vol. 142, pp. 273–277. Springer, Berlin, Heidelberg
10. Hossain, M.S., Hasan, M.M, Ali, M.A, Kabir, M.H., Ali, A.B.M.S.: Automatic detection and recognition of traffic signs. In: IEEE Conference on Robotics Automation and Mechatronics, Singapore, 28–30 June 2010
11. Osigbemeh, M., Onnu, M., Asaolu, O.: Design and development of an improvedtraffic light control system using hybrid lighting system. J. Traffic Trans. Eng. (English Edition) **4**(1), 88–95 (2017)
12. Johanyák, Z.C., Alvarez Gil, R.P.: Fuzzy model for the average delay time on a road ending with a traffic light. In: Kim, K., Kim, H., Baek, N. (eds.), IT Convergence and Security 2017. Lecture Notes in Electrical Engineering, vol. 449. Springer, Singapore (2018)
13. https://www.c40.org/case_studies/led-traffic-lights-reduce-energy-use-in-chicago-by-85
14. Liu, J., Li, J., Zhang, L., Dai, F., Zhang, Y., Meng, X., Shen, J.: Secure intelligent traffic light control using fog computing. Future Gener. Comput. Syst. **78**, 817–824 (2018). ISSN: 0167-739X
15. Dubuc, E., Fan, Y.F, Travernese, L.: Replaceable led light source for an led traffic signal application. US Patent No: US20190011114A1
16. Pogdan M., Ngahane K., Amin M. S. R., "Changing the colour of night in urban streets - LED vs. part-night lighting system", Socio-Economic Planning Sciences, ISSN 0038–0121, 2019
17. Dursun, Y., Almobaied, M., Buyukkinaci, B.: A status of energy efficient LED based traffic lamps in Istanbul. In: Alba, E., Chicano, F., Luque, G. (eds,), Smart Cities. Smart-CT 2016. Lecture Notes in Computer Science, vol. 9704. Springer, Cham (2016). eia.gov/tools/faq.php

A Novel Sentence Scoring Method for Extractive Text Summarization

Kamal Sarkar and Sohini Roy Chowdhury

Abstract Saliency based sentence ranking is a basic step of extractive text summarization. Saliency of a sentence is often measured based on the important words that the sentence contains. One of the drawbacks of such saliency-based sentence extraction method is that it extracts mainly the sentences related to the most common topic in the document. But the input document may contain multiple topics or events and the users may like to see in the summary the salient information for each different topic or event. To alleviate such problem, diversity-based re-ranking approach or sentence clustering-based approach is commonly used. But re-ranking or sentence clustering makes the summarization process slow. In this paper, we propose a novel summarization method that computes the score of a sentence by combining saliency and novelty of the sentence. Without using any re-ranker or clustering of sentences, the proposed approach can automatically take care of the diversity issue while producing a summary. We have evaluated the performance of the system on DUC 2001 and DUC 2002 benchmark single document summarization datasets. Our experiments reveal that it outperforms several existing state-of-the-art extractive summarization approaches.

Keywords Single document summarization · Sentence extraction · Saliency · Novelty · Text mining

1 Introduction

Text summarization is an effective mechanism for managing a large volume of textual data available on the Internet because text summarization presents a document or a cluster of related documents in a condensed form which helps readers in digesting

K. Sarkar (✉) · S. R. Chowdhury
Department of Computer Science and Engineering, Jadavpur University, Kolkata 700032, India
e-mail: jukamal2001@yahoo.com

S. R. Chowdhury
e-mail: sohini.rcd@gmail.com

D. Bhattacharjee et al. (eds.), *Proceedings of International Conference on Frontiers in Computing and Systems*, Advances in Intelligent Systems and Computing 1255,
https://doi.org/10.1007/978-981-15-7834-2_16

the main content of the document(s) very quickly. It has also been applied to many other natural language processing (NLP) applications such as question answering [1], information retrieval [2], indexing [3], document clustering [4]. Though a large number of scientific articles have author-provided summaries, many online documents such as newspaper articles, magazine articles do not come with author-provided summaries. So, there is a need for developing automatic text summarization tools.

Based on the nature of summary produced, text summarization methods are further classified as extractive summarization methods and abstractive summarization methods. Extraction-based summarization produces summary by extracting sentences or text segments from the input, whereas abstraction-based summarization produces summary using some deeper natural language generation techniques. Input to a summarizer can be either single document or multiple documents. In this paper, we focus on single document generic extraction based text summarization.

Automatic text summarization received much attention of the researchers many years ago. In this line, a series of conferences and workshops on automatic text summarization have advanced research on text summarization. Some of the notable workshops and conferences are NTCIR,[1] DUC,[2] special sessions in the reputed conferences such as conference of ACL, COLING, SIGIR, etc.

The most prior research works focused on developing methods for extraction based text summarization [5–10]). Though the researchers have also tried to devise a number of summarization methods that produce summaries in abstractive form, the existing abstractive summarization methods are not yet proven to be successful for producing grammatically correct abstractive summaries which are longer than the very short or ultra-summaries [11–13]. Recently, the deep learning-based summarization approaches have shown relatively better performance compared to the earlier abstractive summarization approaches, but the deep learning-based approaches to generic text summarization have been tested mostly on generating short or headline like summaries [14, 15], and the headline like summaries or the very short summaries do not serve all the purposes of document summarization because they are indicative and less informative. The main reason of popularity of deep learning approaches to short summary generation is that the deep learning approaches need a large training dataset and it is easier to create the large training dataset by automatically downloading short document-headline pairs from online sources than manually creating the large training dataset of document-summary pairs where each summary is relatively longer(100 word or more). In this article, we focus on 100 word summary generation which was the summarization task defined in the shared tasks conducted in conjunction with DUC 2001 and DUC 2002 conferences. Though the DUC 2001 and DUC 2002 summarization datasets are relatively small in size, these are the widely used datasets for 100 word summary generic generation tasks.

Though the fundamental step of extraction based summarization is sentence ranking, diversity-based sentence re-ranking [16], is often used to allow diverse

[1]http://research.nii.ac.jp/ntcir/.

[2]http://duc.nist.gov/.

information in the summary by removing redundant information. One such popular approach is MMR (Maximal Marginal Relevance)-based re-ranking. In this approach, after ranking the sentences of a document based on scores, when the summary is generated, the top ranked sentence is selected first and the next sentence is selected if it is sufficiently dissimilar to the previously selected sentences. This process is repeated until the desired summary length is reached. Since the rank order of the sentences in the summary is further revised at the time of summary generation, this process is generally known as re-ranking. Another approach dealing with redundancy, as well as diversity issue is to cluster sentences, and select representative sentence from each cluster [8]. The main problem with the re-ranking-based method or clustering-based method is that it hampers the speed of the summarization process. In this paper, we propose a sentence ranking method that assigns score to each sentence of the document based on two important factors—(1) saliency-how much salient information contained in the sentence and (2) novelty-how much novel information contained in the sentence. Though our proposed approach uses neither re-ranking nor clustering, it can automatically deal with the redundancy, as well as diversity issue by setting up a trade-off between saliency and novelty of a sentence while scoring the sentence.

Our idea of saliency is implemented using the TF-TDF based term weighting methods [6, 17], which assign score to a sentence based on the number of important words contained in the sentence where importance of a word is measured by the TF-IDF weight of the word. On the other hand, the idea of novelty is implemented using how many novel words contained in the sentence. We define a word is novel to a sentence if it appears first in the sentence, that is, it is the word that was not seen before in the previously encountered sentences of the document. Since we observe that some sentences contain more important novel words compared to other sentences, we distinguish among the novel words present in a sentence based on other features—how many times the novel word is repeated after its first occurrence and whether it is a part of proper name or not. To make our motivation clear, we have shown in Fig. 1, a sample document chosen from DUC dataset and the important novel words have been highlighted in the figure.

According to our assumption, a word which is novel to a sentence X, cannot be novel to other sentences appearing in the document after X. For example, the word "Weather" occurring in the seventh sentence of the document shown in Fig. 1, is novel to this sentence because it appears first in this sentence, it is a proper noun and its global frequency is 2, but when we see that the same word occurs in the eighth sentence of the document, it is no longer novel to the eighth sentence because it is already seen before in the earlier section of the document. With this motivation, we have developed a sentence scoring method which takes into account saliency and novelty both while ranking sentences for generating extractive summary for the document. We observe that, when two sentences have equal or near equal saliency scores, our proposed sentence scoring method gives higher preference to the sentence containing more novel information. The rationale behind such preference over sentences is that a sentence containing more number of novel terms (words) appears most likely at the junction of ending old topic and starting new

"Hurricane Gilbert swept toward the Dominican Republic Sunday, and the Civil Defense alerted its heavily populated south coast to prepare for high winds, heavy rains and high seas.

The storm was approaching from the southeast with sustained winds of 75 mph gusting to 92 mph.

``There is no need for alarm," Civil Defense Director Eugenio Cabral said in a television alert shortly before midnight Saturday.

Cabral said residents of the province of Barahona should closely follow Gilbert's movement. An estimated 100,000 people live in the province, including 70,000 in the city of Barahona, about 125 miles west of Santo Domingo.

Tropical Storm Gilbert formed in the eastern Caribbean and strengthened into a hurricane Saturday night.

The National Hurricane Center in Miami reported its position at 2 a.m. Sunday at latitude 16.1 north, longitude 67.5 west, about 140 miles south of Ponce, Puerto Rico, and 200 miles southeast of Santo Domingo.

The National Weather Service in San Juan, Puerto Rico, said Gilbert was moving westward at 15 mph with a ``broad area of cloudiness and heavy weather" rotating around the center of the storm.

The weather service issued a flash flood watch for Puerto Ricoand the Virgin Islands until at least 6 p.m. Sunday."

Fig. 1 Important novel appearances of the words are highlighted. The novel appearances of the words which occur at least twice are only shown here. The novel words which are part of proper noun are shown in green color

topic, and a term remains novel for the moment it enters first into the discourse. We assume that the important novel words chosen by the writer of the document while transiting from the old topic to a new topic play an important role in extractive summarization. So, by capturing salient and novel information, our proposed summarization approach selects sentences containing important and novel information while creating summary.

The contributions of this paper are given below.

- We propose the new idea of combining saliency and novelty for sentence ranking that improves single document text summarization without using any re-ranker. As a result of not using re-ranker, our proposed summarization method is relatively faster than other summarization methods that use re-ranker [8, 16, 18].

The organization of this paper is as follows: The proposed summarization method has been described in Sect. 2. The experiments and results are described in Sect. 3. In Sect. 4, conclusion and future works are presented.

2 Our Proposed Summarization Method

Our proposed text summarization approach works by ranking sentences based on their scores where the score of a sentence is computed based on two important factors-saliency and novelty. Saliency of a sentence is measured by calculating how much salient information contained in the sentence and novelty of the sentence is measured by calculating how much novel information contained in the sentence. Our idea of combining saliency and novelty for scoring sentences is based on the hypothesis that a salient sentence containing sufficient novel information appearing at the junction of transition from the old topic to the new topic is a potential candidate summary sentence. Our idea of saliency is implemented using the traditional TF-TDF-based term weighting method [6, 17], which compute saliency of a sentence as the sum of the TF-IDF weights of the important terms contained in the sentence. On the other hand, the idea of novelty is implemented using how many novel terms (words) contained in the sentence. We define a word is novel to a sentence if it appears first in the sentence, that is, this word was not seen before in the previously encountered sentences of the document. We observe that a sentence may contain some novel words which contribute more to summary worthiness of the sentence than the other novel words present in the sentence. For example, a novel word which is a part of proper noun is more useful than other words which are not part of any proper noun. So, we distinguish among such novel words based on several features such as how many times the term is repeated after its first occurrence and whether the novel term is a part of proper name or not. We have discussed in this section how to compute saliency and novelty of a sentence. Before computing saliency and novelty score of each sentence, the stop words (a set of highly frequent but less important words and a set of common verbs) are removed from the sentence and then it is stemmed using Porter Algorithm [19].

2.1 Sentence Scoring Based on Saliency

TF-IDF-based Score for a sentence S is

$$\text{TF}-\text{IDF based score (S)} = \sum_{w \in s} \text{TF}-\text{IDF}(w) \tag{1}$$

where TF-IDF(w) is the product of term frequency of the word w and Inverse Document frequency(IDF) of the word w. IDF(w) is computed as log(N/df(w)), where N is corpus size and df(w) is called document frequency of w, that is, how many documents of the corpus contain the word w at least once. To prevent longer sentences from getting the higher TF-IDF-based score, the words with TF-IDF value less than a predefined threshold value are considered as low content words and so they are removed [6], from each sentence.

2.2 Sentence Scoring Based on Novelty

The novelty of a sentence S is measured based on how many novel terms are contained in S

$$\text{Novelty Score (S)} = \sum_{w \in s} \text{Weight}(w) \quad (2)$$

where:

$$\text{Weight}(w) = \begin{cases} \text{TF} - \text{IDF if } w \text{ is novel to } S \text{ and } w \text{ occurs at least twice in the document} \\ \partial * \text{TF} - \text{IDF if } w \text{ is novel to } S \text{ and } w \text{ occurs at least twice in the document} \\ \quad \text{and } w \text{is a part of a proper noun. Here } \partial > 1 \\ 0, \text{ otherwise} \end{cases} \quad (3)$$

We call a word as a novel word for a sentence if it occurs first in the sentence, that is, the word was not contained in the other sentences appeared in the document before the current sentence. A novel term is assigned an additional weight proportional to the TF-IDF weight of the word. The novel word which occurs only once in the document are assigned the weight of 0 because the longer sentences may contain many unimportant novel terms. The rationale behind using such formula given in Eq. (3), for scoring the novel words is to discriminate between the more important novel words and the less important novel words and to maintain uniformity of the novelty score with the scale on which saliency score is computed. This proper scaling is necessary because we finally combine saliency and novelty of a sentence to have a unique score for the sentence. To identify whether a word is a part of a proper noun or not, we simply check the first character of the word; if the initial character of the word is in upper case, we assume that it is likely a part of proper noun. We have used such a simple technique for checking proper noun to avoid using POS tagger or named entity tagger which makes document pre-processing step slower.

In Eq. (3), ∂ is considered as a bosting factor and ∂ is set to a value greater than 1 to assign higher weight to the novel word which is a part of proper name. We observe that setting the value of ∂ to 2 is good enough for our experiments.

2.3 Combining Saliency Score and Novelty Score

Overall score of a sentence $S$is weighted combination of saliency and novelty score.

$$\text{Overall Score (S)} = \text{Saliency Score(S)} + \lambda * \text{Novelty Score (S)} \quad (4)$$

where λ is the weight assigned to the novelty score, in Eq. (4), there is the only one tuning parameter which is tuned trough experimentations for obtaining the best summarization performance. We have obtained the best result by setting $\lambda = 10$.

2.4 Summary Generation

After ranking the sentences of the input document based on the combined score of saliency and novelty, top ranked sentences are selected from the ranked list one by one until the given summary length is reached. We have not used the MMR-based diversity-based re-ranking [16], or its variants while generating summary because our proposed method automatically takes care of the diversity issue.

3 Evaluation

3.1 Experimental Setup

We have conducted our experiments using DUC2001[3] and DUC2002[4] task1 datasets. In DUC2001 and DUC 2002, the task1 was designed to evaluate generic single document summaries where summary length was set to approximately100 words or less. DUC2001 dataset consists of 309 English news articles and DUC2002 dataset consists of 567 English news articles. Each article in the dataset was accompanied by human created model summaries.

We have used the widely used automatic summarization evaluation package named ROUGE toolkit [20], for summary evaluation. It compares the system generated summary with the reference (model) summary and counts overlapping units between them in terms of the *n*-g, word sequences, and word pairs, etc. ROUGE-N is basically an *n*-g based recall score, *n* stands for the length of the *n*-g, for example, unigram (1-g), bigram (2-g), and so on. Among these different scores reported by ROUGE toolkit, the unigram-based ROUGE score (ROUGE-1) has been shown to be correlated with human judgment [20]. We have showed in this paper the experimental results in terms of three different ROUGE metrics: unigram-based (ROUGE-1), bigram-based (ROUGE-2), and skip bigrams (ROUGE-SU4) [21]. We consider in our current study the recall scores for system comparison because we have compared our proposed system with the existing systems that used the older version of ROUGE package that calculates the recall scores for measuring the summary quality [22].

[3] http://www-nlpir.nist.gov/projects/duc/guidelines/2001.html.

[4] http://www-nlpir.nist.gov/projects/duc/guidelines/2002.html.

3.2 System Performance

According to the guidelines of DUC 2001 and DUC 2002, the summary length was set to 100 words. The value of parameter λ mentioned in Eq. (4), has been set to 10 to achieve the best average performance on both DUC datasets.

The results shown in the Tables 1 and 2, are obtained after testing our proposed approach on DUC2001 and DUC 2002 data respectively. We have set to *–m* option for ROUGE-1.5.5 toolkit while evaluating summaries. This means that stemming is done, but stop words are not removed from summaries when system summary and reference summary are compared. As can be seen from the Tables 1 to 2, our proposed summarization system outperforms the summarization system that uses only saliency and MMR-based re-ranking for summary generation. It proves that the novelty score when combined with the saliency score can select salient, as well as diverse sentences for generating summary, and improves single document summarization performance.

It is evident from the results shown in Tables 2 and 3 that novelty score has played an important role in improving summarization performance.

Table 1 ROUGE scores for the systems on DUC 2001 data. 95% confidence intervals are shown in brackets

Systems	R-1	R-2	R-SU4
Our Proposed System (with $\lambda = 10$)	**0.4536** (0.44180 – 0.46535)	0.2010 (0.1882 – 0.2118)	0.2213 (0.2111 – 0.2319)
System with Salience Score only + MMR	0.4443 (0.4325 – 0.4552)	0.1934 (0.1821 – 0.2047)	0.2151 (0.2052 – 0.2251)

Table 2 ROUGE scores for the systems on DUC 2002 data. 95% confidence intervals are shown in brackets

Systems	R-1	R-2	R-SU4
Our Proposed System (with $\lambda =$ 10)	**0.4824** (0.4754 – 0.4897)	0.2255 (0.2181 – 0.2335)	0.2419 (0.2353 – 0.2488)
System with Salience Score only + MMR	0.4528 (0.4449 – 0.4610)	0.1944 (0.1859 – 0.2031)	0.2174 (0.2102 – 0.2248)

Table 3 Comparison of results obtained by our proposed system and two existing systems when they are tested on DUC 2001 and DUC 2002 datasets

	DUC 2001 dataset		DUC 2002 dataset	
	R-1	R-2	R-1	R-2
The system proposed in [8]	**0.4786**	0.1853	0.4669	0.1237
Our proposed summarization system (with $\lambda = 10$)	0.4536	**0.2010**	**0.4824**	**0.2255**
The system with the best method proposed in [18] + MMR	0.4379	0.1614	0.4716	0.2011

3.3 Comparisons with Some Other Existing Systems

We have compared our proposed system with some existing extractive text summarization systems. We have chosen the work proposed in [18] and the work proposed in [8], for comparison with our work because they have been tested on DUC 2001 and DUC 2002 data sets, and do not use sentence position as a feature which is a domain dependent feature. In Table 3, we compare our proposed summarization approach with the two existing summarization approaches. We have considered only ROUGE-1 and ROUGE-2 scores for comparison, because these two kinds of ROUGE scores have been used for evaluation of the systems presented in [8, 18]. Table 3 compares system performance on DUC 2001 data set and DUC 2002 data set.

As we can see from the results presented in the Table 3, on both the data sets, our proposed summarization system outperforms the system proposed by 18]. Table 3 shows that performance of our proposed system on DUC 2001 dataset outperforms the system proposed by Aliguliyev [8], in terms of ROUGE-2 score only. Though, in terms of ROUGE-1 score, our system performs relatively worse on DUC 2001 dataset than the system proposed by [8], our system performs better than the system proposed by Aliguliyev [8], on DUC 2002 dataset in terms of both ROUGE scores. In fact, the system proposed by [8], performs worst on DUC 2002 dataset.

We can see from Table 3 that, on both the data sets, our proposed summarization system outperforms the system proposed in [18]. For DUC 2002 dataset, R-1 score of our system is 2.29%, better than the nearest performing system presented in [18], and R-2 score of our system is 12.13%, better than the system proposed in [18].

The most important difference between our proposed method and the other two existing methods mentioned in this section is that our method does not make use of any separate mechanism for maintaining diversity in the generated summary. This makes our system faster than the other two methods. The other differences are that our system is simpler, but its performance is comparable and sometimes better than the state-of-the-art systems.

4 Conclusion

This paper proposes a novel approach to single document summarization by combining the saliency and novelty of sentences. Our experimental results proves efficacy of the proposed approach for extractive summarization.

In future work, other features can be used to distinguish between the more important novel terms and the less important novel terms present in the sentence. We have a plan to integrate the graph-based centrality based saliency [23, 24], with our proposed novelty-based score for developing a more improved summarization system.

Acknowledgements This research work has received support from the project "JU-RUSA 2.0: research support to faculty members & departmental support towards upgradation in research", funded by Government of India.

References

1. Pande, V., Mukherjee, T., Varma, V.: Summarizing answers for community question answer services. Language Processing and Knowledge in the Web, pp. 16–151. Springer, Berlin, Heidelberg (2013)
2. Kan, M.Y.: Automatic text summarization as applied to information retrieval, Doctoral dissertation, Columbia University (2003). https://www.comp.nus.edu.sg/~kanmy/papers/thesis.pdf. Accessed 2016
3. Sakai, T., Jones, K.S.: Generic summaries for indexing in information retrieval. In: Proceedings of the 24th Annual International ACM SIGIR Conference on Research and Development in Information Retrieval (SIGIR), pp. 190–198 (2001)
4. Wang, X., Shen, D., Zeng, H.-J., Chen, Z., Ma, W.-Y.:. Web page clustering enhanced by summarization. In: Proceedings of the 13th ACM International Conference on Information and Knowledge Management (CIKM), pp. 242–243 (2004)
5. Osborne, M.: Using maximum entropy for sentence extraction. In: Proceedings of the ACL-02, Proceedings of Workshop on Automatic Summarization, (Philadelphia, PA), Annual Meeting of the ACL, Association for Computational Linguistics, Morristown, vol. 4 (2002)
6. Sarkar, K.: Bengali text summarization by sentence extraction (2012). arXiv:1201.2240
7. García-Hernández, R.A., Ledeneva, Y.: Single extractive text summarization based on a genetic algorithm. Pattern Recognition, pp. 374–383. Springer, Berlin, Heidelberg (2013)
8. Aliguliyev, R.M.: A new sentence similarity measure and sentence based extractive technique for automatic text summarization. Expert Syst. Appl. **36**(4), 7764–7772 (2009)
9. Ferreira, R., de Souza Cabral, L., Lins, R. D., e Silva, G. P., Freitas, F., Cavalcanti, G.D., Favaro, L.: Assessing sentence scoring techniques for extractive text summarization. Expert Syst. Appl. **40**(14), 5755–5764 (2013)
10. Sarkar, K.: A keyphrase-based approach to text summarization for english and bengali documents. Int. J. Technol. Diffus. (IJTD) **5**(2), 28–38 (2014)
11. Sarkar, K., Bandyopadhyay, S.: Generating headline summary from a document set. In: International Conference on Intelligent Text Processing and Computational Linguistics, pp. 649–652. Springer, Berlin, Heidelberg (2005)
12. Banko, M., Mittal, V., Witbrock, M.: Headline generation based on statistical translation. In: Proceedings of the 38th Annual Meeting of the Association for Computational Linguistics (ACL-2000), Hong Kong, pp. 318–325 (2000)
13. Zajic, D., Dorr, B., Schwartz, R.: Automatic headline generation for newspaper stories. Workshop on Automatic Summarization, pp. 78–85. Philadelphia, PA (2002)
14. Rush, A., Chopra, S., Weston, J.: A neural attention model for abstractive sentence summarization. In: Proceedings of EMNLP (2015)
15. Nallapati, R., Zhou, B., dos Santos, C., Gulcehre, C., Xiang, B.: Abstractive text summarization using sequence-to-sequence RNNS and beyond. In: The SIGNLL Conference on Computational Natural Language Learning (2016)
16. Carbonell, J.G., Goldstein, J.: The use of MMR, diversity-based re-ranking for reordering documents and producing summaries. In: SIGIR, vol. 98, pp. 335–336, Aug 1998
17. Sarkar, K.: An approach to summarizing Bengali news documents. In: Proceedings of the International Conference on Advances in Computing, Communications and Informatics, pp. 857–862, ACM (2012)
18. Wan, X., Xiao, J.: Exploiting neighbourhood knowledge for single document summarization and keyphrase extraction. ACM Trans. Inf. Syst. **28**(2), Article 8, 8:1–8:34 (2010)
19. Porter, M.F.: An algorithm for suffix stripping. Program **14**(3), 130–137 (1980)
20. Lin, C.Y., Hovy, E.: Automatic evaluation of summaries using n-gram co-occurrence statistics. In: Proceedings of the 2003 Conference of the North American Chapter of the Association for Computational Linguistics on Human Language Technology (NAACL-HLT), pp. 71–78 (2003)
21. Lin, C.Y.: ROUGE: a package for automatic evaluation of summaries. In: Proceedings of the Workshop on Text Summarization Branches Out, July 25–26, Barcelona, Spain (2004)

22. Sarkar, K.: Automatic single document text summarization using key concepts in documents. J. Inf. Process. Syst. **9**(4), 602–620 (2013)
23. Erkan, G., Radev, D.R.: LexRank: graph-based lexical centrality as salience in text summarization. J. Artif. Intell. Res. **22**, 457–479 (2004)
24. Mihalcea, R., Tarau, P.: A language independent algorithm for single and multiple document summarization. In: Proceedings of the 2nd International Joint Conference on Natural Language Processing (IJCNLP): Companion Volume including Posters/Demos and Tutorial Abstracts, pp. 19–24 (2005)

A Novel Approach for Face Recognition Using Modular PCA and MAP–MRF Classifier

Sumit Majumdar, Avijit Bose, and Prasenjit Das

Abstract Automated Face Recognition involves the extraction of features of the facial images of an unknown person and the identification by comparing the features with the available feature set of a known person. Each facial image has been divided into 16 independent sub-blocks which have been used in this work for designing and implementation of the Markov Random Field- (MRF) based classifier which helps to generate the Maximum A Posteriori (MAP) probability. Using the relationship of MRF and Gibbs Distribution, the MAP has been converted into an energy function. A novel approach for defining the energy function has also been proposed in this work.

Keywords Modular PCA · Markov random field · Maximum A posteriori- markov random field (MAP-MRF) · Gibbs distribution

1 Introduction

Automated Face Recognition is one of the biometric solutions available to identify a person by comparing the facial features. Face recognition techniques nowadays are being used in attendance monitoring systems, passport fraud detection, finding missing children, supporting law enforcement by tracking a person, and minimizing identity fraud. It also has a vast application in social media. A computer recognizes faces, by finding out distinguishable features from an unknown face by comparing them to the features of known faces. Efficiency may be affected due to the changes in

S. Majumdar (✉) · A. Bose · P. Das
MCKV Institute of Engineering, Howrah 711204, India
e-mail: sm071985@gmail.com

A. Bose
e-mail: avijit2691@gmail.com

P. Das
e-mail: mr.das.prasenjit@gmail.com

D. Bhattacharjee et al. (eds.), *Proceedings of International Conference on Frontiers in Computing and Systems*, Advances in Intelligent Systems and Computing 1255,
https://doi.org/10.1007/978-981-15-7834-2_17

the illumination and pose. In addition, the system may fail due to aging, the appearance of the beard and spectacles, and other accessories. Researches on face recognition are focused on devising techniques, capable of overcoming these difficulties using more suitable features and powerful classifiers.

In this paper, the authors have mentioned the literature review in Sect. 2. In Sect. 3, a scoop of the present work is briefly discussed. The Feature Extraction method is highlighted in Sect. 4, and in Sect. 5 the proposed classifier has been elaborately discussed. Section 6 concentrates on the yield results and conclusion.

2 History of Face Recognition System

The first automated face recognition was developed in the 1960s, where features of a common reference point were compared to reference data.

In the 1970s, Goldstein, Harmon, and Lesk [1] used 21 specific subjective markers such as hair color and lip thickness to automate recognition.

In 1988, Kirby and Sirovich [2] applied Principal Component Analysis, a standard linear algebra technique, to the face recognition problem. Turk and Pentland [3] in 1991 discovered that while using the Eigenfaces techniques, the residual error could be used to detect faces in images. Mohamed Rizon et al. [6] proposed PCA–ANN based system in 2006.

A statistical approach called Linear Discriminant Analysis (LDA) for face recognition was used by Etemad and Chellappa [4] and the Fisherface method was developed by Belhumeur [5].

In 1994, F. Samaria et al. [7] proposed Hidden Markov Model-based approach and in 1998, Nefian et al. [8] improved that work by reducing the time for training. In 2009, Rui Wang et al. [9] proposed a Markov Random Field-based Bayesian Classifier for face recognition.

3 Brief Idea of the Proposed Work

This work focused on the application of MAP- [9, 10] based classifier using Markov Random Field (MRF) [7–10]. MRF deals with context-dependent constraints of facial images. Contextual relation, i.e. the relation between two neighboring sites, plays a very important role because the face being an array of gray level pixel values, nonlinear transformations like change in illumination conditions, pose variation do not change the pattern of that particular face. Change in the pixel position of the facial image, will distorted the face pattern. Although the histogram of the facial image will not be hampered. The proposed work provides a technique which will use the contextual relation of the different sites in the facial images.

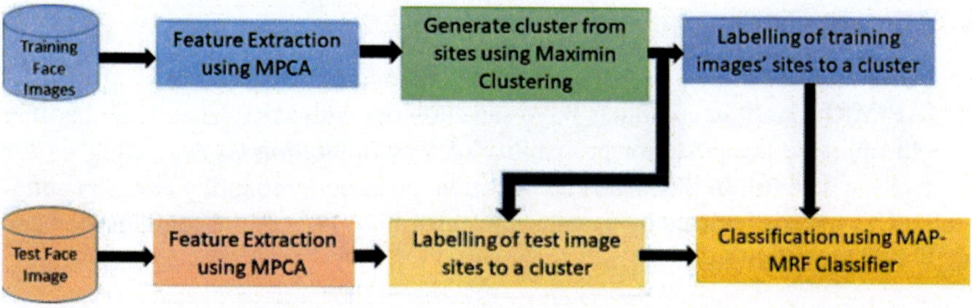

Fig. 1 Stages of the proposed work

This work proposed Modular Principal Component Analysis (MPCA)—[12–14] based feature extraction by applying PCA on the 16 independent sub-blocks (modules) of the facial image.

This 16 independent sub-blocks of each face image have been considered as the sites for the neighborhood system. From the sites of the training images, Maximin Clustering [15] has been used to generate 'g' random clusters. Then each site has been mapped to one of the 'g' labels. Using this random mapping (labeling) methods for calculating the likelihood energy, the One-Site, and the Two-Site clique energies has been proposed. Finally, an MRF-based energy function has been designed by aggregating the likelihood energy, the One-Site, and the Two-Site clique energies. Figure 1 shows the stages of the proposed work.

4 Modular PCA for Feature Extraction

MPCA has been used as a feature extraction technique in this work. It reduces the problem of rotation, transformation, illumination condition, and facial expression.

In case of pose, expression or illumination variation, only some part of the facial image will be affected, keeping rest of the part unchanged. By dividing the face images into independent modules, the computed features of these modules will mostly represent the local information of each module. In this work, faces are divided into 16 independent sub-blocks or modules. Stages of MPCA are given in Fig. 2.

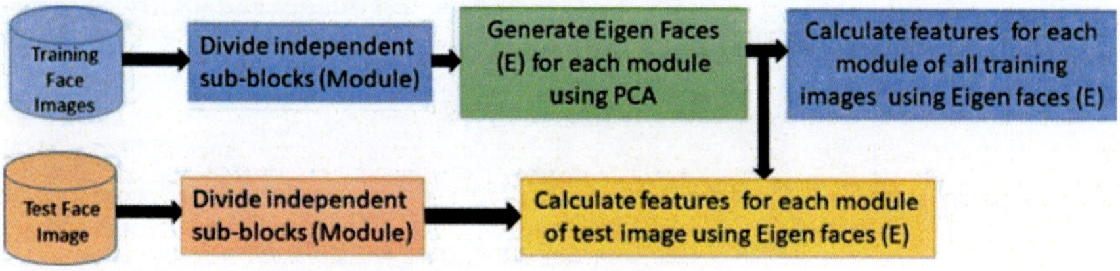

Fig. 2 Stages of feature extraction using Modular PCA

5 Classification Using MAP–MRF

A MAP-MRF classifier combines Bayesian classifier with MRF. The MAP classifier tries to optimize the posterior probability of a configuration (l) depending on the observation data (o). In Bayesian classification, posterior probability of each configuration(f) can be calculated by using prior probability ($P(l)$) of the configuration and likelihood probability ($P(o \mid l)$). It is shown in Eq. 1.

$$l^* = \arg\max_l P(l|o) = \arg\max_l P(l)P(o|l) \tag{1}$$

Considering the configuration(l) as an MRF, which follows Gibbs Distribution [11], the likelihood probability P(o|l) follows an exponential function of the likelihood energy function u(o|l), in the form $e^{-u(o|l)}$; we can write that posterior probability is inversely proportional to posterior energy ($u(l \mid o)$), which is shown in Eq. 2.

$$P(l|\, o) \propto \frac{1}{u(l|\, o)} \tag{2}$$

The MAP estimate can be given by the sum of prior energy (u(l)) and likelihood energy, and the maximization problem becomes a minimization problem of posterior energy, as shown in Eq. 3.

$$l^* = \arg\min_l u(l|o) = \arg\min_l (u(l) + u(o|l)) \tag{3}$$

Now the prior energy is the summation of the energy of all One-Site ($v_1(s_i)$), and Two-Site ($v_2(s_i, s_{i'})$) cliques.

The final posterior energy can be given by Eq. 4.

$$u(l|0) = u(l|0) + \sum_{c \in C_1} v_1(s_i) + \sum_{\{i,i'\} \in C_2} v_2(s_i, s_{i'}) \tag{4}$$

where C_1, C_2 are a set of One-Site cliques and a set of Two-Site cliques, respectively, and $s_i, s_{i'}$ are two neighboring sites.

So, the task classification task has now become, i.e. design of neighborhood system, the labeling of the sites of the training and the test images and finally defining the likelihood and prior energy functions, which are discussed in Sect. 5.3.

5.1 Neighborhood System

In this work, the 16 modules of facial images are considered as a site. 4-neighbor connectivity is considered to design the neighborhood shown in Fig. 3.

Fig. 3 Neighborhood system of 16 sites using 4-neighbor connectivity

The set of One-Site cliques (C_1) = {1, 2, 3, ..., 14, 15, 16}, and the set of Two-Site cliques (C_2) = {{1,2}, {1,5}, {2,1}, {2,3}, {2,6} ..., {13,14}, {14,15}, {15,16}}.

5.2 *Labeling Problem*

Assigning a label to the site is the purpose of the labeling problem. Each site S will be mapped to one of the labels of the set of labels B by the mapping function known as configuration(l).

By combining the feature set of the 16 sites of all the training images (q), a set of 16Xq feature vectors is generated. Maximin Clustering [15] is used to generate 'g' numbers of clusters and their centers. So, the set of labels B has 'g' elements. Euclidean distances of the features of all sites of the training and test images from these 'g' cluster centers have been computed and a site is labeled with the cluster id for which the distance is minimum. A site is labeled as Zero (0) if the minimum distance crosses the threshold value. The threshold is the distance between the cluster to which the site has been labeled and its nearest cluster.

5.3 *Defining Posterior Energy*

From Eq. 4, it can be said that the sum of likelihood energy, One-Site, and Two-Site clique energies is the final posterior energy.

6 Defining Likelihood Energy

The likelihood energy function $u(o \mid l)$ (shown in Eq. 5) does not depend on neighborhood system.

$$u(o|l) = \sum_{i=1}^{\text{no of sites (16)}} u(o(s_i)|l_i) \tag{5}$$

where l_i is the label of si site and $o(s_i)$ is the observation of the si site.

Likelihood energy of the s_i site is given in Eq. 6.

$$u(o(s_i)|l_i) = \begin{cases} D_i \text{ (the distance of } s_i \text{ site from } l_i - \text{th cluster), if } l_i \neq 0 \\ \text{a very large contant}(\vartheta_1)\text{, else} \end{cases} \tag{6}$$

Defining Prior Energy

Let CP be a $g \times g$ matrix, containing cluster center distance from one cluster to all other clusters. CP_{ij} denotes the cluster center distance from i^{th} cluster to j^{th} cluster. This vector has the two following characteristics:

1. $CP_{ii} = 0$, i.e. diagonal elements are Zero.
2. $CP_{ij} = CP_{ji}$, i.e. the CP vector is symmetric.

Now by comparing a site of the unknown image with the corresponding site of each and every known image, the clique potential or prior energy has been calculated.

Lett_i^m represent the label of the ith site of mth training image. Then the One-Site potential of a site with respect to mth training image can be given by

$$v_1^m(s_i) = \begin{cases} CP_{l_i t_i^m}\text{, if } l_i \neq 0 \\ \text{a very large constant } (\vartheta_2)\text{, else} \end{cases} \tag{7}$$

Considering neighborhood system, the Two-Site clique potential of an s_i site and its neighboring site $s_{i'}$, with respect to mth training image, can be given by

$$v_2^m(s_i, s_{i'}) = \begin{cases} CP_{l_i t_i^m} + CP_{l_i l_{i'}} + CP_{l_{i'} t_{i'}^m}, \text{ if } l_i \neq 0 \text{ and } l_{i'} \neq 0 \\ \text{a very large constant}(\vartheta_3), \text{ else} \end{cases} \tag{8}$$

From Eq. 4, the posterior energy of a test image with respect to mth training image can be given by Eq. 9.

$$u^m(l|o) = \sum_{i=1}^{16} u(o(s_i)|l_i) + \sum_{c \in C_1} v_1^m(s_i) + \sum_{\{i,i'\} \in C_2} v_2^m(s_i, s_{i'}) \tag{9}$$

From Eq. 9, for the s_i site of the test image, if $l_i \neq 0$ then likelihood energy will depend upon how close the s_i site and l_i-th cluster center are. The One-Site clique potential will be low when the ith site of the training image and test image are close to each other. For Two-Site clique, the distance between two neighboring site-labels is also taken into consideration, which captures the contextual properties.

If $l_i = 0$, then all three factors in the energy function (Eq. 9) will contribute high value to posterior energy, and if $l_{i'}$ is NULL then the Two-Site clique potential will be high, which will also increase the posterior energy.

Now the class of the training known image has been assigned to the unknown image for which the posterior energy ($u^m(l|o)$) is minimum.

7 Result and Conclusion

Olivetti Research Laboratory (ORL) [16, 17] database has been used in this work. This database consists of 10 images of 40 distinct persons, i.e. 400 images in total.

For finding the optimal number of the features for each imaging site, the number of features is varied from 10 to 60. Considering a 7:3 training to test ration on how the performance of the classifier varies due to the number of features is shown in Fig. 4.

A set of 40 features produces maximum efficiency. After that, the efficiency reduces due to the introduction of less-efficient and noisy features. By identifying a 40 element feature set as the optimal one, recognition performances of the proposed MAP-MRF model are recorded by varying the training and test data sizes. Results are shown in Fig. 5.

The proposed method is tested on the ORL face database. The best accuracy that can be yielded by the proposed work is 99%.

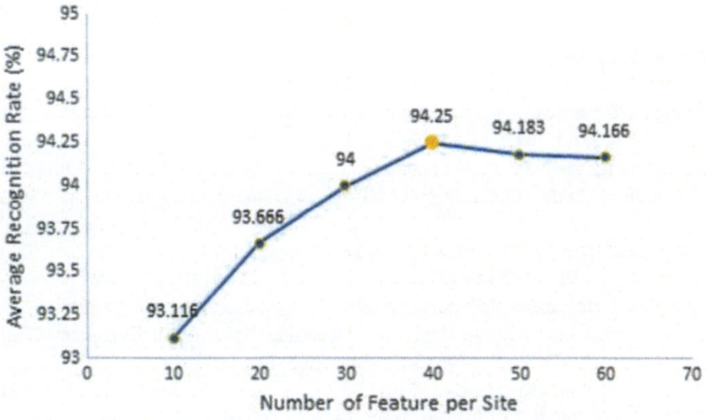

Fig. 4 Variation recognition rate against the number of features per Site

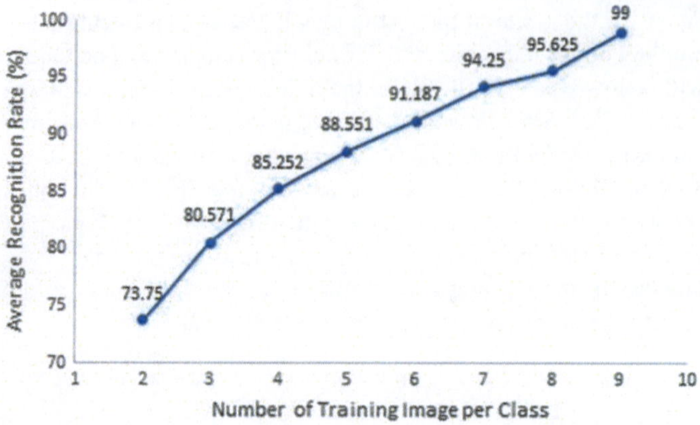

Fig. 5 Recognition rate against the varying number of training images per class

Using a 1 : 1 training to test ration on the same ORL database, the Eigenface method gives a recognition rate of 73%, and the work presented in [7, 8] shows a recognition rate of 84%. The method proposed in this work had achieved an average recognition rate of 88.5% for the same.

To improve this work in the future, pose and illumination invariant filtering can be done in the preprocessing stage. Also, by increasing the number of sites, more pose, facial expression, and illumination invariant features can be extracted. Finally, by considering 8-neighborhood of each site the number of Two-Site cliques can be increased and Three-Site cliques can be produced. This may further improve the recognition performance by exploring more contextual relations among the sites.

References

1. Goldstein, A.J., Harmon, L.D., Lesk, A.B.: Identification of human faces. Proc. IEEE **59**(5), 748–760 (1971)
2. Sirovich, L., Kirby, M.: A Low-Dimensional procedure for the characterization of human faces. J. Opt. Soc. Am. A **4**(3), 519–524 (1987)
3. Turk, M.A., Pentland, A.P.: Face recognition using eigenface. In: Proceedings of the IEEE Computer Society Conference Computer Vision and Pattern Recognition, pp. 586–591. Maui, Hawaii (1991)
4. Etemad, K., Chellappa, R.: Discriminant analysis for recognition of human face images. J. Opt. Soc. Am. A **14**(8), 1724–1733 (1997)
5. Belhumeur, P.N., Hespanha, J.P., Kriegman, D.J.: Eigenfaces versus fisherfaces: recognition using class specific linear projection. In: Proceedings of the 4th European Conference on Computer Vision, ECCV'96, pp. 45–58. Cambridge, UK (1996)
6. Rizon, Md., Mamat, MdR., Hashim, MdH., Shakaff, AYMd., Saad, P., Saad, A.R., Yaacob, S., Desa, H., Karthigayan, M.: Face recognition using eigenfaces and neural networks. Am. J. Appl. Sci. **2**(6), 1872–1875 (2006)
7. Samaria, F., Young, S.: HMM-based architecture for face identification. Image Vis. Comput. **12**, 537–543 (1994)

8. Nefian, A.V., Hayes III, H.M.: Hidden Markov models for face recognition. In: Proceedings of the IEEE International Conference on Acoustics, Speech, and Signal Processing, ICASSP'98, vol. 5, pp. 2721–2724. Seattle, Washington, USA (1998)
9. Wang, R., Lei, Z., Ao, M., Li, S.Z.: Bayesian face recognition based on markov random field modeling. In: Tistarelli, M., Nixon, M.S. (eds.), ICB 2009, LNCS 5558, pp. 42–51 (2009)
10. Huang, R., Pavlovic, V., Metaxas, D.: A hybrid face recognition method using markov random fields. In: Proceedings of International Conference on Pattern Recognition, Cambridge, UK (2004)
11. Liu, Q., Sclabassi, R.J., Li, C.C., Sun, M.: An application of MAP-MRF to change detection in image sequence based on mean field theory. EURASIP J. Appl. Sig n al Process. **13**, 1956–1968 (2005)
12. Wong, S.L.: Markov Random Field, Apr 2002
13. Gottumukkal, R., Asari, V.K.: An improved face recognition technique based on modular PCA approach. Pattern Recogn. Lett. **25**, 429–436 (2004)
14. Pentland, A., Moghaddam, B., Starner, T.: View-Based and modular eigenspaces for face recognition. In: IEEE Conference on Computer Vision and Pattern Recognition (1994)
15. Kumar, V.K.S.: Implementation of maximin algorithm on An S1MD array processor. Master's Thesis submitted to Birla Institute of Technology and Science, Pilani, India (1990)
16. Olivetti & Oracle Research Laboratory: The Olivetti & Oracle Research Laboratory Face
17. Database of Faces. http://www.camrol.co.uk/facedatabse.html

A "Bright-on-Dark, Dark-on-Bright" Approach to Multi-lingual Scene Text Detection

Neelotpal Chakraborty, Ayatullah Faruk Mollah, Subhadip Basu, and Ram Sarkar

Abstract Detecting texts from natural scene images is currently becoming a popular trend in the field of information retrieval. Researchers find it interesting due to the challenges faced while processing an image. In this paper, a relatively simple but effective approach is proposed where bright texts on a dark background and dark texts on a bright background are detected in natural scene images. This approach is based on the fact that there is usually stark contrast between the background and foreground. Hence, K-means clustering algorithm is applied on the gray levels of the image where bright and dark gray level clusters are generated. Each of these clusters are then analyzed to extract the text components. This method proves to be robust compared to the existing methods, giving reasonably satisfactory results when evaluated on Multi-lingual standard datasets like KAIST and MLe2e, and an in-house dataset of images having Multi-lingual texts written in English, Bangla and Hindi.

Keywords Multi-lingual · Scene text · K-means clustering · Morphological operation · Geometric filtering

---

N. Chakraborty (✉) · S. Basu · R. Sarkar
Department of Computer Science and Engineering, Jadavpur University, Kolkata 700032, India
e-mail: neelotpal_chakraborty@yahoo.com

S. Basu
e-mail: subhadip.basu@jadavpuruniversity.in

R. Sarkar
e-mail: raamsarkar@gmail.com

A. F. Mollah
Department of Computer Science and Engineering, Aliah University, Kolkata 700160, India
e-mail: afmollah@aliah.ac.in

D. Bhattacharjee et al. (eds.), *Proceedings of International Conference on Frontiers in Computing and Systems*, Advances in Intelligent Systems and Computing 1255,
https://doi.org/10.1007/978-981-15-7834-2_18

1 Introduction

Detecting texts in the wild is currently gaining more interest within the research fraternity due to its impact on the day-to-day lives of the people since texts are present in signboards, hoardings, name plates, posters, etc. Hence, the demand for newer applications like image to text conversion, tour guidance and vision-based applications [1] has necessitated more research on this domain.

The past few decades have been witnessing the emergence of many techniques for scene text detection. However, these methods [2–7] are observed to be very sensitive to image complexities like presence of noise, blur, uneven intensity and poor lighting conditions. Also, the presence of multi-lingual scene texts [1] needs some versatile processing to retain maximum text components as well as minimum non-text components.

In this paper, we present a simple yet effective solution for multi-lingual scene text detection where "Bright-on-Dark, Dark-on-Bright" (BDDB) strategy is adopted. This strategy is based on the known fact that there exists a good amount of difference in the intensity between a background and its corresponding foreground. In simple terms, bright shade is used for depicting texts on dark-shaded background and vice versa. Here, a color image is taken as input and converted to grayscale. The gray levels of this image are clustered into bright and dark levels where components of both the clusters undergo geometric and morphological filtering to eliminate non-text components and retaining text components. This method is experimentally evaluated on standard multi-lingual datasets like KAIST [8, 9] and MLe2e [10, 11], and an in-house dataset.

2 Literature Study

Recent times have witnessed the emergence of techniques related to scene text detection, that focus on stability and geometric properties of text components in an image. Stability-based techniques focus on detecting Maximally Stable Extremal Region (MSER) where a region is said to be stable if the intensity variation within a certain neighborhood stays minimum. Introduced by Matas et al. in [2], this technique has gained tremendous popularity in scene text detection. However, its high sensitivity to blur affects its efficiency. This problem is tackled in [3] by making it edge based. Still, due to the burden of tuning parameters like region area selection and threshold for intensity variation to suit scene texts of different dimensions, robustness of MSER diminishes. To ameliorate this problem, the works [4, 5] attempt to analyze the text properties and develop relations between MSER parameters and image dimension. An alternative to MSER is developed by Dutta et al. in [6] where the gray levels are binned in a way such that the foreground and background are segregated. The limitation in this method lies in deciding an optimal number of bins for a random image.

Another popular technique, introduced by Epshtein et al. [7], presents Stroke Width Transform (SWT) and the idea that text components usually have strokes of nearly constant width and geometry. This technique has been quite effective in differentiating text from non-text components. However, spurious components resembling textual properties may limit the performance of SWT-based approaches. Some recent methodologies [1, 8, 9] combine MSER and SWT to minimize spurious regions and retain maximum texts.

3 Proposed Methodology

A strategy called "Bright-on-Dark, Dark-on-Bright" (BDDB) is proposed in this paper which is based on the common property of depicting any foreground with a shade that is visibly in contrast to what the background shade is. In other words, a bright foreground is depicted upon a dark background and a dark foreground upon a bright background. In natural scene images, textual contents too exhibit such a property. Initially, a color input image is converted to grayscale and preprocessed to reduce noise, blur and irregular intensity distribution. Then, K-means clustering is applied on all the gray levels in the image so as to cluster them into bright and dark clusters. Finally, the non-text regions are filtered out using a combination of geometric and morphological operations. The overall process is illustrated by a flowchart shown in Fig. 1.

3.1 Image Preprocessing

An image may be hampered by complexities like noise, blur and uneven intensity distribution due to camera quality, movement and lighting conditions. To reduce these, Gaussian filter is, at first, applied on grayscale image to remove noise, and then, the amount of blur is determined using the method presented in [12] and reduced using Richardson-Lucy algorithm [13]. Later, the heterogeneity in intensity distribution is reduced by adaptive contrast enhancement [14]. The step-by-step process of image quality enhancement is shown in Fig. 2.

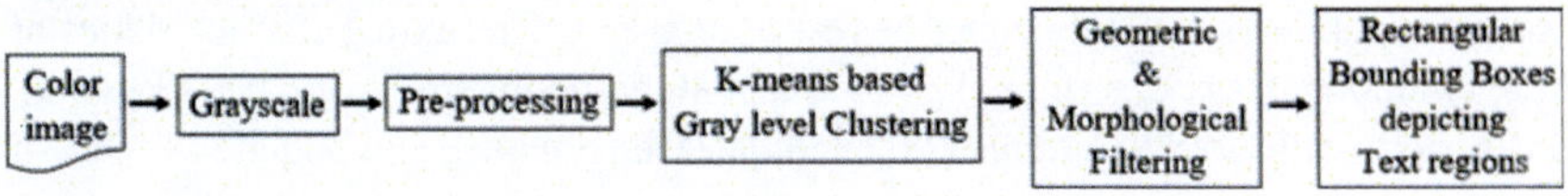

Fig. 1 An overview of the processes involved in BDDB-based scene text detection

Fig. 2 A sample image is taken from in-house dataset. The quality of this image is degraded due to low lighting conditions. Image quality is enhanced in the preprocessing step

Fig. 3 Binary images representing each cluster. The candidate components in both the binary masks are represented in white

3.2 *Gray Level Clustering*

Usually, a gray level value tends to be darker when decreasing and brighter when increasing, with 0 signifying the darkest shade and 255 signifying the brightest. This fact is visually represented in Fig. 3. It becomes difficult to decide the partition point of the scale representing the gray levels, between dark values and bright values. Hence, K-means clustering algorithm is adopted to generate 2 clusters where one cluster represents dark values and the other one bright. Two binary images are generated where each image depicts pixels with values of a particular cluster which is shown in Fig. 3.

3.3 *Filtering of Candidate Components*

The connected components from both the binary images obtained from the previous step, are candidates that may either be text or non-text. The textual characteristics of these components are determined by analyzing their geometric properties like *Area*, *Eccentricity*, *Solidity* and *Extent*. Text candidates are obtained and bounded as shown in Fig. 4.

Area. It is the number of data pixels representing a component

Eccentricity. It is a measure of circularity of the ellipse that approximates the shape of a connected component and is determined by the ratio between the distances of the center from the focus and the vertex of the ellipse.

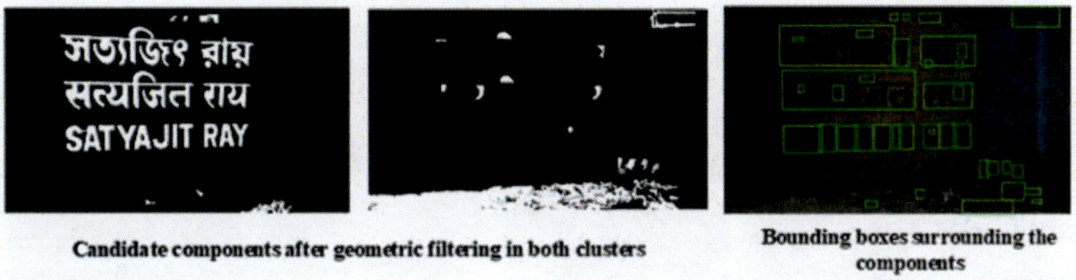

Fig. 4 Components obtained from both the clusters undergo geometric filtering to get text candidates. The bounding boxes obtained undergo morphological processing to further reduce non-texts

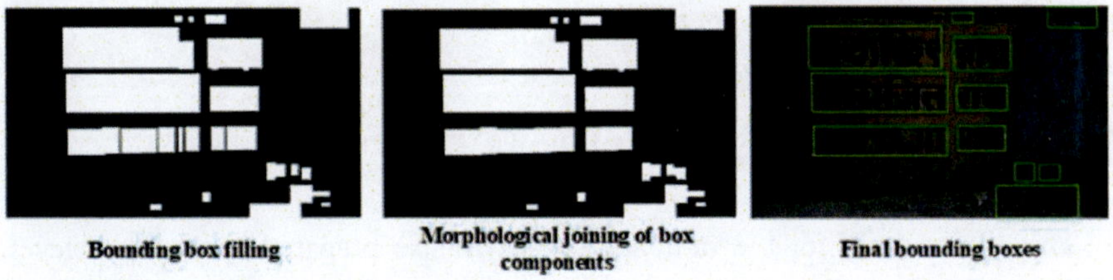

Fig. 5 Morphological closing operation is performed on filled bounding box regions so as to group them at word level. Filtering is also done on the resulting bounding boxes for minimizing non-text regions, thereby generating maximum text regions

Solidity . It is calculated as the ratio between the area of a component and area of a convex hull enclosing the component

Extent. It is the ratio between the area of a component and the area of its corresponding minimum bounding box

There is a possibility of candidate components being broken or surrounded by spurious components with dimensions of a few pixels. Hence, morphological closing with line structuring element is applied to the components, which also serves the purpose of localizing texts at the word level. The resultant bounding boxes of the text regions are shown in Fig. 5.

4 Experimental Results and Discussion

The proposed method is experimentally evaluated on images of different multilingual standard datasets like KAIST having scene images with texts written in Korean and English, and MLe2e having scene images with texts written in Kannada, Chinese, Korean and English. An in-house dataset of 300 scene images is also prepared where texts written in English, Bangla and Hindi are present. Samples from these datasets are shown in Fig. 6.

Fig. 6 Sample images from different datasets showing various types of complexities

The ground truth for the in-house dataset is also prepared at the word level, where each pixel within a rectangular bounding box is considered as true positive as illustrated in Fig. 7.

An input image undergoes the quality enhancement process where a Gaussian filter of size 3×3 is used to remove noise and the image is deblurred if the blur metric is greater than 0.3. Using K-means clustering, the gray levels are clustered into dark cluster (low values) and bright cluster (high values) which, when mapped across the image space, exhibit the contrasting relation between the foregrounds and their corresponding backgrounds. This is quite evident from the cluster maps of images for different datasets as shown in Fig. 8.

This method works efficiently on images from different datasets and exhibits its robustness against the image complexities and multi-lingual environment. The filtering procedure successfully reduces spurious components by exploiting their

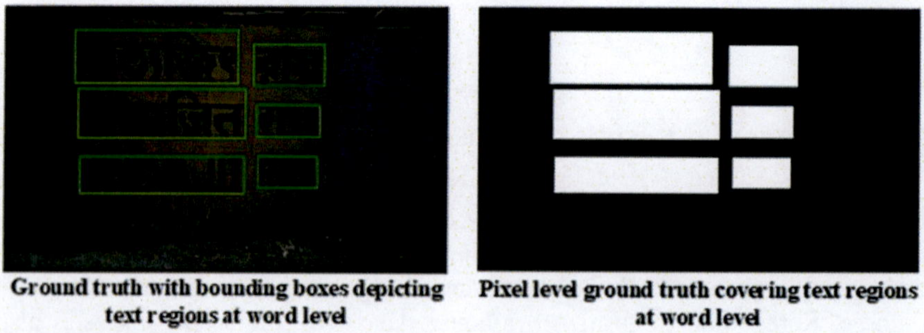

Fig. 7 Ground truth of image from in-house dataset. Rectangular bounding boxes are used for word-level representation. Evaluation is done at pixel level where every pixel within a bounding box is considered as true positive

Fig. 8 Cluster maps of sample images in Fig. 6. The text candidates appear to be distinct and unbroken to a great extent

geometric properties and combining broken texts, if any, or character components into word form using morphological closing operation.

A text component is observed to occupy less than 20% of the image space. Also, based on general observation, it may be stated that eccentricity for a text component is usually less than 1. Solidity and extent of candidate text components are kept less than 0.2 and 0.1, respectively. The final bounding boxes resulting from the methodology are shown in Fig. 9.

Fig. 9 Resultant text bounding boxes in the sample images across different datasets. Some of the bounding boxes highlight either non-text regions or unsuitability of parameter tuning in filtering procedure for some components. This is due to high complexity in images. Still maximum texts are retained with near-accurate bounding boxes

Table 1 Comparison of performance of the proposed method with some state-of-the-art methods for different datasets

Datasets	Methods	Precision	Recall	F-measure
KAIST	Özgen et al. [8]	0.57	0.54	0.55
	Agrawal et al. [9]	0.92	0.34	0.48
	Proposed	**0.93**	**0.64**	**0.76**
MLe2e	Gomes and Karatzas [10]	0.48	0.55	0.51
	Gomes et al. [11]	0.51	0.62	0.56
	Proposed	**0.54**	**0.74**	**0.61**
In-house	Mukhopadhyay et al. [1]	**0.98**	0.41	0.58
	Dutta et al. [6]	0.48	0.84	0.61
	Proposed	0.69	**0.87**	**0.77**

The performance of the overall methodology across the datasets KAIST, MLe2e and in-house is observed to be satisfactory and efficient than some of the existing methods. As evident from Table 1, the proposed BDDB-based method outperforms some state-of-the-art methods.

5 Conclusion

In this paper, a BDDB-based approach is proposed for multi-lingual scene text detection and localization. It applies contrasting relation shared between a foreground and its corresponding background, and is found to be an effective strategy. The filtering procedure successfully reduces spurious regions without sacrificing much the text components. This method is quite robust and it outperforms some of the standard methods based on MSER and SWT. The limitation, however, lies in the thresholding of the geometric properties. Hence, in the future, further improvements shall be aimed to select a threshold criterion of text components.

Acknowledgements This work is partially supported by the CMATER research laboratory of the Computer Science and Engineering Department, Jadavpur University, India, PURSE-II and UPE-II, project. SB is partially funded by DBT grant (BT/PR16356/BID/7/596/2016). RS, SB and AFM are partially funded by DST grant (EMR/2016/007213).

References

1. Mukhopadhyay, A., Kumar, S., Chowdhury, S.R., Chakraborty, N., Mollah, A.F., Basu, S., Sarkar, R.: Multi-Lingual Scene Text Detection Using One-Class Classifier. Int. J. Comput. Vis. Image Proce. (IJCVIP) **9**(2), 48–65 (2019)
2. Matas, J., Chum, O., Urban, M., Pajdla, T.: Robust wide-baseline stereo from maximally stable extremal regions. Image Vis. Comput. **22**(10), 761–767 (2004)
3. Chen, H., Tsai, S.S., Schroth, G., Chen, D.M., Grzeszczuk, R., Girod, B.: Robust text detection in natural images with edge-enhanced maximally stable extremal regions. In: 2011 18th IEEE International Conference on Image Processing, pp. 2609–2612. IEEE, Sep 2011
4. Chakraborty, N., Biswas, S., Mollah, A. F., Basu, S., & Sarkar, R.: Multi-lingual scene text detection by local histogram analysis and selection of optimal area for MSER. In: International Conference on Computational Intelligence, Communications, and Business Analytics, pp. 234–242. Springer, Singapore, July 2018
5. Panda, S., Ash, S., Chakraborty, N., Mollah, A.F., Basu, S., Sarkar, R.: Parameter tuning in MSER for text localization in multi-lingual camera-captured scene text images. In: Das A., Nayak J., Naik B., Pati S., Pelusi D. (eds.), Computational Intelligence in Pattern Recognition. Advances in Intelligent Systems and Computing, vol. 999. Springer, Singapore (2020)
6. Dutta, I.N., Chakraborty, N., Mollah, A.F., Basu, S., Sarkar, R.: Multi-lingual text localization from camera captured images based on foreground homogenity analysis. In: Recent Developments in Machine Learning and Data Analytics, pp. 149–158. Springer, Singapore (2019)
7. Epshtein, B., Ofek, E., Wexler, Y.: Detecting text in natural scenes with stroke width transform. In: 2010 IEEE Computer Society Conference on Computer Vision and Pattern Recognition, pp. 2963–2970. IEEE, June 2010
8. Özgen, A.C., Fasounaki, M., Ekenel, H.K.: Text detection in natural and computer-generated images. In: 2018 26th Signal Processing and Communications Applications Conference (SIU), pp. 1–4. IEEE, May 2018
9. Agrawal, A., Mukherjee, P., Srivastava, S., Lall, B.: Enhanced characterness for text detection in the wild. In: Proceedings of 2nd International Conference on Computer Vision & Image Processing, pp. 359–369. Springer, Singapore (2018)
10. Gomez, L., Karatzas, D.: A fine-grained approach to scene text script identification. In: 2016 12th IAPR Workshop on Document Analysis Systems (DAS), pp. 192–197. IEEE, Apr 2016
11. Gomez, L., Nicolaou, A., Karatzas, D.: Improving patch-based scene text script identification with ensembles of conjoined networks. Pattern Recogn. **67**(X), 85–96 (2017)
12. Crete, F., Dolmiere, T., Ladret, P., Nicolas, M.: The blur effect: perception and estimation with a new no-reference perceptual blur metric. In: Human vision and electronic imaging XII, vol. 6492, p. 64920I. International Society for Optics and Photonics, Feb 2007
13. Fish, D.A., Brinicombe, A.M., Pike, E.R., Walker, J.G.: Blind deconvolution by means of the Richardson-Lucy algorithm. JOSA A **12**(1), 58–65 (1995)
14. Zuiderveld, K.: Contrast limited adaptive histogram equalization. In: Graphics Gems IV, pp. 474–485. Academic Press Professional, Inc, Aug 1994

Pig Breed Detection Using Faster R-CNN

Pritam Ghosh, Subhranil Mustafi, Kaushik Mukherjee, Sanket Dan, Kunal Roy, and Satyendra Nath Mandal

Abstract In this paper, convolutional neural network object detection technology has been used to detect pig breeds with high precision from images captured through mobile cameras. The pretrained model is retrained on several images of 6 different pure breed pigs obtained from organized farms. The Faster R-CNN Inception-ResNet-v2 model has been used in transfer learning fashion for the above task. The training accuracy of this model is 100%, and the testing accuracy of this model is 91% with a confidence level of 94%. From the results achieved, it is noted that this model has produced better results compared to detection accuracy on other datasets like dog dataset, flower dataset, etc.

Keywords Convolutional neural network · Faster R-CNN · Pretrained model · Confidence level

1 Introduction

Pig breeds are classified by their phenotypic and genotypic characterizations. The phenotypic traits of Indigenous pigs like coat color and skin pigmentation, head shape and orientation, ear shape and orientation, tail shape and orientation, body shape, belly type, top line, hoof placement and presence of wattles were observed and recorded by visual observation. If a larger number of pigs carry similar characteristics in some generations, the group of pigs or pig cluster will form a new pig breed. This is time-consuming and laborious work, and some generations have to wait for new pig breed characterizations. Another way of developing new pig breeds is genotyping characterization. If the DNA sequences of a group of pigs are similar and different from existing DNA sequences of registered pig breeds, the group having the new DNA sequences will form a new breed. In genotyping characterization, generally

P. Ghosh (✉) · S. Mustafi · K. Mukherjee · S. Dan · K. Roy · S. N. Mandal
Department of Information Technology, Kalyani Government Engineering College, Kalyani, Nadia 741235, India
e-mail: ghoshpritam25@gmail.com

D. Bhattacharjee et al. (eds.), *Proceedings of International Conference on Frontiers in Computing and Systems*, Advances in Intelligent Systems and Computing 1255,
https://doi.org/10.1007/978-981-15-7834-2_19

blood samples were taken to extract DNA sequences. But, collecting this is a very difficult, painful and time-consuming task. The testing procedure is also very costly. Only eight pig breed names are registered in India but many pig breeds are still non-descriptive, i.e. they have not been characterized till now.

In this paper, six registered pig breeds have been captured through mobile cameras from organized farms, and faster region-based convolutional neural networks (Faster R-CNN) have been used to detect the breeds with a confidence level. The pretrained version of faster region-based convolutional neural networks (Faster R-CNN) known as Inception-ResNet-v2 has been used in transfer learning fashion. The feature extraction part of Faster R-CNN remained intact and top levels have been changed for applying on new image sets [1]. The aim of this paper is the detection of pig breeds from contour pictures only with a confidence level using Inception-ResNet-v2 convolutional neural network. This effort has not been made earlier; this is the motivation behind making this paper.

2 Related Work

AlexNet [2] won the ILSVRC [3] competition in 2012. Since then, multiple Convolutional Network architectures have evolved such as the deep networks by Visual Geometry Group (VGG) and inception module structures by GoogLeNet [4]. However, for scenarios like breed detection, no single architecture or model suits perfectly. Work has also been done on flower [5–8] and bird [9–12] categorization.

3 Faster R-CNN and Transfer Learning

Convolutional Neural Network (CNN) is used to recognize objects from pixel images (Fig. 1). It has two parts, one is feature extraction and the other is classification.

CNN can be used for classification only, but R-CNN [13] not only classifies the object but also localizes the object by drawing a bounding box around it. R-CNN has 3 parts: the first is the region proposal by using selective search, then feature extraction and lastly classification by using SVMs. The R-CNN model had 3 problems: training took multiple stages, was computationally expensive and was extremely slow. As a result, Fast R-CNN [14] was developed, where region proposal is done with the output from the last convolutional layer instead of using the original image. To improve the performance further, instead of using selective search for region proposal, a new network called Region Proposal Network (RPN) was introduced after the last convolutional layer. This modified network is called Faster R-CNN [15]. Incorporating the region proposal step in the main training pipeline helped to train the model much faster (Fig. 2).

This made Faster R-CNN the standard model for object detection programs. In transfer learning, already learned weights and biases acquired by a model trained on

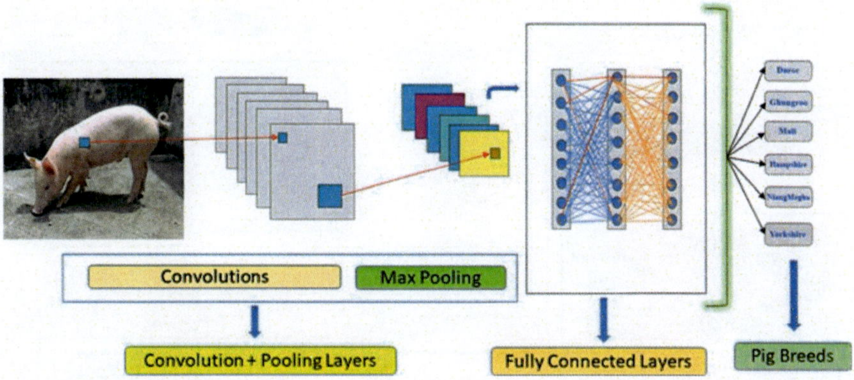

Fig. 1 CNN workflow

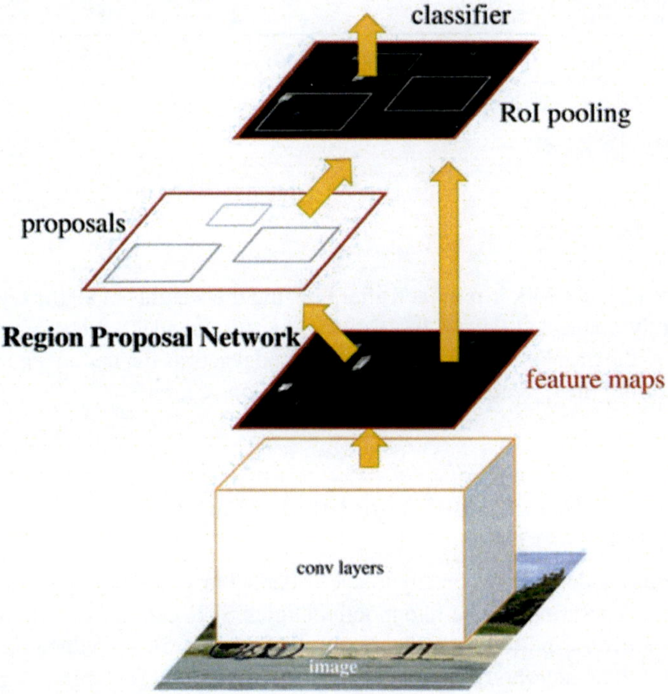

Fig. 2 Faster R-CNN workflow [15]

a large standardized dataset called pretrained model is used. This pretrained model's fully connected layers are then removed and replaced by the required classification layers based on the problem space (Fig. 3).

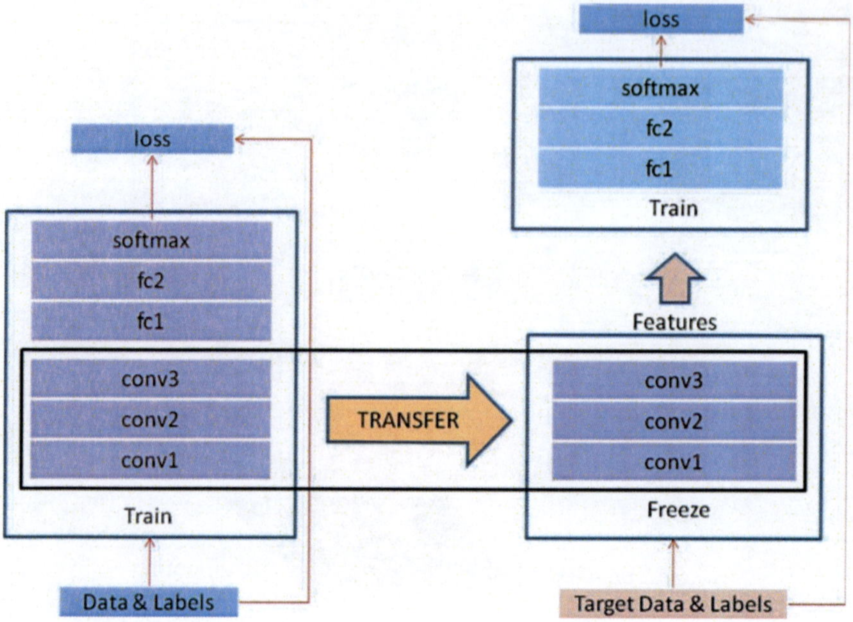

Fig. 3 Transfer learning workflow

Then the new network is retrained after freezing the weights of all the layers other than the fully connected layers. In this way, the fully connected layers get trained according to the problem space where the old model acts as the feature extractor and reduces training time greatly with high classification accuracy.

4 Pig Image Dataset and Annotated Dataset

The proposed model will run on any image collected from the Internet and captured by any image capturing device like mobile, Digital SLR camera, etc. In this paper, images have been captured by mobile with the same profile to reduce recognition complexity. The whole-body contour images of Ghungroo, Yorkshire, Duroc, Mali and Hampshire have been collected from NRC on Pig, Rani, Guwahati, Assam. The images of Niang Megha have been collected from ICAR Research Complex for NEH Region, Barapani, Umiam, Meghalaya. Minimum of two hundred pictures of each breed have been used in this research. The profile of each pig breed is given in Fig. 4.

This annotated dataset has been developed by making a bounding box around each image by labeling software.

Fig. 4 Pig images from the dataset

5 Framework Used and Process Flow

5.1 Framework

The pretrained model that we have used to apply transfer learning on is "Faster R-CNN Inception-ResNet-v2" trained on the MS-COCO large-scale object detection dataset. This model has been taken from the TensorFlow Object Detection API [16]. The overall structure of the network is shown in Fig. 5 [17].

Inception-ResNet is a hybrid of inception-net and residual-net. This network is chosen because currently, Inception-ResNet-v2 is the state-of-the-art network structure on the ImageNet dataset.

5.2 Process Flow of Faster R-CNN Model for Pig Breed Recognition

All convolutional networks will take a lot of time to locate the objects inside the images and classify them. To reduce the execution time, a bounding box has been created over each pig, and labeling the images before training is known as an annotated pig image dataset. The images of this annotated database have been used for training the pretrained Faster R-CNN model. The complete execution steps are given in Fig 6.

The steps are as follows: Create a labeled dataset of pig images using LabelImg (graphical image annotation tool). Create a corresponding .xml file for each image and split the dataset into a training set and a testing set with 9:1 split ratio, respectively; create 2 separate .csv (Comma separated Values) files from the annotated test and

Fig. 5 Overall Inception-ResNet-v2 architecture [17]

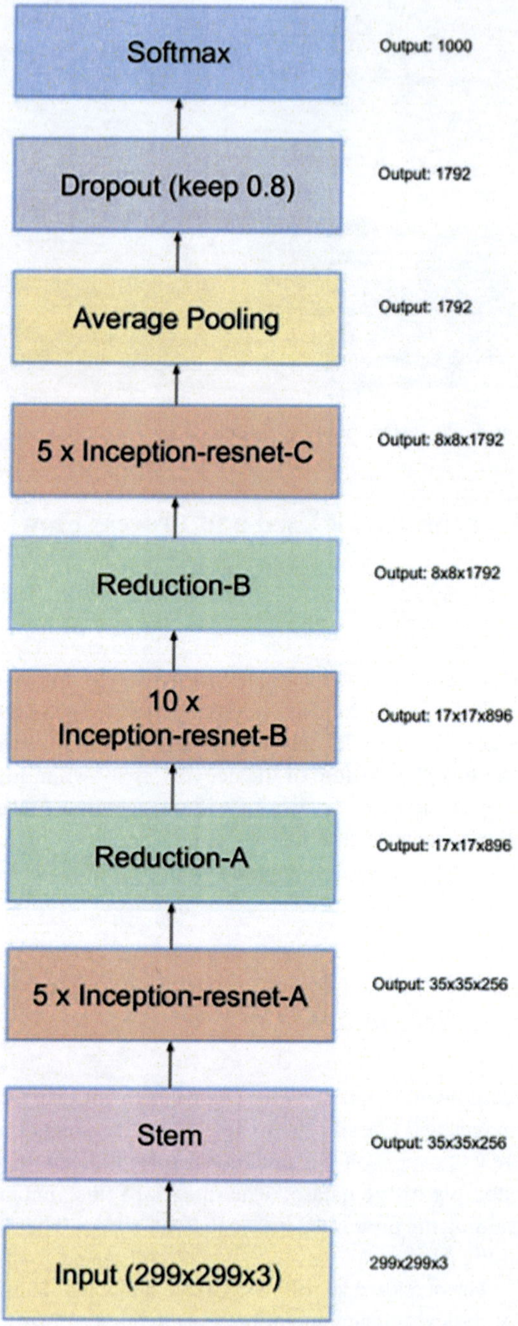

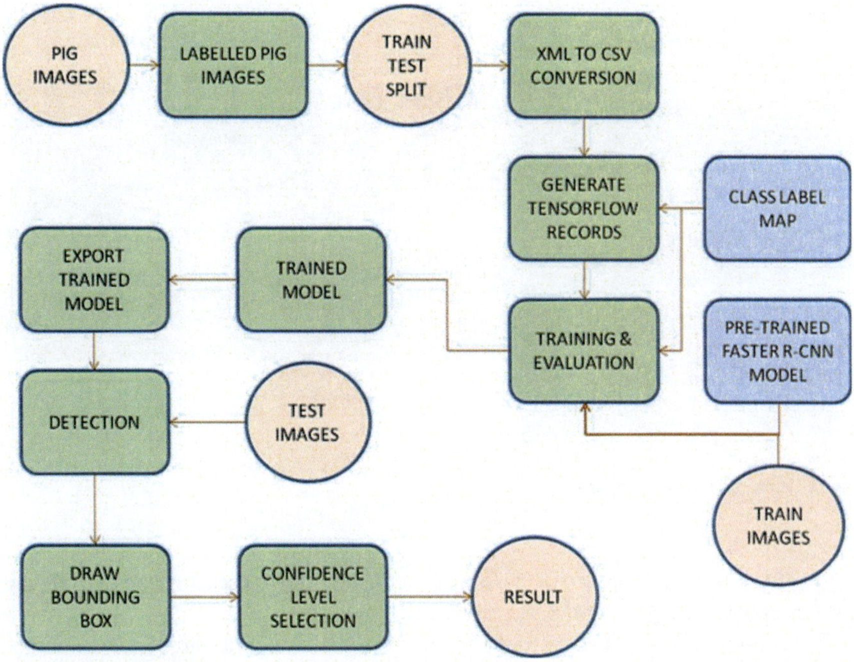

Fig. 6 Process flow diagram

train images sets, create a class-label.pbtxt file which contains all the required class ids and class names for pig breeds. Then, generate TensorFlow. record file with help of the class-label.pbtxt file for test and train sets; choose the pretrained model to be used for transfer learning (Faster R-CNN Inception-ResNet-v2 in our case) and retrain the fully connected layers of the selected model with our own dataset; after completion of training, export the trained model as an inference graph for using it for pig breed detection and apply it on the set of pig images that needs to be detected; feed the raw pig images to the Exported model (implemented in Jupyter Notebook). For each image in the detection set, using matplotlib, an annotation is created on the pig inside the image and its class and detection percentage are mentioned and this final image is stored on the disk for further analysis.

6 Result and Analysis

6.1 Bounding Boxes and Breed Detection

After successful training of the pretrained faster R-CNN model, the numbers of bounding boxes have been bounded on test images with some percentages of each

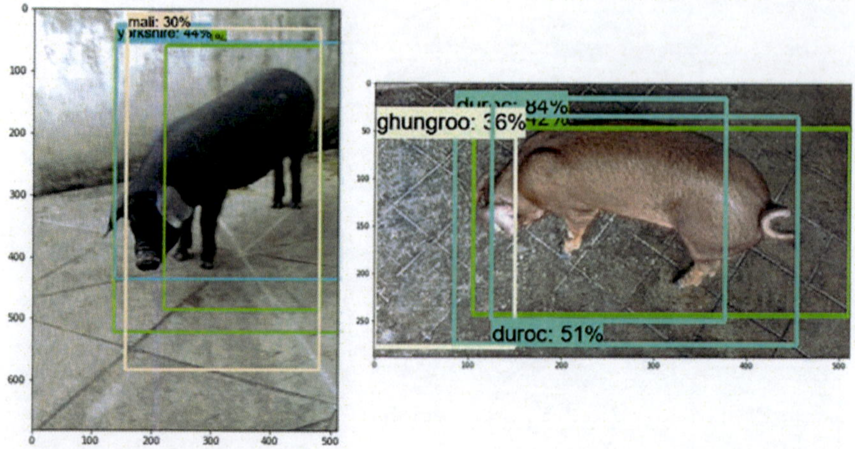

Fig. 7 Multiple bounding boxes

pig breed. The bounding boxes have been reduced after increasing the confidence level as shown in Figs. 7 and 8.

The reason is that with a low confidence level, the model has predicted a number of breeds with different matching percentages but as the confidence level increased, the model has detected the exact breed and only one box has been generated on each pig image with a percentage of matching.

6.2 *Breed Detection in Training and Testing with Fixed Confidence Level*

The annotated images in the pig dataset have been divided into a 9:1 ratio. The model has been tested on both train and test image sets. The model has given 100% accuracy in training and minimum 93% accuracy in all pig breeds with confidence level 0.94 as shown in Table 1.

The reason may be the quality of test images or the number of iterations in training. The accuracy of breed detection in testing may increase if the model is trained for a higher number of iterations.

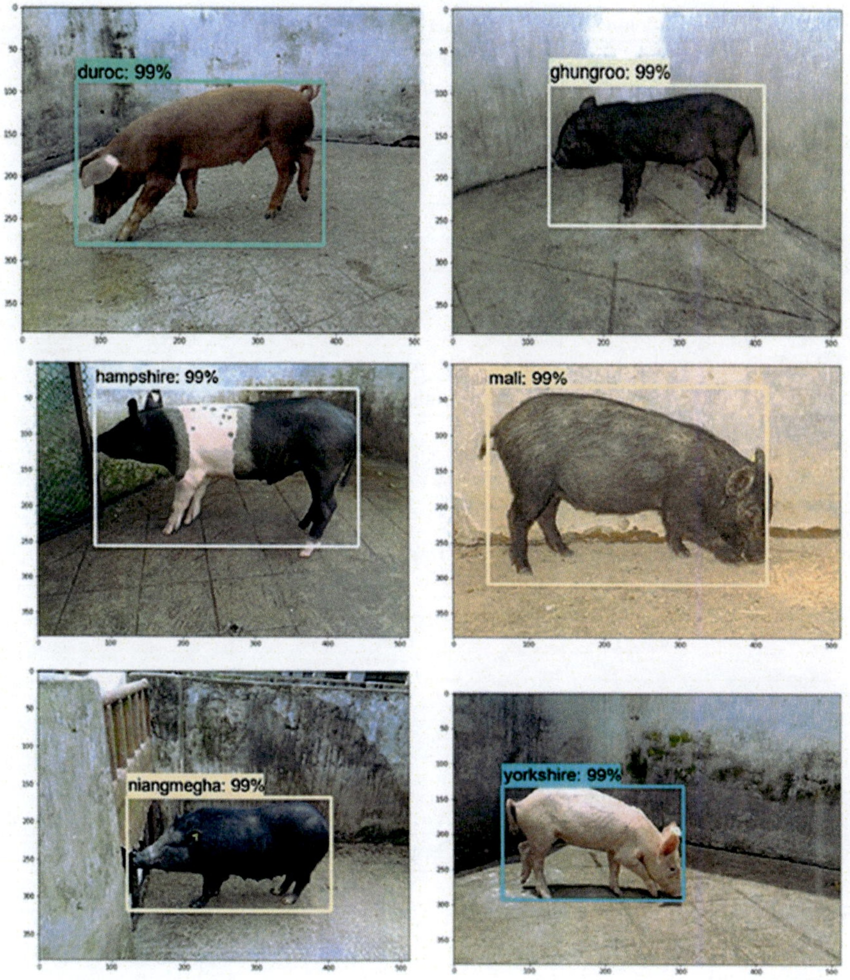

Fig. 8 Single bounding box

6.3 Breed Detection in Training and Testing with Different Confidence Levels

The model has been tested with different confidence levels as shown in Table 2.

32 numbers of pig images from different breeds have been tested from the test set. The model has predicted a higher level of accuracy at a lower confidence level. This may also be due to the quality of test images or due to lower number of training iterations. The confidence level versus prediction accuracy is shown in Fig. 9.

Table 1 Breed detection-wise result confidence level = 0.94

Breed	Total images	Number of testing images	Correctly Identified	Testing accuracy (%)
Duroc	329	33	31	93.93
Ghungroo	1045	105	104	99.04
Hampshire	200	20	18	90.00
Mali	338	34	32	94.11
Yorkshire	216	22	22	100.00
Niang Megha	200	20	19	95.00

Table 2 Confidence level-wise result for all breeds in 32 images from different breeds

Confidence level	Correctly identified	Test accuracy (%)
0.67	32	100.00
0.85	30	93.75
0.90	30	93.75
0.94	29	90.62
0.97	29	90.62
0.99	28	87.50

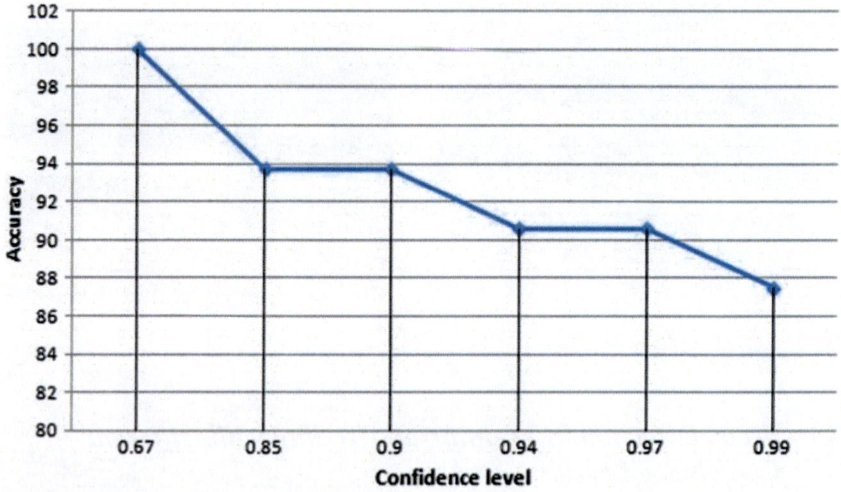

Fig. 9 Confidence versus accuracy in Pig Breed detection

7 Conclusion and Future Work

In this paper, a pretrained Faster R-CNN model named Inception-ResNet-v2 convolutional neural network has been used to detect the six pig breeds. The pig images have been captured in any orientation from organized farms located in different regions in India. The model has been trained in 2,00,000 iterations. The model has given 100% accuracy in 0.94 confidence level in the training set but the accuracy level is decreased in the test set. The accuracy in different confidence levels in test images has been computed. It has been observed that the model has given 100% prediction accuracy in lower confidence levels. The accuracy in higher confidence levels may be increased if the number of iterations is increased during training. Other pretrained models will be used in the future to select the appropriate detection model for pig breed detection. In the future, the pig breed dataset will be modified by including individual images more indigenous pig breeds. Also we propose to implement a new pig breed registration functionality for increasing the versatility of the algorithm by informing the user about unregistered pig breeds.

Acknowledgements The authors would like to thank ITRA-Digital India Corporation (formerly known as Media Lab Asia), Ref. No.: ITRA/15(188)/Ag&Food/ImageIDGP/01 dated 09/11/2016 for funding this research work. The authors would also like to thank Dr. A. Bandopadhyay, Senior consultant, ITRA Ag&Food, Dr. Santanu Banik, Principal scientist, Animal Breeding, NRC on Pig, Assam, Dr. Arnab Sen, Head, Animal Health, ICAR research complex for NEH, Barapani, Dr. Binay Singh, Scientist, ICAR-RC for NEH Region, Tripura Center, Agartala and Dr. Dilip Kumar Hazra, Assistant Professor, Dept. of Agronomy, faculty of agriculture, Uttar Banga Krishi Viswavidyala, Coochbihar, for helping us to implement this research work.

References

1. Torrey, L., Shavlik, J.: Transfer learning. In: Handbook of Research on Machine Learning Applications and Trends: Algorithms, Methods, and Techniques, pp. 242–264. IGI Global (2010)
2. Krizhevsky, A., Sutskever, I., Hinton, G.E.: Imagenet classification with deep convolutional neural networks. In: Advances in Neural Information Processing Systems, pp. 1097–1105 (2012)
3. Russakovsky, O., Deng, J., Su, H., Krause, J., Satheesh, S., Ma, S., Huang, Z., Karpathy, A., Khosla, A., Bernstein, M., et al.: Imagenet large scale visual recognition challenge. Int. J. Comput. Vis. **115**(3), 211–252 (2015)
4. Simonyan, K., Zisserman, A.: Very deep convolutional networks for large-scale image recognition (2014). arXiv:1409.1556
5. Chai, Y., Lempitsky, V., Zisserman, A.: Bicos: a bi-level co-segmentation method for image classification. In: 2011 International Conference on Computer Vision, pp. 2579–2586. IEEE (2011)
6. Nilsback, M.-E., Zisserman, A.: A visual vocabulary for flower classification. In: 2006 IEEE Computer Society Conference on Computer Vision and Pattern Recognition (CVPR'06), vol. 2, pp. 1447–1454. IEEE (2006)
7. Nilsback, M.-E., Zisserman, A.: Automated flower classification over a large number of classes. In: 2008 Sixth Indian Conference on Computer Vision, Graphics & Image Processing, pp. 722–729. IEEE (2008)
8. Varma, M., Ray, D.: Learning the discriminative power-invariance trade-off. In: 2007 IEEE 11th International Conference on Computer Vision, pp. 1–8. IEEE (2007)
9. Branson, S., Wah, C., Schroff, F., Babenko, B., Welinder, P., Perona, P., Belongie, S.: Visual recognition with humans in the loop. In: European Conference on Computer Vision, pp. 438–451. Springer (2010)

10. Khosla, A., Jayadevaprakash, N., Yao, B., Li, F.-F.: Novel dataset for fine-grained image categorization: Stanford dogs. In: Proceedings of the CVPR Workshop on Fine-Grained Visual Categorization (FGVC), vol. 2 (2011)
11. Lampert, C.H., Nickisch, H., Harmeling, S.: Learning to detect unseen object classes by between-class attribute transfer. In: 2009 IEEE Conference on Computer Vision and Pattern Recognition, pp. 951–958. IEEE (2009)
12. Welinder, P., Branson, S., Mita, T., Wah, C., Schroff, F., Belongie, S., Perona, P.: Caltech-ucsd Birds **200** (2010)
13. Girshick, R., Donahue, J., Darrell, T., Malik, J.: Rich feature hierarchies for accurate object detection and semantic segmentation. In: Proceedings of the IEEE Conference on Computer Vision and Pattern Recognition, pp. 580–587 (2014)
14. Girshick, R.: Fast R-CNN. In: Proceedings of the IEEE International Conference on Computer Vision, pp. 1440–1448 (2015)
15. Ren, S., He, K., Girshick, R., Sun, J.: Faster r-cnn: towards real-time object detection with region proposal networks. In: Advances in Neural Information Processing Systems, pp. 91–99 (2015)
16. Huang, J., Rathod, V., Sun, C., Zhu, M., Korattikara, A., Fathi, A., Fischer, I., Wojna, Z., Song, Y., Guadarrama, S. et al.: Speed/accuracy trade-offs for modern convolutional object detectors. In: Proceedings of the IEEE Conference on Computer Vision and Pattern Recognition, pp. 7310–7311 (2017)
17. Szegedy, C., Ioffe, S., Vanhoucke, V., Alemi, A.A.: Inception-v4, inception-resnet and the impact of residual connections on learning. In: Thirty-First AAAI Conference on Artificial Intelligence (2017)

Black Bengal Goat Identification Using Iris Images

Subhojit Roy, Sanket Dan, Kaushik Mukherjee, Satyendra Nath Mandal, Dilip Kumar Hajra, Santanu Banik, and Syamal Naskar

Abstract Animal identification is necessary for records, registration, and proof of ownership. The owner of few Black Bengal Goats can identify his goats by sight but it will create a problem for a larger number of goats as they are looking almost similar. A number of identification tools have been used for Black Bengal Goats like ear tag, tattoo, branding, RFID, etc. The Tattoos are permanent identification marking but inconvenient to read after a few months or years. Most of the farmers and breeders have used ear tags, which contain a number for identification of particular goat but may be lost at the time of grazing. Some organized farmers have placed RFID chips in tags but RFID reader is necessary to read the content of chips. In this paper, an effort has been made to identify individual Black Bengal Goat using their iris image

S. Roy
Gangarampur Government Polytechnic College, Kadighat, Gangarampur 733124, West Bengal, India
e-mail: subhojitroykgec@yahoo.com

S. Dan · K. Mukherjee · S. Nath Mandal (✉)
Department of Information Technology, Kalyani Government Engineering College, Kalyani, Nadia 741235, India
e-mail: satyen_kgec@rediffmail.com

S. Dan
e-mail: sanketdan@gmail.com

K. Mukherjee
e-mail: kaushik8.m@gmail.com

D. K. Hajra
Uttar Bangla Krishi Viswavidyalaya, Pundibari, Cooch Behar 736165, India
e-mail: dhajra@gmail.com

S. Banik
NRC on PIG, Rani, Guwahati 781131, Assam, India
e-mail: sbanik2000@gmail.com

S. Naskar
ICAR-IVRI Eastern Regional station, kolkata, West Bengal, India
e-mail: s_naskar@rediffmail.com

D. Bhattacharjee et al. (eds.), *Proceedings of International Conference on Frontiers in Computing and Systems*, Advances in Intelligent Systems and Computing 1255,
https://doi.org/10.1007/978-981-15-7834-2_20

like a human. The eye images have been captured preprocessed, enhanced, and irises have been segmented. The template has been generated from each segmented iris and stored in the database. The matching has been performed among different segmented iris images from the same goat and also been performed among iris images captured from different goats. It has been observed that the average Hamming distance among iris images captured at different times from the same goat are different from the average hamming distances among iris images from other goats. Finally, the matching threshold has been decided for the identification of Black Bengal Goat.

Keywords Black Bengal goat · Goat identification · Hamming distance · Template matching · Iris scanner

1 Introduction

Animal identification is an important task for understanding disease trajectory, vaccination, production management, and animal ownership assignment. The biometric system provides automatic recognition of an individual based on some sort of unique characteristic of the trait(s) such as visual patterns, nose print, iris, facial features, ear vessel pattern, and the retina. Some external characteristics of animal such as color ring on snake, belly patches in geese or body markings on zebra that gives individual identity. The problem of retina scan is the eye must be dilated and the specialized camera is needed to capture the retinal picture. Iris is highly reliable and unique well-protected organ in any animal in contrast to facial features which may be difficult to identify because of certain limitation are there. The size of iris is fixed throughout the life span of the animal, whereas the size and shape of the face is variable and that's why the iris image is widely used for animal recognition. Iris images can be captured using iris camera, SLR camera, and also mobile camera. The capturing of iris images is easy and not necessary to dilate the eye.

The aim of this paper is to identify individual Black Bengal goat through iris images. The eyes of Black Bengal Goat have been captured from Indian Veterinary Institute animal farm, Kalyani, Nadia(W.B), India using IriShield-USB MK 2102U auto capture iris scanner. The irises have been segmented from eye images; templates have been generated and stored in the iris database. At least ten eye images have been captured from one individual and a minimum of five images have been selected for this experiment in which the irises have been captured as maximum as possible. The matching among the templates from the same goat as well as the different goats has been performed and a threshold has been computed to identify individual goats. This type of effort has not been done to identify individual Black Bengal goat. This is the reason for making this paper.

2 Related Work

From the literature study, it is observed that several research works have been conducted on human recognition. The researchers have used fingerprint, hand vein, iris, retinal, etc., as identification traits. They have captured the traits using different cameras, analyzed these traits, extracted unique features, and finally matching them for human recognition [1–6]. But few research works have been done in animal identification using biometrics such as coat and muzzle pattern, etc. [7–12]. Yue Lu et al. [13] have introduced identification system for cow based on iris analysis and recognition. They have proposed a series of steps for the identification such as iris imaging, iris detection, and recognition process.

3 Existing Method of Identification

Ear tags are a common method of marking cattle, sheep, and goats. Ear tags may be electronic or non-electronic and there are numerous numbers of design, but all entail the piercing of the ear flap(Fig. 1). Ear tattoos are other popular methods of identification marking. This tattoo will be applied to goat kid when it is at least three months old. The tattoo must be placed on the tail or ear, where it can be easily read Figs. 2 and 3.

4 Goat Identification System

A goat recognition system using iris images has two major parts. One is template generation which is composed of capturing eye image, preprocess, segmentation, features extraction, and finally template generation (Figs. 4 and 5). The second part is template matching.

Fig. 1 Ear tagging

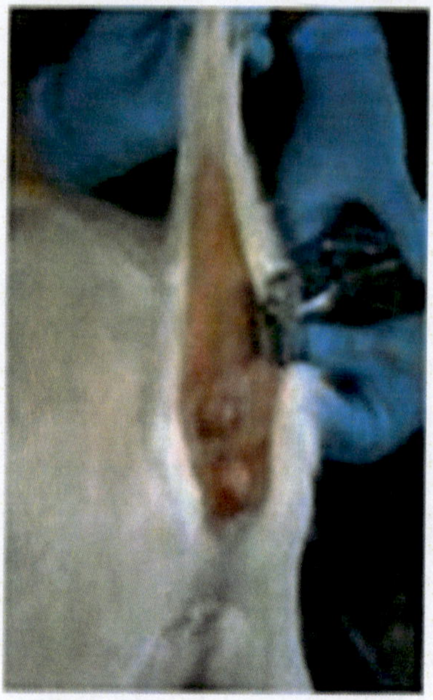

Fig. 2 Applying tattoo on the tail web

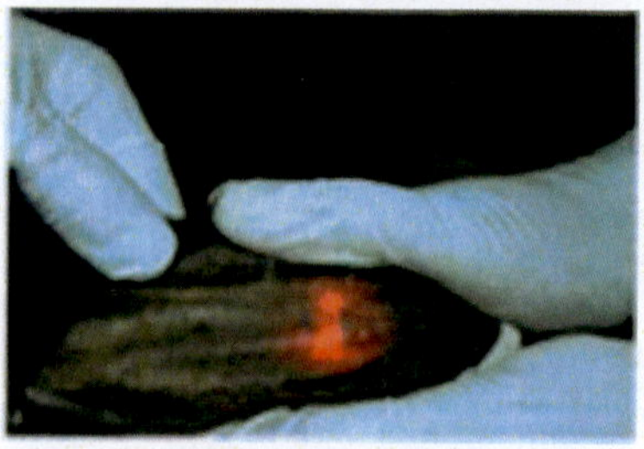

Fig. 3 Reading a tattoo with a flash light providing backlight illumination

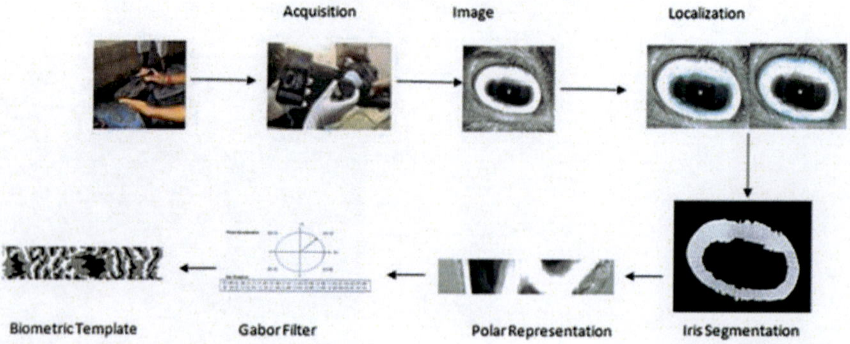

Fig. 4 Template generation

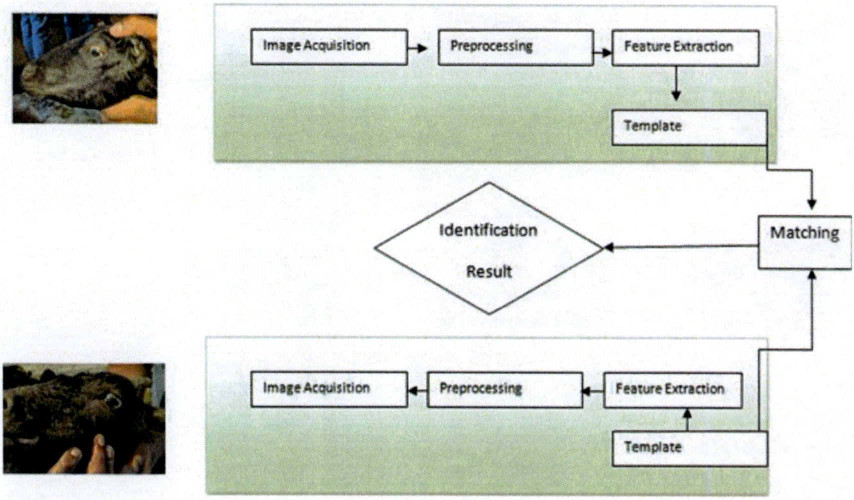

Fig. 5 Template matching

5 Goat Identification Phases

The goat identification using iris images has several phases from capturing of goat eye to template matching is shown in Fig. 6. Each phase has been described in detail.

5.1 Goat Eye Image Acquisition

This is the first step of the identification process and this is considered the most important step in this work, since the quality of the image captures dictates the successful execution of all the subsequent steps. Goat iris images have been captured using specially design iris-shield camera (IriShield MK2120U) developed for human. Here, the specially design hand-held camera which was used to capture the iris image (Fig. 7).

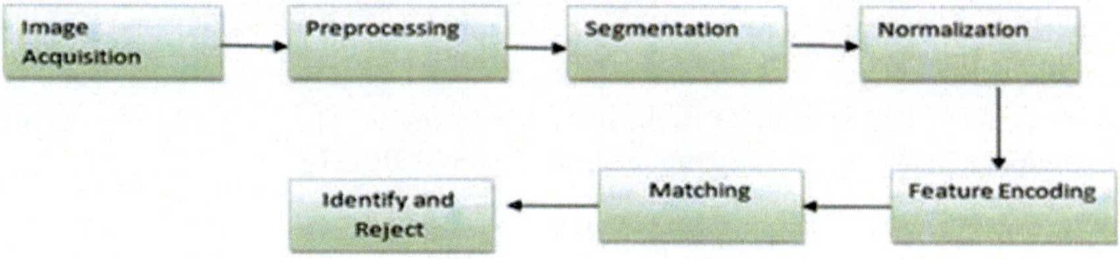

Fig. 6 The phases of goat identification

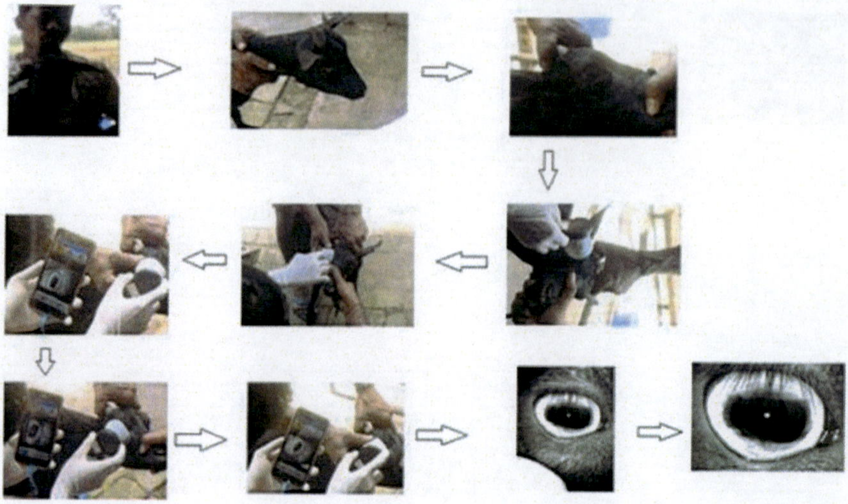

Fig. 7 Capture goat iris images using iris camera

5.2 *Preprocessing*

The iris images captured in the above method has some unnecessary parts such as sclera, eyelids, and pupil. Also, the size of the iris might vary depending upon some factors such as camera-to-eye distance, level of illumination of the eye and amount of reflection on the eye. So this error must be mitigated. Next, eyelid and eyelash must be kept away as much as possible to visualize the whole portion of the iris (Fig. 8). It is better to capture an image where the pupil is in the middle of the eye, the eye is covered with cover and eyelid and eyelash must not influence the image. Finally, images have been resized into 700 × 650.

5.3 *Goat Iris Segmentation*

The shape of the goat pupil is close to rectangular and shape of iris of goat eye is completely arbitrary and these can't be fitted any regular shape. The human iris segmentation algorithm will not work to segment iris from goat eye. At first, perform the complement operation of iris image to make the reflection lighter. The iris image is then filled with holes to darken the reflections. Then this image is again re-complemented to convert the image back to grayscale (Fig. 9).

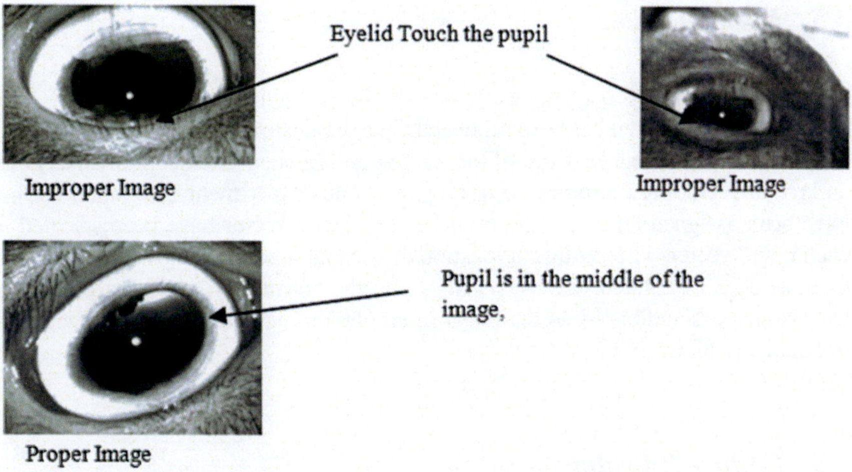

Fig. 8 Goat eye image selection

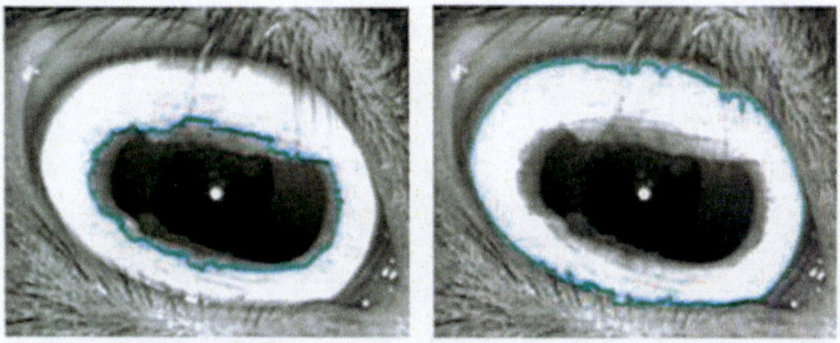

Fig. 9 Detection of Iris/pupil region (Left) and Iris/sclera boundary (Right) after Boundary detection method

5.4 Inner and Outer Goat Iris Boundary Detection

To find the boundary of pupil and iris, at first, grayscale image has been converted into to binary image with a proper threshold value. As the eye image has been stored into matrix form, the searching has been started from the starting element of the matrix and it continued through the image row-wise. The outer boundary has been started from the first nonzero pixel. The Moore neighborhood approach has been used for throughout boundary until the the starting pixel is reached in the clockwise direction.

5.5 Normalization

The normalization has been done after successful segmentation of iris region from a goat eye image so that it has been produced fixed dimension shift and scale invariance iris shown Fig. 10. The centroid of the iris region has been detected known as point and radial vector pass through the iris region. In this experiment, 240 radial vectors have been assigned and 20 data points in each radial vector have been selected in each radial vector. The Cartesian location of each data point along with each radial line has been calculated. Intensity values into the normalized polar representation have been extracted based on the linear interpolation method. The normalized goat iris image is shown in Fig. 11.

5.6 Feature Encoding

Feature encoding has been computed using convolving the normalized iris pattern with 1D LogGabor wavelets. Finally, the encoding process has produced a bit-wise template of size 20 px × 480 px containing some number of bits of information using 1D Gabor filter (Fig. 12).

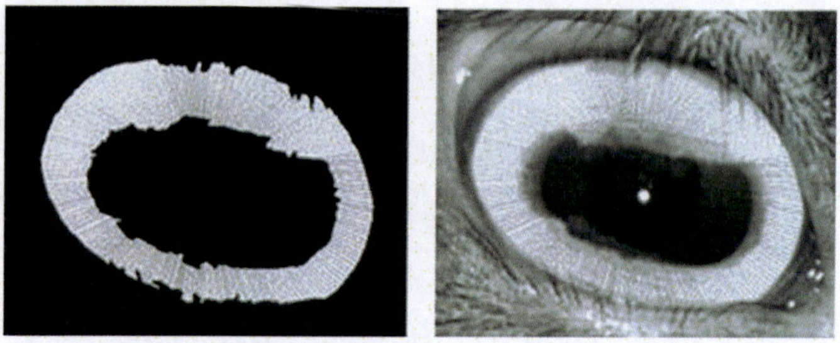

Fig. 10 240 lines are drawn through the center of the iris (Left). Constant number of points are chosen along each radial line (Right)

Fig. 11 Normalized iris image before histogram equalization (Left). Normalized iris image after histogram equalization (Right)

Fig. 12 Biometric template from the normalized iris

5.7 *Matching*

For matching, the Hamming distance of two templates has been calculated, one template is shifted left and right bit-wise and a number of Hamming distance values are calculated from successive shifts using Daugman [14] method. In some cases, some rectification is done for misalignments in normalized iris pattern has been occurred for rotational difference during image capturing. From the calculated Hamming distance values, only the lowest is taken, since this corresponds to the best match between two templates.

6 Result and Discussion

In this experiment, several images have been captured from five Black Bengal Goats. The five eye pictures with maximum coverage of iris have been selected for matching. The iris has been extracted from each image and matching has been performed among the five iris images from the same individual. The matching percentages have been recorded in Table 1. The best eye image has been selected from each individual goat based on maximum coverage of iris in eye image. The five best images have been considered after selection and matching percentages using extracted iris images have been furnished in Table 2. The matching of different iris images from five individuals and matching percentages among the best images from different five individual are furnished in Tables 3 and 4, respectively. The graphical representation of matching of the individual goat is given in Fig. 13.

Table 1 Matching percentage of different goat eye images from same goat based on iris

Sample 1	Sample 2	Matching in %
Image ID-0058_1	Image ID-0058_2	71
Image ID-0058_1	Image ID-0058_3	60
Image ID-0058_1	Image ID-0058_4	58
Image ID-0058_1	Image ID-0058_5	65
Image ID-0058_1	Image ID-0058_6	59
Image ID-0058_1	Image ID-0058_7	59

Table 2 Matching percentage of different goat eye images from different goats based on iris

Sample 1	Sample 2	Matching in %
Image-0035	Image-0058	54
Image-0035	Image-AB34	49.31
Image-0035	Image-NN03	50.81

Table 3 Matching percentages of five different images from each individual

Goat_ID	Id1	Id2	Id3	Id4	Id5	Best pic
0035	100	61	65	62	62	0035_0135
0058	100	71	60	59	65	0058_0184
AB34	100	68	64	62	63	AB34_0089
NN03	100	71	69	68	72	NN03_0112
AB41	100	69	65	70	62	AB41_0210

Table 4 Matching percentages of five best images from five individual

Goat_ID	0035	0058	AB34	NN03	AB41
0035	100	54.16	49.31	50.81	53.21
0058	54.16	100	50.78	54.64	50.71
AB34	49.31	50.78	100	54.19	52.21
NN03	50.81	54.64	54.19	100	53.11
AB41	53.21	50.71	52.21	53.11	100

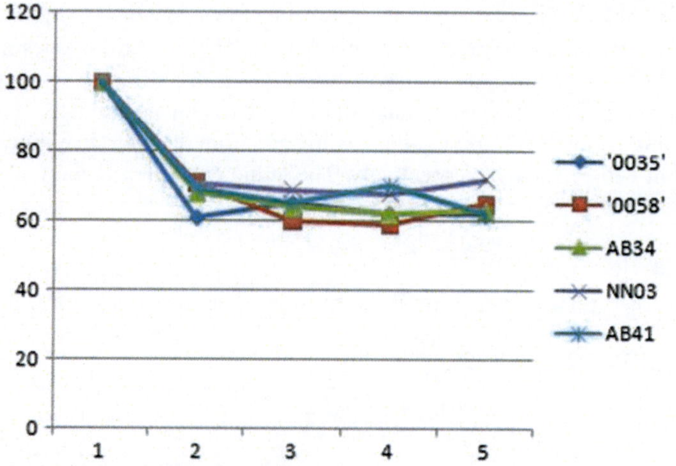

Fig. 13 Matching average of Individual goat

7 Conclusion and Future Work

In this paper, the individual Black Bengal goat has been identified through iris images. The individual Black Bengal goat has been successfully identified by their iris images. This method is purely non-invasive and easy to capture. The image acquisition is difficult as it is not obedient and communicative to human. The minimum matching among iris templates have been found to be 59% and maximum matching found among iris images from different goats is 54%. The gap between matching and mismatching has been found narrow, this is due to image acquisition. The difference will be increased if the goat can be restrained and the images can be captured in a controlled environment. Another bigger problem is IriShield-USB MK 2120U which is a is human iris scanner and the goat iris has been captured using this device. The anatomy of the human eye and the goat's eye are not the same. These are the reasons for a small gap between matching and mismatching. The goat restraining tool will be collected and artificial ambience will be created to capture goat eye images in future and this method will be applied to other animals.

Acknowledgements The authors would like to thank ITRA-Digital India Corporation (formerly known as Media Lab Asia), Ref. No.: ITRA/15(188)/Ag&Food/ImageIDGP/01 dated 09/11/2016 for funding this research work. The authors would also like to thank Dr. A. Bandopadhyay, Senior consultant, ITRA Ag&Food, Dr. Binay Singh, Scientist, ICAR-RC for NEH region, Tripura Center, Agartala, Pritam Ghosh, M.Tech (second year), Subhranil Mustafi, M.Tech (second Year), and Kunal Roy (JRF, DHESTBT project) Kalyani Government Engineering College, Kalyani, Nadia for helping us to implement this research work.

References

1. Ding, Y., Zhuang, D., Wang, K.: A study of hand vein recognition method. In: IEEE International Conference Mechatronics and Automation, 2005, vol. 4, pp. 2106–2110. IEEE (2005)
2. Pan, M., Kang, W.: Palm vein recognition based on three local invariant feature extraction algorithms. In: Chinese Conference on Biometric Recognition, pp. 116–124. Springer (2011)
3. Lu, L., Zhang, X., Zhao, Y., Jia, Y.: Ear recognition based on statistical shape model. In: First International Conference on Innovative Computing, Information and Control-Volume I (ICICIC'06), vol. 3, pp. 353–356. IEEE (2006)
4. Paranjpe, M.J., Kakatkar, M.: Review of methods for diabetic retinopathy detection and severity classification. Int. J. Res. Eng. Technol. **3**(3), 619–24 (2014)
5. Karunanayake, N., Gnanasekera, M., Kodikara, N.: A robust algorithm for retinal blood vessel extraction (2015)
6. Mulyono, D., Jinn, H.S.: A study of finger vein biometric for personal identification. In: 2008 International Symposium on Biometrics and Security Technologies, pp. 1–8. IEEE (2008)
7. Krijger, H., Foster, G., Bangay, S.: Designing a framework for animal identification (2008)
8. Lahiri, M., Tantipathananandh, C., Warungu, R., Rubenstein, D.I., Berger-Wolf, T.Y.: Biometric animal databases from field photographs: identification of individual zebra in the wild. In: Proceedings of the 1st ACM International Conference on Multimedia Retrieval, pp. 1–8 (2011)
9. Noviyanto, A., Arymurthy, A.M.: Automatic cattle identification based on muzzle photo using speed-up robust features approach. In: Proceedings of the 3rd European Conference of Computer Science, ECCS, vol. 110, p. 114 (2012)
10. Stahl, H., Schädler, K., Hartung, E.: Capturing 2d and 3d biometric data of farm animals under real-life conditions. In: Proceedings in International Conference of Agricultural Engineering, SPC03 C, vol. 1034 (2008)
11. Burghardt, T., Campbell, N.: Individual animal identification using visual biometrics on deformable coat patterns. In: International Conference on Computer Vision Systems: Proceedings (2007)

12. Petsatodis, T.S., Diamantis, A., Syrcos, G.P.: A complete algorithm for automatic human recognition based on retina vascular network characteristics
13. Lu, Y., He, X., Wen, Y., Wang, P.S.: A new cow identification system based on iris analysis and recognition. Int. J. Biometr. **6**(1), 18–32 (2014)
14. Daugman, J.: How iris recognition works. In: The Essential Guide to Image Processing, pp. 715–739. Elsevier (2009)

Component-level Script Classification Benchmark with CNN on AUTNT Dataset

Tauseef Khan and Ayatullah Faruk Mollah

Abstract Script identification from camera images is a prerequisite for efficient end-to-end systems in multi-script environment. In recent times, due to the wide usage of digitized multi-lingual documents and images, efficient script identifier becomes an inevitable module in computer vision and pattern recognition applications. Here, a component-level multi-lingual identifier is designed based on CNN to set the benchmark performance on publicly available dataset named AUTNT. The model is evaluated using three different text scripts, viz., Bengali, Latin, and Devanagari. It yields reasonably high accuracy of 92.02% and 89.49% for document and scene component images, respectively, and 92.51% for overall text components irrespective of image sources. This result is first of its kind and it may be convincingly considered as a benchmark for component-level script classification from the said dataset.

Keywords Script identification · Convolutional neural network · Multi-script · AUTNT dataset

1 Introduction

Nowadays, identification of script from camera-captured scene or document-type images is an essential module for end-to-end text detectors. Multi-script text detection and recognition in the wild have gained tremendous attention from past few decades due to its various natural constraints. Script identification is treated as an important component in Optical Character Recognition (OCR) system under scene

T. Khan (✉) · A. F. Mollah (✉)
Department of Computer Science and Engineering, Aliah University, Kolkata 700160, India
e-mail: tauseef.hit2013@gmail.com

A. F. Mollah
e-mail: afmollah@aliah.ac.in

T. Khan
Department of Information Technology, Aliah University, West Bengal Haldia-721657, India

D. Bhattacharjee et al. (eds.), *Proceedings of International Conference on Frontiers in Computing and Systems*, Advances in Intelligent Systems and Computing 1255,
https://doi.org/10.1007/978-981-15-7834-2_21

and document image analysis task. Normally, text present in scene images that carries rich and semantic information needs to be localized accurately for effective information retrieval. Automated text reader from scenes under multi-lingual scenario gains much attention in many potential applications such as automated mobile phone navigator [1], scene text understanding [2], video text recognition [3], automated text translation [4], etc. Multi-lingual script identification is an inevitable step to build a robust end-to-end text detector. However, script identification from scene images is still underexplored. Till now, most of the script identification works are carried out on machine-printed and handwritten texts [5, 6]. Document-level script identification methods may be broadly categorized into four groups, viz., page/paragraph-level script identification [7], text-line-level identification [8], word-level identification [9, 10], and character-level identification [11]. Identification of scripts from well-formatted document layouts is comparatively easier than scene images due to its diverse nature. Several works have been reported on script identification from natural scene images in multi-script environment so far [12–14]. Recently, in deep learning era, identification of scripts from complex images has achieved significant improvement due to several convolutional neural network (CNN) models [15, 16]. But, all these methods have their own limitations in different datasets. Most of the works have been reported for English and other popular scripts, and very few are reported for regional scripts. A high resource environment is often a prerequisite for most of the deep learning approaches.

In this work, a CNN-based multi-script identifier is presented to set the benchmark performance on component-level images of a recently published component-level multi-script dataset i.e. Aliah University Text Non-text (AUTNT) dataset developed by Khan et al. [17]. The dataset comprises of component-level natural scene and complex document images of three different scripts, viz., Latin, Bengali, and Devanagari. Contribution of this paper may thus be stated as: (i) A CNN-based component-level multi-script identifier is designed and used to set benchmark on AUTNT dataset, and (ii) Script-wise separate experiments have been carried out at the component level for document-type, scene-type, and mixed, and extensive results obtained are reported.

2 Related Works

Several methods have been developed so far for document and scene text images. Script of a text is classified based on different spatial and temporal information present in the characters and text lines. So, feature extraction plays an important and critical role in this problem. Features can be broadly categorized as (a) local features and (b) global features [13, 18, 19]. Local features mainly focus on character's intrinsic information. Though handcrafted features are effective for script identification, it is computationally expensive. Also, some features are sensitive to illumination and noise. In the early stage of research, such handcrafted low-level features were used to feed into conventional pattern classifiers to obtain the desired results. Shivakumara et al. [20] proposed directional gradient-based features for script identification of text lines obtained from video. Singh et al. [14] designed novel mid-level features for script identification from scenes. Here, at first, low-level features are extracted from text images, and then these features are grouped in a larger context to represent mid-level features. A novel word-level script identification model using texture-based features for scene image has been reported in [21]. A multi-variant Local Binary pattern (LBP) is used for scene-level script identification method and classified using SVM [13].

Apart from handcrafted features, script identification for scene images has geared up by deep learning approaches in recent times. Deep networks generally extract high-level deep features from texts related to different scripts and process that information to obtain impressive results in a complex scenario. Hasan et al. [15] presented a novel text-line-oriented multi-script identifier using Long Short-Term Memory (LSTM). This model focuses on each character belonging to different scripts rather than an entire text line. Shi et al. [22] used both deep features and mid-level features collectively and fed into a trained discriminative CNN model (DisCNN) for effective discrimination of relatively complex scripts, viz., Chinese and Japanese. Recently, an inter-modality conversion technique is applied for both online and offline handwritten texts for script identification [23]. The model is built using both convolutional and LSTM network, trained using character-level data and identify scripts for both character- and word-level texts. Lu et al. [24] proposed a novel method for script identification task from natural images using both local and global CNN-based frameworks based on ResNet-20. Here, component-level image patches are fed into network to fine-tune the model and finally AdaBoost classifier is used to fuse the extracted features from both the networks which are used for accurate identification of scripts.

3 Proposed Benchmark Model

In this work, a benchmark model is proposed for component-level script identification task on AUTNT dataset [25]. As the primary objective of this work is to set the performance benchmark of multi-script text components of the above dataset, a classical CNN architecture is employed. A detailed architecture of the proposed model is explained below.

3.1 Architecture of Proposed Network

The employed network is deep in terms of convolution layers. The model comprises of three pairs of convolutional and subsampling layers, two dense layers, and final prediction layer. High-level features are extracted using convolutional layers and converted into an 1-D vector in the flattened layer. In every convolutional layer, a

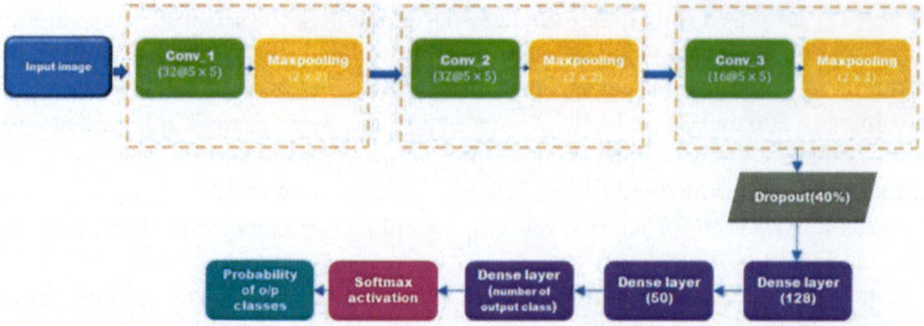

Fig. 1 Detailed architecture of the employed model based on CNN for component-level script identification task

feature map is generated consisting of high-level features and subsequently dimension is reduced by max-pooling. Figure 1 illustrates the overall architecture of the employed network.

Initially, input component-level images are normalized to 50 × 150 dimension using zero-padding bits while maintaining the appropriate aspect ratio. In the first convolutional layer (Conv_1), kernels of size 5 × 5 are slid over the entire image and 32 learnable feature maps are generated. Linear activation function, i.e., ReLU is then applied to increase the non-linearity of feature maps. The generated feature maps of Conv_1 are subsampled using max-pooling technique considering 2 × 2 kernel with a fixed stride size of 2. Each max-pooling layer reduces the feature maps to 50% of actual dimension. However, a number of feature maps, i.e., 32 remains unchanged.

The second convolutional layer (Conv_2) is similar to Conv_1 that generates again 32 feature maps consisting of deeper and high-level features. Similar to Conv_1, max-pooling result is appended to Conv_2 to reduce the dimension of generated feature maps. This dimension reduction of feature maps reduces overall computation cost to train the model. Finally, from the third convolutional layer (Conv_3), a total 16 feature maps are generated with kernels of the same size and subsequent max-pooling layer reduces the size of feature maps further. Later, feature maps of convolutional layers are flattened and fed into dense layers.

Two dense layers mainly work as a linear function, where every neuron of previous layers is fully connected to the next layers. These layers act as normal multi-layer perceptron comprising of 128 and 50 neurons, and ReLU is used as an activation function. Another dense layer is connected which is having a number of output neurons equal to the number of output class label. Finally, softmax activation function is applied that generates class probabilities of each corresponding output class.

3.2 *Training Phase of Network*

The model is trained using AUTNT training dataset containing component-level images. Training images are resized into a fixed dimension as mentioned before. However, a huge amount of training data is required to fine-tune any CNN model. Here, data augmentation technique is applied to increase the overall size of training dataset. In this phase, training images are (a) rotated by 1 and 2 degrees, (b) horizontally flipped, and (c) sharpened. All images are converted to grayscale before feeding into the network.

During training, 10% samples of the total training set are preserved for validating the trained model. The number of epoch is chosen to 25 after several experiments. During compilation of model, categorical cross-entropy is used as a loss function which minimizes the loss between actual output and desired result. Adam optimizer is used which is an extended version of gradient descent algorithm. Learning rate of Adam optimizer is 0.001. Batch size is set to 100 for each epoch. To overcome the overfitting problem, dropout (40%) is applied before passing to dense layers. Table 1 demonstrates the trained sequential CNN model with a number of trainable parameters of each layer for all three scripts, viz., Bengali, Latin, and Devanagari.

Table 1 Proposed network configuration with trainable parameters for three types of scripts

Layer (type)	Output shape	Parameters
Conv2D_1	(150,50,32)	832
ReLU_1	(150,50,32)	0
MaxPooling_1 (stride 2)	(75,25,32)	0
Conv2D_2	(71,21,32)	25,632
ReLU_2	(71,21,32)	0
MaxPooling_2 (stride 2)	(35,10,32)	0
Conv2D_3	(31,6,16)	12,816
ReLU_3	(31,6,16)	0
MaxPooling_3 (stride 2)	(15,3,16)	0
Flatten	720	0
Dense_1	128	92,288
Dense_2	50	6450
Dense_3 (output class)	3	153
Total trainable parameter = 1,38,171		
Non-trainable parameter = 0		

Fig. 2 Sample images of AUTNT dataset. **a–c** Document-type texts with Bengali, Latin, and Devanagari script, respectively, **d–f** Corresponding scene-type text-scripts

4 Experimental Results and Analysis

The proposed model is evaluated using AUTNT standard dataset. In this section, a concise description of the training dataset is included. Then, a set of experiments have been conducted separately for document and scene-type images. Moreover, for each type of images, a different combination of text-scripts are taken for script classification and finally an overall experiment is carried out irrespective of the image source.

4.1 Dataset Outline

AUTNT dataset comprises of multi-script component-level images of both complex document and scenes. It has a total of 10,771 images. Out of total images, 7,890 images are texts belonging to three different scripts, viz., Latin, Bengali, and Devanagari. Images were acquired in the wild with built-in high-resolution camera of cell phones. Most components are complex in nature with uneven lighting, noise, distortion, and other clutters. Figure 2 shows sample text images from the said dataset.

4.2 Evaluation Of Document-Type Text-Scripts

In this section, separate experiments are performed on document-type texts only. Primarily, a combination of two different scripts are classified in one-to-one fashion and finally, all three types of scripts are used for evaluation. Table 2 shows the experimental results of script identification task for document texts. It is worth mentioning that evaluation metrics for three different scripts are computed using one-to-all approach and finally mean value is taken for the final result. It has been

Table 2 Performance measurement of document-type texts for script identification

Scripts	Precision	Recall	F-measure	Error-rate	Accuracy (%)
Bengali (0), Latin (1)	96.35	94.82	95.57	0.037	96.11
Devanagari (0), Latin(1)	97.23	98.40	97.81	0.019	98.05
Bengali (0), Devanagari (1)	91.53	94.82	93.14	0.069	93.01
Bengali (0), Latin (1), Devanagari (2)	92.39	91.57	91.70	0.053	92.02

Note ( ) indicates the class label

Table 3 Performance measurement of scene-type texts for script identification

Scripts	Precision	Recall	F-measure	Error-rate	Accuracy (%)
Bengali (0), Latin (1)	91.57	95.21	93.35	0.049	95.07
Devanagari (0), Latin (1)	82.27	91.54	86.65	0.039	96.08
Bengali (0), Devanagari (1)	86.98	93.62	88.53	0.155	84.47
Bengali (0), Latin (1), Devanagari (2)	81.39	79.4	80.38	0.075	89.49

Note ( ) indicates the class label

noted that script identification accuracy is highest (98.05%) for Latin versus Devanagari. However, overall accuracy falls in the experiments due to the complex nature of Indian regional scripts.

4.3 Evaluation of Scene-Type Text-scripts

Here, the same experiments are conducted for only scene-type texts for script identification task. Table 3 illustrates the performances of each experiment with standard metrics. It may be observed that overall accuracy has slightly decreased due to the complexity of scene images.

4.4 Evaluation of Overall Text-Scripts

The proposed model is now evaluated using overall text-scripts irrespective of image sources. All three types of scripts together are classified for overall text components. Table 4 shows the performance obtained with the CNN-based network.

From Tables 2, 3, and 4, it is observed that script identification accuracy of each script for document texts are slightly higher than scene texts, which is quite obvious due to complex backgrounds, noise, uneven illuminations, etc., associated with scene images. Classification accuracy for both Indian scripts, viz., Bengali and Devanagari for document images are comparatively less due to similarity of font

Table 4 Performance measurement of overall text components for script identification

Scripts	Precision	Recall	F-measure	Error-rate	Accuracy (%)
Bengali (0), Latin (1), Devanagari (2)	91.49	91.93	91.69	0.053	92.51

Note ( ) indicates the class label

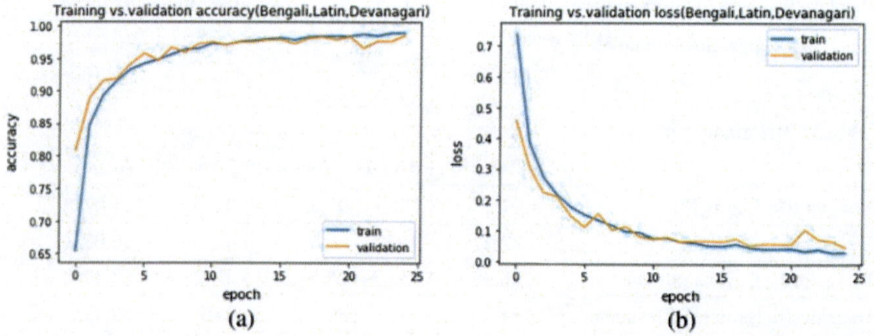

Fig. 3 Epoch wise graphical representation of model performance during training and validation phase. **a** Accuracy graph for training and validation, **b** loss graph for training and validation

style and orientations. Accuracy for Indian scripts of scene images is further slid due to more complexity. However, Latin script has standard orientation compared to Indian scripts, so accuracy is quite high in every case. The proposed model works in a comparatively low-resource environment, unlike other deep learning models. Figure 3 demonstrates the training accuracy and loss for overall text-scripts. It is observed that training accuracy is increasing over validation accuracy and validation loss is higher than training loss with increasing epochs. This scenario demonstrates the effectiveness of the model that is certainly free from overfitting problem. In Fig. 4, some correctly identified scripts for all three types and few false alarms along with predicted output class labels are displayed irrespective of image source. It is observed that Bengali scripts are wrongly identified as Devanagari and vice versa in some cases, whereas Latin scripts are identified correctly in most of the cases.

5 Conclusion

Script identification from complex document and scene images is still considered as an active research area due to certain variability of orientation, style, complex environment, etc. In this work, component-level script classification benchmark is prepared for publicly available AUTNT dataset. Experiments are carried out on a workstation having Intel i5-8400 CPU, 1 TB HDD, 16 GB RAM, NVIDIA GeForce GTX 1060 (6 GB) graphics card. It is observed that script identification accuracy for document-type texts and scene-type texts are 92.02% and 89.49%, respectively,

Fig. 4 Samples of true identified and false alarms. **a** correctly identified Bengali script, **b** corresponding false alarms with predicted class, **c** correctly identified Devanagari scripts, **d** corresponding false alarms, **e** correctly identified Latin scripts, **f** corresponding false alarms

whereas overall accuracy is 92.51% which is slightly better because of the increased number of samples. This result is reasonably satisfactory in the real-world scenario and set as the benchmark performance since it is the first of its kind. In future, more regional and complex scripts can be incorporated in scene environment. Moreover, current method of script identification may be integrated with text localization for the purpose of end-to-end text detection and recognition systems.

Acknowledgements The authors are obliged to Department of Computer Science and Engineering, Aliah University for providing necessary facilities to carry out this research work. The first author is also grateful to University Grant Commission (UGC) for the financial support under the scheme of Maulana Azad National Fellowship.

References

1. Cour, T., Jordan, C., Miltsakaki, E., Taskar, B.: Movie/script: alignment and parsing of video and text transcription. In: European Conference on Computer Vision, pp. 158–171. Springer (2008)
2. Li, L.J., Socher, R., Fei-Fei, L.: Towards total scene understanding: classification, annotation and segmentation in an automatic framework. In: IEEE Conference on Computer Vision and Pattern Recognition, pp. 2036–2043. IEEE (2009).
3. Cheng, Z., Lu, J., Xie, J., Niu, Y., Pu, S., Wu, F.: Efficient Video Scene Text Spotting: Unifying Detection, Tracking, and Recognition (2019). arXiv:1903.03299
4. Bansal, S., Kamper, H., Lopez, A., Goldwater, S.: Towards speech-to-text translation without speech recognition (2017). arXiv:1702.03856
5. Feng, Z., Yang, Z., Jin, L., Huang, S., Sun, J.: Robust shared feature learning for script and handwritten/machine-printed identification. Pattern Recogn. Lett. **100**, 6–13 (2017)

6. Obaidullah, S.M., Halder, C., Santosh, K.C., Das, N., Roy, K.: PHDIndic_11: page-level handwritten document image dataset of 11 official Indic scripts for script identification. Multimed. Tools Appl. **77**(2), 1643–1678 (2018)
7. Singh, P.K., Chatterjee, I., Sarkar, R.: Page-level handwritten script identification using modified log-Gabor filter based features. In: IEEE 2nd International Conference on Recent Trends in Information Systems, pp. 225–230. IEEE (2015)
8. Gomez, L., Nicolaou, A., Karatzas, D.: Improving patch-based scene text script identification with ensembles of conjoined networks. Pattern Recogn. **67**, 85–96 (2017)
9. Sahare, P., Chaudhari, R.E., Dhok, S.B.: Word level multi-script identification using curvelet transform in log-polar domain. IETE J. Res. **65**(3), 410–432 (2019)
10. Jajoo, M., Chakraborty, N., Mollah, A.F., Basu, S., Sarkar, R.: Script identification from camera-captured multi-script scene text components. In: Recent Developments in Machine Learning and Data Analytics, pp. 159–166. Springer (2019)
11. Chaudhari, S., Gulati, R.M.: Script identification using Gabor feature and SVM classifier. Procedia Comput. Sci. **79**, 85–92 (2016)
12. Roy, P.P., Bhunia, A.K., Bhattacharyya, A., Pal, U.: Word searching in scene image and video frame in multi-script scenario using dynamic shape coding. Multimed. Tools Appl. **78**(6), 7767–7801 (2019)
13. Verma, M., Sood, N., Roy, P.P., Raman, B.: Script identification in natural scene images: a dataset and texture-feature based performance evaluation. In: Proceedings of International Conference on Computer Vision and Image Processing, pp. 309–319. Springer (2017)
14. Singh, A.K., Mishra, A., Dabral, P., Jawahar, C.V.: A simple and effective solution for script identification in the wild. In: 12th IAPR Workshop on Document Analysis Systems, pp. 428–433. IEEE (2016)
15. Ul-Hasan, A., Afzal, M.Z., Shafait, F., Liwicki, M., Breuel, T.M.: A sequence learning approach for multiple script identification. In: 13th International Conference on Document Analysis and Recognition, pp. 1046–1050. IEEE (2015)
16. Mei, J., Dai, L., Shi, B., Bai, X.: Scene text script identification with convolutional recurrent neural networks. In: 23rd International Conference on Pattern Recognition, pp. 4053–4058. IEEE (2016)
17. AU Text Non-Text dataset. https://github.com/iilabau/AUTNTdataset
18. Chanda, S., Pal, U.: English, Devanagari and Urdu text identification. In: Proceedings on the International Conference on Document Analysis and Recognition, pp. 538–545 (2005)
19. Chanda, S., Pal, U., Kimura, F.: Identification of Japanese and English script from a single document page. In: 7th IEEE International Conference on Computer and Information Technology, pp. 656–661. IEEE (2007)
20. Shivakumara, P., Sharma, N., Pal, U., Blumenstein, M., Tan, C.L.: Gradient-angular-features for word-wise video script identification. In: 22nd International Conference on Pattern Recognition, pp. 3098–3103. IEEE (2014)
21. Singh, P.K., Sarkar, R., Nasipuri, M.: Offline script identification from multilingual indic-script documents: a state-of-the-art. Comput. Sci. Rev. **15**, 1–28 (2015)
22. Shi, B., Bai, X., Yao, C.: Script identification in the wild via discriminative convolutional neural network. Pattern Recogn. **52**, 448–458 (2016)
23. Bhunia, A.K., Mukherjee, S., Sain, A., Bhattacharyya, A., Bhunia, A.K., Roy, P.P., Pal, U.: Indic handwritten script identification using offline-online multimodal deep network (2018). arXiv:1802.08568
24. Lu, L., Yi, Y., Huang, F., Wang, K., Wang, Q.: Integrating local CNN and global CNN for script identification in natural scene images. IEEE Access **7**, 52669–52679 (2019)
25. Khan, T., Mollah, A.F.: AUTNT-A component level dataset for text non-text classification and benchmarking with novel script invariant feature descriptors and D-CNN. Multimed. Tools Appl. **78**(22), 32159–32186 (2019)

Analysis of Diabetic Retinopathy Abnormalities Detection Techniques

Sudipta Dandapat, Soumil Ghosh, Shukrity Si, and Anisha Datta

Abstract The basic objective of the proposed work is to identify abnormalities caused by Diabetic Retinopathy in human retina. We classified the retinal images into two categories; normal retina and abnormal retina which contains some signs of Diabetic Retinopathy. In this work, we detect these problems using the algorithms: Support Vector Machine (SVM), k-Nearest Neighbour and Convolutional Neural Network (CNN) algorithm. These algorithms are used to build a model and their performances are compared with each other. The result is that the Support Vector Machine (SVM) gives the best accuracy of 96.6% with sensitivity and specificity of 0.66 and 0.95 respectively. Such type of model is very helpful in the early detection and treatment of Diabetic Retinopathy.

Keywords Diabetic retinopathy · Support vector machine · K-nearest neighbour · Convolutional neural network · Sensitivity · Specificity

1 Introduction

Diabetic Retinopathy occurs in many people who are suffering from diabetes. It is a painful disease and sometimes it leads to blindness. It happens when blood sugar levels are too high for a long period of time and it can damage tiny blood vessels that provide blood to the retina. A report from World Health Organization (WHO)

S. Dandapat · S. Ghosh (✉) · S. Si · A. Datta
Department of Computer Science and Enginering, Jalpaiguri Government Engineering College, Jalpaiguri, India
e-mail: soumilghosh16@gmail.com

S. Dandapat
e-mail: dandapatsudipta311@gmail.com

S. Si
e-mail: sukriti.si98@gmail.com

A. Datta
e-mail: dattaanishadatta@gmail.com

D. Bhattacharjee et al. (eds.), *Proceedings of International Conference on Frontiers in Computing and Systems*, Advances in Intelligent Systems and Computing 1255,
https://doi.org/10.1007/978-981-15-7834-2_22

on blindness and vision impairment states that an estimated 253 million people in the world are suffering from eye diseases, out of them 36 million are completely blind and rest 217 million have moderate to severe vision impairment. India has a very large share in these numbers, with around 12 million population who can not see. The main reasons behind these diseases are cataract, glaucoma, Age-related Macular Degeneration (AMD), diabetic retinopathy, childhood blindness, trachoma, and corneal opacities.

These can be detected by proper retinal examination. Fundus photography or Optical coherence Tomography (OCT) are used to take the images of the retina. Optical coherence Tomography is a recent technology in medical imaging. OCT generates images with a 3D profile consisting of different layers of the retina. In Fundus photography, we take photographs of the back of the eyes. Specialized fundus cameras consist of a microscope attached to a flash-enabled camera. Those fundus images are then examined by ophthalmologists, who look for certain patterns and defects in the images to predict diseases.

Exudate is a type of abnormality whose presence in the retina causes eye defects. Exudates are bulges of yellow and white colours appearing in the fundus. They are major indicators of Diabetic Retinopathy and Macular Edoema. Exudates consist of extracellular lipid which is leaked from abnormal retinal capillaries. They are present mainly in the macular region and as the lipids coalesce and extend into the central macula (fovea), which makes eye vision to be severely affected. We focused on Soft exudates in case of Diabetic Retinopathy. They look like greyish-white cotton wool. The main aim of this work is to detect these exudates from fundus images through Image Processing and Machine Learning techniques. Based on these known characteristics and physical features of exudates, we extract features from fundus images and apply pixel clustering. After that, the results of different cluster spaces are combined. The method is very effective in detecting pixels representing exudates in images.

The paper has been structured as follows: Sect. 2 gives a brief review of the literature to understand the problem and to build the model. Section 3 explains the proposed methodology of the model. In Sect. 4, performance of the classifiers is compared with each other in terms of accuracy, sensitivity, specificity, precision, recall, AUROC, F1 score, mean error, RMS error and confusion matrix. Finally, it is concluded with future study in Sect. 4.1.

2 Literature Review

There are several methods for detection and classification of Diabetic Retinopathy. Most of the methods use the techniques based on mathematical morphology, neural networks, pattern recognition, region growing techniques, fuzzy C-means clustering and Gabor filter. The blood vessels present in the retina are detected using 2D matched filter [1]. Automatic identification and the classification of abnormalities present on the vascular network are done by using Gabor filter bank outputs to classify the mild, moderate and severe stages of retinopathy [2]. A method based on recursive region growing and Moat operator is developed by Sinthanayothin which is used to detect Haemorrhages, Microaneurysms, and exudates [3]. There are different methods proposed for detection of optic disc. Principal Component Analysis (PCA) is one of the methods in which the candidate regions for the optic disc are derived by clustering of brighter pixels. Hough Transform is also used for detection of Optic Disc [4]. Microaneurysms and haemorrhages are detected in retinal fundus images using morphological operations with a structuring element and tophat transformation [5]. Mahalanobis classifier is one of the best classifiers which is used in the identification of haemorrhages and microaneurysms [6]. Several techniques of image processing in combination with pattern recognition techniques are used in the detection of microaneurysms and haemorrhages in fundus images [7]. A neural network based approach is used in exudates detection [8]. A fuzzy C-means clustering method [9] and computational intelligence based approach are used for detection of exudates [10]. Acharya et al. [11] classified Diabetic Retinopathy stages as normal, mild, moderate, severe, and proliferative. The feature extraction of the retinal fundus images is done using Higher Order Spectra (HOS). A multi-layer perceptron is used to classify normal and diabetic retinopathy stages [4]. Kahai proposed a model in which a decision system uses Bayes optimality criterion to detect Microaneurysms which enables the early detection of diabetic retinopathy [12]. Area and perimeter calculated from the RGB components of the blood vessels are used as features to classify normal, mild, moderate, severe and proliferative stages of retinopathy using a feed-forward neural network [13]. Automatic classification into normal, mild, moderate, severe, and proliferative classes of Diabetic Retinopathy is done by calculating the areas of several features such as haemorrhages, microaneurysms, exudates, and blood vessels with support vector machine as classifier [14]. Automated diagnosis system is developed to detect retinal blood vessels, and pathologies, such as exudates and microaneurysms together with certain texture properties using image processing techniques. The area of lesions and texture features are then used to construct a feature vector that is input to the multi-class support vector machine (SVM) for classifying images into normal, mild, moderate, severe and proliferative categories [15].

3 Methodology

The important steps in the detection of abnormalities and classification process are as follows:

1. Data Collection.
2. Pre-processing.
3. Histogram equalization.
4. Morphological Operations.
5. Thresholding.
6. Detection of Blood Vessels.

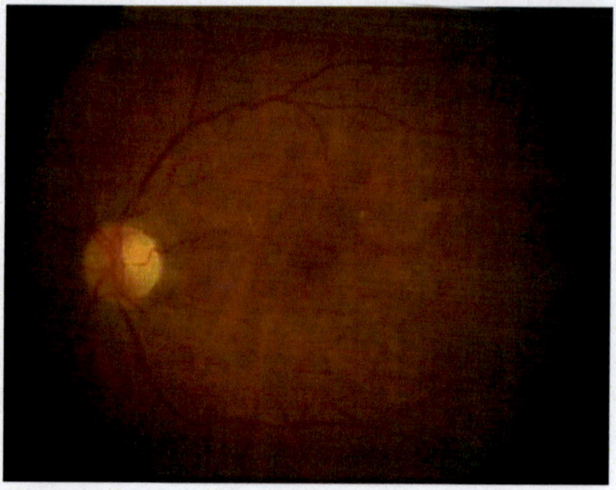

Fig. 1 Normal retina

7. K-means Clustering.
8. Classification.

3.1 Data Collection

To collect the data of retinal fundus images, we use the standard diabetic retinopathy dataset DIARETDB1. The database consists of 89 colour fundus images, of which 84 contain at least mild non-proliferative signs (Micro-aneurysms) of the diabetic retinopathy shown in Figs. 2 and 5 are considered as normal shown in Fig. 1, which do not contain any signs of the diabetic retinopathy according to all experts who participated in the evaluation [16]. Images are captured using the same 50° field-of-view digital fundus camera with varying imaging settings [17]. The data corresponds to a good (not necessarily typical) practical situation, where the images are comparable, and can be used to evaluate the general performance of diagnostic methods. This data set is referred to as "calibration level 1 fundus images" [16]. The image resolution obtained is 1500 by 1152 in 24 bit RGB, i.e. 8 bits (256 intensity levels) for each channel: red, green and blue.

3.2 Preprocessing

The main aim of preprocessing methods is to achieve image normalization by attenuation of intensity variations in the input images. The original images contain non-uniform spatial variations across the image. We convert the original retinal image shown in Fig. 3 into the greyscale image shown in Fig. 4 to extract its features in a better way.

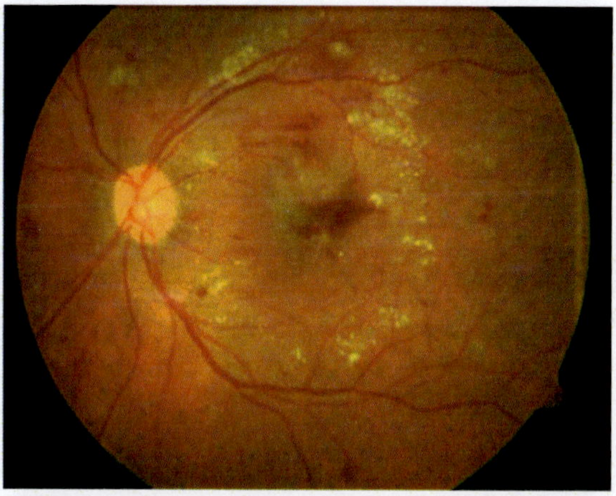

Fig. 2 Abnormal retina

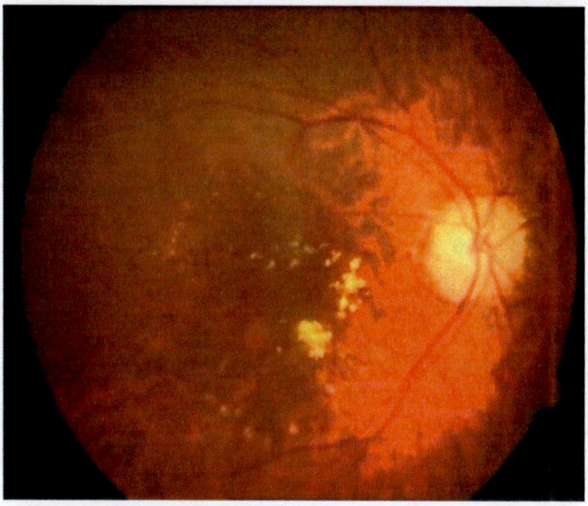

Fig. 3 Original image of retina

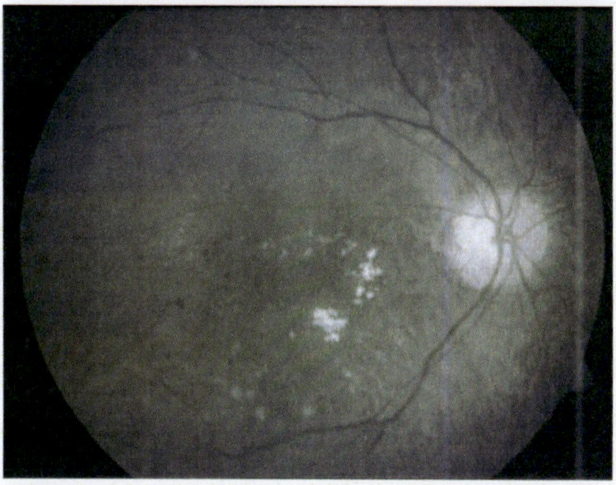

Fig. 4 Greyscale image of retina

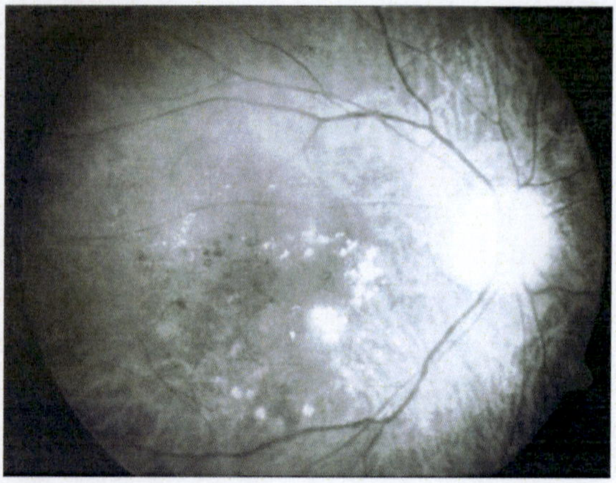

Fig. 5 Contrast image of retina

3.3 *Histogram Equalization*

Histogram equalization is the alternative approach in which we reduce variations by using the redistribution of the grey levels to create a uniform distribution of pixel intensities. In this method, we enhance the contrast of a low contrast image as shown in Fig. 5.

3.4 *Morphological Operation*

In image processing, morphology is the study in which we deal with the shape and its variations. Mathematical morphology gives the mathematical theory for describing shapes using sets. It is used to find the relation between an image and a certain chosen structuring element using the basic operations of erosion and dilation. Dilation, erosion, opening, and closing are the main processes, which involve a special mechanism of combining two sets of pixels in which one set consists of the image being processed and the other constitutes the structuring element or kernel. Opening and closing are two very important transformations. Opening generally smoothens the contour in an image and closing tends to narrow smooth sections of contours, eliminating small holes and filling gaps in contours. Intuitively, dilation expands an image object and erosion shrinks it. The above processes are used in edge detection, noise removal and background removal as well as for finding specific shapes in images as shown in Fig. 6.

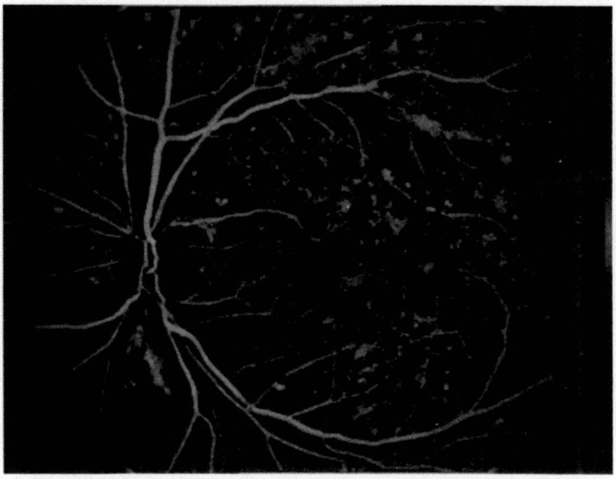

Fig. 6 Applying morphological operations on the image of retina

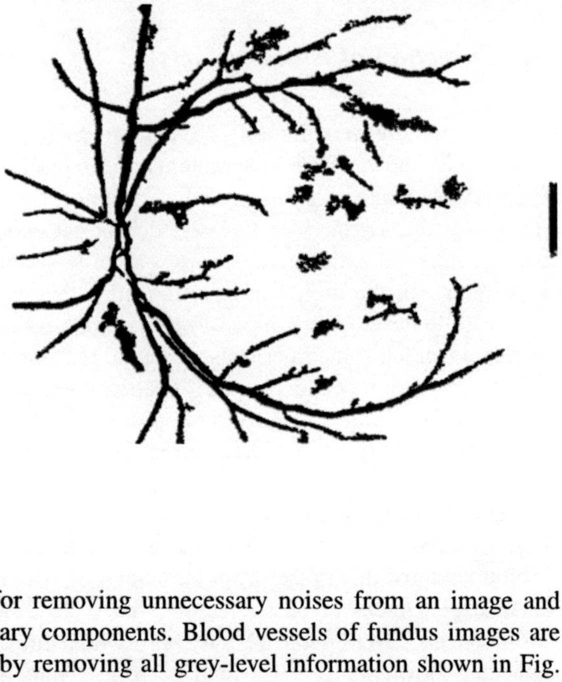

Fig. 7 Applying thresholding on the image of retina

3.5 Thresholding

It is a very useful method for removing unnecessary noises from an image and helps to focus on the necessary components. Blood vessels of fundus images are converted into binary pixels by removing all grey-level information shown in Fig. 7. It is important to differentiate between blood vessels of the foreground from the background information. It is very useful in the image region, which is obscured by similar grey levels. Thus, choosing an appropriate threshold value is important, because a low value may decrease the size of some of the objects. Thresholding helps us in converting image features into numerical data which is then applied on various machine learning algorithms. This provides a channel to convert fundus images into numerical values.

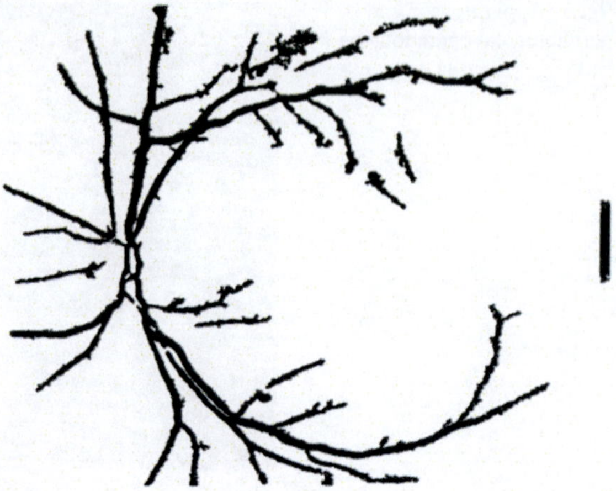

Fig. 8 Segmented blood vessels of retina

3.6 Detection of Blood Vessels

Blood vessels are very important features which help in the examination of the retina, which shows different symptoms of eye diseases. There are swellings in the blood vessels when an eye has diabetic retinopathy. In non-proliferative diabetic retinopathy, the narrow blood vessels do not get enough blood supply because of blockage near the optic nerve. It causes them to get broken down and release fluid in the retina. In the proliferative retinopathy, this gets worse. To counteract this, retina tries to develop new blood vessels. Again these blood vessels are weak and disoriented which sometimes causes them to leak inside vitreous.

So the detection of blood vessels becomes very important. Alternate sequential Filtering (ASF) along with other image processing techniques is used to extract blood vessels. But the presence of haemorrhages and clots make it difficult, thus green channel of the image is extracted for greater contrast. Contrast-Limited Adaptive Histogram Equalization is used to increase contrasts further. Applying ASF on that image gives another image with an average intensity of each region applied over it. Subtraction of this image from the output of CLAHE gives us an image which contains faint traces of blood vessels with optic disk and other things removed. This image is converted into a binary pixel with a threshold T and gets blood vessels segmented shown in Fig. 8. The final image also contains noise and some undesirable elements. Noise is removed by eroding the image. Undesirable elements are removed by taking into account the feature of only blood vessels which are linear in shape.

Fig. 9 A random image of retina after clustering

3.7 K-Means Clustering

Clustering is a method that aims to find similar groups, which are known as clusters. This notion of similarity depends upon the type of distance measure used while clustering shown in Fig. 9.

3.8 Classification

After the features are extracted, according to the extent of features extraction, classification is done using classifiers. The efficiency of the classifier is calculated in terms of its efficiency to classify normal images as normal and abnormal images as abnormal. In our work, we use two classifiers: Support vector machine (SVM) and K Nearest Neighbour classier. Common for these classifiers is that they need to be trained, i.e. they are supervised.

- Support Vector Machine Classification—The classier aims by creating a model that predicts the class labels of unknown data or validation data instances consisting only of attributes, as given in Eq. 1. Kernels of SVM are generally used for mapping non-linearly separable data into higher dimensional feature space instance, which consists of only attributes [18].

$$f(x) = sign\left(\sum_{i=1}^{n} \alpha_i y_i K(x, x_i) + b\right), \tag{1}$$

where $k(x, x_i)$ is a kernel function and the equation for kernel k for Radial Basis function (RBF) as mentioned in Eq. 2.

$$K(x, x_i) = \exp\left(- \frac{||x - x_i||^2}{2\sigma^2} \right), \quad (2)$$

- k-Nearest Neighbour Algorithm—It is a Machine Learning Algorithm based on Euclidean Distance between instances. kNN predicts class labels for different instances by measuring its shortest Euclidean Distance from other instances where the Euclidean Distances are calculated considering all the features or attributes as dimensions as given in the Eq. 3,

$$d(x_i, x'_j) = |||x_i - x_j||^2 = \sum_{k=1}^{d} (x_{ik} - x_{jk})^2, \quad (3)$$

- Convolutional Neural network—Convolutional Neural network is a method of deep learning which is used especially for image classification. The architecture is made by several convolutional layers. The input image is passed through a convolutional layer with ReLU function, then pooling is being done. In our work we use maximum pooling, then it is again passed through a convolutional layer and the process continues. Then it is flattened and softmax function is used in the output layer.
  The model summary of our CNN architecture is shown Fig. 10.

```
Layer (type)                 Output Shape              Param #
=================================================================
conv2d_28 (Conv2D)           (None, 1, 200, 33)        6633
_________________________________________________________________
max_pooling2d_24 (MaxPooling (None, 1, 200, 33)        0
_________________________________________________________________
flatten_24 (Flatten)         (None, 6600)              0
_________________________________________________________________
dense_24 (Dense)             (None, 2)                 13202
=================================================================
Total params: 19,835
Trainable params: 19,835
Non-trainable params: 0
_________________________________________________________________
```

Fig. 10 CNN model summary

4 Results and Analysis

These algorithms are used on the training dataset of digital fundus images to develop the classification model to accurately detect the diabetic retinopathy, for which the accuracy, sensitivity and specificity are given in Table 1. Precision, recall and F1 Score are given in Table 2. AUROC, mean error and RMS error are mentioned in Table 3. The confusion matrices of the SVM and kNN models are given in Table 4. The Area Under ROC curve (AUROC) gives out the predictive performance of these two algorithms shown in Figs. 10, 11, and 12.

By comparing these results, we can state that Support vector machine (SVM) gives better results as compared to the other two models in classifying diabetic retinopathy from digital fundus images.

Table 1 Accuracy, sensitivity and specificity

Classifier	Accuracy	Sensitivity	Specificity
SVM	96.62	0.667	0.954
KNN	94.38	0	0.933
CNN	94.74	–	–

Table 2 Precision, recall and F1 score

Classifier	Precision	Recall	F1 score
SVM	0.97	0.97	0.96
KNN	0.89	0.94	0.92
CNN	0.92	0.92	0.92

Table 3 AUROC, mean error and RMS error

Classifier	AUROC	Mean error	RMS error
SVM	0.70	0.033	0.183
KNN	0.50	0.056	0.237
CNN	0.87	0.10	0.324

Table 4 Confusion matrices

Classifier	Type	Predicted normal	Predicted DR
SVM	Actual normal	2	3
	Actual DR	0	84
KNN	Actual normal	0	5
	Actual DR	0	84

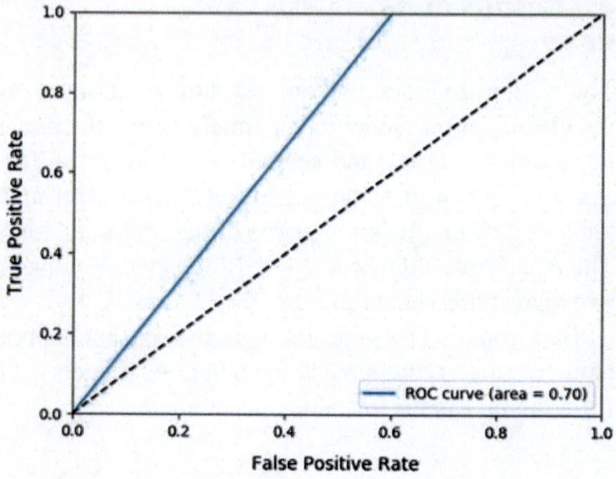

Fig. 11 ROC Curve for SVM

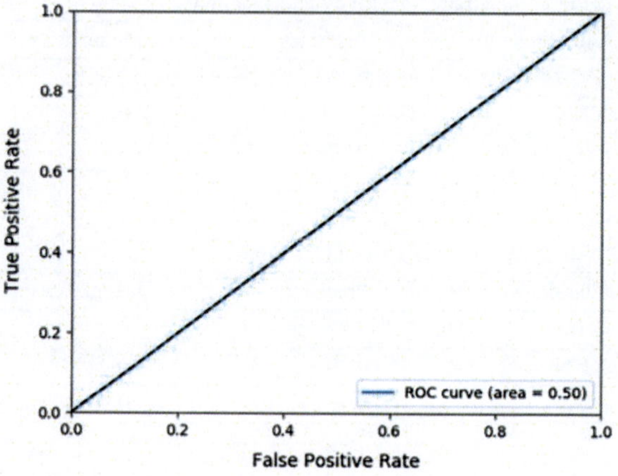

Fig. 12 ROC curve for kNN

4.1 Conclusion

Digital imaging is used for screening of diabetic retinopathy, which provides a high-quality permanent record of the retinal appearance used to monitor the progression or responses to the treatment. Digital images have the potential to be processed by automatic analysis systems.

The system was designed to recognise normal appearances using digital image analysis (optic disc, fovea) and to distinguish abnormal appearances (cotton wool spots, exudates, haemorrhages and microaneurysms) by feature extraction and feeding into a statistical classier for pattern recognition.

Various statistical classifiers like Support vector machine (SVM) and a KNN classier are tested. The system is tested on 89 retinal images. The SVM classier had the best results: with accuracy, sensitivity and specificity of 96.62%, 0.667 and 0.954, respectively.

References

1. Chaudhuri, S., Chatterjee, S., Katz, N., Nelson, M., Goldbaum, M.: Detection of blood vessels in retinal images using two-dimensional matched filters. IEEE Trans. Med. Imag. **8**(3), 263269 (1989)
2. Vallabha, D., Dorairaj, R., Namuduri, K., Thompson, H.: Automated detection and classification of vascular abnormalities in diabetic retinopathy. In: Proceedings of 13th IEEE Signals, Systems and Computers, vol. 2, pp. 1625–1629 (2004)
3. Sinthanayothin, C., Boyce, J., Williamson, T., Cook, H., Mensah, E., Lal, S., Usher, D.: Automated detection of diabetic retinopathy on digital fundus images. Diabet. Med. **19**, 105–112 (2002)
4. Noronha, K., Nayak, J., Bhat, S.: Enhancement of retinal fundus image to highlight the features for detection of abnormal eyes. In: Proceedings of the IEEE Region10 Conference (TENCON2006) (2006)
5. Lay, B., Baudoin, C., Klein, J.-C.: Automatic detection of micro aneurysms in retinopathy fluoro-angiogram. Proc. SPIE **432**, 165 (1983)
6. Ege, B.M., Hejlesen, O.K., Larsen, O.V., Moller, K., Jennings, B., Kerr, D., Cavan, D.A.: Screening for diabetic retinopathy using computer based image analysis and statistical classification. Comput. Meth. Programs Biomed. **62**, 165–175 (2000)
7. Lee, S., Lee, E., Kingsley, R., Wang, Y., Russell, D., Klein, R., Warn, A.: Comparison of diagnosis of early retinal lesions of diabetic retinopathy between a computer and human experts. Arch. Ophthalmol. (2001)
8. Gardner, G., Keating, D., Williamson, T., Elliott, A.: Automated detection of diabetic retinopathy using an artificial neural network: a screening tool. Br. J. Ophthalmol. **86**, 940–944 (1996)
9. Bezdek, J., Pal, M., Keller, J., Krisnapuram, R.: Fuzzy Model and Algorithms for Pattern Recognition and Image Processing. Kluwer Academic Press, London (1999)
10. Osareh, A., Mirmedhi, M., Thomas, B., Markham, R.: Automated identification of diabetic retinal exudates in digital color imaging. Br. J. Ophthalmol. **87**, 1220–1223 (2003)
11. Acharya, U.R., Chua, K.C., Ng, E.Y.K., Wei, W., Chee, C.: Application of higher order spectra for the identification of diabetes retinopathy stages. J. Med. Syst. **32**(6), 481–488 (2008)
12. Kahai, P., Namuduri, K.R., Thompson, H.: A decision support framework for automated screening of diabetic retinopathy. Int. J. Biomed. Imag. **2006**, 18 (2006)
13. Wong, L.Y., Acharya, U.R., Venkatesh, Y.V., Chee, C., Lim, C.M., Ng, E.Y.K.: Identification of different stages of diabetic retinopathy using retinal optical images. Inform. Sci. **178**(1), 106121 (2008)
14. Acharya, U.R., Lim, C.M., Ng, E.Y.K., Chee, C., Tamura, T.: Computer based detection of diabetes retinopathy stages using digital fundus images. J. Eng. Med. **223**(H5), 545553 (2009)
15. Adarsh, P., Jeyakumari, D.: Multiclass SVM-based automated diagnosis of diabetic retinopathy. In: International Conference on Communication and Signal Processing, India (2013)
16. It.lut.fi.: Diaretdb1-standard diabetic retinopathy database (2018). http://www.it.lut.fi/project/imageret/diaretdb1/. Accessed 7 July 2018
17. Kauppi, T., Kalesnykiene, V., Kamarainen, J.-K., Lensu, L., Sorri, I., Raninen A., Voutilainen R., Uusitalo, H., Klviinen, H., Pietil, J.: DIARETDB1 diabetic retinopathy database and evaluation protocol, Technical report
18. Babu, N.R., Mohan, B.J.: Fault classification in power systems using EMD and SVM. Ain Shams Eng. J. (2015)

Supervised Change Detection Technique on Remote Sensing Images Using F-Distribution and MRF Model

Srija Raha, Kasturi Saha, Shreya Sil, and Amiya Halder

Abstract Change detection is a powerful tool used to detect dissimilarities between two images of the same object taken after an interval of time. In the paper, we propose an algorithm for change detection in remote sensing images. This is a supervised technique where we cluster the difference image to obtain approximate training data. Then, segment the difference image on the basis of the training dataset using Markov Random Fields (MRF). F-distribution is used to reduce the segmented image into optimum number of clusters based on the inter-cluster intensities. Experimental results of this proposed method shows more encouraging output than other exiting change detection methods.

Keywords Remote sensing images · F-distribution · MRF model · Change vector analysis

1 Introduction

Remote sensing refers to acquiring images of an object or occurrence without exhibiting any physical touch with it. Remote sensing images are acquired by satellites, weather balloons, drones, etc., over a geographical region. The acquired images are analyzed to determine patterns, anomalies, or changes that have taken place. Change detection is a technique that is used to ascertain the change in attributes of

S. Raha · K. Saha · S. Sil · A. Halder (✉)
Department of CSE, St. Thomas' College of Engineering and Technology, 4 D. H. Road, Kidderpore, Kolkata, India
e-mail: amiya.halder77@gmail.com

S. Raha
e-mail: srijaraha97@gmail.com

K. Saha
e-mail: kasturisaha14@gmail.com

S. Sil
e-mail: shreyas.sil4@gmail.com

D. Bhattacharjee et al. (eds.), *Proceedings of International Conference on Frontiers in Computing and Systems*, Advances in Intelligent Systems and Computing 1255,
https://doi.org/10.1007/978-981-15-7834-2_23

an area over a period of time. In a change detection algorithm, the images of the area captured at different times are used as the input images. The difference image that is calculated from the two images is segmented into two groups of pixels—one group that determines the changing area and the other group that determines the unchanged area. The technique of change detection can be used to determine the changes that have occurred in a geographical area over a period of time by using remotely sensed images of that area. It can detect the changes on the earth's surface after natural calamities like cyclones, earthquakes, landslides, floods, tsunamis, volcanic activities, forest fires, etc., or anthropogenic activities like heavy industrialization, deforestation, urbanization, etc. Over the years, a large number of change detection algorithms have been proposed. Most of these algorithms are based on supervised learning algorithms or unsupervised learning algorithms. The supervised change detection algorithms like K-Nearest Neighbors (K-NN) [1], supervised ant colony optimization [2] and MRF [3, 4] and unsupervised algorithms like SOM [5], bisecting K-Means [6, 7], Gustafson-Kessel Clustering (GKC) [8], modified FCM [9], unsupervised K-NN [10] has been proposed for change detection. Generally, supervised algorithms may need more computation time than unsupervised learning methods. A combination of supervised and unsupervised data may also be used in some techniques like semi-supervised FCM [11, 12]. Also, different existing change detection algebraic techniques like LDD, SCD [13] can also be used. MLP [14] is another extremely popular technique used for change detection. GKC and modified FCM work better for smaller size images but for larger images the complexity increases. In the proposed method, a new efficient change detection algorithm is introduced using the MRF model that use segmentation and applies F-distribution to merge the clusters that belong to the same group. We then compare it with the existing algorithms and analyze the accuracy.

2 Proposed Algorithm

Change vector analysis, commonly known as CVA, is used to detect the changes in two remote sensing images of the same region taken before and after a certain time lapse. In the proposed method, initially, calculate the difference image using change vector analysis. For providing labeled data to the algorithm, then apply the mean-shift clustering technique on the difference image. After that, map the raw data to labeled data obtained from the mean shift clustering. This is done by normalizing the change vector and representing it through Gaussian distribution. The energy of the entire image is calculated through the equation of the MRF model and inter-cluster intensity is found out through F-distribution. Clusters are merged if the inter-cluster intensity is less than the energy of the image. The clusters are merged until all the clusters containing changed pixels and all the clusters containing unchanged pixels are grouped together in different clusters.

2.1 Calculating the Difference Image

Let the two images be $I = \{a_{m \times n}\}$ and $J = \{b_{m \times n}\}$.We calculate the difference image which is obtained by taking the pixel by pixel differences of the two images as shown in Eq. (1)

$$\text{diff}_i = a_i - b_i \, \forall i \in (0, m \times n - 1) \tag{1}$$

The change vector is denoted by a set $D = \{\text{diff}_i\}$.

2.2 Clustering the Difference Image

Then, segment the difference image D into k clusters. For that, assign $C = \{c_j\}$ where C is the set of cluster centers and j is the number of clusters. Depending on the range of i, we must predetermine a window size w. The segmentation has to be done such that for each diff in D, diff belongs in cluster j if the absolute difference between diff and cluster center of j is less than the defined window size as shown in (2)

$$|c_j - \text{diff}| < w \tag{2}$$

For each of these k clusters, new centers are formed calculating the average of all the pixels in c_j. We repeat the procedure $t + 1$ times until the centers converge, which is determined if $c_j^t = c_j^{t+1}$, that is, if the cluster centers of the previous iteration and the current iteration is equal for all k clusters.

2.3 Mean and Standard Deviations of the Clusters

The segmented image set $X = (x_1, x_2, \ldots, x_k)$ is used to train the difference image $D = \{\text{diff}_i\}$ and the cluster means μ_h and standard deviations σ_h are found for all h in $(1, k)$. They are obtained by the formula:

$$\mu_h = \frac{\sum_{i=1}^{\eta_h} \text{diff}_i}{\eta_h} \tag{3}$$

$$\text{and } \sigma_h = \sqrt{\frac{\sum_{i=1}^{\eta_h} (\text{diff}_i - \mu_h)^2}{\eta_h^2}} \tag{4}$$

where η_h is the number of pixels in cluster h.

2.4 Calculating Inter-Cluster Energy

Let ρ_d be the probability of occurrence of the pixel with intensity s in D. From, here we check the inter-cluster energy between each cluster pair(r, s) using Fisher's distribution. The formula for this energy is given by (5), which takes into account the respective means, standard deviations, and number of pixels in cluster pair(r, s).

$$Q = \frac{(\mu_r - \mu_s)(\eta_r + \eta_s - 2)\eta_r\eta_s}{(\eta_r\sigma_r^2 + \eta_s\sigma_s^2)(\eta_r + \eta_s)} \tag{5}$$

The objective is to check whether the inter-cluster intensity is greater than the global energy. If it is, the clusters are kept separated. Else, they are merged.

2.5 Calculating the Global Energy

The global energy is calculated by applying the formula for Gaussian distribution as difference image D is normalized over the clusters C. The global energy is calculated by obtaining a sum of the Gaussian distributed values of each pixel over each cluster as shown in (6)

$$\sum_{h=1}^{k}\sum_{i}^{d}\frac{1}{\sqrt{2\pi}\sigma_h}e^{\frac{(\mathrm{diff}-\mu_k)^2}{2\sigma_h^2}}\rho_i \tag{6}$$

Continuing until no other clusters are left to be merged, we get a segmented image with a few clusters containing changed pixels and a few containing unchanged. The intensity difference between these two groups of clusters is huge. So on performing threshold operation to the output, we get the final output of change detection.

3 Result and Analysis

The performance of the proposed change detection algorithm has been tested and compared against several existing methods. The different coefficients like TP (true positives), TN (true negatives), FP (false positives), FN (false negatives), PCC, JC (Jaccard coefficient), YC (Yule coefficient) between the resultant image and the ground truth image have been calculated and compared on two remote sensing images in Tables 1 and 2. The images used are satellite images of Sardinia and demo fire. The Error, PCC, JC, YC of the algorithm proposed, are compared against those of SOM, K-NN, Bisecting FCM, LDD, SCD, GKC, Modified FCM, unsupervised and supervised Ant Colony Optimization, MRF, etc. The outputs are shown in Figs. 1 and 2. On comparing the coefficients that have been calculated for image 1 and image 2,

Table 1 Comparison of error, PCC, JC, YC of different methods with proposed method for image 1

Methods	TP	FP	TN	FN	Error	PCC	JC	YC
SOM [5]	4906	806	112,938	5663	5.203	0.9480	0.4313	0.8111
K-NN [1]	4377	830	112,914	6192	5.648	0.9435	0.3840	0.7886
Semi-supervised FCM [11]	5336	2017	111,727	5233	5.832	0.9417	0.4240	0.6809
Bisecting FCM [12]	5025	1481	112,263	5544	5.651	0.9435	0.4170	0.7253
Rough C-means [6]	4715	1133	112,611	5854	5.620	0.9438	0.4029	0.7568
GKC [8]	604	13,172	102,446	8091	17.10	0.829	0.028	0.029
Modified FCM [9]	645	13,131	101,087	9450	18.16	0.818	0.028	0.039
Unsupervised ant colony optimization [2]	529	13,247	104,270	6267	15.69	0.843	0.026	0.018
Supervised ant colony optimization [2]	662	13,114	100,541	9996	18.59	0.814	0.028	0.042
MRF [3]	9366	2027	108,510	4410	5.178	0.9482	0.5926	0.7830
Proposed method	8465	423	110,114	5311	4.612	0.9538	0.5961	0.9063

Table 2 Comparison of error, PCC, JC, YC of different methods with proposed method for image 2

Methods	TP	FP	TN	FN	Error	PCC	JC	YC
SOM [5]	490	45	33,063	143	0.5572	0.9944	0.7227	0.9116
K-NN [1]	490	45	33,063	143	0.5572	0.9944	0.7227	0.9116
Bisecting FCM [12]	538	181	32,927	95	0.8180	0.9918	0.6609	0.7454
Rough C-means [6]	490	45	33,063	143	0.5572	0.9944	0.7227	0.9116
Linear dependence detector [13]	257	64	33,044	376	1.3041	0.9870	0.3687	0.7894
GKC [8]	206	386	32,105	158	2.109	0.855	0.298	0.059
Modified FCM [9]	284	286	32,019	234	3.526	0.804	0.369	0.135
Unsupervised ant colony optimization [2]	213	311	32,671	178	2.164	0.864	0.245	0.036
Supervised ant colony optimization [2]	302	314	31,978	259	3.614	0.823	0.336	0.097
MRF [3, 4]	456	26	33,082	177	0.6016	0.9939	0.6919	0.9407
Proposed method	490	42	33,066	143	0.5482	0.9945	0.7259	0.9167

we see that for both the images, the algorithms GKC, Modified FCM, unsupervised and supervised ant colony optimizations do not give effective results. However, it is evident that the proposed algorithm gives more accurate results.

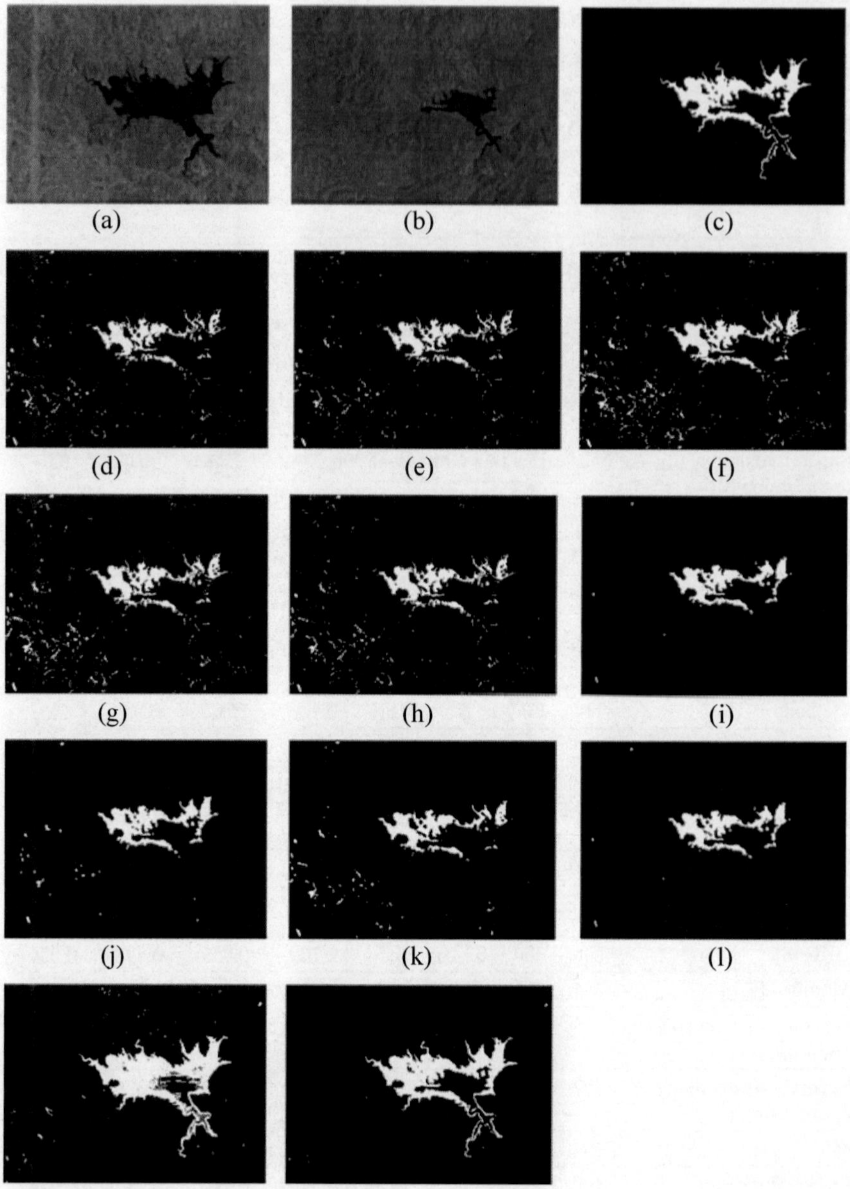

(a) (b) (c) (d) (e) (f) (g) (h) (i) (j) (k) (l) (m) (n)

Fig. 1 Output images of change detection using **d** SOM, **e** K-NN, **f** Semi-supervised FCM, **g** Bisecting K-means, **h** Rough C-means, **i** GKC, **j** Modified FCM, **k** Supervised ACO, **l** Unsupervised ACO, **m** Markov Random Fields, **n** Proposed method, **a**, **b** are two input images and **c** is the ground truth image

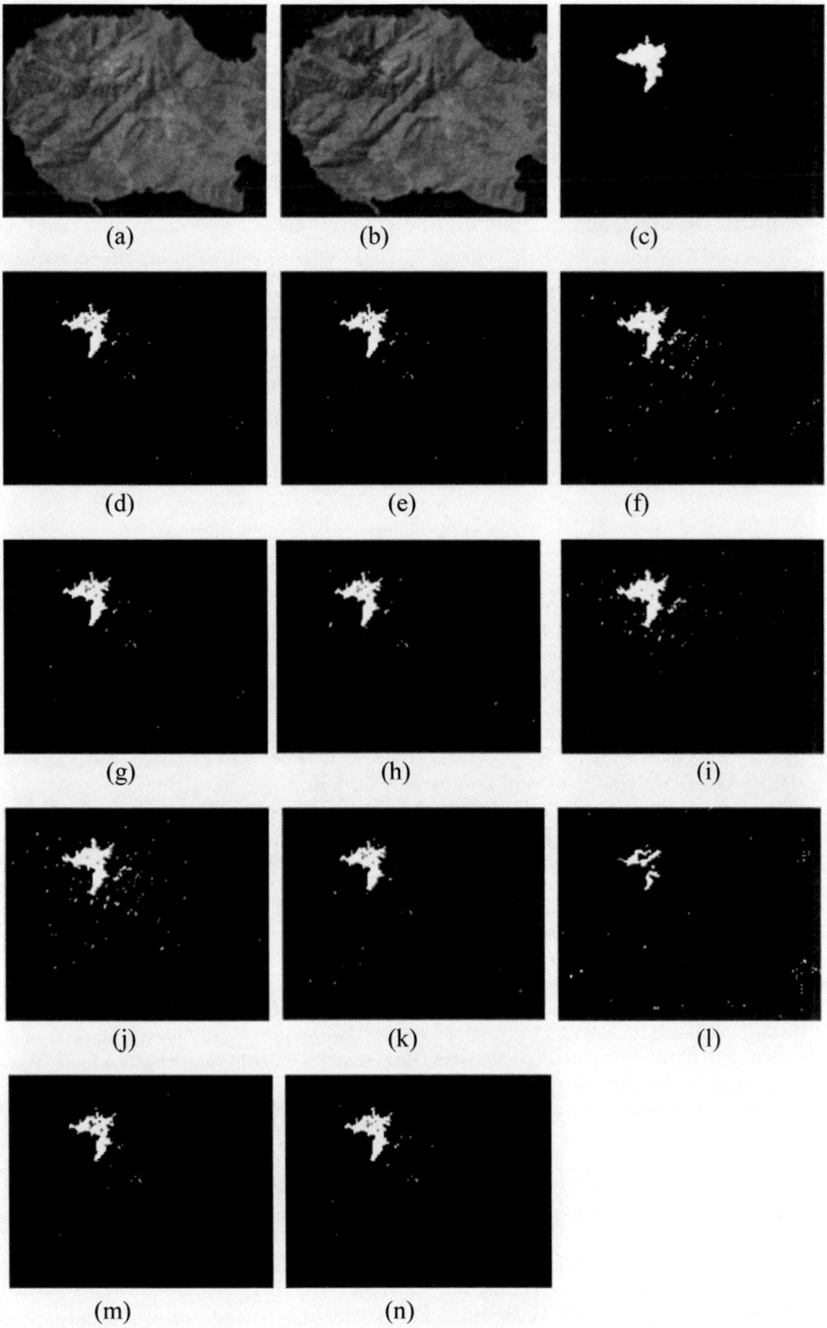

Fig. 2 Output images of change detection using **d** SOM, **e** K-NN, **f** Bisecting K-means, **g** Rough C-means, **h** GKC, **i** Modified FCM, **j** Supervised ACO, **k** Unsupervised ACO, **l** Linear Dependence Detector, **m** Markov Random Fields, **n** F-distribution and MRF model. **a**, **b** are input images and **c** is the ground truth image

4 Conclusion

We present a Mean-shift MRF and F-distribution-based change detection algorithm for change detection in remote sensing images. Based on the Yule, Jaccard, and PCC coefficients, the algorithm gives quite accurate results. The error in the output of the algorithm is less than the error of the outputs obtained through SOM, K-NN, Bisecting FCM, LDD, SCD, GKC, Modified FCM, unsupervised and supervised Ant Colony Optimization, and MRF. The proposed algorithm also gives more accurate results than these algorithms when compared on the basis of PCC, Jaccard, and Yule coefficients. We consider all three parameters along with the error coefficient, because PCC gives misleading estimates when the amount of change is less than 4%. Thus, high ratings can be observed by performing threshold operation on everything. Yule and Jaccard coefficients minimize the elimination effect of large volumes of true negatives.

References

1. Patra, S., Ghosh, S., Ghosh, A.: Change detection of remote sensing images with semi-supervised multilayer perceptron. Fundamenta Informaticae **84**(3), 429–442 (2008)
2. Cihlar, J., Pultz, T.J., Gray, A.L.: Change detection with synthetic aperture radar. Int. J. Remote Sens. **13**(3), 401–414 (1992)
3. Introduction to Markov Random Fields, by Pushmeet Kohli and Andrew Blake
4. Markov Random Fields (A Rough Guide) by Anil C. Kokaram, Electrical and Electronic Engineering Dept., University of Dublin, Trinity College
5. Filippi, A., Dobreva, I., Klein, A., Jensen, J.R.: Self-Organizing Map-Based Applications in Remote Sensing, pp. 231–248 (2010)
6. Jin, X., Han, J.: K-Means Clustering, Encyclopaedia of Machine Learning, Springer US (2010)
7. Abirami, K., Mayilvahanan, P.: Performance analysis of K-means and bisecting-means algorithms in weblog data. Int. J. Emerg. Technol. Eng. Res. (IJETER) **4**(8), 119–124 (2016)
8. Ghosh, A., Mishra, N.S., Ghosh, S.: Fuzzy clustering algorithms for unsupervised change detection in remote sensing images. Inform. Sci., 699–715 (2011)
9. Bezdek, J.C.: Pattern Recognition with Fuzzy Objective Function Algorithms. Springer Publishers, New York (1981)
10. Patra, S., Ghosh, S., Ghosh A.: Semi-supervised Learning with multilayer perceptron for detecting changes of remote sensing images. In: Pattern Recognition and Machine Intelligence, PReMI, vol. 4815. Springer, Berlin, Heidelberg (2007)
11. Ahmed, M.N., Yamany, S.M., Mohamed, N., Farag, A.A, Moriarty, T.: A modified fuzzy C-means algorithm for bias field estimation and segmentation of MRI data. IEEE Trans. Med. Imaging **21**, 193–199 (2002)
12. Chen, S.C., Zhang, D.Q.: Robust image segmentation using FCM with spatial constraints based on new kernel-induced distance measure. IEEE Trans. Syst. Man Cybern. B **34**(4), 1907–1916 (2004)
13. Durucan, E., Ebrahimi, T.: Change detection and background extraction by linear algebra. Proc. IEEE **89**(10), 1368–1381 (2001)
14. Roy, M., Ghosh, S., Ghosh, A.: A novel approach for change detection of remotely sensed images using semi-supervised multiple classifier system. Inf. Sci. **269**, 35–47 (2014)

A New Technique for Estimating Fractal Dimension of Color Images

Chinmaya Panigrahy, Ayan Seal, and Nihar Kumar Mahato

Abstract Fractal dimension (FD) effectively quantifies the roughness of the image surface, which can be applied to many image processing applications. Although a number of methods are available to calculate FD of grayscale images, limited work is done for the color images. In this paper, a new method is proposed to estimate FD of color images in CIE L*a*b* (CIELAB) color space. Firstly, the color image is converted to CIELAB space. Secondly, each of the L, a, and b components of CIELAB space are divided into overlapped grids and n_r value of each corresponding grid at different sizes are obtained using the distance between the minimum and maximum (L, a, b) triplets, where n_r is the number of boxes that constitute the intensity variations over a grid with scale, r. Thirdly, n_r values of the same sized grids are accumulated to obtain the N_r values. Finally, robust least squares regression (RLS) is applied to these N_r values to obtain the FD value. Three latest methods are adopted to differentiate the performance of the proposed method using the synthesized fractal Brownian motion (FBM) and Brodatz image databases. Experimental results show that the proposed method is more accurate in estimating the FD of color images.

1 Introduction

Fractal dimension (FD) efficiently measures the space filled by a self-similar complex object called as a fractal set, which cannot be represented using basic shape primitives like square, cube, etc. FD has many applications in image processing, namely, segmentation [1], classification [10], fusion [12], medical imaging [3], etc.,

C. Panigrahy · A. Seal (✉) · N. K. Mahato
PDPM-Indian Institute of Information Technology, Design and Manufacturing, Jabalpur 482005, MP, India
e-mail: ayan@iiitdmj.ac.in

C. Panigrahy
e-mail: chinmayapanigrahy@iiitdmj.ac.in

N. K. Mahato
e-mail: nihar@iiitdmj.ac.in

D. Bhattacharjee et al. (eds.), *Proceedings of International Conference on Frontiers in Computing and Systems*, Advances in Intelligent Systems and Computing 1255,
https://doi.org/10.1007/978-981-15-7834-2_24

where the terrain surface of an image constituting the intensity values is treated as a fractal set. A plethora of methods are available for computing FD of grayscale images [9], but limited work is done for color images. The differential box counting (DBC) method is one of the favorite methods to estimate FD of grayscale images [6], which is recently used by some researchers to device methods for color images [4, 13, 14]. However, these methods suffer from the perceptually non-uniformity of RGB color space. Simultaneously, DBC uses *quantization-computation* mechanism which overcounts the boxes. In addition to this, it uses nonoverlapped grids that introduce over/under-counting of boxes on the xy-direction. Moreover, it uses linear least squares regression (LLS) for fitting the points, which is sensitive to outliers.

This paper proposes a novel DBC based method to calculate FD of color images in perceptually uniform CIE L*a*b* (CIELAB) color space, which uses an overlapped grid partitioning mechanism to get square grids, *computation* mechanism to estimate the box count, and robust least squares regression (RLS) for fitting the points. Overlapped partitioning mechanism ensures the continuous image surface, *computation* mechanism enhances the box count, and RLS obtains a more accurate line. In this work, we only considered estimating the values of FD for color images in the range [2, 3]. The performance of the proposed method is compared with the three latest methods using two databases and two metrics. Experimental results prove the effectiveness of the proposed method.

The rest of this paper is as follows. Section 2 briefly describes the DBC method, RLS, and CIELAB color space. The proposed method is presented in Sect. 3 whereas Sect. 4 shows the experimental results. Finally, Sect. 5 concludes the paper.

2 Background

This section briefly introduces the DBC method, CIELAB color space, and RLS.

The DBC method proposed by Sarkar et al. [11] first partitions a grayscale image, I, of size $M \times M$ pixels into nonoverlapping square grids of size $s \times s$ pixels such that $s \in \mathbb{Z}^+$, $2 \le s \le \frac{M}{2}$. Let $n_r(i, j)$ be the number of boxes needed to cover the image terrain surface of the (i, j)th grid at scale, r, where $r = \frac{s}{M}$. The $n_r(i, j)$ of a grid can be determined using Eq. 1.

$$n_r(i, j) = \left\lceil \frac{g^{\max}}{h} \right\rceil - \left\lceil \frac{g^{\min}}{h} \right\rceil + 1, \tag{1}$$

where $g^{\min}$ and $g^{\max}$ are the minimum and maximum gray-level intensities over the (i, j)th grid, respectively. Here, $h = \frac{sG}{M}$ represents the box height and G is the number of allowed gray-levels of the image. The total number of n_r values for a particular r, N_r is computed using Eq. 2.

$$N_r = \sum_{i,j} n_r(i, j). \tag{2}$$

Finally, the points $(\log N_r, \log \frac{1}{r}) \forall r$ are fitted using LLS to obtain a line, $y = mx + c$, whose slope, m, corresponds to the FD value. Here, c is the $y-$intercept. The distance error (DE) of the above fitting process is calculated using Eq. 3.

$$DE = \frac{1}{T}\sqrt{\frac{\sum_{i=1}^{N}(mx_i + c - y_i)^2}{1 + m^2}}, \tag{3}$$

where $T = M/2 - 2 + 1$ represents the number of grid sizes used.

In [7], a new RLS mechanism is proposed to enhance the FD values, where the points $(\log N_r, \log \frac{1}{r}) \forall r$ are first grouped into three categories depending on the corresponding s values. First group corresponds to the points for which $s \in [2, \sqrt[3]{M} - 1]$ whereas $s \in [\sqrt[3]{M}, P]$ constitute the second group, where P is the largest integer that assures $\lceil \frac{M}{s} \rceil + 1 \leq \lceil \frac{M}{s-1} \rceil$. The rest of the points are considered as the third group, i.e., $s \in [P + 1, M/2]$. Finally, LLS is applied to all the points of second group, the geometric median (G_M) of first group, and G_M of last group to find FD of the image.

The CIELAB is a perceptually uniform device-independent color space, which is modeled to approximate the human vision. It constitutes all the colors inside the spectrum and outside the human perception [8]. It consists of three components, namely, L, a, and b. L represents the lightness component which greatly resembles the human light perception whereas a and b are the color components representing the red/green and yellow/blue colors, respectively. The extremist values, i.e., $L = 0$, $L = 100$, $a = -100$, $a = 100$, $b = -100$, and $b = 100$ represent the black, white, green, red, blue, and yellow colors, respectively.

3 Proposed Method

In this section, the proposed methodology to compute FD of color images is presented in detail. The proposed method can be described using following steps:

- **Step 1**: The color image, I, of size $M \times M$ pixels, is transformed into CIELAB color space to obtain the L, a, and b components.
- **Step 2**: Each of the L, a, and b components are partitioned into overlapping grids of size $s \times s$ pixels, where two nearby grids will overlap by one row and one column [2]. Here, s varies from 2 to $M/2$ and $s \in \mathbb{Z}^+$.
- **Step 3**: For each grid of size $s \times s$ pixels corresponding to L, a, and b components
  - The (L, a, b) triplets $(g_L^{\min}, g_a^{\min}, g_b^{\min})$ and $(g_L^{\max}, g_a^{\max}, g_b^{\max})$ are estimated, where $g_L^{\min}$ and $g_L^{\max}$ are the minimum and maximum values of the corresponding grid of the L component, respectively. Similarly, $g_a^{\min}$, $g_a^{\max}$, and $g_b^{\min}$, $g_b^{\max}$ are found for the a and b components, respectively.
  - The Euclidean distance between the triplets $(g_L^{\min}, g_a^{\min}, g_b^{\min})$ and $(g_L^{\max}, g_a^{\max}, g_b^{\max})$, d, is computed using Eq. 4.

$$d = \sqrt{(g_L^{\max} - g_L^{\min})^2 + (g_a^{\max} - g_a^{\min})^2 + (g_b^{\max} - g_b^{\min})^2}. \quad (4)$$

– The box height, h, is estimated using Eq. 5.

$$h = \frac{s \times 100}{M}. \quad (5)$$

– The number of boxes needed to wrap the distance d, n_r, is computed using Eq. 6, which uses *computation* mechanism, i.e., no ceiling function is used [9]. Here, r represents the scale of the grid and $r = s/M$.

$$n_r = \frac{d+1}{h}. \quad (6)$$

- **Step 4**: The n_r values of all the grids of size $s \times s$ pixels are accumulated to get N_r using Eq. 2.

$$N_r = \sum_{i,j} n_r(i, j) \quad (2).$$

  Here, i and j correspond to the (i, j)th grid of size $s \times s$ pixels.
- **Step 5**: After finding N_r values $\forall r$, i.e., $\forall s$, RLS [7] is used to fit the points $(\log N_r, \log \frac{1}{r})\forall r$ into a line, $y = mx + c$, where the slope, m, is the FD of I. DE can be estimated using Eq. 3.

4 Experimental Results

This section describes the experimental results along with the databases and two evaluation metrics. Three latest methods, viz., color DBC (CDBC) [4], maximum color distance (MCD) [14], and improved CDBC (ICDBC) [13] are considered to show the effectiveness of the proposed method using the synthesized FBM images and Brodatz database. All the images used in this paper are of size 512 × 512 pixels.

The correctness of the proposed method is shown using 11 color FBM images with known theoretical FD (TFD), which are synthesized using the modified random mid-point displacement algorithm [8]. Figure 1 shows the synthesized FBM images with their TFD values. Twelve images are randomly selected from the Brodatz database [5] to further show the superiority of the proposed method, which are shown in Fig. 2.

The first experiment is performed on the synthesized FBM images to compare different methods. Figure 3 shows the FD values computed by various methods for the images of Fig. 1. It can be seen in Fig. 3 that for maximum FBM images, the proposed method estimates more accurate FD values which are nearest to their TFD values. Furthermore, the proposed technique always obtains larger FD values than other methods. Simultaneously, these FD values are never lower than 2. This shows that the proposed method follows the fractal theory of images [8], which is also

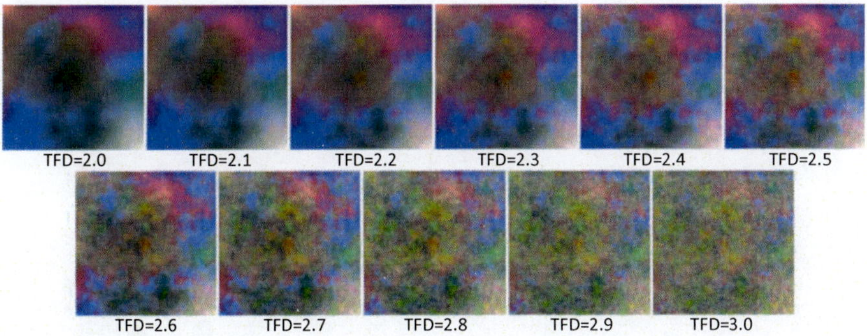

Fig. 1 Synthesized FBM images

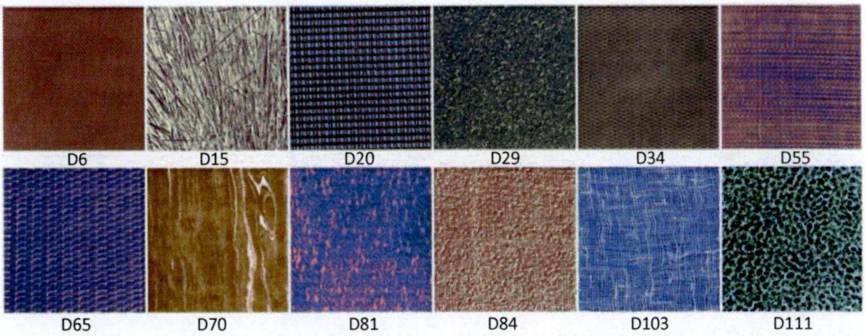

Fig. 2 Twelve images from the Brodatz database [5]

persisted by MCD. CDBC always generates least FD values which are much lower than other methods. Though the FD value of many images computed by ICDBC are higher than other existing methods, it sometimes produces lower FD values which are even lower than 2, which is not appropriate. The DE values of different methods for the FBM images are reported in Table 1. It is evident in the above table that CDBC generates least DE values, however its worst FD values completely overshadow this feature. The proposed method incurs the highest DE values due to RLS, which can be ignored due to better FD values.

In the second experiment, two metrics, namely, average computed FD (ACFD) and average error (AE) [8] are used further to compare the FD values of FBM images computed by various methods. These metrics are only applicable to FBM images due to their known TFD values. ACFD quantifies the extent of computed FD values whereas AE measures the total error in approaching the corresponding TFD values of all the FBM images. The ideal value of ACFD is 2.5. A lower AE value is preferred. Table 2 shows the ACFD and AE values of different methods. The proposed method obtains the highest ACFD value by generating least AE. Thus, the proposed method is best among different methods. In most of the cases, the computed FD value is lower than the respective TFD. Simultaneously, the ACFD value of all the methods

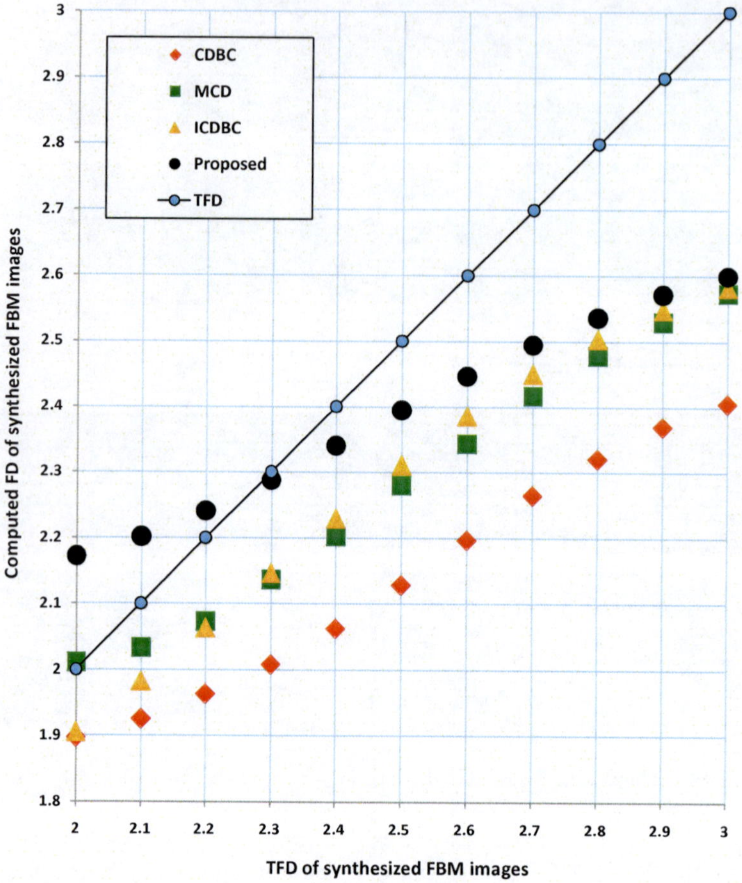

Fig. 3 The FD values of Fig. 1 computed by various methods

is less than 2.5. Therefore, it is likely that the FD values computed by different methods are lower than the corresponding TFDs. The TFD values of images are best approached by the proposed method with higher ACFD value. Hence, a higher FD value is preferred. It can be concluded from the above results that the proposed methodology is most suitable to compute the FD of FBM images.

The images of Fig. 2 are used for the last experiment where the same four methods are considered. Table 3 reports the FD and DE values obtained by these methods for the images of Fig. 2. It is observed in Table 3 that the proposed method always generates higher values of FD than the existing methods. Among the existing methods, ICDBC always produces higher FD values whereas CDBC generates lower FD value for all most all images. The least DE values are incurred by CDBC, but it is likely to compute lower FD values. The proposed technique generates the largest DE value for many images, but these values are sometimes lower than MCD and ICDBC.

Table 1 DE values incurred by different methods for the images of Fig. 1

FBM images		CDBC [4]	MCD [14]	ICDBC [13]	Proposed
H	TFD = 3 − H				
1.0	2.0	0.0044	0.0046	0.0346	0.0519
0.9	2.1	0.0043	0.0061	0.0278	0.0494
0.8	2.2	0.0043	0.0086	0.0234	0.0471
0.7	2.3	0.0044	0.0090	0.0214	0.0451
0.6	2.4	0.0043	0.0091	0.0212	0.0435
0.5	2.5	0.0042	0.0091	0.0219	0.0417
0.4	2.6	0.0040	0.0092	0.0230	0.0397
0.3	2.7	0.0039	0.0095	0.0237	0.0376
0.2	2.8	0.0038	0.0090	0.0241	0.0360
0.1	2.9	0.0036	0.0095	0.0244	0.0344
0.0	3.0	0.0037	0.0102	0.0243	0.0333

Table 2 ACFD and AE values of different methods

	CDBC [4]	MCD [14]	ICDBC [13]	Proposed
ACFD	2.1402	2.2798	2.2827	2.3898
AE	0.3598	0.2222	0.2173	0.1675

Table 3 FD and DE values for the images of Fig. 2

Images [5]	CDBC [4]		MCD [14]		ICDBC [13]		Proposed	
	FD	DE	FD	DE	FD	DE	FD	DE
D6	2.5204	0.0045	2.5133	0.0265	2.6789	0.0387	2.7215	0.0414
D15	2.5626	0.0044	2.5145	0.0443	2.7689	0.0301	2.7951	0.0393
D20	2.5032	0.0049	2.4836	0.0438	2.7155	0.0416	2.7573	0.0380
D29	2.5214	0.0042	2.5395	0.0316	2.6683	0.0350	2.6886	0.0356
D34	2.5212	0.0054	2.5711	0.0395	2.6363	0.0471	2.6672	0.0430
D55	2.5313	0.0044	2.6374	0.0308	2.7366	0.0324	2.7928	0.0401
D65	2.5107	0.0042	2.5822	0.0416	2.6458	0.0415	2.7214	0.0443
D70	2.3554	0.0044	2.4351	0.0273	2.5630	0.0273	2.6448	0.0344
D81	2.5420	0.0045	2.5742	0.0314	2.7789	0.0290	2.7960	0.0463
D84	2.5297	0.0042	2.6442	0.0359	2.7505	0.0334	2.8025	0.0403
D103	2.5845	0.0041	2.6765	0.0369	2.8117	0.0302	2.8827	0.0486
D111	2.6173	0.0050	2.6731	0.0353	2.7856	0.0350	2.8608	0.0411

The results of the first two experiments show that the proposed method generates more accurate FD values when compared with other methods by providing the highest ACFD and lowest AE values. Moreover, different methods are prone to generate a lower FD value for an image, hence a larger FD value is preferred. The last experiment confirmed that the higher FD values among different methods are obtained by the proposed method. However, it incurs higher DE values. Though a lesser DE value is preferred, it may be compromised for a better FD value which has ample significance than DE. Therefore, the proposed method is a better technique to compute FD of color images.

5 Conclusion

This paper proposes a novel method to approximate FD of color images by transforming them into CIELAB color space. The distance between the minimum and maximum points of a grid in CIELAB space is used to compute n_r of that grid, where each grid overlaps one row and one column of its neighboring grids. Finally, RLS is used to obtain the FD value. CIELAB color space discards the non-uniformity of RGB color space whereas RLS prevents outliers to cause distortions. Moreover, the proposed method uses *computation* mechanism to estimate n_r values, hence a more practical FD value is obtained. The effectiveness of the proposed method is shown using the results of three latest methods on the simulated FBM image database and Brodatz database. The TFD values of FBM images are better approached by the proposed method which generates the highest ACFD by incurring least AE value. The proposed method generates a more accurate FD value for almost all images. Nonetheless, it generates quite higher DE values, which needs to be addressed. The n_r values can further be enhanced by considering the triangular partitioning of a grid. The proposed technique can appertain to various image processing applications like texture segmentation, characterization, and classification.

References

1. Chaudhuri, B.B., Sarkar, N.: Texture segmentation using fractal dimension. IEEE Trans. Pattern Anal. Mach. Intell. **17**(1), 72–77 (1995)
2. Li, J., Du, Q., Sun, C.: An improved box-counting method for image fractal dimension estimation. Pattern Recognit. **42**(11), 2460–2469 (2009)
3. Liu, S., Fan, X., Zhang, C., et al.: MR imaging based fractal analysis for differentiating primary CNS lymphoma and glioblastoma. Eur. Radiol. **29**(3), 1348–1354 (2019)
4. Nayak, S.R., Ranganath, A., Mishra, J.: Analysing fractal dimension of color images. In: International Conference on Computational Intelligence and Networks (CINE), pp. 156–159. IEEE (2015)
5. Original Brodatz Texture-Universite de Sherbrooke, 31 August 2019. http://multibandtexture.recherche.usherbrooke.ca/original_brodatz.html

6. Panigrahy, C., Garcia-Pedrero, A., Seal, A., et al.: An approximated box height for differential-box-counting method to estimate fractal dimensions of gray-scale images. Entropy **19**(10), 534 (2017)
7. Panigrahy, C., Seal, A., Mahato, N.K.: Quantitative texture measurement of gray-scale images: fractal dimension using an improved differential box counting method. Measurement (2019). https://doi.org/10.1016/j.measurement.2019.106859
8. Panigrahy, C., Seal, A., Mahato, N.K.: Fractal dimension of synthesized and natural color images in Lab space. Pattern Anal. Appl. (2019). https://doi.org/10.1007/s10044-019-00839-7
9. Panigrahy, C., Seal, A., Mahato, N.K., et al.: Differential box counting methods for estimating fractal dimension of gray-scale images: a survey. Chaos Solitons Fract **126**, 178–202 (2019)
10. Ribas, L.C., Gonçalves, D.N., Silva, J.A., et al.: Fractal dimension of bag-of-visual words. Pattern Anal. Appl. **22**, 89–98 (2019)
11. Sarkar, N., Chaudhuri, B.B.: An efficient differential box-counting approach to compute fractal dimension of image. IEEE Trans. Syst. Man Cybern. **24**(1), 115–120 (1994)
12. Seal, A., Panigrahy, C.: Human authentication based on fusion of thermal and visible face images. Multimed. Tools Appl. **78**(21), 30373–30395 (2019)
13. Zhao, X., Wang, X.: An approach to compute fractal dimension of color images. Fractals **25**(1), 1750007 (2017)
14. Zhao, X., Wang, X.: Fractal dimension estimation of RGB color images using maximum color distance. Fractals **24**(4), 1650040 (2016)

Deep Neural Network for Multivariate Time-Series Forecasting

Samit Bhanja and Abhishek Das

Abstract Recently, Deep Neural Network (DNN) architecture with a deep learning approach has become one of the robust techniques for time-series forecasting. Although DNNs provide fair forecasting results for the time-series prediction, still they are suffering from various challenges. Because most of the time-series data, especially the financial time-series data are multidimensional, dynamic, and nonlinear. Hence, to address these challenges, here, we have proposed a new deep learning model, Stacked Long Short-Term Memory (S-LSTM) model to forecast the multivariate time-series data. The proposed S-LSTM model is constructed by the stacking of multiple Long Short-Term Memory (LSTM) units. In this research work, we have used six different data normalization techniques to normalize the dataset as the preprocessing step of the deep learning methods. Here, to evaluate and analyze the performance of our proposed model S-LSTM, we have used the multivariate financial time-series data, such as stock market data. We have collected these data from two stock exchanges, namely, Bombay Stock Exchange (BSE) and New York Stock Exchange (NYSE). The experimental results show that the prediction performance of the S-LSTM model can be improved with the appropriate selection of the data normalization technique. The results also show that the prediction accuracy of the S-LSTM model is higher than the other well-known methods.

Keywords LSTM · RNN · DNN · Stock market prediction · Data normalization technique.

S. Bhanja
Government General Degree College, Singur 712409, Hooghly, India
e-mail: samitbhanja@gmail.com

A. Das (✉)
Aliah University, New Town, Kolkata 700160, India
e-mail: adas@aliah.ac.in

D. Bhattacharjee et al. (eds.), *Proceedings of International Conference on Frontiers in Computing and Systems*, Advances in Intelligent Systems and Computing 1255,
https://doi.org/10.1007/978-981-15-7834-2_25

1 Introduction

Multivariate time-series data is a set of multidimensional data collected in fixed time intervals. Prediction of the multivariate time-series is an important Time-Series Forecasting (TSF) problem and that has a wide range of application areas, viz., weather, pollution, finance, business, energy, etc. The successful prediction of the multivariate time-series data would definitely smoothen the modern livelihood.

In recent times, the prediction of the financial time-series data, especially the stock market prediction, has become one of the most important application areas for the researchers. The successful prediction of the stock market has a great influence on the socio-economic environment of the county. It also helps the investors to take an early decision whether to sell or buy the stock share or bonds and that reduces the risk of the investment. There are so many complex factors that influence the stock market and the stock market data are high dimensional, volatile, and nonlinear and due to this reason forecasting the stock market data is a highly challenging task.

Data normalization is the primary data preprocessing step for any DNN model for the processing of the time-series data. The time-series data, especially the stock market, is varied over a wide range, so as to produce good quality data from it and to accelerate the learning process, data normalization is essential. The efficiency of any DNN models highly depends on the selection of the proper data normalization technique.

The main focus of this research work is to develop a new DNN model to forecast the multivariate time-series data with a high degree of accuracy. Also to find out the most effective data normalization method for the DNN model.

In this work, we have proposed a DNN model S-LSTM to forecast the stock market as a multivariate time-series forecasting problem. We also analyzed its performance on different data normalization techniques to identify the best suitable data normalization method for the deep learning algorithms to forecast the multivariate time-series data.

We have organized our paper as follows: the literature review is represented in Sect. 2. In Sect. 3, we have provided the basic concepts. Data normalization methods are presented in Sect. 4. The proposed framework and dataset description are represented in Sect. 5. Results and discussion are described in Sect. 6, and finally in Sect. 7, the conclusion is drawn.

2 Literature Review

In the last few years, there are so many sincere efforts which have been made to successfully predict the stock market. These efforts are broadly classified into two categories, viz., statistical approach and soft-computing approach. Support Vector Machine (SVM) [1] and Autoregressive Integrated Moving Average (ARIMA) [2] are the two well-known statistical methods for time-series forecasting. These statistical models can handle the nonlinear time-series data and exhibit a high degree of success for the prediction of the univariate time-series data.

The Artificial Neural Network (ANN)-based models are the most popular soft-computing approach for the time-series prediction. The Artificial Neural Network-based models can perform a large variety of computational tasks faster than the traditional approach [3]. Multilayer Perceptron (MLP) neural network, Back Propagation (BP) [4, 5] neural network, etc. are the popular ANN models. These models are successfully applied to solve the various problems, viz., classification problems, time-series forecasting problems, etc. These ANN models are not suitable for the large volume of highly dynamic, nonlinear, and complex data.

Nowadays, Deep Neural Networks (DNNs) [6–9] exhibit its great success in a wide range of application areas, including multivariate time-series forecasting. The basic difference between the shallow neural networks and the deep neural networks is that the shallow networks have only one hidden layer whereas the deep neural networks have many hidden layers. These multiple hidden layers allow the DNNs to extract the complex features from the large volume of highly dynamic and nonlinear time-series data.

In recent times, Recurrent Neural Networks (RNNs) [10, 11] have become one of the most popular DNN architectures for the time-series classification and forecasting problems. In RNN, output of one time stamp is considered as the input of the next time stamp and for these timestamp concepts, it is most suitable for the processing of the time-series data. But RNNs suffered from the vanishing gradient and exploding gradient problems. For these problems, it cannot represent the long-term dependencies of the historical time-series data. The Long Short-Term Memory (LSTM) [12, 13] is a specialized RNN that overcomes the shortfalls of the traditional RNNs.

3 Basic Concepts

When a neural network has two or more hidden layers, then it becomes a Deep Neural Network (DNN). The most common neural networks, viz., Multilayer Perceptron (MLP) or feedforward neural networks with two or more hidden layers are the representatives of DNN models. DNN models are the basis of any deep learning algorithms. These multiple hidden layers allow the DNN models to capture the complex features from the large volume of the dataset. It also allows the DNN to process nonlinear and highly dynamic information. In recent times, a number of DNN models have been proposed. Out of these large numbers of DNNs, Recurrent Neural Network (RNN) is one of the most popular DNN models to process time-series data.

3.1 Recurrent Neural Networks (RNNs)

RNN [10] is one of the most powerful Deep Neural Network (DNN) models that can process the sequential data. It was first developed in 1986. Since it performs the same set of operations on every element of the sequential data, it is called the recurrent neural network. As per the theory, it can process very long sequences of time-series data, but in reality, it can look only a limited number of steps behind. Figure1 represents the typical architecture of RNN and its expanded form.

In RNN, following equations are used for the computational purpose:

$$h_t = f(Ux_t + Wh_{t-1}) \tag{1}$$

$$O_t = \text{softmax}(Vh_t) \tag{2}$$

where h_t and x_t are, respectively, the hidden sate and the input at the time stamp t, O_t is the output at the time stamp t, and function f is a nonlinear function, viz., $tanh$ or $ReLU$.

The basic difference between the traditional DNNs and the RNN is that RNN uses the same set of parameters (U, V, W as above) for all the steps. This parameter sharing drastically reduces the total number of memorizable parameters of the model.

3.2 Time-Series Data

If a series of data are collected over a fixed time intervals, then that dataset is called the time-series data. If every data points of time-series dataset, in spite of a single value, it is a set of values, then that type of time-series data is called the multivariate time-series data. There are numerous application areas present where multivariate time-series data are present, viz., weather, pollution, sales, stocks, etc. and these data can be analyzed for the forecasting purpose [14, 15]. The general format of the time-series data is as follows:

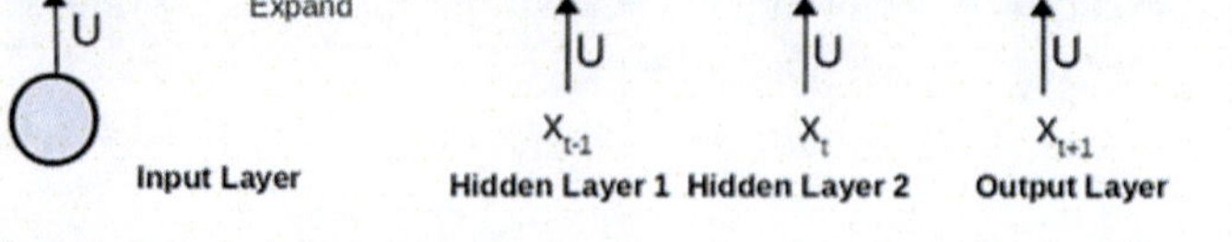

Fig. 1 A typical RNN and its expanded architecture

$$X = \{x(1), x(2), ..., x(t)\} \quad (3)$$

where $x(t)$ is current value and $x(1)$ is the oldest value. If X is multivariate time-series data then every data point $x(i)$ will be a vector of a fixed-length k. So, $x(i) = \{x_{i,1}, x_{i,2}, ..., x_{i,k}\}$.

4 Data Normalization Methods

The efficiency of any DNN models is heavily dependent on the normalization methods [16]. The main objective of the data normalization is to generate quality data for the DNN model. The nonlinear time-series data, especially the stock market data fluctuates over a large scale. So, the data normalization is essential to scale down the data to a smaller range to accelerate the learning process of the DNN models. Although there are different numbers of data normalization techniques are available, in all of these techniques, each input value a of each attribute A of the multivariate time-series data is converted to a_{norm} to the range [low, high]. Some of the well-known data normalization techniques are described below.

4.1 Min-Max Normalization

Here, the data scale down to a range of [0, 1] or [−1, 1]. The formulae for this method are as follows:

$$a_{\text{norm}} = \frac{(\text{high} - \text{low}) * (a - \min A)}{\max A - \min A} \quad (4)$$

where $\min A$ and $\max A$ are, respectively, the smallest and the largest values of the attribute A.

4.2 Decimal Scaling Normalization

In this method, all the values of each attribute are converted to the complete fractional number by moving the decimal points of each value. And this decimal point movement is done based on the maximum value of each attribute.

$$a_{\text{norm}} = \frac{a}{10^d} \quad (5)$$

where d is the number of digits present in the integer part of the biggest number of each attribute A.

4.3 Z-Score Normalization

In this normalization method, all the values of each attribute A are scaled down to a common range of 0 and standard deviation of that attribute. The formulae are as follows:

$$a_{\text{norm}} = \frac{a - \mu(A)}{\delta(A)} \tag{6}$$

where $\mu(A)$ and $\delta(X)$ are, respectively, the mean value and the standard deviation of the attribute A.

4.4 Median Normalization

In this method, all the values of each attribute A is normalized by the following formulae:

$$a_{\text{norm}} = \frac{a}{\text{meadian}\,(A)} \tag{7}$$

4.5 Sigmoid Normalization

In this technique, the sigmoid function is used to normalize all the values of each attribute A. The formulae are as follows:

$$a_{\text{norm}} = \frac{1}{1 - e^{-a}} \tag{8}$$

4.6 Tanh Estimators

This method is developed by Hample. Here, the data normalization is done by the following formulae:

$$a_{\text{norm}} = 0.5\left[\tanh\left[\frac{0.01 * (a - \mu)}{\delta}\right] + 1\right] \tag{9}$$

where μ is the mean value of the attribute A and δ is the standard deviation of the attribute A.

5 Proposed Framework and Dataset Description

In this section, we have described the overall architecture of our proposed DNN model, named stacked Long Short-Term Memory (S-LSTM) model. Figure 2 shows the detailed architecture of our proposed model S-LSTM. The basic building blocks of the S-LSTM model is the LSTM unit. The main reason for the selection of the LSTM unit over RNN is that RNN suffers from the vanishing gradient and exploding gradient problem and due to these problems RNN is not capable to learn the features from the long sequences of the historical time-series data. On the contrary, the LSTM unit has the gated structure and due to this gated structure, it can extract the features from the long sequences of the historical data. The key part of the LSTM unit is its memory cell (cell state). This memory cell comprises three gates, viz., input gate, forget gate, and output gate. The basic gated structure of the LSTM unit is shown in Fig. 3 [9].

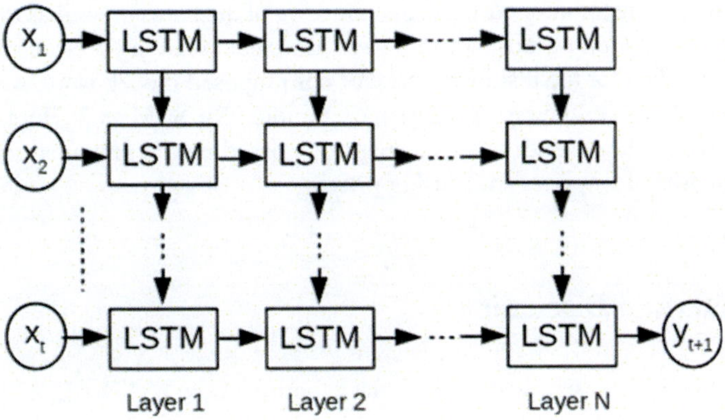

Fig. 2 Proposed forecasting model

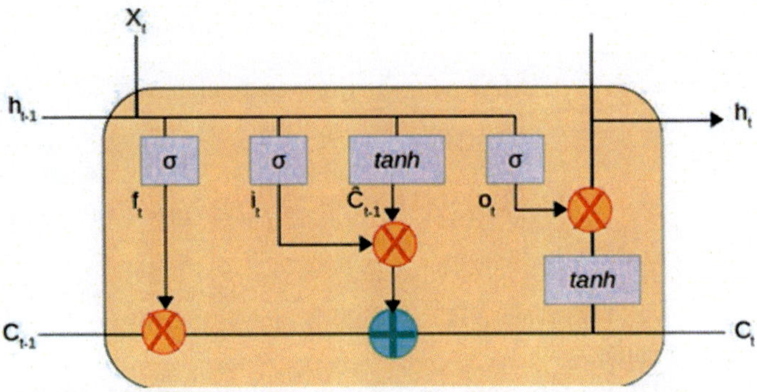

Fig. 3 Gated structure of LSTM

We develop our proposed S-LSTM model by stacking of N number of LSTM layers [13] as shown in Fig. 2 and each layer can be expanded to t number of LSTM units. Here the number of time stamps is represented by t. For example, if we want to forecast the fifth-day closing price of the stock market based on the previous 4 d data, then each LSTM layer must be expanded to the four numbers of LSTM units.

In this research work, we have taken the historical stock market data as an example of the multivariate time-series data. Here, we have used this stock market data to evaluate and analyze the performance of our proposed model S-LSTM and to test the effectiveness of each data normalization methods on our proposed model. Although the stock market data has a large number of parameters, in this work, we have only considered the four important parameters, viz., opening price, low price, high price, and closing price. We have taken the historical stock market data from two stock exchanges, viz., Bombay Stock Exchange (BSE) [17] and New York Stock Exchange (NYSE) [18]. Here, we have collected a total of 1717 d BSE data and a total of 1759 d NYSE data from January 1, 2012 to December 31, 2018. We have used the first 70% of the total data (1202 d BSE data and 1232 d NYSE data) as the labeled data for the training purpose. Second 15% of the data are used as the labeled data for the validation purpose and we have used the remaining 15% of the data as the unlabeled data for the testing purpose of our proposed model. Here, we have set the number of stacked layers of our proposed model S-LSTM as 3. Here, we have tried to forecast the seventh day's closing price based on the previous 6 d opening price, high price, low price, and closing price.

6 Results and Discussion

In this research work, we have done all the experiments by MATLAB R2016b with Neural Network Toolbox.

Here, as the performance metric, we have used the Mean Absolute Error (MAE) and Mean Squared Error (MSE). The formulae for calculating these errors are as follows:

$$\text{MAE} = \frac{1}{k}\sum_{i=1}^{k}(|o_i - p_i|) \tag{10}$$

$$\text{MSE} = \frac{1}{k}\sum_{i=1}^{k}(o_i - p_i)^2 \tag{11}$$

where the number of observation is k. o_i and p_i are, respectively, the actual value and the predicted value.

In Tables 1 and 2, we represent the different prediction errors (MSE and MAE) of the proposed model for each data normalization method of BSE and NYSE data, respectively. Figures 4 and 5 graphically show the foretasted closing price of BSE

Table 1 Forecasting errors of BSE dataset

Normalization method	MSE	MAE
Min-Max	3.1579e–05	0.0046
Decimal scaling	1.3143e–07	2.7651e–04
Z-Score	3.0571e–04	0.0161
Sigmoid	3.1234e–08	1.3581e–04
Tanh Estimator	**1.2439e–08**	**8.3422e–05**
Median	2.3169e–06	0.0017

Table 2 Forecasting errors of NYSE dataset

Normalization Method	MSE	MAE
Min-Max	2.1031e–05	0.0041
Decimal scaling	1.7521e–06	9.8705e–04
Z-Score	2.3471e–04	0.0215
Sigmoid	7.8731e–08	2.0161e–04
Tanh Estimator	**1.6359e–08**	**9.3741e–05**
Median	1.5125e–06	8.8129e–04

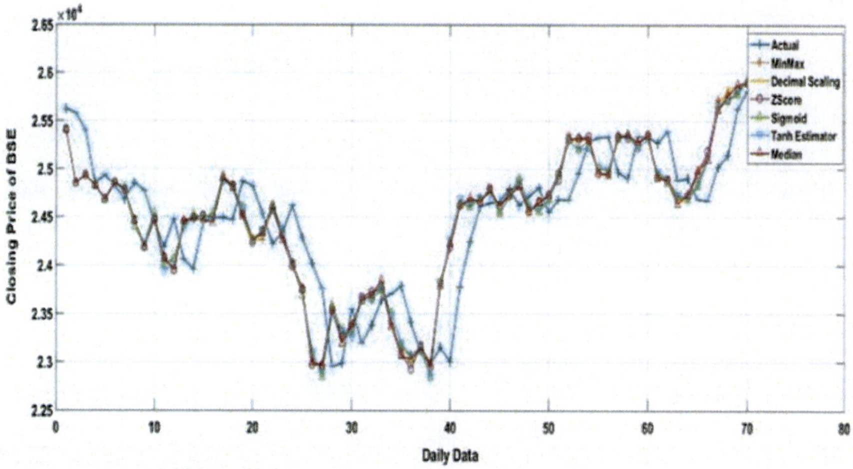

Fig. 4 Forecasting results of BSE

and NYSE for each data normalization technique. In Table 3, we have compared our proposed model S-LSTM with the other popular models concerning with their prediction errors (MSE and MAE).

From Tables 1 and 2, we can observe that the prediction errors are varied with the different normalization methods and the Tanh estimator produces lower prediction errors for both the prediction of BSE and NYSE indices compared to the other

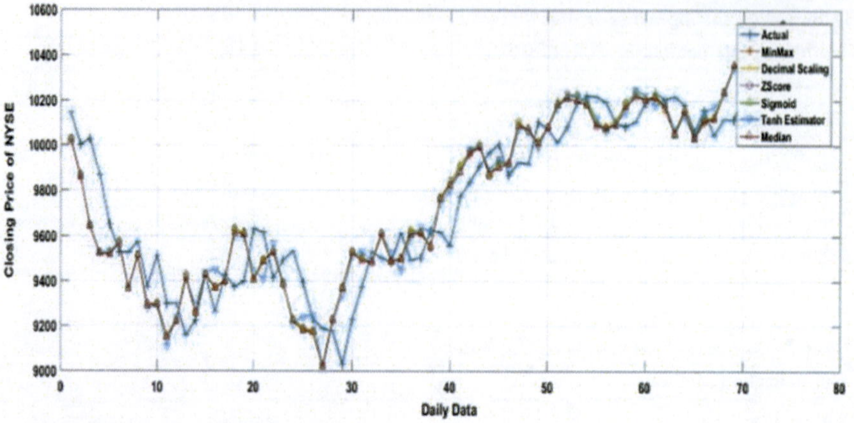

Fig. 5 Forecasting results of NYSE

Table 3 Forecasting errors of different models for BSE dataset

Model	MSE	MAE
SVM	3.9652e–02	6.3119e–1
ARIMA	2.6943e–05	9.1638e–3
S-LSTM	**1.2439e–08**	**8.3422e–05**
RNN	2.8139e–07	5.7621e–04
LSTM	1.0429e–07	3.0538e–04

normalization methods. Figures 4 and 5 also show that the Tanh estimator data normalization method produces better forecasting results. It is quite clear from Table 3 that our proposed model (S-LSTM) exhibits the smallest forecasting errors (MSE and MAE) compared to other well-known models.

7 Conclusion

In this work, we have proposed a deep neural network model S-LSTM for forecasting the multivariate time-series data. Moreover, we have also tried to find out the most suitable data normalization method for the deep neural network models. Here, as a case study, we have used BSE and NYSE historical time-series data for multivariate time-series forecasting purposes. From Tables 1 and 2 and also from Figs. 4 and 5, we can conclude that the Tanh estimator data normalization method is the best normalization method for deep neural network models. From all these observations, we can draw the conclusion that our proposed deep neural network model S-LSTM has outperformed all other well-known models for the forecasting of the BSE and NYSE data.

In the future, we also want to analyze our proposed model for the forecasting of different multivariate time-series data, such as weather, pollution, etc.

References

1. Meesad, P., Rasel, R.I.: Predicting stock market price using support vector regression. In: 2013 International Conference on Informatics, Electronics and Vision (ICIEV), pp. 1–6. IEEE (2013)
2. Rodriguez, G.: Time series forecasting in turning processes using arima model. Intell. Distrib. Comput. XII **798**, 157 (2018)
3. Sulaiman, J., Wahab, S.H.: Heavy rainfall forecasting model using artificial neural network for flood prone area. In: IT Convergence and Security 2017, pp. 68–76. Springer (2018)
4. Werbos, P.J., et al.: Backpropagation through time: what it does and how to do it. Proc. IEEE **78**(10), 1550–1560 (1990)
5. Lee, T.S., Chen, N.J.: Investigating the information content of non-cash-trading index futures using neural networks. Expert Syst. Appl. **22**(3), 225–234 (2002)
6. Schmidhuber, J.: Deep learning in neural networks: an overview. Neural Netw. **61**, 85–117 (2015)
7. Du, S., Li, T., Horng, S.J.: Time series forecasting using sequence-to-sequence deep learning framework. In: 2018 9th International Symposium on Parallel Architectures, Algorithms and Programming (PAAP), pp. 171–176. IEEE (2018)
8. Cirstea, R.G., Micu, D.V., Muresan, G.M., Guo, C., Yang, B.: Correlated time series forecasting using multi-task deep neural networks. In: Proceedings of the 27th ACM International Conference on Information and Knowledge Management, pp. 1527–1530. ACM (2018)
9. Bhanja, S., Das, A.: Deep learning-based integrated stacked model for the stock market prediction. Int. J. Eng. Adv. Technol. **9**(1), 5167–5174 (2019). October
10. Williams, R.J., Zipser, D.: A learning algorithm for continually running fully recurrent neural networks. Neural Comput. **1**(2), 270–280 (1989)
11. Shih, S.Y., Sun, F.K., Lee, H.Y.: Temporal pattern attention for multivariate time series forecasting. Mach. Learn. **108**(8–9), 1421–1441 (2019)
12. Bengio, Y., Simard, P., Frasconi, P., et al.: Learning long-term dependencies with gradient descent is difficult. IEEE Trans. Neural Netw. **5**(2), 157–166 (1994)
13. Sagheer, A., Kotb, M.: Time series forecasting of petroleum production using deep lstm recurrent networks. Neurocomputing **323**, 203–213 (2019)
14. Hsu, C.M.: Forecasting stock/futures prices by using neural networks with feature selection. In: 2011 6th IEEE Joint International Information Technology and Artificial Intelligence Conference, vol. 1, pp. 1–7. IEEE (2011)
15. Tang, Q., Gu, D.: Day-ahead electricity prices forecasting using artificial neural networks. In: 2009 International Conference on Artificial Intelligence and Computational Intelligence, vol. 2, pp. 511–514. IEEE (2009)
16. Nayak, S., Misra, B., Behera, H.: Impact of data normalization on stock index forecasting. Int. J. Comput. Inform. Syst. Ind. Manag. Appl. **6**(2014), 257–269 (2014)
17. Yahoo! finance (June 2019). https://in.finance.yahoo.com/quote/%5EBSESN/history?p=%5EBSESN
18. Yahoo! finance (June 2019). https://finance.yahoo.com/quote/%5ENYA/history/

Study on Information Diffusion in Online Social Network

Sutapa Bhattacharya and Dhrubasish Sarkar

Abstract Nowadays, Online Social Network (OSN) is very trendy in business, politics, and health care. This one can have the wider range of accessibility of information diffusion. The Online Social Network (OSN) is very significant as it provides interaction platform to the users across the globe. The impact joins both human attributes alongside community relationships. There are a group of individuals who has very strong connections to a range of social networks. These networks are capable of forwarding more information. So, it gives much better performance for single connection rather than multiple connections within a single network. Social influence plays a very important role in information diffusion. That's why information diffusion is the methodology where information transmits through certain target nodes over time among them. In this paper, some methodologies related to information diffusion, features, and limitations have been discussed. Delivering a detailed analysis, giving thoughtful social activities, and providing user's views are the main goals of this paper.

Keywords Information diffusion · Social influence · Online social network · Social network analysis

1 Introduction

In 1954, Barnes [1] has referred as "Social Networks" in the journal named as Human Relations. SixDegrees.com was the earliest Online Social Network (OSN) which was launched in 1997. After that, Facebook and Flickr in 2004, YouTube in 2005, Twitter in 2006, and Sina Micro-blog in 2009 came one by one.

S. Bhattacharya
Siliguri Institute of Technology, Siliguri, India
e-mail: sutapa2007@gmail.com

D. Sarkar (✉)
Amity University Kolkata, Kolkata, India
e-mail: dhrubasish@inbox.com

D. Bhattacharjee et al. (eds.), *Proceedings of International Conference on Frontiers in Computing and Systems*, Advances in Intelligent Systems and Computing 1255,
https://doi.org/10.1007/978-981-15-7834-2_26

Social Network Analysis (SNA) mainly emphasizes on the structure of the interactions among inhabitants, organizations, states, and many other's social entities. The techniques of social network analysis [2] are that in which way to represent and evaluate the relationships and flows among citizens, groups, organizations, animals, computers or other information or knowledge processing entities. A tie [2] joins a couple of users by single or multiple associations. Pairs might sustain a knot based under just single next of kin. Suppose the members of identical society where they share information, give financial supports, and attend meetings mutually.

Furthermore, ties vary in substance, route, and potency. As of now, information diffusion has participated in a main role in Online Social Networks (OSN). Ties play an important role in message propagation in information diffusion [3].Strong ties are individually influential for message transfer but weak ties are also significant for the dissemination of information in OSN. There is a nice explanation of information diffusion [3].

From social media, information have been extracted and stored. The user can be influenced by human's emotions, behavior, and relationship in same community or different communities. After that prediction can be needed for follower–following relationship through tweeter. This is the total scenario which has been seen in information diffusion process [4]. A real-life scenario has been noticed in this case also. When someone watches a movie and gives rating point of that movie, then his other neighbors or friends may be influenced and judge that movie in same logic and give the rating point same also. In fact, it will happen in some times but not always. So user actions are also main factors in these cases. Happiness, obesity, and yawning are all contagious.

The rest of the paper is organized as follows. In Sect. 2, some common terms and concepts have been discussed. In Sect. 3, information diffusion-related work which describes the features of Online Social Network (OSN) has been discussed. Last but not the least part contains conclusion and future scopes.

2 Common Terms and Concepts

Social networks are a set of actors knotted by one or more types of relationships. The representation of social networks has been done by graph theory concepts [2].

There are several concepts about the social structure of network and the arrangements of interactions among members have been described here. Some of the terms have been presented using diagrams in Fig. 1. There are common terms named as follows:

Network or graph: A representation of connections among a set of items, where items mean nodes or vertices and connections are edges or links. That means, a set of objects (nodes) with interconnections (edges).

Node: "Actor" or people on which relationships act. Nodes represent the entities of social network like users, pages, or groups.

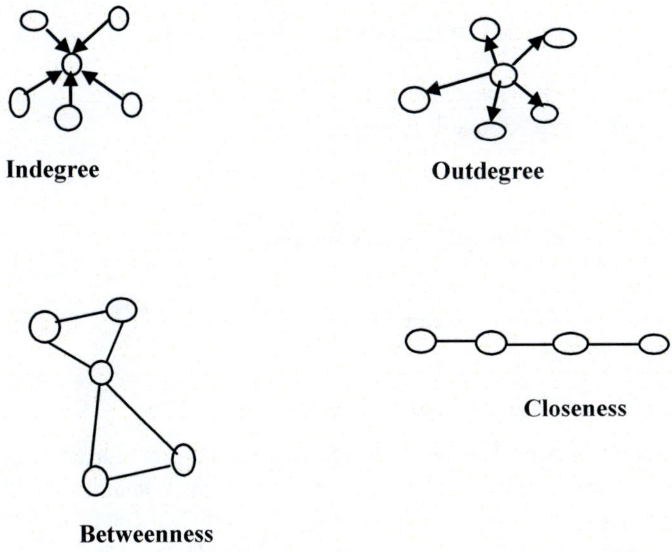

Fig. 1 Different properties of social networks [6]

Edge: Relationship connecting nodes which can be directional or un-directional.

Degree: A number of associations or edges that have to bond with other nodes [2].

Adjacent nodes: Any two nodes connected by an edge.

Community: Group of similar people.

Path: Path is a track where after traversing graph vertices or edges cannot be repeated.

Centrality (group or individual measure): Centrality [5] deals with the individual or group level. It is the way to find out the most central or important node within the graph. Four types of centrality have been seen.

Degree Centrality (undirected networks): It is a calculation of the total number of nearest neighbors [5]. If we consider a graph $G = (V, E)$ with V vertices and E number of edges, then degree centrality is

$$C_{Deg}(V) = \frac{d_y}{|N| - 1} \tag{1}$$

where N represents number of nodes in graph and d_y is the degree of node V.

Degree Centrality (directed networks): There are two different divisions named as in degree and out degree [5].

In degree: It is the calculation of sum of total ties incoming to a node, most influential. Suppose, a graph $G = (V, E)$ with N number of nodes, then in-degree centrality is

$$C_{\text{indeg}}(V) = \frac{d_{\text{yin}}}{|N| - 1} \tag{2}$$

where, d_{vin} is in degree of node V in graph.

Out degree: It is the sum of the total ties outgoing from a node, which determines which node disseminates information. The out-degree centrality can be determined for a graph $G = (V, E)$ having V vertices and E edges,

$$C_{\text{outdeg}}(V) = \frac{d_{\text{yout}}}{|N| - 1} \tag{3}$$

where d_{vout} is the out degree of node V in graph.

Betweenness Centrality: It is lying between each other sets of actors and quickly approachable. Considering a graph $G = (V, E)$, which has V number of vertices and E number of edges, then betweenness centrality for vertex V can be denoted by C_B (V), which could be computed for each pair of vertices (s, t) [5] as

$$C_B(V) = \frac{\sum \partial_{st}(V)}{\partial_{st}} \tag{4}$$

where, ∂_{st} denotes the number of shortest paths from s to t, and $\partial_{st}(V)$ is numeral of shortest or straight paths from s to t that pass through a vertex V [5].

Closeness Centrality: It helps to determine the closeness of a node to all other nodes in the system. The average length of shortest path, i.e., mean geodesic distance between one vertex V to all other vertices can be measured here [5]. For a graph $G = (V, E)$, the closeness centrality $Cc(V)$ can be calculated as

$$C_C(V) = \frac{\sum d_G(v, t)}{|N| - 1} \tag{5}$$

where $N >= 2$ is the dimension of the networks and N is the number of nodes.

In other words, the reciprocal of the sum of geodesic distance $d_G(v, t)$ to all other vertices of V, which means

$$C_C(V) = \frac{1}{\sum d_G(v, t)} t \in \frac{V}{v} \tag{6}$$

Eigenvector Centrality: It allocates comparative scores to all the nodes in the network. One norm is maintained here that links to higher score nodes which supplies additional scores where that node has equal connections but gives lower score nodes

[2]. The application of this technique is like Google's Page Rank which ranks web pages [5].

Cohesion (group measure): It deals with connectedness of each node couples at network level.

Density (group measure): Density measures structural strength of network.

$$\text{Density} = \frac{\text{The number of edges}}{\text{The total number of possible edges}} \tag{7}$$

Cut vertex or articulation point: If a vertex of graph G has been detached then graph G increases the number of components. That vertex is called cut vertex or articulation point.

Bridge or cut edge: If the removal of an edge between pair of vertices comes the graph G will be partitioned into two components or sub-graphs, then the edge is called cut edge or bridge of a graph G.

Connected: It is path or track from one vertex to any other vertex of a graph G.

Clustering coefficient: The nodes in the graph tend to be clustered together. Using clustering coefficient a degree to which nodes in a network are clustered to each other can be measured [2].

Local clustering coefficient: Fraction of pairs of the node's friends that are friends with each other.

The following diagrams [6] have shown the properties of social network.

Tie: The connectivity of a pair of actors based on single or numerous relationship [2] which always uphold a knot based. For example, employees of the similar organization keep up a multiplex tie, depending on many relations.

Weak tie: Weak ties are rarely maintained, non-intimate relationships, e.g., between colleagues who do not share common tasks or friendship.

Strong tie: Strong ties consist of emotional intensity; self-disclosure; intimacy; mutual understanding; kinship; and regular contact as between classmates, friends, or colleagues.

Triadic closure: The tendency for people who share connections in a social network to become connected.

Homophile: It is the characteristics of actors who have similar features on the basis of status and beliefs. Homophile ties can be strong or weak.

3 Related Work

In social network analysis, every person is capable of influencing the power of thoughts or actions of others. Influential node and user choice identification plays a major role in information diffusion. In Twitter, we can identify some influential users those have seen in sociocentric (which includes whole network) and egocentric network (for individual and all other connected persons). A model [7] was designed by such a way to recognize influential nodes for a user in egocentric online social networks that detects the movement or behaviors of each and every person. With the help of using sentiment analysis and hashtag term analysis, investigate the personality or qualities of both the user and the influencers.

In paper [8], the modified k-shell decomposition algorithm explained the process of computing user influence on Twitter. It clearly explains two things that K-shell decomposition algorithm is rectified by assigning logarithmic values to users and it makes as a surprisingly well-distributed bell curve to spot and takes away peering interaction from the network to other distinguished users [8]. The authors in [9] mention the broad range of concepts to explain when people interacts to each other then their each action is also affected by others people. So, social influence is a usual phenomenon, but it should be also noted that for applying it person's mindset and attitude [10] can be changed. Current social media has shown how contents which were shared by many people has been affected by each and every life's positivity and negativity. Social influence worked in several types of fields such as review on marketing and business, politics, and leadership. In a nutshell behind a person's every positive and negative actions social influence plays a huge role.

Twitter with geo-tagging system [11] can be very helpful to identify user location such as home and work places. Now the question remains why these two information are so important? Answer is simple. A person spends more time in these two locations only according to user's routine.

Now let us describe the novel page rank-based approach [12] which is clearly based on study, analysis, and understanding of user's preference, that is, what one likes, what one dislikes, what one hides, and, of course, what one shares often in Facebook. The second interesting phenomena can be described as follower–following relation. Now if the follower can be described with one node in an algorithm then the actual figure of the neighbor nodes [13] to categorize nodes in bunch center can be calculated and then among same four-layer neighbors, the coefficient of local centrality based on decreasing function for the local clustering coefficient of nodes can be determined [14].

After discussion about the page rank approach, now there is its extension version also. As per extension version, the similarity in between user and structured link [15] has a significant impact with homophiles classification. It is clear that people with similar interest most often generated the following–follower relation. Now the primary target for this approach would be to find out the seed nodes. It is quite clear that other researchers will significantly consider three explicit aspects of nodes [4] such as degree centrality, closeness centrality, and betweenness centrality.

Apart from all the abovementioned nodes there is also an implicit one and certainly that is not degree-centrality nodes. In paper [16], the authors elaborate that how people's herd and collective behavior adoption in information diffusion with different timestamps have been changed normally. Now the entire phenomenon is important to understand which node will actually influence the social network. This is also important to understand which will maximize the influence [17] among advertising and stake holder influence. Competitive influence has a significant impact with advertising information. The result can be positive or negative influence. In fact, there is a way to classify these positive and negative analogies to determine the current situation on users, emotional expression, and deep sentiment [17].

Predictive and explanatory models are used in information diffusion [4]. Predictive models can be classified as game theory model, linear threshold model, and independent cascade model. Explanatory models can be categorized as epidemic models and influence models. So, it is very important to understand how the model of information diffusion process works in the social networks. The authors in [18] describe the way to identify influential spreaders in various approaches. Predictive models can be graph based or non-graph based. Explanatory models may be static or dynamic network [18]. A generic model [19] named as T-Basic explains the estimation of its parameters from user's behaviors through machine learning techniques. This is main issue of predicting the temporal dynamics of the diffusion process.

There is a short description among all techniques in Table 1.

4 Conclusion and Future Scopes

The importance of information diffusion of social network is to extract information from user. Information present in users' news feeds and others sharing behavior on facebook.com are observed. External events can be followed up in Facebook and may enlighten their sharing behavior.

Two major conscripts build a sequential bridge between the extraction and the influence research which affects information diffusion. So, in detail, extraction of information leads to description as well as affects factor analysis which is further narrowed down to information diffusion researchers. It helps us to understand the prediction of user's behavior. But as the research progresses the limitations that emerge is briefly categorized as follows: study based on sentiment or emotion and predictions based on group status combination which limits to identify individuals' personality and psychometric needs. There are two scenarios which have been seen in information diffusion process [20]—contagion-based diffusion (can be influenced by its neighbors) and homophile-based diffusion (taken by features of nodes present in social network).

In conclusion, in spite of having so many problems in information diffusion, the information of weak nodes have been collected as a user-centric and shared goals in recommendation system in future. The recent approaches of recommendation system can be incorporated in information diffusion model to achieve the better result.

Table 1 Models of information diffusion in OSN

Model	Techniques/Methods
Modified K-shell decomposition [8]	The basic algorithm produces highly skewed k-shell levels and takes n times for execution. But this method gives the nodes which prepares statistical observations those are hard. So this algorithm can be modified with the help of a logarithmic mapping. Here each k-shell level represents approximately the log value algorithm only needed to iterate $\log_2 n$ times
Finding key locations with the help of Geo-tagged Twitter Activity [11]	Firstly, take out the details from individual user-profile data entry for home location. Secondly, compare the outcomes with openly presented socio-economic data from Twitter dataset. These two methods are for detecting home. Now for the workplace locations, it can be identified both from LinkedIn and Twitter data
Novel Page Rank-based approach [12]	The common centrality deals with activity graph. It has also combined concepts from research on people's connectivity and users' contact activity that basically incorporates communication movement, the strength of user's links. In a nutshell, Page Rank-based approach has been used to recognize key users
Evaluation of social influence analysis using microscopic and macroscopic models [16]	Microscopic models are used for human communication purpose and macroscopic models consider unique spread probability and identical influential power for all nodes
Micro- and macro-models in information diffusion [17]	According to user's online and offline modes, discrete-time bi-probability-independent cascade model has been used
A generic T-Basic model [19]	The model can be calculated from user's behavior parameters which can be taken from the real Twitter datasets
Friendship Paradox-based models and statistical interference [20]	Find out states of network (statistical inference or dynamic model), interactions among friends during message transfer in information diffusion
Two heterogeneous nonlinear models used for cascading trees [21]	WeChat application can be used for providing information which can be forwarded or shared by users
Competitive diffusion model [22]	For analyzing the stability of the diffusion infection-free equilibrium and learning the numerical simulations
Two novel information diffusion models using independent cascade model [23]	For extracting influential nodes, ranking the sets of influential nodes

(continued)

Table 1 (continued)

Model	Techniques/Methods
Content similarity model [24]	Using learning-based framework and binary classification problem to detect active edges in diffusion of message in network

Reference

1. Barnes, J.A.: Class and committees in a Norwegian island parish, human relations. Hum. Relat., 39–58 (1954)
2. https://www.archiv.politaktiv.org/documents/10157/29141/SocNet_TheoryApp.pdf. Accessed on 30 November 2019
3. Bakshy, E., Rosenn. I.: The role of social networks in information diffusion. In: IW3C2 (2012)
4. Li, M., Wang, X., Gao, K., Zhang, S.: A survey on information diffusion in online social networks: models and methods**8**(4), 1–21 (2017)
5. https://www2.unb.ca/~ddu/6634/Lecture_notes/Lecture_4_centrality_measure.pdf. Accessed on 30 November 2019
6. Sarkar, D., Kole, D. K., Jana, P.: Survey of influential nodes identification in online social networks. Int. J. Virtual Commun. Soc. Netw. **8**(4), 57–67 (2016)
7. Sarkar, D., Debnath, S., Kole, D.K., Jana, P.: Influential nodes identification based on activity behaviors and network structure with personality analysis in egocentric online social networks. Int. J. Ambient Comput. Intell. **10**(4), 1–24 (2019)
8. Brown, P.E., Feng, J.: Measuring user influence on twitter using modified K-shell decomposition. In: Association for the Advancement of Artificial Intelligence, pp. 18–23 (2011)
9. Hillmann, R., Trier, M.: Influence and dissemination of sentiments in social network communication patterns. In: Proceedings of the 21st European Conference on Information Systems (2013)
10. Snijders, R., Helms, R.W.: Analyzing social influence through social media, a structured literature review. In: 7th IADIS International Conference on Information Systems, Spain (2014)
11. Efstathiades, H., Antoniadis, D., Pallis, G., Dikaiakos, M.D.: Identification of key locations based on online social network activity. In: IEEE/ACM International Conference on Advances in Social Networks Analysis and Mining, Cyprus, pp. 218–225 (2015)
12. Heidemann, J., Klier, M., Probst, F.: Identifying key users in online social networks: a pagerank based approach. Completed Research Paper, pp. 1–21 (2010)
13. Zhao, X., Liu, F., Wang, J., Li, T.: Evaluating influential nodes in social networks by local centrality with a coefficient. Int. J. Geo-Inf. **6**(2), 1–11 (2017)
14. Razis, G., Anagnostopoulos, I., Zeadally, S.: Modeling Influence with Semantics in Social Networks: A Survey, pp. 1–61 (2018)
15. Sarkar, D., Roy, S., Giri, C., Kole, D.K.: A statistical model to determine the behavior adoption in different timestamps on online social network. Int. J. Knowl. Syst. Sci. **10**(4), 1–17 (2019)
16. Althoff, T., Jindal, P., Leskovec, J.: Online actions with offline impact: how online social networks influence online and offline user behavior. In: Proceedings of 10th ACM International Conference on Web Search and Data Mining (2017)
17. Li, K., Zhang, L., Huang, H.: Social influence analysis: models, methods, and evaluation **4**(1), 40–46 (2018)
18. Guille, A., Hacid, H., Favre, C., Zighed, D.A.: Information diffusion in online social networks: a survey. SIGMOD Rec. **42**(2), 17–28 (2013)
19. Guille, A., Hacid, H., Favre, C.: Predicting the temporal dynamics of information diffusion in social networks (2013)
20. Krishnamurthy, V., Nettasinghe, B.: Information diffusion in social networks: friendship paradox based models and statistical inference, pp. 1–37 (2018)

21. Liu, L.,BinChen, B., Hanjalic, A., Wang, H.: Modelling of information diffusion on social networks with applications to WeChat, vol. 496, pp. 318–329 (2018)
22. Sun, Q., Li, Y., Hu, H., Cheng, S.: A model for competing information diffusion in social networks. IEEE Access (2019)
23. Kimura, M., Saito, K.: Tractable Models for Information Diffusion in Social Networks, pp. 259–271. Springer (2006)
24. Varshney, D., Kumar, S., Gupta, V.: Modeling information diffusion in social networks using latent topic information. In: International Conference on Intelligent Computing, Switzerland, pp. 137–148 (2014)

A Multi-layer Content Filtration of Textual Data for Periodic Report Generation in Post-disaster Scenario

Sudakshina Dasgupta, Indrajit Bhattacharya, and Tamal Mondal

Abstract Data filtration has been considered as a research area in various real-time applications like packet routing, data stream processing, etc. from various heterogeneous sources. Due to huge overhead of information in these real-time applications, filtration approaches have been designed and considered as significant feature that needs to be incorporated. In any disaster response situation, the volunteers (doctors, nurses, police, army, etc.) of various government or non-government organizations serve in affected regions in order to reduce number of fatalities. Besides, local group of people might also engage themselves in relief works along with the volunteers. The volunteers, local groups, and victims of affected regions are capable of generating and exchanging/transferring situational data regarding resource requirements, number of fatalities, etc. through their smartphones. The generated data are then dumped into Information Drop-boxes (IDBs) situated at various Shelter Points (SPs) whenever mobile nodes come in contact with them. As a result, large volume of situational data might be accumulated at IDBs. This information overhead might lead to two distinct issues that need to be addressed: a) data redundancy and b) data inconsistency. In order to deal with such issues, it is not only the unique objective to eliminate redundancies, noise, etc. from messages, but also the information format must be made comprehensible to the policy-makers. In the proposed work, such an approach has been adopted by designing a layered filtration approach in order to refine various contents of transferred messages, i.e., location, message body, and frequently changing information (number of fatalities, etc.). The refined information can be reorganized date-time-wise by constructing one or more sentences with an objective to produce periodic reports. From evaluation perspective, it has been

S. Dasgupta (✉)
Government College of Engineering and Textile Technology, Serampore, West Bengal, India
e-mail: sudakshinadasgupta@yahoo.com

I. Bhattacharya · T. Mondal
Kalyani Government Engineering College, Kalyani, Nadia 741235, West Bengal, India
e-mail: indra51276@gmail.com

T. Mondal
e-mail: tamalkalyanigov@gmail.com

D. Bhattacharjee et al. (eds.), *Proceedings of International Conference on Frontiers in Computing and Systems*, Advances in Intelligent Systems and Computing 1255,
https://doi.org/10.1007/978-981-15-7834-2_27

observed that the proposed approach is much effective and can be incorporated as a feature for IDBs.

Keywords Filtration · Data · Location · Content · Numeric

1 Background

In any post-disaster situation, various government and non-government organizations work across the affected regions in order to reduce number of fatalities. In past studies [1, 2], it has been observed that existing communication infrastructures like GSM, Internet, etc. damaged completely across all affected regions. In order to deal with communication challenges, hybrid opportunistic network infrastructure has been proposed [3, 4]. Various mobile sensor nodes generate situational information using application [5] under DTN [6] connectivity. On contact, the mobile nodes exchange their summery vectors of information with other nodes. Information Dropboxes (IDBs) situated at various permanent Shelter Points (SPs) (*like government buildings, schools,* etc.) that act as data repository from various mobile nodes whenever any of them comes in contact. A variety of data mules (cars, UAVs, etc.) act as data collector which collect data from various IDBs situated at different SPs using any predefined path planning approach [7]. Finally, the collected information is received by master control station (MCS) situated at any urban region, where the real-time policies to serve various affected regions are carried out. Any disaster-affected region primarily comprises three distinct types of people carrying smartphones (called as mobile nodes). These nodes are scattered across affected locations for their distinct or equivalent purposes. The categorizations of these nodes are *Volunteers, Local People, and Victims.* Each of these nodes is capable of generating data, transferring situational data to any other mobile node on contact [3] through their smartphones, and storing data into their buffers. It may be noted that the mobile nodes can provide real-time localized views across different locations of any affected region. Whenever any mobile node reaches within the vicinity of nearest IDB, the stored messages are seamlessly transferred to that IDB. Therefore, IDBs can provide a global view of the situation and damage of an affected region. However, at any point of time during disaster response situation, large volume of situational data might be generated that are being transferred by various mobile nodes. Large data volume at any time instance might affect two main parameters that need to be addressed: *(a) The buffers at IDBs are limited. As a result, important messages might be unnecessarily discarded due to insufficient storage space. (b) Data inconsistencies might exist due to noise and redundancies in stored data.* In order to address such challenges, real-time refinement of situational data might require discarding duplicate or unwanted data from IDBs. Information should be filtered in such a way that (a) the buffer space can be utilized efficiently and (b) significant information can effectively be preserved at any time stamp. Furthermore, the information should be stored in structured format in order to generate periodic reports about various facts (*number of fatalities, resource requirement,* etc.) in post-disaster situation. In a disaster response situation, large volume of situational data are generated by the mobile nodes in real time. In order to process those real-time data, the IDBs should be featured in such a way that, at any time stamp only important and significant messages are transferred to various data mules (*cars, boats, UAVs,* etc.) in a summarized form within a very short time interval, as contact time between IDBs and data mules is limited [7]. On the other hand, the buffer spaces of IDBs should be effectively utilized to accumulate periodic situational messages.

2 Contribution

In this section, the contribution of the proposed work has been introduced. It has been depicted in Sect. 1 that huge volume of situational data might affect the performance of the system with respect to different evaluation parameters like *buffer management and data consistency.* Due to those issues, the concept of three layer filtration technique has been designed for the IDBs. It has been considered that, any generated textual message primarily contains three parts as follows: *Location:* latitude and longitude of generated situational text data, *Content:* disaster- and non-disaster-related content words that construct the body of the message, and *Numeric Digits:* numeric part in the message that describes number of fatalities (dead, injured, etc.). Apart from such information, generated text messages must contain *Date and Time* that depicts the date and time of message creation. In order to preserve such consideration, the generated messages have been parsed in the format as shown in Fig. 1. The primary task would be to convert messages originating from heterogeneous sources to a specified format as described in Fig. 1. The mobile nodes periodically generate textual messages. The messages are stored in the buffer of the respective nodes. Whenever the mobile nodes come in contact with any of the nearest IDBs situated at various SPs, the messages stored in buffer (summary vectors) are seamlessly transferred to the IDBs. At IDBs, the messages are parsed conforming the specified format. Then *Location clustering* is performed on text data in order to classify the GPS trails of generated messages. However, it might be possible that within a cluster, same type of messages might be circulated number of times for a particular date–time combination. Diffusion of such type of information should be eliminated in order to reduce inconsistency. Therefore, *content filtration* has been performed for each cluster of messages. Within each cluster it might be possible that messages contain information that changes frequently with time. As an example, messages describe fatality-related information (*dead, damaged, injured,* etc.). The variation in the messages that report different numbers associated with fatalities often leads

Date and Time	GPS Location	Body of Message

Fig. 1 Generated message format as per proposed work

to inconsistency. Such type of unpredictability needs to be addressed in order to assess number of fatalities time to time. The aftermath of each layer of proposed filtration technique obtains (a) various clusters containing number of GPS locations of generated messages. It gives a proper real-time visualization of various affected regions at IDBs, (b) various content words/phrases along with cardinal digits that can be reorganized in order to construct group of meaningful sentences from filtered messages for a particular date–time combination. The procedure will be repeated time to time for each cluster, and (c) reports related to fatalities can be generated for each cluster through which number of casualties can be predicted for any particular time window. The rest of the paper has been organized as follows.

Various developed modules of proposed approach, i.e., *Location Clustering, Content Filtration, and Numeric Filtration* have been rigorously described in Sect. 3. In Sect. 4, the performance of the modules has been analyzed for datasets obtained from heterogeneous sources (disaster mock drills carried at past in various places of West Bengal, India, and OSNs). Finally, Sect. 5 concludes the paper.

3 Proposed Filtration Approach

In this section, the proposed data filtration approach has been discussed in greater details. It has already been discussed in Sect. 1, IDBs deployed at different SPs can play a major role in post-disaster situation analysis. Through periodic collection of situational crowd sourced data, it assists the policy-makers by providing a proper visualization of various events. Thus, it is much more necessary to incorporate or improve data processing capabilities of IDBs with more perfection. These kinds of improvements not only affect buffer space but also preserve information consistency. In proposed work, such features have been adopted especially for IDBs in terms of filtration. Through deliberate processing of message contents, the proposed approach not only reduces information overhead but also wipes out inconsistencies. Note, in proposed technique, it has been assumed that messages can be generated from heterogeneous sources that have different data format. Therefore, messages coming from various sources to different IDBs must be parsed to the required format depicted in Fig. 1. For effective application of proposed filtration approach, the structural content of messages must be preserved to a common format. Though from previous studies [4, 5], it can be observed that IDBs situated at SPs act as data repository. But, due to incorporation of additional features such as *filtration* at IDBs, common information format must be maintained. Therefore, in proposed work, the incoming messages are parsed to preserve the message format as shown in Fig. 1. After successful parsing, the messages are preserved to a defined format as shown in Fig. 1.

3.1 Location Clustering of Messages

Clustering refers to the categorization/grouping of similar types of objects [8]. In proposed work, we simply overlooked the detailed analysis of various types of clustering approaches. Rather we were much interested on such clustering type that is mostly suitable for our problem. In post-situation, a large number of volunteers of various distinct organizations are working across affected regions. The volunteers can generate situational messages through their smartphones. Most of the time, it might be possible that messages have been generated within a small precision of GPS locations. As a result, it is more likely to cluster those messages as they might contain similar type of situational information regarding *casualties, resource requirements,* etc. Therefore, irrespective of processing each message manually, visualization of message clusters might improve the quality of service in identification of event location. In proposed work, one of the partition-based clustering approach has been used, i.e., k-means [9] for classifying the messages based on their generated locations. Note, in disaster response situation real-time data processing took place. The data processing time will be limited in IDBs. Data must be transferred to MCS through mules as early as possible in order to define real-time policies. In such dynamic scenarios, the performance parameter of clustering approach should be measured in terms of *execution time* and *simplicity*. Therefore, for clustering of messages for proposed technique we are simply relaxing the parameter *accuracy* which might be considered as important parameter for other type of problems. Now, the question lies on the optimal value of k, i.e., *how many clusters should be formed from the set of GPS locations?* From literature, it has been observed that for evaluation of cluster numbers various popular methods are available. In proposed work, *Silhouette* technique has been adopted in order to evaluate the optimal value k. This technique measures the quality of clustering, i.e., belongingness of each object with its cluster.

3.2 Content Filtration

In previous subsection, it has been clearly discussed how the generated messages are clustered based on their GPS trails. Now, the content filtration has been performed for each cluster of messages. In disaster response scenarios, generated situational text messages from heterogeneous sources mainly contain *disaster-related content words* (recue, dead, damaged, etc.) along with some *non-disaster content words* (building, bridge, boats, etc.). In some past studies, it has also been observed that situational messages with limited set of content words circulate for a particular time window. Hence, within a cluster there might be possibility that those situational messages containing disaster-related content words are repeated for a particular date–time combination. Furthermore, in disaster response situation, volume of situational message generated is also large. Therefore, within minute IDBs may encounter with numerous such messages. Through effective clustering of generated messages, in proposed work, it has been assumed that there might be some uniformity in messages on the events they are based on. That is, within a cluster, messages which have been encountered within a shorter time interval (Say 1 min) might have a likeness about any event. Hence, there must be some mechanism through which such repetitions can be eliminated. In proposed work, such a technique has been adopted. First of all, in each cluster for a time window of a particular date, all the messages are merged. In proposed solution, the time window taken is as 1 minute. Suppose there are six messages generated in every 5 s from 09:01:00 to 09:01:30 on 2015-04-10. Then the content of the six messages will be merged and the time format will be 09:01 and this will be stored as a single message according to the format described in Fig. 1. Here some redundancies are eliminated by storing single occurrence of repetitive date, time, and GPS location of centroid. Now, for each date–time combination, the words listed in NLTK STOPWORDS for ENGLISH language and punctuations were removed. Note, it has been considered that removal of stop words has no impact on disaster-related information. Thus, it can be removed to make the content part more informational. A corpus list L of disaster-related content words has been prepared from www.dictionary.com1 that contains all types of disaster-specific content words. Now, for each date–time combination within a cluster, (a) each content word or its synsets within a message group will be checked with L to verify whether it is the list of disaster-related content words or not, (b) number of occurrences of that content word or its synsets within the message group are stored in vector *V*, and (c) the message group is updated by eliminating multiple occurrences of those content words. After content filtration, the message groups are obtained for distinct date–time combination. Now, real-time reports can be generated through effective arrangement of words or phrases as discussed in [10].

3.3 Numeric Filtration

As it has been mentioned in Sect. 1, it might be possible that within a very short time interval reported number of dead or injured person changes frequently. Within any cluster, it might be possible that numeric values related to fatalities change time to time. This type of situation might create confusion about actual number of fatalities within a period of time. In proposed work, such scenarios have also been taken care of by using *exponential smoothing* approach. Using a set of numeric values related to fatalities, we want to predict the actual number which can be placed with keyword related to fatalities within the message group. In fact in past, the authors have shown that during Hyderabad Blast, reported number of victims or injured persons changed during the period of only 7 minutes. Thus, such type of information must also be effectively taken care of. In proposed work, we consider disaster-specific content words like "dead," "kill," "injured," etc. or their synsets in each cluster of messages along with numeric content attached to them. If such content words along with numeric digits are found, then the words are segregated with numeric digits and date–time based on their type (i.e., whether it related to dead or injury). But the major challenge lies in the extraction of appropriate numeral associated with fatality-related keyword. It might not be the case all the time that numerals are directly associated with the keywords. In order to deal with such challenge, in proposed work, the following algorithm has been proposed.

Algorithm 2: Extracting casualty numbers with keywords from messages

1. Create a list of casualty related content words from various authorized disaster related corpuses.
2. **for** each message do
    Find and convert (if any) numerical words to numerical digits. e.g. Three becomes 3
    Tokenize the content of each message using nltk_word_tokenize
    for each item X from the list do
        if(X in word_tokenize_list)
        L1: Find the location of occurrence of X. Search for numeric digits from prefix and suffix of the occurrence of X upto three places i.e. search between previous three words and next three words. If any numeric values found, extract it with X along with Date/ time.
            If there are none found repeat with next value of X
        if (X not in word_tokenize_list)
            Check whether any of the synonym of X is present.
            if (found) then goto L1
    end
end
3. Stop

Now, after the segregation of fatality-related content words along with their date–time and numeric values using Algorithm 4, the next task is to predict appropriate numeral values which will be associated with these content words for that particular time window. That is, we need to know, which value to consider. Moreover, the input values may contain noise, due to which the results might be faulty. In order to solve this issue, we predict the actual value of numerals associated with content words using exponential smoothing technique.

4 Results

It can be observed from Fig. 2, based on the GPS locations of generated messages for three datasets, that the number of cluster requirements has been evaluated. Based on the distribution of GPS locations it has been observed that for kandi mock drill the number of cluster requirement has been evaluated are 4. Whereas for Nepal

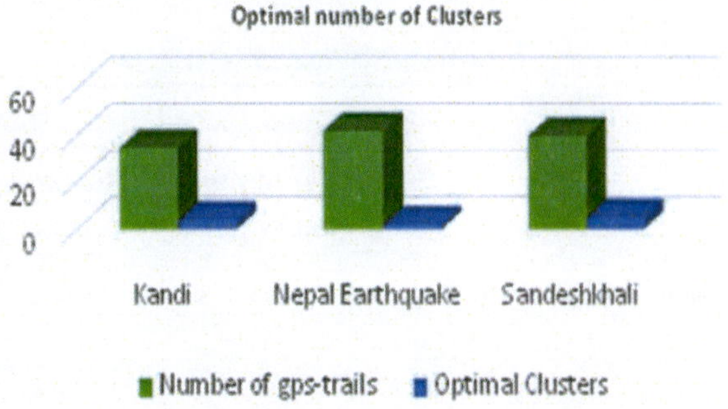

Fig. 2 Optimal number of clusters for three datasets after calculating silhouette score

earthquake and sandeshkhali mock drill, the numbers are evaluated as 2 and 5, respectively. After successful evaluation of silhouette score, the incoming messages have been clustered using k-means clustering approach as discussed in Sect. 3 (see Fig. 3). The diamond shapes in the clusters are the cluster centroid. After successful clustering of three test data available, content filtration approach has been carried out as shown in Fig. 4. Note, here the evaluation has been carried out in terms of reduction of phrases from messages. It can be observed from Fig. 4, a total of 7800 content words were in Nepal earthquake test data. For sandeshkhali and kandi data, the incoming messages contain 3051 and 3466 content words, respectively. However, after performing content filtration a total of 4442 content words remained in two clusters for Nepal earthquake, 1813 content words remain in four clusters for sandeshkhali mock drill, and 1987 content words remain in five clusters for kandi mock drill. It can be observed from Fig. 5 that messages containing fatality-related keywords are extracted along with numeric values associated with them using Algorithm 2. After that, same type of keywords or their synsets are segregated along with their numeric values. Then through exponential smoothing, values are predicted for each type of fatality-related keyword.

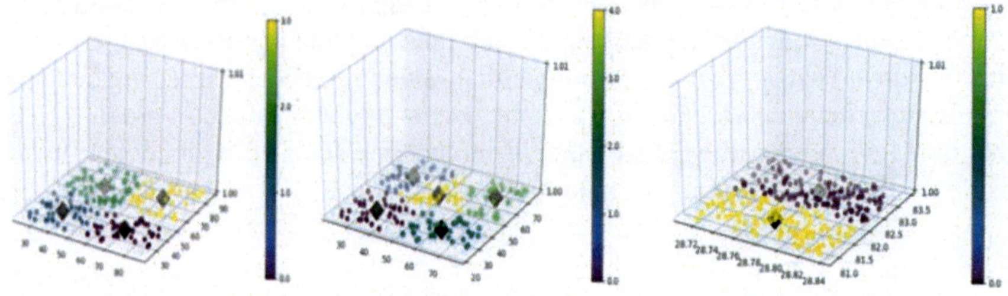

(a) Kandi Data (b) Sandeshkhali Data (c) Nepal Earthquake

Fig. 3 3d Scatterplot of message clustering based on GPS locations

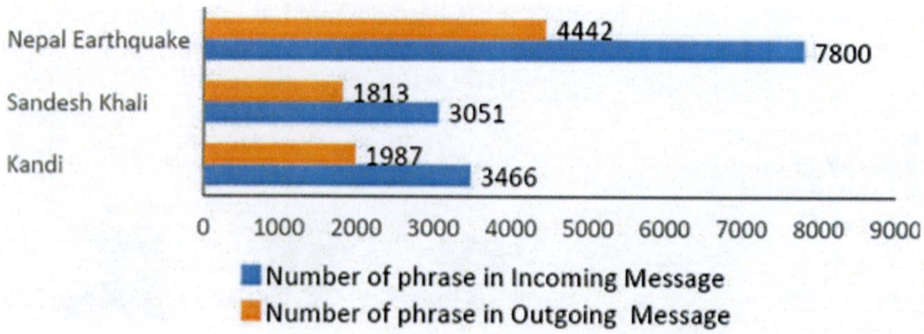

Fig. 4 Total phrase traffic comparisons in incoming and outgoing messages

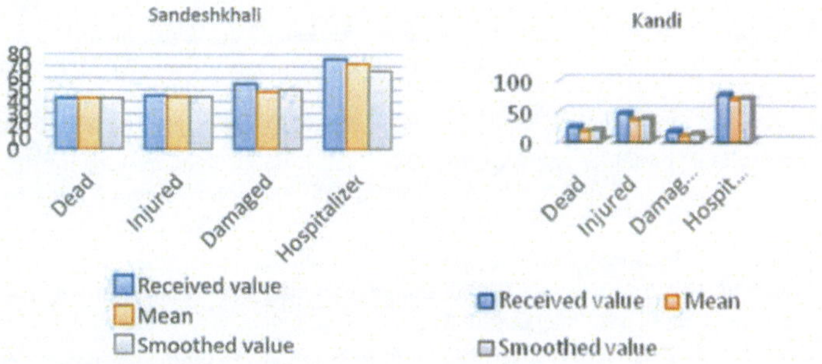

Fig. 5 Bar graph representation of original, mean, and smoothed values

5 Discussion

The proposed work mainly focuses on developing a layered filtration approach through which overhead of incoming messages can be minimized and buffer space can be preserved for IDBs. Various attributes of the generated messages have been addressed in order to wipe out redundancies and inconsistencies. By implementing each distinct module of filtration approach in IDBs, it is not only the fact that information overhead can be minimized but also the policy-makers can obtain various different snapshots of post-disaster situation. Through incorporation of *Location Clustering,* a real-time global snapshot of various affected regions can be obtained where the affected regions can be marked as clusters based on the generation of messages. On the other hand, by incorporating *Content Filtration* approach, messages are grouped and multiple occurrences of identical disaster-related content words are wiped out for every distinct date–time combination within every cluster. Such type of arrangement leads the policy-makers to apply sentence construction technique that might be used to generate summary reports. Through *Numeric Filtration,* numbers of fatalities are predicted in real time for every cluster. Such prediction also assists the policy-makers to obtain number fatalities in any post-disaster situation that often helps in generating fatality reports time to time.

References

1. Uddin, M.Y.S., Nicol, D.M., Abdelzaher, T.F., Kravets, R.H.: A post-disaster mobility model for delay tolerant networking. In: Winter Simulation Conference, pp. 2785–2796 (2009)
2. Gupta, A.K., Mandal, J.K., Bhattacharya, I., Mondal, T., Shaw, S.S.: CTMR-collaborative time-stamp based multicast routing for delay tolerant networks in post disaster scenario. Peer-to-Peer Netw. Appl. **11**(1), 162–180 (2018)
3. Saha, S., Shah, V.K., Verma, R., Mandal, R., Nandi, S.: Is it worth taking a planned approach to design ad hoc infrastructure for post disaster communication? In: Proceedings of the Seventh ACM International Workshop on Challenged Networks, pp. 87–90. ACM (2012)

4. Saha, S., Nandi, S., Paul, P.S., Shah, V.K., Roy, A., Das, S.K.: Designing delay constrained hybrid ad hoc network infrastructure for post-disaster communication. Ad Hoc Netw. **25**, 406–429 (2015)
5. Paul, P.S., Ghosh, B.C., De, K., Saha, S., Nandi, S., Saha, S., Bhattacharya, I., Chakraborty, S.: On design and implementation of a scalable and reliable Sync system for delay tolerant challenged networks. In: 2016 8th International Conference on Communication Systems and Networks (COMSNETS), pp. 1–8. IEEE (2016)
6. Fall, K.: A delay-tolerant network architecture for challenged internets. In: Proceedings of the 2003 Conference on Applications, Technologies, Architectures, and Protocols for Computer Communications, pp. 27–34. ACM (2003)
7. Mondal, T., Roy, J., Bhattacharya, I., Chakraborty, S., Saha, A., Saha, S.: Smart navigation and dynamic path planning of a micro-jet in a post disaster scenario. In: Proceedings of the Second ACM SIGSPATIAL International Workshop on the Use of GIS in Emergency Management, p. 14. ACM (2016)
8. Berkhin, P.: A survey of clustering data mining techniques. In: Grouping Multidimensional Data, pp. 25–71. Springer, Berlin, Heidelberg (2006)
9. Hartigan, J.A., Wong, M.A.: Algorithm AS 136: a k-means clustering algorithm. J. R. Stat. Soc. Ser. C (Applied Statistics) **28**(1), 100–108 (1979)
10. Mondal, T., Pramanik, P., Bhattacharya, I., Saha, A., Boral, N.: Towards development of FOPL based tweet summarization technique in a post disaster scenario: from survey to solution. In: 2017 51st Annual Conference on Information Sciences and Systems (CISS), pp. 1–6. IEEE (2017)

Categorization of Videos Based on Text Using Multinomial Naïve Bayes Classifier

Arghyadip Sinha and Jayjeet Ganguly

Abstract After the dawn of the Internet, we are surrounded by a world filled with multimedia content, especially video. Two type of videos every Indian loves and watches a lot are News and Cricket. News and cricket are two very different types of category of videos and thus helping to annotate and categorize these two type of videos will be valuable. We have presented a method to categorize videos on the basis of textual content in them and as the text in a Cricket Video is generally quite significantly different from that of a News Video, the categorization of videos is done in two parts, the first being Text Detection and Extraction, where Edge Detection, Thresholding, Dilation, Contour Detection, and Optical Character Recognition (OCR) are used for the exposure of the part of Videos which has text in them and extract the text present in it. The second stage is that of Text Categorization, using Term Frequency (TF) and Inverse Document Frequency (IDF) and applying Multinomial Naive Bayes Classifier to classify the text into a category.

Keywords Sobel edge detection · Otsu thresholding · Morphological Dilation · Optical character recognition · Term frequency · Inverse document frequency · Multinomial Naive Bayes

1 Introduction

Owing to the expansion of Internet and advances made recently concerning compression technology a lot of video data is present over the Internet and various other places. Every passing day a huge amount of video data is uploaded via the Internet. This has created a need for categorizing the videos so that they can be efficiently indexed, browsed, searched, and relevant materials can be retrieved.

A. Sinha · J. Ganguly (✉)
Jalpaiguri Government Engineering College, Jalpaiguri, West Bengal, India
e-mail: jayjeetganguly@gmail.com

A. Sinha
e-mail: adsinha09@gmail.com

D. Bhattacharjee et al. (eds.), *Proceedings of International Conference on Frontiers in Computing and Systems*, Advances in Intelligent Systems and Computing 1255,
https://doi.org/10.1007/978-981-15-7834-2_28

The content present in the videos will be required for indexing them. This content is often strongly related to the textual information appearing in them [1]. In this paper, categorization of videos has been done on the basis of text present in them. We have considered Cricket and News videos, both being very popular in India. Textual content in a video conveys a lot of information about the video. Text present in a cricket video will significantly be different than that of a news video, in general. This provides a suitable ground for differentiating videos based on the text present in them. Video Categorization has been achieved through two major steps, which are Text Detection and Extraction from video, and text categorization.

Text Detection refers to the problem of detecting the areas in a video frame, where textual contents are located. Once the text is located, we need to recognize the text present in that area and extract the text. Text detection and text recognition in images has seen a lot of development and work been done on it recently [2, 3]. But in spite of various sources for video data, less advancement has been made on the field of text detection and text recognition in the Video Domain.

Text Categorization is the process of grouping text files and documents in different categories based on the content present in them [4]. We use the bag-of-words representation and apply supervised machine learning algorithm of Multinomial Naïve Bayes Classifier to classify the text files. Thus, a video is first converted to a text file containing the text present in it and then that text file is categorized, thus categorizing the Video as a whole.

2 Text Detection and Extraction

The first step for Video Categorization is, in a video frame we focus on the text region and then use Optical Character Recognition to extract the textual content in that frame. Since many successive video frames will contain the same textual information, we assume that treating every hundredth video frame for text detection will suffice.

Most text regions have edges between its boundaries and the background. Thus, some detection of edges is necessary to find the text regions. To detect these edges, we use the Sobel edge detector on the video frames. Now we need to differentiate the regions where edges are detected from regions where it is not. We use thresholding to do so as it segments the edge-detected video frame based on intensity. We use Otsu thresholding to achieve it. After the edge detection, we need to connect them to form a single block. We use Morphological dilation to achieve it as it bonds the edges into clusters.

After Morphological Dilation, we need to find and extract the contours, which are the borders of a shape with matching intensity. These contours are basically the probable regions where the texts are present and thus they provide the location of the text regions. We apply certain geometrical constraints so that we derive a better result. The text regions are then used to extract the text by using Optical Character Recognition (OCR) on it.

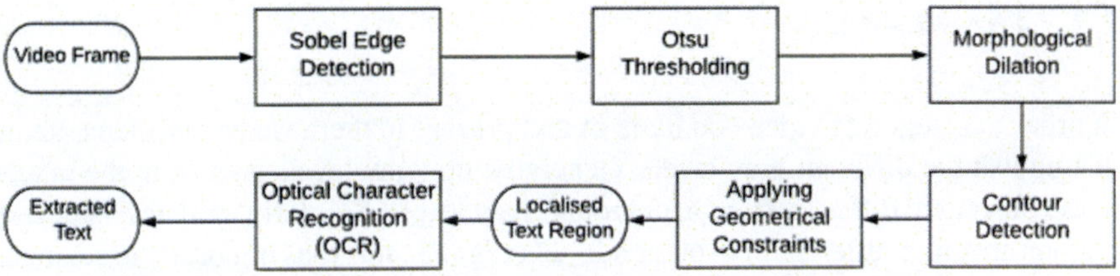

Fig. 1 Flowchart of our proposed methodology for text detection and extraction

The flow diagram of our planned methodology for text detection and text extraction has been shown in Fig. 1.

2.1 Edge Detection

In an image, we used edge detection to find discontinuities in intensity. The main purpose of detection of discontinuities within image intensity is the identification of important events as well as changes in properties of the world. Text regions generally have higher contrast than their local neighbors. Therefore, all pixels in the text region as well as some pixels not in the text region which have similar high color contrast are registered in edge image [1]. The pattern of the texture of text in the video frames can be thought of as a group of short horizontal and vertical edges mixed together [5]. These text edges can be in various orientation and are connected to each other. To detect these edges, we use the Sobel edge detector on the video frames.

2.1.1 Sobel Edge Detection

A very common operator for doing edge detection is a Sobel operator. When one region is darker or is brighter than the other side it will give a response. The sign of the output doesn't matter. We also need to compute for the y-direction. The Sobel operator is a well- known edge detector [6]. It is an approximation to the derivative of an image. So there it is separate in the y- and x-direction. We have mainly used Sobel operator due to its advantages as in the average factor as it smoothens the random noise in the image. So it enhances the edge elements on both sides. Thickness property may be seen in the edge. Its emphasis on the regions, mainly of high spatial frequency corresponding to the edges is also a reason for using it. The main equation is presented as follows:

$$t_1 = \sqrt{4\sum_{i=1}^{h-1}\sum_{j=1}^{w-1}(G_x^2(i,j) + G_y^2(i,j))/(h-1)(w-1)} \quad (1)$$

2.2 *Thresholding*

In order to categorize videos on basis of text present in them we need to separate or distinguish the different gray levels. Generally, the gray levels present in the pixels in the object are different from background gray levels. Similarly, gray levels of text regions are quite different than other non-text regions. It is thus important to segment the two regions in a video frame. Thresholding is very simple to use where segmentation of binary images is created from a grayscale image. Thresholding comprises of a certain region corresponding to the various regions and can be classified by using range function applied to the pixel intensity values of the image or video frame [6]. We have applied Otsu thresholding on the edge-detected video frame.

2.2.1 Otsu Thresholding

Our main purpose is to use clusters to reduce the graylevel image into binary form with the help of thresholding and so the best method to do this in image processing is generally performing Otsu's thresholding method where the algorithm follows that in images there are pixels in only two classes that are classified as background pixels and the other is foreground images which form a bimodal histogram. In Otsu thresholding, the threshold that minimizes the combined spread or intra-class variance or the threshold which maximizes the inter-class variance is exhaustively searched [6].

2.3 *Morphological Dilation*

We have used morphological dilation for the purpose of grouping operations which are non-linear and the video frames or images of which have a relation to the shape. Morphological operations being dependent only on pixels relative ordering and not on pixel's numerical values and so are suitable for binary image processing. Two most important data are required by these operations, first being the image, which in this case will be the video frames that are inputted which will be a binary image. The other is the structuring element. The structuring component is a relatively small matrix consisting of pixels, every pixel being either zero or one. The dimensions of the matrix are used to specify the structuring Element's size. The structuring elements consist of coordinates of numbers of distinct points having relation to the origin. The pattern comprising of ones and zeroes are used for the specification of the shape of the defined structuring component or element. As the important texts that will convey some information about the video are generally aligned horizontally, morphological dilation is being performed on the defined horizontal direction and thus a structuring element of rectangular shape having 10 pixels horizontally and 2 pixels vertically is used for morphological dilation. Morphological dilation is then used for the addition of pixels required to the boundaries of objects on the image or grows or thickens images in the binary image. The edges of text characters when calculated, the number is dependent on structuring elements size and is also dependent on its shape. It helps to connect the edges of text characters together forming a block so that contours of it can be extracted later [7].

2.4 Contour Detection

A Contour can be defined as a closed curve which joins each continuous point (along the same boundary), which have matching intensity or color. Contours are then used for detecting and recognizing objects. Contours are then used for analyzing the shape of objects. Finding contours is similar to differentiating black background containing white objects. Here objects are considered as white and background is to be taken as black. After morphological dilation, edges of text characters are then used for combination forming white blocks which then are used for finding contours by separating the white block from the black background. Each contour is a bounding box and is rectangle in shape. Contour detection will give us the probable regions in the video frame that will contain text in them. There are three arguments needed to find the contours. The first one being the source image or the source or the video frame, the second being retrieval mode of the contour and third being approximation method. The objects boundary points which are converted into coordinates and arranged in an array are the individual contours. For the contour retrieval mode, we use the external mode to retrieve only the extreme outer contours. Also, we don't need to store the coordinates of all the points on the boundary. For the Contour Approximation method, we use ChainApproxSimple. It is helpful in the removal of redundant points compressing vertical, horizontal, and also diagonal segments as well as leaving endpoints only.

2.5 Geometrical Constraints

After the morphological processing and contour detection, we will extract the bounding rectangles over the contours. A lot of contours will be falsely identified as regions of text. In addition to that, as we are interested in horizontally aligned text as contextual text information in both Cricket and News videos are found horizontally, we need to ignore the vertical bounding rectangles and concentrate on the horizontal bounding rectangles. Thus, in order to obtain more accurate results we need to impose a few geometrical constraints over the bounding rectangles (Fig. 2).

There are three types of Geometrical Constraints that has been used. The first one is the width of the bounding rectangle. The width has been kept at a minimum of 25 pixels as any bounding rectangle with smaller width will not provide sufficient

Fig. 2 Before application of OCR

information. The next Geometrical Constraint is the height of the bounding rectangle, which has been kept between the range of 5–100 pixels. The last Geometrical Constraint is the bounding rectangles aspect ratio.

2.6 Optical Character Recognition (OCR)

After applying optical character recognition on the contours, we extract the text in machine-readable form. This text is important as for text categorization to be performed, in order to analyze the text and train machine learning algorithms on it, it must be in machine-readable form.

3 Text Categorization

Text categorization or text classification consists of automated grouping of the natural languages texts to predefined categorization on the basis of their content [8]. As we need to categorize the videos as to whether they are a News or Cricket video according to the text present in them, therefore after extracting the text in the videos we need to classify the text in the two given categories so that the entire video can be categorized as a whole.

Applying Machine Learning our objective is to automatically categorize any extracted text after learning example texts which have already been assigned to a category. This is a supervised learning problem. After the text is extracted from a video, it is important for us to classify that under which category it must be assigned to. The text from videos are stored as a dataset and is divided into dataset—training and testing. Supervised Machine learning is implemented to learn from the training dataset in order to classify the text in the test dataset. Since only two classes are involved in this categorization, it is a binary classification.

There are two major steps for Text Categorization. The first being preprocessing where we convert the original text file into a form on which Machine Learning can be

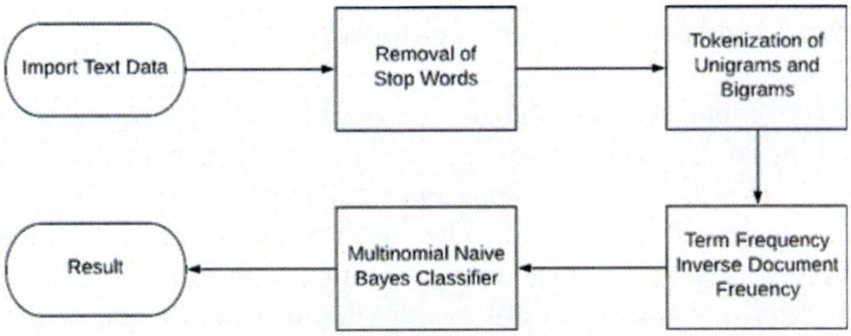

Fig. 3 Flowchart of our proposed methodology for Text Categorization

implemented. It is done by using stop word removal, unigram and bigram tokenization, Term Frequencies (TF), and Inverse Document Frequencies (IDF). The second step is to implement Multinomial Naive Bayes algorithm for the classification of the text files. The flow diagram of our proposed methodology for Text Categorization is given in Fig. 3.

3.1 Text Preprocessing

The text preprocessing step converts the original text data, where the most significant features of the text are served in order to classify between text-categories. Here text preprocessing is done on the text files in the text dataset before applying it to the Multinomial Naive Bayes classifier. Text preprocessing consists of stop word Removal, which removes words which are commonly used such as prepositions which carry no useful information. Next step is tokenization of unigrams and bigrams. The third step is to use TF and IDF.

3.1.1 Tokenization of Unigrams and Bigrams

In a text, we find all the *n*-grams present which is the sequence of n consecutive characters and their frequency is calculated [9]. Thus tokenization of consecutive sequence of *n*-words is called *n*-grams. When we set the range for *n* to (1, 2), we are considering both unigram and bigram for further operation. A single word is considered to be a unigram whereas two consecutive words integrated as a single token is called bigram. Bigrams are also helpful because they capture some partial information about local word order [10].

3.1.2 Term Frequency and Inverse Document Frequency

After stop word removal and tokenizing the words into unigrams and Bigrams, we create a bag-of-words containing all the tokens. Term Frequency (TF) summarizes how often a given word or a token appears within a text file.

Inverse Document Frequency downscales the word which appears a lot in the text file. If a word or token appears in many text files or documents, then its Inverse Document Frequency is low and if it appears in only one document, then it is the highest.

3.2 *Multinomial Naive Bayes*

Naive Bayes befalls a family of algorithms, which is established on applying Bayes theorem alongside a simple assumption which states that every component is independent of the others to predict the category which the sample is under. They happen to be probabilistic classifiers. The probability of each class will be calculated from which the class with the highest probability will be considered as output.

In the Multinomial Naive Bayes classifier, each text file is viewed as a collection of words but the Order in which the words appear is regarded irrelevant. The probability of a class value c given a test text file d is estimated as

$$P(c|d) = \frac{P(c)\prod_{w \varepsilon d} P(w|c)^{n_{wd}}}{P(d)} \tag{2}$$

Using this, we have also used the following equation:

$$P(w|c) = \frac{1 + \sum_{d \varepsilon D_c} n_{wd}}{k + \sum_{w'} \sum_{d \varepsilon D_c} n_{w'd}} \tag{3}$$

where D_c is the collection of all training text documents in class c, and k is the size of the vocabulary (i.e., the number of distinct words in all training documents).

4 Experiments and Result

We have created a dataset of 500 videos each of which were converted to text files based on the text present in them. The text files were then divided between training and testing dataset and Multinomial Naive Bayes was applied to categorize the text files, thus categorizing the videos themselves. The accuracy achieved in the Text Categorization phase was 0.94. Precision, Recall and F1-Score for Cricket and News Videos are given in Table 1 (Fig. 4).

Table 1 Experiment result

Video category	Precision	Recall	F1-Score
News	0.95	0.91	0.93
Cricket	0.93	0.96	0.95

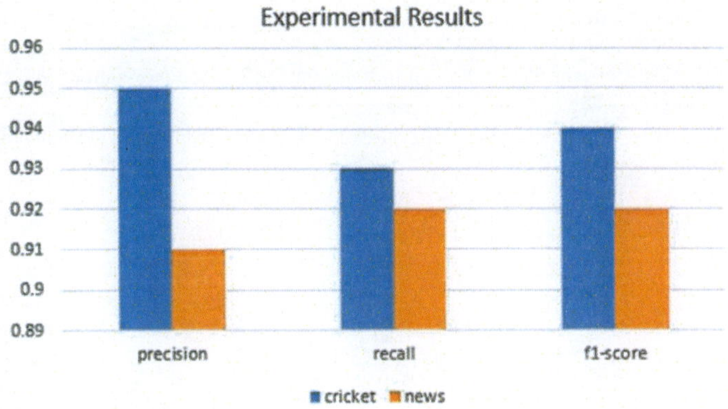

Fig. 4 Experimental results

5 Conclusion

In this paper, we have proposed a novel method for the Categorization of Videos using the text in it using two major phases—Text detection and extraction, and Text categorization. The final classified result which gave an accuracy of 0.94 was achieved due to the coordinated working of the two major phases.

Video Categorization can be very useful for categorizing large video Databases, where the textual content in the videos provide valuable information about the video.

References

1. Gllavata, J., Ewerth, R., Freisleben, B.: A robust algorithm for text detection in images In: 3rd International Symposium on Image and Signal Processing and Analysis. ISPA 2003. Proceedings of the, IEEE, vol. 2, pp. 611–616 (2003)
2. Mishra, A., Alahari, K., Jawahar, C.: Top-down and bottom-up cues for scene text recognition. In: 2012 IEEE Conference on Computer Vision and Pattern Recognition, pp. 2687–2694. IEEE (2012)
3. Mishra, A., Alahari, K., Jawahar, C.: Scene text recognition using higher order language priors. In: BMVC-British Machine Vision Conference, BMVA (2012)
4. Joachims, T.: A probabilistic analysis of the rocchio algorithm for text categorization. Technical report. Carnegie-mellon univ pittsburgh pa dept of computer science (1996)

5. Chen, D., Bourlard, H., Thiran, J.P.: Text in complex background using SVM. In: Proceedings of the 2001 IEEE Computer Society Conference on Computer Vision and Pattern Recognition. CVPR 2001, vol. 2, p. II. IEEE (2001)
6. Sahoo, P.K., Soltani, S., Wong, A.K.: A survey of thresholding techniques. Comput. Vis. Gr. Image Proc. **41**(2), 233–260 (1988)
7. Wolf, C., Jolion, J.M.: Extraction and recognition of text in multimedia documents. Formal Pattern Anal. Appl. **6**(4), 309–326 (2004)
8. Sebastiani, F.: Machine learning in automated text categorisation: a survey. Technical report IEI-B4-31-1999 (1999)
9. Rahmoun, A., Elberrichi, Z.: Experimenting *n*-grams in text categorization. Int. Arab J. Inf. Technol. **4**(4), 377–385 (2007)
10. Joulin, A., Grave, E., Bojanowski, P., Mikolov, T.: Bag of tricks for text classification (2016)

Improved Multi-scale Opening Algorithm Using Fuzzy Distance Transform Based Geodesic Path Propagation

Nirmal Das, Indranil Guha, Punam K. Saha, and Subhadip Basu

Abstract Vessel tree segmentation from CT scan angiogram images of the human brain is a challenging task. The complex geometry, interconnections, and fusion with soft tissues and bones make the segmentation process harder. The segmented cerebrovasculature plays a major role in the fast analysis of vascular geometry leading to an effective diagnosis of the diseased cerebrovascular segment. The present work proposes a geodesic path propagation based on fuzzy distance transform to improve the multi-scale opening algorithm for effective segmentation of carotid vessel with less user intervention. The geodesic path is estimated between a pair of vessel seeds given by the user. The points on the paths are used as the initial pure vessel seeds during the multi-scale opening of the vascular tree from other fused conjoint components (bone, soft tissue etc.) in a shared intensity space. We developed a 2D/3D user interface to mark user-specified vessel/bone seeds or separators on the input images. Experiments on three patients' CTA images show significant qualitative improvement in segmentation results with much lesser user intervention.

Keywords Cerebrovasculature · Carotid vasculature · Fuzzy distance transformation · 3D rendering · Geodesic paths · Multi-scale opening

N. Das (✉) · S. Basu
Department of Computer Science and Engineering, Jadavpur University, Kolkata, India
e-mail: das.nirmaljis@gmail.com

S. Basu
e-mail: subhadip.basu@jadavpuruniversity.in

I. Guha · P. K. Saha
Department of ECE and Radiology, University of Iowa, Iowa City, IA 52242, USA
e-mail: indranil.guha21@gmail.com

P. K. Saha
e-mail: pksaha@engineering.uiowa.edu

D. Bhattacharjee et al. (eds.), *Proceedings of International Conference on Frontiers in Computing and Systems*, Advances in Intelligent Systems and Computing 1255,
https://doi.org/10.1007/978-981-15-7834-2_29

1 Introduction

This segmentation of objects embedded in 2D/3D images has drawn great attention of the image processing research community over several decades [8, 19, 20, 22, 25]. In this work, we will discuss the segmentation of objects in 3D images. Although, a significant research has been carried out on the image segmentation algorithm [11, 14, 17, 23] over the years, researchers are still unable to figure out a universal segmentation approach. Often we face challenges in segmentation as the existing segmentation algorithms fail to perform efficiently. The main challenge of the proposed segmentation method is the separation of conjoint objects, fused at different locations and scales in a shared intensity band. Please refer to the following articles to know about the segmentation problem mentioned above and the possible solutions [12, 16, 18, 21]. The background and area of applications, current state of the art, and the contribution of the proposed method are discussed in the following.

The present article describes the proposed segmentation algorithm in the context of one medical imaging application: segmentation of vascular tree in the CT scan images of the human brain. Patient-specific cerebrovascular geometry helps clinicians to better understand the patient's cerebrovascular health and to detect/localize the diseased vascular segment as in case of aneurysm [7, 24]. Segmentation of the complete vascular tree from CT images of the human brain is the foremost step for further analysis of the cerebrovascular geometry and hemodynamic simulation in the underlying vasculature [10]. Analysis of various parameters of blood flow dynamics like velocity, wall shear stress help in the diagnosis of several vascular diseases. The hemodynamic analysis also helps to predict the disease-prone vascular segments.

In CTA images, bones take higher intensity range and soft tissues, skin takes lower intensity range and the intensity of the vascular tree falls in an intensity band in between [1]. Coupling of bone and vessel at different locations and scales, especially at near carotid sinus and nasal regions makes the segmentation problem harder.

Several methods can be found in the literature to segment cerebrovasculature as well as aneurysm from brain images of various modalities. In [2], Banerjee et al. reported a Bezier curve based approximate phantom building method and later they extended the method to segment cerebrovasculature in the human brain CTA images. But the segmentation approach is entirely based on interactive human vascular seed selection and vulnerable to human errors. Steinman et al. [26] reported a semiautomatic segmentation method for segmentation of giant intracranial aneurysm on Posterior Communicating Artery (PCA) using the method of discrete dynamic contouring. Cebral and colleagues presented level sets and deformable models based 3D reconstruction algorithm to segment a cerebral aneurysm [9]. Frangi and his colleagues applied a region-based surface modeling method to achieve the segmentation task [28]. However, the above-mentioned algorithms are mainly focused on segmentation of aneurysm and the small attached vessel ignoring the complete cerebrovasculature. Segmentation of aneurysm with the entire vascular tree has some important additional advantages. Segmentation of complete cerebrovasculature allows hemodynamic analyses with more contextual information and as a result, improves the analytic result. Also, segmentation of only aneurysm in the close proximity of the skull and neighboring vessels are difficult. These unavoidable issues have not been comprehensively addressed in the above-mentioned studies. Saha and his colleagues in [18, 21] have reported a multi-scale opening (MSO) algorithm to differentiate objects conjoined at different locations with varying scales and complex geometry at isointensity space as well as shared intensity space. The method successfully applied to separate vessel-bone in non-contrast CT images of human lung and in the subsequent extension of the work, it is also used to bone-vessel in shared intensity band in human brain CT images [4]. In [18], the MSO algorithm is theoretically established and separation/segmentation of conjoined objects is made independent of contrast protocol used in CT images. The existing MSO algorithm often requires a huge number of user-given seed points in low-resolution regime.

The algorithm presented in this paper reduces the user interventions during the initial seeding process in the multi-scale opening process. The proposed algorithm uses fuzzy distance transform (FDT) based geodesic path propagation to improve the MSO algorithm. The points on the geodesic paths will be used as pure vessel seeds by the MSO algorithm. Theoretical fundamentals of the MSO-based segmentation algorithm has already been discussed in [13]. The experimental results show that the performance of the MSO algorithm improves significantly with the use of FDT-based geodesic paths as the initial vascular seeds.

In the following sections, first, we briefly introduced the theory and notations related to the proposed algorithm then the methodology and experimental results.

2 Theory and Method

In this section, the basic definitions and notations related to 3D digital image space are described first. Then the algorithm of FDT-based geodesic path propagation is explained followed by a brief discussion on the existing MSO algorithm [18] and the improved MSO algorithm.

2.1 Basic Definitions and Notations

A 3D cubic grid is denoted by $\{Z^3 | Z$ is the set of positive integers$\}$. A point $p \in Z^3$ is called a voxel denoted by (x_1, x_2, x_3). Two voxels $p = (x_1, x_2, x_3)$ and $q = (y_1, y_2, y_3) \in Z^3$ are adjacent iff $\max(|x_i - y_i|) \leq 1 | 1 \leq i \leq 3$, where $| \cdot |$ means the absolute value. Two adjacent voxels are called neighbors of each other. 26 neighbors of a voxel p without itself is denoted as $N^*(p)$. An object O is a fuzzy set $(p, \mu_{f_O}(p)) | p \in Z^3$, where $\mu_{f_O} : Z^3 \rightarrow [0, 1]$ is the fuzzy membership function. The voxels with membership value > 0 defines the support of the object O, $\theta(O) = p | p \in Z^3 \wedge \mu_O(p) \neq 0$. $\bar{\theta}(O) = Z^3 - \theta(O)$ is the set of background voxels. Let$S$ is the set of object voxels. A path π in S from $p \in S$ to $q \in S$ is a sequence of successive adjacent voxels, i.e., $< p = p_0, p_1, \ldots, p_l = q >$. A path between two adjacent voxels p and q is called a link. The path length $\pi =< p_0, p_1, \ldots, p_l >$ in O, denoted by $\Pi_O(\pi)$ is sum of the length of all links along the path, i.e.,

$$\prod_O = \sum_0^{l-1} \frac{1}{2}(\mu_{fo}(p_i) + \mu_{fo}(p_{i+1}))\, |p_i - p_{i+1}|$$

The fuzzy distance between two voxels p, q in object O expressed as $D_{fo}(p, q)$ is the length of one of the shortest paths from p to q, i.e.,

$$D_{fo}(p, q) = \min_{\pi \in P(p,q)} \Pi_O(\pi) | P(p, q)$$

where $P(p, q)$ is the set of all paths from p to q. The FDT of an object O is an image $\{(p, FDT_O(p)) | p \in Z^3\}$, where $FDT_O : Z^3 \rightarrow \Re^+ | \Re^+$ is the set of all real numbers including zero, is the fuzzy distance from the background.

$$FDT_O(p) = \min_{\bar{\theta}(O)} D_{fo}(p, q)$$

Local maxima of object O are the set of locally deepest voxels. Let, $L_{\max} \subset \theta(O)$ be the set of locally deepest voxels i.e.,

$$L_{\max} = \{p | p \in \theta(O) \Lambda \forall q \in N_l(p), FDT_O(q) \leq FDT_O(p)\}$$

where $N_x(p)$ is the $(2x + 1)^3$ neighborhood of p. So a point may have more than one number of local maxima point in its neighborhood. If there are multiple local maxima points, we will choose the voxel with larger FDT value.

2.2 *FDT Based Geodesic Path*

The minimum cost shortest path between two points is called geodesic path. An image can be represented as an undirected graph $G = (V, E)$, here V denotes set of vertices, $V = \{P | P \in O\}$ and $E = \{(p, q) | p, q$ are adjacent voxels$\}$. In this article, we have used Dijkstra's shortest path algorithm [14] to calculate the shortest path between two points in the FDT image in 3D with the restriction that shortest path will always run through the nearest local maxima point if it exists, otherwise it will choose the neighbor with highest FDT value. A voxel p may have more than one local maxima or may not have any local maxima in $N_x(p)$ i.e $(2x + 1)^3$ neighborhood of p. We have taken $x = 1$ to avoid noisy local maxima. To pass the geodesic path from a point to its nearest local maxima, the edge-weight between them should be minimum. Let point $p = (x_1, x_2, x_3) \in Z^3$ and the nearest local maxima point be $q = (y_1, y_2, y_3)$. The edge-weight δ between them is given by the below cost function:

$$\delta(p, q) = \left(\frac{2}{DT(p) + DT(q)}\right) \times D_c(p, q) \tag{1}$$

$D_c(p, q)$ denotes Chamfer distance [6] between the voxels.

Dijkstra's algorithm follows the working of the greedy algorithm. After termination of the algorithm, we will get a connected set of approximate axial points.

2.3 MSO Algorithm and Its Improvement

Now, we will briefly describe the basic notations and definitions used in the MSO algorithm. *Fuzzy morpho-connectivity* (*FMC*) strength of a path $\pi =< p_1, p_2, ..., p_l >$ in a fuzzy object O, expressed as $\Gamma(\pi)$, is the minimum FDT value along the path.

$$\Gamma_O(\pi) = \min_{1 \leq i \leq l} FDT_O(p_i)$$

FMC between two voxels $p, q \in Z^3$, is the strength of the strongest morphological path between them. *FMC* strength between the two voxels can be denoted by

$$\gamma_O(p, q) = \max_{\pi \in P(p,q)} \Gamma_O(\pi)$$

Optimum erosion of a fuzzy object A represented by the set of seeds S_A with respect to its rival object B represented by the set of seed voxels S_B and a set of common separator S_S is the set of all voxels p such that there exists an erosion scale that disconnects p from B while leaving it connected to A, i.e.,

$$R_{A,0} = \{p | \max_{a \in S_A} \gamma_{A,0}(a, p) > \max_{b \in S_B} \gamma_{B,0}(b, p)\}$$

where the *FMC* functions $\gamma_{A,0}$ and $\gamma_{B,0}$ are defined from the FDT maps $D_{f_{A,0}}$ and $D_{f_{B,0}}$, respectively. The optimum erosion $R_{B,0}$ of the object B is defined similarly. As *FMC* strength depends on the FDT value of the voxels along the path, hence in optimum erosion the region through which vessel passes plays an important role in the segmentation process. Specifically, in shared intensity regions where the bone dominates vessel. As two separated regions only capture an eroded version of the target object, hence constraint dilation has been used to further improve the data. Constrained dilation of $R_{A,0}$ with respect to its co-object $R_{B,0}$ within the fuzzy object O, denoted as $M_{A,0}$ is the set of all voxels $p \in N_O(R_{A,0})$ which are strictly closer to $R_{A,0}$ than $R_{B,0}$, i.e.,

$$M_{A,0} = \{p | p \in N_O(R_{A,0}) \Lambda \min_{a \in R_{A,0}} D_{f_O}(a, p) < \min_{b \in R_{B,0}} D_{f_O}(b, p)\}$$

where $N_{O(R_{A,0})}$ is morphological neighborhood of a set of voxels $p \in \theta(O)$ such that $\exists q \in R_{A,0}$ for which $D_{f_O}(p, q) < D_{f_O}(q)$ and p is connected to q by a path π of monotonically increasing FDT values. As p is connected to q by monotonically increasing FDT values, hence in a shared intensity region where bone dominates vessel and sufficient vessel seed points are not given by the user, MSO algorithm fails to segment the vessel as the algorithm favors the bone seeds to grow more than the vessel seeds.

But using FDT-based geodesic path, intermediate approximate axial points are generated as the set of initial seed points S_{vessel}. This prevents the rapid growth of bone seeds, S_{bone}, in the shared intensity region where the pure vessel points are only marked by user seeds. The developed method connects the user-given vessel seed points by the FDT-based geodesic path. Therefore, the number of pure vessel points increase significantly in the shared intensity space with minimal user intervention. If required, additional manual seeds may be introduced to correct the intermediate points on the path.

The MSO algorithm starts from the initial set of vessel points generated by the geodesic path. The algorithm separate the objects at a specific scale and iteratively propagates to finer scales. For each of the two objects, the FDT values are set to zero over the region currently acquired by its rival object. It puts an artificial wall at the periphery of each object already separated stopping paths from one object to pass through the region already occupied by the rival object. Specifically, after each iteration, the FDT image of object A is updated as follows:

$$\text{FDT}_{O_{A,i}}(p) = \begin{cases} 0, & \text{if } p \in N_O(R_{B,i-1} - M_{A,i-1}) \\ \text{FDT}_{O_{A,i-1}}(p) & \text{otherwise} \end{cases}$$

The FDT of the co-object is updated similarly. The seed voxels for the two objects are replaced by $M_{A_{i-1}}$ and $M_{B_{i-1}}$, respectively. With this configuration, the algorithm enters into the next iteration and the morphological separation of M_{A_i} and M_{B_i} is derived using the Equations.

3 Experimental Result

The performance of the proposed method is assessed in terms of qualitative results on 3 patients' CT Scan angiogram images of the brain. This dataset specification has already been published in a conference paper by Basu et al. [4]. Two image intensity threshold, one lower and another higher were manually selected from 3 randomly chosen data from CT Angiogram dataset. The averages of the two thresholds were, respectively, assigned to $I_{\min}$ and $I_{\max}$. In our dataset, the lower threshold was set to $I_{\min} = 100$ Hu and the higher threshold was set to $I_{\max} = 450$ Hu. The experimental methods are facilitated with the help of the developed graphical user interface. The user-given seeds/separators are taken with the help of the GUI. The purpose of the proposed improvement of the MSO algorithm is to separate the vascular tree

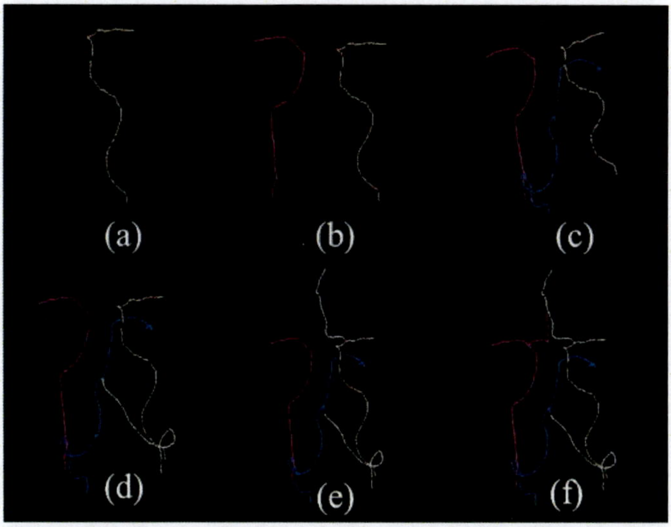

Fig. 1 Generation of geodesic path between pair of seeds in a sample image. **a–f** different steps of the FDT-based geodesic path formation

from soft bones as well as bones in their overlapping intensity zone. The proposed algorithm of FDT-based geodesic path propagation between two given vessel seed points generates a connected set of intermediate local maxima voxels/voxels with maximum FDT value. These automatically generated intermediate points serve the requirement of vessel seeds, especially in the region where the number of bone voxels is very, low scale, low resolution, and limited SNR value. Another advantage of the proposed method is that it overcomes the vulnerability of inappropriate seed points placed by the user. Figure 1 describes the steps of FDT-based shortest path algorithm on the segmented phantom of major arteries from a real patient CT scan angiogram image. The algorithm uses a total of 10 seed points to generate a discrete set of skeletal points. Two different kinds of seed points are used. The first kind of seed points are placed in pair as start and endpoint and the second kind of seed point is placed alone as 'joining point'. The qualitative improvement of the results compared to the existing MSO algorithm can easily be observed in Fig. 2a–c, where segmentation results are shown on 3 patients' CT angiogram images. The quantity of total generated vasculature is much higher in all of the 3 segmented vasculatures in Fig. 2a Data id-3029 (b) Data id -2005 (c) Data id-2008. Table 1 gives a comparison of the no. of seed points required to segment the vasculature on 3 patients' CT scan angiogram images of the human brain. In all of the 3 data, the requirement of user-specified seed points has been reduced significantly.

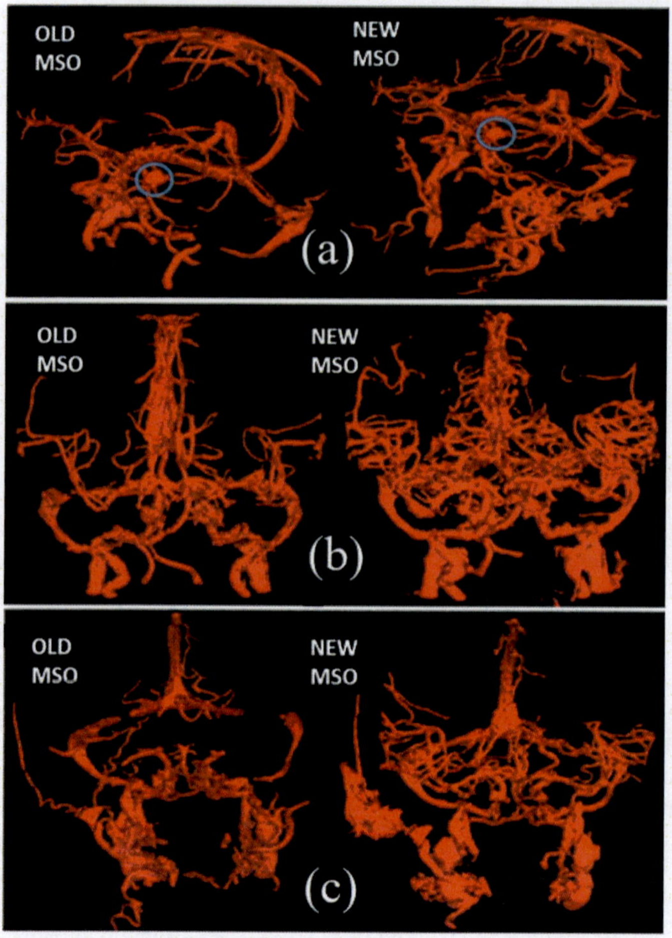

Fig. 2 Qualitative comparison of improved performance of the MSO algorithm on three image samples using the intermediate approximate axial points **a** Data id-3029 (encircled area shows an aneurysm) **b** Data id-2005, **c** Data id-2008

Table 1 Comparison of number of seeds used to segment of vascular tree in patients' CT Scan angiogram images of human brain in the existing MSO and the proposed algorithm

Image id	Number of seeds required		
	No. of vessel seeds used in [18]	Total no. of seeds used in [18]	Total no. of seeds in proposed method
3029	13	41	14
2005	28	50	13
2008	16	34	14

4 Conclusion

The proposed work is an indispensable step to reduce the requirement of user-given vessel seeds and improve the existing MSO algorithm. This method has adopted FDT-based geodesic path propagation approach described in [13]. The generated geodesic path increases the number of vessel seeds in the area where the number of pure bone seeds is high. We argue that the proposed algorithm successfully improved the existing MSO algorithm for better segmentation results. Segmented vasculature is essential in the study of vascular bends, branching locations, joins and in the simulation of digital fluid flows in underlying structure[10]. The freely available software ITK-SNAP [27] is used for 3D visualization of the segmented vascular tree. This segmented tree can be used for structural analysis and hemodynamic simulation in human cerebrovasculature. FDT-based geodesic path propagation may also be applied to other biomedical applications in 2D/3D [3, 5, 15].

Acknowledgements The authors acknowledge Dr. Robert E. Harbaugh, Penn State Hershey Medical Center and Prof. Madhavan L. Raghavan, Department of Biomedical Engineering, University of Iowa, for providing the CT Scan Angiogram datasets used in this study. This project is partially supported by the CMATER research laboratory of the CSE Department, Jadavpur University, India; Department of Biotechnology grant (BT/PR16356/BID/7/596/2016), Govt. of India, and CSIR SRF DIRECT Fellowship (File No. 09|096(0921)2K18 EMR-I), CSIR-HRDG, Government of India.

References

1. Abrahams, J.M., Saha, P.K., Hurst, R.W., LeRoux, P.D., Udupa, J.K.: Three-dimensional bone-free rendering of the cerebral circulation by use of computed tomographic angiography and fuzzy connectedness. Neurosurgery **51**(1), 264–269 (2002)
2. Banerjee, A., Dey, S., Parui, S., Nasipuri, M., Basu, S.: Synthetic reconstruction of human carotid vasculature using a 2-d/3-d interface. In: 2013 International Conference on Advances in Computing, Communications and Informatics (ICACCI), pp. 60–65. IEEE (2013)
3. Basu, S., Plewczynski, D., Saha, S., Roszkowska, M., Magnowska, M., Baczynska, E., Wlodarczyk, J.: 2dSpAn: semiautomated 2-d segmentation, classification and analysis of hippocampal dendritic spine plasticity. Bioinformatics **32**(16), 2490–2498 (2016)
4. Basu, S., Raghavan, M.L., Hoffman, E.A., Saha, P.K.: Multi-scale opening of conjoined structures with shared intensities: methods and applications. In: 2011 International Conference on Intelligent Computation and Bio-Medical Instrumentation, pp. 128–131. IEEE (2011)
5. Basu, S., Saha, P.K., Roszkowska, M., Magnowska, M., Baczynska, E., Das, N., Plewczynski, D., Wlodarczyk, J.: Quantitative 3-d morphometric analysis of individual dendritic spines. Sci. Rep. **8**(1), 3545 (2018)
6. Borgefors, G.: Distance transformations in arbitrary dimensions. Comput. Vis. Gr. Image Proc. **27**(3), 321–345 (1984)
7. Brisman, J.L., Song, J.K., Newell, D.W.: Cerebral aneurysms. New Engl. J. Med. **355**(9), 928–939 (2006)
8. Bushberg, J.T., Boone, J.M.: The essential physics of medical imaging. Lippincott Williams & Wilkins (2011)
9. Cebral, J.R., Castro, M.A., Appanaboyina, S., Putman, C.M., Millan, D., Frangi, A.F.: Efficient pipeline for image-based patient-specific analysis of cerebral aneurysm hemodynamics: technique and sensitivity. IEEE Trans. Med. Imaging **24**(4), 457–467 (2005)

10. Das, N., Rakshit, P., Nasipuri, M., Basu, S.: 3-d digital flows in cerebrovascular phantoms. In: 2017 Ninth International Conference on Advances in Pattern Recognition (ICAPR), pp. 1–5. IEEE (2017)
11. Falcão, A.X., Stolfi, J., de Alencar Lotufo, R.: The image foresting transform: theory, algorithms, and applications. IEEE Trans. Pattern Anal. Mach. Intell. **26**(1), 19–29 (2004)
12. Gao, Z., Grout, R.W., Holtze, C., Hoffman, E.A., Saha, P.: A new paradigm of interactive artery/vein separation in noncontrast pulmonary ct imaging using multiscale topomorphologic opening. IEEE Trans. Biomed. Eng. **59**(11), 3016–3027 (2012)
13. Guha, I., Das, N., Rakshit, P., Nasipuri, M., Saha, P.K., Basu, S.: Design of cerebrovascular phantoms using fuzzy distance transform-based geodesic paths. In: Progress in Intelligent Computing Techniques: Theory, Practice, and Applications, pp. 359–367. Springer (2018)
14. Heimann, T., Meinzer, H.P.: Statistical shape models for 3d medical image segmentation: a review. Medical Image Anal. **13**(4), 543–563 (2009)
15. Krzystyniak, A., Baczynska, E., Magnowska, M., Antoniuk, S., Roszkowska, M., Zareba-Koziol, M., Das, N., Basu, S., Pikula, M., Wlodarczyk, J.: Prophylactic ketamine treatment promotes resilience to chronic stress and accelerates recovery: correlation with changes in synaptic plasticity in the ca3 subregion of the hippocampus. Int. J. Mol. Sci. **20**(7), 1726 (2019)
16. Lei, T., Udupa, J.K., Saha, P.K., Odhner, D.: Artery-vein separation via MRA-an image processing approach. IEEE Trans. Med. Imaging **20**(8), 689–703 (2001)
17. Pham, D.L., Xu, C., Prince, J.L.: Current methods in medical image segmentation. Ann. Rev. Biomed. Eng. **2**(1), 315–337 (2000)
18. Saha, P.K., Basu, S., Hoffman, E.A.: Multiscale opening of conjoined fuzzy objects: theory and applications. IEEE Trans. Fuzzy Syst. **24**(5), 1121–1133 (2015)
19. Saha, P.K., Borgefors, G., di Baja, G.S.: A survey on skeletonization algorithms and their applications. Pattern Recognit. Lett. **76**, 3–12 (2016)
20. Saha, P.K., Chaudhuri, B.B., Majumder, D.D.: A new shape preserving parallel thinning algorithm for 3d digital images. Pattern Recognit. **30**(12), 1939–1955 (1997)
21. Saha, P.K., Gao, Z., Alford, S.K., Sonka, M., Hoffman, E.A.: Topomorphologic separation of fused isointensity objects via multiscale opening: separating arteries and veins in 3-d pulmonary CT. IEEE Trans. Med. Imaging **29**(3), 840–851 (2010)
22. Saha, P.K., Strand, R., Borgefors, G.: Digital topology and geometry in medical imaging: a survey. IEEE Trans. Med. Imaging **34**(9), 1940–1964 (2015)
23. Saha, P.K., Udupa, J.K., Odhner, D.: Scale-based fuzzy connected image segmentation: theory, algorithms, and validation. Computer Vis. Image Underst. **77**(2), 145–174 (2000)
24. Saleem, M.A., Macdonald, R.L.: Cerebral aneurysm presenting with aseptic meningitis: a case report. J. Med. Case Rep. **7**(1), 244 (2013)
25. Sonka, M., Hlavac, V., Boyle, R.: Image processing, analysis, and machine vision. Cengage Learn. (2014)
26. Steinman, D.A., Milner, J.S., Norley, C.J., Lownie, S.P., Holdsworth, D.W.: Image-based computational simulation of flow dynamics in a giant intracranial aneurysm. Am. J. Neuroradiol. **24**(4), 559–566 (2003)
27. Yushkevich, P.A., Piven, J., Hazlett, H.C., Smith, R.G., Ho, S., Gee, J.C., Gerig, G.: User-guided 3d active contour segmentation of anatomical structures: significantly improved efficiency and reliability. Neuroimage **31**(3), 1116–1128 (2006)
28. Zhang, C., Villa-Uriol, M.C., De Craene, M., Pozo, J.M., Frangi, A.F.: Morphodynamic analysis of cerebral aneurysm pulsation from time-resolved rotational angiography. IEEE Trans. Med. Imaging **28**(7), 1105–1116 (2009)

Voice-Based Railway Station Identification Using LSTM Approach

Bachchu Paul, Somnath Bera, Tanushree Dey, and Santanu Phadikar

Abstract Enormous research is going on Automatic Speech Recognition (ASR) in the past decade. Human-Computer Interaction (HCI) will become more efficient and hands-free through voice-based commands. In our proposed work of speech recognition, we have taken a list of ten major railway stations in South Eastern Railway (SER) from Howrah (One major railway station in West Bengal, India) to Medinipur (A station in West Bengal, India). We have chosen the ten important stations where most number of passengers travels through the local train. The passengers spent a huge amount of time in the long queue for collecting the tickets. We have created a small speech corpus, where 20 people have uttered these stations ten times; a total of 2000 audio samples. We have done a prepocessing phase, followed by a Mel-Frequency Cepstral Coefficients (MFCC), Δ MFCC and Δ Δ MFCC feature extraction method and finally a Long Short-Term Memory (LSTM) sequence classification has been used for correct identification of the station's name and obtained the highest training accuracy of 96.87% for the different hyperparameters discussed in Sect. 5.

Keywords Short time energy · ASR · Zero crossing · FFT · MFCC · Deep learning · LSTM · Mini-batch

B. Paul (✉) · S. Bera · T. Dey
Department of Computer Science, Vidyasagar University, Midnapore 721102, West Bengal, India
e-mail: ableb.paul@gmail.com

S. Bera
e-mail: somnathh.beraa@gmail.com

T. Dey
e-mail: deytanu87@gmail.com

S. Phadikar
Department of Computer Science and Engineering, Maulana Abul Kalam Azad University of Technology, Kolkata 700064, India
e-mail: sphadikar@yahoo.com

D. Bhattacharjee et al. (eds.), *Proceedings of International Conference on Frontiers in Computing and Systems*, Advances in Intelligent Systems and Computing 1255,
https://doi.org/10.1007/978-981-15-7834-2_30

1 Introduction

Human can communicate through speech which is one of the natural ways to communicate with each other). For a long period of time, human languages have been taken as a subject of research to make computer understandable [1, 2]. There are several research areas in speech technology such as speech recognition, speech enhancement and coding, speaker recognition and verification. Speech recognition is the process of transforming speech signal into text which is then converted into machine-readable format [3]. Speech recognition has various application areas such as dictating computers instead of manually typing to help handicapped people, smart room, etc. To arrive at possible phonemes distribution, classification algorithms are used in speech recognition model for every frame. MFCC can be used as a classification algorithm whereas the Hidden Markov Model(HMM) is used in decoding phase and also used to find a sequence of phones which maps to output words [4]. HMM is concerned with a pre-trained language model. In the form of a multiple layered model, Deep Neural Network (DNN) is used to extract specific feature and information. To enhance the capabilities of the computer, so that it understands the human activities, deep learning algorithm has been mostly used. It also used in different applications of speech recognition like isolated word recognition, speech to text conversion, audio-visual speech identification, speaker recognition and speaker adaptation [9].

We focused on a real application of speech command interaction between man and machine. We have observed that most of the passengers travel through local train in SER route, are not adequate to handle a smart phone. They till now collect the ticket from railway counters by waiting 30–45 min during the morning on every working day. So a voice-based system can help the passengers to buy a ticket more quickly. In this paper, we have worked on speech recognition in Deep Neural Network, where MFCC, Δ MFCC and Δ Δ MFCC have been introduced for feature extraction and LSTM network can be used to train data.

2 Literature Review

Petridis et al. [5] worked on an end-to-end visual speech recognition system with the help of LSTM networks. This system learns feature extraction and classification stages jointly. In this model, features were directly extracted from pixels. This model considered two streams: one of which was used for static information encoding and another one for encoding local temporal dynamics. This model could be extended to multiple streams so that audio streams are easily added and performance could be evaluated on audiovisual speech recognition tasks.

Amodei et al. [6] worked on an end-to-end deep learning, which can be used to identify English and Mandarin Chinese speech. End-to-end learning has the advantage of handling a diverse variety of speech in noisy environments. Here Batch Dispatch with GPU technique was used to show that the system can be inexpensively used in an online setting. This system is RNN which consists of one or more convolutional input layers. Multiple recurrent (uni or bidirectional) layers follow this. The whole systems were trained for either 20 epochs of full English dataset or full Mandarin dataset and stochastic gradient descent with Nesterov momentum was used along with mini-batch of 512 utterances. Finally, they tested noisy English speech using the test sets from third CHiME challenge. The CHiME audio provided improvement of performance using all 6 channels.

Gaikwad and Gawali [3] provided an overall technological development technique in every stage of speech recognition for Human-Computer Interface (HCI). For future extraction of speech, MFCC is used in many development techniques. So many modeling technique, i.e., Acoustic phonetic method, Pattern recognition technique, Template-based method, DTW, Knowledge-based policy, Statistical-based algorithm, etc., along with their merits and demerits were discussed in [3]. From the explanations given in [3], it is observed that GHM and HMM provided better performance than the others, hence in our paper these two models have been considered.

Kenny et al. [7] illustrated the use of Deep Neural Network in their work where they extracted Baum-Welch statistical method and find i-vector oriented text-independent speaker recognition. Their proposed model focused on phonetic events, rather than the usual acoustic ones. Here also it is seen that the proposed method has a comparable performance with baseline. This method performed better in low-false alarm area. The authors in [7] demonstrated how the outputs of DNN were used to train i-vector extractor and extract i-vectors.

Noda et al. [8] proposed an AVSR system based on deep learning for audio and visual parameter extraction and a multi-stream HMM for multimodal feature integration to recognize isolated words. They introduced a connectionist HMM for noise-robust AVSR and they used a CNN to extract visual features from raw mouth area images. The authors have shown that deep de-noising auto-encoder is able to separate noise effect on original clean audio inputs that were obtained de-noised audio feature and this could achieve remarkable noise robustness in separated word recognition job.

Nassif et al. [9] provided a statistical analysis of the use of deep learning in speech-related application. It was seen that various works had been carried out on speech recognition, whereas a few works had been performed on speech enhancement, speaker identification, speech emotion recognition, and speech transcription. Here also it is seen that for feature extraction in DNN, MFCC was extensively used to extract feature from speech. However, Linear Discriminate Analysis (LDA), Short-Time Fourier Transform (STFT), maximum likelihood linear transform (MLLT), Perceptual Linear Predictive (PLP), etc., have been fewly used for the same. MFCC is mostly used in classical classifiers such as HMM, GMM. The authors in [9] have also demonstrated that Recurrent Neural Network (RNN) can be used in speech recognition.

But, our research more effectively and efficiently can recognize an isolated word speech, since we obtained the highest accuracy of 96.8%. The methodology used is very simple to recognize a word.

The proposed method is given in the schematic diagram in Fig. 1.

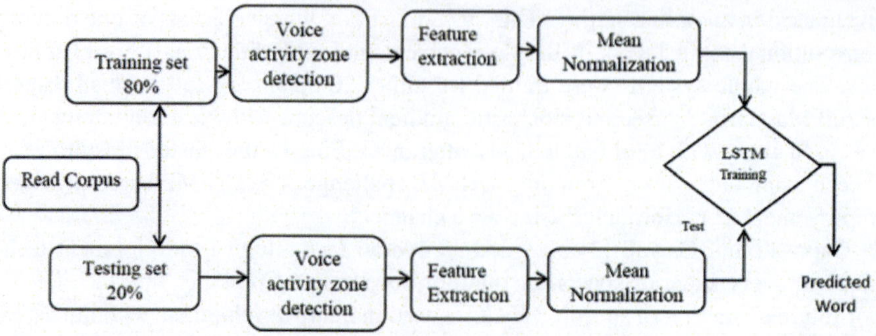

Fig. 1 Schematic diagram of the proposed method

Our paper is structured as: Sect. 3 explains the dataset used and the preprocessing phase, Sect. 4 explains the feature extraction, Sect. 5 describes the network structure and result of the method, and finally Sect. 6 discusses the conclusion.

3 Dataset and Preprocessing

3.1 Dataset Used

In our proposed work of speech recognition, we recorded the name of 10 stations of South Eastern Railway of Indian Railways. We have chosen 20 people among them 14 are male and 6 are female to record the words. We have taken most of the data are in a normal room environment and very few records with a little noise of raining outside or running fan. We used the Audacity software [10] to record the word with a sampling frequency 16 KHz with .wav format. For each station 200 samples, a total of 2000 speech sample has been taken for the proposed model. Among them, 1600 samples for training and 400 samples for testing.

3.2 Preprocessing

In this stage, the voice activity zone is detected from each of the uttered word. This is done by framing the signal of 25 ms with 50% overlapping. Then for each of the frame the average energy and average zero crossing has been computed by the formula given in Eqs. 1 and 2, respectively. The energy of a frame calculates how much information it holds and zero crossing takes decision for a noise or noiseless frame with some threshold [11].

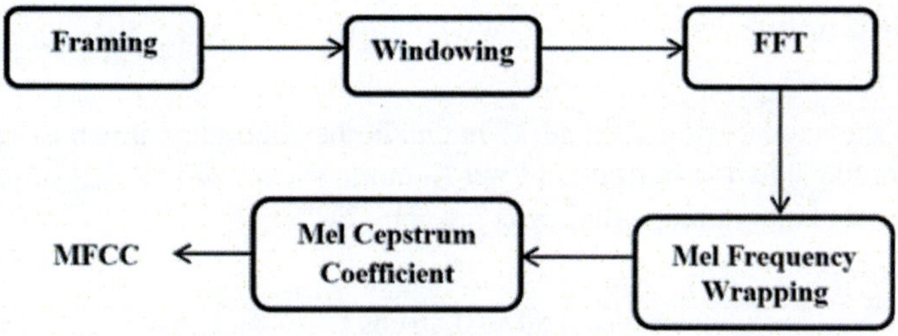

Fig. 2 Steps taken to find MFCC

$$E_n = \sum_{m=-\infty}^{\infty} [X(m) - W(n-m)]^2 \tag{1}$$

where $X(.)$ is the frame and $W(.)$ is the windowing function.

$$\text{ZCR} = \frac{1}{2N} \sum_{j=i-N+1}^{i} |sgn(x(j) - sgn(x(j-1))|w(i-j) \tag{2}$$

where

$$sgn(x(j)) = \begin{cases} 1, & \text{if } (x(j) \geq 0. \\ 0, & \text{if } x(j) < 0. \end{cases}$$

4 Feature Extraction

For each of the voice activity zone, we have computed the 14 Mel-Frequency Cepstral Coefficients (MFCC) and 14 ΔMFCC and 14 Δ Δ MFCC as our feature. The MFCC is computed in the following steps given in Fig. 2.

To find MFCC from the speech signal, we used the following steps given in Fig. 2.

4.1 Framing

The voiced section for each file detected in Sect. 3 is segmented into 25 ms frame with 50% overlap. A single frame contains 400 samples, i.e., 80 frames per second.

4.2 Windowing

Since speech is an aperiodic signal, to maintain the continuity at two extreme ends of a frame, the signal is multiplied by a Hamming window [11, 12] of same size. The equation of a hamming window is given by Eq. 3.

$$w(n) = 0.54 - 0.46 \cos\left(\frac{2\pi n}{N-1}\right) \tag{3}$$

4.3 Fast Fourier Transform (FFT)

The time domain into frequency domain is converted using the FFT [13] to measure the energy distribution over frequencies. The FFT is calculated using the Discrete Fourier Transform (DFT) formula given in Eq. 4.

$$S_i(k) = \sum_{n=1}^{N} s_i(n)e^{-\frac{j2\pi kn}{N}} \qquad 1 \le k \le K \tag{4}$$

K is the DFT length.

4.4 Mel-Frequency Wrapping

In this step, the power spectrum is mapped onto mel scale using 20 number of triangular bandpass filters. The relationship between frequency (f) and mel (m) is given in Eq. 5.

$$m = 2595 log_{10}(1 + \frac{f}{700}) \tag{5}$$

4.5 Mel Cepstrum Coefficient

The frequency domain into time domain of the signal is converted by Discrete Cosine Transform (DCT) using Eq. 6.

$$C_m = \sum_{k=1}^{M} \cos\left[m\left(k - \frac{1}{2}\right)\frac{\pi}{M}\right]E_k \tag{6}$$

here M is the number of filter bank and 20 in our case, $1 \le m \le L$ is the number of MFCC coefficients.

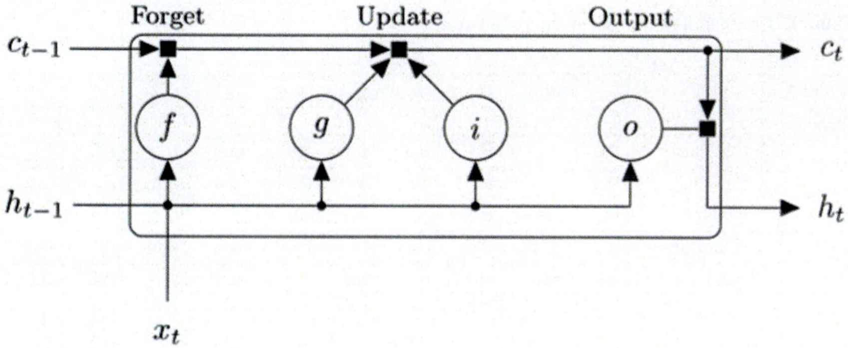

Fig. 3 A long short-term memory network

The first 14 coefficients are the MFCC as our primary feature. The ΔMFCC and Δ ΔMFCC are computed from MFCC parameters. Thus, for a single frame, the 42 number of features are used as our feature vector.

5 Result and Discussion

The Schematic diagram of an LSTM network is given in Fig. 3.

Where x_t is the input, C_t is the cell state, h_t is the hidden state, f is forget gate, g is memory cell, I is input gate, and o is the output gate. The advantages of using the LSTM model are better for classification in sequential data and avoid vanishing gradients with respect to vanilla RNN [14].

From each of the uttered word, we obtained a 42-dimensional feature with a variable number of frames, since it varies from person to person and word to word. All frames from a single record are grouped together with their categorical class label and are extracted both for training and testing dataset. The training set contains a total of 1600 sample and testing set of 400 samples. Then both training and testing sets are mean normalized to convert varying values of feature into a close interval. The training dataset is trained with the different hyperparameters of the proposed method of LSTM model as follows:

Number of input units: **42**-dimensional features
Number of LSTM units: **200** hidden units
Number of output units: **10** fully connected layer
Maximum epoch: **100**
Mini-batch size: **256**

We obtained the highest accuracy of 96.87% for the training set of 1600 sample. The confusion matrix for the only misclassification is given in Table 1.

The prediction accuracy depends on the dataset, the recording devices and the environment used. From the confusion matrix, we observed that some of the mis-

Table 1 Confusion matrix for misclassification

	Actual label										
Predicted label	Andul	–	0.6	0	0	0.4	0.4	0.3	0.2	0.2	0.5
	Bagnan	0.8	–	0	0.5	0.6	0	0.4	0	0.7	0.4
	Balichak	0.6	1	–	0	0.8	0	1.4	0.2	0	0.5
	Howrah	0	0	0.4	–	0.3	0	0	0.3	0.5	0
	Kharagpur	0	0.5	0.3	0	–	0.1	0.6	0.7	0	0.3
	Mecheda	0.8	0.2	0.1	0	0.5	–	0.7	0	0.8	0.6
	Medinipur	0.3	0	0.6	0.5	0.6	0.4	–	0	0.1	0
	Panskura	0	0.5	1	0.6	0	0.6	0	–	0.2	0
	Santragachi	1	0.2	0	0	0.2	1	1	0.6	–	1
	Uluberia	0.6	0.8	0.5	0.2	0	0.4	0.3	0	0.7	–

classification occurs because of the pronunciation styles and duration of voice of the same word depends on person to person as every human being a different vocal structure and different instruments used during recording. We also compared with different machine learning based classifier for the same dataset and obtained the highest training accuracy of 91.53%. The comparison of accuracy for 10 uttered words in machine learning and deep learning based classifier is given in Fig. 4. The testing accuracy obtained for the LSTM model was 87.6% for 400 utterances. The corresponding machine learning based highest testing accuracy was 80.5% using Support Vector Machine (SVM) classifier.

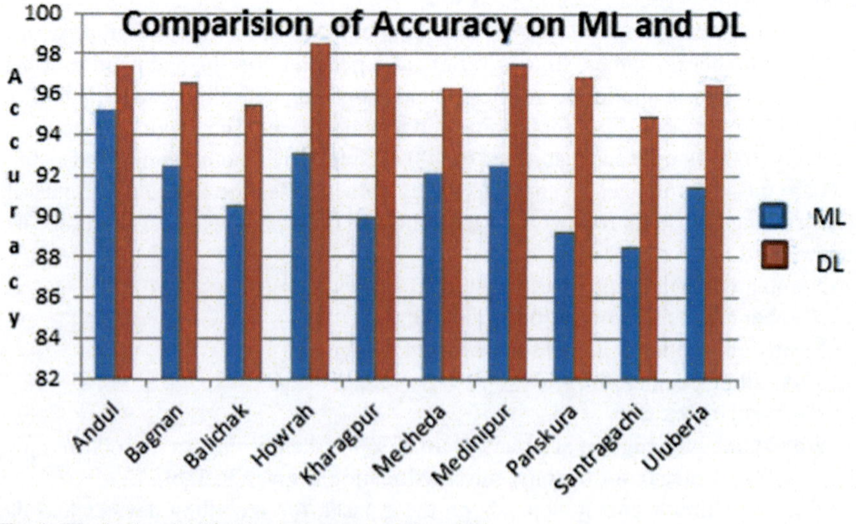

Fig. 4 Word-wise classification accuracy on ML and DL

With the analysis of the confusion matrix and Fig. 4 for sequence type of classification problem with time varies the data, the LSTM model works fine indeed of the traditional machine learning model.

6 Conclusion

In this paper, we have discussed the technique which is developed in every stage of the speech recognition system. Here we see that MFCC, ΔMFCC, and Δ ΔMFCC are widely used for feature extraction and LSTM network are used for data training where data is taken as almost noiseless data. But in future, we will work on noisy data for speech recognition in Deep Neural Network. As sequence classification for the recognition of the uttered word is almost correct using the LSTM model. Our dataset is very small as limited to a number of words. In our future work, we will emphasis on a big corpus of all the railway stations in South Eastern Railway of India and will take the record from the passengers in the platform area. So we have to give more importance to filtering the signal for noise cancelation. As a feature, some other audio parameters such as Linear Predictive Coefficients (LPC), Vector Quantization will be used in our future work.

References

1. Abdullah-al-Mamun, M.D., Mahmud, F., Islam, A., Zuhori, S.T.: Automatic Speaker Recognition System Using Mel Frequency Cepstral Coefficients (MFCC) and Vector Quantization (VQ) Approach
2. Selvaraj, S.P.P., Konam, S.: Deep Learning for Speaker Recognition (2017). Accessed April 19
3. Gaikwad, S.K., Gawali, B.W., Yannawar, P.: A review on speech recognition technique. Int. J. Comput. Appl. **10**(3), 16–24 (2010)
4. https://hub.packtpub.com/how-deep-neural-networks-can-improve-speech-recognitionand-generation/
5. Petridis, S., Li, Z., Pantic, M.: End-to-end visual speech recognition with LSTMs. In: 2017 IEEE International Conference on Acoustics, Speech and Signal Processing (ICASSP), pp. 2592–2596. IEEE (March 2017)
6. Amodei, D., Anubhai, R., Battenberg, E., Case, C., Casper, J., Catanzaro, B., Chen, J., et al.: Deep speech 2: end-to-end speech recognition in English and mandarin. In International conference on machine learning, pp. 173–182 (June 2016)
7. Kenny, P., Stafylakis, T., Ouellet, P., Gupta, V., Alam, M.J.: Deep neural networks for extracting baum-welch statistics for speaker recognition. In: Odyssey, vol. 2014, pp. 293–298 (June 2014)
8. Noda, K., Yamaguchi, Y., Nakadai, K., Okuno, H.G., Ogata, T.: Audio-visual speech recognition using deep learning. Appl. Intell. **42**(4), 722–737 (2015)
9. Nassif, A.B., Shahin, I., Attili, I., Azzeh, M., Shaalan, K.: Speech recognition using deep neural networks: a systematic review. IEEE Access **7**, 19143–19165 (2019)
10. Mazzoni, D., Dannenberg, R.: Audacity [software]. The Audacity Team, Pittsburg, PA, USA (2000)

11. Bachu, R.G., Kopparthi, S., Adapa, B., Barkana, B.D.: Separation of voiced and unvoiced using zero crossing rate and energy of the speech signal. In: American Society for Engineering Education (ASEE) Zone Conference Proceedings, pp. 1–7 (June 2008)
12. Scarr, R.: Zero crossings as a means of obtaining spectral information in speech analysis. IEEE Trans. Audio Electroacoust. **16**(2), 247–255 (1968)
13. Palia, N., Kant, S., Dev, A.: Performance evaluation of speaker recognition system. J. Discrete Math. Sci. Cryptogr. **22**(2), 203–218 (2019)
14. Sak, H., Senior, A., Beaufays, F.: Long short-term memory recurrent neural network architectures for large scale acoustic modeling. In: Fifteenth Annual Conference of the International Speech Communication Association (2014)

Voting in Watts-Strogatz Small-World Network

Soujanya Ray, Kingshuk Chatterjee, Ritaji Majumdar, and Debayan Ganguly

Abstract Social network science and social interaction science is an interesting and steadily growing field of research in current time. Social network influences the lives of millions, and therefore the propagation of influence in such a network deserves study. In this paper, we have studied voting in a Watts-Strogatz small-world network for two parties. In our model, each node has an initial bias towards one of the parties (or can be neutral) and are influenced by their neighbours to vote for a particular party. We show via simulation that (i) for linear of logarithmic voting function, the small-scale variation is minimum, but the majority of the nodes tend to align towards one of the parties in the long term, (ii) for periodic voting function, the small-scale variations are sharp and oscillating, but in the long term the number of voters in each party roughly remains the same, (iii) the degree of the graph does not seem to have a strong influence on the result, and (iv) networks with higher small-world probability tend to resist the alignment of voters towards a particular party for a longer time as compared to networks with lower small-world probability. This majority voting model seems to efficiently capture the potential behaviour of voters on small-world network over a campaigning period.

S. Ray
Department of Computer Science and Engineering, Netaji Subhash Engineering College, Kolkata, India
e-mail: shaun.ray.1996@gmail.com

K. Chatterjee
Department of Computer Science, Government College of Engineering and Ceramic Technology, Kolkata, India
e-mail: kingshukchaterjee@gmail.com

R. Majumdar
Advanced Computing and Microelectronics Unit, Indian Statistical Institute, Kolkata, India
e-mail: majumdar.ritajit@gmail.com

D. Ganguly (✉)
Department of Computer Science, Government College of Engineering and Leather Technology, Kolkata, India
e-mail: debayan3737@gmail.com

D. Bhattacharjee et al. (eds.), *Proceedings of International Conference on Frontiers in Computing and Systems*, Advances in Intelligent Systems and Computing 1255,
https://doi.org/10.1007/978-981-15-7834-2_31

Keywords Social network analysis · Majority voting model · Small-world network

1 Introduction

Influence propagation over an underlying network has been widely studied in solid-state physics [1, 2], where the electron spins are arranged in an underlying topology and the entire system [3] tries to attain a minimum energy configuration. Each electron tries to maintain or flip the spin direction of its neighbouring electrons in order to reduce the tension between them. Such an interaction is termed as nearest neighbour interaction. However, other systems, where each electron can affect other electrons which may be some k hops away, have also been studied. Such a system is termed as a Voting Model [4] since it captures the essence of voters trying to influence others towards their own party.

In this paper, we study campaigning or voting in a social network. A majority voting model refers to a pool of population selecting and voting in favour of one of many candidates; the candidate with the maximum number of votes is declared to be the winner. This model differs from the previous voting models in that each voter may have varying inclination towards one of the parties, or may be completely neutral. Each voter tries to influence its nearest neighbour towards his own party. However, this interaction is not only a function of the bias of the receiver, but also the weight of the edge between them, which captures the influence of the voter to the receiver.

The rest of the paper is arranged as follows: In Sect. 2, we discuss our model and the underlying topology in detail. The methodology of our study and the results are discussed in Sect. 3 and Sect. 4, respectively. We conclude in Sect. 5.

2 The Voting Model and Its Underlying Structure

A social network can be modelled as a connected undirected graph where nodes represent individuals and edges denote the connection between them. This kind of graph, which captures social media, is often referred to as a small-world network [5, 6]. Different graphical models have been proposed to study a small-world network. The first random graph was the Erdős and Rényi [7] random graph model. However, this model of graph lacks two important properties observed in the real-world networks—(i) they do not generate local clustering and triadic closures, and (ii) they do not account for the hubs and do not show the existence of power law. For our study, we have considered a Watt-Strogatz network [8]. Watts-Strogatz small-world network is a k regular connected graph and is characterized by p, the probability with which a node connects to its nearest node. If $p \sim 0$ then all the nodes connect to their nearest neighbour, whereas when $p \sim 1$, the nodes connect to their farthest neighbour. These properties make Watts-Strogatz network ideal to model a real-world social network.

For the rest of this paper, small world or small-world network or network will always imply the Watts-Strogatz small-world network unless otherwise stated.

The network can be identified with three characteristics—the voters, their bias and their influence. For a network with n nodes, each node (i.e. voter) is assigned a unique number between 0 and $n-1$. This number can be considered as the identification of the voter [9]. For this paper, we have limited ourselves to a model with two parties A and B. Each voter can either be biased towards party A or party B, or can be neutral. Bias is represented as an integer where positive implies inclination towards party A and negative implies inclination towards party B. The absolute value of the bias is an indication of the voter's loyalty for the party. For example, if two nodes i and j have biases 3 and 4, respectively, then i is biased towards party A, j towards party B and the loyalty of j is more towards his party than that of A. Each node has some influence towards its neighbour. If two nodes i and j are neighbours, then the undirected edge between them can be considered as two directed edges, where $w(i, j)$ implies the influence of i on j and $w(j, i)$ implies the influence of j on i. In general $w(i, j) \neq w(j, i)$.

In the campaigning phase, each voter tries to influence its neighbour towards his party. In other words, if b is the bias of a node j, then after a single round of campaigning, the new bias of that node is

$$b_{\text{new}} = b + \sum \text{sign}(i) * f(w(i, j))$$

where $\text{sign}(i) = +1$ if the ith neighbour is biased towards party A, -1 if he is biased towards party B, and 0 otherwise. $f(w(i, j))$ is some function of the influence of i on j, and the summation is over all neighbours of the node j. Therefore, depending on the number of influences coming for each party, the bias or endurance of a particular voter towards a party either increases or decreases. In this paper, we have considered linear, logarithmic and periodic weight functions. Each of the functions clearly shows different effects on the network.

A single run of the campaigning function [10, 11] denotes that each of the nodes (voter) has interacted with its neighbouring nodes (voters) only once. In real life, campaigning is carried out over some time period. In this paper, we start with a random number of voters for party A and party B, while some of the voters are neutral. We repeat the campaigning step for $t > 1$ iterations and denote the final number of voters for each party.

In the next section, we describe our simulation methodology explicitly.

3 Methodology

The entire simulation process can be broadly categorized into two steps: (i) generation of the Watts-Strogatz network and (ii) the campaigning process. We discuss the steps of each of these processes henceforth.

3.1 Generation of the Network

The network is generated physically in the following two steps:

1. Construct a ring lattice graph with N nodes each connected to k neighbours, $k/2$ on each side, with an edge (i, j) present between two nodes i and j if and only if

$$0 < |i - j| \mod (N - k/2) \leq k/2$$

2. For every node $i = 0,\ldots, N-1$, take every edge connecting i to its $k/2$ rightmost neighbours, i.e. every edge $(i, j \; mod \; N)$ with $i < j \leq i + k/2$, and rewrite it with probability β.

Once the network is generated, the logical properties of the nodes are as follows:

1. Node identification number: $0,\ldots, N-1$.
2. Bias/endurance: For this paper, only two party voting schemes are considered. Each voter is assigned a random number in the range $(-b, b)$ where positive (negative) value implies that the voter is biased towards party A(B). The higher the absolute value of bias, the more is the voter inclined towards that party.
3. Influence: Each edge (i, j) is assigned with two values $w(i, j)$ and $w(j, i)$. These values are randomly generated to take value in the range (0, 1).

3.2 Campaigning Phase

As discussed before, the bias of each voter is affected by each campaigning step as follows:

$$b_{\text{new}} = b + \sum \text{sign}(i) * f(w(i, j))$$

We have considered three voting functions:

1. Linear: $f((w(i, j)) = w(i, j)$
2. Logarithmic: $f((w(i, j)) = \log(w(i, j))$ where the logarithm is with respect to base 2
3. Periodic: $f((w(i, j)) = \sin(w(i, j))$

4 Results and Discussions

The parameters used for simulation are as follows:

(i) Number of nodes (N) = 10,000
(ii) Number of campaigning phases = 100.

The nodes were randomly assigned bias values, and the bias determines the initial number of voters for party A and party B, i.e. if the bias is >1 (<1), then the node is assigned to party A(B). Some nodes are neutral. We show the variation in the number of voters for each party when the three types of voting functions are used.

Linear Campaigning

The results shown in Figs. 1, 2 and 3 have used a linear voting function, degree 5 for each node and the small-world probability of the graphs are 0.1, 0.5 and 0.9, respectively. As the number of iterations increases, the nodes tend to get aligned to the same party, and the process of alignment towards a single party is somewhat linear. Moreover, it is also notable that for lower probability, the alignment is visible

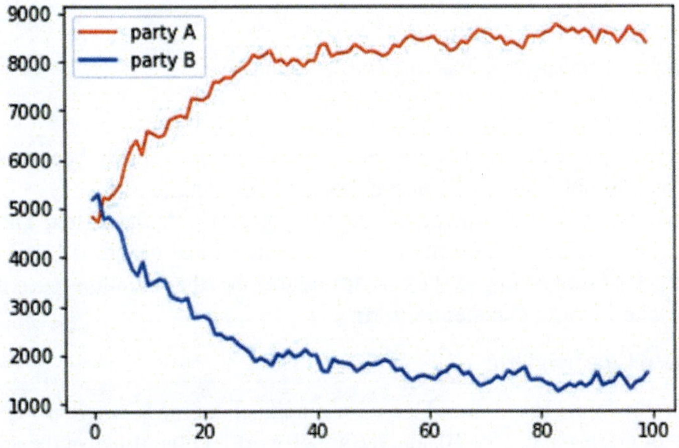

Fig. 1 Distribution of voters with a small-world probability 0.1

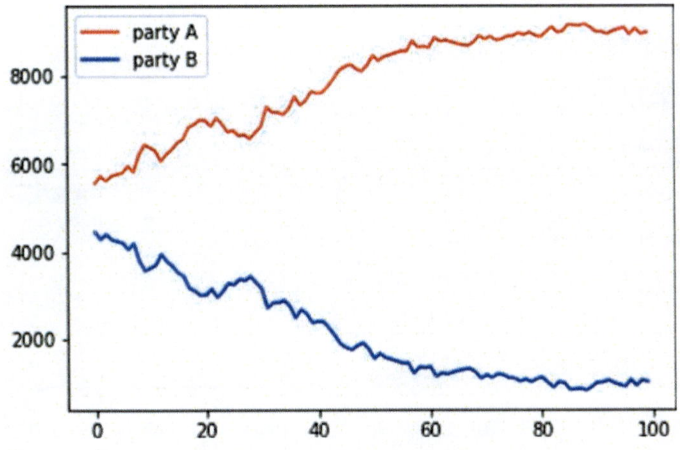

Fig. 2 Distribution of voters with a small-world probability 0.5

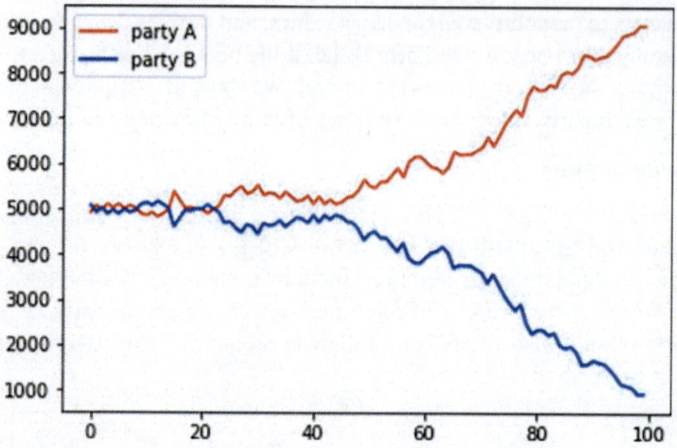

Fig. 3 Distribution of voters with a small-world probability 0.9

within a few steps of iterations only. However, as the probability becomes higher, the nodes tend to hold on to their original bias a little longer.

The same exercise has been carried out for degree 10 regular graph. The images of Figs. 4, 5 and 6 show the results for small-world probability of 0.1, 0.5 and 0.9, respectively. It is interesting to see that for degree 10, the variation loses its linear nature and takes a more exponential form.

Logarithmic Campaigning

The results shown in Figs. 7, 8 and 9 are for graphs of degree 5, whereas Figs. 10, 11 and 12 are for graphs of degree 10, and the small-world probabilities of the graphs are 0.1, 0.5 and 0.9, respectively. The voting function used for this step is logarithmic.

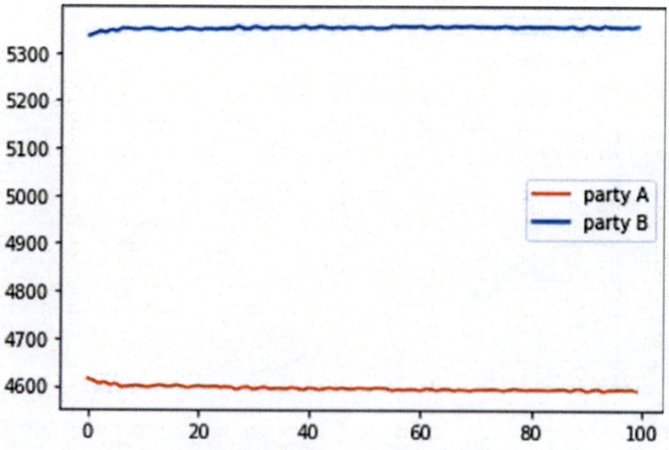

Fig. 4 Distribution of voters with a small-world probability 0.1

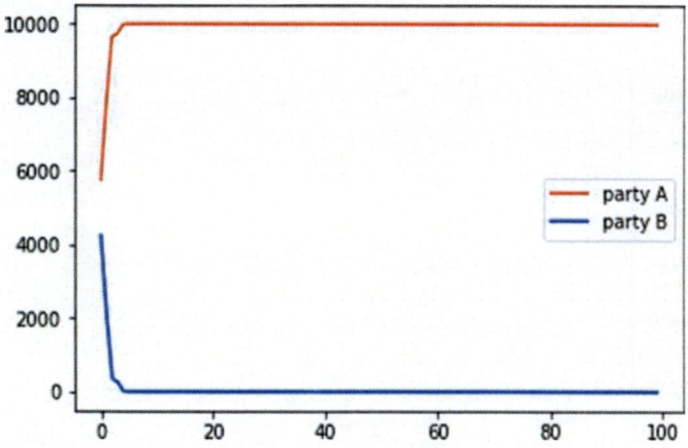

Fig. 5 Distribution of voters with a small-world probability 0.5

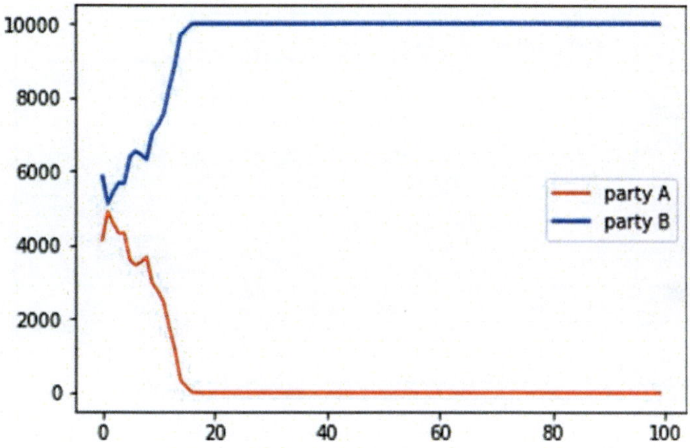

Fig. 6 Distribution of voters with a small-world probability 0.9

Similar to linear function, one of the parties captures almost all the nodes as the number of iterations is increased. In accordance with the voting function, variation has a logarithmic nature.

Periodic Campaigning

The results shown in Figs. 13, 14 and 15 are for graphs of degree 5, whereas Figs. 16, 17 and 18 are for graphs of degree 10, and the small-world probabilities of the graphs are 0.1, 0.5 and 0.9, respectively. The voting function used for this step is periodic.

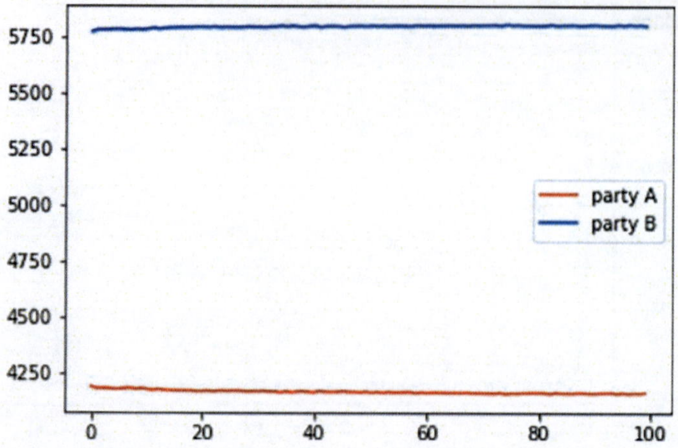

Fig. 7 Distribution of voters with a small-world probability 0.1

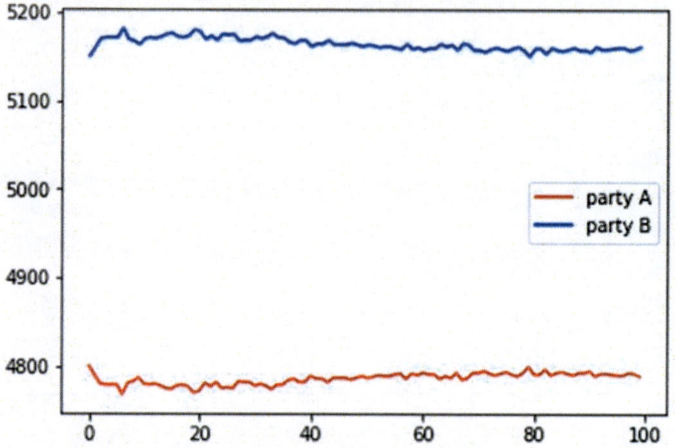

Fig. 8 Distribution of voters with a small-world probability 0.5

Interestingly, this is the only function of the three where one of the parties does not end up influencing all the nodes. The number of supporters for each party oscillates throughout the number of steps, and in certain cases (Figs. 15 and 16), at the end of 100 rounds of influence (i.e. 100 iterations), the number of voters in each party is almost equal. Even when the number of voters is significantly different, an oscillating pattern is noticeable in each step of campaigning.

Discussion

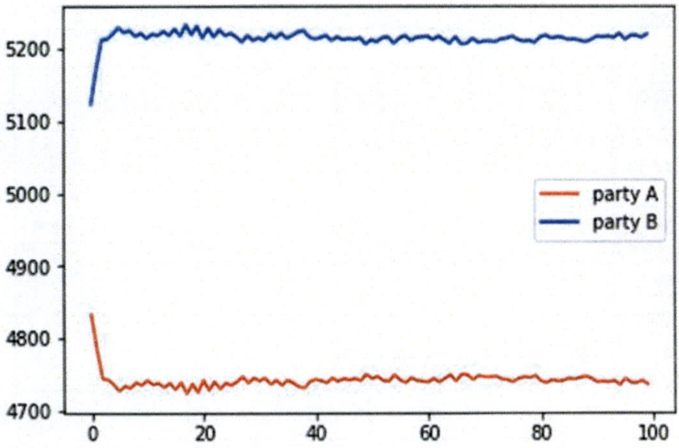

Fig. 9 Distribution of voters with a small-world probability 0.9

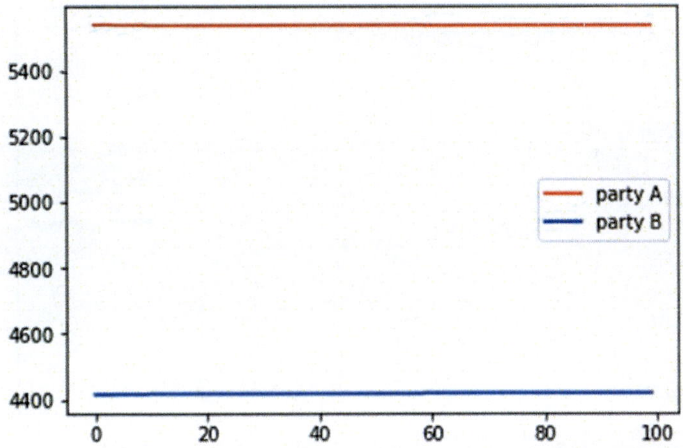

Fig. 10 Distribution of voters with a small distribution of voters with a small-world probability 0.1

In this section, we have shown the simulation results of campaigning on Wattz-Strogatz network with various small-world probabilities and two different degrees. One of the inferences that seems to be drawable from these simulation results is that for linear and logarithmic campaigning functions, one of the two parties tends to influence all the nodes of the network, irrespective of the degree. The variation in the number of supporters for each party with increasing steps follows linear and logarithmic slope for those two voting functions. However, for periodic voting function, the situation is slightly different. Although this function injects an oscillating nature in the number of voters for each party, this voting function does not seem to align all the voters towards one of the parties. More simulation and mathematical studies are necessary in order to predict the number of voters for each party if the initial number

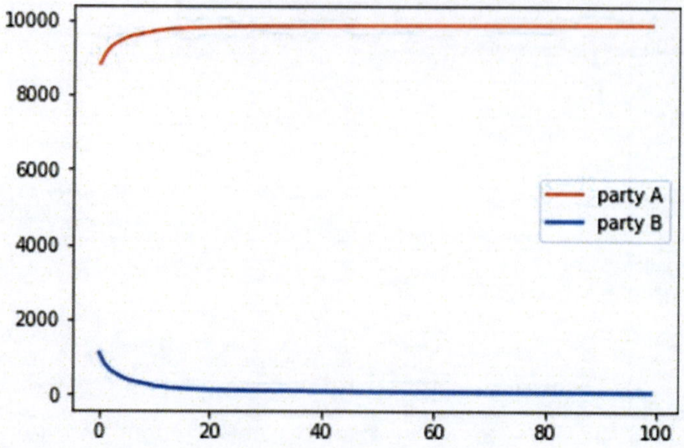

Fig. 11 Distribution of voters with a small-world probability 0.5

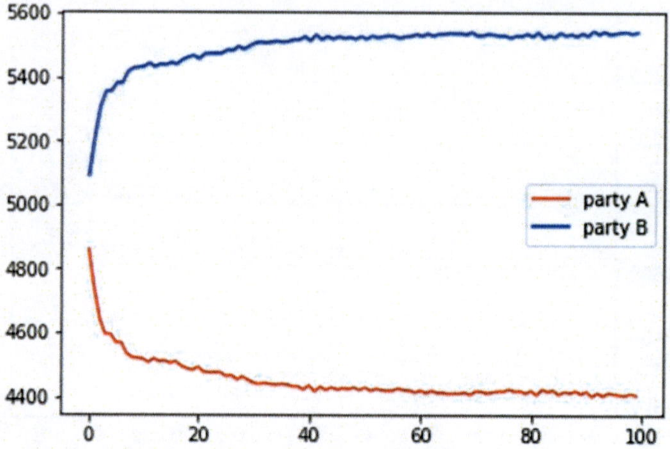

Fig. 12 Distribution of voters with a small-world probability 0.9

of voters, degree of graph, small-world probability and the number of iterations is provided.

It is easy to extend this study with more parties. In such a scenario, the bias, and each party can be considered as vectors, and the scalar product of the instantaneous bias with the party will provide its instantaneous alignment.

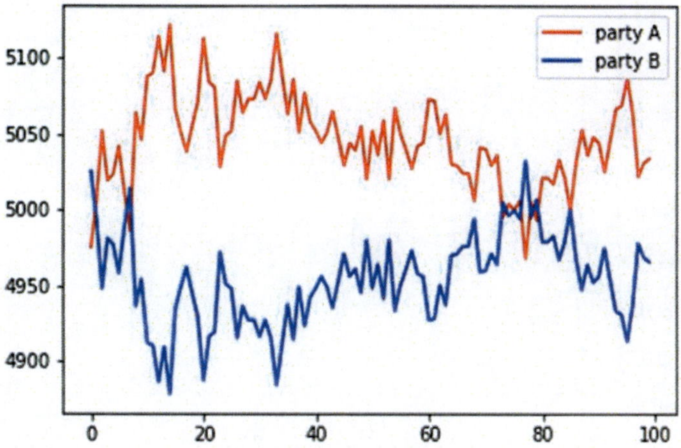

Fig. 13 Distribution of voters with a small-world probability 0.1

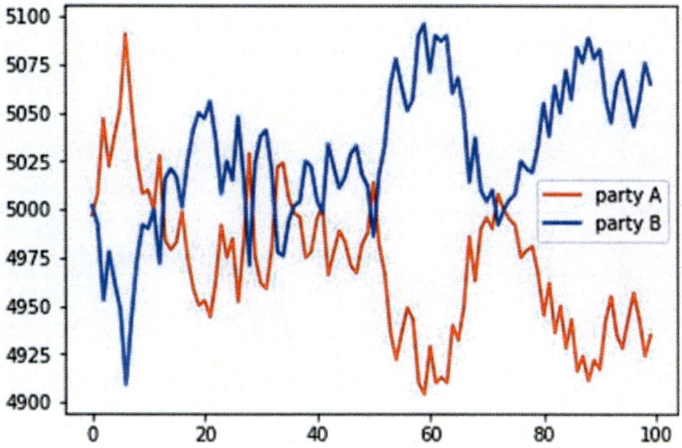

Fig. 14 Distribution of voters with a small-world probability 0.5

5 Conclusion

In this paper, we have studied numerically the distribution of the alignment of voters over a Watts-Strogatz network. We have shown the results for different small-world probabilities, different degrees of graph and different campaigning functions. The inference of this simulation is that linear and logarithmic function quickly aligns most of the voters towards one of the parties. On the other hand, periodic function does not have such a drastic effect in long range. But in short range, this function oscillates the number of voters rigorously. Therefore, a hypothesis may be that a small-term

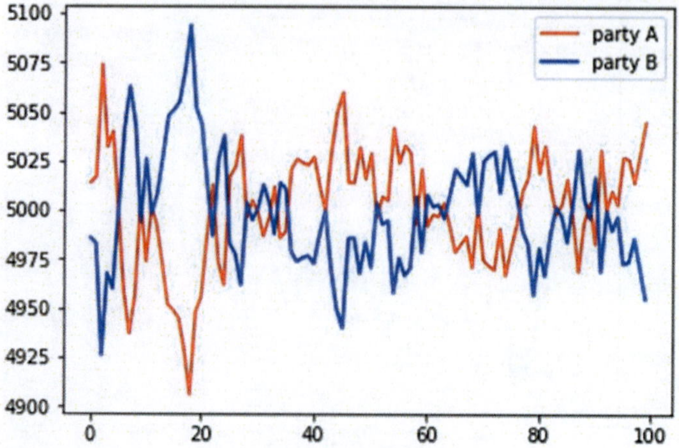

Fig. 15 Distribution of voters with a small-world probability 0.9

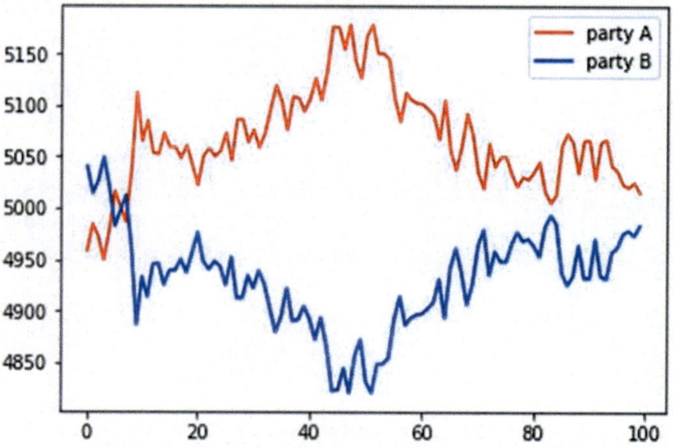

Fig. 16 Distribution of voters with a small distribution of voters with a small-world probability 0.1

sudden change in the number of voters is governed by the periodic function, whereas slow, long time change is governed by linear and logarithmic functions.

This model simulation can be further improved by inducing sudden change and anomaly in bias due to adverse events and studying the impact on the whole network. A future scope of this work is to try to predict the number of voters for each party if the initial number of voters, the small-world probability, the degree of graph and the campaigning function is provided.

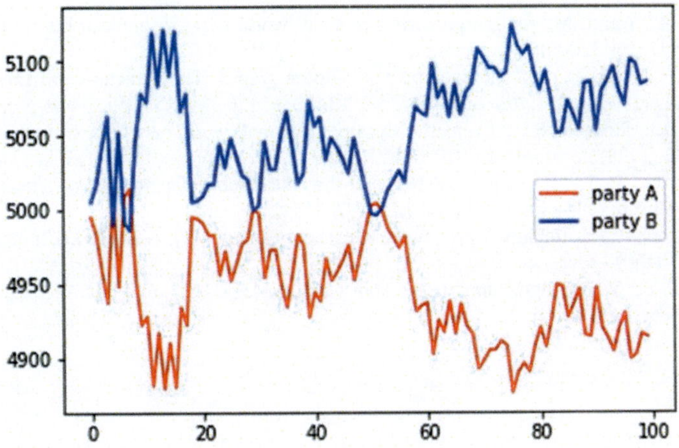

Fig. 17 Distribution of voters with a small-world probability 0.5

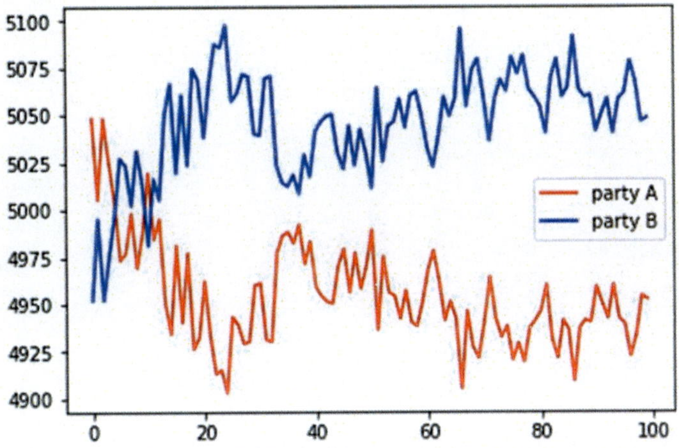

Fig. 18 Distribution of voters with a small-world probability 0.9

References

1. Glauber, R.J.: Time-dependent statistics of the Ising model. J. Math. Phys. **4**(2), 294–307 (1963). https://doi.org/10.1063/1.1703954
2. Cicera, M.V.A., Lima, F.W.S.: Majority-vote model in one-dimensional on directed small-world networks. Phys. Astron. Int. J. **2**(6) (2018). https://doi.org/10.15406/paij.2018.02.00140
3. Luz, E.M.S., Lima, F.W.S.: Majority vote on a directed small world network. Int. J. Mod. Phys. C **18**(08), 1251–1261 (2007). https://doi.org/10.1142/s0129183107011297
4. Tullock, G.: Problems of majority voting. J. Polit. Econ. **67**(6), 571–579 (1959). https://doi.org/10.1086/258244
5. Amarla, L.A.N., Scala, A., Barthelemy, M., Stanley, H.E.: Classes of small-world networks. Proc. Nat. Acad. Sci. **97**(21), 11149–11152 (2000). https://doi.org/10.1073/pnas.200327197

6. Barrat, A., Weigt, M.: On the properties of small-world network models. Europ. Phys. J. B **13**(3), 547–560 (2000). https://doi.org/10.1007/s100510050067
7. Erdos, P., Renyi, A.: On the evolution of random graphs. The Structure and Dynamics of Networks (1960). https://doi.org/10.1515/9781400841356.38
8. Watts, D.J., Strogatz, S.H.: Collective dynamics of small world networks. Nature **393**(6684), 440–442 (1998). https://doi.org/10.1038/30918
9. Gudgin, G., Taylor, P.J.: Electoral bias and the distribution of party voters. Trans. Inst. Br. Geogr. **63**, 53 (1974). https://doi.org/10.2307/621532
10. Gardenfors, P.: Positionalist voting functions. Theory Decis. **4**(1), 1–24 (1973). https://doi.org/10.1007/bf00133396
11. Parhami, B.: Voting algorithms. IEEE Trans. Reliab. **43**(4), 617–629 (1994). https://doi.org/10.1109/24.370218

Transfer Learning in Skin Lesion Classification

Samrat Mukherjee and Debayan Ganguly

Abstract The number of patients diagnosed with Melanoma, a fatal type of skin cancer, is increasing day by day. Convolutional Neural Networks (CNNs) have shown promising results in image classification problems. Transfer learning is, using the weights of already trained models on a different dataset, and using that model to classify images into respective classes. It is quite useful when there is the lack of training data. Using a custom model, trained on cat-dog dataset, we got a classification sensitivity of 0.90.

Keywords Transfer learning · Melanoma · Sensitivity

1 Introduction

Melanoma is described as a type of cancer which could be life-threatening. Dermatologists can detect it through visual inspection with an accuracy of 65–66% [1]. Though the death rate due to melanoma cancer is very high, it can be cured if it is detected earlier [2]. To test whether or not the interested lesion is melanoma or benign, physicians use methods like ABCD (Asymmetry, Border irregularity, Color patterns, and Diameter) rule [3] and seven-point checklist [4]. Deep learning-based approach of extracting features from data has outperformed systems with hard-coded rules in domains like computer vision, natural language processing, etc. In our case, we used Convolutional Neural Networks (CNNs) for extracting and learning features from the training images required classifying the images (Fig. 1).

We used the trained model, in this paper, for the purpose of classification of lesions into their respective classes. At first, we train our model with the dataset of

S. Mukherjee · D. Ganguly (✉)
Government College of Engineering and Leather Technology, Kolkata 700098, India
e-mail: debayan3737@gmail.com

S. Mukherjee
e-mail: samratpisv123@gmail.com

D. Bhattacharjee et al. (eds.), *Proceedings of International Conference on Frontiers in Computing and Systems*, Advances in Intelligent Systems and Computing 1255,
https://doi.org/10.1007/978-981-15-7834-2_32

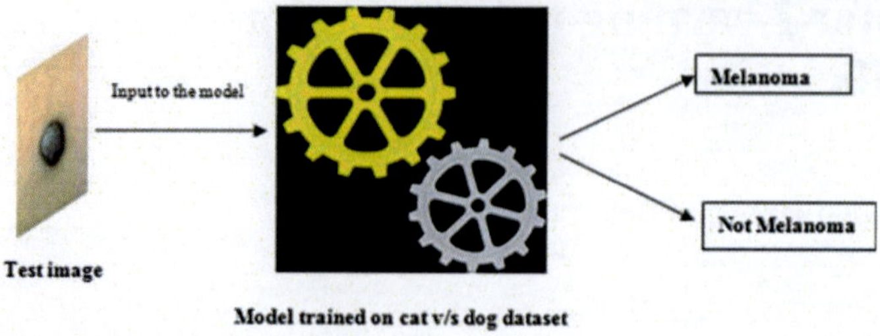

Fig. 1 Working of the proposed system during testing time

cat versus dogs [5]. Then, we fetch the data [6], for the purpose of testing classification of lesions.

Transfer Learning can be defined as a method to improve the learning in a new task T_t on the target domain D_t by transferring the knowledge from a related task T_s from source domain D_s. This methodology of learning improves the performance of predictive function $f_T(\cdot)$ for learning task T_t by discovering and transferring latent knowledge from D_s and T_s, where D_s $!=$ D_t and/or T_s $!=$ T_t. Also, in most of the cases, the size of D_s is much larger than the size of D_t.

The way we applied transfer learning can be termed as heterogeneous transfer learning. We term it so because, although the training and testing data are all image data, the source and target domain don't share the label space and belong to different feature spaces.

Our proposed model achieved classification sensitivity better than the state-of-the-art.

2 Related Work

Nasr-Esfahani et al. [7] used data from [6] for their methodology. They created a custom CNN model with two convolutional layers each followed by a MaxPool layer. Each image was pre-processed by correcting the distortion due to illumination, and Gaussian Filter (with $\sigma = 2$) was applied on some parts of the image based on the mask generated previously, using k-means classifier. The images were rotated by 90, 180, 270, and 360 degrees, hence creating four copies each. Images were cropped from the corners as described in the paper, creating eight copies excluding the original. Hence, total number of images got multiplied by 36. The result obtained has been described in Table 1, in Sect. 4.

Zagrouba and Barhoumi [8] attempted to classify skin images as melanoma or non-melanoma by an artificial neural network classifier using image processing techniques. They applied median filter on each of the images, for noise removal. Following that, they applied a PCA-like analysis for better edge detection. The

Table 1 Quantitative comparison of various proposed methods in literature, highlighted represents best results

Methods	Metrics				
	Sensitivity	Specificity	PPV	NPV	Accuracy
Zagrouba and Barhoumi [8]	0.46	**0.87**	0.7	0.71	0.7
Munteanu and Cooclea [9]	0.82	0.57	0.56	0.83	0.67
Nasr-Esfahani et al. [7]	0.81	0.8	**0.75**	**0.86**	**0.81**
Proposed method (evaluated on test set)	**0.9**	0.34	0.56	0.78	0.6

lesion was extracted by the use of region growing method, succeeded by a pre-processing step based on fuzzy sets. Then, after studying several existing attributes which described the lesion malignity, a set of nine features were formulated to distinguish melanoma from benign lesions. For the purpose of classification of the lesions into melanoma or non-melanoma, a mono-layer perceptron was trained with those attributes discussed.

3 Experiment

3.1 Dataset

We started with the data available at [5], for the cat-dog classification problem. The dataset contains roughly equal number of cat and dog images (~ 12,500 images). The data for skin lesion type was obtained from [6] which contain a total of 179 images of two types of lesion, melanoma and non-melanoma. 86 images are of the not melanoma class and the rest are melanoma. The dataset also contains corresponding segmentation masks for the images with lesions. But we didn't use it for our proposed method.

3.2 Methodology

The sizes of the input were made uniform 256 * 256 * 3, where 3 represent the number of color channels. We used Convolutional Neural Networks (CNNs) to extract the attributes from the training images of cats and dogs, and hence classify them into respective classes. We then use the trained model to classify images of skin lesions. CNNs require large amount of training data so as to satisfactorily extract features from the images for the purpose of classification. Due to the lack of availability of data, we used transfer learning concept to tackle the problem.

The proposed CNN model has four convolutional layers. Each convolutional layer is followed by a layer of Batch Normalization (BN) and MaxPooling2D with a pool size of (2, 2). Batch normalization is commonly used layer in neural networks to normalize the input to the next layer.

The general formula for batch normalization is

$$x' = \frac{x - E(x)}{\mathrm{sqrt}(\mathrm{var}(x))}$$

where sqrt and var being square root and variance, respectively.

First two blocks are succeeded by a dropout layer with probability of 0.25 to minimize the risk of over-fitting. The convolving kernel used for this paper is of size (3, 3). The first convolutional layer has 32 filters followed by 64, 128, and 256 filters in the following convolutional layers. 'ReLU' being the most commonly used activation function has been applied to the convolutional layers. The output of the fourth convolutional block is fed into a fully connected layer with 128 nodes. It was again followed by a batch normalization and dropout layer with the same probability. We added the final dense layer with two nodes for the classification of images, with softmax activation. Figure 2 describes the said model in a pictorial form.

The images of the training set were fed into the model after applying some random rotation, random zoom-in.

3.3 Training

We used Kaggle kernels to test our proposed methodology. The described model was trained for 50 epochs, using Adam optimizer while minimizing categorical cross-entropy loss. Although this is a binary classification problem, we opted for categorical cross-entropy loss, instead of binary-cross-entropy, as the latter is a special case of the former. The equation for cross-entropy loss is given by

$$L(\theta) = -\frac{1}{n}\sum_{i=1}^{n}\sum_{j=1}^{m} y_{ij}\log\left(p_{ij}\right)$$

where i denotes number of samples and j indicates number of classes. For binary classification problem $m = 2$.

The model was ~95.5% accurate on the validation set. Figure 3 describes the training process.

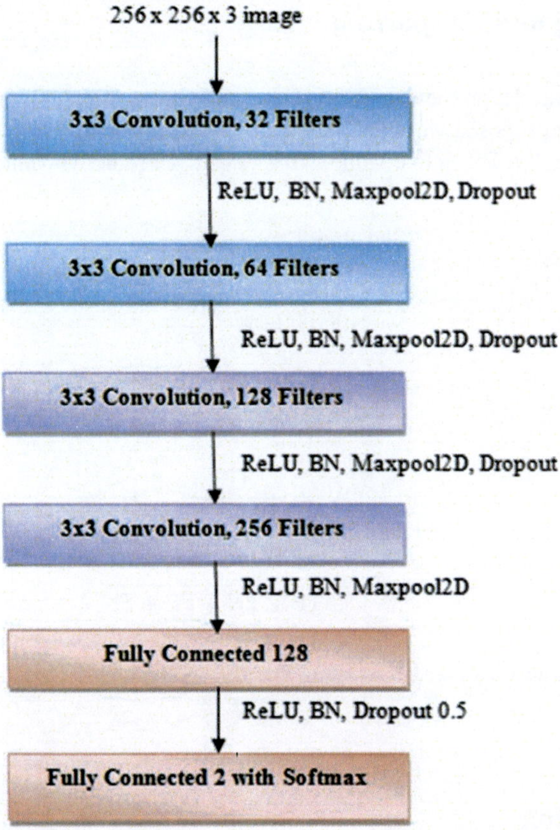

Fig. 2 Network architecture used

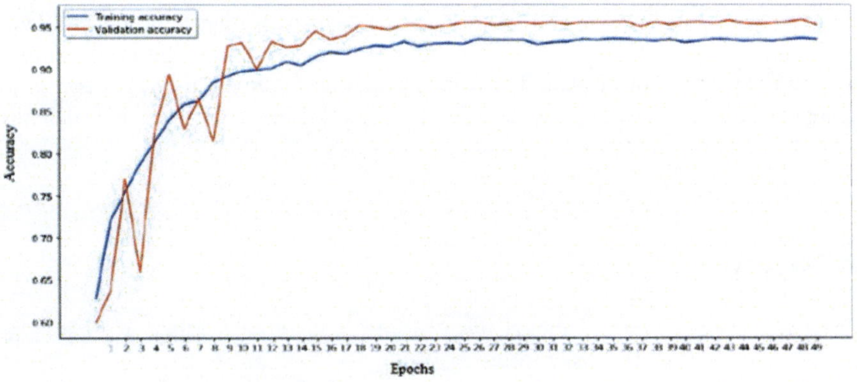

Fig. 3 Accuracy plot during training procedure

3.4 Testing and Comparison

We tested our model on the data from [6], to know how efficient it is to explore the problem in the proposed way. The metrics we used for comparison are sensitivity, specificity, Negative Predictive Value (NPV), Positive Predictive Value (PPV), where

$$\text{Sensitivity} = \frac{\text{TP}}{\text{TP} + \text{FN}} \tag{1}$$

$$\text{Specificity} = \frac{\text{TN}}{\text{TN} + \text{FP}} \tag{2}$$

$$\text{NPV} = \frac{\text{TN}}{\text{TN} + \text{FN}} \tag{3}$$

$$\text{PPV} = \frac{\text{TP}}{\text{TP} + \text{FP}} \tag{4}$$

$$\text{Accuracy} = \frac{\text{TP} + \text{TN}}{\text{TP} + \text{TN} + \text{FP} + \text{FN}} \tag{5}$$

where TP = True Positive, FP = False Positive, TN = True Negative, and FN = False Negative.

4 Conclusion

Proper and early diagnosis of melanoma skin cancer will save a lot of precious human lives. Though we are far from state-of-the-art performance in this problem, the proposed methodology might be helpful in cases where we don't have access to a sufficient amount of data. The accuracy can be improved if we get access to more data. Also, if we can get dataset from similar domain to train on, then also it is expected to perform better.

References

1. Jayasekara Pathiranage, A.: Convolutional neural networks for predicting skin lesions of melanoma. All Regis University Theses, 843 (2017)
2. American Cancer Society: Cancer Facts & Figures 2016. American Cancer Society, Atlanta, GA, USA (2016)
3. Nachbar, F., Stolz, W., Merkle, T., Cognetta, A.B., Vogt, T., Landthaler, M., Bilek, P., Falco, O.B., Plewig, G.: The ABCD rule of dermatoscopy: high prospective value in the diagnosis of doubtful melanocytic skin lesions. J. Am. Acad. Dermatol. **30**(4), 551–559 (1994)

4. Argenziano, G., Fabbrocini, G., Carli, P., Giorgi, V., De, S.E., Delfino, M.: Epiluminescence microscopy for the diagnosis of doubtful melanocytic skin lesions: comparison of the ABCD rule of dermatoscopy and a new 7-point checklist based on pattern analysis. Arch. Dermatol. **134**(12), 1563–1570 (1998)
5. Kaggle Home-page. kaggle.com
6. Skin cancer detection, Vision image processing labs. https://uwaterloo.ca/vision-image-processing-lab/research-demos/skin-cancer-detection
7. Nasr-Esfahani, E,. Samavi, S, Karimi, N., Soroushmehr, S.M.R., Jafari, M.H., Ward, K., Najarian, K.: Melanoma detection by analysis of clinical images using convolutional neural network. In: 38th Annual International Conference of the IEEE Engineering in Medicine and Biology Society (EMBC), Orlando, USA (2016)
8. Zagrouba, E., Barhoumi, W.: A preliminary approach for the automated recognition of malignant melanoma. Image Anal. Stereol. **23**(2), 121–135 (2004)
9. Munteanu, C., Cooclea, S.: Spotmole-melanoma control system (2009). http://spotmole.com

Knowledge-Based Expert System for Diagnosis of Agricultural Crops

Subhankar Halder and Sourav Kumar Singh

Abstract Agricultural crops are the most essential entity to sustain human life. A huge chunk of these crops get damaged every year due to the infestation of several diseases, which poses a serious threat to global food security. Detecting these diseases at an early stage is a necessity in order to treat them and prevent any damage. This requires the need of someone skilled to diagnose the anomaly correctly and provide the methods of treatment. Expert Systems can easily help to detect these diseases and provide the appropriate methods of treatment to the user. The advancements in the fields of computer vision methodologies and deep learning have enabled the development of models that can accurately detect the diseases and pave the course for smartphone-enabled detection system. Utilizing a public datatset of 43429 healthy and diseases crop leaves accumulated in controlled condtions, we trained a deep convolutional neural network to recognize 14 types of crops and 38 types of illness (or absence of). The model that is trained attains the accuracy of 99.7% on a test set that is held-out. The model assists the expert system to correctly recognize the disease that the crop is suffering from and this approach to train models of deep learning on huge image datasets provides a comprehensible path toward more efficient crop disease identification and treatment on a universal scale.

Keywords Deep learning · Expert systems · Crop diseases

1 Introduction

Humans have always been dependent on edible plants to survive and the first civilizations thrived after the practices of agriculture came under prominence. Modern advancements in science and technology have enabled human beings to be able to

S. Halder (✉) · S. Kumar Singh
Jalpaiguri Government Engineering College, Jalpaiguri 735102, India
e-mail: sh1944@cse.jgec.ac.in

S. Kumar Singh
e-mail: sks1927@cse.jgec.ac.in

D. Bhattacharjee et al. (eds.), *Proceedings of International Conference on Frontiers in Computing and Systems*, Advances in Intelligent Systems and Computing 1255,
https://doi.org/10.1007/978-981-15-7834-2_33

produce sufficient enough food to satisfy the need of about 7 billion people. However, a variety of factors including plant diseases [1] and climate change [2] pose an immense amount of threat to global food security. Crop diseases not only cause a great deal of threat to food security overall, but they also cause catastrophic problems for small-scale farmers whose entire means of support rely on these crops. In the developing world, excess of about 80% of the overall agricultural production is generated by small-scale farmers [3], and records of the loss of production of about 50% of the crops are due to diseases and pests [4]. Moreover, the considerable fraction of people engrossed with hunger (50%) resides in households dependent on smallholder farming [5], leading to smallholder farmers being the section of people that is majorly endangered with food supply disrupted by pathogens.

Diagnosis of the sickness of crops is not an effortless job. It requires expertise and knowledge regarding diseases that affect crops. People can rely on a smart system that has the knowledge and expertise (expert systems) to provide them the ability to identify and diagnose any type of disease, and choosing appropriate treatment for that. A number of systems have been invented to reduce the crop loss that occurs mostly because of diseases. The recognition of illness accurately when it affects first is a major step towards efficient disease control. Organizations working in the field of agriculture and clinics working for the improvement of crop health have historically provided the means of the identification of diseases that the crops suffer from. However, recently the information provided online with respect to the diagnosis of diseases of crops has also helped with the cause, the extension of the growth of Internet anchoring it worldwide. Recent developments in smartphones, equipped with advanced HD camera and associated technologies has resulted in a condition where correct identification of disease leveraged by automated computer vision techniques, can enable their implementation at great magnitude. In the paper, we display the technical practicality of these techniques with the help of a deep learning system that uses 43429 images of 14 species of crops that suffer from overall 38 diseases (or healthy). These images have been provided openly through the PlantVillage project [6]. Figure 1 lists down all the pairs of crops and diseases with an example of each.

2 Literature Review

Riley et al. [7] came up with an expert system that can identify diseases that crops suffer from with the help of perceptible symptoms that are showed by the diseased plants. A coordinated expert system for the managing of cucumber plants was developed by El-Dessoki et al. [8]. Rajkishore et al. [9] created a system for the identification of pest infestation, illness and any anomalies with the Indian Mango. Lopez-Morales et al. [10] created JAPIEST: a cohesive unit that is intelligent enough for the identification of and limiting the illness of tomatoes and pest infestation that they face in hydroponics greenhouses. An expert system for the identification of diseases and pests in olive crops was developed by Gonzalez-Andujaret [11]. All of these expert

Fig. 1 Example of images of crops and corresponding diseases represented in the PlantVillage dataset. (1) Apple Scab (2) Apple Black Rot (3) Apple Cedar Rust (4) Apple healthy (5) Cherry healthy (6) Cherry Powdery Mildew (7) Corn Gray Leaf Spot (8) Corn Common Rust (9) Corn healthy (10) Corn Northern Leaf Blight (11) Grape Black Rot (12) Grape Black Measles (13) Grape Healthy (14) Grape Leaf Blight (15) Peach Bacterial Spot (16) Peach healthy (17) Bell Pepper Bacterial Spot (18) Bell Pepper healthy (19) Potato Early Blight (20) Potato healthy (21) Potato Late Blight (22) Strawberry Healthy (23) Strawberry Leaf Scorch (24) Tomato Bacterial Spot (25) Tomato Early Blight (26) Tomato Late Blight (27) Tomato Leaf Mold (28) Tomato Septoria Leaf Spot (29) Tomato Two Spotted Spider Mite (30) Tomato Target Spot (31) Tomato Mosaic Virus (32) Tomato Yellow Leaf Curl Virus (33) Tomato healthy

systems discussed actually focus on a single species of plants and employ only one process of identification. In the study discussed in this paper, a variety of agricultural crops have been covered along with a number of diseases associated with them.

A number of developments have come up so far for examining and identification of illness of crops, that comprises of RGB imaging, hyperspectral and multispectral technologies [12], etc. Macedo-Cruz et al. [13] developed a system that measured the impact because of frost in case of oats. The first step that they performed was the changing of images in RGB form to L*a*b space of color. Three separate thresholding plans of action were used by authors: Otsu's Technique, Isodata procedure, and fuzzy thresholding. Yao et al. [14] developed a system that was able to diagnose and differentiate three separate categories of illness that the considered rice crops were suffering from. Here as well, Otsu's Technique was employed, which resulted in the isolation of the regions affected by the corresponding diseases. Features with textures, outline and the color were extracted, the former one originating out of HSV color space. Ultimately, the details are provided to a system that performs the final classification with the help of Support Vector Machine.

3 Proposed Method

3.1 Expert System

3.1.1 Knowledge Acquisition

Various sources are used for collecting the knowledge and the data that are employed by the expert systems. Primarily, the source of data is engineers and scientists working in the field of agriculture and farmers. Secondarily, specialized online websites and corresponding databases are used for source of information. After collection of knowledge, the information is then coded into a language known as CLIPS (mostly used in case of expert systems). Here, we have used the PyKnow module, which is a Python library that is mostly employed for coding expert systems as it is very heavily inspired from the CLIPS language.

3.1.2 Knowledge Representation

Rule-based systems are systems that provide results based on deduction of symbols, which here need the changing of information specific to crop diseases to a symbolic statement. Expert systems discussed here are the same. The rules and facts have been coded using PyKnow, which is strongly based on CLIPS. CLIPS is an Expert System shell that is pattern marching and forward reasoning. The inference engine technique provided here automatically compares the facts against the patterns present in rules and then with the help of this it decides which of the rules would be applicable.

Subsequently, the actions associated with the particular rules that are applicable are executed. The executed actions might lead to new facts kicking in which in turn would lead to new rules being applicable. The same process continues till there are not more suitable rules present which satisfy the required conditions. The format of the rules is in the form of If-Then statements, which means that the action is performed only when the condition mention under the If statement is satisfied to be true.

3.1.3 Knowledge Inference

The inference engine of a expert system provides the final result of a expert system. The inference predicted by the model, i.e., the disease that the crop is suffering from is provided to the inference engine which then treats this as a fact. The engine matches this fact with the patterns in the rules which are in the knowledge base, to find the rules which are applicable in this scenario. The rule applicable for execution, the rule whose pattern matches with the required fact is the one which provides with the advice or the steps to be taken after the diagnosis of the disease. The PyKnow module helped to create facts and rules, which when executed by the input from the model, provided the further steps to be taken in case of successful diagnosis of the crop disease.

3.2 Transfer Learning

It should be noted here that the quantity of images that we can use to learn the required information is limited. A useful way in this scenario would be Transfer Learning, which is used to develop strong networks based on classification using limited data, by tuning the parameters of an already trained network on some large dataset, in our case ImageNet [15]. Even though our task of classification of diseases is a targeted classification task for image category with fine grains in comparison to ImageNet, the function of the low lying layers is to only encode features that are simple and hence can be used further for most tasks related to the comparison. An example would be the foremost layer represents color and direction. Even if initially the model is not trained on the PlantVillage dataset of images, the same model can be used to be activated for the spots on the leaves that are symptoms of diseases. ResNet architecture yielded very good results in the 2015 ILSVRC. The ResNet involves layers with different weights and portrays a considerable improvement with respect to previous configurations. The convolutional filters used are very small.

3.3 Transfer Learning of the Residual Neural Network

The ResNet-50 that comes pre-trained is a CNN which uses about a million images from ImageNet databse [15]. The depth of the network is 50 layers and it is capable of classifying images to about 1000 various categories, including that of objects, living beings, etc. Hence, the network comes learned with rich feature representations with respect to a varied category of images. The input size of the network is 224-by-224.

At first we trained Resnet-50 model on segmented version of the dataset with default values (RMSProp Optimizer). The graph didn't quite converge as it is supposed to, rather it showed some weird behavior. After Fine-Tuning the model and changing some hyperparameters such as the learning rate and also changing the optimizer to Adam optimizer, it did started to converge but it still didn't produce the best result. After adding Learning rate scheduler the train Val graph converged perfectly and also produced the best result.

As a summary, we can say that our experimental configurations are based on tuning the following criteria:

1. Architecture for Deep Learning
   - ResNet-50
2. Training mechanism choice
   - Transfer Learning
3. Dataset Type Choice
   - Color
4. Distribution of Training-Testing set
   - Train: 80%, Test: 20%.

An attempt for hyperparameters standardization was also made by us across our experiments in order to achieve fair comparison. The following settings for the hyperparameters were used for our experiments:

- Base Learning rate: 0.0001
- Optimizer: Adam
- Learning rate scheduler: multiplies by 0.3 after every 4th
- Train Batch Size: 32
- Val Batch Size: 32.

All the models were created in keras which is a Python open-source neural network library.

Table 1 Leaf disease detection techniques comparison

References	Applied technique	Plant diseases	Classification accuracy (%)
This work	Resnet-50 with color images	All 14 crop species	98.48
Mohanty et al. [16]	GoogLeNet with color images	All 14 crop species	99.35
Monzurul Islam et al. [17]	SVM with RGB imaging	Potato	95
Hu et al. [18]	Hyperspectral imaging	Potato	95
Asfarian et al. [19]	Texture analysis and PNN	Rice	83
Xu et al. [20]	K-nearest neighbor	Tomato	82

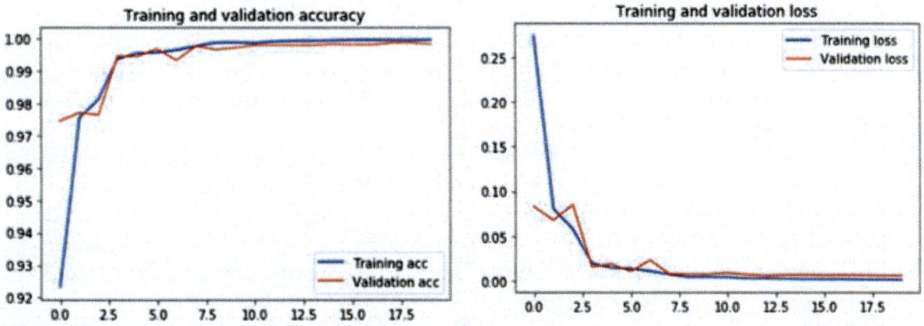

Fig. 2 Accuracy and Loss that the Resnet-50 model comes across during training and validation

4 Results and Discussion

The ResNet-50 model is applied on the datatset with 43429 images with 38 classes, after fine-tuning the hyperparameters. During the experiment, the dataset was split into three divisions: the training set (80%), the validation set (10%), and the testing set (10%). Test Accuracy of 99.78% was achieved in the Resnet-50 model. This work has been compared to other proposed works in Table 1.

Figure 2 portrays the accuracy and loss that the Resnet-50 model comes across during training and validation.

5 Conclusion

In this paper, deep learning CNN models are used to develop a system that is automated and can be easily availed. The most common illness in 14 agricultural crops is diagnosed with the help of computational powers and the expert system provides the steps to be taken further so that the global food security can be sustained. Our system will provide farmers an efficient and practical method of disease diagnosis and will inform them of the steps to take after that. We plan to work on integrating this expert system to a smartphone assisted environment where the image taken from a smartphone can help a farmer directly access the help of the expert system.

References

1. Strange, Richard N., and Peter R. Scott. Plant disease: a threat to global food security. Annual review of phytopathology 43 (2005)
2. Tai, A.P., Martin, M.V., Heald, C.L.: Threat to future global food security from climate change and ozone air pollution. Nat. Clim. Change **4**(9), 817–821 (2014)
3. IFAD, UNEP. Smallholders, food security and the environment. Rome: International Fund for Agricultural Development (2013)
4. Harvey, C.A., et al.: Extreme vulnerability of smallholder farmers to agricultural risks and climate change in Madagascar. Philos. Trans. Royal Soc. Lond. B: Biol. Sci. **369**(1639) (2014)
5. Sanchez, Pedro A., and Monkombu S. Swaminathan. Cutting world hunger in half. Science 307, no. 5708 (2005): 357-359
6. Hughes, D.P., Salathé, M.: An open access repository of images on plant health to enable the development of mobile disease diagnostics (2015). CoRR abs/1511.08060
7. Riley, M.B., Williamson, M.R., Maloy, O.: Plant disease diagnosis. Plant Health Instr (2002). https://doi.org/10.1094/PHI-I-2002-1021-01
8. El-Dessoki, A., Edrees, S., El-Azahry, S.: CUPTEX: An Integrated Expert System for Crop Management of Cucumber (ESADW-93), p. 1993. Molar, Cairo-Egypt, May 1993
9. Rajkishore, P., Ranjan, K., Sinha, A.K.: AMRAPALIKA: an expert system for the diagnosis of pests, diseases and disorders in Indian mango. Knowl.-Based Syst. **19**, 9–21 (2006)
10. Lopez-Morales, V., Lopez-Ortega, O., Ramos-Fernandez, J., Munoz, L.B.: JAPIEST: an integral intelligent system for the diagnosis and control of tomatoes diseases and pests in hydroponic greenhouses. Expert Syst. Appl. **35**, 1506–1512 (2008)
11. Gonzalez-Andujar, J.L.: Expert system for pests, diseases and weeds identification in olive crops. Appl. Expert Syst. (2008). https://doi.org/10.1016/j.eswa.2008.01.007
12. Wahabzada, M., Mahlein, A.-K., Bauckhage, C., Steiner, U., Oerke, E.C., Kersting, K.: Plant phenotyping using probabilistic topic models: uncovering the hyperspectral language of plants. Sci. Rep. 6, Nature, Article number: 22482 (2016)
13. Macedo-Cruz, A., Pajares, G., Santos, M., Villegas-Romero, I.: Digital image sensor-based assessment of the status of oat (Avena sativa L.) crops after frost damage. Sensors **11**(6), 6015–6036 (2011)
14. Yao, Q., Guan, Z., Zhou, Y., Tang, J., Hu, Y., Yang, B.: Application of support vector machine for detecting rice diseases using shape and color texture features. In: 2009 International Conference on Engineering Computation, IEEE, Hong Kong, pp. 79–83 (2009)
15. Mehdipour Ghazi, M., Yanikoglu, B., Aptoula, E.: Plant identification using deep neural networks via optimization of transfer learning parameters. Neurocomputing **235**, 228–235 (2017)
16. Mohanty, S.P., Hughes, D.P., Salathe, M.: Using deep learning for image-based plant disease detection. Front Plant Sci. **7**, 1419 (2016)
17. Monzurul Islam, A., Dinh, A., Wahid, K., Bhowmik, P.: Detection of potato diseases using image segmentation and multiclass support vector machine. CCECE.2017.7946594
18. Hu, Y.H., Ping, X.W., Xu, M.Z., Shan, W.X., He, Y.: Detection of late blight disease on potato leaves using hyperspectral imaging technique. PubMed **36**(2), 515–519 (2016)
19. Asfarian, A., Herdiyeni, Y., Rauf, A., Mutaqin, K.M.: Paddy diseases identification with texture analysis using fractal descriptors based on fourier spectrum. In: Proceeding of International

Conference on Computer, Control, Informatics and Its Applications, pp. 77–81. Jakarta. IEEE (2013)
20. Xu, G., Zhang, F., Shah, S.G., Ye, Y., Mao, H.: Use of leaf color images to identify nitrogen and potassium deficient tomatoes. Pattern Recogn Lett **32**, 1584–1590 (2011)

An Effective Hybrid Statistical and Learning Based Approach to Confined Domain of a Web Document from Corpus

Amit Dutta

Abstract Textual documents are unstructured by nature. In order to organize a large archive of these documents, it is required a way to identify these document by using some key information. Keyword extraction is one such important method for text summarization. Getting a suitable domain of document from a large archive of documents is one of prime challenge today. Categorization of unstructured documents for faster retrieval; preferences of individual based on their likings, information could be retrieve from an information hub. In this work, Web documents undergoes web noise elimination prior to keywords extraction using term frequency (tf) and inverse document frequency (idf) method to create document corpus. This corpus is used to generate ontological representation of domain based on the keyword contain in the documents. This method produces domain representative document. The document corpus is classified by using multiclass classification technique. An exhaustive testing is done for predicting the domain of the document and obtained an accuracy of 88–93%.

Keywords Domain · Keyword · Ontology · Unstructured document

1 Introduction

Textual document management system like Knowledge Management System, Digital Library System, and Information Retrieval System requires a major concept of text summarization for storage and retrieval of unstructured textual document. The document content identification process involve retaining the representative words of the document and eliminating less important words like stop word and others which are generally used to make the document content grammatically right. There are various statistical models that are popularly used in finding keyword automatically. Such models are like Bayesian, tf-idf, K-Nearest Neighbor.

A. Dutta (✉)
Department of Information Technology, St. Thomas College of Engineering and Technology, Kolkata, India
e-mail: to.dutta@gmail.com

D. Bhattacharjee et al. (eds.), *Proceedings of International Conference on Frontiers in Computing and Systems*, Advances in Intelligent Systems and Computing 1255,
https://doi.org/10.1007/978-981-15-7834-2_34

For the purpose of the said work, documents are collected from web, require elimination of noise mainly HTML tags and control characters. Web content mining played a significant role as the original documents are collected from web. It is an important work in searching useful knowledge or information from Web [1]. Web mining broadly is of two types, content mining and structure mining [2]. A novel work of Noise Elimination from Web Page Based on Regular Expressions for Web Content Mining is described in [3]. Noise eliminated document undergoes for keyword extraction. There are several methods used for finding keywords from a document. A Keyword Extraction algorithm based on scaled conjugate gradient backpropagation in 2017, is described in [4]. In 1998, Tseng applied the schemes of selecting multilingual keywords [5]. In 2000, D. Freitag proposed the combination of schemes of generating keywords and important information from the given document [6]. In 2009, Wu et al., proposed a unique method of Data Preprocessing in SVM-Based Keywords Extraction from Scientific Documents [7]. A support vector machine approach for keyword extraction is proposed in 2006, by Kuo Zhang [8], and others. In 2000, Freitag proposed [6], the combination of schemes of generating keywords and important information from the given document. Maron [9] describe a Probabilistic Indexing and Information Retrieval method [9]. Information retrieval on internet using meta-search engines is proposed [10], by Manoj et al., in 2008. A domain specific ontology for web crawler is initiated in [11]. Another approach is to find key words from noisy documents is initiated in [12]. Filtering of web document for key phrase is an approach of web mining is explained in [13]. Using Neighborhood Knowledge finding, single document key phrase is proposed in [14]. Document classification is done using feature clustering method is be described in [15].

Five categories of documents are collected from e-journal/authentic website. Preprocessing of web documents are done in two stages, the document set undergoes web noise elimination process; here mainly the HTML tags and control characters are filtered out. In the next stage, stop words and other less significant characters with respect to the keyword are eliminated. The keyword extraction method tf-idf is applied to get document represented by key phrases only. This data set is used for the proposed method. In this paper, Sects. 2 and 3, are used to describe preliminaries and proposed method whereas Sects. 4 and 5, represent experimental results and conclusion, respectively.

2 Preliminaries

The documents are collected from web. Preprocessing of document involves elimination of web noise, stop words, and other less significant word or character with respective to key phrase before applying keyword extraction method tf-idf. The processed document corpus is taken as input to proposed method.

Noise in a Web Document: In a document, anything apart from document content is a noise. Web document usually contains two types of noises, local and global. Local or Intra-page noise are blocks of information within a single webpage, like banner, advertisements, navigational guides, decoration pictures etc. Global noise or Inter-page noise is usually no smaller than individual pages like, mirror sites, legal/illegal duplicated web page, old versioned web page, etc. Applying web noise elimination method as discussed in [3], the pre-pre-processed data is obtained.

Term Frequency or tf: The term frequency of a term or candidate denotes the number of times that candidate appears throughout the test document. Candidates which are keywords are found to be repeated more number of times compared to the non-keywords. Thus, a candidate is more likely to be a keyword, if it appears more throughout the document, i.e., if the term frequency is high, it is more likely to be a keyword.

Inverse Document Frequency or idf: To obtain the document frequency of a term, a corpus of documents is needed. The document frequency of a term denotes the number of documents in which the term appears (at least once). It has been found that the words (or group of consecutive words), which are found in more number of documents are more likely to be common with little significance to the contents of the document. Thus, candidates who have high document frequency are less likely to be keywords. In other words, the candidates with a lower document frequency value are more likely to be keywords, so we need to obtain inverse document frequency from the document frequency using the formula

$$\text{idf}_{\text{term}} = \log 10 * \frac{N}{1 + \text{df}_{\text{term}}} \tag{1}$$

where N, is the number of document in domain corpus. df_{term} and idf_{term} are respectively document frequency and inverse document frequency.

Calculate Term Frequency-Inverse Document Frequency or tf-idf: The term frequency (tf) and inverse document frequency, together represent how the frequencies of the candidates in the corpus and in the document itself are related to the relevance of the candidate in the document. Greater the value of tf, greater is the chance of being keyword, while smaller the value of document frequency (df), greater is the chance of being keyword. So, the product of the two values (td and idf), together would denote the likelihood of a candidate being a keyword. The value tf-idf is directly related to the likelihood of a keyword, thus, greater the value of tf-idf, greater the chance of the candidate being a keyword. The value of tf-idf for each candidate must be calculated for each document using expression (2).

$$\text{tf\_idf}_{\text{term}} = \text{tf}_{\text{term}} * \text{idf}_{\text{term}} \tag{2}$$

3 Proposed Method

In the commencement phase five categories of documents are collected from e-journal/authentic website. These domains are, respectively, "artificial intelligence", "banking", "cricket", "image processing", "computer networking". Preprocessing of these web documents are done in two stages, the document set undergoes web noise elimination process; mainly the HTML tags, control characters and others [3], are filtered out. In the consequent stage, stop words and other less significant characters with respect to the keyword are eliminated from the initially processed document set using [4]. Further, we apply keyword extraction method tf-idf to get the index value of all candidate keywords of the respective document in the document corpus. In the subsequent phase, respective domain representative document RD_i (i = 1–5) is created by combining all the keywords of each document in the respective domain. Here duplicate words are not considered. We have considered 200 authentic documents spread on said five fields. The said method is applied on half of the total number of document in the corpus to find domain representative document. Eventually, five domain representative documents are created as five reference classes. Now using these reference classes, we test the entire data set in threefold, obtained result is discussed in the experimental result section.

Step 1: For each document in the archive checks for web noise. If the web noise is present it is eliminated by using [3].

Step 2: Input of step1 undergoes the following stages as shown in Fig. 1 below to obtain document with candidate keywords, first 20 candidates with highest tf-idf score are taken only to reduce computational overhead.

Step 3: Let A be the archive of half number of documents from corpus, equally distributed in five domains. Considering AD_i (i = 1–5) is the individual domain document corpus, containing one-fifth of document in A.

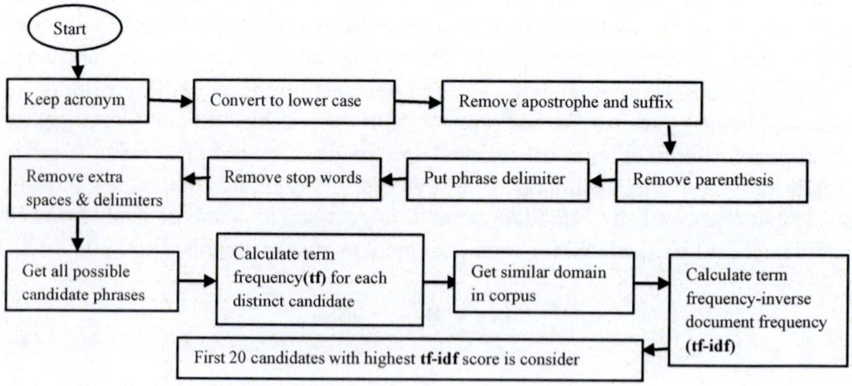

Fig. 1 Steps to preprocessing of document

Fig. 2 A partial view of ontological representation of the domain 'computer networking'

OSI Model ----- ARP ---- Modem
TCP/IP Model ----- Broadcast ----- Multicast
UTP ---- DNS ----- NFS
STP ---- dial-up line ----- POP
PPP ---- EGP ----- Router
LAN ---- FTP ----- RPC
WAN ---- HTTP ----- X11
Crosstalk ---- Ibgp ----- SMTP
PSTN ---- IMAP ----- Socket
Analog Transmission ---- IP ----- Switch
Digital Transmission ---- ISO ----- Telnet
Access Point ---- Modem ----- X.25

Step 4: Using each document of the domain AD_i ($i = 1$–5), an ontological representation of the domain containing unique keywords are prepared by combining all the candidate keywords of each document in the respective domain. Duplicate words are not considered.

Step 5: Thus, *domain representative document* RD_i ($i = 1$–5) forms and gives the sense of the domain.

Step 6: Classification of testing documents.

6.1 Applying step1 & step2 consecutively to obtain the candidate keywords of the test document. Now apply similarity matching of candidate keywords with domain representative document RD_i ($i = 1$–5).
6.2 If N_i ($i = 1$–5) be the matching number for each document with domain representative document RD_i ($i = 1$–5).
6.3 For each document Max(N_i) ($i = 1$–5) is the predicted domain RD_i.

A partial view of ontological representation of the domain "computer networking" is presented in the following Fig. 2. Terms shown here are related to each other in some way.

4 Experimental Results

The experiment is conducted on the basis of documents collected from e-journal/authentic website. The five categories of documents are, respectively, "artificial intelligence", "banking", "cricket", "image processing", "computer networking". For the purpose of collecting document, Google search engine is used. A document corpus is formed with size 200. We consider equal number of documents in each domain category.

The experiment is carried out using PYTHON (3.6), in 32-bit Windows 7 Professional with Service Pack 1. The processor used is Core i3, 3.70 GHz and 4.00 GB RAM. Experiment is done in threefold. In first fold, 100% data (here document) is used. In the next two consecutive phase, 75 and 50% data are used for testing purpose.

Table 1 Actual and predicted result

Data size		Domain D1	Domain D2	Domain D3	Domain D4	Domain D5
100% of document corpus	Actual	40	40	40	40	40
	Predicted	37	37	37	35	38
75% of document corpus	Actual	30	30	30	30	30
	Predicted	27	27	27	27	28
50% of document corpus	Actual	20	20	20	20	20
	Predicted	18	18	18	17	18

Table 2 Accuracy in true prediction of domain

	Domain D1	Domain D2	Domain D3	Domain D4	Domain D5
Fold 1	92.50	92.50	92.50	87.50	95.00
Fold 2	90.00	90.00	90.00	90.00	93.33
Fold 3	90.00	90.00	90.00	85.00	90.00
Average accuracy	90.83	90.83	90.83	87.50	92.78

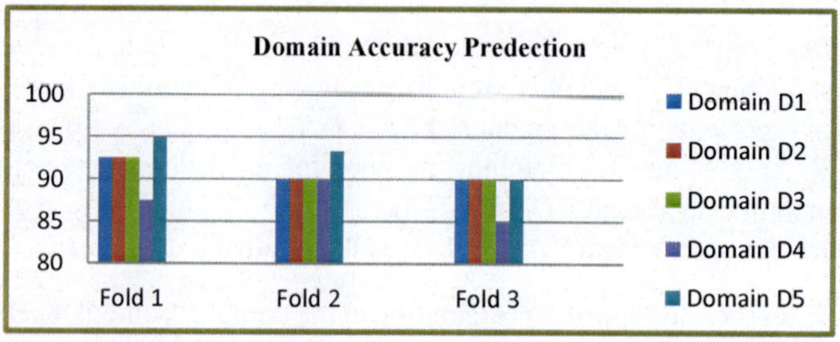

Fig. 3 Chart shows comparative fold result

In the first row of Table 1, 40 samples to each of the 5 domains are consider as fold1 and predicted values are 37, 37, 37, 35, 38 for each domain respectively shown in next row.

In the following Table 2 accuracy in true prediction of domain is given (Figs. 3 and 4).

5 Conclusion

Majority of the real-world textual documents is an unstructured in nature. Unstructured document arrangement and its categorization for organized storage and faster retrieval is one of the prime challenge today. The proposed method for finding domain of a document using domain representative document is able to predict domains of a

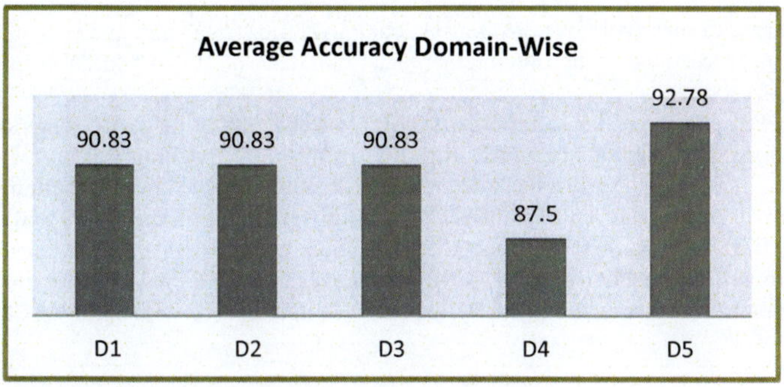

Fig. 4 Chart shows domain-wise average accuracy of prediction

test documents with descent accuracy level of 88–93%, which is effective and efficient compared with other methods like feature clustering method described in [15]. The result may produce better level of accuracy if the sample size is increased to few thousand. For this purpose, other statistical measurement tools like recall and precision could be useful.

References

1. Han, J., Chang, K.C.-C.: Data mining for web intelligence. IEEE Comput. **35**(11), 64–70 (2002)
2. Srivastava, J., Desikan, P., Kumar, V.: Web mining—concepts, applications, and research directions. In: Chu, W., Lin, T.Y. (eds.) Foundations and Advances in Data Mining. STUDFUZZ, vol. 180, pp. 275–307. Springer, Heidelberg (2005)
3. Dutta, A., Paria, S., Golui, T., Kole, D.K.: Noise elimination from web page based on regular expressions for web content mining. In: ICACNI-2014, Kolkata, vol. 27, pp. 545–554, June 2014. Springer
4. Aich, A., Dutta, A., Chakraborty, A.: A scaled conjugate gradient backpropagation algorithm for keyword extraction. In: LNCS Conference Proceedings, INDIA—2017, VietNam, 15–17 June 2017. Springer
5. Tseng, Y.: Multilingual keyword extraction for term suggestion. Proc. SIGIR **98**, 377–378 (1998)
6. Freitag, D.: Machine learning for information extraction in informal domains. Mach. Learn. **39**, 169–202 (2000)
7. Wu, C., Marchese, M., Wang, Y., Krapivin, M., Wang, C., Li, X., Liang, Y.: Data preprocessing in SVM-based keywords extraction from scientific documents. In: ICICIC'09, pp. 810–813. IEEE Computer Society, Washington, DC, USA (2009)
8. Zhang, K., Xu, H., Tang, J., Li, J.: Keyword extraction using support vector machine. In: WAIM'06, pp. 85–96. Springer, Berlin, Heidelberg (2006)
9. Maron, M.E.: Probabilistic indexing and information retrieval. In: Sparck, K., Willett, P. (eds.) Readings in Information Retrieval. Readings in Information Retrieval, pp. 39–46 (1997)
10. Manoj, M., Elizabeth, J.: Information retrieval on Internet using meta-search engines: a review. CSIR 739–746 (2008)
11. Mukhopadhyay, D., Biswas, A., Sinha, S.: A new approach to design domain specific ontology based web crawler. In: 10th International Conference on Information Technology, pp. 289–291 (2007)

12. Grineva, M., Grinev, M., Lizorkin, D.: Extracting key terms from noisy and multi-theme documents. In: Proceedings of the 18th International Conference on World Wide Web, pp. 661–670, NY, USA (2009)
13. Turney, P.D.: Coherent key phrase extraction via web mining. In: Proceedings of the 18th International Joint Conference on AI, IJCAI'03, pp. 434–439, San Francisco, CA, USA (2003)
14. Wan, X., Xiao, J.: Single document key phrase extraction using neighborhood knowledge. In: Proceedings of the 23rd AAAI Conference on Artificial Intelligence, pp. 855–860 (2008)
15. Harish, B.S., Udayasri, B.: Document classification: an approach using feature clustering. In: Thampi, S., Abraham, A., Pal, S., Rodriguez, J. (eds.) Recent Advances in Intelligent Informatics. Advances in Intelligent Systems and Computing, vol. 235. Springer, Cham (2014)

Face Recognition Using Siamese Network

Srinibas Rana and Dakshina Ranjan Kisku

Abstract Face recognition has become very popular in biometrics in recent time, especially after the availability of GPU-based processing technology and ease of implementing deep learning methodologies. Still, it is treated as computationally expensive when the training database is large and efficiency is in stake when little data is available for training. One such example is one-shot learning in which prediction has to be made correctly given only a single image for training. Though human with a good memory can recognize a face fairly in difficult conditions, it is challenging for a machine to identify if little data is available for training. In this paper, a framework, which leverages the convolutional neural network in Siamese architecture, is proposed to solve the problem. This network learns a mapping of input image from image space to a compact 4096-dimensional vector space where distance among points ranks similarity. Once the model is trained with an eye on optimization function's behaviour it can utilize its discriminating potential to recognize the faces.

Keywords Siamese neural network · Classification · Face recognition · Unconstrained face

1 Introduction

Face recognition [1] is a well-researched biometric trait used for recognition of individuals. However, due to changing environments and facial dynamics face recognition becomes difficult to use mostly in unconstrained scenarios. Unconstrained faces

S. Rana (✉)
Department of Computer Science and Engineering, Jalpaiguri Government Engineering College, Jalpaiguri 735102, West Bengal, India
e-mail: srinibas.rana@cse.jgec.ac.in

D. R. Kisku
Department of Computer Science and Engineering, National Institute of Technology Durgapur, Mahatma Gandhi Road, A-Zone, Durgapur 713209, West Bengal, India
e-mail: drkisku@cse.nitdgp.ac.in

D. Bhattacharjee et al. (eds.), *Proceedings of International Conference on Frontiers in Computing and Systems*, Advances in Intelligent Systems and Computing 1255,
https://doi.org/10.1007/978-981-15-7834-2_35

often make recognition system unable to identify an individual correctly. It happens due to the changes in illumination, head pose, occlusion, clutter, facial expression, background, facial make-up, scale, etc. To address these issues in an unconstrained environment, traditional face recognition approaches [1], as well as deep learning models [2], are made available. However, intra-class variations of faces in such an environment make recognition work non-viable. Many deep networks are being used to capture the variations, while a large number of examples are used for training. However, with less number of training examples, it becomes difficult to make prediction correctly.

To work with the challenge of using less number of training examples, this paper uses a deep learning variant known as Siamese network which makes use of two convolutional neural networks together to get rid of the problem to some extent for recognition of faces in an unconstrained environment.

Siamese network was first coined by Bromley et al. [3] where it was used to match two images of signatures. Siamese network is built using two identical convolutional neural networks; each shares the same architecture as well as the same parameters. These twin networks are connected at the last layer to compute an energy function based on the differences of the two distinct input images fed into two networks as depicted in the Fig. 1. In contrast, networks in traditional classification try to learn some inherent feature in the input that recognizes the input as the class it belongs to (e.g. dog/elephant/zebra, etc.). In Siamese, twin network tries to learn what features make the two input same or different. The motivation of using twin networks is if the input image pair is very similar, then they will be mapped by their corresponding network to close locations with each other in feature space as both the networks possess same weight vector, same other parameters and compute the same objective function. Also, networks are symmetric in the sense that if same inputs are fed into the networks, interchanging their choice between the twins will give same result in distance metric computation.

Another motivation is the requirement of less number of samples per class for training. Generally, in order to achieve a standard accuracy in traditional classification, a huge number of training samples per class is required. Here, few sample images (even a single image per class) is needed to train.

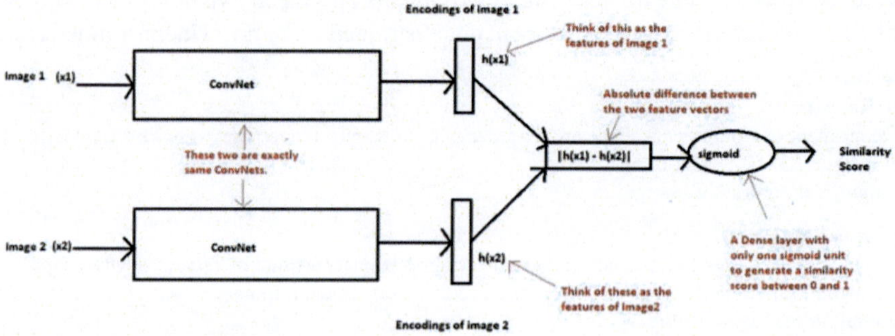

Fig. 1 Siamese framework

This paper is organized as follows: Sect. 2 discusses some existing state-of-the-artworks. Section 3 illustrates the proposed methodology. Section 4 reports and compares experimental results with other methods. Section 5 draws conclusion.

2 Literature Review

In the early 2000s, the world has seen gradual improvements in accuracies of face recognition using engineered features [4, 5]. But the performances of these techniques greatly reduce when applied to unconstrained face databases. Afterwards, the focus transferred from the engineered feature-based approach to deep learning models. Deep learning approaches experienced a rapid growth when AlexNet won ImageNet Large-Scale Visual Recognition Challenge (ILSVRC 2012) [6]. Gradually, efficient deep convolutional neural network architectures like Inception [7], Resnet [8] and others [9–11] cropped up and become the state-of-the-art in image classification applications.

The work on one-shot image classification started when Fe-Fei et al. [12] developed a Bayesian network to categorize objects in an unsupervised manner. Koch et al. [13] proposed a Siamese network based on the standard convolutional architecture and successfully applied on Omniglot dataset [14].

The deep model architectures that exhibit the best performance in many image classification problems follow a common trait: multiple convolutional layers before the feature map-flattening layer followed by fully connected layers with softmax or sigmoid layer at the end.

3 Proposed Methodology

3.1 Model

In the proposed approach, the convolutional neural network has been used. The architecture [13] consists of a series of three convolution max pooling pairs after that one convolution layer and flattening as in the Fig. 2.

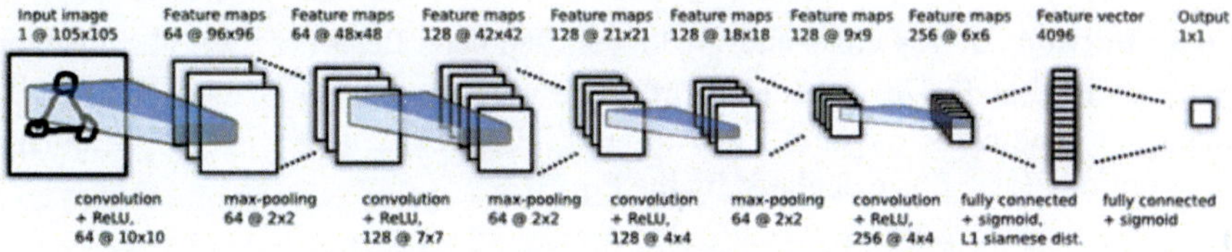

Fig. 2 Detailed convolutional architecture used in each of the Siamese twin

In these convolutional layers, ReLU (Rectified Linear Unit) activation function is used. Max pooling layer has a filter of size 2 and stride 2. Thus, if $\mathcal{W}_{l-1,l}$ be the 3D tensor of weights between layer l–1 to l, $\mathcal{P}_{l-1}$ is the feature map of $(l-1)$th layer and b_l is the bias of lth layer then lth layer feature map after ReLU activation is given by the Eq. (1).

$$\mathcal{P}_l = \text{maxpool}(\max(0, \mathcal{W}_{l-1,l} * \mathcal{P}_{l-1} + b_l), 2) \tag{1}$$

where '*' represents convolution operation.

The flattened feature map produces 4096-dimensional vectors when passed through fully connected layer and are named h_u and h_l, respectively, got from upper and lower Siamese twin. The L1 distance of h_u and h_l are then pushed into a fully connected layer of single-output neuron with sigmoid activation function at the end to produce similarity score. Therefore, prediction p is calculated by Eq. (2)

$$p = \sigma\left(\sum_j \alpha_j |h_u - h_l|\right) \tag{2}$$

where α_js are additional parameters of the hidden layers to be learned.

3.2 Method

Given the model architecture, a batch of input image pairs with known targets (0 for similar and 1 for dissimilar) is fed into the twin networks. If batch size is k, then first $k/2$ of the batch is prepared randomly selecting dissimilar face images (i.e. images of a pair belong to different classes) and no images are repeated. So if there are I instances in each of C classes, then $\binom{I}{2}$ pairs of same class are generated resulting to $\binom{I}{2} * \text{C}$ possible pairs of same class overall. But the total number of possible pairs (both same and different) is given by $\binom{I*C}{2}$. As the possible pairs are huge in number, training the network with all of them is very hard. So, practically, pairs of similar and dissimilar classes are randomly chosen in the ratio of 1:1 for training the model.

3.3 Training

Loss Function: Binary cross-entropy loss is used here. Let $Y(x_1^{(i)}, x_2^{(i)})$ be a vector of length M (batch size), each element of the vector is a label. When $x_1^{(i)}$ and $x_2^{(i)}$ belong to same class, target is 1 and when different class, target is 0. Now, binary cross entropy loss, L is calculated by the Eq. (3).

$$L(x_1^{(i)}, x_2^{(i)}) = Y(x_1^{(i)}, x_2^{(i)}) \log f(x_1^{(i)}, x_2^{(i)}) + (1 - Y(x_1^{(i)}, x_2^{(i)})) \log (1 - f(x_1^{(i)}, x_2^{(i)})) \tag{3}$$

where

$$f(s) = \frac{1}{1 + e^{-s}}$$

Optimization: Adaptive moment estimation (Adam) optimization is used to update network weights when gradient is additive across Siamese networks with a uniform learning rate for all layers.

Weight Initialization: Network weights for all convolutional layers are initialized from normal distribution with zero mean and standard deviation 10^{-2}, bias, when used were initialized with mean 0.5 and standard deviation 10^{-2}.

3.4 Evaluation

Trained and optimized network can be used for verification. The similarity score resulted from the last sigmoid layer has a value between 0 and 1. But just observing the numerical score, it is difficult to assess the performance of the model. So M-way k-random one-shot evaluating protocol, as described in the Algorithm 1 is used for performance measure. The average accuracy of all the images of the test set is finally reported.

Result: Percentage of correct predictions
M = 40; (M way one shot)
k = 250; (k is the number of random trials)
while *any image of the test set is left unexplored* **do**
    Select an unexplored image $\hat{x}$ from class C_i of test set.
    while *k number of trials is not over* **do**
        1. Build a support set of N images in the following manner
        (i) Select N-1 images at random from the gallery with their label. S={ $(x_1,y_1),(x_2,y_2),\ldots,(x_{N-1},y_{N-1})$ } such that $x_p \notin C_i$.
        (ii) Select a single image with it's label (x_N,y_N) from C_i in gallery at random.
        2. Generate the input pairs $(\hat{x},x_q)$, where $x_q \in$ S.
        3. The pairs from step 2 are now fed into the model in batches to get the similarity scores.
        4. If maximum score happened to be found in the pair which was deliberately created from the same class and placed in the support set, then it is a correct prediction, otherwise wrongly classified.
    end
    Calculate percentage of correct prediction of k random trials as
    $P_{correct} = \frac{n}{k} * 100$, when n times the model predicted correctly
end

Algorithm 1: M-way k random one-shot classification

4 Experimental Results

The proposed face recognition method is evaluated with the unconstrained Facial Images UFI [15] database. The database has two versions, viz. *Cropped images* and *Large images*. Some of the samples from that dataset are depicted in Fig. 3.

The proposed model is tested on a cropped version of the dataset. The challenging issues of this dataset are (a) unconstrained and (b) variable number of images per class. This set contains images of 605 people with an average of 7.1 images per class in the directory called 'train' and one exclusive image per class in the directory called 'test'. Every image size is uniform (128 × 128 pixels).

The training is continued for 16000 epochs and the observed training loss is depicted in Fig. 4. The Table 1 compares the results of the proposed method reporting two variations of N and k values with some baseline approaches reported in [15].

Fig. 3 Some sample images from UFI cropped version are shown

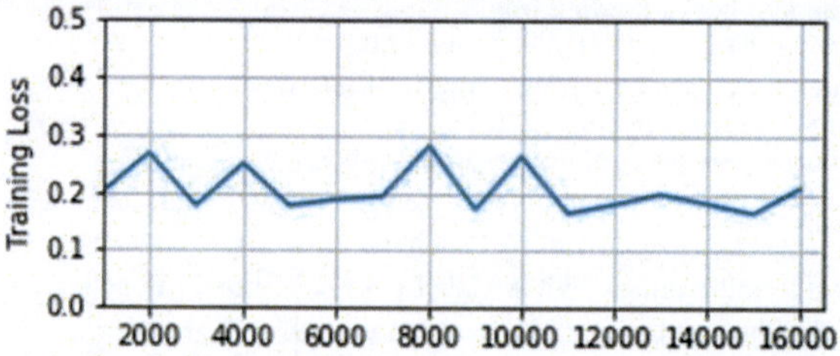

Fig. 4 Training loss variations with epochs

Table 1 Comparison of the proposed method with some baseline method on *Cropped* partition

Method	Accuracy in %
LBPHS [4]	55.04
LDPHS [16]	50.25
POEMHS [17]	67.11
FS-LBP [18]	63.31
Proposed approach (40-way 250 random)	**69.64**
Proposed approach (20-way 50 random)	**77.67**

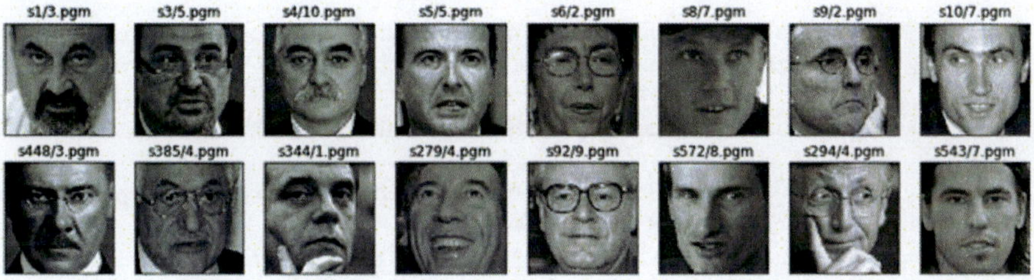

Fig. 5 Some misclassified images with class label

It is also observed that some inter-class samples are so close that even human can experience difficulty in recognizing them. Also, pose and appearance are misleading for the machine to find distinguishing features for classification. Some misclassified samples are depicted in Fig. 5.

5 Conclusion

This paper has studied Siamese convolutional neural network to identify what makes two images similar or dissimilar, and further, it has been applied on unconstrained face recognition. Experimental results show improved performance compared to histogram-based approaches. Also in M-way k-random classification, it has been observed that reduced value of M and k can lead to higher accuracy. In future, model architecture can be explored to find the best-performing architecture.

References

1. Jain, A.K., Li, S.Z.: Handbook of Face Recognition. Springer (2011)
2. Wang, M., Deng, W.: Deep face recognition: a survey (2018). arXiv preprint arXiv:1804.06655
3. Bromley, J., Guyon, I., LeCun, Y., Säckinger, E., Shah, R.: Signature verification using a "siamese" time delay neural network. In: Advances in Neural Information Processing Systems, pp. 737–744 (1994)
4. Ahonen, T., Hadid, A., Pietikainen, M.: Face description with local binary patterns: application to face recognition. IEEE Trans. Pattern Anal. Mach. Intell. **12**, 2037–2041 (2006)
5. Lowe, D.G.: Distinctive image features from scale-invariant keypoints. Int. J. Comput. Vis. **60**(2), 91–110 (2004)
6. Krizhevsky, A., Sutskever, I., Hinton, G.E.: Imagenet classification with deep convolutional neural networks. In: Advances in Neural Information Processing Systems, pp. 1097–1105 (2012)
7. Szegedy, C., Liu, W., Jia, Y., Sermanet, P., Reed, S., Anguelov, D., Erhan, D., Vanhoucke, V., Rabinovich, A.: Going deeper with convolutions. In: Proceedings of the IEEE Conference on Computer Vision and Pattern Recognition, pp. 1–9 (2015)

8. He, K., Zhang, X., Ren, S., Sun, J.: Deep residual learning for image recognition. In: Proceedings of the IEEE Conference on Computer Vision and Pattern Recognition, pp. 770–778 (2016)
9. Bengio, Y. et al.: Learning deep architectures for AI. Found. Trends® Mach. Learn. **2**(1), 1–127 (2009)
10. Srivastava, N., Salakhutdinov, R.R., Hinton, G.E.: Modeling documents with deep boltzmann machines (2013). arXiv preprint arXiv:1309.6865
11. Iandola, F.N., Han, S., Moskewicz, M.W., Ashraf, K., Dally, W.J., Keutzer, K.: Squeezenet: Alexnet-level accuracy with 50x fewer parameters and <0.5 MB model size (2016). arXiv preprint arXiv:1602.07360
12. Fe-Fei, L., et al.: A bayesian approach to unsupervised one-shot learning of object categories. In: Proceedings Ninth IEEE International Conference on Computer Vision, pp. 1134–1141. IEEE (2003)
13. Koch, G., Zemel, R., Salakhutdinov, R.: Siamese neural networks for one-shot image recognition. In: ICML Deep Learning Workshop, vol. 2 (2015)
14. Lake, B.M., Salakhutdinov, R., Tenenbaum, J.B.: Human-level concept learning through probabilistic program induction. Science **350**(6266), 1332–1338 (2015)
15. Lenc, L., Král, P.: Unconstrained facial images: database for face recognition under real-world conditions. In: Mexican International Conference on Artificial Intelligence, pp. 349–361. Springer (2015)
16. Zhang, B., Gao, Y., Zhao, S., Liu, J.: Local derivative pattern versus local binary pattern: face recognition with high-order local pattern descriptor. IEEE Trans. Image Process. **19**(2), 533–544 (2009)
17. Vu, N.-S., Dee, H.M., Caplier, A.: Face recognition using the poem descriptor. Pattern Recogn. **45**(7), 2478–2488 (2012)
18. Lenc, L., Král, P.: Automatically detected feature positions for LBP based face recognition. In: IFIP International Conference on Artificial Intelligence Applications and Innovations, pp. 246–255. Springer (2014)

Contrast Enhancement Algorithm Using Definite Integration Mathematical Method Trapezoidal Rule

Amiya Halder and Nikita Shah

Abstract In this paper, we propose a novel contrast enhancement algorithm based on discrete integration mathematical Trapezoidal formula. It improves the contrast of the low contrast images. Various novel algorithms have been already proposed for contrast enhancement. But these methods mainly depend on histogram equalization or modified histogram equalization. Our method is simple and easy to understand that only applies the trapezoidal rule and it is clearly shown that it works very efficiently on low contrast images. In comparison to other different exiting algorithms, the proposed method works efficiently in case of different standard low contrast images.

Keywords Low contrast image · Contrast enhancement · Peak-Signal-to-Noise-Ratio (PSNR) · Trapezoidal rule

1 Introduction

Image enhancement is one of the essential operations in image processing techniques. It improves the quality of the images for human viewers for further image processing techniques. Contrast enhancement is one of the basic operations of the image enhancement technique. Contrast mainly represents the brightness or luminance of the image and it can be defined as the difference in the pixel intensities of the highest and lowest pixel intensities of the image [6, 13]. The contrast of images is the most important feature to judge the quality of images that can be good or poor [8].

A. Halder (✉) · N. Shah
Department of CSE STCET, 4 D. H. Road, Kidderpore 700023, Kolkata, India
e-mail: amiya.halder77@gmail.com

N. Shah
e-mail: nikitashah9714@gmail.com

D. Bhattacharjee et al. (eds.), *Proceedings of International Conference on Frontiers in Computing and Systems*, Advances in Intelligent Systems and Computing 1255,
https://doi.org/10.1007/978-981-15-7834-2_36

A lot of essential information cannot be retrieved in a dark or a low contrast image and hence it is required to establish efficient contrast enhancement algorithms for processing the images that needed in the field of digital image processing, computer vision, and pattern recognition [5].

To increase the image quality, various contrast enhancement methods like genetic algorithms (GA) based image enhancement [12] method, adaptive histogram equalization (AHE) [18], local histogram equalization based technique [7], contrast limited adaptive histogram equalization (CLAHE) [14] have been proposed.

In this paper, different methods for contrast enhancement have been presented (Dynamic histogram equalization (HE) [1], Modified adaptive histogram equalization (AHE) [10], weighted threshold histogram equalization with improved switching median filtering technique (WTHEISMF) [16], Image histogram adjustment (IM) [9], adaptive gamma correction with weighting distribution (AGCWD) [5]). AHE method improves the contrast of the image without any loss of information but fails when dominating portions appear in the image. IM algorithm is a modified histogram equalization that performs automatically and AGCWD has enhanced the brightness of low contrast images using gamma correction but fails when the image lacks bright pixels.

To further improve the contrast of the images, histogram equalization(HE) [15, 17], sigmoid function [6], histogram equalization(HE) [15, 17], local mean method (GLM) [4], piecewise-constant interpolation (InterPWC) and local max-min interpolation (InterMaxMin) [2, 19], arc the most commonly used spatial domain contrast enhancement techniques. These methods are successfully increase the contrast of the images.

We present a mathematical formula numerical integration method Trapezoidal rule-based contrast enhancement algorithms. In this paper, Trapezoidal rule is applied on a three points formula in the specified interval with the degree of precision being one. The proposed method is very simple and easy to implement. To evaluate the performance of the proposed method using different quantitative measures (PSNR, RMSE, and SSIM). The efficiency of the proposed method with different exiting contrast enhancement algorithms is demonstrated on a set of different standard images.

2 Background

This section presents contrast enhancement method using mathematical integration formula Trapezoidal rule [3, 11]. In numerical analysis, we can evaluate definite integration using the Trapezoidal rule.

$$\int_a^b f(\omega)d\omega \tag{1}$$

It is one of the statistical methods to get the value of definite integration using subinterval between end-points a and b. Let us consider (ω) to be a set of points of $[a, b]$ such that $a = \omega_0 < \omega_1 < \omega_2 < \omega_3 < \ldots\ldots < \omega_{n-1} < \omega_n = b$ and $\Delta\omega_k$ is the length of k subinterval, then

$$\int_a^b f(\omega)d\omega \simeq \sum_{k=1}^{n} \frac{f(\omega_{k-1}) + f(\omega_k)}{2} \Delta\omega_k \tag{2}$$

$$= \frac{\Delta\omega}{2}(f(\omega_0) + 2f(\omega_1) + 2f(\omega_2) + 2f(\omega_3) + ... + 2f(\omega_{n-1}) + f(\omega_n)) \tag{3}$$

For example, to evaluate integration using a four points Trapezoidal formula in the interval $[a, b]$. Considering $\Delta\omega_k = (b - a) = h$ in Eq. (3), we get Eq. (4) as

$$I = \frac{(b-a)}{2}[\vartheta_0 + 2 * \vartheta_1 + 2 * \vartheta_2 + \vartheta_3] = \frac{h}{2}[\vartheta_0 + 2 * \vartheta_1 + 2 * \vartheta_2 + \vartheta_3] \tag{4}$$

The final form of the consecutive four points ϑ_0, ϑ_1, ϑ_2 and ϑ_3 Trapezoidal formula is shown in Eq. (4).

3 Proposed Method

In the proposed method, we consider ϑ_0, ϑ_1 and ϑ_2 are the consecutive pixel intensities of the low contrast image. The method enhances the contrast of the images using three pixels of neighborhood operation based on mathematical integration. In the proposed method, Trapezoidal rule is three points formula in the interval [a, b]. The above groundwork is used in the Algorithm 1 for improving the contrast enhancement of low contrast images.

4 Experimental Results

The proposed algorithm (PA) is compared with several existing methods namely, InterMaxMin, InterPWC, WTHEISMF, GA, HE, AGCWD, GLM, AHE, IM methods. All the methods are implemented in MATLAB and C language and run in WINDOWS Operating System, machine configuration is Intel Core i3 CPU 4010U@1.70 GHz and 4 GB memory. Different quantitative parametric quantity PSNR, RMSE, and SSIM are used to compare the performance of the different algorithms with the proposed method. For demonstrate purpose, we are testing different

Algorithm 1: Proposed Algorithm using Trapezoidal Rule

Input: Input image $\varphi(P, Q)$
Output: Output image $\chi(P, Q)$
for $\rho = 1 \ldots P - 1$ **do**
  for $\theta = 1 \ldots Q - 1$ **do**
    $\psi(\rho, \theta) = \frac{h}{2}[\varphi(\rho - 1, \theta - 1) + 2 * \varphi(\rho, \theta) + \varphi(\rho + 1, \theta + 1)]$
    if $\psi(\rho, \theta) < 0$ **then**
      $\chi(\rho, \theta) = 0$
    else if $\psi(\rho, \theta) > 255$ **then**
      $\chi(\rho, \theta) = 255$
    else
      $\chi(\rho, \theta) = \psi(\rho, \theta)$

standard images `Lena`, `Baloon`, `Cameraman`, `Barbara`, `Pepper`, `Donald`, and `Mandril`. The calculated RMSE values are shown in Table 1, for the standard images `Lena`, `Baloon`, `Cameraman`, `Barbara`, `Pepper`, `Donald`, and `Mandril` using different methods. From the Table. 1, it is clear that the RMSE value of the proposed algorithm is minimum,that is, proposed method gives better result than other exiting methods.

Table 1 RMSE values using InterMaxMin, InterPWC, WTHEISMF, GA, HE, AGCWD, GLM, AHE, IM, and Proposed algorithm (different low contrast images)

Methods	Lena	Barbara	Pepper	Donald	Cameraman	Mandril	Baloon
InterPWC	48.14	25.52	24.40	45.09	31.09	34.34	40.80
InterMaxMin	45.12	28.21	30.02	44.40	37.50	136.27	43.02
WTHEISMF	46.21	29.56	24.37	39.41	31.00	43.27	42.43
GA	54.31	54.91	56.23	63.84	65.84	74.31	55.76
HE	45.44	24.76	23.97	39.30	27.81	35.08	39.98
AGCWD	53.09	45.65	47.01	31.96	44.54	58.66	58.74
GLM	78.85	96.41	100.97	58.73	101.39	95.73	90.45
AHE	24.09	32.81	39.23	47.83	64.42	31.40	44.18
IM	23.80	16.21	23.62	34.81	21.45	27.75	32.63
PA	15.75	17.12	21.83	14.24	20.82	26.28	19.03

Table 2 PSNR values using InterMaxMin, InterPWC, WTHEISMF, GA, HE, AGCWD, GLM, AHE, IM, and Proposed algorithm (different low contrast images)

Methods	Lena	Barbara	Pepper	Donald	Cameraman	Mandril	Baloon
InterPWC	14.48	19.99	20.38	15.04	18.27	17.41	15.91
InterMaxMin	15.04	19.12	18.57	15.18	16.64	5.44	15.45
WTHEISMF	14.83	18.71	20.39	16.21	18.30	15.40	15.57
GA	13.43	13.34	13.13	12.02	11.76	10.71	13.20
HE	14.98	20.25	20.54	16.24	19.25	17.23	16.09
AGCWD	13.63	14.94	14.68	18.03	15.15	12.76	12.75
GLM	10.19	8.44	8.04	12.75	8.01	8.50	9.0
AHE	20.49	17.81	16.26	14.53	11.95	18.19	15.23
IM	20.59	23.93	20.66	17.29	21.50	19.26	17.85
PA	24.18	23.45	21.34	25.05	21.75	19.73	22.53

Table 3 SSIM values using InterMaxMin, InterPWC, WTHEISMF, GA, HE, AGCWD, GLM, AHE, IM, and Proposed algorithm (different low contrast images)

Methods	Lena	Barbara	Pepper	Donald	Cameraman	Mandril	Baloon
InterPWC	0.8015	0.9071	0.8827	0.5444	0.7667	0.8110	0.7496
InterMaxMin	0.6748	0.7343	0.7444	0.4935	0.6216 0	0.4712	0.6786
WTHEISMF	0.7178	0.7210	0.8276	0.5698	0.7108	0.5258	0.7002
GA	0.6878	0.6684	0.7178	0.8212	0.6882	0.6007	0.7493
HE	0.7789	0.8814	0.8579	0.5450	0.7729	0.7949	0.7118
AGCWD	0.8371	0.8914	0.9007	0.9086	0.9298	0.8696	0.8353
GLM	0.4035	0.3688	0.3691	0.7904	0.4447	0.4385	0.4027
AHE	0.8961	0.8428	0.8675	0.8110	0.7286	0.8852	0.8428
IM	0.8365	0.9129	0.948	0 0.8149	0.9279	0.9085	0.7710
PA	0.9749	0.9782	0.9725	0.9916	0.9699	0.8894	0.9708

PSNR values are calculated using different exiting methods with the proposed algorithm that are tabulated in Table 2. The highest PSNR value represents the high-quality image and from Table 2, it is noted that the proposed method has better PSNR values than other methods. All SSIM values for different standard images are shown in Table 3.

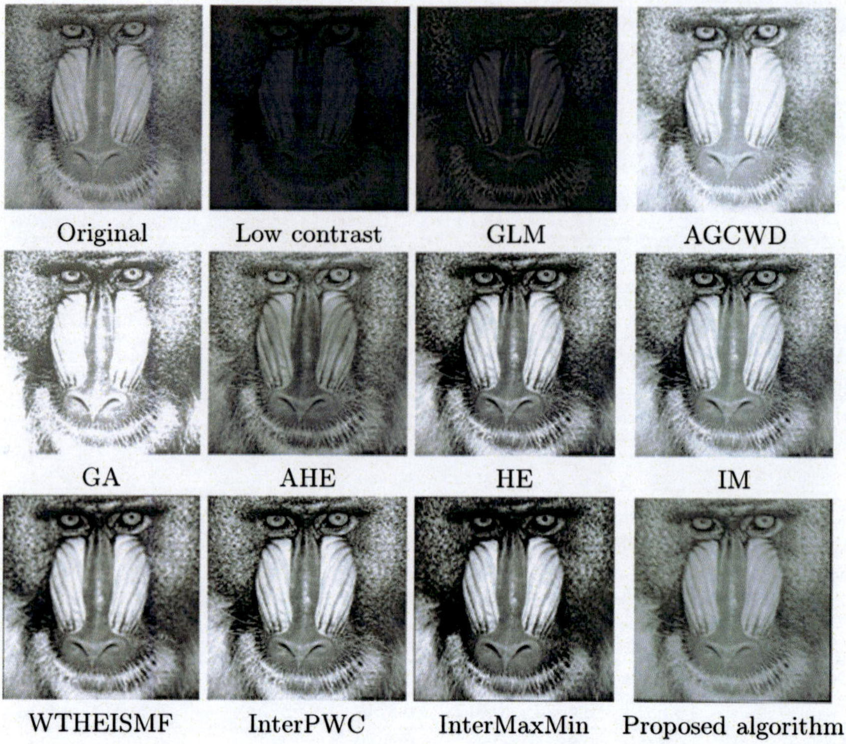

Fig. 1 Compare of the output images for low contrast `Mandril` image using different methods

Figures 1 and 2 gives the original, low contrast, and output images using different methods: InterMaxMin, InterPWC, WTHEISMF, GA, HE, AGCWD, GLM, AHE, IM methods on the `Mandril`, `cameraman` and `Peppers` images. From the results, it is noted that the proposed method shows better performance than the other existing contrast enhancement methods.

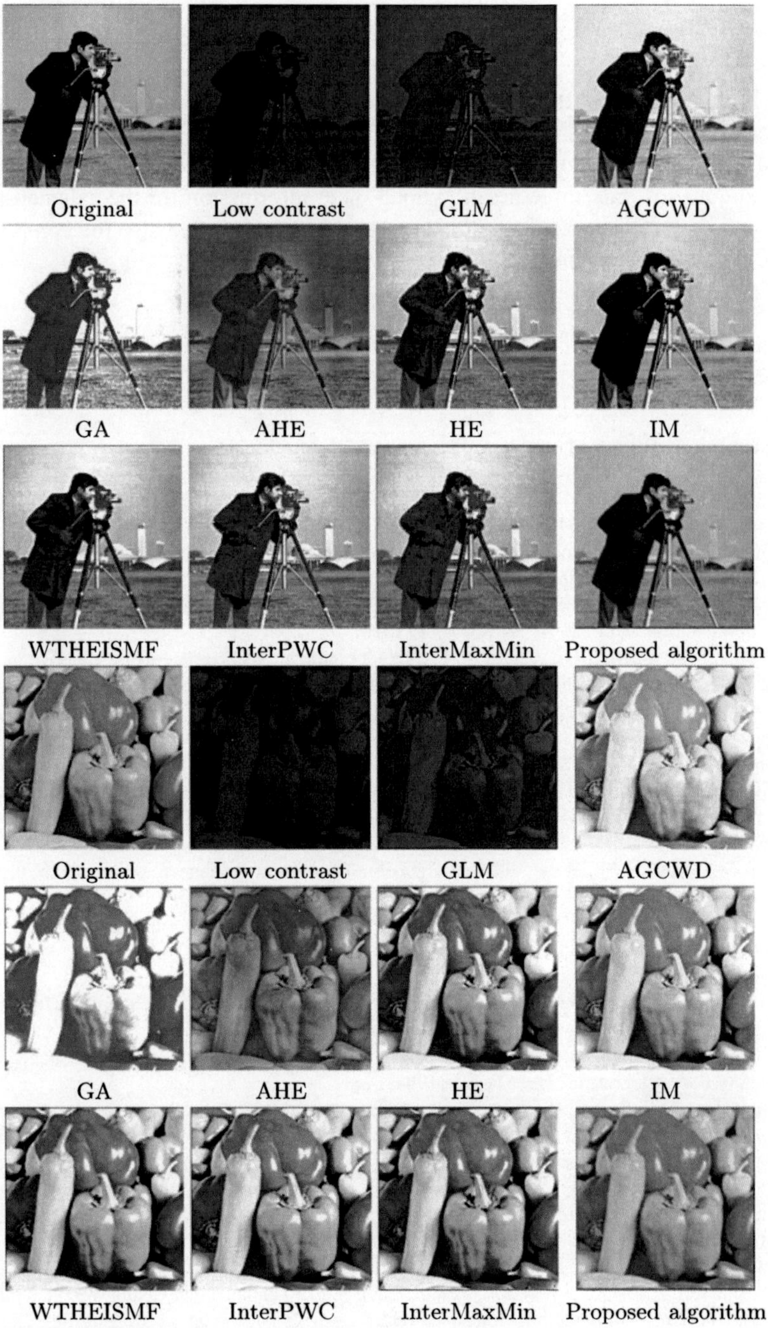

Fig. 2 Compare of the output images for low contrast `Cameraman` and `Pepper` image using different methods

5 Conclusion

The proposed method introduces a novel algorithm to increase the contrast of dark low contrast images using simple, easy mathematical integration. The Trapezoidal rule-based proposed method has several advantages as compared to the standard existing methods. It uses only three neighborhood pixel information for the calculation. It gives more prominent results than other existing methods and also the complexity of the proposed algorithm is less.

References

1. Abdullah-Al-Wadud, M., Kabir, M.H., Dewan, M.A.A., Chae, O.: A dynamic histogram equalization for image contrast enhancement. IEEE Trans. Consum. Electr. **53**(2), 593–600 (2007)
2. Agarwal, M., Mahajan, R.: Medical image contrast enhancement using range limited weighted histogram equalization. Procedia Comput. Sci. **125**, 149–156 (2018)
3. Datta, N., Jana, R.: Introductory Numerical Analysis. Shreedhar Prakashani (2010)
4. Gautam, G., Mukhopadhyay, S.: Efficient contrast enhancement based on local–global image statistics and multiscale morphological filtering. In: Advanced Computational and Communication Paradigms, pp. 229–238. Springer (2018)
5. Huang, S.C., Cheng, F.C., Chiu, Y.S.: Efficient contrast enhancement using adaptive gamma correction with weighting distribution. IEEE Trans. Image Process. **22**(3), 1032–1041 (2012)
6. Ismail, N.H.B., Der Chen, S., Ng, L.S., Ramli, A.R.: An analysis of image quality assessment algorithm to detect the presence of unnatural contrast enhancement. J. Theor. Appl. Inform. Technol. **83**(3), 415–422 (2016)
7. Kim, J.Y., Kim, L.S., Hwang, S.H.: An advanced contrast enhancement using partially overlapped sub-block histogram equalization. IEEE Trans. Circ. Syst. Video Technol. **11**(4), 475–484 (2001)
8. Lal, S., Chandra, M., et al.: Efficient algorithm for contrast enhancement of natural images. Int. Arab J. Inf. Technol. **11**(1), 95–102 (2014)
9. Mittal, N.: Automatic contrast enhancement of low contrast images using Matlab. Int. J. Adv. Res. Comput. Sci. **3**(1) (2012)
10. Moldvai, C.: Adaptive contrast enhancement. US Patent 7,760,961, July 20, 2020
11. Mollah, S.A.: Numerical Analysis and Computational Procedures. ArunabhaSen Books And Allied (P) Ltd. (2012)
12. Pal, S.K., Bhandari, D., Kundu, M.K.: Genetic algorithms for optimal image enhancement. Pattern Recogn. Lett. **15**(3), 261–271 (1994)
13. Gonzalez, R.C., Woods, R.E.: Digital Image Processing. Prentice Hall (2002)
14. Reza, A.M.: Realization of the contrast limited adaptive histogram equalization (CLAHE) for real-time image enhancement. J. VLSI Signal Process. Syst. Signal Image Video technology **38**(1), 35–44 (2004)
15. Senthilkumaran, N., Thimmiaraja, J.: Histogram equalization for image enhancement using MRI brain images. In: 2014 World Congress on Computing and Communication Technologies (WCCCT), pp. 80–83. IEEE (2014)
16. Sharmila, R., Uma, R.: A new approach to image contrast enhancement using weighted threshold histogram equalization with improved switching median filter. Int. J. Adv. Eng. Sci. Technol. **7**(2), 208–211 (2011)
17. Singh, R.P., Dixit, M.: Histogram equalization: a strong technique for image enhancement. Int. J. Signal Process. Image Process. Pattern Recogn. **8**(8), 345–352 (2015)

18. Stark, J.A.: Adaptive image contrast enhancement using generalizations of histogram equalization. IEEE Trans. Image Process. **9**(5), 889–896 (2000)
19. Yu, H., Inoue, K., Hara, K., Urahama, K.: Histogram interpolation methods for image contrast enhancement. J. Inst. Indus. Appl. Eng. **5**(1), 7–14 (2017)

Self Organizing Map-Based Strategic Placement and Task Assignment for a Multi-agent System

Mukund Subhash Ghole, Arabinda Ghosh, Arindam Singha, Chinmay Das, and Anjan Kumar Ray

Abstract Strategic placement and task assignment for a multi-agent system are important research topics in various cooperative working environment. In this article, authors proposed a procedure to attend demands from various areas of workspace by multi-agent system. Agents are trained through self organizing map (SOM) to locate the strategic position. Two modes of operations are proposed. One is operating multi-agent system from base camp through task assignment. Another is placing the multi-agent system to a strategic location to address the demands of attending task. Extensive simulations are presented to validate the proposed work through a simulated environment and integrating real locations from Google map. A comparative study is presented to highlight the effectiveness of the individual modes of operation.

Keywords Task assignment · Self organizing Map (SOM) · Google map · Surveillance · Multi-agent system

---

M. S. Ghole (✉) · A. Ghosh · A. Singha · C. Das · A. K. Ray (✉)
Electrical and Electronics Engineering, National Institute of Technology Sikkim, Ravangla 737139, Sikkim, India
e-mail: mukund.ghole@gmail.com

A. K. Ray
e-mail: akray.nits@gmail.com

A. Ghosh
e-mail: phee16023@nitsikkim.ac.in

A. Singha
e-mail: arindamsingha008@gmail.com

C. Das
e-mail: m180038@nitsikkim.ac.in

D. Bhattacharjee et al. (eds.), *Proceedings of International Conference on Frontiers in Computing and Systems*, Advances in Intelligent Systems and Computing 1255,
https://doi.org/10.1007/978-981-15-7834-2_37

1 Introduction

Natural or human A few examples include tsunami in Sumatra, Indonesia in 2004 [1], earthquake in Wenchuan, China [2], in 2008, in Haiti in 2010 [3] and in Nepal in 2015 [4]. Such disasters around the world result in loss of human lives. Further, pre and post disaster emergency responses have raised challenges regarding relief and rescue operation, and overall surveillance of the areas of interest. Multi-agent System (MAS) is the combination of two or more agents which achieve the group objectives, through coordination. A MAS is used in many domains including traffic control and transportation system [5]. Task assignment is a process to assign the tasks among the available resources (e.g. humans, computers, drones, etc.). Task assignment(TA) has applications in tracking dynamic targets [7], in intelligent logistic system using Unmanned Aerial Vehicles (UAV) [8], in task tour assignment of UAV [9], in cooperative task allocation under resource constraints [10], TA in post earthquake scenario [11], post disaster rescue using aerial drones [12], cooperative working environment of Autonomous Underwater Vehicle (AUV) [13].

The coordination among agents in a MAS can be achieved through incorporation of machine learning. In this work, the paradigm of self organizing map (SOM) is used which was introduced by Kohonen in 1980 [6]. It is inspired by the characteristics of the human brain. In a competitive learning environment, agents are trained for the topology of the input space. In [19], SOM based TA of a swarm of robots in a 3-D dynamic environment is proposed.

Some related works from the literature regarding TA and/or SP of multi-agent system is discussed as follows. In [15], cellular learning automata approach is developed to allocate spatial-temporal tasks in MAS. A multi-agent architecture is developed to envisage communication and cooperation mechanisms among agents in [16]. A negotiation-based algorithm is proposed in [17], for TA at smaller negotiation cost and is implemented on an UAV testbed. In [18], SP of drones is used to deliver a defibrillator during the crisis of myocardial arrest to improve the survival rate in Sweden.

In this work, authors proposed two different procedures of attending tasks coming from defined locations. First, sending multi-agent system to the tasks directly. Second, the tasks were attended strategically following the self organizing map paradigm. Extensive simulations are presented to validate the proposed work and it is also integrated with real locations from google map. A comparative study between the above mentioned procedures is also presented.

The paper is arranged as follows: in Sect. 2, the proposed approach is presented. In Sect. 3, the results are presented to validate the proposed work including simulated environment and the actual environment through Google map. The overall conclusion is made at Sect. 4.

Table 1 Nomenclature used

N	Total number of agents	$d_{j,k}$	Lattice distance between winner and jth agent
M	Total number of demands/tasks	$h_{j,k(x)}$	Neighborhood function
t	Sample number	$\sigma(t)$	Width of the neighborhood function
$t_{\max}$	Maximum number of samples	σ_0	Initial value of width function
x	Input position	$\eta(t)$	Learning rate
ω	Weights of an agent	η_0	Initial value of learning rate
m	Dimension of the search space	τ_1	Time constant of width function
$k(x)$	Index of the winner agent	τ_2	Time constant of learning rate
A	Set of all agents	T	Set of all tasks
T_r	Set of all remaining tasks	d_{ijr}	Distance overhead for A_i from task T_{jr}
dia	Distance overhead for assigned tasks for ith agent	p	Number of tasks assigned in each loop of task assignment
M_r	Number of remaining tasks	d_i	Closest task distance for agent A_i
d_{jr}	Closest agent distance from T_{jr}		

2 Methodology

In this work, a multi-agent multi-task assignment problem based on the task to agent distances is addressed. Consider a situation like a natural or man-made disaster such as an earthquake or a landslide. Such an area is considered as the workspace and a set of areas are defined as shelter areas or area of interest. A base camp is defined nearby the considered workspace where the agents reside initially. The agents are trained to place themselves to their strategic locations to serve the task coming from the workspace and/or areas on interest. The work is divided into the TA and SP of agent for quick responses. The nomenclature used in the proposed method is tabulated in Table 1.

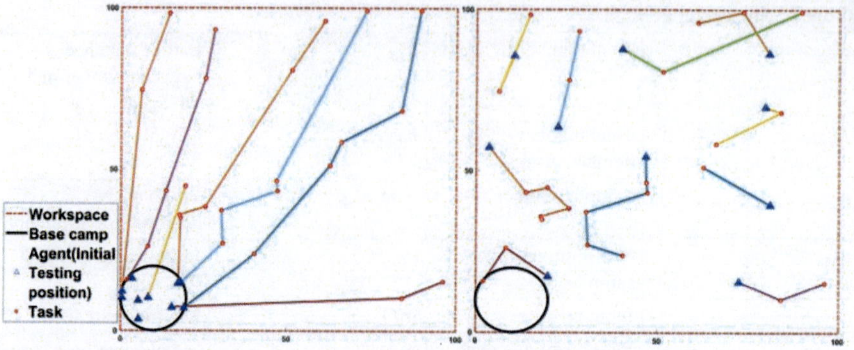

Fig. 1 (Case I) Agents attending tasks coming from the whole workspace; base camp is at a corner place; (left) TA only, (right) SP and TA

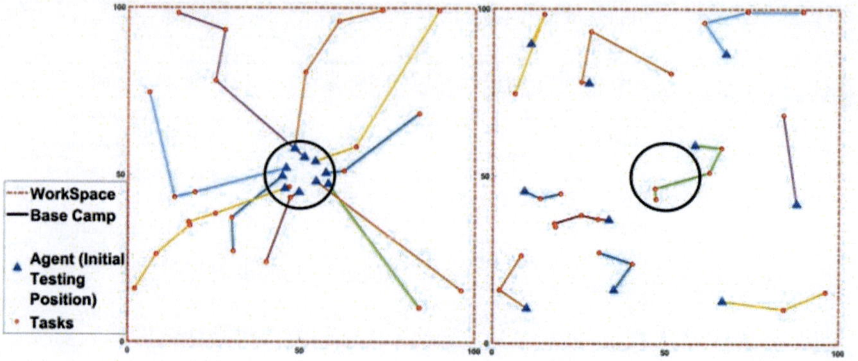

Fig. 2 (Case II) Agents attending tasks coming from the whole workspace; base camp is at a center place; (left) TA only, (right) SP and TA

2.1 Selection of the Workspace and Areas of Interest

In this work, simulation environments and Google map environments are considered to represent this work. For example, a rectangular workspace is considered in simulated environment (Figs. 1 and 2), and quadrilateral environment is considered in Google map (Figs. 4 and 5). Moreover, the areas of interest of different shapes within the workspace are also considered (Figs. 3 and 6).

2.2 Multi-agent Task Assignment

– For modeling of multi-agent task assignment, the following relation are considered

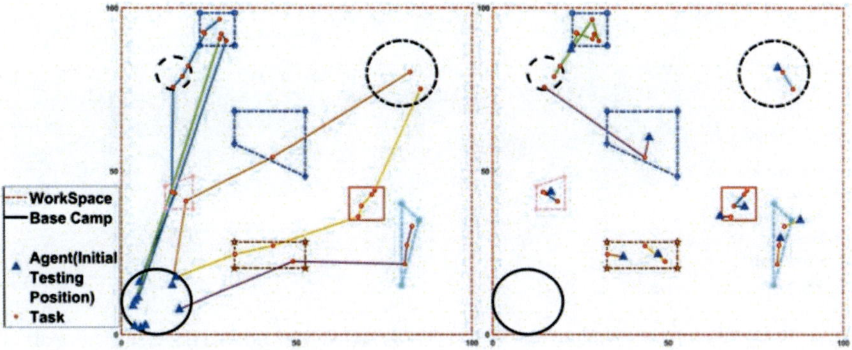

Fig. 3 (Case III) Agents attending tasks coming from the areas of interest; base camp is at a corner place; (left) TA only, (right) SP and TA

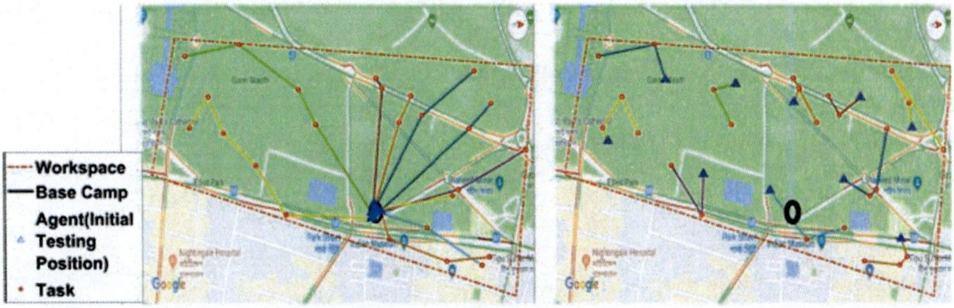

Fig. 4 (Case I) Agents attending task demands from the Kolkata maidan (Google map) workspace; base camp is at a corner place; (left) TA only, (right) SP and TA

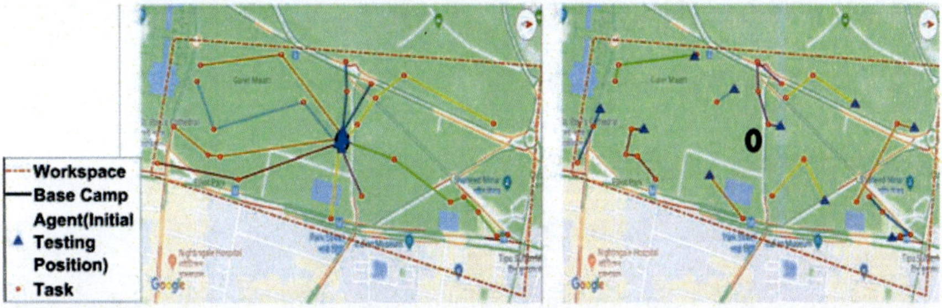

Fig. 5 (Case II) Agents attending task demands from the Kolkata maidan (Google map) workspace; base camp is at a center place; (left) TA only, (right) SP and TA

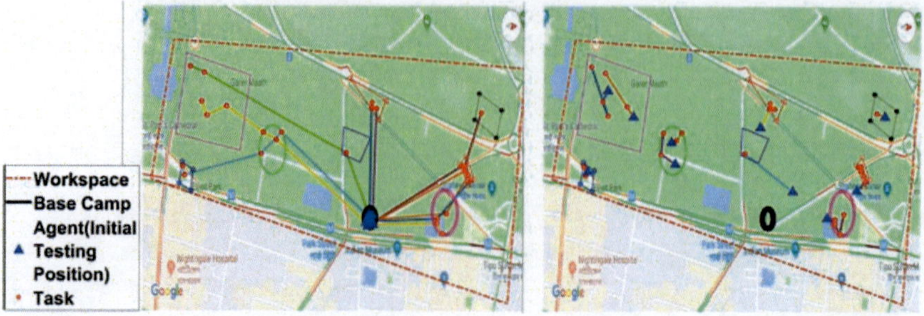

Fig. 6 (Case III) Agents attending task demands from the Kolkata maidan (Google map) areas of interest; base camp is at a corner place; (left) TA only, (right) SP and TA

$$A_i \in A, \forall\, i = 1, 2, \ldots, N,\ T_j \in T,\ \forall\, j, T_{j_r} \in T_r,\ \forall\, j_r \tag{1}$$

where, j refers to the ID of M number of tasks, j_r refers to the M_r number of remaining unassigned tasks. Initially $M = M_r$, $T_j = T_{jr}$, dia $= 0$.

- Find the task to agent distances as

$$d_{ijr} = ||A_i, T_{jr}|| + \text{dia},\ \forall\, A_i,\ \forall\, T_{jr} \tag{2}$$

- Find the closest task for each agent

$$d_i = \underset{i}{\text{argmin}}(d_{ijr})\ \forall\, A_i,\ \forall\, j_r \tag{3}$$

- Find the closest agent for each task

$$d_{jr} = \underset{j_r}{\text{argmin}}(d_{ijr})\ \forall\, T_{jr},\ i = 1, 2, \ldots,\ N \tag{4}$$

- Now for a pair of agent A_i and T_{jr}, iff

$$d_i = d_{jr}\ \forall\, jr,\ i = 1, 2, \ldots, N \tag{5}$$

  then assign the task T_{jr} to the agent A_i which are henceforth referred as $A_{(i,\,\text{assigned})}$ and $T_{(jr,\,\text{assigned})}$, for all such 'p' number of tasks $T_{(jr,\,\text{assigned})}$ satisfying the Eq. (5). Exclude the tasks to form the $T_{jr} \in T_r$.
- Update the remaining tasks and the position of agents to the already assigned tasks as follows

$$\begin{aligned} M_r &= M_r - p \\ A_{(i,\,\text{assigned})} &\Leftarrow T_{(jr,\,\text{assigned})} \\ \text{dia} &= ||A_{(i,\,\text{assigned})}, T_{(jr,\,\text{assigned})}|| + \text{dia} \end{aligned} \tag{6}$$

– Repeat the process from Eqs. (2)–(6), till all the tasks are assigned.

2.3 Strategic Placement

For applications like area surveillance, impending threat detections, quick responses to tasks demanded from multiple areas of interest, strategic placement of a multi-agent system plays a vital role. In the following sections, the framework to utilize the SOM paradigm is proposed. This includes integration of real locations from the Google map.

2.4 Lattice Structure of Agents and Their Weights

Each agent is considered as a neuron in the SOM. The agents are placed as a single dimensional lattice. For any demand from the workspace or the area of interest, these agents will move through the search space and try to place themselves near the demand. Every agent has a weight associated with it represented by ω. Let us consider, in a m dimensional search space, the weight of each agent can be represented by

$$\omega_i = [\omega_{i1},\ \omega_{i2}, \ldots,\ \omega_{im}] \quad \text{for} \quad i = 1,\ 2, \ldots,\ N \tag{7}$$

In this paper, the weights of the agents represent the coordinates in the base camp or the workspace. Thus, m is 2 and N is the total number of agents. The initial weights of the agents are selected from the base camp.

2.5 Generation of Input Points

2.5.1 On Simulated Environment

The demands of attending tasks may come from any place in the workspace or the areas of interest. In a 2 dimensional search space, an input can be defined as

$$x(t) = [x_1,\ x_2] \tag{8}$$

where t is the sample number. In this paper, an input is generated from either the workspace or areas of interest using the Eqs. (10)–(15).

2.5.2 Using Google Map

Google Map uses GPS coordinates given in degrees, minutes, and seconds (DMS) format. The format is changed to a Decimal degrees (D.d) using the following equation [14]

$$D^\circ M' S'' = D + \frac{M'}{60} + \frac{S''}{3600} = D.d \tag{9}$$

The Eqs. (10)–(15) are used to generate an input point x for the workspace or the areas of interest.

For a circular shaped area, let the coordinates of the center of the circle be (c_1, c_2) and the coordinate of a point on circumference is (a_1, b_1). The radius, r of the circle is

$$r = \sqrt{(c_1 - a_1)^2 + (c_2 - b_1)^2} \tag{10}$$

The input point $(x_1, x_2) \in x$ is considered to be inside or on the circle if

$$(x_1 - c_1)^2 + (x_2 - c_2)^2 \leq r^2 \tag{11}$$

For a triangular shaped area, let the coordinates of the vertices be (a_1, b_1), (a_2, b_2), and (a_3, b_3). The area A_T is given by

$$A_T = \frac{1}{2}\,|a_1(b_2 - b_3) + a_2(b_3 - b_1) + a_3(b_1 - b_2)| \tag{12}$$

To check a point $(x_1, x_2) \in x$ to be inside, the area of the triangle is calculated with the help of Eq. (12), by taking 2 vertices of the triangle and that point. A point $(x_1, x_2) \in x$ is inside the triangle with coordinate (a_1, b_1), (a_2, b_2), and (a_3, b_3) iff,

$$A_{T1} + A_{T2} + A_{T3} = A_T \tag{13}$$

For a quadrilateral shaped area, let the coordinates of the four vertices of the quadrilateral be (a_1, b_1), (a_2, b_2), (a_3, b_3), and (a_4, b_4). The total area of the quadrilateral is

$$A_Q = \frac{1}{2}|(a_1b_2 - a_2b_1) + (a_2b_3 - a_3b_2) + (a_3b_4 - a_4b_3) + (a_4b_1 - a_1b_4)| \tag{14}$$

To check a point $(x_1, x_2) \in x$ to be inside a quadrilateral, the area of the triangle is calculated with the help of Eq. (12), by taking 2 adjacent vertices of quadrilateral in cyclic order and that point, by naming it as A_{QT1}. This process is repeated to get A_{QTi} for i = 1, 2, 3, 4.

A point $(x_1, x_2) \in x$ is inside the quadrilateral with coordinate (a_1, b_1), (a_2, b_2), (a_3, b_3), and (a_4, b_4) iff,

$$A_{QT1} + A_{QT2} + A_{QT3} + A_{QT4} = A_Q \tag{15}$$

2.6 Training of Agents Using SOM

The euclidean distance between the weights of each agent and the input is calculated to decide a winner agent. The index of the winner agent is [6]

$$k(x) = \text{index}\left[\min_i \|x(t) - \omega_i(t)\|\right] \text{ for } i = 1, 2, \ldots, N \tag{16}$$

The winner agent will influence its neighborhood to move toward the input point. In this paper, a Gaussian neighborhood function is considered and is given by

$$h_{j,k(x)}(t) = e^{-\left(\frac{d_{j,k}^2}{2\sigma^2(t)}\right)} \text{ for } t = 0, 1, \ldots, t_{\max} \tag{17}$$

In Eq. (17), $d_{j,k}$ is the distance between the lattice position of the winner agent and lattice position of other jth agent, $\sigma(t)$ represents the width of the neighborhood and $t_{\max}$ is the maximum number of samples. As training of the agents proceed, the influence of the winner on its neighborhood will gradually decrease. Therefore, $\sigma(t)$ is given as

$$\sigma(t) = \sigma_0 e^{-\left(\frac{t}{\tau_1}\right)} \tag{18}$$

σ_0 and τ_1 are the initial value and time constant of the width function, respectively. The weights of the winner agent and its neighborhood is updated in each sample to move them toward the input. The change in weight of each agent in each sample is given by

$$\Delta\omega_i(t) = \eta(t) h_{j,k(x)}(t)(x(t) - \omega_i(t)) \ \forall \ i \tag{19}$$

In Eq. (19), $\eta(t)$ is the learning rate parameter and is given by

$$\eta(t) = \eta_0 e^{-\left(\frac{t}{\tau_2}\right)} \tag{20}$$

η_0 and τ_2 are the initial value and time constant of the learning parameter. The value of the updated weight can be given by

$$\omega_i(t+1) = \omega_i(t) + \Delta\omega_i(t) \text{ for } i = 1, 2, \ldots, N \tag{21}$$

3 Results

The proposed method is tested under different shape and sizes of the workspace and areas of interest. Also, the base camp locations are altered within the same workspace. The effects of the task assignment and strategic placement of a multi-agent system are presented under simulated and Google map environments. For the training with SOM, the parameters are $t_{max} = 1000, \sigma_0 = 5, \eta_0 = 5, \tau_1 = 5, \tau_2 = 40$. In this work, $N = 10$ and $M = 25$ are considered.

3.1 *Case I and Case II*

In Case I, the base camp is at the corner of the workspace and in Case II, the base camp is at the center of the workspace as shown in Fig. 1 and Fig. 2, respectively. The demands are generated simultaneously from random places within the workspace. Two modes of operations are considered in each case.

- In TA only, the agents are attending the demands with the help of the proposed task assignment method. This is demonstrated through Figs. 1 (left) and 2 (left), for the Case I and Case II, respectively.
- In SP and TA, the agents are strategically placed in the workspace to attend any demand quickly. This is achieved by a two- step process. First, to place the agents at proper locations, agents are trained by the data points from the whole workspace by SOM. Second, for quick response, agents are attending the demands using the proposed task assignment procedure from their respective strategically placed positions. The results are shown in Figs. 1 (right) and 2 (right), for the Case I and Case II, respectively.

3.2 *Case III*

In Case III, base camp is considered in the corner of the workspace. Task inputs for training and testing are generated from specific areas of interest.

- In TA only, agents are attending the task from the specific base camp utilizing the proposed task assignment process. The is demonstrated in Fig. 3 (left).
- In SP and TA, the agents are first placed into strategic positions in the workspace using SOM. The result is shown in Fig. 3 (right). As demands of attending tasks come, each agent responds from its strategic position.

Initial base camp positioning of agents and the task positions within the workspace or areas of interest is taken to be the same for each mode.

Table 2 Comparative results for strategic placement and task assignment

Agent	Case I		Case II		Case III	
	TA (units)	SP and TA (units)	TA (units)	SP and TA (units)	TA (units)	SP and TA (units)
A1	0	21.7896	29.3352	19.6739	87.7229	7.2063
A2	87.7806	27.5807	53.9068	21.5073	0	4.5639
A3	0	25.0475	58.5406	30.8793	98.4824	10.3676
A4	83.7403	25.9360	64.23	26.8263	79.2767	41.9580
A5	0	55.5508	45.3212	35.3536	83.8363	42.2896
A6	107.7086	30.0736	65.8489	40.1660	0	7.9374
A7	81.677	26.4477	23.6157	17.3807	0	3.1504
A8	122.1279	49.4847	32.6596	11.1131	85.0589	9.2573
A9	97.0193	41.2465	52.8365	43.9342	101.1903	8.0127
A10	0.3624	36.8788	59.5015	34.6871	0	4.9132
Total distance	616.296	340.363	485.796	281.522	535.567	139.657

The analysis is also extended to real area using Google map. An area of Kolkata maidan is taken for this [14]. This is demonstrated in Fig. 4 through to Fig. 6. Here, all the above mentioned cases and modes of operations are also demonstrated.

Depending on the requirement, the proposed method can be utilized under the task assignment mode only and the strategic placement along with the task assignment mode. The comparative information of task assignment with distance overhead of both the modes of operations is tabulated in Table 2, with respect to the simulated environment. It is observed that the combination of strategic placement along with the task assignment results in less distance overhead. This allows a quick response to the demand for attending a set of tasks. From Table 2, it is also observed that under TA mode, a few agents were not assigned with any task. Also, the individual agent needs to travel more distance to attend the given tasks than the SP and TA mode. So, during urgency, the SP and TA mode is preferable.

4 Conclusion

In this work, SOM based strategic placement and task assignment of multi-agent system is presented. From the results, it is observed that agents which are already at their strategic locations respond quickly against demands. These agents travel lesser distances to attend the demands. A better workload balance is achieved through SP and TA mode. Incorporation of Google map exhibits the possibility of extending this work for realistic scenarios of surveillance. This phase of the work is focussed on the task assignment for agents. In future, path planning of these agents including, obstacle avoidance and interagent collision avoidance will be considered.

Acknowledgments This work is supported by the Visvesvaraya Ph.D. Scheme, Digital India Corporation (formerly known as the Media Lab Asia) for the project entitled "Intelligent Networked Robotic Systems".

References

1. Saatcioglu, M., Ghobarah, A., Nistor, I.: Performance of structures in Indonesia during the December 2004 great Sumatra earthquake and Indian Ocean Tsunami. Earthq Spectra **22**(S3), 295–319 (2006)
2. Han, Q., Du, X., Liu, J., Li, Z., Li, L., Zhao, J.: Seismic damage of highway bridges during the 2008 Wenchuan earthquake. Earthq. Eng. Eng. Vibr. Springer **8**(2), 263–273 (2009)
3. Cavallo, E., Powell, A., Becerra, O.: Estimating the direct economic damages of the earthquake in Haiti. Econ. J. **120**(546), F298–F312. Oxford Unversity Press, Oxford, UK (2010)
4. Goda, K., Kiyota, T., Pokhrel, R.M., Chiaro, G., Katagiri, T., Sharma, K., Wilkinson, S.: The 2015 Gorkha Nepal earthquake: insights from earthquake damage survey. Front. Built Environ. Front. **1**(8) (2015)
5. Burmeister, B., Haddadi, A., Matylis, G.: Application of multi-agent systems in traffic and transportation. IEE Proc. Softw. **144**(1), 51–60 (1997)
6. Kohonen, T.: The self-organizing map. Proc. IEEE **78**(9), 1464–1480 (1990)
7. Zhong, L., Luo, Q., Wen, D., Qiao, S., Shi, J., Zhang, W.: A task assignment algorithm for multiple aerial vehicles to attack targets with dynamic values. IEEE Trans. Intell. Transp. Syst **14**(1), 236–248 (2013)
8. Kuru, K., Ansell, D., Khan, W., Yetgin, H.: Analysis and optimization of unmanned aerial vehicle swarms in logistics: an intelligent Delivery Platform. IEEE Access **7**, 15804–15831 (2019)
9. Shima, T., Rasmussen, S., Gross, D.: Assigning micro UAVs to task tours in an urban terrain. IEEE Trans. Control Syst. Technol. **15**(4), 601–612 (2007)
10. Huang, L., Qu, H., Zuo, L.: Multi-type UAVs cooperative task allocation under resource constraints. IEEE Access **6**, 17841–17850 (2018)
11. Zhu, M., Du, X., Zhang, X., Luo, H., Wang, G.: Multi-UAV rapid-assessment task-assignment problem in a post-earthquake scenario. IEEE Access **7**, 74542–74557 (2019)
12. Rivera, A.J.A., Villalobos, A.D.C., Monje, J.C.N., Mariñas, J.A.G., Oppus, C.M.: Post-disaster rescue facility: Human detection and geolocation using aerial drones. In 2016 IEEE Region 10 Conference, pp. 384–386 (2016)
13. Chen, M., Zhu, D.: A workload balanced algorithm for task assignment and path planning of inhomogeneous autonomous underwater vehicle system. IEEE Trans. Cogn. Devel. Syst. (2018)
14. Kolkata Maidaan, lat 22.554779 (deg) and lon 88.347090 (deg), Google Earth. Accessed 17 Sept 2019
15. Khani, M., Ahmadi, A., Hajary, H.: Distributed task allocation in multi-agent environments using cellular learning automata. Soft Comput. **23**(4), 1199–1218 (2019)
16. Vallejo, D., Castro-Schez, J.J., Glez-Morcillo, C., Albusac, J.: Multi-agent architecture for information retrieval and intelligent monitoring by UAVs in known environments affected by catastrophes. Eng. Appl. Artif. Intell. **87**, 103243 (2020)
17. Moon, S., Oh, E., Shim, D.H.: An integral framework of task assignment and path planning for multiple unmanned aerial vehicles in dynamic environments. J. Intell. Robot. Syst. **70**(1–4), 303–313 (2013)

18. Lennartsson, J.: Strategic placement of ambulance drones for delivering defibrillators to out of hospital cardiac arrest victims. Thesis, KTH Royal Institute of Technology, Sweden (2015)
19. Yi, X., Zhu, A., Yang, S.X., Luo, C.: A bio-inspired approach to task assignment of swarm robots in 3-D dynamic environments. IEEE Trans. Cybern. **47**(4), 974–983 (2017)

Classification of Indian Languages Through Audio

Samim Raja, Suchibrota Dutta, Debanjan Banerjee, and Arijit Ghosal

Abstract Discrimination of different Indian languages is a very demanding job. This finds its importance in the domains of dialect identification, speech to text conversion, as well as in other domains. Indian languages vary in the point of view of irregularity, energy variation, and also the number of times they pass through the Zero axis which can be best represented by Skewness based perceptual audio feature; Short Time Energy, in short STE and Zero Crossing Rate, in short ZCR based time domain audio features, respectively. Hence, this advised approach banks on the said facets. For studying the characteristics of ZCR and STE accurately, statistical features based on their co-occurrence matrices are considered. As Mel Frequency Cepstral Coefficient (MFCC) is an overused aural facet, advised approach will also act as an alternative to it. Supervised classifiers are used for categorization purpose. Obtained classification accuracy has been compared with earlier research work to establish potency of advised facet set.

Keywords Language categorization · Skewness · ZCR · STE · Co-occurrence matrix

---

S. Raja (✉)
Cognizant Technology Solutions, Kolkata 743502, West Bengal, India
e-mail: samimr591@gmail.com

S. Dutta (✉)
Department of Information Technology and Mathematics, Royal Thimphu College, Thimphu, Bhutan
e-mail: suchibrota@gmail.com

D. Banerjee
Department of Management Information Systems, Sarva Siksha Mission Kolkata, Kolkata 700 042, West Bengal, India
e-mail: debanjanbanerjee2009@gmail.com

A. Ghosal (✉)
Department of Information Technology, St. Thomas College of Engineering and Technology, Kolkata 700023, West Bengal, India
e-mail: ghosal.arijit@yahoo.com

D. Bhattacharjee et al. (eds.), *Proceedings of International Conference on Frontiers in Computing and Systems*, Advances in Intelligent Systems and Computing 1255,
https://doi.org/10.1007/978-981-15-7834-2_38

1 Introduction

India is such type of nation where language has huge diversity. There are several language groups existing in Indian language mainly Indo-Aryan, Dravidian, and Austroasiatic. Indo-Aryan language group includes Bengali & Hindi language, Dravidian language group includes Tamil & Telugu language, and Austroasiatic language group includes Santhali & Munda languages. Discrimination of language especially Indian language is a motivating research area because of this huge diversity of Indian languages. It is impossible for a single person to learn all the languages of India. Still, just by listening, one can at least get an idea about whether the language is a south Indian language or north Indian language. Here perception plays a great role in these understanding. This job aims to identify the category of the spoken language through machine learning. It will simply classify input speech into three Indian language groups—Indo-Aryan, Dravidian and Austroasiatic, based on perceptual field and time field aural facets which are able to pigeonhole them. Mel Frequency Cepstral Coefficient (MFCC) is extremely utilized aural facet in auditory related work and because of this MFCC has not been employed in this exertion.

To convert speech into text during runtime, preliminarily the language of the spoken words need to be identified. Otherwise it will be tough to convert speech into text. This work will act as the preliminary step toward this task. There exist several dialects also for every language. Dialect identification is also an important and promising research work. Speech to text conversion task will be more perfect if dialect is also taken into account. This work can be treated as a fundamental step toward dialect identification, as language identification is the initial step for dialect identification. Discrimination of speech also finds its importance in the domain of data mining. If a digital language archive is required to be made, all speech signals need to be stored language wise so that in future they can be retrieved in a very easy way. This is a kind of audio indexing. There are several fields where audio indexing is very helpful. This work will pay some important contribution in this domain also.

In this effort, concise depiction in relation to foregoing research works is explained in Sect. 2. Suggested methodology for discriminating Indian languages is described in next Sect. 3. Investigational outcomes with contrast with previous work is made clear in Sect. 4, followed by the section describing conclusions.

2 Previous Works

Vocal speech identification necessitates a chain of signal dealing out stairs and education algorithms to sculpt discriminating distinctiveness of diverse speech. Most of the preceding efforts are found complex, whereas uncomplicated proposal is always wished. Yasminet al. [1] included the classification of three commonly spoken verbal communications in India—English, Hindi in addition to Bengali though English is not an Indian language. The scheme was based on Pitch, Zero Crossing Rate (ZCR),

Mel Frequency Coefficients, in short MFCC, and the numerical textural facets by analyzing co-occurrence matrix structured from MFCC. In the work of Gwon et al. [2], the authors presented a thin distinguishing facet education skeleton for verbal communication detection. Itrat et al. [3] presented an automatic language identification system that differentiates between two different spoken utterances in Urdu and Sindhi. A thorough cram on habitual verbal communication detection was accomplished and provided in the work of Karpagavalli and Chandra [4], that covered the structural design, verbal communication parameterization, tactics, individuality, topics, records, gizmos, and submissions. Lakhani and Mahadev [5] compared the performance of using audio spectrum in the log scale and using Polyphonic sound sequences from raw audio samples to train the neural network and to classify speech as either English or Spanish.

Kotsakis et al. [6] focused on the differentiation of aural substance, originating since radio constructions, depending on verbal idiom. Santhi and Sekar [7] dealt with the efficiency of an LID system for English and Hindi. Sigappi and Palanivel [8] outlined a line of attack for identifing a chosen lexis of terms spoken in Tamil verbal communication. The fundamental thinking was to extort the facets making use of Mel Frequency Cepstral Coefficients or MFCC since the vocal lexis that were made use of as envoy facets of the vocalizations to produce sculpts that aided in identification. Suo et al. [9] used a duo wise posterior probability estimation or PPPE rule to attune the yield of apiece classifier. Intendd loomes were estimated on 2005, National Institute of Standards and Technology (NIST). Hegde and Murthy [10] investigated the exercise of facets drawn from Fourier alter period for putting into practice LID systems. Statistical feature extraction method was proposed by Haralick [11].

3 Proposed Methodology

Previous research works were mostly involved in discriminating speech into different non-Indian languages. As geographically India is a vast country there are different types of language groups mainly Indo-Aryan, Dravidian, and Austroasiatic. Paces of the intended approach is depicted in Fig. 1. The input to the system is mono speech

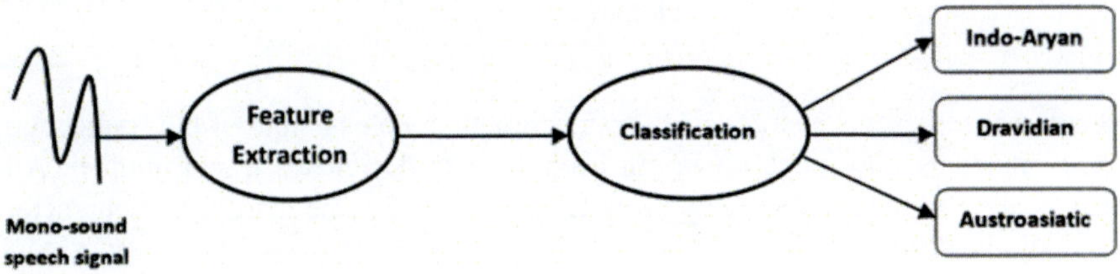

Fig. 1 Steps of the proposed approach

files and the output will be the language family to which these speech files belong to.

Some of the previous works involved Mel Frequency Cepstral Coefficients, in short, MFCC which is a very good aural facet and also excessively used. This exertion aims to put forward a simple acoustic facet group which does not include MFCC.

3.1 Computation of Features

When people speak in different languages, it certainly varies on the point of view of irregularity which is well represented by Skewness based features which is a perceptual feature. Moreover, energy variation, as well as the amount of times for a prearranged occasion distance the verbal communication signal passes through a rate of zero differs widely for different languages and this can be measured in time domain. This observation has motivated to employ Short Time Energy or STE, Zero Crossing Rate or ZCR depended time domain features as well as Skewness based perceptual features.

3.1.1 Features Supported on Short Time Energy (STE)

Verbal communication is constructed from an instance diverging voiced tract arrangement with instance diverging stimulation. As a result the speech signal is nonstationary in nature. The approach taken in this work is to analyze vigor of a speech signal and the signal is split into a number of small sections called frames, and it can be assumed within all fairness that signal remains stationary for analysis within these frames. It is true that level of energy is not the same for all Indian languages rather it varies. But at the same time it has been observed that energy level belonging to the same language group does not vary much, whereas it varies widely between two different language groups. This phenomenon is best represented by Short Time Energy (STE). STE is calculated for any input aural signal divided into Q frames $\{x_i(m): 1 \leq i \leq Q\}$ by Eq. (1), as

$$\text{ste}_n = \frac{1}{\text{length}} * \sum_{m=0}^{n-1} [x_n(sa)]^2 \tag{1}$$

where, ste_n represents energy of nth frame, length denotes length of the frame, and $x_n(sa)$ indicates sath mock-up in nth frame and m designates frame number. All casings are 50% overlied to keep away from failure of some margin line character for a definite casing.

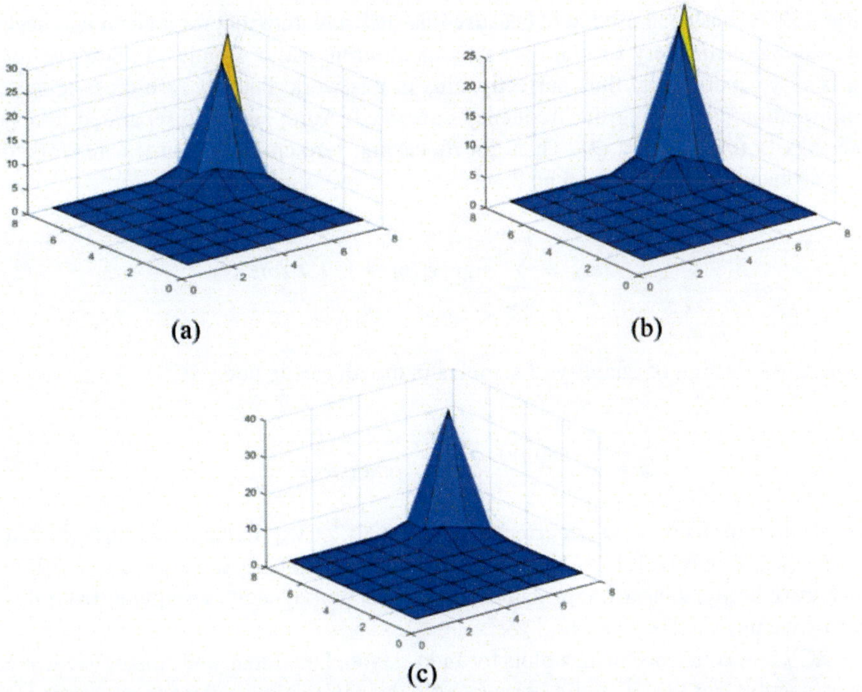

Fig. 2 Co-occurrence nature of STE plot for **a** Indo-Aryan, **b** Dravidian, and **c** Austroasiatic language group

As mean, in addition to standard deviation bestows only general thought, to precisely study the characteristics of STE, a co-occurrence matrix of STE is necessitated. Numerical facets proposed by Haralick [11], are evaluated from the matrix. Co-occurrence matrix is considered because it is capable to reflect the recurring distinctiveness of a feature. A co-occurrence matrix CO_ST_m having dimension $R \times R$ (where, $R = \max\{ste_i\} + 1$) is outlined. An element in the matrix $CO\_ST_m (p, q)$ reflects the occurrence number of STE p and q in consecutive time instances. Numerical features [11] like energy, entropy, contrast, correlation, and homogeneity are then computed from co-occurrence matrix.

STE co-occurrence matrix plots for Indo-Aryan, Dravidian, and Austroasiatic are depicted in Fig. 2a, b along with Fig. 2c correspondingly. These plots reflect that occurrence nature of STE varies for Indo-Aryan, Dravidian, and Austroasiatic language group evidently.

3.1.2 Feature Based on Zero Crossing Rate (ZCR)

The zero-crossing pace is the pace of changing of signs down a signal, means it is the pace at which signal modifies from upbeat to negative or else reverse. Zero Crossing

Rate (ZCR) has been used here because it is such a feature that for a given language its value remains very similar over various different audio samples. In other words it is very accurate and thus perfectly suits in this work. Pace of zero cross reflects information concerning the frequency contented. Aural signal is separated into Q casings $\{x_i(m)$: $1 \leq i \leq Q\}$. Then, for ith casing, zero crossing charges are worked out as mentioned in equation no 3:

$$zcr_n = \sum_{m=1}^{n-1} \text{sign}[x_i(m-1) * x_i(m) \tag{2}$$

where, n be a sign of quantity of samples in the ith casing and

$$\text{sign}[v] = \begin{cases} 1, & \text{if } v > 0 \\ 0, & \text{otherwise} \end{cases} \tag{3}$$

Similar to STE a co-occurrence matrix CO_ZCR_m having measurement M $\times$ M (M $= \max\{zcr_i\} + 1$) is formed and then numerical attributes [11] like energy, entropy, contrast, correlation along with homogeneity computed from this co-happening matrix.

ZCR co-occurrence matrix plots for Indo-Aryan, Dravidian, and Austroasiatic are portrayed in Fig. 3a, b, and c correspondingly. These plots reflect that occurrence nature of ZCR varies for Indo-Aryan, Dravidian, and Austroasiatic language group noticeably.

3.1.3 Feature Based on Skewness

Skewness is a perceptual facet. It quantifies the unevenness or indiscretion existing for a particular normal circulation compared to its mean location. Indo-Aryan, Dravidian, and Austroasiatic language group at all times maintain a quantity of indiscretion in their normal circulation. Skewness *ske* is calculated using subsequent Eq. (4)

$$sk = \left[E(d-\mu)^3\right]/\sigma^3 \tag{4}$$

where μ indicates mean of example information d, standard variation of d is denoted by σ, and the expected assessment of magnitude y is symbolized by $E(y)$.

3.2 *Classification*

A classifier distinguishes between various classes of data and assigns the testing data labels (class) based on minimum number of facets given to it aiming to decrease

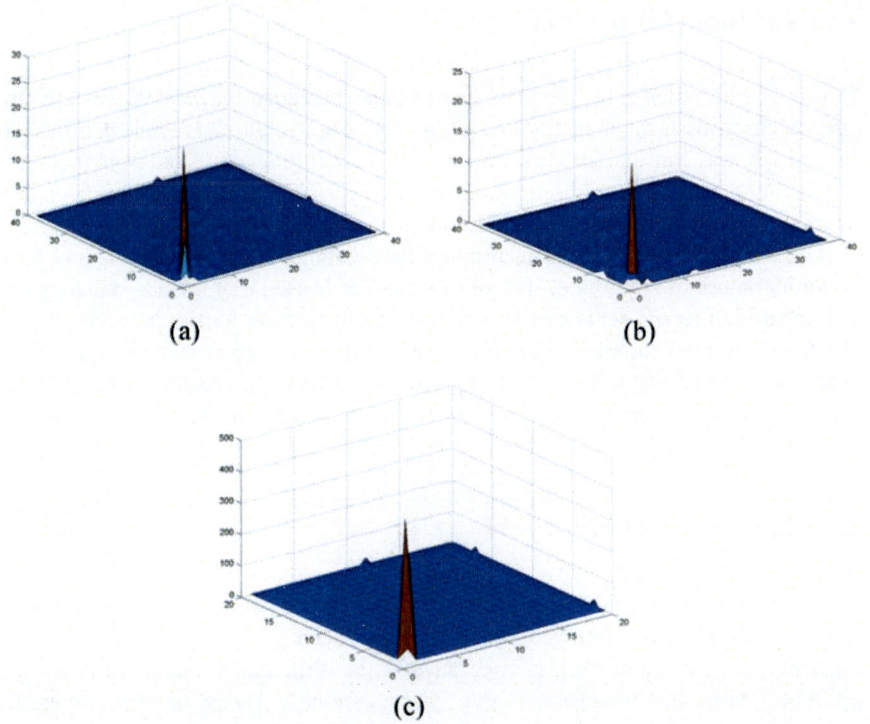

Fig. 3 ZCR co-happening plot for **a** Indo-Aryan, **b** Dravidian, and **c** Austroasiatic language group

computational complexity while maintaining the accurateness of language categorization good. Discriminating power of the proposed 11 dimensional facet set (5 numerical facets dependant on STE co-happening matrix + 5 numerical facets dependant on ZCR co-happening matrix + 1 skewness based facet) is tested by feeding it to Neural Network (NN), Random Forest, and Naïve Bayes. Multi Layer Perceptron or MLP has been employed to put into service Neural Network or NN. An aural data set has been prepared consisting of 600 audio files – 200 files each for Indo-Aryan (Hindi & Bengali 100 each), Dravidian (Tamil & Telegu 100 each), and Austroasiatic (Santhali & Munda 100 each) language group. The whole dataset is segregated into two equal fractions—training records set and testing records set. Multi Layer Perceptron or MLP model is considered having 11 neurons in input stratum as a sign of 11 facets. This model has 3 neurons in the output stratum denoting three categories of Indian language groups. There are 7 neurons in the concealed stratum of this MLP model. Concept of decision tree classifier has been implemented through Random Forest. To apply Naïve Bayes classifier, tenfold cross-validation is considered.

4 Experimental Results

The entire audio files of the record set are of mono category and have 90 s extent. To offer extensive assortment in this data set, speech of both male and female of different aged people was considered. These speech aural cases are accumulated from footages of CD/DVD, aural footages of assorted be alive performances in addition from the Internet. A number of these verbal communication audio cases are boisterous also.

All these speech files are broken into 50% overlapped frames to avoid any loss of border nature of any frame. Half of this data set is used as guidance statistics set and left behind record set is considered as testing record set for the supervised type classifier. After the completion of classification job, training and testing data set has been reversed and classification job is performed once again. Average of these two classification job is viewed as final cataloging outcome and is tabulated in Table 1.

4.1 Relative Analysis

Cataloging power of the put forward facet assembly has been evaluated with times of yore work. Dataset employed in this work has been used to employ the scheme suggested by Yasmin et al. [1]. They have to use Principal Component Analysis (PCA) to shrink the dimension of the feature assembly. Usage of PCA increases time complexity of the work, whereas this work is not suffering from that problem. Proposed feature set yields better accuracy compared to that work and this is tabulated in Table 2.

Table 1 Indian language classification accuracy

Cataloging proposal	Cataloging exactness (in %) meant for suggested feature set
Neural network	95.33
Random forest	97.33
Naïve Bayes	98.33

Table 2 Relative presentation of submited exertion amid erstwhile work in % exactness

Erstwhile technique	Classification accuracy (in %)
Yasmin et al. [1]	97.33
Proposed method (Naïve bayes)	98.33

5 Conclusion

This research discloses that by employing proposed facets a fine level of cataloging exactness is obtained. This feature set do confirm an assure exactness at a level of 98.33% while classifying speech into different category of Indian languages. The facet group length is shaped without exercising PCA to coach classifiers for making difference between the three divisions of Indian language groups. But also at the same time there stays a large amount possibility for progress and this may be stated that this exertion endeavors to boost the stoutness of the present verbal communication recognition manners.

References

1. Yasmin, G., Das Gupta, I., Das, A.K.: Language discrimination from speech signal using perceptual and physical features. In: Computational Intelligence in Data Mining, pp. 357–367. Springer, Singapore (2019)
2. Gwon, Y.L., Campbell, W.M., Sturim, D.E., Kung, H.T.: Language recognition via sparse coding. In: Interspeech 2016 (2017)
3. Itrat, M., Ali, S.A., Asif, R., Khanzada, K., Rathi, M.K.: Automatic language identification for languages of Pakistan. Int. J. Comput. Sci. Netw. Secur. (IJCSNS) **17**(2), 161 (2017)
4. Karpagavalli, S., Chandra, E.: A review on automatic speech recognition architecture and approaches. Int. J. Signal Process. Image Process. Pattern Recogn. **9**(4), 393–404 (2016)
5. Lakhani, V.A., Mahadev, R.: Multi-language identification using convolutional recurrent neural network (2016). arXiv preprint arXiv:1611.04010
6. Kotsakis, R., Mislow, A., Kalliris, G., Matsiola, M.: Feature-based language discrimination in radio productions via artificial neural training. In: Proceedings of the Audio Mostly 2015 on Interaction With Sound, p. 22, October 2015. ACM (2015)
7. Santhi, S., Sekar, R.: An automatic language identification using audio features. Int. J. Emerg. Technol. Adv. Eng. **3**(1), 358–364 (2013)
8. Sigappi, A.N., Palanivel, S.: Spoken word recognition strategy for Tamil language. Int. J. Comput. Sci. Iss. (IJCSI) **9**(1), 227 (2012)
9. Suo, H., Li, M., Lu, P., Yan, Y.: Using SVM as back-end classifier for language identification. EURASIP J. Audio Speech Music Process. **2008**(1), 674859 (2008)
10. Hegde, R.M., Murthy, H.A.: Automatic language identification and discrimination using the modified group delay feature. In: Proceedings of 2005 International Conference on Intelligent Sensing and Information Processing, pp. 395–399. IEEE (2005)
11. Haralick, R.M., Shapiro, L.G.: Computer and Robot Vision, vol. I (1992)

Neural Dynamics-based Complete Coverage of Grid Environment by Mobile Robots

Arindam Singha, Anjan Kumar Ray, and Arun Baran Samaddar

Abstract In this work, an algorithm is presented for complete coverage of a grid cell-based environment by mobile robots. The proposed paradigm consists of two biologically inspired neural network models. These allow mobile robots to navigate through collision-free paths and overcome dead-end situations. In this work, inter grid cell diagonal movement is restricted to enhance safety and prevent collision with obstacles. It also ensures inter robot collision-free navigation. The simulation results have shown the effectiveness of the proposed method for a single mobile robot and a multiple mobile robot system. A comparative study is also presented which showcases improvement of the proposed work over the existing literature.

Keywords Complete grid coverage · Autonomous multi-robot system · Bioinspired neural network · Dead-end situation · Path planning

1 Introduction

Complete grid coverage (CGC) of single and multiple mobile robots have a wide area of applications. The cooperative complete grid coverage approach has a certain privilege over single robot-based applications. There are many application areas of cooperative complete grid coverage; like ship cleaning task, terrain-covering, region full feeling, cooperative cleaning, and many more. Different methods are presented in the literature by several researchers to cope up with complete area coverage complete. The methods of CGC are classified as neural network-based approach [1–3], cell decomposition method [4, 5], graph-based method and recursive search-based

A. Singha (✉) · A. K. Ray · A. B. Samaddar
National Institute of Technology Sikkim, Ravangla 737139, Sikkim, India
e-mail: arindamsingha008@gmail.com

A. K. Ray
e-mail: akray.nits@gmail.com

A. B. Samaddar
e-mail: absamaddar@yahoo.com

D. Bhattacharjee et al. (eds.), *Proceedings of International Conference on Frontiers in Computing and Systems*, Advances in Intelligent Systems and Computing 1255,
https://doi.org/10.1007/978-981-15-7834-2_39

approach, and decentralized sweep coverage algorithm [6, 7], etc. Simultaneous allocation and complete coverage of multiple autonomous robots were presented in [8]. A biologically inspired self-organizing map was presented in [9], for the application of task assignment of swarm robots. The cooperative multi-agent sweep algorithm is presented in [10], for irregular shape of the environment. Ergodicity-based cooperative multi-agent area coverage was described in [11]. In [12], authors presented a complete area path planning algorithm to cope up with deadlock situations in a dynamic environment. They had used a global backtracking method for the robot to find an unvisited position when the robot faces any deadlock situation. Complete coverage path planning for an aerial vehicle is proposed in [13]. In this article, following aspects have been incorporated based on the drawbacks in the aforementioned article for navigation of mobile robots in a grid-based environment.

1. In the research articles [2, 3], the mobile robots can move through diagonal grid cells. This assumption may lead to a collision with the obstacles in a practical situation. This work ensures such collision avoidance by restricting inter grid cells' diagonal movement of mobile robots. The mobile robots can move only to their horizontal and vertical grid cells.
2. In [12] authors had used the backtracking method to overcome from deadlock or dead-end situations. In this article, an intelligent neural network-based paradigm introduces to cope up with the dead-end situation.

The grid-based environment is described in Sect. 2. The neural dynamics-based movement of mobile robots in a free grid cell environment is presented in Sect. 3. The avoidance of the dead-end situation is addressed in Sect. 4. The complete grid coverage for both the situations is presented in Sect. 5. The simulation results and comparative studies with existing works for a single mobile robot and multiple mobile robots are demonstrated in Sect. 6. In Sect. 7, the conclusion and the future direction of the work is presented.

2 Grid-Based Environment

The proposed model has applied to a grid-based environment. The entire working environment W is a 2-D grid cell environment. The current position of the mobile robot is denoted as p_c, where $p_c = [x_c\ y_c]$. The neighboring grid cell of the current mobile robot position is described in Eq. (1) [16]

$$\begin{aligned} p_n =&[x_n\ y_n] \in P_n \ \&\ P_n \in W \\ &\forall x_n \in (x_c - 1, x_c + 1), y_n \in (y_c - 1, y_c + 1)(x_c, y_c) \neq (x_n, y_n) \end{aligned} \tag{1}$$

p_n represents nth neighboring grid cell and P_n is the set of neighboring grid cells with respect to the current robot position p_c. The movement of the mobile robot is restricted to its neighboring horizontal or vertical free grid cell. So, the possible next positions p_{np} of the mobile robot is given by

$$p_{np} = \|p_c - p_n\| = 1 \,\&\, I_n = E \tag{2}$$

where I_n is the external input associated with p_n and is given by

$$I_n = \begin{cases} E, & \text{Unvisited} \\ -E, & \text{Obstacle} \\ 0, & \text{Visited} \end{cases} \tag{3}$$

and E is assumed to be a large positive constant number.

3 Movement into a Free Cell Using Bioinspired Neural Network

In free space, the mobile robot will use the biologically inspired neural network model for determining its next position. The proposed model is inspired by the electrical equivalent circuit of the biological model of cell membrane potential as explained in [14]. The neural dynamic equation can be represented as [2]

$$\frac{d\psi_n}{dt} = -A\psi_n + (B - \psi_n)(s_n^e) + (D - \psi_n)(s_n^i) \tag{4}$$

where, ψ_n represents the neural activity of nth neighboring grid cell, A, B, and D are positive constant values. S_n^e is excitatory input and S_n^i is inhibitory input to the system. Excitatory input gives information about free grid cells in its neighborhood and inhibitory input provides information about the neighboring grid cells containing obstacles. The excitatory (S_n^e) and inhibitory (S_n^i) inputs are designed as [15]

$$S_n^e = \left([I_n]^+ + \sum_{j=1}^{k} w_{nj}[\psi_j]^+ \right) \tag{5}$$

$$S_n^i = \left([I_n]^- + \sum_{j=1}^{k} w_{nj}[\psi_j]^- \right) \tag{6}$$

When the value of $I_n = E$, then $[I_n]^+ = I_n$ and $[I_n]^- = 0$. When $I_n = -E$, then $[I_n]^+ = 0$ and $[I_n]^- = I_n$. k represents total number of neighboring grid cells. The unvisited areas with the highest external activity attract the mobile robot and obstacle positions with the lowest external input repeals the mobile robot. It can be concluded from the previous statement that the unvisited positions have a global impact on the mobile robot, where the obstacle positions only have a local impact. Similar to current mobile robot position, every neighboring grid cell has a local lateral connection with jth number of neighboring connection. So, ψ_j denotes the neural activity of the jth neighboring grid cell of nth grid cell. w_{nj} is weight between nth grid cell to its jth neighboring grid cell. w_{nj} is calculated as

$$w_{nj} = \|p_n - p_j\| \tag{7}$$

The weights between two neurons are symmetric, by means that $w_{nj} = w_{jn}$. The next position of the mobile robot is determined by maximum neural activity among the possible next positions of neighboring grid cells $\psi_{p_{np}}$. So, the next position of the mobile robot will be $n_p \leftarrow max(\psi_{p_{np}})$.

4 Avoidance of Dead-End Situation

The neural network model defined in this article is inspired by [16]. Dead-end situation occurs for the mobile robot, when all the possible next positions are either visited or having obstacles. To overcome from the dead-end situation, the mobile robot will choose an unvisited position as its target position ($p_t = [x_t\ y_t]$) and move towards it. The neural activity of neighboring grid cells, while the mobile robot is in the dead-end situation, will be determined by the following Eq. (8)

$$\psi_n = g(w^d \psi_c + I_n) \tag{8}$$

g is a function, which is a combination of neural activity of the current position of mobile robot ψ_c, external input of the neighboring grid cell I_n and weights between current to neighboring grid cell for dead-end situation denoted as w^d. Weights between two grid cells are calculated by the following Eq. (9)

$$w^d = e^{\gamma \|p_c - p_n\|} \tag{9}$$

γ is a positive constant. Interpretation of p_c, p_n, and I_n are same as before. Function g is defined as

$$g(x) = \begin{cases} 0, & x \leq 0 \\ \beta x, & x \in (0\ 1] \\ 1, & x > 1 \end{cases} \tag{10}$$

β is a positive constant number. The details about selection of next position (n_p) for dead-end situation is described elaborately in [16].

5 Complete Coverage

The complete coverage of the grid cell environment requires avoidance of dead-end situations and free space movement. The free space movement of the mobile robot is addressed in Sect. 3, and overcoming from dead-end situation is addressed in Sect. 4. The overall neural activity calculation for both free space and dead-end situation is proposed as

$$\psi_n = RP + (1 - R)Q \tag{11}$$

where R is a binary number which signifies whether the robot is in free space or in a dead-end situation. The parameter R is defined as

$$R = \begin{cases} 1, & if \text{ not dead-end;} \\ 0, & if \text{ dead-end;} \end{cases} \tag{12}$$

P is the neural activity achieved from Eq. (4) and Q is the neural activity achieved from Eq. (8) for the dead-end situation.

6 Simulation Results

In this section, the proposed neural dynamic model is validated through different situations for single and multiple mobile robots system. A comparative analysis is also given in order to showcase the effectiveness of the proposed model. To simulate the proposed model the parameter values are taken as $A = 80$, $B = 1$, $D = 1$, $E = 200$, $\beta = 1$, and $\gamma = 3$.

6.1 Complete Grid Coverage by a Single Mobile Robot

The effectiveness of the proposed algorithm on a single mobile robot system is verified with two situations. Along with that, the proposed algorithm is also compared with the model proposed in [2]. Each grid cell is initialized with zero neural activity. In both the situations the environment is equally distributed in (16×11) number of grid cells. Though the initial position of the mobile robot is different. It is observed from Figs. 1 and 2, that in both the situations the mobile robot is successfully visiting

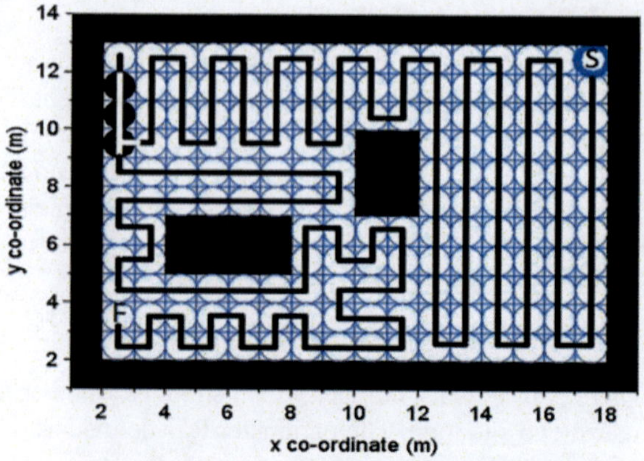

Fig. 1 Single mobile robot starting from top right grid cell

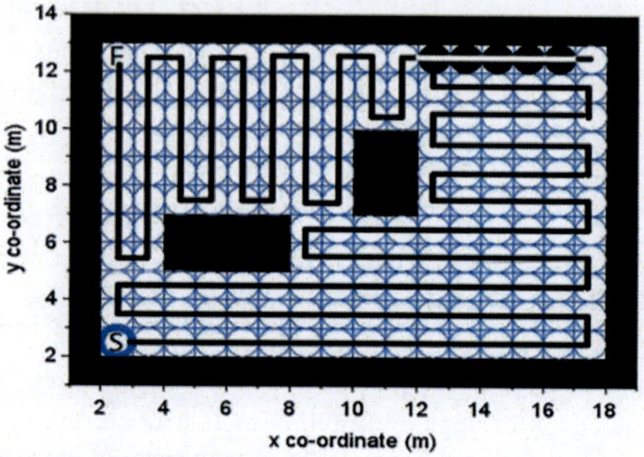

Fig. 2 Single mobile robot starting from bottom left grid cell

Table 1 Studies of single robot system for different initial position

Robot initial position	Steps	Turns	Overlaps
Top right corner	165	65	3
Bottom left corner	167	40	5

all the grid cells, though the number of turns and overlapped grid cells are different. The overlapped grid cells are more in situation 2, though the mobile robot took fewer turns to completely cover all the grid cells. A detailed comparison result is given in Table 1, for both the situations.

6.2 *Comparative Studies on Complete Grid Coverage by Single Mobile Robot*

The proposed algorithm is compared with Chaomin's model [2], for a single mobile system. The environment of is divided into (11 × 8) number equally organized grid cells. The mobile robot is starting from the bottom left grid cell position in both the situations. When compared to Chaomin's model [2], for the same initial position and environmental characteristics the mobile robot took fewer turns to completely cover all the unvisited grid cells. The detailed comparison results are given in Table 2. It can be concluded that for this given condition and environmental characteristics, our proposed algorithm works better than the algorithm proposed in [2]. Figures 3 and 4, depict complete grid coverage by the mobile robot using the algorithm proposed in [2], and in this article, respectively.

6.3 *Cooperative Grid Coverage by Multiple Mobile Robots*

A group of mobile robots is expected to be more effective than a single mobile robot system. The proposed algorithm is applied in a decentralized manner on multiple mobile robots systems, for two different situations. The proposed algorithm is applied to two different environments of size (14 × 10) and (16 × 12), respectively. The initial and final positions of the mobile robots are denoted as $S1$, $S2$, and $F1$, $F2$, respectively. It is depicted in Figs. 5 and 6, that in both the situations, the mobile

Table 2 Comparison studies for a single robot system

Model	Steps	Turns	Overlaps
Chaomin's model	80	30	0
Proposed model	80	25	0

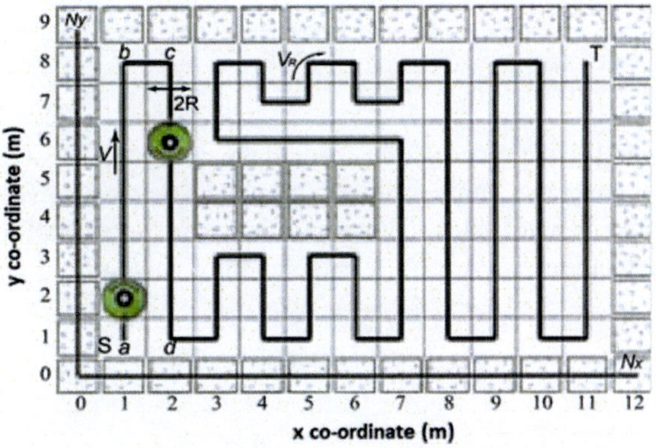

Fig. 3 Comparative result using Chaomin's model [2]

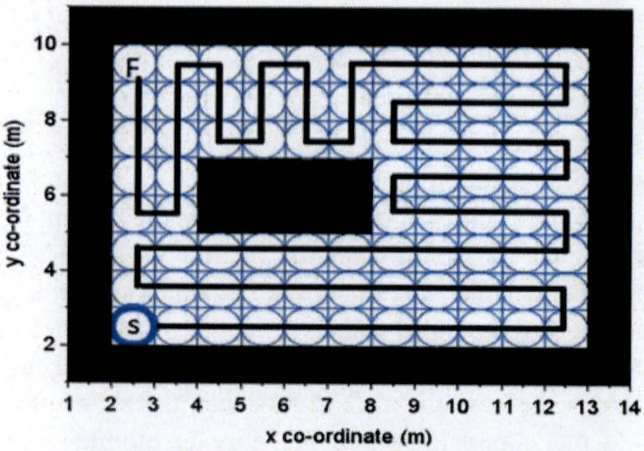

Fig. 4 Comparative result using proposed model

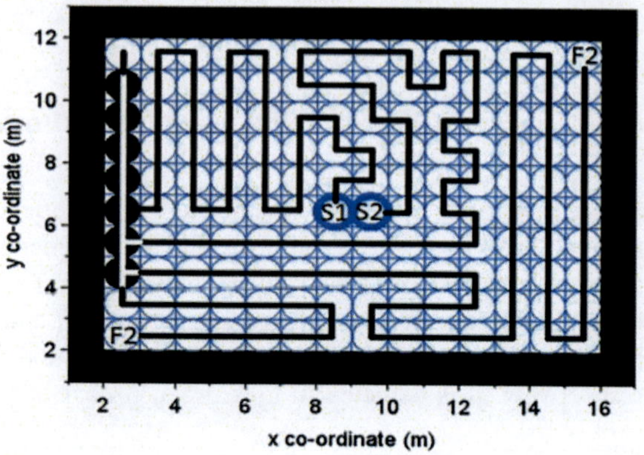

Fig. 5 Multiple mobile robots moving in no obstacle situation

Table 3 Studies on multiple robot system for different sets of obstacle and initial position

Number of obstacles	Coverage (%)		Turns		Overlaps	
	1st robot	2nd robot	1st robot	2nd robot	1st robot	2nd robot
No obstacle	41.50	58.50	19	31	7	0
Two obstacles	55.73	44.27	29	27	7	2

robots have successfully covered the whole workspace. The details about percentage of coverage, number of turns taken and grid cells overlapped for both the situations are given in Table 3. This signifies that the initial position of the mobile robot doesn't affect the neural activity calculation of neighboring grid cells.

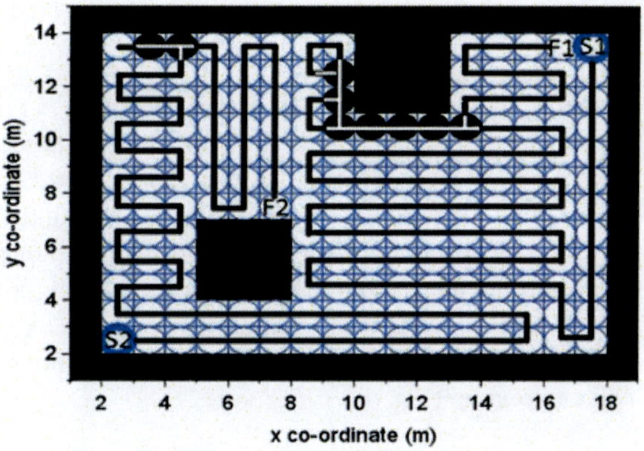

Fig. 6 Multiple mobile robots moving in presence of obstacles

6.4 Comparative Studies on Cooperative Grid Coverage by Multiple Mobile Robots

The proposed algorithm is compared with the model presented in [3], for multiple mobile robots system. The environment consists of (10 × 10) number of grid cells. A detailed comparison result is also given in Table 4. In [3], the inter grid cell diagonal movement is allowed, which eventually escalating the probability of collision with other grid cells containing obstacles, as shown in Fig. 7. In this work, the diagonal movement is restricted (Fig. 8), which results in 1 extra overlapped grid cell movement, but it ensures a collision-free path with obstacles.

Table 4 Comparison studies on multiple robot system

Model	Coverage (%)		Overlaps		Diagonal movement
	1st robot	2nd robot	1st robot	2nd robot	
Bin's model [3]	46.35	53.65	0	6	Allowed
Proposed model	50.60	49.40	3	4	Restricted

7 Conclusion

In this paper, a neural network-based algorithm is proposed for complete grid coverage. The simulation results show that mobile robots can successfully cover every grid cell in the environment. It is also demonstrated that mobile robots are capable of overcoming the dead-end situation. By restricting the inter grid cell diagonal movement the chances of collision with obstacles are removed. A detailed comparative study showed the effectiveness of the proposed algorithms over other algorithms for both a single and a multiple mobile robot system. A potential scope of improvement of this work is to incorporate task assignments among multiple robots within the grid cell environment.

Acknowledgments This work is supported by the Visvesvaraya Ph.D. Scheme, Digital India Corporation (formerly known as the Media Lab Asia) for the project entitled "Intelligent Networked Robotic Systems".

Fig. 7 Comparative result using Bin's model [3]

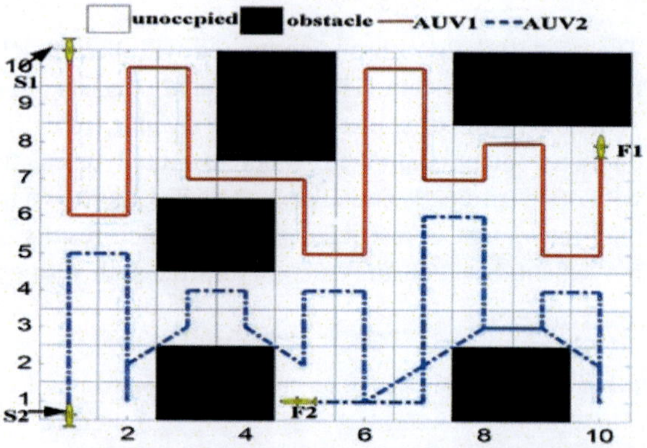

Fig. 8 Comparative result using Proposed model

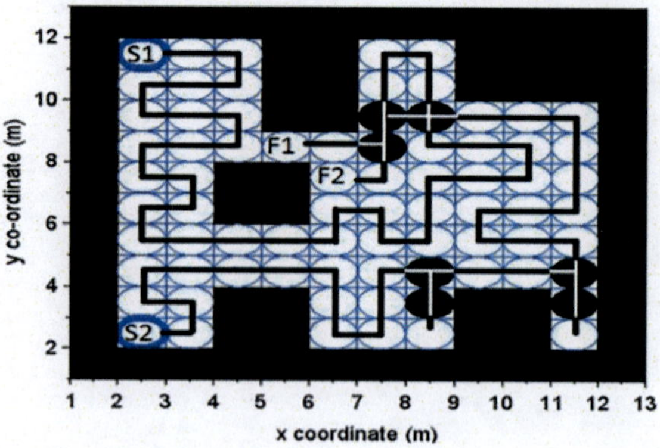

References

1. Yang, S.X., Luo, C.: A neural network approach to complete coverage path planning. IEEE Trans. Syst. Man Cybern. Part B (Cybern.) **34**(1), 718–724 (2004)
2. Luo, C., Yang, S.X., Li, X., Meng, M.Q.H.: Neural-dynamics-driven complete area coverage navigation through cooperation of multiple mobile robots. IEEE Trans. Indus. Electr. **64**(1), 750–760 (2017)
3. Sun, B., Zhu, D., Tian, C., Luo, C.: Complete coverage autonomous underwater vehicles path planning based on Glasius bio-inspired neural network algorithm for discrete and centralized programming. IEEE Trans. Cogn. Devel. Syst. **11**(1), 73–84 (2018)
4. Oh, J.S., Choi, Y.H., Park, J.B., Zheng, Y.F.: Complete coverage navigation of cleaning robots using triangular-cell-based map. IEEE Trans. Indus. Electr. **51**(3), 718–726 (2004)
5. Li, Y., Chen, H., Er, M.J., Wang, X.: Coverage path planning for UAVs based on enhanced exact cellular decomposition method. Mechatronics **21**(5), 876–885 (2011)
6. Cheng, T.M., Savkin, A.V., Javed, F.: Decentralized control of a group of mobile robots for deployment in sweep coverage. Robot. Autono. Syst. **59**(7–8), 497–507 (2011)
7. Zhai, C., Hong, Y.: Decentralized sweep coverage algorithm for multi-agent systems with workload uncertainties. Automatica **49**(7), 2154–2159 (2013)
8. Hassan, M., Liu, D.: Simultaneous area partitioning and allocation for complete coverage by multiple autonomous industrial robots. Auton. Robot. **41**(8), 1609–1628 (2017)
9. Yi, X., Zhu, A., Yang, S.X., Luo, C.: A bio-inspired approach to task assignment of swarm robots in 3-D dynamic environments. IEEE Trans. Cybern. **47**(4), 974–983 (2017)
10. Shi, M., Qin, K., Liu, J.: Cooperative multi-agent sweep coverage control for unknown areas of irregular shape. IET Control Theory Appl. **12**(14), 1983–1994 (2018)
11. Ivic, S., Crnkovic, B., Mezic, I.: Ergodicity-based cooperative multiagent area coverage via a potential field. IEEE Trans. Cybern. **47**(8), 1983–1993 (2017)
12. Liu, H., Ma, J., Huang, W.: Sensor-based complete coverage path planning in dynamic environment for cleaning robot. CAAI Trans. Intell. Technol. **3**(1), 65–72 (2018)
13. Guastella, D.C., Cantelli, L., Giammello, G., Melita, C.D., Spatino, G., Muscato, G.: Complete coverage path planning for aerial vehicle flocks deployed in outdoor environments. Comput. Electr. Eng. **75**, 189–201 (2019)
14. Hodgkin, A.L., Huxley, A.F.: A quantitative description of membrane current and its application to conduction and excitation in nerve. J. Physiol. **117**(4), 500–544 (1952)
15. Grossberg, S.: Nonlinear neural networks: principles, mechanisms, and architectures. Neural Netw. **1**(1), 17–61 (1988)
16. Singha, A., Ray, A.K., Samaddar, A.B.: Navigation of mobile robot in a grid-based environment using local and target weighted neural networks. In: 8th International Conference on Computing, Communication and Networking Technologies (ICCCNT), pp. 1–6 (2017)

Solving Student Project Allocation with Preference Through Weights

Juwesh Binong

Abstract Student project allocation is a common problem for many university departments. In this work, a solution is proposed for allocating projects to students with their preferences. No preference for the faculties was considered. Students with better academic performance were provided a higher prospect for their preferences. The proposed approach has been implemented in Microsoft Excel and applied to allocate projects to the final year students of Bachelor of Engineering in the North-Eastern Hill University. The simplicity and ease of the method, makes it suitable for anyone to use in allocating projects.

Keywords Student Project Allocation (SPA) · SPA problem · Preference · Excel

1 Introduction

Student Project Allocation (SPA) is a common problem found in many departments of an educational institution [1, 3, 4]. Students in their higher semester are required to complete projects as a part of their academic program. Generally, a project is assigned to a student and a faculty is assigned as a project guide to the student. Usually, a wide range of projects are made available to assign each student a project of his/her choice. A faculty also offers a range of projects of his/her area of specialization(s), and hence a teaching department of an educational institution like university offers a wide range of projects for students as each department can have many faculties.

A student has preferences over the available projects and can choose a project of his/her choice. Many times faculties, too, can have preferences over students that he/she wants to guide; but a faculty may not have preferences for students depending on the department. Also, there may be upper bounds on the number of students that can be assigned to a particular project, and the number of students that a faculty can

J. Binong (✉)
Department of Electronics and Communication Engineering, North-Eastern Hill University, Shillong, India
e-mail: binong.j@gmail.com

D. Bhattacharjee et al. (eds.), *Proceedings of International Conference on Frontiers in Computing and Systems*, Advances in Intelligent Systems and Computing 1255,
https://doi.org/10.1007/978-981-15-7834-2_40

supervise. In this work, an algorithm is presented to allocate students a project of their choice among the available ones and the capacity of the faculties, but based on their academic performance.

SPA is considered an example of a two-sided matching [12], which is an important research area in the field of operational research and decision analysis. The two-sided matching decision problem is derived from Gale and Shapley's research on stable marriage matching and college admission problem [8]. Allocating projects manually is tedious, hence different automated techniques are reported in the literature. In [2], authors developed an integrated program for the SPA problem which is solved using a solver. In [14], defined a linear program for some special case of SPA problem. A heuristic-based approach like the Genetic algorithm is also found in literature [15]. In [9], a genetic algorithm is presented as an aid for assigning projects to students, where students can choose from a list of projects by indicating their preference in advances. In [5], Manlove has shown the complexity of the SPA problem, and he proposed an approximation algorithm for the same. Works, where faculties can have preferences for students are also found in literature [7]. The stable marriage problem which is a one-one matching problem, also applied to the SPA problem [6].

In this paper, an algorithm is presented for allocating students' projects of their preferences for a given instance of SPA. This algorithm is student-oriented and an academically good student can obtain the project of his/her choice. This motivates students to perform well in the lower semesters. Another advantage of the proposed approach is that it is very simple, easy to understand, and can be implemented easily in any spreadsheet including Microsoft Excel [11]. No knowledge of new software is required to use the proposed method, only some basic knowledge of Microsoft Excel is sufficient to take the benefit of the proposed method. The algorithm can be implemented in data science approach too and this let to take the advantages of powerful data science tools.

The remainder of the paper is structured as follows. In Sect. 2, the proposed problem is described; followed by the problem formulation in Sect. 3. Then, in Sect. 4, an algorithm for the SPA problem is presented. In Sect. 5, results are presented and discussed. Finally, concluded in Sect. 6.

2 Problem Description

During the final year of the Bachelor of Technology in the Electronics and Communication Engineering at the North-Eastern Hill University (NEHU), students have to take a minor project during the odd semester and a major project during the even semester, as a part of the course requirement, each selected from the lists of available projects. Each faculty of the department of Electronics and Communication Engineering offers projects based on his/her areas of specialization(s). Students are allowed to opt a set of project guides or supervisors, i.e., faculties who are offering the preferring projects among the available ones. Usually, faculties are not given preferences over the students, and each faculty assigned a number of students based

Table 1 An instance of the SPA problem

Student_ID	Students' average CGPA	Students' preferences
s_1	p_1	f_1, f_2, f_4, f_7
s_2	p_2	f_n, f_2, f_8, f_5
...	......	.
s_n	p_n	f_8, f_2, f_4, f_7

on his/her capacity. A project (i.e. a faculty member) guide may be chosen by many students, and the number of students preferring the project guide may outnumber the capacity of the faculty offering a certain number of projects; and in such case, some students may not assign the preferred project (see Table 1).

3 Problem Formulation

The Student Project Allocation (SPA) problem of NEHU, may be formulated as follows. Let $S = \{s_1, s_2, \ldots, s_n\}$ be a set of students, let $F = \{f_1, f_2, \ldots, f_m\}$ be a set of faculties. Each student s_i provides a preference list taking from F in order of preference. Each s_i have to provide equal number of preferences, as asked by the project coordinator of the department.

4 Proposed Methodology

4.1 Preference Scales

According to the Oxford online dictionary, preference is a greater liking for one alternative over another or others [10]. It is a qualitative term. A preference is a technical term in psychology, economics, and philosophy usually used in relation to choosing between alternatives. For example, someone prefers A over B if they would rather choose A than B [16].

In the analytic hierarchy process (AHP) [13], preferences are measured with preference scales. Students' preference for faculty is an important step in decision-making. Here preferences are measured in term of scale, the lowest preference carrying a value of 2, and the second-lowest a value of 3, and the highest preference carrying the highest value.

An example of such a preference scale is shown in Table 2, where "n" is any positive integer specifying the total number of preferences allowed.

Allocation of a project guide f_j to a student s_i is done based on the academic performance of the student for the semesters completed. The academic performance

Table 2 The preference scale

Preferences	Scale
First preference	$n + 1$
Second preference	n
Third preference	$n - 1$
…………..	…
Last preference	2

of a student s_i is measured in terms of average Cumulative Grade Point Average (CGPA) p_i which is the average of the CGPA of the all the semester the student has completed. A student s_i is allocated a project supervision f_j of its preferences based on its academic performance p_i. If a project supervision f_j is chosen by more than one student, the project supervision f_j is assigned to the student with the highest academic performance. But in such an allocation process, some students lose all the preferences of their choice. In that case, a score called the *preference score* q_i of the project guides whose capacity is still vacant are calculated and the project guides are ranked based on the preference scores.

The rank r_j of the academic performance of the remaining students are also calculated, and each student assigned a project guide that matches the rank of the student. A student ranked r_j is allocated the project guide with score q_j. The rank of a project guide is calculated as follows:

$$q_j = \frac{\text{Number of } s_i \text{ preferring } f_j}{\sum s_i} \tag{1}$$

4.2 *Data Preparation Method*

The data is collected and a table is prepared in the format as shown in Table 3. The weight of a student for a preference is calculated by multiplying his/her average CGPA with the preference scale.

4.3 *The Proposed Algorithm*

The time required to solve the SPA problem increases exponentially with the size of instances, i.e., the SPA problem is an NP-hard problem [5]. The SPA problem mentioned in this work can be solved by the algorithm described in Algorithm 1.

All the students are allocated project guide of their choice, but a student with poor academic performance may not be allocated guide of his/her choice.

Algorithm 1: The proposed Student Project Allocation Algorithm

```
   Input: Students Preference Dataset
   Output: Allocated list of students based on preference
1  Sort the dataset in descending order based on weight.
2  Assign rank to each row of the dataset with the highest weight as rank '1'
3  forall the preference k = 1 to q do
4      Assign 'k' for the row with preference 'k'.
5      forall the preference k do
6          Remove the allocated students if the preference (k − 1) ≠ 0.
7          Arrange students s_i based on project guide f_j.
8          Sort the students in descending order of p_i.
9          Allocate the students to the preferred guide to his/her capacity as follows and store
           the allocated list of students:
10             (a) If numbers of students s_i preferring a project guide f_j is more his/her capacity,
           allocate students only to him/her full capacity;
11             (b) Else, allocate all the students to the guide.
12     end
13 end
14 If one or more student losses all it preferences, allocate a project guide as follows:
15     (a) Compute the rank for all the guides using Eq. (1).
16     (b) Obtain the average CGPA of the unallocated students and assigned ranks, making top
   student as rank 1.
17     (c) Allocate the highest rank student to the project guide with highest rank.
```

4.3.1 Complexity Analysis

The time complexity of the algorithm depends on the loops, as the dataset can be made sorted one. If n is the numbers of students and q is the numbers of preferences allowed to each student, then the total number of records is $n \times q$. For a sorted dataset, the students can be arranged in groups based on project guides in n operations. Sorting of students under all the guides can be done approximately in $(n + 1)q/2$ operations. Steps 14–17 can be done in constant time. Moreover, these steps can be pre-calculated from the dataset. All other steps have a negligible number of operations comparatively. Hence, the number of total operations is approximately nq. But the maximum value of q can not exceed r, the number of project guides. The maximum number of preferences that can be given to each student is equal to the number of project guides, i.e., $q = r$. If $n = r$ and all the students are allowed to select all the guides, in worst-case, the time complexity turns out to be $O(n^2)$.

The SPA problem mentioned in this work can be implemented easily in any spreadsheet like Microsoft Excel or data science approach using python and its associated libraries including pandas, keras, numpy, etc.

5 Results and Discussion

All of the 20 students of the Bachelor of Technology in the Department of the Electronics & Communication Engineering in the North-Eastern Hill University were asked to provide their preferences for project guides whom they would like to be supervised for their final year project.

The names of the four supervisors were asked to be provided in order of preference in a Google sheet. CGPAs of all the students were also collected through the same google sheet. The average CGPA of each student was calculated. The data were recorded in a spreadsheet consisting of the columns as shown in Table 3.

The preferences of the students for faculties are shown in Fig. 1; where *T#x*, x being an integer, represents the students' preference for faculty "x"; and 1st–4th represent preferences 1st–4th respectively.

Out of 20 students, 11 students could obtain the supervisor of their first preference, 4 students got their second preference, 2 students got their third preference, and no student was left with their fourth preference. But the remaining 3 students lost all their preferences, and hence were allocated project guide based on *preference score*. That is, the number of students allocated in the first preference is 55%; 20% in the second preference; just 10% in third preference; and only 15% through the *preference score*.

Table 3 Data preparation table

Student_ID	Average CGPA	Guide_ID	Preference_No	Preference scale	Weight
S_1	$CGPA_1$	f_w	First preference	$n+1$	$CGPA_1 * (n+1)$
S_1	$CGPA_1$	f_x	Second preference	n	$CGPA_1 * n$
S_1	$CGPA_1$	f_y	Third preference	$n-1$	$CGPA_1 * (n-1)$
……	……	……	……	………	…………
S_1	$CGPA_2$	f_k	Last preference	$n+1$	$CGPA_1 * 2$
S_2	$CGPA_2$	f_k	First preference	$n+1$	$CGPA_1 * (n+1)$
S_2	$CGPA_2$	f_z	Second preference	n	$CGPA_1 * n$
S_2	$CGPA_2$	f_x	Third preference	$n-1$	$CGPA_1 * (n-1)$
……	……	……	……	………	…………
S_2	$CGPA_2$	f_y	Last preference	$n+1$	$CGPA_1 * 2$
……	……	……	……	………	…………
S_n	$CGPA_n$	f_x	Last preference	$n+1$	$CGPA_n * 2$

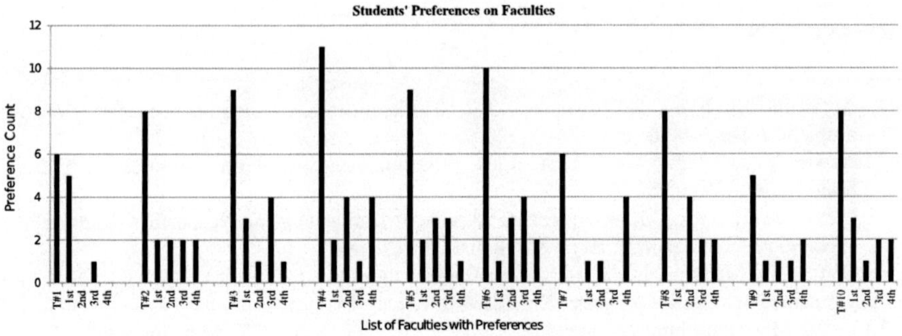

Fig. 1 Preferences of 20 students for 10 faculty members

Fig. 2 Result of the allocation process based on Algorithm 1

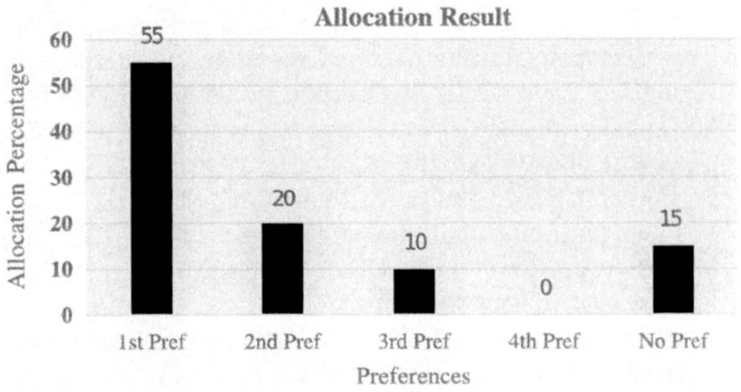

The result of the project allocation process is shown in Fig. 2. As only 15% of the students were allocated project through the *preference score*, this implies the effectiveness of the proposed algorithm.

6 Conclusion

In this paper, an algorithm for solving the SPA problem is presented. The algorithm can be implemented easily in Excel. The Excel-based approach may be handy for most of the project coordinator. People comfortable with python scripts and its associated libraries for manipulating data frame can use python-based data science techniques to get more benefit out of the proposed algorithm.

Some future works on the proposed algorithm may be listed as follows:

- The proposed approach may be extended to the case where faculties too can have their preference for students.
- The proposed algorithm needs to be tested on some benchmark dataset.
- The performance of the proposed method still remains to compare with the existing works.

References

1. Abraham, D.J., Irving, R.W., Manlove, D.F.: Two algorithms for the student-project allocation problem. J. Disc. Algor. **5**(1), 73–90 (2007)
2. Anwar, A.A., Bahaj, A.S.: Student project allocation using integer programming. IEEE Trans. Educ. **46**(3), 359–367 (2003)
3. Chown, A.H., Cook, C.J., Wilding, N.B.: A simulated annealing approach to the student-project allocation problem. Am. J. Phys. **86**(9), 701–708 (2018)
4. Cooper, F., Manlove, D.: A 3/2-approximation algorithm for the student-project allocation problem (2018). arXiv preprint arXiv:1804.02731
5. David, M.: Algorithmics of Matching Under Preferences, vol. 2. World Scientific (2013)
6. Dye, J.: A constraint logic programming approach to the stable marriage problem and its application to student-project allocation. B.Sc Honours project report, University of York, Department of Computer Science (2001)
7. El-Atta, A.H.A., Moussa, M.I.: Student project allocation with preference lists over (student, project) pairs. In 2009 Second International Conference on Computer and Electrical Engineering, vol. 1, pp. 375–379, December 2009. IEEE (2009)
8. Gale, D., Shapley, L.S.: College admissions and the stability of marriage. Am. Math. Month. **120**(5), 386–391 (2013)
9. Harper, P.R., de Senna, V., Vieira, I.T., Shahani, A.K.: A genetic algorithm for the project assignment problem. Comput. Oper. Res. **32**(5), 1255–1265 (2005)
10. Lexico, powered by OXFORD. Preference. https://www.lexico.com/en/definition/preference. Accessed 16 Sept 2019
11. Ragsdale, C.T., Zobel, C.W.: A simple approach to implementing and training neural networks in excel. Decis. Sci. J. Innov. Educ. **8**(1), 143–149 (2010)
12. Roth, A.E., Sotomayor, M.: Two-sided matching. In: Handbook of Game Theory with Economic Applications, vol. 1, pp. 485–541 (1992)
13. Saaty, T.L.: What is the analytic hierarchy process? In: Mathematical Models for Decision Support, pp. 109–121. Springer, Berlin, Heidelberg (1988)
14. Saber, H.M., Ghosh, J.B.: Assigning students to academic majors. Omega **29**(6), 513–523 (2001)
15. Salami, H.O., Mamman, E.Y.: A genetic algorithm for allocating project supervisors to students. Int. J. Intell. Syst. Appl. **8**(10), 51 (2016)
16. Wikipedia, The Free Encyclopedia. Preference. https://en.wikipedia.org/wiki/Preference. Accessed 16 Sept 2019

Computational Biology Track

Deep Learning-Based Automated Detection of Age-Related Macular Degeneration from Retinal Fundus Images

Rivu Chakraborty and Ankita Pramanik

Abstract The early and intermediate stages of age-related macular degeneration (AMD) are often asymptomatic and may lead to a neovascular form, which ends up causing blindness. The existing works on the detection of AMD make use of image processing and manual feature extraction methods. These methods detect drusen properties and use decision-making algorithms to obtain the desired results. The proposed work is a novel solution for the problem of AMD detection using a deep learning approach. The proposed method screens retinal images for detecting direct evidence of AMD. As deep learning model calculates features and learns on its own, there is less chance of neglecting any important feature which may happen in the existing methods. The proposed approach is applied to check for the presence of AMD on a dataset of healthy and diseased cases and a detection accuracy of 84% is obtained.

Keywords Deep learning · Convolutional neural network · Age-related macular degeneration · Medical imaging · Retinal image analysis

1 Introduction

Age-related macular degeneration (AMD) is a degenerative disorder in the macular region. Most of the adults above the age of 50 years lose their eyesight due to AMD [1, 2]. The incidence rate of AMD is higher than diabetic retinopathy in elderly people [3]. AMD is spread in women at an earlier age than men. Among all the races, caucasians have a higher tendency to be affected by AMD. The early and

R. Chakraborty (✉) · A. Pramanik
Indian Institute of Engineering Science and Technology, Shibpur, India
e-mail: rivuchakraborty17@gmail.com

A. Pramanik
e-mail: ankita@telecom.iiests.ac.in

D. Bhattacharjee et al. (eds.), *Proceedings of International Conference on Frontiers in Computing and Systems*, Advances in Intelligent Systems and Computing 1255,
https://doi.org/10.1007/978-981-15-7834-2_41

(a) (b)

Fig. 1 Comparison between **a** normal vision, and **b** simulated vision of AMD (*Source*: National Eye Institute; https://medialibrary.nei.nih.gov/)

intermediate stages of AMD are usually asymptomatic. But when it forwards to the advanced stage, usually a considerable amount of central vision loss sets in, which later leads to blindness. A comparison between normal vision and simulated AMD vision is shown in Fig. 1.

The actual cause of AMD is not properly determined, but it can be associated with several factors like chronic photodestruction effect, genetics, nutritional disorder, etc. A major indication of AMD is the appearance of drusen [4], small yellow excrescence of fatty proteins (lipids) accumulated between the retinal pigment epithelium (RPE) basement membrane and the remaining part of the Bruch membrane. The indication of intermediate stage of AMD consists of at least 1 large druse ($\geq$125 μ) or multiple medium-sized drusen (63–125 μ) or geographic atrophy (GA) of the RPE that excludes the fovea [5].

AMD is categorized into dry and wet AMD. Dry AMD, also known as nonexudative AMD, is not neovascular and generally can be observed in intermediate and advanced stages under ophthalmoscopy. It is identified by progressive atrophy of RPE [6]. Wet AMD, also known as exudative or neovascular AMD, is identified by active neovascularization under RPE [5], eventually leading to exudation, hemorrhage, and scar. As a result, daily activities like recognizing objects and reading get affected. Ultimately, it will damage the photoreceptors and cause an abrupt loss of vision, which cannot be reversed, if left untreated in the early and intermediate stages.

The existing algorithms for automated retinal image analysis are mostly dependent on conventional techniques that used engineered image features (e.g., histogram of oriented gradients, wavelets, scale-invariant feature transform, etc., [7, 8]), which were selected manually. Those features were then applied in a classifier (e.g., support vector machines, logistic regression, random forests, boosting, etc., [9–12]). By contrast, deep learning methods absorb task-specific image features with multiple levels of abstraction being independent of manual feature engineering. Latest developments in deep learning have enhanced performance levels rapidly for several image analysis tasks. Lately, deep learning has been applied for performing various retinal image analysis studies, such as the classification of diabetic retinopathy, myopia, etc. A deep convolutional neural network (DCNN) architecture is built from scratch for the automated classification of AMD and non-AMD retinal fundus images in the proposed work.

Section 2 of this paper explains the dataset, data partitioning, and the proposed DCNN architecture adopted in this study. Section 3 discusses on the obtained results using the proposed method and comparison with existing works, respectively. Lastly, this paper is concluded in Sect. 4, with a discussion on limitations, possible improvements, and future scope of works.

2 Methodology

This work is targeted to solve a binary classification problem, i.e., classifying non-AMD versus AMD retinal fundus images. In traditional machine learning algorithms, the features are calculated manually by researchers. As deep learning model calculates its features with multiple levels of abstraction and learns on its own, there is less chance of neglecting any important feature which may happen in the existing traditional machine learning algorithm. Recent researches have shown great improvements using deep learning methods. So a DCNN approach is taken to solve this problem. It uses the convolutional layers to calculate spatial features of the images and trains on its own. The aim of this work is to review the comparison between the performance of the proposed deep learning method and a human clinician.

2.1 Dataset

The retinal fundus images are downloaded from Baidu Research Open Access Datasets (BROAD) for iChallenge-AMD [13]. This dataset consists of a total of annotated 400 retinal fundus images from both AMD patients (22.25%) and non-AMD subjects (77.75%). All the images are color images. Labels of AMD/non-AMD, disc boundaries, and fovea locations, as well as masks and boundaries of kinds of lesions, are provided to train models for automated AMD assessment. The reference standard for AMD presence obtained from the health records is not only based on retinal fundus image, but also take optical coherence tomography (OCT), visual field, and other factors into consideration. For training images, AMD, and non-AMD labels (the reference standard) are reflected in the image file names. Sample retinal fundus images with AMD and non-AMD included in the dataset are shown in Fig. 2.

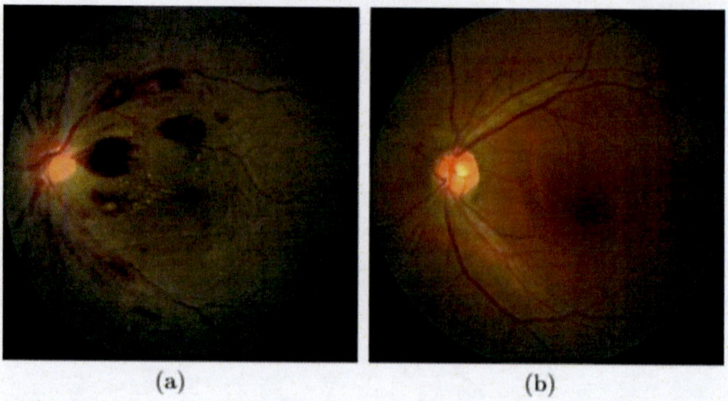

Fig. 2 Sample color retinal fundus images **a** with AMD, and **b** without AMD in the dataset [13]

2.2 *Data Partitioning*

In any machine learning algorithm, the given data is randomly partitioned into two types of datasets namely—training and test sets. The model is trained using the training set which contains data and labels. The test set is then used to validate the performance of the underlying model. The purpose of validation is to draw conclusions on the model in the presence of a similar type of unknown data.

The proposed method was validated using four-fold cross-validation. As shown in Fig. 3, the images are taken from the dataset were randomly partitioned into 4 equal-sized subsets. Among them, one of the subsets was retained as a test set for validation purpose and the rest of the subsets were used for training the proposed DCNN architecture. This method was iterated for 4 times with different test sets reserved for validation each time and the rest of them were used for training. The results obtained from each validation iteration were analyzed to get a better understanding of the model in this given dataset.

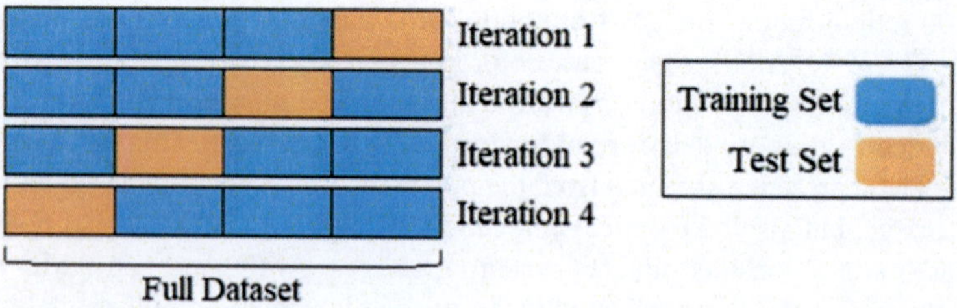

Fig. 3 Test/train split in four-fold cross-validation

2.3 Proposed DCNN Architecture

When Deep Learning is applied in the image processing [14] field the initial layers of all the networks detect the basic things in the image like horizontal lines, vertical lines, texture, etc. Thus, it is not necessary to train the whole network every time a set of new images are taken. Thereby, the computational time is reduced significantly.

The model simulation process was done in a PC manufactured by HP with Intel Core i5-8500 CPU @ 3.00GHz and 4Gb RAM and no GPU. So to work in such a low-end PC some preprocessing was necessary. The images were resized to 400 x 400 pixels as the higher dimension crashed the memory of the computer. A simple convolutional neural network was modeled with 5 convolutional layers with ReLU function and 5 Maxpool layers alternatively, a fully connected layer with ReLU function with dropout and the final layer is built with a fully connected layer with Softmax function as shown in the block diagram in Fig. 4. The kernels used in all the convolutional and maxpool layers have a receptive field of 5 x 5 and 2 x 2 pixels, respectively. The Adam optimization technique was chosen and the loss was calculated by categorical cross-entropy. This network has been trained by feeding the training images with mini-batch size 8 for 50 epochs with a 0.001 learning rate. It took almost 45 min to build the prediction model. This trained network gives an output that consists of the probabilities of the classes—AMD and non-AMD. The final output was selected having the higher probabilities. This network was built, trained, and tested in Python using TFLearn with TensorFlow as backend.

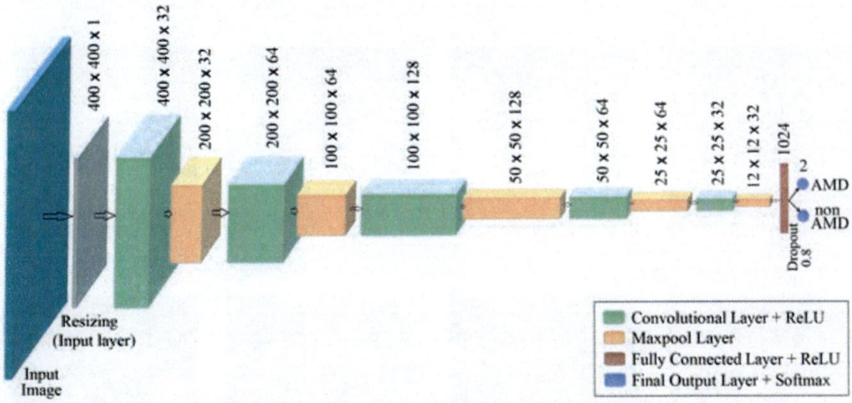

Fig. 4 Proposed CNN architecture for classification of AMD and non-AMD images

3 Result Obtained and Comparison

The proposed network was validated using four-fold cross-validation technique. The best performance obtained from each validation fold were then averaged to determine the final result. It gives a validation accuracy of 84% on the grayscale images. The performance achieved in each fold is listed in Table 1. A sample of randomly selected correctly and incorrectly classified images and corresponding predicted classes are shown in Fig. 5.

A considerable amount of work has been assigned to model the automated detectors for particular retinal pathologies, e.g., diabetic retinopathy [15], myopia, etc. However, a negligible amount of work has been done for AMD in spite of its prevalence. A work on classifying the 4 stages of AMD using a deep learning approach has been taken up in [16]. In the work, a differentiation between the stages was done so that the detection of the early stages can be achieved timely. OverFeat features [17],

Table 1 Results obtained using four-fold cross-validation technique

Fold	Accuracy
1	0.83
2	0.86
3	0.84
4	0.83
Average	0.84

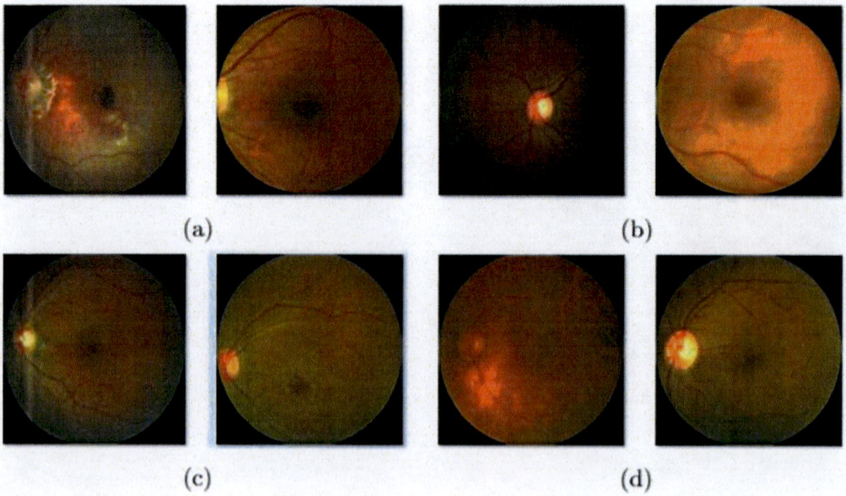

Fig. 5 Classification of retinal fundus images. **a** Correctly classified AMD images, **b** correctly classified non-AMD images, **c** incorrectly classified AMD images, **d** incorrectly classified non-AMD images

Table 2 Comparison of accuracy of existing works [19] with the proposed method

Methods	Accuracy (%)
Human	90.2
DCNN-A (WS)	91.6
DCNN-A (NSG)	90.7
DCNN-A (NS)	90.0
DCNN-U (WS)	83.7
DCNN-U (NSG)	83.9
DCNN-U (NS)	83.2
Proposed method (Grayscale images)	84.0

WS: With stereo pairs; NS: No stereo; NSG: No stereo, graded

resulting from a specific pretrained DCNN on ImageNet dataset [18] (consisting of one thousand general-purpose nonmedical image classes with more than 500 images in each class) have been considered. The image classes consist of animals, plants, foods, instrumentations, etc. The resulting feature vector obtained from the preprocessed image is applied as input to a linear support vector machine (LSVM) classifier. Later in the same year, they reported a better result in the automated grading of AMD from color fundus images using DCNN [19]. They implemented transfer learning to model different DCNN architectures (DCNN-A, DCNN-U) for finding a solution to the referable AMD classification problem. The weights of all layers of the AlexNet [20] and OverFeat [17] are optimized via training to model the previously mentioned architectures. A comparison between the performance of humans, their work, and the proposed method is represented in Table 2. The proposed method surpasses all the DCNN-U methods in [19], but a better accuracy was achieved in their DCNN-A methods as shown in Table 2.

4 Conclusion

A novel deep learning model using CNN was adopted in the proposed model to distinguish AMD from others using fundus images as a binary classification task. This custom-designed DCNN architecture gives an average accuracy of 84% using grayscale images as input by applying a four-fold cross-validation technique. In the proposed method, as the images were resized to 400 x 400 pixels, a huge part of the information is lost. If a complex deeper network can be modeled it could give a better result. Also, a neural network is a data-hungry system. The dataset consists of only 400 images out of which only 300 images were used for training. So it could have given a better result if trained with more images. In this dataset, only 22.25% of total images are of AMD class. So the prediction of non-AMD class is more likely when an unknown data is tested.

The proposed method can be improved by developing an efficient method with proper tuning of hyperparameters. Proper preprocessing of images may contribute to the modification of the existing network. Segmentation and analysis of the optic disc, fovea, and 4 typical kinds of lesions related to AMD (drusen, exudate, hemorrhage, scar) and others in retinal fundus images could help the system to classify the severity of the disease. So, if the improved model can achieve higher accuracy than the human, then the automated detection of AMD would be helpful to the physicians to prescribe better suggestions.

References

1. Bressler, N.M., Bressler, S.B., Congdon, N.G., Friedman, D., Klein, R., Lindblad, A., Milton, R., Seddon, J., et al.: Potential public health impact of age-related eye disease study results: Areds report no. 11. Archives of ophthalmology (Chicago, 1960) **121**(11), 1621–1624 (2003)
2. Bressler, N.M.: Age-related macular degeneration is the leading cause of blindness. JAMA **291**(15), 1900–1901 (2004)
3. Abràmoff, M.D., Garvin, M.K., Sonka, M.: Retinal imaging and image analysis. IEEE Rev. Biomed. Eng. **3**, 169–208 (2010)
4. Gass, J.D.M.: Drusen and disciform macular detachment and degeneration. Arch. Ophthalmol. **90**(3), 206–217 (1973)
5. Bressler, N.M., Bressler, S.B., Fine, S.L.: Age-related macular degeneration. Surv. Ophthalmol. **32**(6), 375–413 (1988)
6. Bird, A., Bressler, N.M., Bressler, S.B., Chisholm, I., Coscas, G., Davis, M., De Jong, P., Klaver, C., Klein, B., Klein, R., et al.: An international classification and grading system for age-related maculopathy and age-related macular degeneration. Surv. Ophthalmol. **39**(5), 367–374 (1995)
7. Dalal, N., Triggs, B.: Histograms of oriented gradients for human detection (2005)
8. Lowe, D.G.: Distinctive image features from scale-invariant keypoints. Int. J. Comput. Vis. **60**(2), 91–110 (2004)
9. Burlina, P., Freund, D.E., Dupas, B., Bressler, N.: Automatic screening of age-related macular degeneration and retinal abnormalities. In: 2011 Annual International Conference of the IEEE Engineering in Medicine and Biology Society, pp. 3962–3966. IEEE (2011)
10. Freund, D. E., Bressler, N., Burlina, P.: Automated detection of drusen in the macula. In: 2009 IEEE International Symposium on Biomedical Imaging: From Nano to Macrol, pp. 61–64. IEEE (2009)
11. Trucco, E., Ruggeri, A., Karnowski, T., Giancardo, L., Chaum, E., Hubschman, J.P., Al-Diri, B., Cheung, C.Y., Wong, D., Abramoff, M., et al.: Validating retinal fundus image analysis algorithms: issues and a proposal. Investig. Ophthalmol. Vis. Sci. **54**(5), 3546–3559 (2013)
12. Pacheco, K.D., Wolfson, Y., Burlina, P., Freund, D.E., Feeny, A., Joshi, N., Bressler, N.M.: Evaluation of automated drusen detection system for fundus photographs of patients with age-related macular degeneration. Investig. Ophthalmol. Vis. Sci. **57**(12), 1611–1611 (2016)
13. Baidu road: Research open-access dataset. Ichallenge-amd dataset. http://ai.baidu.com/broad/subordinate?dataset=amd (2019). Last accessed 15 Oct 2019
14. Goodfellow, I., Bengio, Y., Courville, A.: Deep learning. MIT press (2016)
15. Gulshan, V., Peng, L., Coram, M., Stumpe, M.C., Wu, D., Narayanaswamy, A., Venugopalan, S., Widner, K., Madams, T., Cuadros, J., et al.: Development and validation of a deep learning algorithm for detection of diabetic retinopathy in retinal fundus photographs. JAMA **316**(22), 2402–2410 (2016)
16. Burlina, P., Pacheco, K.D., Joshi, N., Freund, D.E., Bressler, N.M.: Comparing humans and deep learning performance for grading AMD: a study in using universal deep features and transfer learning for automated amd analysis. Comput. Biol. Med. **82**, 80–86 (2017)

17. Sermanet, P., Eigen, D., Zhang, X., Mathieu, M., Fergus, R., LeCun, Y.: Overfeat: integrated recognition, localization and detection using convolutional networks. arXiv preprint arXiv:1312.6229 (2013)
18. Deng, J., Dong, W., Socher, R., Li, L.J., Li, K., Fei-Fei, L.: Imagenet: a large-scale hierarchical image database. In: 2009 IEEE Conference on Computer Vision and Pattern Recognitionl, pp. 248–255. IEEE (2009)
19. Burlina, P.M., Joshi, N., Pekala, M., Pacheco, K.D., Freund, D.E., Bressler, N.M.: Automated grading of age-related macular degeneration from color fundus images using deep convolutional neural networks. JAMA Ophthalmol. **135**(11), 1170–1176 (2017)
20. Krizhevsky, A., Sutskever, I., Hinton, G.E.: Imagenet classification with deep convolutional neural networks. In: Advances in Neural Information Processing Systems, pp. 1097–1105 (2012)

An Artificial Bee Colony Inspired Density-Based Approach for Clustering with New Index Measure

Ankita Bose and Kalyani Mali

Abstract This article suggests a clustering approach that is inspired by the random searching of the Artificial Bee Colony optimization. It considers the density distribution of the objects which gives an insight into identifying the cluster structures. The unknown density distribution has been approximated by using Kernel Density Estimation, it eventually helps in identifying the thresholds that discriminate the core objects from the noise and intercluster density transition. We have introduced a new index measure that serves the purpose of cluster evaluation. The superiority of the proposed approach has been drawn from the experimental analysis over different data sets.

Keywords Clustering · Bee colony optimization · Density estimation · Index measure

1 Introduction

In this era of technological evolution, huge amount of data is stored every day. Preprocessing of this data is necessary to reduce the computational burden and the processing time. Cluster analysis, specifically density-based cluster analysis plays an important role in determining cluster structures from an unknown data set. A density-based approach considers the clusters as a dense region, that is nonspherical and of arbitrary shape. Density connectivity is an implicit way of understanding the underlying cluster structure. Like the most algorithms, selection of input parameters is a big hurdle for density-based approaches and cluster structures are sensitive to these parameters. These parameters are mainly the radius (ε) of the surrounding area of an object and the minimum number of points (MinPts) within this area. Domain

A. Bose (✉) · K. Mali
Compuetr Science and Engineering, University of Kalyani, Kalyani 741235, India
e-mail: boseankita8@gmail.com

K. Mali
e-mail: kalyanimali1992@gmail.com

D. Bhattacharjee et al. (eds.), *Proceedings of International Conference on Frontiers in Computing and Systems*, Advances in Intelligent Systems and Computing 1255,
https://doi.org/10.1007/978-981-15-7834-2_42

knowledge may help a little, but for an unknown domain, the situation will be more difficult.

The earliest algorithm that introduces the notion of density-based clustering is the DBSCAN. In 1996, Easter el al. [1], developed A density-Based Clustering for Discovering Clusters (DBSCAN). DBSCAN can determine clusters of arbitrary shape, automatically determines the number of clusters, not sensitive toward the ordering of the data points, but it is difficult to determine the input parameters (ε and MinPts) [2], and DBSCAN is not able to identify varying densities. The problem of varying densities has been solved with the introduction of VDBSCAN [3]. It does not perform well on high dimensional data set. There is another variation of DBSCAN namely DVBSCAN [4], it can identify clusters with varying densities in large spatial data sets, but it is also sensitive to the density parameters. DBCLASD [5] and ST-DBSCAN [6] are also not able to process varying densities. Another important algorithm is DENCLUE [7], it is a gird based approach that builds a tree structure and process only those cells that contain data. It requires to specify the size of the grid and a global density threshold. Next comes OPTICS [8], introduced by Ankerst et al. in 1999. It arranges the ordering of the data points based on MinPts, provided by the user and it requires visual interaction to identify the cluster structures, which is not very suitable for large spatial data sets. DBCURE and DBCURE-MR [9] show better performance than OPTICS and also capable of identifying varying densities but they are also parameter dependent. There is another density clustering approach [10], that categorize elements depending on their similarity.

There are some existing clustering algorithms that consider the hybridization of the random search technique of the evolutionary algorithms with the density-based approaches. The commonly used evolutionary algorithms are, Particle Swarm Optimization (PSO) [11], Ant Colony Optimization [12], Artificial Bee Colony Optimization (ABC) [13], Genetic Algorithm (GA) [14], and Differential Evolution (DE) [15], Radial Basis Function and k-means based clustering [16], these rich set of approaches provides randomized searching with the iterative nature of minimizing the error and reach the optimum, make them efficient to randomly search the density parameters. Genetic Algorithm with a Density-Based Approach for Clustering (GADAC) [17], is one such technique, although it does not directly follow the steps of GA. Karami et al. in 2014 [18], developed a DE inspired density-based approach, but it suggests only supervised clustering strategy. A PSO inspired density-based approach is suggested by Guan et al. which uses several index measures as fitness function.

In this article, we have introduced a density-based approach for clustering, which is inspired by Artificial Bee Colony Optimization. We choose ABC because of its first convergence, less parametric complexity, clustering accuracy. We named the proposed method as ABCDAC. The overall procedure is based on the ordering of the data points, that requires MinPts and an initial object position which is drawn randomly by ABC. It is important to notice that the ordering of the data points may vary depending on the initial object position. This ordering suggests an unknown density distribution. We approximate this distribution using Kernel density estimation [19], and we choose Gaussian kernel as a reference. This approximation helps in determining the thresholds for core, noise, and intercluster density transition. We have suggested an index measure as a measure of fitness for ABC. It is basically the ratio of cohesiveness of the cluster elements with the between cluster separation.

The overall organization is as follows. Section 1 gives a brief introduction, Sect. 2 gives the background study, Sect. 3 shows the outline of the proposed ABCDAC, Sect. 4 contains the experimental analysis, Sect. 5 is devoted to conclusion and future scope.

2 Background Study

2.1 Effect of Noise Over Density Distribution

Density distribution gives us a way to represent the density connectivity of the objects of a data set, hence discovering the cluster structures. To obtain accurate density distribution it is, therefore, important to follow an ordering while processing the objects. We start with an initial object (I_P), then we choose the nearest objects, then we process these objects to find the nearest objects to them, and so on. So, in this processing, we must maintain an order so that we can have better density distribution. While analyzing the distribution curve (Fig. 1), we have noticed some interesting facts such that, there will always be a transition between clusters and it shows higher ε value but there must not be any such transition for the nearest MinPts values. This type of transition is the density transition between clusters. When we encounter noises, it will give at least two successive transitions for a single noise. From the figure below, we can see that for a single noise there are two successive high transitions.

2.2 Kernel Density Estimation (KDE)

In reality, it is difficult to understand the density distribution of an unknown data set. So, we try to assume the functional form to determine its parameters. Parameter estimation becomes a difficult task for an unknown distribution, but we are lucky enough as there are some nonparametric approaches to estimate the parameters of an unknown density distribution. Kernel Density Estimation (KDE) is one such approach. It is suggested by Parzen in 1962 [19]. KDE suggests a class of estimators. We can have Triangular and Gaussian kernel functions as some of the commonly used kernels. In ABCDAC we choose Gaussian kernel as a rule of thumb. Let there are independent and identically distributed n samples, x_1, x_2, x_3.....x_n. This set of samples is drawn randomly from an unknown distribution g_x(y) and y∈(x-h,x+h). Here, h represents a bandwidth and g_x(y) suggests that for some y, g(y) is dependent on x. We can define the estimate for g_x(y) as $\hat{f}_n(\text{y})=\frac{1}{n}\sum_{i=1}^{n} g_x(\text{y})$. In ABCDAC we choose Gaussian distribution as a rule of thumb and estimate the density distribution,

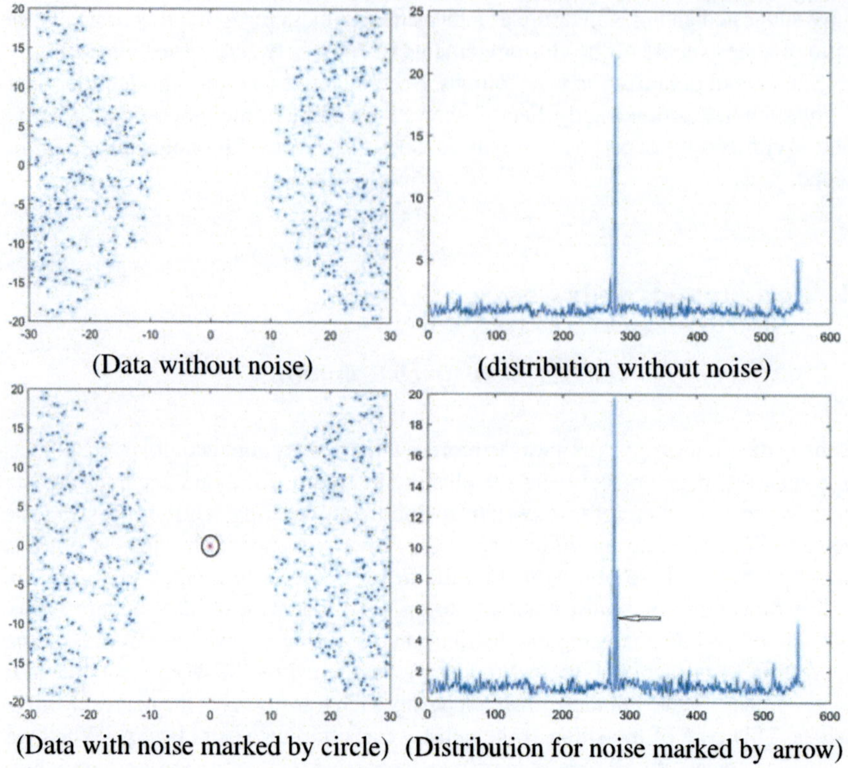

Fig. 1 Effect of noise over density distribution

which is basically the orientation of the data points obtained for each pair of MinPts and initial object position (I_P). To choose the density threshold (Fig. 2) we take the maximum of the second derivative of the KDE, that lies on the right side of the Gaussian curve. We can follow ref [20] for bandwidth (h) selection.

3 Proposed ABCDAC

Artificial Bee Colony Optimization (ABC) is inspired by the biological profile of the honey bees. The overall random searching is divided into three phases, namely employed bee phase, onlooker bee phase, and the scout bee phase. The employed bees explore the search space. The investigation starts with the random initialization of a number of food sources.

Algorithm 1

Input: Database (DB)
Output: Labeled Database, MinPts, varying ε (ε_C) [Here C is the optimum number of clusters]

1. n_i=MinPts, I_P [I_P is the initial object position]
2. Label←C
3. EPS←DensityDistribution(MinPts,DB)
4. f← KDE(EPS)
5. SD← SecondDerivative(f)
6. CTh← MAX(SD)[choose the max(SD) that lies in the right side of the Gaussian curve of f]
7. TH ←EPS(EPS>CTh)[TH contains density value of the noises and inter cluster density transition values]
8. for k ← 1 to Sizeof(EPS)
9. if EPS(k)< MIN(TH)
10. Label(k) ←C
11. else 12. ESS ← EPS(k+1 to k+MinPts)
13. if ESS>TH
14. C← Next Label [When there is no more transitions within the MinPts range of an object, it denotes a inter cluster density transition]
15. else
16. Label(k)←Noise
17. end
18end [Once we obtained the labeled matrix, if we take the average of the epsilon values for their corresponding label, it will give the ε of the corresponding cluster]

19. DensityDistribution(Minpts,DB)
20. consider eps as a Density matrix
21. for each object P in DB
22. eps(P)←AVG(nearest MinPts points of P)
23. Neighbor←nearest MinPts points of P
24. for each object P1 in Neighbor
25. eps(P1)←AVG(nearest MinPts points of P1)
26. Neighbor1←nearest MinPts points of P1
27. Neighbor←SORT(Neighbor ∪ Neighbor1)

Initialization Initially we choose $N = \{n_1, n_2...n_{10}\}$, 10 food sources. Each food source is a set {MinPts, I_p} containing the cluster parameters to be optimized. Each parameter is randomly initialized within its pre specified upper (x_{High}) and lower (x_{Low}) bound. For each parameter x_j ($j = 1, 2$) of n_i ($i = 1, 2...10$). We can randomly choose them as x_j=x_{Low} + rand (0,1)*(x_{High}−x_{Low}). We consider the objective function as

$$J_i = F - \text{measure} + Di$$

The objective function is a measure of summation of both F-measure [21] and Di (Density index proposed in Sect. 4.1). We can use F-measure when we have ground truth available for the data set and for an unknown data set, we can use Di to measure

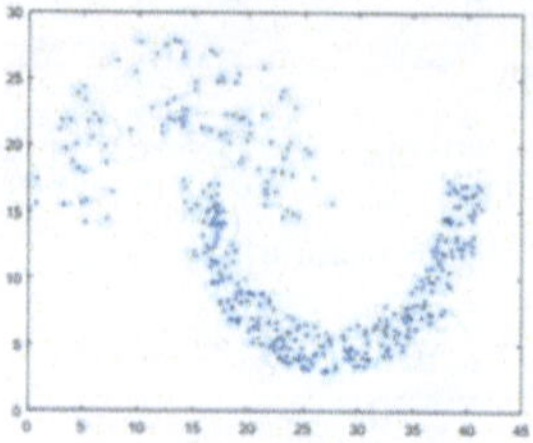

(Synthetic data set)

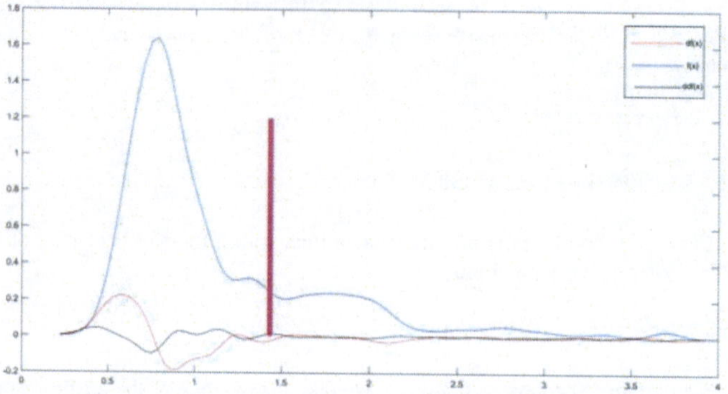

(The blue curve, red curve and black curve represents the KDE plot, first derivative of KDE and second derivative of KDE respectively, Threshold is represented by bar plot)

Fig. 2 Threshold selection

the cohesiveness among the objects of a cluster. Algorithm 1 describes how we can obtain clusters of objects showing density-wise connectivity.

Searching by Employed Bees: In this phase, selection is done based on the fitness and some previously selected threshold. If fit_{n_i} is the fitness of an object and fit_{n_k} is the fitness of it's neighbor and $fit_{n_i} > fit_{n_j}$, then the object will be selected, otherwise the neighbor will be selected as the food source. We can obtain the neighbor as n_i=x_{ij}+P*($x_{ij} - x_{kj}$), here P is a random number in the range of [−1,1] and k = {1, 2...10} and j = 1, 2, and i≠k. The scale factor P is there to adjust the parameter of the newly specified neighbor within its predetermined bounds. If the fitness of the neighbor is better then the food source will be updated by its neighbor.

$$fit_{n_i} = \begin{cases} \frac{1}{1+J_i}, & \text{if } J_i > 0 \\ 1 + abs(J_i), & \text{otherwise} \end{cases}$$

Selection by Onlooker Bees: After the employed bee phase, each food source (n_i) and it's neighbor (n_k) is compared against some threshold T_R, if both fails to satisfy the threshold, i.e., T_R>Pr(n_i) then it will be considered as the failure for that food source n_i and the onlooker bee will search for the next food source as Pr(n_i)=$\frac{fit_{n_i}}{\sum_{i=1}^{10} fit_{n_i}}$

Selection by Scout Bees: At this stage, scout bee selects the food source with maximum failure to satisfy the threshold T_R. If the failure is greater than some prespecified threshold the food source will be replaced by a random food source in its neighbor. At the end, the food source with maximum fitness will be selected as optimum.

4 Experimental Analysis

The overall analysis is done by considering 4 synthetic data sets with arbitrary shape and known cluster label. Beside this, we choose 6 real data sets (Table 1). For the evaluation of the clustering results, we consider the F-measure statistics for the synthetic data sets and we have suggested an index measure for the evaluation of the clustering results of real data sets. We have compared the performance of ABCDAC with evolutionary approaches, namely Particle Swarm Optimization (PSO), Genetic Algorithm (GA) Radial Basis Function, and k-means based algorithm (RBFk), and considered two density-based approaches namely OPTICS, DBSCAN and density clustering (Denclust).

4.1 Index Measure

Indices perform an important role in the evaluation of the clustering result. There are several indices that have been developed so far. Most of them suggest a centroid based approach such as Davies–Bouldin, Xie–Beni, Dunn, etc. For density-based approaches we can never assume a cluster prototype or centroid, therefore, we are not able to apply those commonly used validity indices. The evaluation of a density-based approach should be in terms of density connectivity. Therefore, measurement of cohesiveness of the cluster elements is important. We have suggested one such index measure and named it as D-index. Here is the expression for D-index. In Eq. 1 A_{x_i} represents the average intra cluster distance of cluster x_i and Tran(x_i,x_k) represents the density transition between cluster x_i and x_k and N is the number of objects in the database. So, the equation below is the ratio of the compactness of the cluster elements to its separation from the other clusters.

$$Di = \frac{\sum_{i=1}^{c} A_{x_i}}{N \sum \mathrm{Tran}_{i \neq k}(x_i, x_k)} \quad (1)$$

Therefore, more compact cluster with the maximum separation will give the better result, i.e., lower value of Di shows the better performance

It can be noticed from the D-index measure of Tables 2 and 3, that except Iris data, ABCDAC performs well for the other 5 data sets. The increase in data dimension does not affect the efficiency of ABCDAC, in case of DBSCAN, we are not able

Table 1 Data set description

Data sets	Type	Instance	Attribute	Data sets	Type	Instance	Attribute
Data Set1	Synthetic	373	2	Iris	Real	150	4
Outlier	Synthetic	600	2	Glass	Real	214	10
Half kernel	Synthetic	1000	2	Wine	Real	178	13
Two spiral	Synthetic	2000	2	Seed	Real	210	7
				Heart	Real	270	13
				User knowl-edge modeling	Real	516	5

Table 2 Performance of ABCDAC over synthetic data sets

Data sets	MinPts	Varying ε	F-measure	D-index (Di)
Data Set1	7	0.1472, 0.3208	0.83	1.6990
Outlier	5	0.9976, 0.5391, 1.0562, 0.8491	1	4.8523
Half kernel	10	0.9994, 0.7640	1	3.1622
Two spiral	10	0.1738, 0.1706	1	1.7800

Table 3 D-index measurement for real data sets

Data sets	MinPts	ABCDAC	GA	PSO	RBFk	OPTICS	Denclust
Iris	9	3.3851	2.5900	1.8289	2.8062	2.6740	3.7836
Glass	6	1.8980	6.8985	6.6035	4.3988	3.1559	2.3725
Wine	9	2.5312	3.3560	3.7560	6.6049	7.3269	3.6134
Seed	8	1.5291	7.2772	7.8127	5.1257	1.6937	8.7144
Heart	10	1.8792	8.9246	8.9220	4.3215	2.1653	2.4931
User knowl-edge modeling	5	3.7320	4.7283	3.9261	7.4225	5.6635	4.1477

to find any suitable result (Table 4), for wine and heart data set. Figure 3 shows the clustering by ABCDAC for synthetic data set, we have observed that there are more variations in the density curves for ABCDAC in comparison with OPTICS and Denclust, that helps in identifying the density threshold from the KDE for these density curves. One important thing is, due to the less parametric and robustness of ABC, the convergence takes less than 50 iteration for the real data sets and for synthetic data sets it is less than 10. It is clearly depicted in Fig. 4.

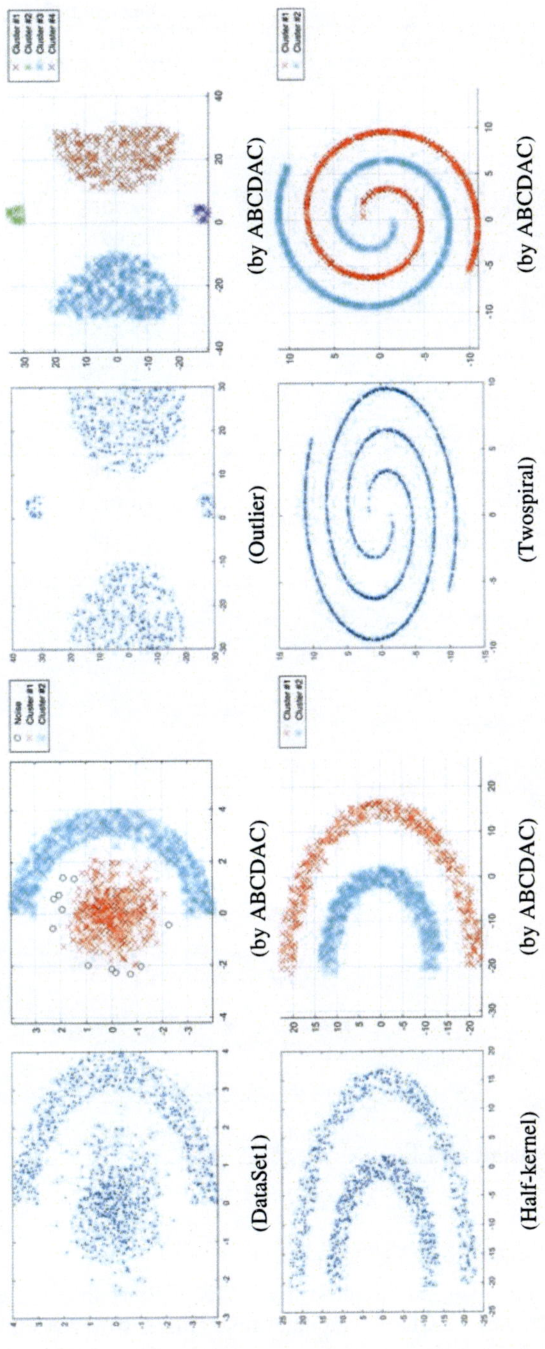

Fig. 3 Synthetic data sets and there clustering results by ABCDAC

Table 4 D-index measurement for real data sets by DBSCAN

Data sets	ε	D-index (Di)
Iris	0.4	2.0098
	0.5	2.5943
	1	8.8736
	1.5	8.8736
Glass	0.5	1.8965
	1	2.7011
	1.5	3.9921
	2	9.8870
Wine	0.5	9.9738
Seed	0.5	5.0072
	1	11.3200
	1.5	11.6646
	0.8	2.7002
Heart	1	8.7949
User knowledge modeling	0.2	8.1020
	0.25	13.5000
	0.3	9.5033
	0.4	6.7680

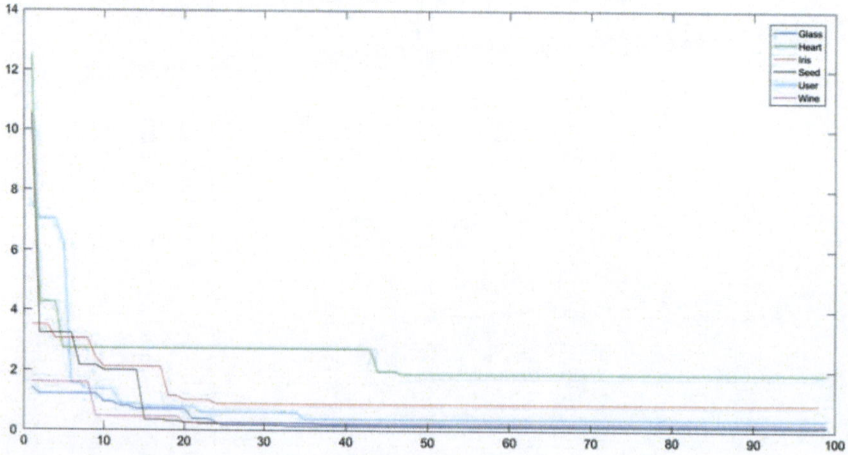

Fig. 4 Convergence Plot for all data sets by ABCDAC

5 Conclusion

In this article, we have suggested an Artificial Bee Colony Inspired density-based approach for clustering. ABCDAC can determine cluster structures that are density connected, it can identify clusters with varying densities and does not require any input parameter. Hence, provide a fully automated density-based clustering technique. From the experimental analysis, we can understand the robustness of the proposed ABCDAC over GA, PSO, RBFk, DBSCAN, OPTICS, and Denclust. For the synthetic data sets with arbitrary cluster structure, it is expected that density-based algorithms like DBSCAN and OPTICS will show better result, but from the measurement statics, it is clear that ABCDAC performs well over the other algorithms. Similarly, for the real data sets, ABCDAC shows better performance over GA, PSO, RBFk, OPTICS, DNSCAN, and Denclust. The random searching strategy of ABC along with the approximation by KDE and introduction of measurement index that measures the cohesiveness of the objects, all these factors make ABCDAC less complex, flexible, and efficient approach for density-based clustering.

References

1. Ester, M., Kriegel, H.-P., Sander, J., Xiaowei, X., et al.: A density-based algorithm for discovering clusters in large spatial databases with noise. KDD **96**, 226–231 (1996)
2. Schubert, E., Sander, J., Ester, M., Kriegel, H.-P., Xiaowei, X.: Dbscan revisited, revisited: why and how you should (still) use dbscan. ACM Trans. Database Syst. (TODS) **42**(3), 19 (2017)
3. Rasheduzzaman Chowdhury, A.K.M., Mollah, Md.E., Rahman, Md.A.: An efficient method for subjectively choosing parameter 'k' automatically in vdbscan (varied density based spatial clustering of applications with noise) algorithm. In: 2010 The 2nd International Conference on Computer and Automation Engineering (ICCAE), vol. 1, pp. 38–41. IEEE (2010)
4. Ram, A., Jalal, S., Jalal, A.S., Kumar, M.: A density based algorithm for discovering density varied clusters in large spatial databases. Int. J. Comput. Appl. **3**(6), 1–4 (2010)
5. Xu, X., Ester, M., Kriegel, H.-P., Sander, J.: A distribution-based clustering algorithm for mining in large spatial databases. In: Proceedings 14th International Conference on Data Engineering, pp. 324–331. IEEE (1998)
6. Birant, D., Kut, Alp: St-dbscan: an algorithm for clustering spatial-temporal data. Data Knowl. Eng. **60**(1), 208–221 (2007)
7. Hinneburg, A., Keim,, D.A., et al.: An efficient approach to clustering in large multimedia databases with noise. KDD **98**, 58–65 (1998)
8. Ankerst, M., Breunig, M.M., Kriegel, H.-P., Sander, J.: Optics: ordering points to identify the clustering structure. In: ACM Sigmod record, vol. 28, pp. 49–60. ACM (1999)
9. Kim, Y., Shim, K., Kim, M.-S., Lee, J.S.: Dbcure-mr: an efficient density-based clustering algorithm for large data using mapreduce. Inf. Syst. **42**, 15–35 (2014)
10. Rodriguez, A., Laio, A.: Clustering by fast search and find of density peaks. Science **344**(6191), 1492–1496 (2014)
11. Eberhart, R., Kennedy, J.: Particle swarm optimization. In: Proceedings of the IEEE international conference on neural networks, vol. 4, pp. 1942–1948. Citeseer (1995)
12. Dorigo, Marco., Maniezzo, Vittorio., Colorni, Alberto., et al.: Ant system: optimization by a colony of cooperating agents. IEEE Trans. Syst. Man Cybern. Part B Cybern. **26**(1), 29–41 (1996)
13. Karaboga, D.: An idea based on honey bee swarm for numerical optimization. Technical report, Technical report-tr06, Erciyes University, engineering faculty, computer (2005)
14. Goldberg, D.E., Holland, J.H.: Genetic algorithms and machine learning (1988)
15. Storn, Rainer, Price, Kenneth: Differential evolution-a simple and efficient heuristic for global optimization over continuous spaces. J. Glob. Optim. **11**(4), 341–359 (1997)
16. Dubey, A.D.: K-means based radial basis function neural networks for rainfall prediction. In: 2015 International Conference on Trends in Automation, Communications and Computing Technology (I-TACT-15), pp. 1–6. IEEE (2015)

17. Lin, Chih-Yang., Chang, Chin-Chen, Lin, Chia-Chen: A new density-based scheme for clustering based on genetic algorithm. Fundamenta Informaticae **68**(4), 315–331 (2005)
18. Karami, A., Johansson, Ronnie: Choosing dbscan parameters automatically using differential evolution. Int. J. Comput. Appl. **91**(7), 1–11 (2014)
19. Parzen, E.: On estimation of a probability density function and mode. Annals Math. Stat. **33**(3), 1065–1076 (1962)
20. Rosenblatt, M.: Remarks on some nonparametric estimates of a density function. Annals Math. Stat. 832–837 (1956)
21. Sasaki, Yutaka., et al.: The truth of the f-measure. Teach Tutor Mater **1**(5), 1–5 (2007)

An Investigation of Accelerometer Signals in the 0.5–4 Hz Range in Parkinson's Disease and Essential Tremor Patients

Olga S. Sushkova, Alexei A. Morozov, Alexandra V. Gabova, Alexei V. Karabanov, and Larisa A. Chigaleychik

Abstract An investigation of the 0.5–4 Hz little-studied frequency range of acceleration signals (ACC) was performed in patients with Parkinson's disease (PD) and essential tremor (ET). In this frequency range, new neurophysiological regularities were revealed. A new method was used to analyze the wave train electrical activity of the muscles based on the analysis of Morlet wavelet spectrograms and ROC curves. The method idea is to find local extrema (named "wave trains") in the Morlet wavelet spectrogram and to calculate various parameters describing these wave trains: the number of wave trains per second, the duration of the wave trains in periods, the leading frequency of the wave trains, the width of the wave train frequency band. The AUC functional dependence on the values of the bounds of the ranges of these parameters is investigated. This method is aimed at studying changes in the time-frequency parameters (the shape) of signals including changes that are not related to the power spectral density of the signal.

Keywords Parkinson's disease · Essential tremor · Trembling hyperkinesis · Accelerometer · ACC · Tremor · Wavelet spectrogram · Wave trains

---

O. S. Sushkova (✉) · A. A. Morozov
Kotel'nikov Institute of Radio Engineering and Electronics of RAS, Russian Federation, 125009 Moscow, Russia
e-mail: o.sushkova@mail.ru

A. A. Morozov
e-mail: morozov@cplire.ru
URL: http://www.fullvision.ru

A. V. Gabova
Institute of Higher Nervous Activity and Neurophysiology of RAS, Russian Federation, 117485 Moscow, Russia
e-mail: agabova@yandex.ru

A. V. Karabanov · L. A. Chigaleychik
FSBI Research Center of Neurology, Russian Federation, 125367 Moscow, Russia
e-mail: doctor.karabanov@mail.ru

L. A. Chigaleychik
e-mail: chigalei4ick.lar@yandex.ru

D. Bhattacharjee et al. (eds.), *Proceedings of International Conference on Frontiers in Computing and Systems*, Advances in Intelligent Systems and Computing 1255,
https://doi.org/10.1007/978-981-15-7834-2_43

1 Introduction

Preclinical diagnosis is very important for the treatment of Parkinson's disease (PD) and essential tremor (ET). Although PD and ET have been intensively studied in recent decades, the preclinical indicators of these movement disorders remain to be discovered. Several approaches were proposed but the definitive method is still lacking [4]. The diagnosis can be done by motion sensors (accelerometers) [8]. Measurements of tremor of the limbs by ACC can be used for an objective and quantitative analysis of neuromuscular function and movement disorders in PD and may provide useful information to characterize preclinical PD [7]. Common mathematical approaches used for the analysis of acceleration signals in PD are based on spectral methods (average frequency, power fraction in certain frequency bands) [9]. At that, the frequency range below 4 Hz is usually not investigated in acceleration signals, since it is considered that it is impossible to find statistically significant differences between groups of patients and healthy subjects in this range [3]. For example, in paper [7], it is stated that the 0–3.5 Hz frequency range in ACC is not to be considered because of this opinion.

Earlier, we have developed a method for investigation of the electrical activity of the cerebral cortex. The method is based on the analysis of Morlet wavelet spectrograms and ROC curves [11, 13]. The method idea is in that an electroencephalogram (EEG) is considered as a wave train set [12]. We consider the wave train as an EEG typical component, but not as a special kind of EEG signals. In contrast to works on the finding of the electrical activity of one or two specific types, such as sleep spindles [6] and alpha spindles [2], we investigate any type of the wave train electrical activity in the cerebral cortex over a wide frequency range. Previously, such an approach was proposed in [5, 20]. In this paper, we have adopted and used the developed method for ACC signals analysis. The adopted method is based on the wavelet spectrogram statistical analysis, visualizing the results of statistical analysis, and improved wave train detection algorithm.

The method of signal analysis is considered in Sect. 2. In the Sect. 3, the experimental setting used for the verification of the wave train analysis method is described. The results of the data analysis are discussed in Sect. 4.

2 The Method of Signal Analysis

For the study of ACC, the analysis method of the wave train electrical activity of the muscles was used. The method is based on the analysis of Morlet wavelet spectrograms and ROC curves [11, 13]. The method idea is to search for local extrema (named "wave trains") in the Morlet spectrograms and computation of different parameters of these extrema: the number of wave trains per second, the duration of the wave trains in periods, the leading frequency of the wave trains, the width of the wave train frequency band. The degree of difference in the group of patients with

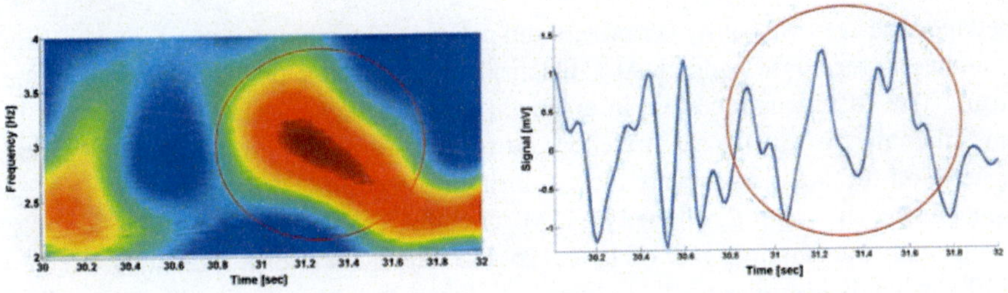

Fig. 1 A wave train in the wavelet spectrogram of ACC (left). A wave train in ACC (right). The wave train is indicated by the red circle

PD and ET from the control group of subjects in the space of these parameters is analyzed. For this, ROC curves are used. The functional dependence of AUC (the area under the ROC curve) on the values of the bounds of the ranges of the parameters under consideration is investigated. This method is aimed at studying changes in the time-frequency characteristics (the shape) of signals, including those which are not related to the changes in signal power spectral density.

The tremor in patients in the first stage of PD studied in this work appears only on one side of the body. In contrast to PD, in patients with ET, tremor appears on both sides of the body at once. Further, we will conditionally call the PD patient's hand, on which the tremor manifests, as "tremor" hand and we will conventionally call the opposite hand as "healthy" hand.

Let us consider a wave train example; this wave train is detected in the Morlet wavelet spectrogram of ACC on the patient's "healthy" left-hand (see Fig. 1, left), using our method. The patient has the tremor of the right side of the body. The central frequency of the wave train is 3.1 Hz, the signal is well-localized in time and frequency. The ACC, which was used to calculate the spectrogram in Fig. 1 (left), is represented in Fig. 1 (right). On the signal, one can see 3 periods of the wave train. In this paper, we study the number of wave trains (per second) in the 0.5–4 Hz frequency range in ACC of patients with PD and ET. The number of wave trains is compared with the healthy subject data using special AUC diagrams and the Mann-Whitney non-parametric statistical test. A detailed description of the AUC diagrams is given in [10, 12–18].

3 Experimental Setting

Data of patients with ET and PD in the early stages were compared with the data of healthy people. All patients were untreated; they did not take specific medicine previously. The PD patients group included patients at the first stage of PD on the Hoehn and Yahr scale. ? left-hand tremor was founded in 9 patients. A right-hand tremor was founded in 11 patients. ET was founded in 13 patients. The group of healthy volunteers included 8 people. There are no statistically significant differences between the ages of healthy volunteers and patients. Note that all patients and healthy volunteers were right-handed. ACC electrodes were located on the outer sides of the arms. The ACC was recorded in a special pose of the subject. The subject sat in an armchair, his arms were stretched out right in front of him, and his legs stood quietly on the floor. Eyes were closed during all recordings. To record the ACC, a special accelerometer developed by E.M. Timanin (IAP RAS) [19], was used. ACC signals have a sampling rate of 1378.125 Hz. The Butterworth filter with the 0.1–240 Hz bandwidth was applied. Then, the decimation with factor 20 was applied. The duration of each record was approximately two minutes. Records were analyzed as is, without selecting special areas in the signal.

4 Data Analysis and Results

Patients with the right-hand tremor and left-hand tremor were investigated separately. We have calculated wave trains in the frequency range from 0.5 to 4 Hz for ACC in each PD patient, ET patient, and healthy volunteer. AUC values were calculated for different frequency ranges in the interval from 0.5 to 4 Hz. We observe differences in ACC both in the "healthy" hands and in the "tremor" hands of the PD patients and both hands of the ET patients in the 0.5–4 Hz frequency range. But the frequency AUC diagrams (see, for instance, Fig. 2, left) indicate that there are poor differences in ACC in this frequency range. Let us investigate the reasons for these poor differences and make the differences more strong.

Let us consider the frequency AUC diagram of the "healthy" left-hands of the PD patients (Fig. 2, left). The blue color means that the wave train quantity in the patients is less than in the healthy subjects. The red color means that the wave trains quantity in the patients is greater than in the healthy subjects. One can see pronounced regularities (red and blue areas) in the frequency ranges 1.7–2.9 Hz (a blue spot) and 0.5–1.8 Hz (a red spot). We have calculated the periods, amplitude, and bandwidth AUC diagrams for these frequency ranges. Using these AUC diagrams we show that the differences in the frequency AUC diagram became more strong not only in the "tremor" hands but also in the "healthy" hands in ACC when we consider separated wave trains corresponding to special period, amplitude, and bandwidth ranges.

Let us consider the amplitude AUC diagram for the 0.5–1.8 Hz frequency range (a red spot) for the left "healthy" hands of patients with the right side tremor (Fig. 2, right). It can be seen that the 60 μV threshold can be chosen (AUC = 0.69) to separate the wave trains that are characteristic of the PD patients but not of the healthy people. The similar threshold values can be calculated for both bandwidth and duration in periods parameters of the wave trains. After the selection of threshold values (the amplitude of the wave trains is from 0 to 60 μV, the bandwidth of the wave trains is from 0 to 1 Hz, and the full width on the square root of 1/2 height of the wave trains is from 0 to 1 period), the considered frequency range was corrected; it became from 1 to 4 Hz (AUC=0.81). Thus, the following frequency ranges and wave train parameters were selected (see Table 1, for the blue area and Table 2, for the red area). These

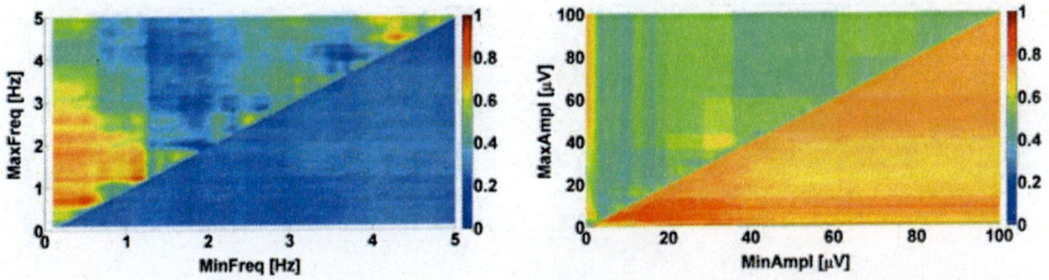

Fig. 2 At left: a frequency AUC diagram of ACC of the left "healthy" hands of the PD patients. The ordinate is the upper bound of the frequency ranges. The abscissa is the lower bound of the frequency ranges. The frequencies from 0.5 to 5 Hz with the 0.1 Hz step are considered. At right: an amplitude AUC diagram of ACC of the left "healthy" hands of the PD patients. The ordinate is the upper bound of the amplitude ranges. The abscissa is the lower bound of the amplitude ranges. The amplitudes are considered in the range 0–100 μV

Table 1 Wave train parameters for the blue area

No.	Group	Frequency	Amplitude	Periods	Bandwidth	AUC	*p*-value
1	Left-LH	0.5–4	0–100	1–2	0–2	0	0.00008
2	Right-LH	1.9–2.4	0–160	1–3	0–2	0.06	0.002
3	ET-LH	1.3–3.3	0–140	1–3	0–2	0.21	0.03
4	Left-RH	2.3–2.9	10–70	0–2	0–2	0.06	0.001
5	Right-RH	1.9–2.7	0–100	0–3	1–3	0.05	0.0007
6	ET-RH	2.1–2.8	0–110	0–2	1–2	0.15	0.01

Table 2 Wave train parameters for the red area

No.	Group	Frequency	Amplitude	Periods	Bandwidth	AUC	*p*-value
1	Left-LH	1–4	0–100	0–1	0–1	1	0.00008
2	Right-LH	1–4	0–60	0–1	0–1	0.81	0.02
3	Left-RH	1.1–2	0–40	0–3	0–2	0.92	0.002
4	Right-RH	1–1.7	0–220	0–1	0–2	0.90	0.01

parameters enable to separate the wave trains that are typical for the patients but not for the healthy people. In Table 1, the first row is the left-hand tremor PD patients, electrode LH (left-hand). The second row is the right-hand tremor PD patients, LH. The third row is the ET patients, LH. The fourth row is the left-hand tremor PD patients, RH (right-hand). The fifth row is the right-hand tremor PD patients, RH. The sixth row is the ET patients, RH. In Table 2, the first row is the left-hand tremor PD patients, LH. The second row is the right-hand tremor PD patients, LH. The third row is the left-hand tremor PD patients, RH. The fourth row is the right-hand tremor PD patients, RH. The selected frequency is given in Hz, amplitude in μV, duration in periods, and bandwidth in Hz. Characteristics of wave trains, AUC values, and the Mann-Whitney test p-values are reported in the tables.

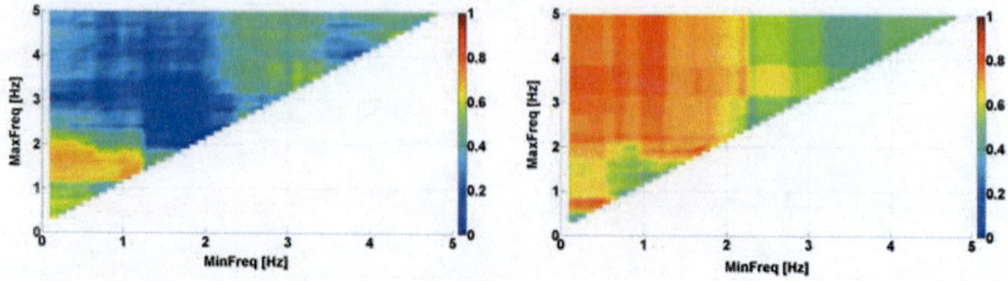

Fig. 3 Diagram of frequency values of AUC calculated for various bands of frequencies with selected thresholds. At left: the AUC diagram with selected thresholds for the blue area. At right: the AUC diagram with selected thresholds for the red area. Frequencies from 0.5 to 5 Hz in 0.1 Hz step are considered

Using the Mann-Whitney statistical test to detect significant differences in the parameters of electrical "wave train" activity of muscles, one can demonstrate that statistically significant differences exist in both "tremor" and "healthy" hands of the PD patients and in both hands of the ET patients in comparison with the healthy volunteers. Thus, the set of selected thresholds make the found regularities in frequency AUC diagrams stronger.

Let us consider the frequency AUC diagram for the left "healthy" hands of patients with the tremor of the right side (Fig. 3), with selected thresholds for amplitude, duration in periods, and bandwidth. The frequency AUC diagram with selected thresholds for the blue area is presented in the left figure; the frequency AUC diagram with selected thresholds for the red area is presented in the right figure. It can be seen that after narrowing the space of the wave trains under consideration (applying the thresholds), the red and blue regions on the frequency AUC diagrams are well-separated.

Thus, after the application of selected thresholds for amplitude, duration in periods, and bandwidth, the significant differences between the patients with PD (both "tremor" and "healthy" hands) and healthy volunteers, and between the ET patients (both hands) and healthy volunteers in ACC were found in the blue area. Also, the significant differences between the patients with PD (both "tremor" and "healthy" hands) and healthy volunteers in ACC were found in the red area.

5 Conclusions

A new method of exploratory data analysis was developed. This method involves calculating AUC values and non-parametric testing of statistical hypotheses to detect significant differences in the parameters of "wave train" electrical activity of muscles. A detailed analysis of the data of ET and PD patients in the poorly studied frequency range 0.5–4 Hz was carried out. Statistically significant differences from the healthy group of subjects were found both in the "tremor" hands and "healthy" hands of the patients with PD and in both hands of the ET patients. The found regularities in the blue area in frequency AUC diagrams can be a prospective for early diagnosis of PD and ET. The found regularities on the "healthy" hands of the PD patients are of considerable interest for early diagnosis of PD because the "healthy" hands of the PD patients can be used as a model of the processes occurring at the preclinical stages of PD [1]. The existence of the red area in frequency AUC diagrams in patients with PD and the absence of this area in the ET patients are of considerable interest for differential diagnosis of PD and ET patients. It can be assumed that wave trains in the frequency range 0.5–4 Hz reflect the increased electrical activity of the muscle fiber groups that make up the muscles. The obtained results indicate an individual picture of frequency parameters in the 0.5–4 Hz frequency range for specific diseases.

Parkinson's disease is a systemic disease, and manifestations of this disease include impaired muscle tone, both in the shaking hands and non-shaking hands. The reason for this is that, due to the disease, the reciprocal (cross) connections and the downward effect of the extrapyramidal system on the segmental level of tone control (alpha motoneurons, etc.) of the muscles are disrupted. Identification of these changes allows determining the degree of decompensation from the "healthy", intact side, and also allows predicting the clinical manifestation of focal neurological symptoms. The monitoring of these changes can be used as a promising prognostic parameter of decompensation and an assessment of the effectiveness of a specific treatment.

The application of the developed signal analysis method allowed us to identify new regularities in the ACC in the 0.5–4 Hz frequency range that previously could not be detected using standard spectral methods based on the power spectral density analysis.

Acknowledgements The authors are grateful to the corresponding member of RAS Sergey N. Illarioshkin for helping with the work. The research was supported by the Russian Foundation for Basic Research, project No. 18-37-20021. The research was carried out within the state task framework. The authors are grateful to the Scholarship of the President of Russia to young scientists and post-graduate students, grant No. SP-5247.2018.4.

References

1. Andreeva, Y., Khutorskaya, O.: EMGs spectral analysis method for the objective diagnosis of different clinical forms of Parkinson's disease. J. Electromyogr. Clin. Neurophys. **36**(3), 187–192 (1996)
2. Lawhern, V., Kerick, S., Robbins, K.A.: Detecting alpha spindle events in EEG time series using adaptive autoregressive models. BMC Neurosci. **14**, 101 (2013). http://www.biomedcentral.com/1471-2202/14/101
3. Lyons, K.E., Pahwa, R.: Handbook of Essential Tremor and Other Tremor Disorders. CRC Press (2005)
4. Meigal, A.Y., Rissanen, S.M., Tarvainen, M.P., Airaksinen, O., Kankaanpää, M., Karjalainen, P.A.: Non-linear EMG parameters for differential and early diagnostics of Parkinson's disease. Front. Neurol. **4**, 135 (2013)

5. Obukhov, Y.V., Korolev, M.S., Gabova, A.V., Kuznetsova, G.D., Ugrumov, M.V.: Patent no. 2484766 Russian Federation. Method of early encephalographic diagnostics of Parkinson disease (2013). 20 June 2013 (2013)
6. O'Reilly, C., Nielsen, T.: Automatic sleep spindle detection: benchmarking with fine temporal resolution using open science tools. Front. Hum. Neurosci. **9**, 353 (2015). http://doi.org/10.3389/fnhum.2015.00353
7. Palmerini, L., Rocchi, L., Mellone, S., Valzania, F., Chiari, L.: Feature selection for accelerometer-based posture analysis in Parkinson's disease. IEEE Trans. Inf. Technol. Biomed. **15**(3), 481–490 (2011)
8. Rissanen, S.M., Kankaanpää, M., Meigal, A., Tarvainen, M.P., Nuutinen, J., Tarkka, I.M., Airaksinen, O., Karjalainen, P.A.: Surface EMG and acceleration signals in Parkinson's disease: feature extraction and cluster analysis. Med. Biol. Eng. Comput. **46**(9), 849–858 (2008)
9. Robichaud, J.A., Pfann, K.D., Vaillancourt, D.E., Comella, C.L., Corcos, D.M.: Force control and disease severity in Parkinson's disease. Mov. Disord. **20**(4), 441–450 (2005)
10. Sushkova, O.S., Morozov, A.A., Gabova, A.V.: Development of a method of analysis of EEG wave packets in early stages of Parkinson's disease. In: Proceedings of the International conference Information Technology and Nanotechnology (ITNT 2016, Samara, Russia, May 17–19, 2016), CEUR, Samara, pp. 681–690 (2016a). http://ceur-ws.org/Vol-1638/Paper82.pdf
11. Sushkova, O.S., Morozov, A.A., Gabova, A.V.: A method of analysis of EEG wave trains in early stages of Parkinson's disease. In: International Conference on Bioinformatics and Systems Biology (BSB-2016). IEEE, pp. 1–4 (2016b)
12. Sushkova, O.S., Morozov, A.A., Gabova, A.V., Karabanov, A.V.: Data mining in EEG wave trains in early stages of Parkinson's disease. In: Proceedings of the 12th Russian-German Conference on Biomedical Engineering, pp. 80–84 (2016c)
13. Sushkova, O.S., Morozov, A.A., Gabova, A.V.: Data mining in EEG wave trains in early stages of Parkinson's disease. In: Advances in Soft Computing MICAI 2016 Lecture Notes in Computer Science **10062**, 403–412 (2017a)
14. Sushkova, O.S., Morozov, A.A., Gabova, A.V.: Investigation of specificity of Parkinson's disease features obtained using the method of cerebral cortex electrical activity analysis based on wave trains. In: SITIS. IEEE, pp. 168–172 (2017b)
15. Sushkova, O.S., Morozov, A.A., Gabova, A.V., Karabanov, A.V.: Application of brain electrical activity burst analysis method for detection of EEG characteristics in the early stage of Parkinson's disease. SS Korsakov J. Neurol. Psychiatry **118**(7), 45–48 (2018a)
16. Sushkova, O.S., Morozov, A.A., Gabova, A.V., Karabanov, A.V.: Investigation of surface EMG and acceleration signals of limbs' tremor in Parkinson's disease patients using the method of electrical activity analysis based on wave trains. In: Simari, G., Eduardo, F., Gutierrez, S.F., Melquiades, J.R. (eds.) Advances in Artificial Intelligence: 16th Ibero-American Conference on AI, pp. 253–264. Springer, Cham (2018b)
17. Sushkova, O.S., Morozov, A.A., Gabova, A.V., Karabanov, A.V.: An investigation of the specificity of features of early stages of Parkinson's disease obtained using the method of cortex electrical activity analysis based on wave trains. J. Phys. Conf. Ser. **1096**(1), 012, 078 (2018c)
18. Sushkova, O.S., Morozov, A.A., Gabova, A.V., Karabanov, A.V.: Investigation of the multiple comparisons problem in the analysis of the wave train electrical activity of muscles in Parkinson's disease patients. In: Proceedings of ITNT-2019: V International Conference and Youth School Information Technologies and Nanotechnologies, pp. 329–337 (2019)
19. Timanin, E.M., Gustov, A.V., Eremin, E.V.: Patent no. 2483676. Russian Federation. Device for complex analysis of different types of human tremor (2013). 10 June 2013
20. Zhirmunskaya, E.A.: Klinicheskaya elektroentsefalografiya (tsifry, gistogrammy, illyustratsii) (In Russian). Mezhotraslevoy nauchno-issledovatel'skiy inzhenerno-tekhnologicheskiy tsentr "Skan", Moscow

Simulation of Action Potential Duration and Its Dependence on $[K]_O$ and $[Na]_I$ in the Luo-Rudy Phase I Model

Ursa Maity, Anindita Ganguly, and Aparajita Sengupta

Abstract Cardiac arrhythmia is a major group of heart diseases, that are caused due to irregular heartbeats. There are several risk factors for lethal cardiac disrhythmia that have been discovered over the years and the primary methods of diagnosis have been based on models of cardiac action potential in mammalian ventricular cells. This paper aims towards giving an insight on the effect that ionic concentrations have on the Action Potential Duration (referred to as APD) curves (Luo-Rudy Phase I model and Faber Rudy Model) through simulations using SIMULINK. Abnormalities in K^+, Na^+ and Ca^{+2} concentrations across the cell membranes may cause fatal cardiac deaths due to arrhythmic beats.

Keywords Cardiac arrhythmia · Action potential · Sudden death · Electrophysiology

1 Introduction

Several models of membrane action potential have been developed over the course of time in the last few decades. The very first model by Hodgkin and Huxley et al. [1, 2], represents the electrical behaviour of the membrane modelled using ionic currents (Sodium currents, Potassium currents and leakage currents) and membrane capacitance. The electrical activity in Purkinje fibres which are specialized conducting fibres cardiac myocytes, was modelled by McAllister et al. [3].

U. Maity (✉) · A. Sengupta
Department of Electrical Engineering, Indian Institute of Engineering Science and Technology, Howrah, Shibpur, India
e-mail: ursa5597@student.ubc.ca

A. Sengupta
e-mail: sgaparajita@gmail.com

A. Ganguly
Department of Electrical Engineering, Guru Nanak Institute of Technology, Kolkata, India
e-mail: aninditaganguly80@gmail.com

D. Bhattacharjee et al. (eds.), *Proceedings of International Conference on Frontiers in Computing and Systems*, Advances in Intelligent Systems and Computing 1255,
https://doi.org/10.1007/978-981-15-7834-2_44

Beeler and Reuter in [4] using the Hodgkin and Huxley equations as the framework and using voltage clamp techniques, developed a mathematical model that introduced the fast Sodium current (i_{Na}), the time-dependent Potassium current (i_K), time-independent Potassium current (i_{K1}), and slow inward current due to Calcium ions (i_{si}) and ODEs of the different ionic gating variables and the primary governing ODE of the membrane voltage, (V_m). Luo and Rudy modified the Beeler-Reuter model and developed the Luo-Rudy Phase I model in [5] and the phase II model was published in 1994 [6].

As suggested by Levi et al. in [7], in cardiac tissues, the imbalances in ionic concentrations can be considered as antecedent to the onset of arrhythmia. The ionic concentration of Calcium is used in a dynamic equation and is directly involved in the generation of APDR curves. A process called Calcium-induced Calcium release is facilitated by Calcium ions. This was discussed by Livshitz et al. [8], using calmodulin-dependent protein kinase II (CaMKII), whose activity determines the nature of CaT alternans in different beat frequencies. However, the basic framework of the mathematical model of the ventricular action potential in mammalian cardiac myocytes still holds and has been used as the basis of the Simulink Model drawn in MATLAB for this review. A hardware implementation of a similar kind has been carried out by Othman et al. [9], using HDL coder and FPGA for solving the ODEs and generating APDR curves.

The present work aims to exhibit the Luo-Rudy Phase I model in a purely mathematical framework omitting the physiological details to attract the attention of the less trained. All the equations are realized in a complete SIMULINK model. This additionally incorporates the ionic concentrations as variable parameters which can be entered directly. The corresponding changes in the APDR curves are recorded using SIMULINK Data Inspector. The results obtained are consistent with the experimental findings, as well as simulations carried out in existing literature. No work appears to be available in the literature which contains a similar simulation model to show the effect of variations in ion concentration on the APDR. The fact that ionic imbalances is a major cause of irregular heartbeats or Heart Rate Variability (HRV) can be further investigated for the purpose of estimating APDR curves of heart patients and analyzing the results for diagnosis and treatment using pacing protocols such as deterministic and stochastic pacing [10].

Organization of this paper is as follows: The mathematical formulation and the Simulink model is given in Sect. 2. The results are given in Sect. 3.1 and the conclusions are drawn in Sect. 3.2.

2 Methods

2.1 Theory and Mathematical Formulation

The membrane voltage (V_m) is represented as an ODE depending on the total current (i_{total}), the stimulus current (i_{stim}) and the membrane capacitance (C_m) as in (1) [1].

$$\frac{\mathrm{dV}_m}{\mathrm{d}t} = \frac{-(i_{\mathrm{total}} + i_{\mathrm{stim}})}{C_m} \tag{1}$$

The current i_{total} is represented as the sum of the fast sodium current (i_{Na}), slow inward calcium current (i_{Ca}), time-dependent potassium current (i_{K}), time-independent potassium current (i_{Ki}), the plateau potassium current(i_{plat}) and the background current(i_{bg}) as shown in (2) [5].

$$i_{\mathrm{total}} = i_{\mathrm{Na}} + i_{\mathrm{Ca}} + i_{\mathrm{K}} + i_{\mathrm{Ki}} + i_{\mathrm{plat}} + i_{\mathrm{bg}} \tag{2}$$

The currents in (2), are determined by a set of gate variables and corresponding voltage-rate dependent constants. The formulations for the respective currents are shown below [1–5], in (3)–(8), respectively.

$$i_{\mathrm{Na}} = g_{\mathrm{Na}} \cdot m^3 \cdot h \cdot j \cdot (\mathrm{V}_m - E_{\mathrm{Na}}) \tag{3}$$

$$i_{\mathrm{Ca}} = g_{\mathrm{Ca}} \cdot d \cdot f \cdot (\mathrm{V}_m - E_{\mathrm{Ca}}) \tag{4}$$

$$i_K = g_K \cdot x \cdot x_{\mathrm{inac}} \cdot (\mathrm{V}_m - E_K) \tag{5}$$

$$i_{\mathrm{Ki}} = g_{\mathrm{Ki}} \cdot k_{\mathrm{inac}} \cdot (\mathrm{V}_m - E_{\mathrm{Ki}}) \tag{6}$$

$$i_{\mathrm{plat}} = g_{\mathrm{plat}} \cdot k_{\mathrm{plat}} \cdot \left(\mathrm{V}_m - E_{\mathrm{plat}}\right) \tag{7}$$

$$i_{\mathrm{bg}} = 0.03921 \cdot (V_m + 59.87) \tag{8}$$

The constants in the above equations are defined as follows in (9)–(17)

$$E_{\mathrm{Na}} = \frac{RT}{F} \ln\left(\frac{[\mathrm{Na}_o]}{[\mathrm{Na}_\mathrm{i}]}\right) \tag{9}$$

$$E_{\mathrm{Ca}} = 7.7 - 13.0287 \ln[\mathrm{Ca}_i] \tag{10}$$

$$E_{\mathrm{K}} = \frac{RT}{F} \ln\left(\frac{[\mathrm{K}_o] + pr_{\mathrm{NaK}}[\mathrm{Na}_o]}{[\mathrm{K}_i] + pr_{\mathrm{NaK}}[\mathrm{Na}_i]}\right) \tag{11}$$

$$E_{\mathrm{Ki}} = E_{\mathrm{plat}} = \frac{RT}{F} \ln\left(\frac{[K_o]}{[\mathrm{K}_i]}\right) \tag{12}$$

$$g_{\mathrm{K}} = 0.282\sqrt{\frac{[\mathrm{K}_0]}{5.4}} \tag{13}$$

$$g_{Ki} = 0.6047\sqrt{\frac{[K_0]}{5.4}} \tag{14}$$

$$k_{plat} = \frac{1}{1 + e^{(7.488 - V_m)/5.98}} \tag{15}$$

$$x_{inac} = 2.387 \cdot \frac{e^{(0.04V_m + 3.04)} - 1}{(V_m + 77) \cdot e^{(0.04V_m + 1.4)}} \tag{16}$$

$$\frac{d[Ca_i]}{dt} = 0.07[Ca_i] - 10^{-4}(i_{Ca} - 0.07) \tag{17}$$

Here, g_{ion} represents the maximum conductance of the corresponding ionic channel ($mScm^{-2}$), E_{ion} is the reversal potential of the corresponding ion (mV), $[E_o]$ and $[E_i]$ are the extracellular and intracellular concentrations of ion E (mM) and k_{-plat} is the inactivation gate for i_{Kplat}. The value of g_{Na} is taken $23mScm^{-2}$ [4], g_{si} is $0.09mScm^{-2}$ and g_{Kplat} is $0.0183mScm^{-2}$. R is the gas constant with value 8.135 J K^{-1} mol^{-1}J, T is the ambient temperature in K and F is Faraday's constant with value 9.684×10^4 C mol^{-1}.

The variables m, h, j, x, d and f is the respective gate variables for the corresponding ionic channels. Their values range between 0 and 1. The details are listed in Table 1.

$$\frac{d\gamma}{dt} = \theta_\gamma - \gamma(\theta_\gamma + \mu_\gamma) \tag{18\}$$

$$\theta_m = \frac{0.032 \cdot (V_m + 47.13)}{1 + e^{(-4.713 - 0.1V_m)}} \tag{19}$$

$$\mu_m = 0.008 \cdot e^{(-V_m/11)} \tag{20}$$

$$\theta_h = 0.135 \cdot e^{\frac{-(V_m + 80)}{6.8}} \tag{21}$$

Table 1 Description of the gate variables

Gate variable	Purpose	Current
m	Activation	i_{Na}
h	Slow inactivation	i_{Na}
j	Fast inactivation	i_{Na}
x	Activation	i_K
x_{inac}	Inactivation	i_K
k_{inac}	Inactivation	i_{Ki}
d	Activation	i_{Ca}
f	Inactivation	i_{Ca}

$$\mu_h = 3.56 \cdot e^{0.079\mathrm{V}_m} + 3.1 \times 10^5 \cdot e^{0.079\mathrm{V}_m} \tag{22}$$

$$\theta_j = \frac{\{-1.2714 \cdot e^{0.24\mathrm{V}_m} + 3.1 \times 10^5 e^{0.044\mathrm{V}_m}\}(\mathrm{V}_m + 37.78)}{1 + e^{(24.64+0.311\mathrm{V}_m)}} \tag{23}$$

$$\mu_j = \frac{0.1212 \cdot e^{-0.01052\mathrm{V}_m}}{1 + e^{-5.53-0.311\mathrm{V}_m}} \tag{24}$$

$$\theta_x = \frac{0.0005 \cdot e^{4.15+0.083\mathrm{V}_m}}{1 + e^{2.85+0.057\mathrm{V}_m}} \tag{25}$$

$$\mu_x = \frac{0.00013 \cdot e^{-1.2-0.06\mathrm{V}_m}}{1 + e^{0.8+0.04\mathrm{V}_m}} \tag{26}$$

$$\theta_{\mathrm{Kinac}} = \frac{1.02}{1 + e^{0.2385(\mathrm{V}_m - E_{\mathrm{Ki}}) - 14.12}} \tag{27}$$

$$\mu_{\mathrm{Kinac}} = \frac{0.4912 e^{\{0.08032(\mathrm{V}_m - E_{\mathrm{Ki}} - 5.476)\}} + e^{\{-.06175(\mathrm{V}_m - E_{\mathrm{ki}}) - 594.31\}}}{1 + e^{0.2385(\mathrm{V}_m - E_{\mathrm{Ki}}) - 14.12}} \tag{28}$$

$$\theta_{\mathrm{d}} = \frac{0.095 \cdot e^{(0.05-0.01\mathrm{V}_m)}}{1 + e^{(0.144+0.072\mathrm{V}_m)}} \tag{29}$$

$$\mu_{\mathrm{d}} = \frac{0.07 \cdot e^{(-0.748-0.0017\mathrm{V}_m)}}{1 + e^{(2.2+0.05\mathrm{V}_m)}} \tag{30}$$

$$\theta_f = \frac{0.012 \cdot e^{(-0.224-0.008\mathrm{V}_m)}}{1 + e^{(4.2+0.15\mathrm{V}_m)}} \tag{31}$$

$$\mu_f = \frac{0.0065 \cdot e^{(-0.6-0.002\mathrm{V}_m)}}{1 + e^{(6-0.2\mathrm{V}_m)}} \tag{32}$$

The gate variables and the related rate constants are calculated as shown in (18–32).

Here, the ODE in (18), holds for any gate variable γ and having rate constants θ_γ and μ_γ (msec^{-1}) (22)–(24) hold for values of $\mathrm{V_m} < -40$ mV.

The SIMULINK model is created using the above mathematical framework and is improved to include an additional dependence of potassium current on sodium concentration using the relationship given by Faber et al. [11]. The formulation for Na^+ activated K^+ current is given in (35), combining Na^+ dependence given by Kameyama et al. [12] and voltage dependence by Sanguinetti [13], in (33) and (34), respectively.

$$p_V = 0.8 - \frac{0.65}{1 + e^{(V_m+125)/15}} \tag{33}$$

$$p_{\mathrm{Na}} = \frac{0.85}{1 + \left(\frac{66}{[\mathrm{Na}_i]}\right)^{2.8}} \tag{34}$$

$$i_{\mathrm{K(Na)}} = g_{\mathrm{K(Na)}} \cdot p_V \cdot p_{\mathrm{Na}} \cdot (V_m - E_{\mathrm{K}}) \tag{35}$$

Here, $g_{K(Na)}$ is maximum conductance of $i_{K(Na)}$ with value 0.12848 $mScm^{-2}$ [11].

2.2 *SIMULINK Model*

The SIMULINK Model is made using MATLAB functions and blocks. The output is obtained for the various parameters. For the ODEs, integrator blocks were used, and the solver was 'ode45' with a fixed-step and error tolerance set at *auto*. The stimulus current i_{stim} is applied as a pulse of duration 1 ms in a period of 500 ms. The first pulse appears at 100 ms, the next pulse at 600 ms and it continues as such till 2100 ms. The value of the current pulse is set at −80 μAcm^{-2}. The output was recorded using the Simulink Data Inspector for varying concentrations of potassium and sodium.

A schematic representation of the SIMULINK Model created using the equations in Sect. 2.1, is given in Fig. 1.

The SIMULINK Model in Fig. 1, uses MATLAB function blocks to calculate the values of the rate constants and the integrator blocks (shown in deep blue) for creating the differential equations. The blocks that calculate the ionic currents (shown in yellow) take the variables such as electric potentials and conductances as described in Sect. 2.1, respectively, as the inputs. The summation blocks (shown in green) are used for summing up all the ionic currents and also for adding the stimulus current.

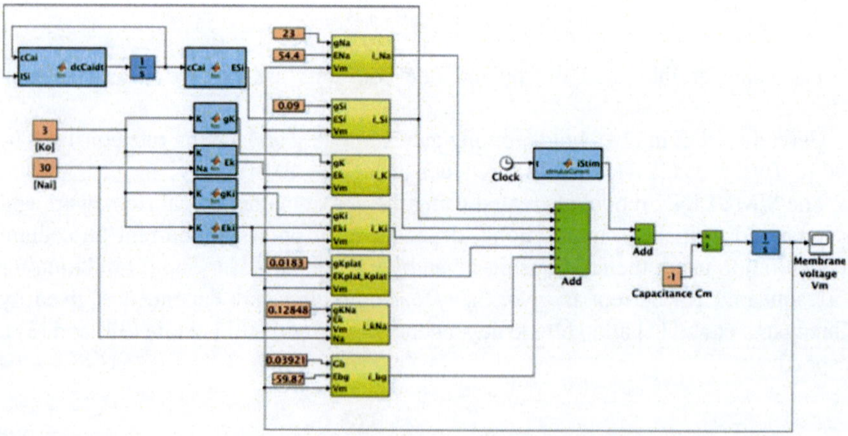

Fig. 1 SIMULINK model of Luo-Rudy PHASE I model with Faber Rudy modification

Table 2 Initial values taken in the Simulink Model

Parameters	Initial value
m	0.00167502
h	0.983166
j	0.989356
x	0.00567105
d	0.00298536
f	0.999665
$[Ca]_i$	0.000178550

The constant blocks (shown in light orange) act as inputs to the various blocks, for example, the values of $[K_o]$ and $[Na_i]$ are entered and varied by changing the values in the constant blocks. For the stimulus current, a clock is used as an input and then a MATLAB function block that calculates and provides the stimulus current (iStim) as the output. The final output obtained is the membrane voltage (Vm). As the equations are coupled, the membrane voltage is fed as input to the current blocks. It is to be noted that all the MATLAB function blocks (shown in cyan) are not visible externally as some of them are a part of the subsytems of the current blocks.

The initial values of the gate variables are chosen according to the standard database for Luo-Rudy model [5]. The same have been used for simulation of various other Luo-Rudy MATLAB codes. They are provided as the initial conditions of the integrator blocks corresponding to the calculation of the different gate variables and calcium ion concentration, respecively. These values are listed in Table 2.

3 Discussions

3.1 Results

The extracellular concentration of potassium determines the length of the Action Potential Duration (APD) as discussed by Luo and Rudy [5], the APD shortens with an increase in concentration. A similar result is achieved when the input values of $[K]_o$ are increased in the SIMULINK model (Fig. 1). The action potential curves for five pulses taking $[K_o]$ as 5.4 mM and $[Na]_i$ to be 18 mM is shown in Fig. 2, in order to demonstrate how Action Potential curves are generated one after the other by the periodic application of stimulus current. Three values of $[K]_o$ are entered and the corresponding curves are recorded as shown in Fig. 3. The values taken respectively are 3 mM, 5.4 mM and 7 mM keeping $[Na]_i$ constant at 18 mM. Another set of curves is obtained by keeping $[K]_o$ constant at 5.4 mM and setting the values of $[Na]_i$ at 6 mM, 18 mM and 30 mM respectively. The results obtained are shown in Fig. 4.

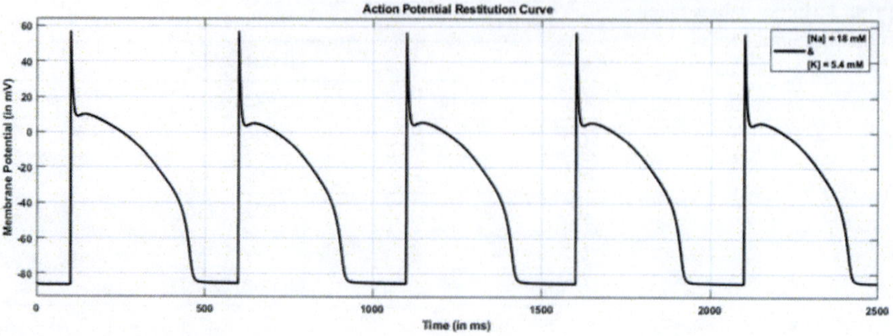

Fig. 2 The action potential curves for $[K]_o = 5.4$ mM and $[Na]_i = 18$ mM

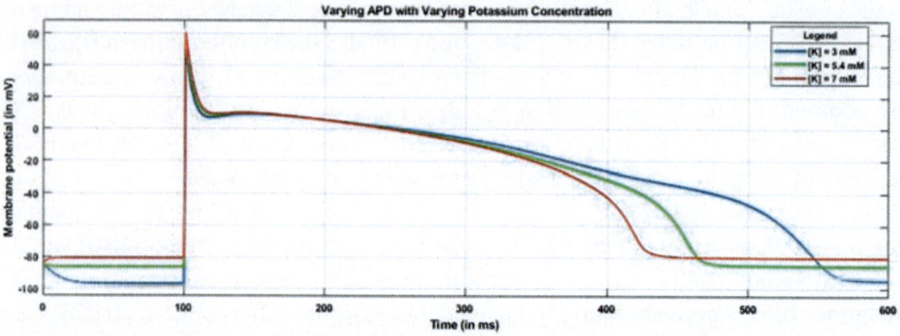

Fig. 3 Action Potential curves at $[K]_o = 3$, 5.4 and 7 mM respectively, $[Na]_i = 18$ mM

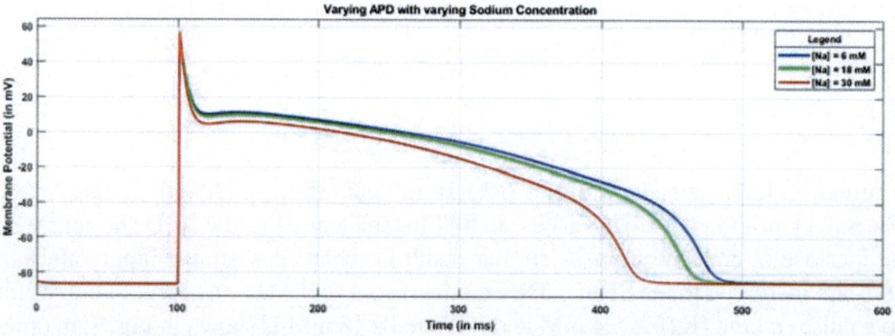

Fig. 4 Action Potential curves at $[Na]_i = 6$, 18 and 30 mM respectively, $[K]_o = 5.4$ mM

3.2 Conclusion

This work develops a detailed SIMULINK model for the Action Potential in mammals. The Luo-Rudy model [5], is one of the fundamental models that provide a detailed understanding of the generation of the action potential of the heart. The

SIMULINK model used here may be improved further to accommodate the details of the more advanced models [10]. This will recast the problem of curbing arrhythmia as an Engineering one on which the effect of controlling pacing protocols may be examined. As evident from the graphs obtained as shown in Figs. 3 and 4, it can be concluded that the concentrations of the ions play an important role in determining the length of APD alternans. According to Faber et al. [11], sodium overload may cause serious problems in the regular rhythm of the heart. In a similar way, the deviation of levels of concentration of potassium from normal values may also cause arrhythmia. Fluctuations in electrolyte concentrations in our cardiac tissues directly cause irregularities in heartbeats and may often lead to fatal conditions if not diagonised and treated in time.

3.3 Future Scope of Work

The computational study of the effect of varying ion concentrations in the Luo-Rudy [5], model may be helpful in understanding the phenomenon of arrhythmia in mathematical models for generation of cardiac action potential. Subsequently this may help in developing control strategies to mitigate its effects. A future direction of research may be oriented to estimate APD values from real time data and correlating the same with varying ion concentrations. Treatment of arrhythmia is dependent on estimation of irregularities in the APD patterns. Furthermore, having an estimation of the irregularities shall also help determine an anti-arrhythmic pacing protocol for future research

Acknowledgements The authors are grateful to Prof. Jan Kucera, Department of Physiology, University of Bern, for exposing them to further investigation on the Luo-Rudy model. This work is funded by the TARE scheme of Scientific and Engineering Board (SERB), Government of India.

References

1. Hodgkın, A.L., Huxley, A.F.: The components of membrane conductance in the giant axon of Loligo. J. Physiol. **116**, 473–496 (1952a)
2. Hodgkin, A.L., Huxley, A.F.: A quantitative description of membrane current and its application to conduction and excitation in nerve. J. Physiol. (Lond) **117**, 500–544 (1952b)
3. McAllister, R.E., Noble, D., Tsien, R.W.: Reconstruction of the electrical activity of cardiac Purkinje fibres. J. Physiol. (Lond) **251**, 1–59 (1975)
4. Beeler, G.W., Reuter, H.: Reconstruction of the action potential of ventricular myocardial fibres. J. Physiol. (Lond) **268**, 177–210 (1977)
5. Luo, C.H., Rudy, Y.: A model of the ventricular cardiac action potential. Depolarization, repolarization, and their interaction. Circ. Res. **68**(6), 1501–1526 (1991)
6. Luo, C.H., Rudy, Y.: A dynamic model of the cardiac ventricular action potential. I. Simulations of ionic currents and concentration changes. Circ. Res. **74**, 1071–1096 (1994)

7. Levi, A.J., Dalton, G.R., Hancox, J.C., Mitcheson, J.S., Issberner, J., Bates, J.A., Evans, S.J., Howarth, F.C., Hobai, I.A., Jones, J.V.: Role of intracellular sodium overload in the genesis of cardiac arrhythmias. J. Cardiovasc. Electrophysiol. **8**, 700–721 (1997)
8. Livshitz, L.M., Rudy, Y.: Regulation of Ca2 + and electrical alternans in cardiac myocytes: role of CaMKII and repolarizing currents. Am. J. Physiol. Heart Circ. Physiol. **292**(6), H2854–H2866 (2007)
9. Othman, N., Jabbar, M.H., Mahamad, A.K., Mahmud, F.: Luo Rudy phase i excitation modeling towards HDL coder implementation for real-time simulation. 978-1-4799-4653-2/14/©. IEEE (2014)
10. Dvir, H., Zlochiver, S.: The interrelations among stochastic pacing, stability, and memory in the heart. Biophys. J. **107**, 1023–1034 (2014)
11. Faber, G.M., Rudy, Y.: Action potential and contractility changes in $[Na^+]_i$ overloaded cardiac myocytes: a simulation study. Biophys. J. **78**, 2392–2404 (2000)
12. Kameyama, M., Kakei, M., Sato, R., Shibasaki, T., Matsuda, H., Irisawa, H.: Intracellular Na^+ activates a K^+ channel in mammalian cardiac cells. Nature **309**, 354–356 (1984)
13. Sanguinetti, M.C.: In: Colatsky, T.J. (ed) Na^+-activated and ATP-sensitive K^+ channels in the heart in potassium channels: basic function and therapeutic aspects, pp. 85–109. Alan R. Liss Inc., New York (1990)

Automated Classification and Detection of Malaria Cell Using Computer Vision

Subhrasankar Chatterjee and Pritha Majumder

Abstract With the advancement of technology, its involvement in medical science is inevitable. Rapid progress has been made in the fields of automated disease detection and diagnosis. The same concept has been extended to the field of malaria cell analysis in this research. From the image of a cell, this model is capable of determining whether it is uninfected or parasitized with malaria infection. The method discussed here is lightweighted and easily implementable in terms of time and space complexity. Also, it has a training accuracy of 97.98% and a validation accuracy of 97.6%. A new method called "Information Padding" has been used and discussed in the methodology section.

Keywords Malaria · Parasites · Deep learning · Convolutional neural network · Information paddin

1 Introduction

Plasmodium parasites are the main cause of malaria. Female mosquitoes are the prime vector of this parasite. The human body is affected by five variants of Plasmodium species. They can be listed as P. falciparum, P. vivax, P. malariae, P. ovale, and P. knowlesi. Out of which P. falciparum and P. vivax, the two most common species, are responsible for most malaria-related deaths globally. There are over 400 diverse species of *Anopheles* mosquito. Among them, about 30 are malaria vectors. People infected with the parasite usually feel very sick. The main symptoms are fever, headache, and chill but it is preventable and curable. As per the most recent ***World***

S. Chatterjee (✉)
Computer Science Engineering Department, Indian Institute of Technology, Kharagpur 721302, West Bengal, India
e-mail: sc1933@cse.jgec.ac.in

P. Majumder
Department of Information Technology, Jalpaiguri Government Engineering College, Jalpaiguri 735102, West Bengal, India

D. Bhattacharjee et al. (eds.), *Proceedings of International Conference on Frontiers in Computing and Systems*, Advances in Intelligent Systems and Computing 1255,
https://doi.org/10.1007/978-981-15-7834-2_45

Malaria Report, in 2017 approximately 4,35,000 lifes were lost due to malaria. The major victims of malaria are the children under 5 years of age [1].

Early diagnosis can help in the reduction of malarial disease. There are various ways to diagnose malaria such as Clinical Diagnosis, Microscopic Diagnosis, Rapid Diagnostic Tests (RDT), Polymerase Chain Reaction (PCR), and many more. ***Clinical diagnosis*** is based on the patient's symptoms and on physical evidence inferred from the examination. Clinical findings should always be backed up by a laboratory test for malaria which is tedious and also error-prone due to human inconsistency. The current method for diagnosis of malaria in the field is **light microscopy** of blood films. It is also known as "**gold standards**". For the identification of malaria parasites under a microscope, a glass slide is used and a drop of the patient's blood is applied upon, which is then absorbed in a staining solution to make parasites more easily observable under a conventional light microscope, generally with a 100 × oil objective. There are two variants of blood smears that are prepared for diagnosing malaria, namely thick smears and thin smears. **Thick Smears** are used to perceive the presence of parasites in a drop of blood and **Thin Smears** are used to detect the type of malaria species and to indentify the parasite stages. Thin and thick smears microscopy methods are used equally nowadays. **RDT**s detect substantiations of malaria parasites (antigens) and take approximately 10–15 min to process but it doesn't give any information about parasite count in the blood. RDTs are currently costlier than a microscopic diagnosis. **PCR** is considered the most precise among all tests. It can trace very low parasite concentrations in the blood and can differentiate among species but it is very much costly method [2].

The main goal of this paper is to discuss an automated method that can identify the presence or absence of malaria parasite in thin blood film smears. Moreover, it detects whether a blood cell is parasitized or uninfected. Here, we are using **computer vision** technique to diagnose and recognize malaria.

The Introduction in Sect. 1 is followed by Literature Survey in Sect. 2, Proposed Methodology in Sect. 3, Results in Sect. 4, Conclusion in Sect. 5, and References at the end.

2 Literature Survey

Several attempts have been made on Malaria Cell Detection using machine learning, deep learning, and many more other approaches. Some approaches are given below:

- Andrea Loddo et al. proposed "The Malaria Parasite Image Database for Image Processing and Analysis" [3].
- In this paper, image processing and pattern matching technique have been used to improve computerized analysis of malaria parasites.
- Itishree Mohanty et al. proposed "Automatic Detection of Malaria Parasites Using Unsupervised Techniques" [4].

- The methodology followed by them is based on unsupervised learning technique. They achieved a maximum accuracy of 87.5%.
- Komal B. Rode et al. proposed "Automatic Segmentation of Malaria Affected Erythrocyte in Thin Blood Films" [5].
- In this work, they have been used as a conventional technique following the triangle's thresholding technique, HSV segmentation.
- Prakriti Aggarwal et al. proposed "An Intensity Threshold based Image Segmentation of Malaria-Infected Cells" [6].
- In this research, image processing based robust algorithm has been used which achieved an accuracy of 93%.
- Kristofer Delas Peñas et al. proposed "Analysis of Convolutional Neural Networks and Shape Features for Detection and Identification of Malaria Parasites on Thin Blood Smears" [7].
- In this paper, the deep learning method has been used which achieved accuracies of 92.4% trained from scratch and 93.60% using transfer learning.
- Allisson Dantas Oliveira et al. proposed "An Automatic System for Computing Malaria Parasite Density in Thin Blood Films" [8].
- The methodology used in this paper is based on image processing and achieved an accuracy of 89.3%.
- Kishor Roy et al. proposed "Detection of Malaria Parasite in Giemsa Blood Sample Using Image Processing" [9].
- Various segmentation techniques have been used in this paper such as watershed segmentation, HSV segmentation.

3 Proposed Methodology

3.1 The Dataset

The Malaria Dataset from the U.S. National Library of Medicine has been used for the purpose of training and testing the model. The dataset has a total of 27,558 cell images in which equal instances of parasitized and uninfected cells are present. The images of the parasitized cells were studied from Olympus and Motif microscope models. 13779 images were labeled as parasitized and the rest as uninfected [10]. Some sample cell images are shown in Fig. 1.

3.2 The Pre-processing

The following preprocessing was used (Fig. 2).

Grayscale Conversion. All the images were converted into Grayscale using weighted grayscale conversion technique. The RGB intensity values were multiplied

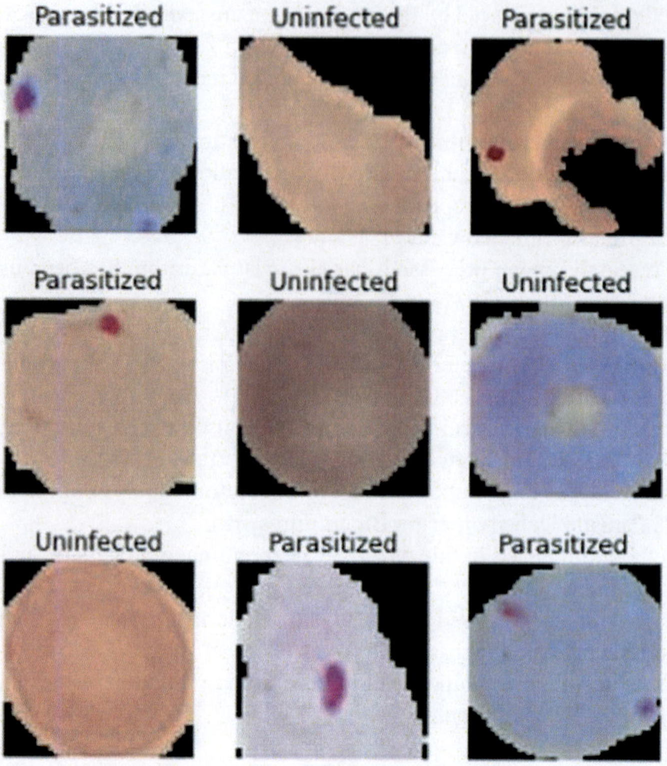

Fig. 1 Dataset

Fig. 2 Grayscale conversion

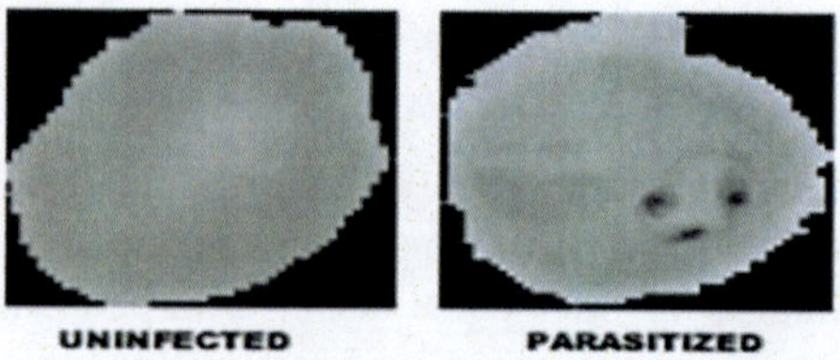

to (0.299, 0.587, 0.114), respectively. Here, 'R' stands for R-channel, 'G' stands for G-channel, and 'B' stands for B-channel.

$$Y = (0.299 * R + 0.587 * G + 0.114 * B)$$

Information Padding. A colored image contains three channels, namely RGB. But for the sake of this experiment, the number of channels is increased to four and, the last channel is fed with information directly from the image after the grayscale conversion. So, now the images have four channels, RGB and gray. This step is crucial to the network model we used in the later section.

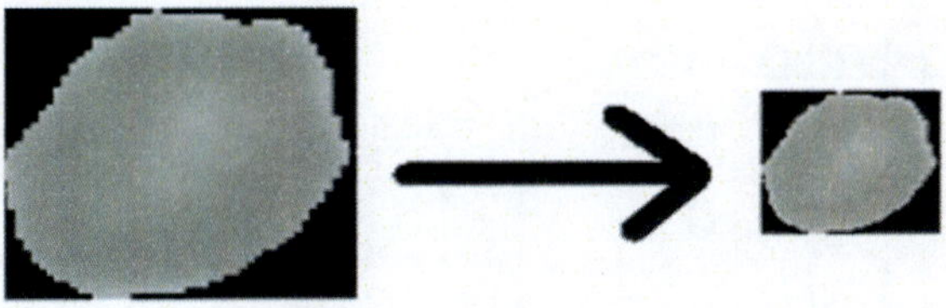

Fig. 3 Resizing

Image Resizing. The outputs are re-sized to a constant dimension of (50 X 50) pixels. Fig. 3 demonstrates the resizing operation.

Pixel Scaling. All the pixels in every image is mapped to a range of 0–1 by dividing with 255.

3.3 *The Network Model*

The peculiarities in the images are extracted using a three-layer Convolutional Operation. The grayscale channel obtained after information padding undergoes the same convolutional operation but with separate training kernels. So, the model has combined capabilities of both the RGB training and the grayscale training.

Convolution Operation. Convolution can be defined mathematically as the combined integration of two functions which depicts the influence of one function over the other. The equation can be given as

$$(f * g)(t) = \int_{-\infty}^{+\infty} f(x)g(t - x)dx$$

The number of kernels for convolution in the three layers is 64, 128, and 256, respectively. All the kernels are of size (3 X 3) and stride (1 X 1). Fig. 4 shows an example of the convolution operation.

Rectified Linear Unit (ReLU): The ReLU is an activation function which is mathematically defined as follows. ReLU is used to introduce non-linearity in the feature map.

Figure 5 shows a graph of the ReLU function.

Max Pooling: For a given (N X N) kernel placed over the ReLU-activated feature map, the max pooling operation selects the maximum value among the (N X N) options and passes it on to the output matrix. Max Pooling is typically used to conserve the features in the spatial domain of the image. After every ReLU-activated Convolution operation a (2 X 2) max pooling was used. Fig. 6 explains the concept of max pooling clearly.

The output from the final ConvPool pair is then flattened, i.e., the feature map is converted into a single feature vector.

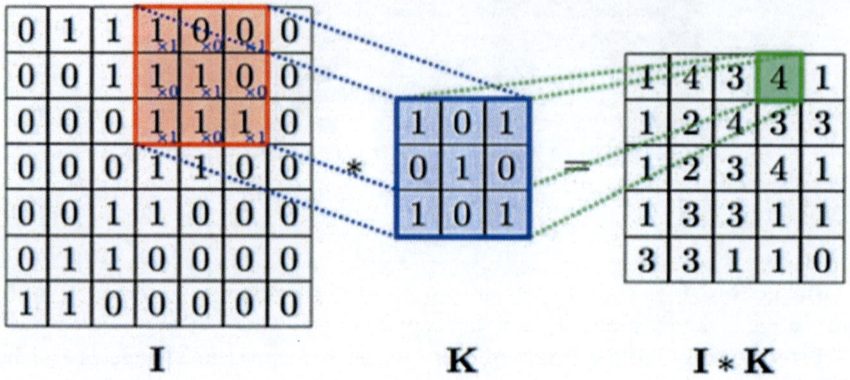

Fig. 4 Convolution operation

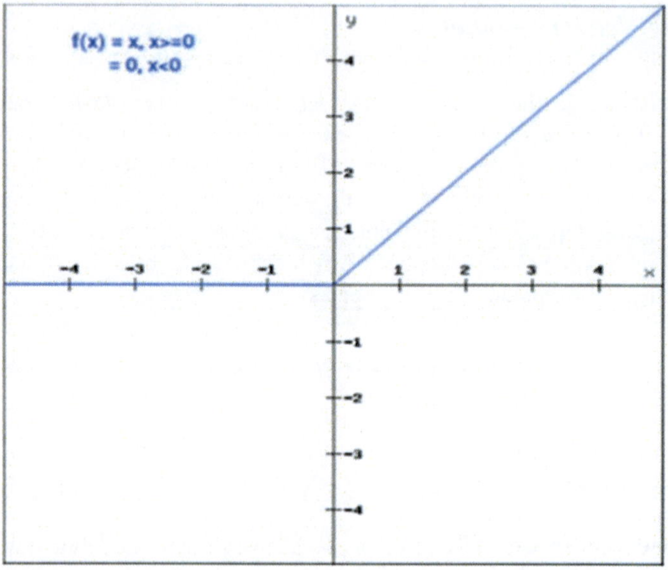

Fig. 5 Rectified linear unit

Fully Connected Layer: The single feature vector serves as the input to the fully connected layer. There is only one hidden layer with 128 nodes, which takes in values from the input layer via a ReLU activation through forward propagation. The output layer contains only one node which has a sigmoid activation. If the output value is greater than 0.5 the cell is parasitized, uninfected otherwise. The CNN is trained using back propagation, where binary cross-entropy function is used to calculate the loss.

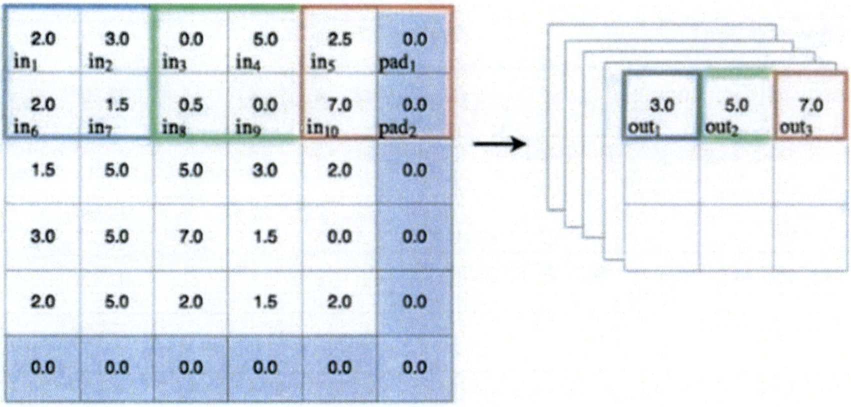

Fig. 6 Max pooling

$$L(w) = -\left(\frac{1}{N}\right)\sum_{n=1}^{N}[y_n \log(y_n^{pred}) + (1 - y_n)\log\left(1 - y_n^{pred}\right)]$$

The specifics of the model are shown via the diagram as follows (Fig. 7)

3.4 *Training the Model*

Out of the 27558, 16534 images are set aside for the training set (in which 50% is parasitized and the rest is uninfected). The rest of the 11024 instances are divided into two equal parts one for the testing set and the other for the validation set.

3.5 *Details of the Learning Process*

The Adam Optimizer was used for the purpose of training the model. The Adam Optimizer was used for the purpose of optimization of the gradient descent algorithm. The batch size used was 256 images per batch, for mini batch gradient descent. The network was trained for 15 epochs.

4 Results

The metrics on which Performance Analysis are done are:

Training Accuracy refers to the accuracy achieved on the Training Set.

Layer (type)	Output Shape	Param #
conv2d_10 (Conv2D)	(None, 48, 48, 64)	2368
max_pooling2d_10 (MaxPooling	(None, 24, 24, 64)	0
conv2d_11 (Conv2D)	(None, 22, 22, 128)	73856
max_pooling2d_11 (MaxPooling	(None, 11, 11, 128)	0
conv2d_12 (Conv2D)	(None, 9, 9, 256)	295168
max_pooling2d_12 (MaxPooling	(None, 4, 4, 256)	0
flatten_4 (Flatten)	(None, 4096)	0
dense_7 (Dense)	(None, 256)	1048832
dense_8 (Dense)	(None, 1)	257

Total params: 1,420,481
Trainable params: 1,420,481
Non-trainable params: 0

Fig. 7 Model summery

Validation Accuracy refers to the accuracy achieved on the Validation Set. The convergence graph of the Loss and Accuracy is shown in Figs. 8 and 9. It is evident from the graph that the model is good and has almost converged.

5 Conclusions

This paper is shared light on two things in particular—A non-conventional architecture of a deep learning model and its accuracy for a standard Malaria Classification problem which is 97.98%. The research done in this paper is a target for the automated classification and detection of cells affected by the malaria parasite. The future scope of this model includes ensemble architecture to yield even higher accuracy.

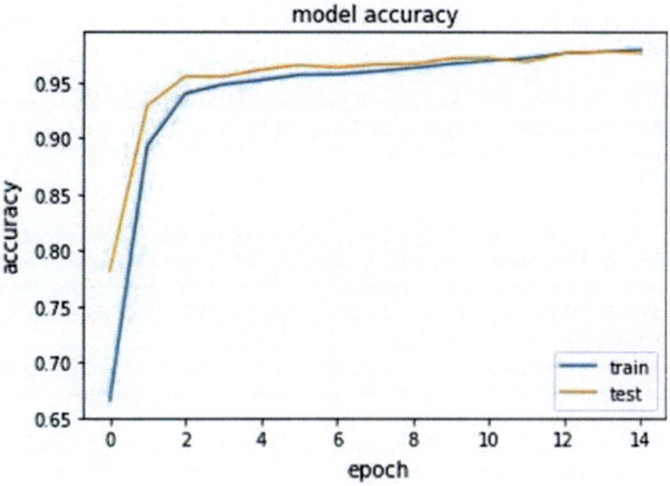

Fig. 8 Accuracy curve

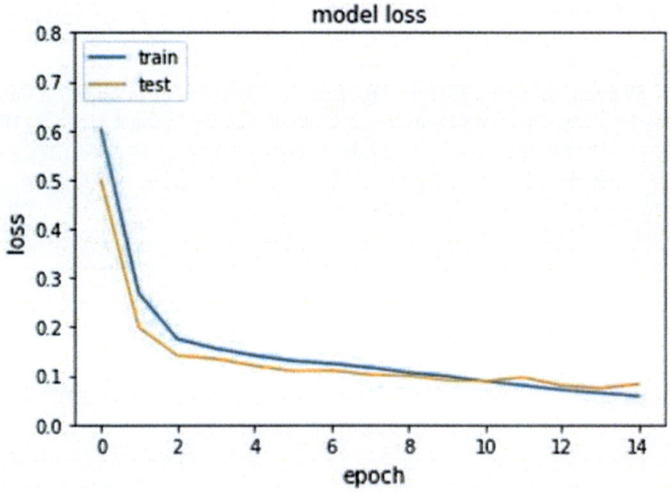

Fig. 9 Loss convergence crve

References

1. World Health Organization (Official Website) last accessed 13 Jan 2020, https://www.who.int/newsroom/factsheets/detail/malaria
2. Centers for Disease Control and Prevention, last accessed 13 Jan 2020. https://www.cdc.gov/malaria/diagnosis_treatment/diagnosis.html

3. Loddo, A., Di Ruberto, C., Kocher, M., Prod'Hom, G.: MP-IDB: the malaria parasite image database for image processing and analysis (2019). https://doi.org/10.1007/978-3-030-13835-6_7
4. Mohanty, I., Pattanaik, P.A., Swarnkar, T.: Automatic detection of malaria parasites using unsupervised techniques. In: Pandian, D., Fernando, X., Baig, Z., Shi, F. (eds.) Proceedings of the International Conference on ISMAC in Computational Vision and Bio-Engineering 2018 (ISMAC-CVB). ISMAC 2018. Lecture Notes in Computational Vision and Biomechanics, vol 30. Springer, Cham (2018)
5. Rode, K.B., Bharkad, S.D.: Automatic segmentation of malaria affected erythrocyte in thin blood films. In: Pandian, D., Fernando, X., Baig, Z., Shi, F. (eds.) Proceedings of the International Conference on ISMAC in Computational Vision and Bio-Engineering 2018 (ISMAC-CVB). ISMAC 2018. Lecture Notes in Computational Vision and Biomechanics, vol 30. Springer, Cham (2018)
6. Aggarwal, P., Khatter, A., Vyas, G.: An intensity threshold based image segmentation of malaria-infected cells. In: 2018 Second International Conference on Computing Methodologies and Communication (ICCMC) (2018)
7. Peñas, K.D., Rivera, P.T., Naval, P.C.Jr.: Analysis of convolutional neural networks and shape features for detection and identification of malaria parasites on thin blood smears In: Nguyen, N., Hoang, D., Hong, TP., Pham, H., Trawiński, B. (eds.) Intelligent Information and Database Systems. ACIIDS 2018. Lecture Notes in Computer Science, vol 10752. Springer, Cham (2018)
8. Oliveira, A.D., Carvalho, B.M., Prats, C., Espasa, M., Gomez i Prat, J., Codina, D.L., Albuquerque, J.: An automatic system for computing malaria parasite density in thin blood films. In: Mendoza, M., Velastín, S. (eds.) Progress in Pattern Recognition, Image Analysis, Computer Vision, and Applications. CIARP 2017. Lecture Notes in Computer Science, vol. 10657. Springer, Cham (2017)
9. Roy, K., Sharmin, S., Mufiz Mukta, R.B., Sen, A.: Detection of malaria parasite in giemsa blood sample using image processing. Int. J. Comput. Sci. Inf. Technol. (IJCSIT) **10**(1) (2018)
10. The Malaria Dataset from U.S. National Library of Medicine https://ceb.nlm.nih.gov/repositories/malaria-datasets/

A Novel Approach to 3D Face Registration for Biometric Analysis Using RCompute_ICP

Parama Bagchi, Debotosh Bhattacharjee, and Mita Nasipuri

Abstract In this present work, 3D face registration with ICP (Iterative Closest Point) algorithm on 3D face manifolds has been presented. This present work takes an unknown 3D face manifold as input and then registers it using a pre-initialized version of ICP algorithm named "RCompute_ICP." The present work first assumes that there are many source models and one target model which has been given as input. We have to predict the pose of the unknown source model and find out the registration performance of the source model. Our present algorithm uses ICP for the prediction of pose and registration performance using a one to many mapping technique by mapping each source model to each of the target models present. Finally, the target model which gives the least error after mapping of the source model is finally designated to be the probable pose of the unknown source model which is then subsequently registered to the corresponding target model. We synthesize our results, through the experiments conducted on 3D face manifolds from three different databases, namely Frav3D, GavabDB, and KinectJUDeiTy3DK Database (our new 3D face database). The least error obtained after registration was as low as 0.0032 which subsequently depicts the robustness of our present method.

Keywords Registration · ICP · Manifold

P. Bagchi (✉)
Department of Computer Science and Engineering, RCC Institute of Information Technology, Kolkata, India
e-mail: paramabagchi@gmail.com

D. Bhattacharjee · M. Nasipuri
Department of Computer Science and Engineering, Jadavpur University, Kolkata, India
e-mail: debotosh@ieee.org

M. Nasipuri
e-mail: mitanasipuri@gmail.com

D. Bhattacharjee et al. (eds.), *Proceedings of International Conference on Frontiers in Computing and Systems*, Advances in Intelligent Systems and Computing 1255,
https://doi.org/10.1007/978-981-15-7834-2_46

1 Introduction

Biometric security systems recognize an individual based on his features, e.g., handwriting, finger prints, palm print, face, iris, etc. In the earlier days, handwriting recognition, fingerprint recognition were highly popular, but those systems can easily be forged. Slowly as technology improved, 2D face recognition systems started to replace the traditional handwriting recognition systems because there is little chance for a human face to be forged. Another advantage of using face as a biometric trait is that, unlike other traits, a face can be acquired even without active participation of selected persons. In comparison to 2D biometric systems, 3D systems found a significant place in the field of biometric security because of their enormous capability to handle the effects caused due to occlusion, and change in illumination, pose, and expressions and as well as a combination of all of these. This robustness is obtained as 3D manifolds models work only with depth information.

However, variations in poses, expressions, etc., in the face images captured at different time instants need to be addressed. This is done by finding a suitable transformation through which a face image can be aligned to a target template. This process is known as face registration. In this work, a 3D face registration algorithm which uses point cloud-based ICP (Iterative Closest Point) technique termed as "RCompute_ICP" has been proposed. This algorithm takes as input a source and a target point cloud and registers them using ICP and predicts the pose of the unknown model.

Section 2 describes the related works on 3D face registration. Section 3 describes the proposed work related to "RCompute_ICP" algorithm. Section 4 discusses the experimental results. Finally, comparative analysis has been drawn in Sect. 5. Conclusion and future scope is discussed in Sect. 6.

2 Related Work

The task of a registration algorithm is to find out a transformation, which would minimize the alignment error between two different manifolds. 3D face registration techniques could be subdivided into

(a) Rigid face registration.
(b) Non-Rigid face registration.

Rigid registration is the simplest registration technique which uses only rotation and translation to register an object.

Non-Rigid registration includes non-linear deformations such as local shear and stretching. It includes these additional transformations in addition to the translation and rotation. It is important to note that rigid registration in the presence of occlusions, expression, and pose still remains an area to be explored.

Several works have been performed mostly on non-rigid registration where face manifold registration or point cloud registration have been performed mostly by locating 3D face features on 3D face registration. Chen et al. [1, 2] attempted a non-rigid facial registration technique by aligning the facial point clouds using a coarse registration process at first and then applied ICP to bring the surfaces closely aligned to each other. Experiments were performed mainly on frontal profiles, facial expressions, and two side profiles up to 10° yaw rotation. Experiments were performed on ECL—IV2 3D face database and the recognition rate obtained after registration was 92.68%. Berretti et al. [3] proposed approaches of non-rigid registration of 3D frames on the Florence Superface v2.0 dataset with poses from −25° to +25°. Here, a holistic-based registration was performed using ICP which was applied one by one on several frames and finally a super-resolution registered model was created. The ICP was done using an initialization process by registering the second frame with the first frame and those parameters were further passed on to the corresponding frames. Nair et al. [4] proposed a 3D point-based distribution model where at first several feature points were first localized, then a coarse registration process was computed by taking distances between feature points into account after that a fine registration was accomplished using ICP. Experiments were performed on 3D face manifolds taken from GavabDB and NTU face dataset. Gokberk et al. [5] recognized faces by first registering all 3D face manifolds to an Average Face Model.

Several notable works were performed on rigid 3D face registration in presence of occlusions, expressions, and pose. Yaniv et al. [6] discussed various types of ICP-based registration [7] techniques. Another registration [8, 9] technique was mentioned by Masuda et al. [10]. In this rigid registration technique, closest feature points matched using PCA (principal component analysis), and then the outliers was discarded using the RANSAC algorithm. Rusinkiewicz et al. [11] proposed the various variants of ICP. Lee et al. [12] presented a registration technique based on an efficient anisotropic ICP registration technique, which was applied on two large-scale RGB-D object databases. The initial poses were perturbed with noise, and sampling intervals were applied. The convergence stage was declared when the error threshold was tiny. Yan et al. [13], attempted an ICP-based ear registration technique where a pre-computed voxel of closest neighbors procedure was adopted. Experiments were performed on a 3D ear biometric database. Chang et al. [14] performed a recognition technique on a database of 4000 scans by attempting to register the nose region by using ICP. It is obvious from the above discussion that not much work on rigid 3D face registration in the presence of pose, expressions as well as a combination of pose and expression had been performed. It would be important to address the above problems in the present work on 3D face registration. It would be an added advantage to find how effectively rigid registration performs under poses up to +45° to −45° along with a combination of expressions.

The present work concentrates on the registration [15] of 3D face manifolds based on the initialization of ICP and has been inspired by Besl et al. [16]. Listed below are the major contributions of the present work:

- Proposing a mathematical model of 3D face registration "RCompute_ICP" which uses a unique one to many mapping from source to the target models and thereby predicting the orientation of the source point cloud.

- Proposing an efficient method of point cloud-based registration formed by initializing ICP with the principal components of the source and target point clouds across pose, expressions as well as a combination of pose and expressions.

3 Proposed Work

The proposed algorithm takes as input an unknown 3D face manifold of a person. The assumption is that there are several known 3D face point clouds present across pose orientations of yaw, pitch, and roll. We need to find the pose of the unknown models with respect to the known models and register it with the known models. In the present work, three databases have been considered, namely:

1. GavabDB [17]
2. Frav3D [18] and
3. A new 3D face database (KinectJUDeiTy3DK) developed by us.

Following are the steps of our algorithm:

(a) 3D Data Acquisition: 3D raw data is first acquired using an acquisition device such as scanner which captures data in point cloud form. Frav3D and GavabDB databases have been acquired using David Laser Scanners. KinectJUDeiTy3DK face database has been developed using Kinect Camera. The raw point cloud to be of size nx3 where n stands for the number of points in the point cloud.
(b) Construction of the target models: After data has been acquired in point cloud form, the next step is the construction of the target models to which the source models will be registered. Figure 1 shows different target models in various poses that we have taken for experimentation.
(c) Construction of the source models: After the target models have been created, the next step is the creation of the source model. A source model is also a point cloud constructed in a particular pose and is input to the system through the user. Now, a source point cloud having a different size k in a particular pose say $s_{k\_pose\_source}$, is to be registered to each target model $t^{i}_{pose\_target}$. Now, we want to detect the pose of this unknown source model and register it to the desired target model. Figure 2 shows a source model being input to the system.

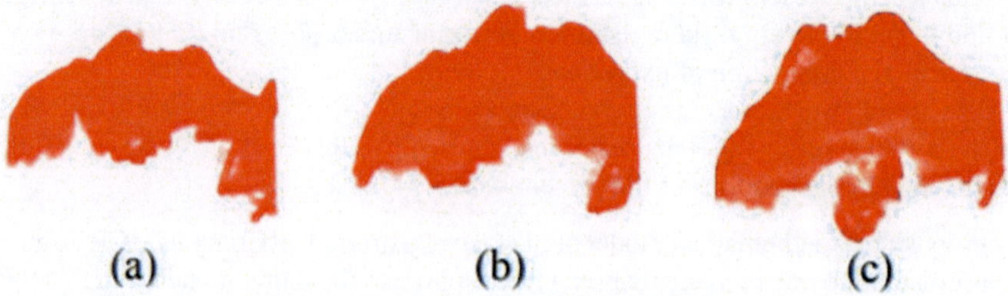

Fig. 1 Target models in various poses (Yaw-left, Roll-left, pitch up)

Fig. 2 Source model in a frontal pose

(d) Alignment Detection and Registration of source to target model: After the source and the target models have been created, the last and final step is the alignment and registration. Here, we divide the work into two parts:

   (a) In the alignment detection step, the correct pose of the source model is determined.
   (b) In registration step, the source is aligned to the target. Here, the target models are in frontal pose.

The above tasks are performed using our present rigid registration algorithm named "RCompute_ICP" using a unique one to many mapping technique. The mathematical implementation for part (a) is given below:

$$\mathrm{T1} : \mathrm{s}_{\mathrm{k\_pose\_source}} \rightarrow \mathrm{t}^{\mathrm{i}}_{\mathrm{p}_{\mathrm{pose}_{\mathrm{target}}}} \tag{1}$$

where T1 is the rigid transformation that minimizes the distance between source and target point clouds using the proposed algorithm.

For detecting the correct alignment, we used "RCompute_ICP'" algorithm to minimize the error of misalignment between source and target models using the minimum distance metric RMSE (Root mean square error). This can be depicted as follows:

$$dist\left(s_{k\_\mathrm{pose\_source}}, t^{i}_{p\_\mathrm{pose\_target}}\right) = \sum_{ms_{k\_\mathrm{pose\_source}}} \sum_{nt^{i}_{p\_\mathrm{pose\_target}}} (m - n)^2 \tag{2}$$

$$\text{Correct Alignment} = \text{minimum}(dist) \tag{3}$$

The model which gives minimum value of dist is said to give the correct pose alignment for $s_{k\_pose\_source}$. The mathematical implementation for part (b) is as follows:

$$\text{T2: } s_{k\_pose\_source} \underset{\text{Register}}{\rightarrow} t^{i}_{p\_pose\_target} \quad (4)$$

In this transformation, the posed source model is registered, i.e., rendered to frontal position by aligning it to a frontal target model. Both parts (a) and (b) formulate our proposed "RCompute_ICP" algorithm. The key feature of the algorithm is that we are initializing ICP algorithm using differently sized point clouds. Here, we are using KDtree version of ICP. ICP is a global rigid registration algorithm which uses two point clouds (one is called "source" and the other is called "target'") and tries to minimize the distance between them using a minimum distance transformation parameter. The present version of ICP, renamed as "'RCompute_ICP'" algorithm, proposes a registration cum alignment detection algorithm which initializes ICP with point clouds of different sizes. There are several target models present and one source model. Using the present algorithm, the source needs to be aligned to each target and then separately registered in order to render it to a frontal position. For this reason, let us consider different point clouds consisting of one source S and several target models T1, T2, T3,... Tm. The algorithm, as described below, uses matched pairs to calculate the minimum distance metric.

RCompute_ICP algorithm

Start of Algorithm

Step (1) Initialize ICP parameters, namely number of iterations, matching technique, edge rejection.

Repeat Step 2–6 till RMSE converges

Step (2) For each point of S find it's matched pair in T1, T2, T3,..Tm using nearest neighbor search and find the minimum distance between each corresponding pair in S and T1, T2, T3,...Tm.

Step (3) Remove edge vertices because they do not contribute to registration.

Step (4) Estimate the rotational and translational parameters (R, t) by centralizing S and each of the target point clouds.

Step (5) Rotate and translate the source point cloud as shown below: S = R*S + t

Step 6) Find RMSE between corresponding pairs 'P' in S and 'O.'

End of Algorithm.

Figure 3 shows a snapshot of the convergence using the present algorithm.

Point matching technique using KDtree [19] is $O(\log n)$ which is a fast process. Rcompute_ICP is an optimal global algorithm, where there are two point clouds, one is called the target is fixed and the source is used to match the reference. The algorithm uses translation and rotation to minimize error metric and uses this transformation to map the source to the target point cloud.

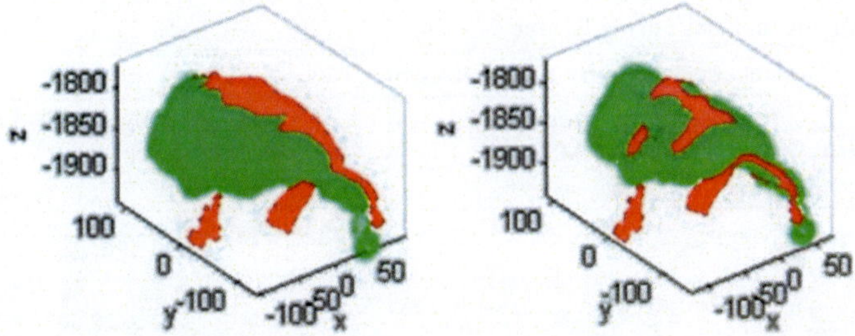

Fig. 3 A snapshot of the registration process by RCompute_ICP

(e). Evaluation of registration: After pose detection and registration the performance is measured by RMSE (Root Mean Square Error).

4 Experimental Results

Here, a full explanation using has been given for the results obtained for facial alignment and registration on Frav3D Database. In Frav3D database, 100 images consisting of Yaw with +15°, Yaw with +25°, Yaw with −15°, Yaw with -25°, Pitch up with +10°, Pitch down with −10°, Roll left with +25°, Roll right with −25°, and two frontal profiles have been selected to test the present method.

4.1 Results of Alignment Detection

For alignment detection, we randomly selected two models from one of the ten classes and checked its alignment with all the eight models numbered 1–8 of Class 1. The number of iterations was kept fixed at 100. The frontal model was excluded because they were separately kept only for evaluation of registration.

As evident from Table 1, if two models after alignment have very close RMSE error, then the correct orientation could be found by the model with which the unknown model has the least RMSE error. The "known" model with which after alignment, the RMSE error between the "known" and the "unknown" model was minimum served as the correct alignment.

4.2 Results of Registration

The results of registration are shown in Table 2. There are two frontal models in Class 1 against whom all the other models from Class 2 to Class 10 have been registered.

Table 1 Alignment results using KDtree

Name of the known models	Alignment detection By KDtree matching						
	2_05(Across left Yaw)	Starting RMSE	……..	……	Ending RMSE	Aligned 3D face	Number of iterations
With 1_05	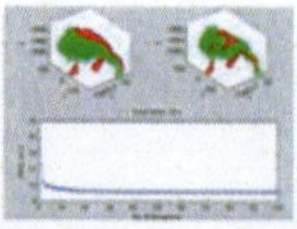	16.70	12.63	10.43	9.15	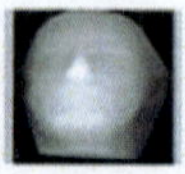	99 iterations
With 1_06	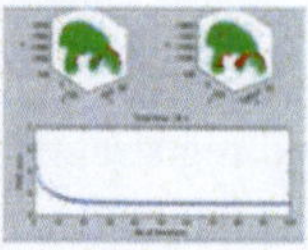	6.27	5.20	4.56	3.55	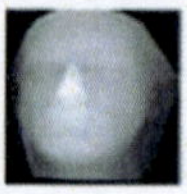	70 iterations

Table 2 Registration performance using "RCompute_ICP"

Name of the known models and with which it is registered	Performance of registration By KDtree matching						
	Name of the unknown models for alignment detection: 2_05 (across left Yaw)	Starting RMSE	…….	……	Ending RMSE	Registered 3D face	Number of iterations
2_06 with 1_01	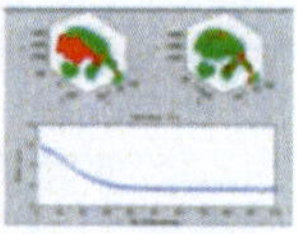	11.87	10.26	9.86	5.64		85 iterations needed to converge
2_06 with 1_02	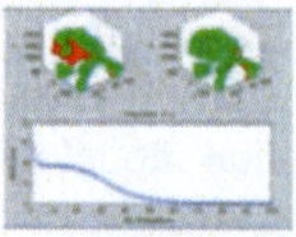	10.55	9.41	8.58	4.11		Did not converge even in 100 iterations

After registration, the registration errors obtained showed robust registration because it went down to 4.11. But since the maximum number of iterations was set to 100, it was found that even till 120 iterations there was no substantial reduction in RMSE.

5 Comparative Analysis

In this section, a comparison of the present registration algorithm "RCompute_ICP" over the already prevailing registration algorithms has been enlisted in Table 3.

Table 3 Comparative analysis

Sl. No	Registration method	Database	Size of the database		Performance of registration
ICP using initialization					
1	Fast registration algorithm using point to plane technique [20]	1. The wave object having 2 surfaces with phase difference of 10° 2. Angel object from Ohio State University 1. Potato-head	Wave object	Points: 22500 Triangles: 44 402	RMSE error: 14 mm
			Angel object	Points: 14089 Triangles: 27213	RMSE error: 0.73 mm
			Potato head	Points: 13827 Triangles: 27144	RMSE error: 0.63 mm
2	Fast point to plane registration [21]	322 faces from 3DMD face database	3D Faces	100,000 triangles 50,000 points	Mostly frontal faces with mean error below 4 mm
ICP without initialization (coarse followed by fine alignment)					
					Profiles / **RMSE**
3	Multistep alignment scheme [22]	GavabDB database	3D Faces	50,000 points	Neutral1: 0.0 Neutral: 0.0 Random: 1.6 Laugh: 1.6 Smile: 0.0 Look up: 13.1 Look down: 0.0
4	Automatic landmark annotation and dense correspondence registration (non rigid) [22]	115 Chinese (56 males, 59 females) and 124 (48 males, 76 females) database	3D faces	30,000 vertices per surface on average	RMSE error of ~1.7 mm on absolutely frontal poses with expressions only
5	Present work	Poses like yaw, pitch and roll as well as poses up to 45° along with expressions	3D faces	Vertices: approx. 30,000	Least RMSE error attained was as low as 0.0032

6 Conclusion

The present work depicts that proper initialization is essential for face registration. This has been used for ill posed subjects. In this case, "RCompute_ICP" is used to register ill-posed subjects, and the present work is used to register extreme poses. For future work, 3D face recognition is proposed as a part of future work.

References

1. Ben Amor, B., Ardabilian, M., Chen L.: New experiments on ICP-based 3D face recognition and authentication. International Conference on Pattern Recognition, 3, pp. 1195-1199, ISSN 1051-4651 (2006)
2. Ouji, K., Amor, B.B., Ardabilian, M., Ghorbel, F., Chen, L.: 3D face recognition using ICP and geodesic computation coupled approach. Signal Process. Image Enhanc. Multimed. Process. **31**, 141–151 (2006)
3. Berretti, S., Pala, P., Bimbo, A.D.: Face recognition by super-resolved 3D models from consumer depth cameras. IEEE Trans. Inf. Forensics Secur. **9**(9) (2014)
4. Nair, P., Cavallaro, A.: 3D face detection, landmark localization, and registration using a point distribution model. IEEE Trans. Multimed. **11**(4), 611–23 (2009)
5. Alyuz, N., Gokberk, B., Akarun, L.: Regional registration for expression resistant 3D face recognition. IEEE Trans. Inf. Forensics Secur. **5**(3) (2010)
6. ZivYaniv: Rigid Registration. Image-Guided Interventions, pp. 159–192
7. Chua, C., Jarvis, R.: 3-d free-form surface registration and object recognition. Int. J. Comput. Vis. **17**(1), 77–99 (1996)
8. Gelfand, N., Mitra, N.J., Guibas, L.J., Pottmann, H.: Robust global registration. In: Symposium on Geometry Processing, pp. 197–206 (2005)
9. Johnson, A., Hebert, M.: Using spin images for efficient object recognition in cluttered 3d scenes. IEEE Trans. Pattern Anal. Mach. Intell. **21**(5), 433–449 (1999)
10. Masuda, T.: Multiple range image registration by matching local log-polar range images. In the 7th Asian Conference on Computer Vision, pp. 948–957 (2006)
11. Rusinkiewicz, S., Levoy, M.: Efficient variants of the ICP algorithm. In Proc. 3DIM, pp. 145–152 (2001)
12. Lee, B., Lee, D.D.: Learning anisotropic ICP (LA-ICP) for robust and efficient 3D registration. In: IEEE International Conference on Robotics and Automation (ICRA) (2016)
13. Yan, P., Bowyer, K.W.: A fast algorithm for ICP-based 3D shape biometrics. In: Fourth IEEE Workshop on Automatic Identification Advanced Technologies, pp. 213– 218 (2005)
14. Chang, K.I., Bowyer, K.W., Flynn, P.J.: Multiple nose region matching for 3D face recognition under varying facial expression. IEEE Trans. Pattern Anal. Mach. Intell. **28**(10) (2006)
15. Lu, X., Jain, A.K., Colbry, D.: Matching 2.5D face scans to 3D models. IEEE Trans. Pattern Anal. Mach. Intell **28**(1) (2006)
16. Besl, P., Mckay, N.: A method for registration of 3-D shapes. In: PAMI (1992)
17. Bagchi, P., Bhattacharjee, D., Nasipuri, M., Basu, D.K.: A novel approach for nose tip detection using smoothing by weighted median filtering applied to 3D face images in variant poses. In: International Conference on Pattern Recognition, Informatics and Medical Engineering (PRIME) (2012). http://gavab.escet.urjc.es/recursos_en.html
18. Chanop, S., Hartley, R.: Optimised KD-trees for fast image descriptor matching. In: IEEE Conference on Computer Vision and Pattern Recognition (2008)
19. Park, S.Y., Subbarao, M.: An accurate and fast point-to-plane registration technique. In: Pattern Recognition Letters (2003)

20. Koppen, W.P., Chan, C.H., Christmas, W.J., Kittler, J.: An intrinsic coordinate system for 3D face registration. In: 21st International Conference on Pattern Recognition (ICPR) (2012)
21. Stormer, A., Rigoll, G.: A multi-step alignment scheme for face recognition in range images. In: 15th IEEE International Conference on Image Processing (2008)
22. Guo, J., Mei, X., Tang, K.: Automatic landmark annotation and dense correspondence registration for 3D human facial images. In: BMC Bioinformatics (2013)

Sequence Characterization of Glutamate Receptor Genes of Rat (Vertebrate) and Arabidopsis Thaliana (Plant)

Antara Sengupta, Pabitra Pal Choudhury, and Subhadip Chakraborty

Abstract iGluR gene family of a vertebrate (rat) and glutamate-like receptor (AtGLR) gene family of plant (Arabidopsis thaliana) by Darwin and Darwin (in The Power of Movement in Plants, 1880) perform few similar kind of functionalities in neurotransmission. These have been compared quantitatively depending upon the biochemical characteristics of 20 amino acids comprising the amino acid sequences of the aforesaid genes. 19 AtGLR genes and 16 iGluR genes have been taken as datasets. Thus, we detected the commonalities (conserved elements) which plants and animals have got from a common ancestor during the long evolution by Darwin and Darwin (in The Power of Movement in Plants, 1880). Eight different conserved regions have been found based on individual amino acids. Different conserved regions are also found, which are based on chemical groups of amino acids. We have tried to find out different possible patterns which are common throughout the dataset taken. Nine such patterns have been found with size varying from 3 to 5 amino acids at different regions in each primary protein sequence. Phylogenetic trees of AtGLR and iGluR families have also been constructed. This approach is likely to shed lights on the long course of evolution.

Keywords Glutamate receptors · Arabidopsis thaliana · Chemical properties · Directed graph · Phylogenetic tree

---

A. Sengupta (✉)
MCKV Institute of Engineering, Howrah 711204, India
e-mail: antara.sngpt@gmail.com

P. P. Choudhury
Indian Statistical Institute, Kolkata, India
e-mail: pabitrapalchoudhury@gmail.com

S. Chakraborty
Nabadwip Vidyasagar College, Nabadwip 741302, India
e-mail: subhadipiicb@gmail.com

D. Bhattacharjee et al. (eds.), *Proceedings of International Conference on Frontiers in Computing and Systems*, Advances in Intelligent Systems and Computing 1255,
https://doi.org/10.1007/978-981-15-7834-2_47

1 Introduction

Plants grow in silence but react to wounding. Specific genes are there in plants which make control over such functionalities in neurotransmission [2]. On the other hand, animals have well-developed nervous systems [3]. Earlier it was believed that plants do not have anything like a nervous system, they do not respond to stimuli in the way animals do. But according to the great pioneer Scientist Sir Acharya J.C. Bose, [4] plant also has an alternative sort of sensitive nervous system, not like that of animals but can respond to various external stimuli [1]. It has been reported that like glutamate receptors of vertebrate nervous system similar receptors are also there in the plant equivalent to it and have so close similarities that they must have evolved from a common ancestor [5]. Arabidopsis thaliana has a large family of Glutamate Receptor-like genes (AtGLR) in them like ionotropic glutamate receptor (iGluR) gene family of animals. GLRs in Arabidopsis thaliana are organisms those contribute in neurotransmission. Ionotropic glutamate receptors are essential for mammalian central nervous system. They intervene in chemical synaptic transmission at the clear majority of excitatory synapses [6]. Mutations in both genes can lead to neurological and psychiatric disorder. In rat, ionotropic glutamate receptor delta-1 is expressed in the caudate putamen and delta-2 is expressed in Purkinje cells of the cerebellum [7]. Amino acids are organic compounds containing specific biochemical compositions that instruct primary protein sequences to have specific structure, to perform specific functions and to get stability. Quantitative analysis of genes at molecular level and investigation of evolutionary relationships existing among various species through similarity or dissimilarity analysis are possible to be performed through mathematical parameters which evaluates some data and verify facts [8]. Graphical representation of primary protein sequence gives alignment-free method which provides visual inspection of the sequence and instant provision of analyzing similarities with others [9]. Numerous authors [10–15] proposed several 2D and 3D graphical representations of DNA as well as amino acid sequences. Some research papers are reported, where evolutionary relationships among various species are investigated based on multiplet structure of amino acid sequences [13]. It is attempted to find out conserved areas of amino acid sequences of organisms based on chemical properties of amino acids, which are invariant [8, 10, 16]. Several works have also been done with PPIs, where it is tried to find the nature of distribution of chemical properties of amino acids in any hub gene and the genes in a network and proximity among them [17]. In our paper, methodologies are discussed through different sub-sections of Sect. 2. In first subsection, the 20 amino acids have been classified depending upon the chemical properties they have and the methodologies have been described thoroughly in next sub-sections. As our primary objective is to make quantitative comparison between iGluR gene family of mammal (Rat) and AtGLR of plant (Arabidopsis thaliana) based on their chemical features and investigate evolutionary relationships among them, in this context, the datasets on which the experiments are carried out have been specified in the third section, and application and results are applied on dataset taken. The last section is kept for conclusion. Thus, the novelty lies in the mathematical modeling to establish the fact that like glutamate receptor gene family in mammal, plant also has some genes which may have come from a common ancestry.

2 Methodology

2.1 Classification of Amino Acids

To analyze the structure and chemical properties of protein, it is necessary to understand the chemical properties of amino acids. It has been reported that amino acid properties are not only decisive in determining protein structure and functions performed by them, but also play a role in molecular evolution [8, 10, 16].

Hence, it is aimed to understand how the biochemical nature of each amino1 acid sequence figure out evolutionary conserved regions among glutamate genes of mammal and plant during the long process of evolution based on eight distinct chemical properties of amino acids. 20 amino acids are classified accordingly as shown in Table .

2.2 Numerical Representation of Dataset Taken as Per Classification

Quantitative understanding of a primary protein sequence demands numerical representation of the amino acids it contains. Let $P = P_1, P_1, \ldots, P_k$, is an arbitrary primary protein sequence, where for any P, P_i represents a single amino acid from the set, such that $P_j \in \{A, C, D, E, F, G, H, 1, K, L, M, N, F, Q, R, S, T, V, W, Y\}$. Let 20 amino acids are classified into a set of classes C. Then it is possible to read each amino acid of set P by its corresponding class C, where $C \in \{1,2,3,4,5,6,7,8\}$. As an example, according to Table 1, a chunk of amino acid sequence 'MEALT' will be represented as '61447' after the encoding into the numeric representation.

Table 1 Classification of 20 amino acids as per their chemical properties

Class no.	Class name	Amino acids
1	Acidic	Aspartate, Glutamate
2	Basic	Histidine, Arginine, Lysine
3	Aromatic	Tryptophan, Tyrosine, Phenylalanine
4	Aliphatic	Isoleucine, Glycine Leucine, Valine, Alanine,
5	Cyclic	Proline
6	Sulfur containing	Methionine, Cysteine
7	Hydroxyl containing	Threonine, Serine
8	Acidic amide	Glutamine, Asparagine

2.3 Calculate Percent-Wise Presence of Each Amino Acid Individually and Class-Wise in Each Amino Acid Sequence

Presence of all 20 amino acid sequences is not the same for all species. Even the rate of frequency varies sequence to sequence. In this subsection, rate of occurrence, i.e., frequency of each amino acid present as well as the percentage of each group of amino acid present in each amino acid sequences is calculated as amino acid sequence of a protein that contains the necessary information to determine the three-dimensional structure of that protein and its stability. Sometimes amino acid encodes some information which are inherited from the previous stage (before translation). Amino acids which belong to basic and acidic chemical properties contribute in beta sheet formation [6], whereas aromatic amino acids provide a negative electrostatic potential surface that leads to cation–pi interaction, which has significant contribution in overall stability of proteins.

2.4 Common Patterns/Blocks Finding and Identifying Conserved Regions

Conserved amino acid sequence regions are extremely useful to investigate, identify, and study functional and structural importance of those regions. Here, in this part of the manuscript, firstly conserved regions are detected. Given an input of pattern length D ($2 < D < 8$), every possible pattern (made up of any combination of numeric value 1–8) of length D is investigated among the sequences. If the pattern is found in every sequences, then the pattern is stored along with its location, else it is discarded.

2.5 Investigating Evolutionary Relationships Between Two Species

Let P = {P1, P2,… Pn} be a set of amino acid sequences, where each sequence Pi for i = {1,2,…n} can be represented through weighted directed multigraph Gm. Gm = {V,A} is ,say, a multigraph, where set of vertices V = {Vi, V_2,V_3, V_4, V_5, V_6, V_7, V_8} represent the elements of class C, where $C = \{1, 2, 3, 4, 5, 6, 7, 8\}$. Arcs A represent the possible parallel arcs present between any two vertices. For any amino acid sequence, say, Vi and Vj are any two vertices for which the graph may have parallel arcs from Vi to Vj. Define the weight of each arc from Vi to Vj as i, so that for every arc from Vi to Vj, total weight of the arcs of graph Gm is

$$w_m(V_i, V_j) = \sum w(V_i, V_j) \quad (1)$$

Thus, we will get an (8X8) adjacency matrix M for each pair of vertices, which gives us 64-dimensional vectors in row order, say M, where M = (w(1,1), w(1,2), w(1,8),…, w(8,1),… w(8,8)). Now the directed multigraph can be drawn through adjacency matrix. Now let for any two amino acid sequences, say, P1 and P2, the corresponding set of vectors are Q = {Q1, Q2,…, Q64} and R = {R1, R2,….R64}, respectively. The weight deviation (WD) between P1 and P2 can be represented as stated in Eq. 2.

$$WD(P_1, P_2) = \frac{\sum_{i=1}^{64} |Q_i - R_i|}{64} \quad (2)$$

3 Application and Results

3.1 Dataset Specification

To carry out the entire experiment, 19 of such sequences have been selected to carry out the investigations. i6 amino acid sequences of iGluR family of Rat have also been taken for studies which are reported in NCBI and shown in Table 2.

3.2 Calculate Percentage of Each Group of Amino Acids Present in Each Sequence

It has been discussed before that the chemical properties of primary protein sequences have significant role in protein folding and protein structure making. It can be observed that in both the tables (Tables 3 and 4) group 6, i.e., amino acids of aliphatic group have remarkably major contributions to make primary protein sequences in both species, whose hydropathy profile is hydrophobic.

3.3 Common Pattern Finding in All Primary Protein Sequences

20 amino acids are classified based on their chemical properties, and each of them is identified with specific numeric value, so when common pattern finding procedure has been made for all the primary protein sequences, some numerical patterns are found which are common in all the primary protein sequences taken. In this section, it has been tried to make microscopic view of those patterns. Table 4 contains patterns of length of three amino acids (AA) for which same amino acids are contributed in

Table 2 Dataset 19 AtGLR gene family in Arabidopsis and 16 iGluR gene family in rat

(Rat) iGhiR gene family				(Arabidopsis) AtGLR gene family					
Gene name	Accession No.	Gene name	Accession No.	Gene name	Accession No.	Gene name	Accession No.	Gene name	Accession No.
UluRA	PI9490	KA-2	Q63273	AtGLRl. 1	AAF26802.1	AtGLR2.5	CAB96656.1	AtGLR3.5	AAC69939.1
3taRB	P10491	Delta 1	Q(52(540	AtGLRl.2	BAAseseo.i	AtGLR2.6	CAB96653.1	AtGLR3.6	CAB63C12.1
jluRC	AAA41245.1	Delta 2	Q63226	AtOLR1.3	BAA96961.2	AtGLR2.7	AAC33239.1	AtGLR3.7	AAC69938.1
UluRD	AAA41246.1	NMDA Zetal	P35439	AtGLRl.4	AAF02156.1	AtGLR2.8	AAC33237.1		
31uR5	AAA02873.1	NMDA2A	AAC03565.1	AtGLR2.1	A AB61068.1	AtGLR2.9	AAC33236.1		
3 MW	P42260	NMDA2B	AAA41714.1	AtGLR2.2	AAD26895.1	AtGLR3.2	CAA18740.1		
3hR7	AACS0577.1	NMDA2C	AAA41713.1	AtGLR2.3	AAD26894.1	AtGLR3.3	AAG51316.1		
KA-1	AAA17830.1	MDA(NR2D)	AAC37647.1	AtGLR2.4	CAA19752.1	AtGLR3.4	A AB71458.1		

Table 3 Presence of each group of amino acids in each sequence of rat in percentage

Rat		GluRA	GluRB	GluRC	GluRD	GluR5	GluR6	GluR7	KA1	KA2	Delta1	Delta2	NMDAZ1	NMDA2A	NMDA2B	NMDA2C	NMDA2D
Group	1	10.5	11.1	10.9	10.7	11	10.8	11.2	10.5	10.8	10.4	10.1	8.01	10.7	11.5	11.8	9.15
	2	12	11.4	13	12.6	13.2	12	12.3	12.6	12.6	11.5	12.8	12.2	14.6	13.6	13.8	12.6
	3	8.13	9.73	10.4	10.5	9.91	11.1	11.1	11.8	10.8	9.1	8.17	8.49	8.96	9.7	9.78	8.92
	4	36.2	35.7	35.2	35.7	36.2	36.4	36.9	35.3	37.3	37.1	37.4	38.3	35.3	31.1	31.8	39.5
	5	4.56	4.07	4.02	4.07	4.28	3.64	2.94	3.6	3.88	4.5	5.92	8.41	4.16	4.92	4.99	11.5
	6	4.26	4.07	4.46	4.74	5.18	4.3	3.74	4.17	4.32	5.65	5.11	4.2	4.48	4.51	4.79	3.48
	7	15.7	14.4	14	14	12.7	12.9	13.8	13.1	13.2	13.1	13.5	13.8	13.4	15.1	14.8	10.9
	8	1.72	9.53	7.93	8.72	7.55	8.82	7.92	9.01	7.2	8.68	7.05	6.55	8.42	9.63	8.3	4.08

Table 4 Presence of each group of amino acids in each sequence of arabidopsis in percentage

Arabidopsis		AtGLR 2.1	AtGLR 2.2	AtGLR 2.3	AtGLR 2.4	AtGLR 2.5	AtGLR 2.6	AtGLR 2.7	AtGLR 2.8	AtGLR 2.9	AtGLR 3.2	AtGLR 3.3	AtGLR 3.4	AtGLR 3.5	AtGLR 3.6	AtGLR 3.7	AtGLR 1.1
Amino acid group	1	8.85	9.67	10.2	9.69	12.2	11.3	10.5	10.5	10.9	9.73	9.73	9.46	10.6	9.33	10.7	10.4
	2	9.49	11	11.2	11.3	12.3	12.7	11.8	12.5	12.2	13.1	13.1	11.3	12.1	11.7	13.3	13.1
	3	11.5	11.6	12.1	10.6	10.9	9.6	10.6	11.7	11.5	10.5	10.5	10.5	11.4	9.82	10.1	9.65
	4	37.7	36.4	36.5	37.1	36.2	35.8	35.1	34.5	34	36.1	36.1	35.8	36.2	40	38.2	38
	5	4.8	5.11	5.25	5.46	4.22	4.86	4.28	4.38	4.68	4.54	4.54	4.78	5.03	4.62	4.45	2.23
	6	3.3	3.7	3.8	3.56	3.38	3.64	3.64	3.24	3.51	3.24	3.24	3.43	2.23	3.37	3.15	3.09
	7	14.6	14.4	14.1	15	14.1	14.6	15.2	14.7	14.8	15	15	15.8	14.3	12.1	13.5	15.7
	8	8.7	8.7	6.92	7.35	6.75	7.62	8.89	8.46	8.41	7.79	7.79	8.94	8.16	8.95	6.63	7.8

Table 5 Common pattern finding (length of 3 AA) for which same amino acids are contributed in same order as per classification

Pattern-'444'					
Gene name	Position	AA	Gene name	Position	AA
ATGLR2.1	564	VVA	ATGLR3.5	55	VVA
ATGLR2.5	28	VVA		125	VVA
ATGLR2.6	101	VVA		238	VVA
ATGLR2.7	148	VVA		529	VVA
ATGLR2.8	169	VVA	ATGLR3.7	96	VVA
ATGLR2.9	165	VVA		414	VVA
ATGLR3.2	584	VVA	ATGLR1.1	17	VVA
ATGLR3.3	584	VVA	GluR5	75	VVA
ATGLR3.4	115	VVA	Delta1	91	VVA
	298	VVA	NMDAZ1	566	VVA
	548	VVA	NMDA2A	287	VVA
	609	VVA		572	VVA
ATGLR3.6	283	VVA	NMDA2B	69	VVA
	339	VVA	NMDA2D	323	VVA
	731	VVA		593	VVA

same order, whereas Table 5 contains patterns of length of 3 AA and Table 6 contains patterns of length of 4 AA and 5 AA, respectively. It is remarkable that in all cases amino acids from group 4 (aliphatic) has major contribution in pattern forming. Aliphatic group of chemical properties is hydrophobic, and they are nonpolar too. Hence, they have major contribution in protein folding [6].

It has been observed in Table 5 that 'VVA' is a block of length 3 AA which is found in majority of primary protein sequences of Arabidopsis and six of rat. In Table 6, blocks are having lengths of 4 AA. 'FIVL', 'FLVL', 'FVVL', 'EIAK', and 'DLLK' are the five blocks which are found in some primary protein sequences of both the species.

Table 6 also shows common block 'VLFLV' of 5 AA length, which is found in AtGLR2.1 of Arabidopsis and in GluR6, GluR7, KA1, KA2, and Delta1 of rat.

Table 7 shows regions in each primary protein sequences, where the regions specified are conserved depending upon the classification of amino acids. The regions are varying from 1 AA to 2 AA long. It can be easily observed that the conserved regions majorly belong to aliphatic group of chemical properties. The regions which are conserved based on individual amino acid are shown in Table 8 along with the regions of conservations. G, I, R, F, P, W, G, and W are those amino acids. These types of conservations depict regional importance of those patterns from functional and structural point of view to make 3D protein structures of each primary protein sequence.

Table 6 Common pattern finding (length of 4 AA and 5 AA) for which same amino acids are contributed in same order as per classification

Pattern-'3444'									Pattern-'1442'						Pattern '44344'		
Name	Position	AA	Name	Position	AA	Gene Name	Position	AA	Name	Position	AA	Name	Position	AA	Name	Pos ition	AA
ATGLR2.9	11	FIVL	ATGLR2.2	650	FLVL	ATGLR3.4	698	FVVL	ATGLR1.1	719	EIAK	ATGLR1.1	88	DLLK	ATGLR2.1	863	VLFLV
ATGLR3.2	11	FIVL	ATGLR2.5	22	FLVL		618	FVVL	ATGLR1.2	752	EIAK	ATGLR1.3	100	DLLK	GluR6	553	VLFLV
ATGLR3.3	11	FIVL	KA1	11	FLVL		658	FVVL	GluR6	445	EIAK	ATGLR2.4	236	DLLK	GluR7	560	VLFLV
GluRC	839	FIVL	NMDA2A	654	FLVL	Delta2	808	FVVL	GluR57	452	EIAK	GluRC	481	DLLK	KA1	563	VLFLV
GluRD	824	FIVL							KA1	4 55	EIAK	GluR5	437	DLLK	KA2	561	VLFLV
GluR5	794	FIVL							KA2	4 53	EIAK	GluR57	246	DLLK	Delta1	56 1	VLFLV

Table 7 Regions conserved based on chemical groups of amino acids

Gene names	Group-wise conservation of amino acids																			
Positions	347	359	428	483	510	596	641	661	675	773	834	836–837	922	932	934	937	967	1234	1246	1269	1277–1278
ATGLR37	I	V	L	I	W	I	L	F	I	I	F	ID	A	D	S	Y	L	I	I	L	LY
ATGLR32	L	L	L	I	W	I	F	F	I	I	Y	ID	A	D	T	Y	M	I	I	L	LF
ATGLR34	I	L	L	V	W	V	F	F	V	I	Y	ID	T	D	T	F	M	I	I	L	LF
ATGLR35	I	L	L	V	W	V	F	F	I	I	F	ID	T	D	T	F	M	I	I	I	LF
ATGLR33	I	L	L	V	W	L	L	F	I	I	F	ID	A	D	T	Y	M	I	I	L	LF
ATGLR36	V	L	L	V	W	L	F	F	V	I	F	VD	T	D	T	Y	M	I	I	L	LF
ATGLR11	L	I	L	V	W	L	L	I	I	V	F	VE	T	D	T	Y	L	I	M	F	LF
ATGLR14	L	I	L	V	W	L	L	F	I	V	Y	ID	T	D	T	F	L	I	I	L	LF
ATGLR12	L	I	M	V	W	L	L	L	I	V	F	IE	T	D	T	Y	L	I	M	I	LF
ATGLR13	L	I	L	V	W	L	L	L	I	V	F	IE	T	D	T	F	L	I	M	I	LF
ATGLR24	V	I	L	V	W	L	L	F	I	I	F	ID	T	D	T	Y	L	i	L	L	LF
ATGLR21	M	V	L	V	W	L	L	F	I	I	F	ID	T	D	S	Y	L	I	L	L	LF
ATGLR22	L	V	L	I	W	L	L	F	I	I	F	ID	T	D	T	F	L	I	L	L	LF
ATGLR23	L	V	L	L	W	L	L	F	I	I	F	ID	T	D	T	F	L	I	L	L	LF
ATGLR25	L	I	L	I	W	L	L	L	I	I	F	ID	T	D	A	Y	L	I	I	L	LF
ATGLR26	L	I	L	I	W	I	L	L	I	I	F	ID	T	D	A	Y	L	I	I	L	LF
ATGLR27	M	I	L	V	W	L	L	L	V	V	Y	IE	T	D	T	Y	L	I		L	LF
ATGLR28	M	I	L	V	W	L	L	L	I	V	Y	ID	T	D	T	Y	L	I	I	L	LF
ATGLR29	M	I	L	V	W	L	L	L	I	V	Y	IE	T	D	T	F	L	I	I	L	LF
NMDA2A	L	L	I	L	W	L	L	F	V	I	F	ID	T	D	S	F	V	L	L	L	VF

(continued)

Table 7 (continued)

NMDA2B	L	L	I	L	W	I	L	F	I	I	F	ID	T	D	S	F	V	I	L	L	LF
DELTA2	L	L	V	V	W	L	F	F	V	V	F	ID	T	D	S	F	V	L	L	L	VF
NMDA2C	L	L	V	V	W	L	F	F	V	V	F	ID	T	D	S	F	V	L	L	L	VF
NMDA2D	L	V	I	L	W	V	L	F	V	V	F	ID	T	D	S	F	V	L	L	L	VF
NMDAZ1	V	L	L	I	W	V	F	F	I	I	F	ID	T	E	S	F	L	I	L	L	VF
GLURA	L	L	L	I	W	L	M	F	I	V	F	ID	T	D	S	Y	V	I	L	L	VF
GLURB	L	L	L	I	W	L	M	F	I	V	F	ID	T	D	T	Y	L	I	L	L	VF
GLUR6	L	I	L	V	Y	M	I	F	V	V	Y	VE	T	D	S	F	I	V	L	L	VF
GLUR7	I	I	L	I	Y	M	I	F	I	V	Y	VD	T	D	S	F	I	V	L	L	VF
KA1	L	V	M	L	Y	I	I	F	V	V	Y	VD	T	D	S	F	I	V	L	L	VF
KA2	L	I	M	V	Y	M	M	F	V	V	Y	VD	T	D	S	F	I	V	L	L	VF
DELTA1	V	I	V	I	Y	V	L	F	I	V	F	VD	T	D	S	F	V	I	L	L	IF
GLURC	V	I	I	I	Y	V	F	F	I	V	Y	LD	T	D	S	F	I	I	M	L	IF
GLURD	V	I	I	I	Y	V	F	F	V	V	Y	ID	A	D	S	F	I	I	M	L	IF
GLUR5	V	I	I	I	F	V	F	F	I	V	Y	ID	T	D	S	F	I	I	M	L	IF

Table 8 Conserved regions of dataset taken based on individual amino acid (AA)

Sr. no.	Position	AA in conserved regions
1	821	G
2	913	I
3	917	R
4	923	F
5	952	P
6	958	W
7	1205	G
8	1240	W

3.4 *Investigating Evolutionary Relationships Between Two Species*

To investigate and identify evolutionary relationships between the datasets taken, firstly, Eq. 1 has been used to derive 8X8 weight matrix for each AA sequence. As example, weight matrices for AtGLR1.1 of Arabidopsis have been stated in Table 9, whereas, Fig. 1 shows the directed graph showing the pictorial representation of it. Equation 2 mentioned in previous section is used to get weight deviation (WD) between each pair of amino acid sequences taken.

Thus, a distance matrix, say dissimilarity matrix, has been formed to analyze evolutionary relationships existing between them. Figure 2a, b is the phylogenetic trees, which are being constructed from the dissimilarity matrix derived using Eq. 2. In Fig. 2a, all the 16 primary protein sequences of rat and 19 primary sequences of Arabidopsis have participated.

Here AtGLR2.1 to AtGLR 2.9 and AtGLR3.2 to AtGLR 3.7 genes of Arabidopsis thaliana except AtGLR2.5 and AtGLR3.6 have close evolutionary relationships with GluRA and GluRB of rat, whereas AtGLR3.6 is close to NMDAZ1, NMDA2B, NMDA2C, and NMDA2D of Rat. In Fig. 2b, phylogenetic tree has been constructed

Table 9 Distance matrix of AtGLR1.1

Vertices	1	2	3	4	5	6	7	8
1	11	17	9	31	2	3	9	8
2	20	10	16	41	1	1	13	7
3	7	10	8	32	4	4	13	9
4	30	37	33	122	8	8	59	22
5	0	2	4	11	0	3	2	3
6	3	4	4	12	1	2	4	1
7	15	19	8	48	5	6	25	12
8	4	11	5	22	4	3	13	5

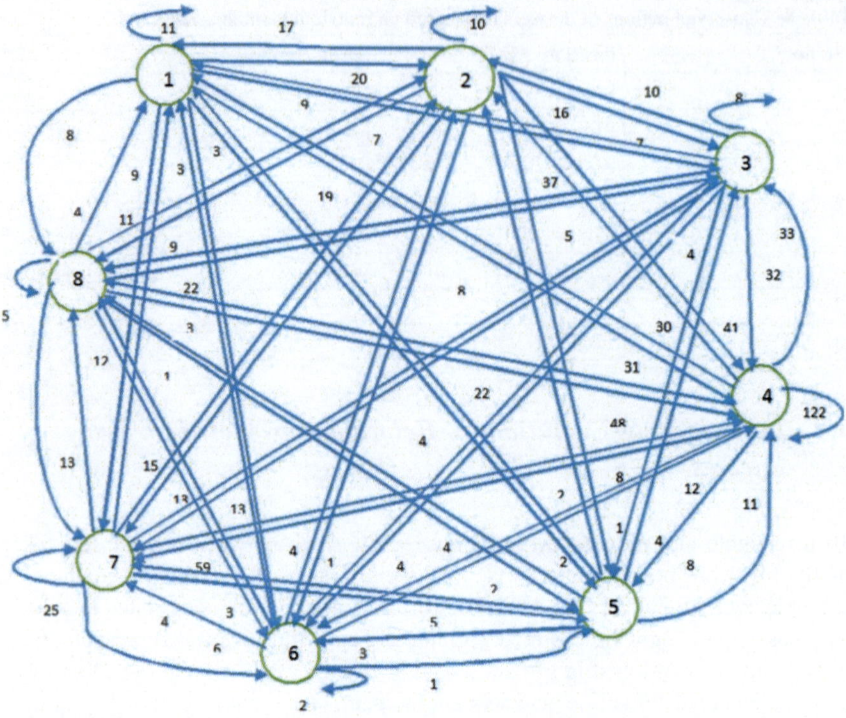

Fig. 1 Corresponding weighted directed multigraphical representation of AtGLR1. 1

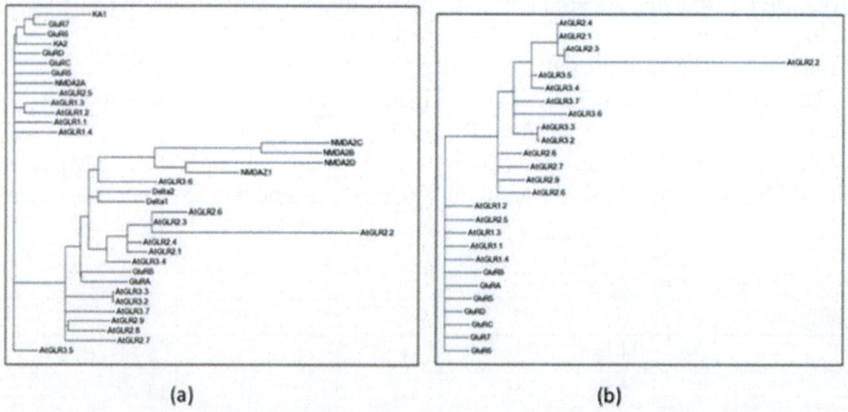

Fig. 2 **a** Phylogenetic tree constructed for total datasets taken. **b** Phylogenetic tree constructed for all the 19 AtGLR genes of Arabidopsis thaliana and 8 iGluR genes (GluRA, GluRB, GluRC, GluRD, GluR5, GluR6, GluR7, and GluR8)

with all the 19 AtGLR genes of Arabidopsis thaliana and 8 iGluR genes (GluRA, GluRB, GluRC, GluRD, GluR5, GluR6, GluR7, and GluR8) of rat where we got more finer view about evolutionary relationships between glutamate gene families of rat and Arabidopsis. They have similarities in chemical properties among them. The observation thus depicts the evolutionary relationships of ionotropic glutamate receptor (iGluR) genes of vertebrates with AtGLR genes in Arabidopsis.

4 Conclusion and Discussion

The analysis throughout the paper confirms in silico the recent biological claim that the iGluR family of the rat, a mammal, and AtGLR family of Arabidopsis, a plant, have some common functionalities in neurotransmission and have been conserved during the long process of evolution from a common ancestor. The graph theoretic approach which is applied to investigate evolutionary relationship among all the primary protein sequences taken is an alignment-free method and hence the time complexity is directly proportional to the sequence length N, that is, O(N).

Acknowledgements The authors are grateful to Professor R. L. Brahmachary, Dr. Jayanta Kr. Das, and Dr. Santanu Sen for their valuable suggestions.

References

1. Darwin, C., Darwin, F.: The Power of Movement in Plants (1880)
2. Anthony, T.: Plant Behaviour and Intelligence, 304 pp. Oxford University Press (2014). illus. $90.20 (ISBN: 0199539545)
3. Alexander, C., Erwin G.: Plant Biology: Electric Defence (2013)
4. Life Movements in Plants, vol. 1. First Published 1918, Reprinted 1985
5. Mousavi Seyed, A.R., Adeline, C., Franjois, P., Stephan, K., Farmer, E.E.: Glutamate receptor-like genes mediate leaf-to-leaf wound signaling. https://doi.org/10.1038/nature12478
6. Randic, M.: Graphical representations of DNA as 2-D map. Chem. Phys. Lett. **386**, 468–71 (2004)
7. Szpirer, C., Molne, M., Antonacci, R., Jenkins, N.A., Finelli, P., Szpirer, J., Riviere, M.M., Rocchi, M., Gilbert, D.J., Copeland, N.G.: The genes encoding the glutamate receptor subunits KA1 and KA2 (GRIK4 and GRIK5) are located on separate chromosomes in human, mouse, and rat
8. Das, J.K., Das, P., Ray, K.K., Choudhury, P.P., Jana, S.S.: Mathematical characterization of protein sequences using patterns as chemical group combinations of amino acids. PLoS One **11**(12), e0167651 (2016). https://doi.org/10.1371/journal.pone.0167651
9. Blaisdell, B.E.: A measure of the similarity of sets of sequences not requiring sequence alignment. Proc. Natl. Acad. Sci. USA **83**, 5155–9 (1986)
10. Basak, P., Maitra-Majee, S., Das J.K., Mukherjee, A., Ghosh Dastidar, S., Pal Choudhury, P., Lahiri Majumder, A.: An evolutionary analysis identifies a conserved pentapeptide stretch containing the two essential lysine residues for rice L-myo-inositol 1-phosphate synthase catalytic activity. https://doi.org/10.1371/journal.pone.0185351

11. Guo, X., Nandy, A.: Numerical characterization of DNA sequences in a 2-D graphical representation scheme of low degeneracy. Chem. Phys. Lett. **369**, 361–366 (2003)
12. Liao, B., Wang, T.: New 2D graphical representation of DNA sequences. J. Comput. Chem. **25**(11), 1364–1368 (2004)
13. Sengupta, A., Das, J., Pal Choudhury, P.: Investigating evolutionary relationships between species through the light of graph theory based on the multiplet structure of the genetic code. Electronic ISSN: 2473-3571. https://doi.org/10.1109/iacc. 2017.0175
14. Sengupta, A., Hassan, S., Pal Choudhury, C.: Article: quantitative understanding of breast cancer genes. In: IJCA Proceedings on National Conference cum Workshop on Bioinformatics and Computational Biology NCWBCB(1), 15–17 (2014)
15. Qi, Z.H., Qi, X.Q.: Novel 2D graphical representation of DNA sequence based on dual nucleotides. Chem. Phys. Lett. **440**, 139–144 (2007)
16. Das, J.K., Pal Choudhury, P.: Chemical property based sequence characterization of PpcA and its homolog proteins PpcB-E: a mathematical approach. PLOS One (2017). https://doi.org/10.1371/journal.pone.0175031
17. Sengupta, A., Pal Choudhury, P., Manners, H.N., Guzzi, P.H., Roy, S.: Chemical characterization of interacting genes in few subnetworks of alzheimer's disease. https://doi.org/10.1109/bibm.2018.8621335

The Estimation of Inter-Channel Phase Synchronization of EEG Signals in Patients with Traumatic Brain Injury Before and Post the Rehabilitation

Renata A. Tolmacheva, Yury V. Obukhov, and Ludmila A. Zhavoronkova

Abstract The identical inter-channel phase coherency of electroencephalogram (EEG) signals is determined for control volunteers during cognitive and motor tests. EEG signal phase is evaluated at the points of ridge of their wavelet spectrogram. Inter-channel EEG phase coherency for patients with traumatic brain injury (TBI) is represented. Phase-connective pairs of EEG channels obtained from the results of EEG records of patients with TBI before and post the rehabilitation are considered.

Keywords EEG · Wavelet spectrogram · Ridges · Phase synchronization

1 Introduction

In neurophysiology, coherency $Coh_{xy}(f)$ is determined by the normalized complex cross- correlation $C_{xy}(f)$ of signals $x(t)$ and $y(t)$ [1–3]:

$$Coh_{xy}(f) = \langle | C_{xy}(f) | \rangle, C_{xy}(f) = \frac{S_{xy}(f)}{\left(S_{xx}(f)S_{yy}(f)\right)^{1/2}}. \tag{1}$$

In coherency, $Coh_{xy}(f)$ is averaged in different time intervals and in a certain frequency interval that is determined using neurophysiological data. Normally, such intervals correspond to the delta (2–4 Hz), theta (4–8 Hz), and alpha (8–12 Hz)

R. A. Tolmacheva (✉) · Y. V. Obukhov
Kotel'nikov Institute of Radio Engineering and Electronics of RAS, Russian Federation, 125009 Moscow, Russia
e-mail: tolmatcheva@ya.ru

Y. V. Obukhov
e-mail: yuvobukhov@mail.ru

L. A. Zhavoronkova
Institute of Higher Nervous Activity and Neurophysiology of RAS, Russian Federation, 117485 Moscow, Russia
e-mail: lzhavoronkova@hotmail.com

D. Bhattacharjee et al. (eds.), *Proceedings of International Conference on Frontiers in Computing and Systems*, Advances in Intelligent Systems and Computing 1255,
https://doi.org/10.1007/978-981-15-7834-2_48

EEG rhythms. Such averages and the presence of the threshold level of coherence that is used to select the phase-connective pairs of signals are disadvantages of the coherent analysis that lead to instability in the determination of the inter-channel phase synchronization of EEG signals. These disadvantages are considered in detail in [4]. The validity of the coherent analysis of essentially non-stationary EEG signals is questioned [2, 4]. Analytical signal $x^*(t) = x(t) + iH(x(t))$ also can be used for the assessment of phase connectivity, where $H(x(t))$ is the Hilbert transform [5]. The phase synchronization of two signals takes place when [6]:

$$\left|\Phi_{x,y}(t)\right| = \left|n\Phi_x(t) - m\Phi_y(t)\right| \le const, \tag{2}$$

where Φ_x, Φ_y are phases of the signals $x(t)$ and $y(t)$, and n, m are integers. We will use below $n = m = 1$. We proposed an approach to the assessment of the inter-channel phase connectivity of the EEG based on calculation and comparison of phases of signals at the points of the ridges of their wavelet spectrograms [7]. Under this approach, phase-connective pairs of EEG signals for the group of control volunteers and for the group of patients with TBI during two cognitive and motor tests are considered. The definition of phase-connective pairs of EEG signals can be useful for analyzing EEG records of people with TBI before and after rehabilitation. This can determine the positive or negative dynamics of treatment.

2 Methods

The ridge of the signal Morlet wavelet spectrum (4) is defined as the points of the stationary phase [8] that is in them $\frac{d\Phi}{dt} = \omega$. In [7, 9], it was shown that for the time-asymptotic signal $x(t) = A(t)\exp(i\Phi(t))$ under condition (3) the amplitude and phase of the signal are corresponding to (5):

$$\left|\frac{d\Phi(t)}{dt}\right| >> \left|\frac{1}{A(t)}\frac{dA(t)}{dt}\right|, \left|\frac{1}{A(t)}\frac{dA(t)}{dt}\right| << \left|\frac{1}{\psi(t)}\frac{d\psi(t)}{dt}\right|, \tag{3}$$

$$W(\tau, f) = \sqrt{f}\int x(t)\psi^*\left((t-\tau)f\right)dt, \psi(\eta) = \frac{1}{\sqrt{\pi F_b}}e^{2\pi F_c\eta}e^{-\frac{\eta^2}{F_b}}, \tag{4}$$

where usually $F_b = F_c = 1$ is used.

$$A(t) \approx \max|W(t, f)| \equiv |W(t, f_r)| \text{ and} \Phi(t) \approx \arctan\left(\frac{\mathrm{Im}W(t, f_r)}{\mathrm{Re}W(t, f_r)}\right). \tag{5}$$

In this case, the non-stationary phase points of wavelet spectra are not included in the assessment of the phase of signals. When the ridge points do not satisfy the condition for asymptotic ridges (3), we obtain errors in the calculations of the phase difference. However, we assume that such an error is significantly less than the errors related to the averaging of the phase in (1). The international system of the location of scalp electrodes ≪ 10 − 20% ≫ is used to the EEG record. The records of 19-channel EEG were analyzed for the group 8 control volunteers and for patients with TBI. These are 5 patients with TBI without rehabilitation and 3 patients with TBI, who had EEG records before and after rehabilitation. Figure 1 shows an example of the power spectral density of Morlet wavelet spectrograms in a pair of EEG channels with selected points of the maximum spectral power density (ridge). Figure 2 shows the dependences of phases on time of two EEG signals $x(t)$ and $y(t)$ at the points of the ridges t_k of their wavelet spectrograms. Then, we plot the histograms of values of the phase difference of two signals $x(t)$ and $y(t)$ as in [6]. Figure 3 shows normalized histograms $\rho_{x,y} = n_{x,y}/N$, where $n_{x,y}$ is number of reference points of ridges with the discretization of the histogram $\left|\Delta\Phi_{x,y}(t_k)\right| < 0.05\pi$, N is a total number of EEG signal reference points during the test. The first pair of channels Fp1-Fp2 (see Fig. 3a) can correspond to a phase-connective pair, because the histogram has sharp pick with an excellent signal-to-noise ratio. Another pair Fp1-O2 (see Fig. 3b) can correspond to a phase-unconnected pair. Figure 3a shows that $\rho_{x,y} < 0.1$ can be considered as a background. We consider the threshold $\rho_{x,y}^{thr} = 0.1$ and we will assume that the values of the portions of the ridge points above it will correspond to the phase-connective pairs of channels. Note that the phase difference of two EEG channels (e.g., Fp1–Fp2) can be used to characterize the phase-locking dynamics. Distributions graphs of the maximum values of portions max $\rho_{x,y}$ over pairs of EEG channels, based on the histograms obtained (see Fig. 3), were arranged in the ascending order max $\rho_{x,y}$ for

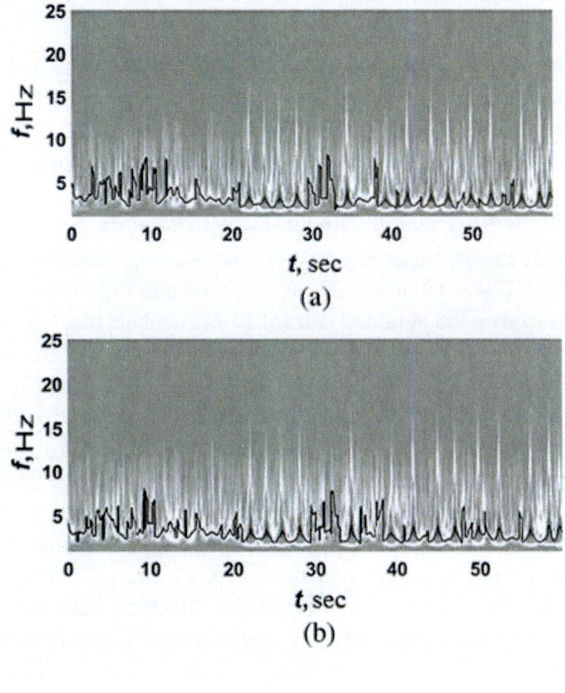

Fig. 1 Morlet wavelet spectrograms of the signal for a pair of the EEG channels with the selected ridge: **a** first channel; **b** second channel

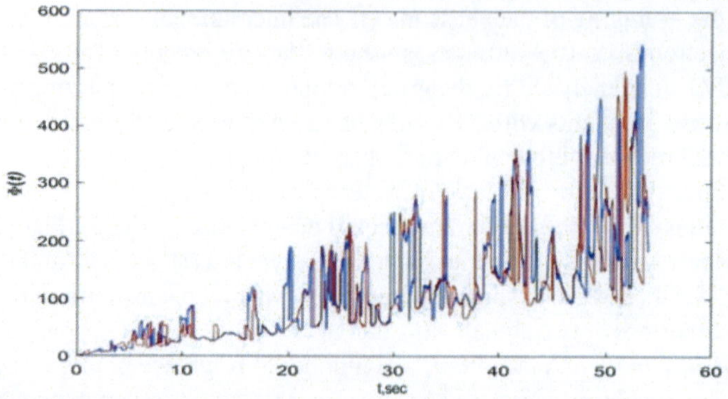

Fig. 2 Dependences of phases on time of two EEG channels. Red line is the phase of EEG signal of FP1 channel. Blue line is the phase of EEG signal of FP2 channel

EEG records without tests, with the cognitive calculation-logical test (CT1), with the cognitive spatial-pattern test (CT2), and with the motor test (MT1) are represented in Fig. 4. The records of 19-channel EEG were analyzed, and therefore the number of pairs of channels is 171. In Fig. 4, a certain abscissa corresponds to a pair of the EEG channels.

3 Results

By the new approach of assessment, the inter-channel EEG and the phase synchronization during two cognitive tests (CT1 and CT2) and MT1 test for the group 8 control volunteers and for the group patients with TBI are considered [10]. From the dependences of Fig. 4, it can be distinguished phase-connective pairs of the EEG channels that occur only during the test, by comparing the phase-connective pairs of the EEG record without the test and at the time of the EEG records during one of the tests. Phase-connective pairs that are absent in EEG records without tests and that appear only during tests are shown in Fig. 5 for the group of control volunteers. Based on the obtained pairs of EEG channels (see Fig. 4), it is possible to find phase-connective pairs of the EEG channels that occur in a group of 8 control volunteers during the CT1 test, the CT2 test, and MT1 (see Fig. 5). Figure 5 shows that the greater number of the phase-connective pairs of the EEG channels that occur in the group of control volunteers during the CT1 test is in the left hemisphere. The greater number of the phase-connective pairs of the EEG channels that occur in the group of control volunteers during the CT2 test is in the right hemisphere. In accordance with published works, the prefrontal regions of the left and right hemispheres are predominantly activated with the CT1 and CT2 tests, respectively, for control volunteers [11]. Our method confirms this. Phase-connective pairs of EEG channels for each patient are represented by a certain color. Patients with TBI have not phase-

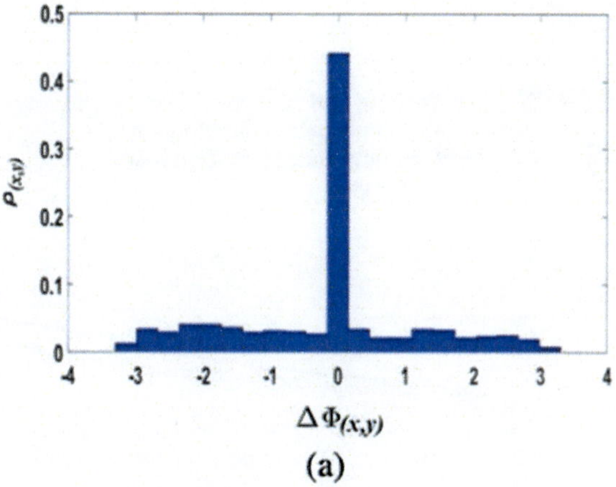

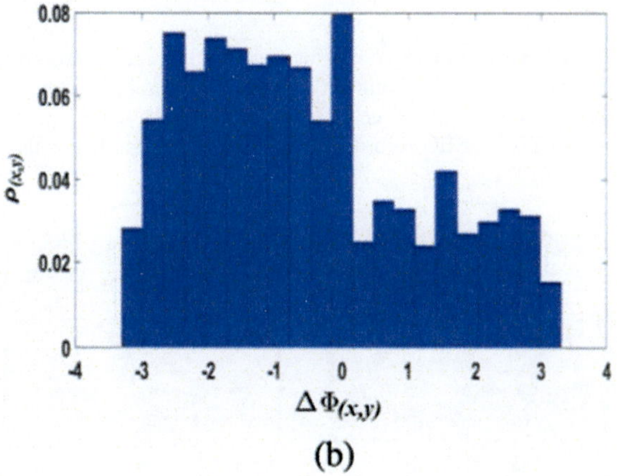

Fig. 3 Histograms of portions $\rho_{x,y}$ for the phase difference at ridge points of their wavelet spectrograms for two EEG channels: **a** phase-connective pair of EEG channels; **b** unconnected pair of EEG channels

connective pairs that occur in most people in the group of control. The activation of the left hemisphere during of the CT1 test is absent by patients with TBI (see Fig. 6b). Also, the activation of the right hemisphere during the CT2 test is absent by patients with TBI (see Fig. 6c).

Next, we consider phase-connective pairs of EEG channels obtained from the results of EEG records of three patients with TBI who had repeated EEG records. The first investigation was performed about 1–4 months after the injury, after the initial treatment. The second investigation was performed after the rehabilitation. The rehabilitation is the training using dual tasks (cognitive and motor), different from the test tasks, which are used in research during EEG records. The rehabilitation was performed during 1–2 months for 40–45 min for two times a week by two patients with TBI. Patients had to perform 2 tasks simultaneously. The 2nd investigation of the patient № 1 was performed after 1 month the rehabilitation. The number of phase-connective pairs of EEG channels, which occur only during the MT1 (see Fig. 7a),

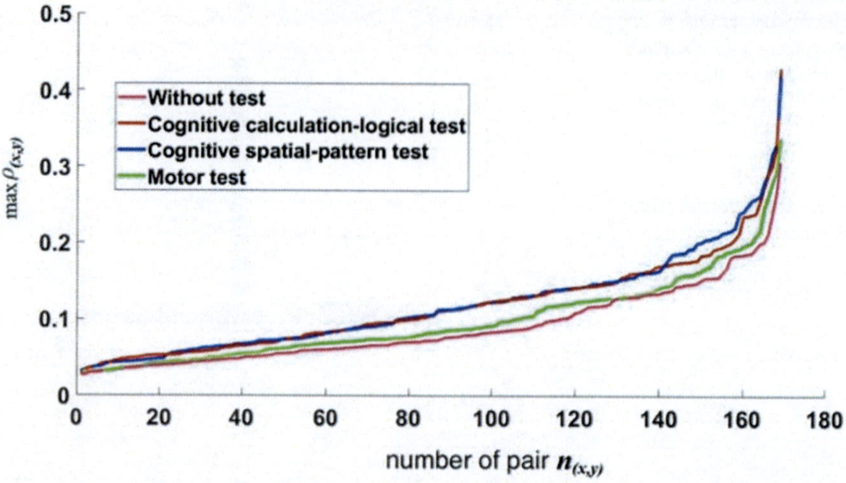

Fig. 4 Distributions graphs of the maximum values of portions max $\rho_{x,y}$ over pairs of EEG channels arranged in the ascending order max $\rho_{x,y}$. Numbers of pairs EEG channels are plotted on the abscissa. Purple line is the EEG record without test; red line is the EEG record with the CT1 test; blue line is the EEG record with the CT2 test; green line is the EEG record with the MT1

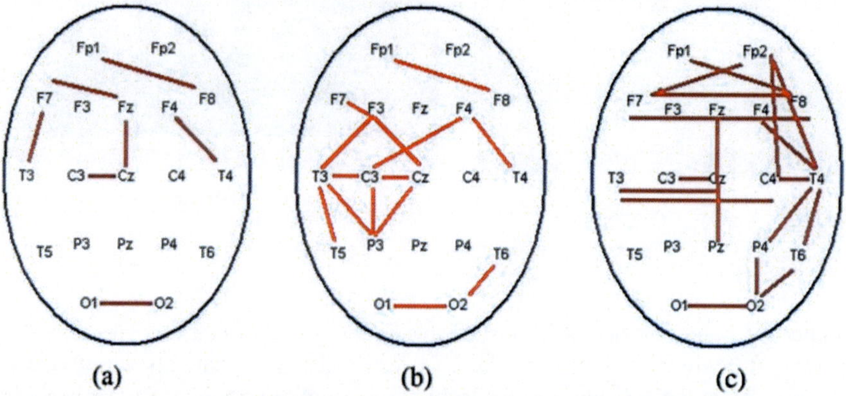

Fig. 5 Phase-connective pairs that occur in the group of 8 control volunteers during CT1, CT2, and MT1 tests. **a** MT1; **b** CT1; **c** CT2

in the second investigation compared with the first investigation was decreased. Phase-connective pairs are less compared to the control (see Fig. 5a). The number of phase-connective pairs of EEG channels, which occur only during the CT1 test (see Fig. 7b), in the second investigation compared with the first investigation was increased. The number of inter-hemispheric phase-connective pairs of EEG channels was increased. More phase-connective pairs appeared in the left hemisphere. These results do not contradict the results of the control during the CT1 test (see Fig. 5b). The number of phase-connective pairs of EEG channels, which occur only during the

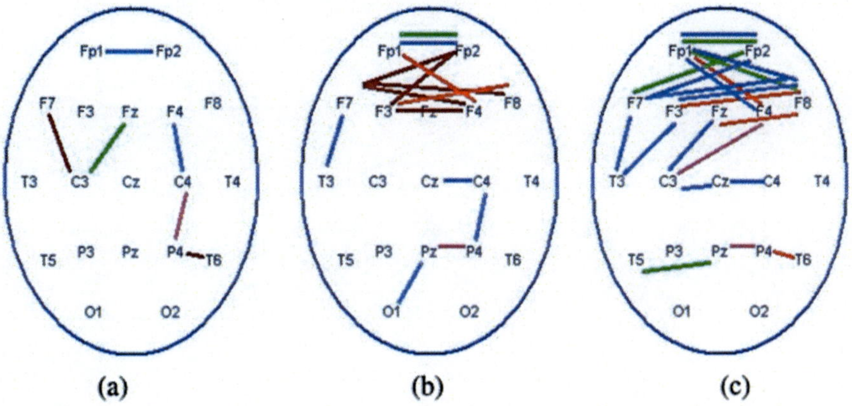

Fig. 6 Phase-connective pairs that occur for patients with TBI during CT1, CT2, and MT1 tests. **a** MT1; **b** CT1; **c** CT2. Phase-connective pairs of EEG channels for each patient are shown by certain color

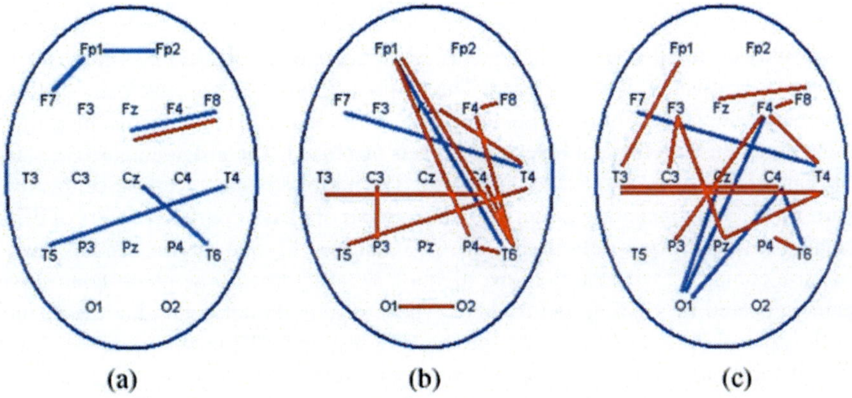

Fig. 7 Phase-connective pairs of EEG channels that occur during MT1 (**a**), CT1 test (**b**), and CT2 test (**c**) in a patient № 1 with TBI. Blue lines are phase-connective pairs of EEG channels after the first investigation. Red lines are phase-connective pairs of EEG channels after the second investigation

CT2 test (see Fig. 7c), in the second investigation compared with the first investigation was increased. The inter-hemispheric phase-connective pairs of EEG channels were observed. The phase-connective pairs of channels appeared in the right hemisphere. These results do not contradict the results of the control during the CT2 test (see Fig. 5c). After the rehabilitation on the basis of EEG record during the MT1 there is a negative dynamics in motor function, during the CT1 and CT2 tests there is a positive dynamics in cognitive function.

The second investigation of the patient № 2 was performed after 2 months after the rehabilitation. The number of phase-connective pairs of EEG channels, which

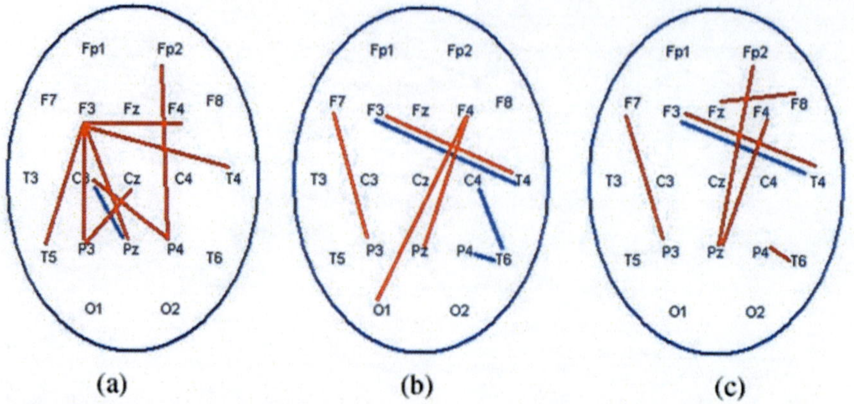

Fig. 8 Phase-connective pairs of EEG channels that occur during MT1 (**a**), CT1 test (**b**) and CT2 test (**c**) in a patient № 2 with TBI. Blue lines are phase-connective pairs of EEG channels after the first investigation. Red lines are phase-connective pairs of EEG channels after the second investigation

occur only during the MT1 (see Fig. 8a), in the second investigation compared with the first investigation was increased. The number of phase-connective pairs of EEG channels, which occur only during the CT1 test (see Fig. 8b), in the second investigation compared with the first investigation was increased. The phase-connective pairs of channels appeared in the left hemisphere. The inter-hemispheric phase-connective pairs of EEG channels were observed. The number of phase-connective pairs of EEG channels, which occur only during the CT2 test (see Fig. 8c), in the second investigation compared with the first investigation was increased. The phase-connective pairs appeared in the right hemisphere. These results do not contradict the results of the control during the MT1, during the CT1 test and CT2 test (see Fig. 5). After the rehabilitation on the basis of re-EEG record during the MT1, there is a positive dynamics of a motor function, during the CT1 test and during the CT2 test, there is a positive dynamics in a cognitive function.

The patient № 3 had the dynamic investigation without the rehabilitation. The second investigation was performed after 1,5 5 month. The number of phase-connective pairs of EEG channels, which occur only during the MT1 (see Fig. 9a), in the second investigation compared with the first investigation was increased. These results do not contradict the results of the control during the MT1 test (see Fig. 5a). The number of phase-connective pairs of EEG channels, which occur only during the CT1 test (see Fig. 9b), in the second investigation compared with the first investigation was decreased. The number of phase-connective pairs of EEG channels, which occur during the CT2 test (see Fig. 9c), in the second investigation compared with the first investigation was increased. The number of inter-hemispheric phase-connective pairs

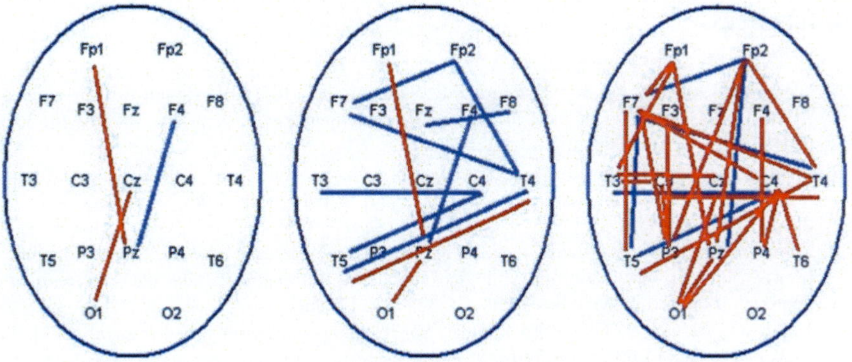

(a) (b) (c)

Fig. 9 Phase-connective pairs of EEG channels that occur during MT1 (**a**), CT1 test (**b**) and CT2 test (**c**) in a patient № 3 with TBI. Blue lines are phase-connective pairs of EEG channels after the first investigation. Red lines are phase-connective pairs of EEG channels after the second investigation

of EEG channels was increased. More phase-connective pairs of channels appeared in the right hemisphere, the frontal region became involved too. These results do not contradict the results of the control during CT2 test (see Fig. 5c). After the rehabilitation on the basis of re-EEG record during the MT1 there is a positive dynamics of a motor function, during CT2 test there is a positive dynamics in a cognitive function, during the CT1 there is a negative dynamics.

4 Conclusions

The new approach to the assessment of the inter-channel phase connectivity of EEG signals based on the calculation and comparison of the phases at points of the ridge their wavelet spectrograms is proposed. Phase-connective pairs of EEG signals were determined for 8 control volunteers during motor and cognitive tests. Areas of the left hemisphere are predominantly activated with the calculation-logical cognitive test. Areas of the right hemisphere are predominantly activated with the cognitive spatial-pattern test. Patients with TBI have not phase-connective pairs that appear in most people in the group of control. The definition of phase-connective pairs of EEG signals can be useful for analyzing EEG records of patients with TBI before and after rehabilitation. This can determine the positive or negative dynamics of treatment. This is necessary for monitoring the treatment of patients with TBI.

Acknowledgements The work was carried out within the framework of the state task and partially was supported by the Russian Foundation for Basic Research, project No. 18-07-00609.

References

1. Wending, F., Ansari-Asl, K., Bartolomei, F., Senhadji, L.: From EEG signals to brain connectivity: a model based evaluation of interdependence measures. J. Neurosci. Methods **183**(1), 9–18 (2009)
2. Nolte, G., Bai, O., Wheaton, L., Mari, Z., Vorbach, S., Hallett, M.: Identifying true brain interaction from eeg data using the imaginary part of coherency. Clin. Neurophys. **115**(10), 2292–2307 (2004)
3. Zhan, Y., Halliday, D., Jiang, P., Liu, X., Feng, J.: Detecting time-dependent coherence between non stationary electrophysiological signals. A combined statistical and time-frequency approach. J. Neurosci. Methods **156**(1–2), 322–332 (2006)
4. Kulaichev, A.: (2009) Ob informativnosti kogerentnogo analiza v issledovaniyah EEG. Zh Vyssh Nerv Deiat I P Pavlova **59**(6), 757–767
5. Rosenblum, M., Pikovsky, A., Kurths, J., Schafer, C., Tass, P.: Handbook of Biological Physics, vol. 4. In: Phase Synchronization: From Theory to Data Analysis, pp. 279–321 (2001)
6. Tass, P., Rosenblum, M., Weule, J., Kurths, J., Pikovsky, A., Volkmann, J., Schnitzler, A., Freund, H.-J.: Detection of n:m phase locking from noisy data: application to magnetoencephalography. Phys. Rev. Lett. **81**(15), 3291–3294 (1998)
7. Tolmacheva, R.A., Obukhov, YuV., Polupanov, A.F., Zhavoronkova, L.A.: New approach to estimation of interchannel phase coupling of electroencephalograms. J. Commun. Technol. Electron. **63**(9), 1070–1075 (2018)
8. Mallat, S.: A Wavelet Tour of Signal Processing. The Sparse Way, Academic, Amsterdam (2009)
9. Guilleemain, P., Kronland-martinet, R.: Characterization of Acoustic Signals Through Continuous Linear Time-Frequency Representations. Proc. IEEE **84**, 561–585 (1996)
10. Obukhov, Y.V., Tolmacheva, R.A., Polupanov, A.F., Zhavoronkova, L.A.: On an approach to assessing the inter-channel phase synchronization of electroencephalogram signals. In: Proceedings of ITNT-2018: IV International Conference and Youth School Information Technologies and Nanotechnologies (ITNT 2018, Samara, Russia, April 24–27, 2018), CEUR, Samara, pp. 84–88
11. Zhavoronkova, L., Shevtsova, T., Maksakova, O.: How Does Human Brain Simultaneously Solve Two Problems?. LAP LAMBERT Academic, Saarbryuken (2017)

Identification of Differentially Expressed Genes Using Deep Learning in Bioinformatics

Srirupa Dasgupta, Supriti Mondal, Abhinanadan Khan, Rajat Kumar Pal, and Goutam Saha

Abstract Bioinformatics data can be used for the ultimate prediction of diseases in different organisms. The microarray technology is a special form of 2D representation of genomic data characterized by an enormous number of genes across a handful of samples. The actual analysis of this data involves extraction or selection of the relevant genes from this vast amount of irrelevant and redundant data. These genes can be further used to predict classes of unknown samples. In this work, we have implemented two popular deep learning segmentation architectures, namely, SegNet and U-Net. These techniques have been applied to the microarray dataset of colon cancer (typically containing tumour and normal tissue samples) to extract the culprit/responsible gene. The performance of the reduced set formed from these genes has been compared across different classifiers using different existing methods of feature selection. It is found that both deep learning based approaches outperform the other methods. Lastly, the biological significance of the genes has also been verified using ontological tools, and the results are significant.

Keywords Deep learning · Gene expression · Microarray data · SegNet · U-Net

S. Dasgupta (✉)
Government College of Engineering and Leather Technology, LB-11, Sector-III, Salt Lake, Kolkata 700106, India
e-mail: srirupa.dasguptam@gmail.com

S. Mondal · A. Khan · R. K. Pal
University of Calcutta, Acharya Prafulla Chandra Roy Shiksha Prangan, JD-2, Sector-III, Salt Lake, Kolkata 700106, India
e-mail: mondaltubai5@gmail.com

A. Khan
e-mail: khan.abhinandan@gmail.com

R. K. Pal
e-mail: pal.rajatk@gmail.com

G. Saha
North Eastern Hill University, Umshing Mawkynroh, Shillong 793022, Meghalaya, India
e-mail: dr.goutamsaha@gmail.com

D. Bhattacharjee et al. (eds.), *Proceedings of International Conference on Frontiers in Computing and Systems*, Advances in Intelligent Systems and Computing 1255,
https://doi.org/10.1007/978-981-15-7834-2_49

1 Introduction

In the twentieth century, scientists came up with many ways to study genes such as analysing the proteins they encode. Human beings have around 20,000 genes. It would take a very long time to investigate each human gene, one at a time. DNA microarray technology [4] allows us to study thousands of gene in an organism all at once. It will enable us to investigate the levels of gene expression under various experimental conditions. A single experiment provides us with information regarding the functional relationships between the cellular processes of biological organisms. Cancer identification by analysing microarray dataset is a very delicate, time-consuming, and non-scalable task. Cancer starts when cells in a particular organ begin to grow out of control. These cells usually form a tumour that can often be seen on an X-ray or felt as a lump. Identification of the genes responsible helps in early diagnosis of cancer and increases the chances of survival, significantly. A tumour is considered to be malignant if the cells can grow into surrounding tissues or spread to distant areas of the body.

1.1 Analogy Between the Human Brain and Neural Networks

In the human brain, millions of neurons, that are connected amongst each other via axons, form a network in a hierarchy. Synapses pass electrical signals between layers. This is how the human brain learns things. Whenever we see, hear, feel, or think about something, a synapse (an electrical impulse) is fired from one neuron to another in a networked pattern, which causes us to understand, interpret, and memorize phenomena. It extracts patterns by exploiting the unknown structure in the input data at multiple layers. Deep learning models have proven successful in modelling the hierarchical structures of system biology and predicting bio-molecule interaction. Driven by the power of computational capacity and improved algorithms, enormous amounts of data generated in the biological field can now be handled very efficiently.

1.2 Shallow Neural Nets to Deep Neural Nets

The neural network is basically a function that is used for mapping [3, 9]. Each subnetwork is also a function that does the same thing containing components of a transformation $w * x$ (linear) and a transformation $\delta(z)$ (non-linear). By including several function blocks into a single layer and forming a stack of such layers, we obtain a network, which expresses a non-linear mapping function with a complexity that is relatively high.

Deep learning can successfully handle sequential data [6, 15]. When trained with backpropagation, stochastic gradient descent, and perceptron, deep convolution neural networks can detect underlying features in sequential data. Convolutional neural networks having 1D filters are used to handle such data. Deep learning works wonders in handling 2D data and data, which is similar to a tensor, like images (biomedical) and gene expression profiles. Using consecutive layers of convolutional and pooling, such networks can efficiently extract the underlying relationships in the original data in several ways and correlate the result to an intrinsically obtained concealed area, where the detailed representation is not only meaningful but also fit for supervised learning. It handles raw data very efficiently because it carries out both extraction of features, as well as classification in a wholesome manner. This brings out the significant high-level features smoothly with careful tuning of the hyper-parameters that are involved.

1.3 Model Parameters

In a neural network, many factors can influence model performance. Hidden layers in the network that express the complex relationship are an important factor. More the number of nodes included in a neural network, the higher is the complexity in the model. If there are few nodes, the model would be too plain to bring out the elaborate relation amongst the input and the output, which stands out as *under-fitting*. Using a huge number of nodes, the model turns out to be so complicated that even little noise present in the input will be exaggerated in the output, which stands out as *over-fitting*. We need to find a method which optimizes the number of hidden layers. In the training process, we have to select efficiently the loss function, optimizer, and learning rate to be used. The widely used functions for calculating loss are cross-entropy [7] for classification. The optimizer includes stochastic gradient descent (SGD) [14], Adam [5], etc.

1.4 Deep Learning Segmentation Architectures

In our experiments, we have explored two popular segmentation architectures in deep learning: (i) SegNet [2] and (ii) U-Net [12] on microarray datasets. SegNet [2] is generally used for semantic segmentation but has not been used much on microarray datasets. SegNet is designed to be an architecture for pixel-wise semantic segmentation. Training parameters such as rate of learning, size of the filter, filter counts, and epochs counts are varied to observe the effects on performance. Such networks are faster than fully connected convolution neural networks [16] as only the pooling indices are passed to the up-sampling layers which are relatively smaller in number.

The other model, U-Net [12], was initially published for biomedical segmentation. The utility of the network is its capacity to learn from very little data. U-Net obtains both coarse and fine information by using a multi-scale method which skips connections at required segments in the layers of deconvolutional. U-Net is slower because its up-sampling involves learning, and hence it requires more parameters for the purpose and is more time-consuming to train than SegNet. However, SegNet does not bring out multi-scale information as efficiently as the U-Net. These are the strengths and weaknesses of the two models. It performs better classification than other popular machine learning algorithms.

2 Background

The underlying information needs to be extracted from biological datasets to understand the correlation between input features, such as genes, explicitly. The convolution neural network (CNN) [16] has two specialties: *local connectivity* and *weight sharing*. The data vector goes through several hidden layers (i.e. layers of convolution and layers of pooling), at the end of which the vector length decreases whereas the channel number increases. Here, different channels personify the input sequence on aspects that are different. This pooling layer helps the network to extract not only higher level but also wide-range of properties from the vector at the input. With these two layers, this architecture captures the input properties in different aspects, including both motifs that are local and interactions which are long-ranged, with relatively high complexity. Here, in this work, we have trained two encoder–decoder convolutional architectures.

2.1 SegNet

In SegNet architecture, there is no previously trained model. It is simply built by stacking one layer after another. This work is focused on analysing the underlying SegNet architecture (Fig. 1) by varying the parameters used to train.

Each layer is represented using layer parameters, such as the size of the input, size of the filter, number of filters, etc. In the encoder, the convolutional (CONV) layer is followed by the rectified linear unit (ReLU) [10] and pooling (Pool) segments. Output feature map from the pooling layer is down-sampled by a factor that is preferred. After going through many convolution-ReLU-pooling layers, the length of the input is reduced to a large extent. To obtain the original length of the input, decoders are used for unpooling. For every decoding unit, there is an associated encoding unit. Decoders are applied one after another for up-sampling. The resized volume is finally provided to the Softmax [13] layer for classification from the convolution layer having a filter size of 1×1.

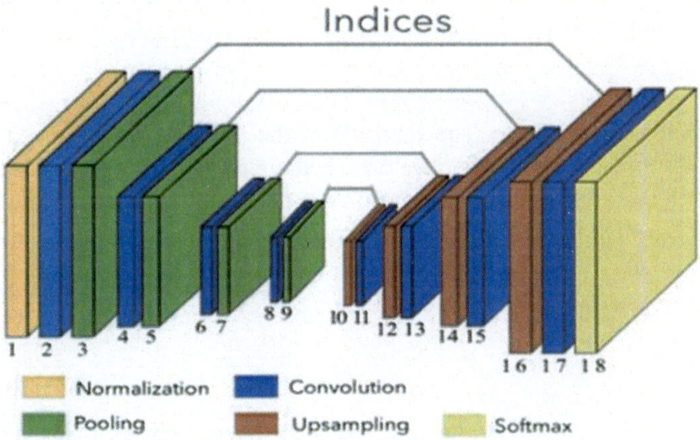

Fig. 1 The SegNet architecture

Encoders and decoders are joined in such a way so as to share a common feature map size and indices used for pooling. There are two types of SegNet obtained using two- or four-encoder architecture. In this work, we have used a two-encoder architecture. In the architecture that uses two encoders, there are eight layers of convolution, two layers of pooling, and two unpool layers. This architecture using four-encoders has 16 number of convolution layers, 4 number of pooling layers, and 4 unpool layers. At the convolution layer, the filter has been moved all over the input. The feature obtained from the output of the neuron is given by

$$O_j = f\left(\sum W_i x_i + b\right), \tag{1}$$

where x_i are the values of pixel in filter area; w_i are the filter weights; b is a bias; O_j is the output from the neuron; and f is the activation function. Options of training, like momentum, has been set to a value of 0.9 as the default value and the learning rate initially has been set to 0.01. Optimisation algorithm and the cost function are chosen as SGD and cross-entropy has been set as the loss function. The cost function has been calculated using

$$C(w) = \frac{1}{n}\sum_{i=1}^{n} C_p(w), \tag{2}$$

where p is the output class number. The cost function has been evaluated in the very last level. It measures the difference between the original and the observed data label obtained in the training phase. The difference has been back propagated, and the weights in the further layers have been readjusted to bring down this value to zero. The weights have been updated by

$$w' = w - \eta \nabla C(w) = w - \eta \sum_{i=1}^{n} \nabla C_p(w)/n, \quad (3)$$

where η is the learning rate. The schedule of the learning rate has been set to be *adaptive*. As a result of this adaptive nature, the rate of learning has been changed after a scheduled time with a preset factor. All the samples have been shuffled after every epoch, which overcomes bias in the process of training the architecture.

2.2 U-Net

The model of the U-Net architecture has a U-shape with attributes of an input dataset learned using different levels using a series of consecutive convolutional layer and pooling layers. The feature maps passed through deconvolutional layers (up-sampling)to obtain the position of the genes in the microarray dataset, which are responsible for a disease. To produce coarse as well as fine information, feature maps of same size input obtained from down-sampling layers as well as up-sampling layers are joined during up-sampling.

The basic network structure consists of two parts: convolutional encoding unit and decoding unit. In both parts of the network, the basic convolution operations are performed followed by the ReLU [10] activation. For down-sampling in the encoder, 2×2 max-pooling operations are performed. Deconvolution or convolution transpose was done in the decoding phase to up-sample the feature maps. The early version of U-Net was used to crop and copy feature maps from encoder to decoder units. Although the number of samples provided is very few, yet high performance is obtained using the model. It has been shown in Fig. 2.

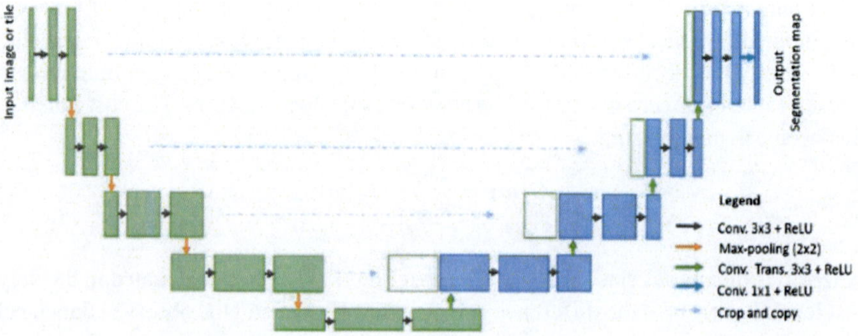

Fig. 2 The U-Net architecture

3 Methodology

3.1 Preprocessing Using Min-Max Normalization

Min-Max is a technique that helps to normalize the data that scales the data within [0, 1]. This normalization helps the model to understand the data easily.

$$v' = \frac{v - v_{\min}}{v_{\max} - v_{\min}}, \tag{4}$$

where v is the respective value of the attribute. Before applying the SegNet and U-Net models on the microarray datasets, we have normalized the data using Eq. (4). From both networks, the last layer feature maps have been extracted and stored in a NumPy array. From each feature map, we get the gene location, which helps us to find the gene ID. Then, we can check the gene ID in the DAVID [8] for the biological significance of the identified genes.

3.2 Performance Evaluation

To evaluate the performance, the accuracy of the result has been calculated according to the following relation:

$$\text{accuracy } (ACC) = \frac{\text{correctly predicted data}}{\text{total testing data}} \tag{5}$$

3.3 Frameworks

Several deep learning frameworks exist and are being continuously upgraded and maintained, such as Tensorflow, Keras, PyTorch, etc. [11]. It is very convenient and user-friendly but does not incorporate the adaptiveness provided by Tensorflow, which can be used to control every single detail. After preprocessing, we have converted our dataset into a tensor, using the Torch package. Torch is a tensor library like NumPy, which uses GPU (graphical processing unit) support. Tensor is a matrix-like data structure, which is an unavoidable component used in deep learning libraries for efficient computation. GPUs are used primarily for calculating operations that involve tensors, and this has spurred the tremendous impetus in deep learning capability in recent times. "torch.nn" modules have used for neural network functions like convolutional layers, pooling layers, up-sample layers, etc.

4 Results and Discussion

The results for the colon dataset used have been presented below. The colon cancer dataset obtained from [1] contains 2000 genes across 62 samples, belonging to two classes. It has been trained using 80% of the samples. Initially using SegNet, the number of hidden layers and the number of iterations have been varied and, as high as, 69% accuracy has been obtained. The corresponding p-value of the genes reported for this case was very significant and of the order of minus seven, as shown in Table 1.

In the second approach using SegNet, the different parameters like the number of layers, kernel sizes, epochs count or iterations, filter numbers per convolution, and the initial rate of learning have been considered. Varying these five parameters, four different cases have been produced, which has been presented in Table 2. The performance of SegNet on different indices has been given in Table 3. However, when trained with U-Net using the same combination, the highest accuracy of 100% has been obtained within 100 iterations as shown in Table 4 for an iteration of 100 and kernel size of five. As the performance was the highest, no further investigations have been done based on the U-Net framework.

Table 1 Accuracy and p-value reported using SegNet

No. of iterations	No. of hidden layers	Accuracy	p-value
100	4	0.44	$1.3e^{-2}$
50	19	0.20	$1.2e^{-3}$
100	19	0.69	$6.3e^{-7}$

Table 2 The different training parameters considered

Parameters	Case 1	Case 2	Case 3	Case 4
No. of layers	8	8	8	8
Filter size	3	5	3	5
No. of epochs	20	20	50	50
No. of filters per convolution	1, 64, 128, 256, 512			
Initial learning rate	0.001			

Table 3 Evaluation results for the test dataset for all the cases

Parameters	Case 1	Case 2	Case 3	Case 4
Highest accuracy	0.69	0.64	0.66	0.69
Mean accuracy	0.69	0.61	0.62	0.68
Time elapsed	21 min 50 s	22 min 04 s	55 min 34 s	55 min 01 s

Table 4 Accuracy and p-value reported using U-Net

No. of iterations	No. of hidden layers	Accuracy	p-value
50	19	0.8	$1.4e^{-4}$
100	19	1.0	$2.0e^{-5}$

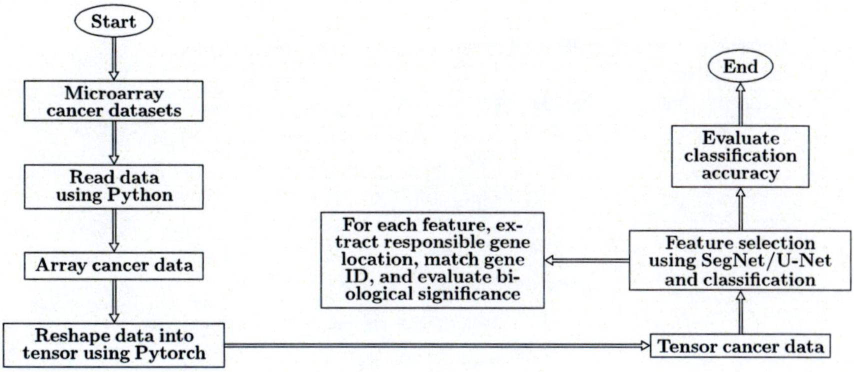

Fig. 3 Proposed methodology

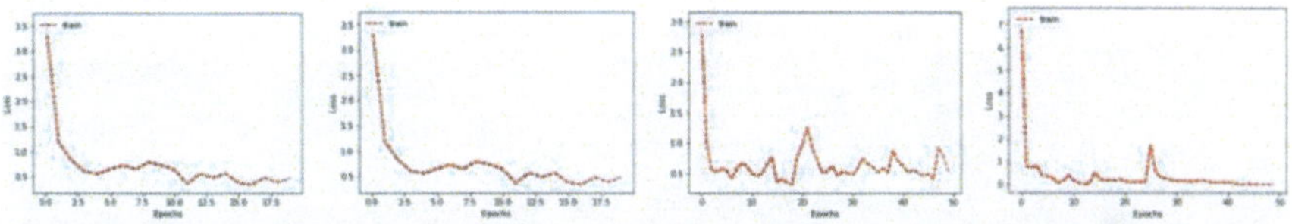

(a) kernel size = 3 (b) kernel size = 5 (c) kernel size = 5 (d) kernel size = 3

Fig. 4 Learning ability of the proposed classifier measured in terms of loss with different kernel sizes and number of epochs. The first two represent an epoch size of 20, while the next two signify an epoch size of 50

Figure 3 shows the entire methodology for the machine learning model. In the graphs shown in Fig. 4, the loss or the amount of the wrong prediction has been determined for all the four cases considered above. It has been found that this value becomes more or less constant after a certain number of iterations, which indicates that the model was neither over-fitted nor under-fitted. Although the highest accuracy has been obtained using U-Net (Fig. 5), yet the biological significance of the genes reported by SegNet is higher. Top five genes reported using SegNet having an accuracy of 0.69 and biological significance of $6.3e^{-7}$ have further been presented in Fig. 6. The genes were reported from DAVID Ontology.

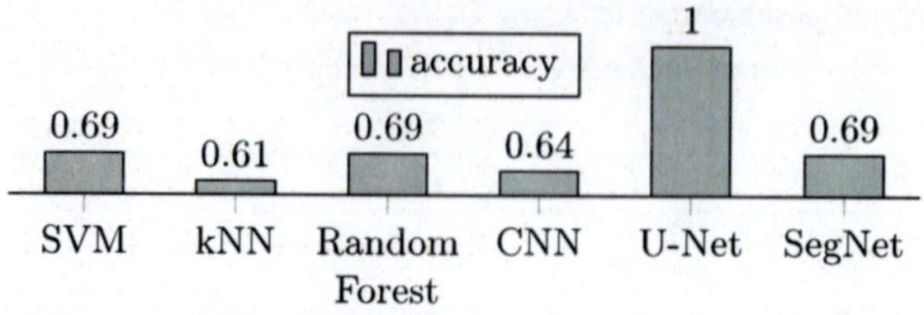

Fig. 5 Comparative results of different classifiers for the colon cancer dataset

From	To	Species	David Gene Name
U30498	7805	Homo sapiens	lysosomal protein transmembrane 5(LAPTM5)
M57732	6927	Homo sapiens	HNF1 homeobox A(HNF1A)
K02566	3827	Homo sapiens	kininogen 1(KNG1)
M69043	4792	Homo sapiens	NFKB inhibitor alpha(NFKBIA)
U20158	3937	Homo sapiens	lymphocyte cytosolic protein 2(LCP2)
M36981	4831	Homo sapiens	NME/NM23 nucleoside diphosphate kinase 2(NME2)
D13639	894	Homo sapiens	cyclin D2(CCND2)
M74491	377	Homo sapiens	ADP ribosylation factor 3(ARF3)
L08187	10148	Homo sapiens	Epstein-Barr virus induced 3(EBI3)
X68194	6856	Homo sapiens	synaptophysin like 1(SYPL1)
M24470	2766	Homo sapiens	guanosine monophosphate reductase(GMPR)
U30872	1063	Homo sapiens	centromere protein F(CENPF)
U31248	7727	Homo sapiens	zinc finger protein 174(ZNF174)
M55268	1459	Homo sapiens	casein kinase 2 alpha 2(CSNK2A2)
D21094	10989	Homo sapiens	inner membrane mitochondrial protein(IMMT)
J03239	7067	Homo sapiens	thyroid hormone receptor, alpha(THRA)
M28373	351	Homo sapiens	amyloid beta precursor protein(APP)

Fig. 6 A list of the top significant genes

5 Conclusion

The experimental results of the proposed work showed that the cancer dataset provides higher accuracy based on the U-Net model compared to SegNet. Although deep learning with its huge amount of complexity was primarily designed for large image segmentation and identification of essential characteristics from the images, proper selection of parameters can make such architecture very efficient in handling microarray datasets. As a result, we have obtained very high accuracy when using U-Net with an epoch number of 50 and kernel size of 5, giving an accuracy of 100%. U-Net performed better than SegNet. Also, the prominent genes obtained from the microarray dataset have been verified for their significance using the DAVID website. We have achieved a very low p-value, which proves the biological relevance of the proposed methodology.

The work can be further extended to test the performance on other microarray datasets to achieve some crucial findings. Moreover, other parameters of deep learning could also be explored in the future to obtain more convincing results.

References

1. Alon, U., Barkai, N., Notterman, D.A., Gish, K., Ybarra, S., Mack, D., Levine, A.J.: Broad patterns of gene expression revealed by clustering analysis of tumor and normal colon tissues probed by oligonucleotide arrays. Proc. Natl. Acad. Sci. **96**(12), 6745–6750 (1999)
2. Badrinarayanan, V., Kendall, A., Cipolla, R.: SegNet: A deep convolutional encoder-decoder architecture for image segmentation. IEEE Trans. Pattern Anal. Mach. Intell. **39**(12), 2481–2495 (2017). https://doi.org/10.1109/TPAMI.2016.2644615
3. Bianchini, M., Scarselli, F.: On the complexity of neural network classifiers: a comparison between shallow and deep architectures. IEEE Trans. Neural Netw. Learn. Syst. **25**(8), 1553–1565 (2014). https://doi.org/10.1109/TNNLS.2013.2293637
4. Bolón-Canedo, V., Sánchez-Maroño, N., Alonso-Betanzos, A., Benítez, J.M., Herrera, F.: A review of microarray datasets and applied feature selection methods. Inf. Sci. **282**, 111–135 (2014). https://doi.org/10.1016/j.ins.2014.05.042
5. Browniee, J.: Gentle introduction to the Adam optimization algorithm for deep learning (2017), https://machinelearningmastery.com/adam-optimization-algorithm-for-deep-learning/
6. Chen, Y., Li, Y., Narayan, R., Subramanian, A., Xie, X.: Gene expression inference with deep learning. Bioinformatics **32**(12), 1832–1839 (2016). https://doi.org/10.1093/bioinformatics/btw074
7. De Boer, P.T., Kroese, D.P., Mannor, S., Rubinstein, R.Y.: A tutorial on the cross-entropy method. Annals Oper. Res. **134**(1), 19–67 (2005). https://doi.org/10.1007/s10479-005-5724-z
8. Huang, D.W., Sherman, B.T., Lempicki, R.A.: Systematic and integrative analysis of large gene lists using david bioinformatics resources. Nat. Protoc. **4**(1), 44–57 (2009). https://doi.org/10.1038/nprot.2008.211
9. Li, Y., Huang, C., Ding, L., Li, Z., Pan, Y., Gao, X.: Deep learning in bioinformatics: introduction, application, and perspective in the big data era. Methods (2019). https://doi.org/10.1016/j.ymeth.2019.04.008
10. Nair, V., Hinton, G.E.: Rectified linear units improve Restricted Boltzmann machines. In: ICML 2010—proceedings, 27th international conference on machine learning, pp. 807–814 (2010). http://citeseerx.ist.psu.edu/viewdoc/summary?doi=10.1.1.165.6419
11. PyTorch: Welcome to PyTorch tutorials—PyTorch tutorials 0.3.1.post2 documentation (2017). http://pytorch.org/tutorials/
12. Ronneberger, O., Fischer, P., Brox, T.: U-net: convolutional networks for biomedical image segmentation. In: Lecture Notes in Computer Science (including subseries Lecture Notes in Artificial Intelligence and Lecture Notes in Bioinformatics), vol. 9351, pp. 234–241. Springer (2015). https://doi.org/10.1007/978-3-319-24574-4_28
13. Stanford University: Unsupervised feature learning and deep learning tutorial—Softmax regression (2013). http://ufldl.stanford.edu/tutorial/supervised/SoftmaxRegression
14. Stanford University: Unsupervised feature learning and deep learning tutorial optimization: Stochastic gradient descent overview (2015). http://ufldl.stanford.edu/tutorial/supervised/OptimizationStochasticGradientDescent

15. Stanford University Course cs231n: CS231n convolutional neural networks for visual recognition (2018). http://cs231n.github.io/convolutional-networks/
16. Zeebaree, D.Q., Haron, H., Abdulazeez, A.M.: Gene selection and classification of microarray data using convolutional neural network. In: ICOASE 2018—international conference on advanced science and engineering, pp. 145–150. Institute of Electrical and Electronics Engineers Inc (2018). https://doi.org/10.1109/ICOASE.2018.8548836

An Efficient Region of Interest Detection and Segmentation in MRI Images Using Optimal ANFIS Network

K. Amal Thomas, S.P. Kaarmukilan, Sucheta Biswas, and Soumyajit Poddar

Abstract The detection of tumour regions in Glioma brain images is a time-consuming task. This paper discusses the algorithm for efficient detection of tumour using Optimal Adaptive Neuro-Fuzzy Inference System (OANFIS). The proposed methodology consists of five modules: pre-processing, feature extraction, selection, classification and finally segmentation. The database are pre-processed with the help of median filter. Gray Level Co-occurrence Matrix (GLCM) features are extracted from both the images. To avoid the complexity, important features are selected using Crow Search Optimization (CSO) algorithm. In the next step, the shortlisted features are given to the ANFIS classifier to make classify an image as an abnormal or normal image. The ROI region of the abnormal images is separated with PFCM (Probabilistic Fuzzy C-means Clustering). The performance of the proposed methodology is analyzed in terms of sensitivity, accuracy and specificity.

Keywords Brain tumour · Region of interest · PFCM · ANFIS · CSO · Neural network · Fuzzy logic system and feature selection

1 Introduction

Brain tumour is an intracranial abnormal growth of cells within the central spinal canal or in the brain region of our human body. Brain tumour is one of the most common and deadly diseases in the present scenario. So the of treatment based

K. Amal Thomas (✉) · S.P. Kaarmukilan · S. Biswas · S. Poddar
Indian Institute of Information Technology, Guwahati, Assam, India
e-mail: amal.kakkassery@outlook.com

S.P. Kaarmukilan
e-mail: kaarmukilansp@gmail.com

S. Biswas
e-mail: suchetab1093@gmail.com

S. Poddar
e-mail: poddar18@gmail.com

D. Bhattacharjee et al. (eds.), *Proceedings of International Conference on Frontiers in Computing and Systems*, Advances in Intelligent Systems and Computing 1255,
https://doi.org/10.1007/978-981-15-7834-2_50

on detection of the tumour in its initial stages. There are different types of brain tumours in medical science which make the decisions very much complicated for the detection [1]. Hence, brain tumour classification is very much important. A good classification leads to the right decision and provides the right treatment [2]. In most cases, physician gives the treatment for the strokes rather than the treatment for the brain tumour. Therefore, detection of the tumour is a very essential task. The lifetime of the person affected by the brain tumour will improve if it is detected early [3]. In general, early stage tumour diagnoses mainly includes Computed Tomography scan, Magnetic Resonance Imaging scan, Biopsy and Nerve test. [4]. Presently, with the rapid growth of the Artificial Intelligence development in computer-aided diagnosis, Biomedical and Magnetic Resonance Imaging attracts lots of attention [5]. Cause of these particular kind of cancers is termed as malignant tumour in brain. Not all brain tumours are malignant . Some types of brain tumours are of benign. Brain cancer is also known as glioma [6]. This is why feature extraction and classification are important steps in brain tumour detection. In a feature set, it should have good discriminating features. The particular feature set will reduce the dismissal of features to neglect the issues related to dimensionality problem [5].

2 Literature Review

In the literature survey, a couple of methods have been proposed for brain tumour classification and segmentation in image processing. Most recently published works are presented as follows:

Meiyan Huang et al. [7] have explained the Brain tumour Segmentation Based on Local Projection Independent Classification. In the local independent projection-based classification or LPIC method was used to classify every single voxel into different classes. In the local independent projection-based classification or LPIC method was used to classify every single voxel into different classes. A novel classification was derived by introducing the LPIC into the classification model.

Kailash D.Kharat et al. [8] explained the two Neural Network techniques for the classification of the MRI images of the brain. The Neural Network technique consists of three stages. Feature extraction, dimensionality reduction and finally classification. In the primary stage, they obtained the features from MRI images using Discrete Wavelet Transform (DWT).

Zhan-Li Sun et al. [9] have explained the tumour classification using Eigen-gene based classifier learning algorithm. Eigen-gene extracted using Independent Component Analysis (ICA) was one kind of effective feature for brain tumour classification procedure. They used Eigen-gene and support vector machine-based optimal Classifier Committee Learning (CCL) algorithm.

A.P. Nanthagopal and R.Sukanesh [10] have explained the wavelet statistical texture features-based segmentation and classification of brain CT images. The authors present a method to select both the dominant run length, as well as co-occurrence features of wavelet approximation ROI region of each slice to be segmented by using a Support Vector Machine (SVM).

3 Proposed Model

The basic aim of our proposed approach is to detect the ROI region (tumour part) from the MRI images. The proposed brain tumour classification includes four major modules such as pre-processing, feature extraction, tumour detection process and segmentation. Initially, the images are pre-processed with the help of enhancement technique. After the pre-processing process, GLCM features are extracted from the input image. After the feature extraction step, the important features are filtered from the total number of features using the CSO algorithm. Then, the duly selected features are given to the ANFIS classifier to classify an image as abnormal or normal. ROI part is segmented from the abnormal image with the help of PFCM. The overall architecture of the proposed methodology is given in Fig. 1.

3.1 Pre-processing

The pre-processing method is performed to avoid impurities or noisy contents within the image to decrease the processing complexity by enhancing the performance of the image. Median filter is conducted here to decrease the processing complexity. Typically, the median filter is using for reduction of noise in an image. It's a non-linear filter that calculates median of the pixel that falls within the mask. It is more robust and does not create a new realistic pixel value, which preserves edge blurring

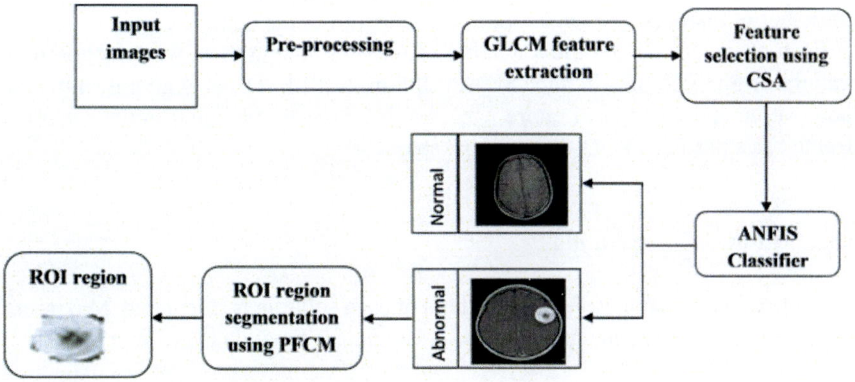

Fig. 1 Proposed brain tumour detection model

and loss of image detail. It preserves the sharp high frequency details. It is well suited to eliminate salt and pepper noise from the image.

3.2 *Feature Extraction*

After the pre-processing process, GLCM features are extracted from each image. Totally, 22 features are extracted from each image. Table 1 shows the extracted values of GLCM for both abnormal and normal MRI images of the brain. Here, the *p(i, j)* implies for the elements in the GLCM matrix, Also, G^L is the Number of gray levels in the image as specified under Quantization. *i, j* represents the intensities between 0 and G^L-1. So the sample intensity of the specified parameters can be assigned to discrete gray levels for the calculation of features. In G^L value, each gray level representing a range of sample intensity figures.

3.3 *Features Selection Using Crow Search Algorithm*

After finding the feature of each image present in the database D^D, we have to reduce the dimension of features for improving the accuracy. Because higher number of features are one of the great obstacles for classification technique.

3.3.1 Solution Representation

To optimize the features, the CSO algorithm initially creates a population of solution [12]. In an optimization algorithm, solution creation is an important step in helping to identify the optimal solution in less time consumption. Each of the images having *n* number of features. We select optimal features from the complete list.

Each of images having the *n* number of features among them we select optimal features. In Self Learning Particle Swarm Optimizer (SLPSO), we randomly initialize the position and velocity of entire particles. Here, each individual in the swarm is deemed as a particle in a D-dimensional space.

3.3.2 Fitness Computation

After the initial solution generation, fitness of each solution is calculated. Maximum fitness is considered as the good solution. The fitness function is given in (1).

$$\text{Fitness} = \max(A) \tag{1}$$

where,

Table 1 Extracted Features

Feature descriptor	Features name	Computation
FD^1	Angular second moment	$\sum_{x=0}^{G^L-1}\sum_{y=0}^{G^L-1} p(x,y)^2$
FD^2	Contrast	$\sum_{n=0}^{G^L-1} n^2 \sum_{x=0}^{G^L-1}\sum_{y=0}^{G^L-1} p(x,y), \mid x-y \mid = n$
FD^3	Inverse difference moment	$\sum_{x=0}^{G^L-1}\sum_{y=0}^{G^L-1} \frac{1}{1+(x-y)^2} p(x,y)$
FD^4	Entropy	$-\sum_{x=0}^{G^L-1}\sum_{y=0}^{G^L-1} p(x,y) * log(p(x,y))$
FD^5	Correlation	$\sum_{x=0}^{G^L-1}\sum_{y=0}^{G^L-1} \frac{(x,y)p(x,y)-\mu_i*\mu_y}{\sigma_i*\sigma_y}$
FD^6	Variance	$\sum_{x=0}^{G^L-1}\sum_{y=0}^{G^L-1} (x-\mu)^2 p(x,y)$
FD^7	Sum average	$\sum_{x=0}^{2G^L-1} x p_{i+j}(x)$
FD^8	Sum variance	$\sum_{x=0}^{2G^L-1} (x-\text{Average})^2 p_{i+j}(x)$
FD^9	Sum entropy	$-\sum_{x=0}^{2G^L-2} p_{i+j}(x) log(p_{i+j}(x))$
FD^{10}	Difference entropy	$-\sum_{x=0}^{G^L-1} p_{i+j}(x) log(p_{i+j}(x))$
FD^{11}	Inertia	$\sum_{x=0}^{G^L-1}\sum_{y=0}^{G^L-1} (x-y) * p(x,y)$
FD^{12}	Cluster shade	$\sum_{x=0}^{G^L-1}\sum_{y=0}^{G^L-1} (x+y-\mu_i-\mu_y^3 p(x,y)$
FD^{13}	Cluster prominence	$\sum_{x=0}^{G^L-1}\sum_{y=0}^{G^L-1} (x+y-\mu_i-\mu_y^4 p(x,y)$
FD^{14}	Dissimilarity	$\sum_{x,y} \mid x-y \mid p(x,y)$
FD^{15}	Homogeneity	$\sum_{x,y} \frac{1}{1-(x-y)^2} p(x,y)$
FD^{16}	Energy	$\sum_{x,y} p(x,y)^2$
FD^{17}	Auto correlation	$\sum_{x=0}^{G^L-1}\sum_{y=0}^{G^L-1} (p_i-\mu_x)(p_j-\mu_y)/\sigma_x\sigma_y$
FD^{18}	Maximum probability	$\max(x,y)p(x,y)$
FD^{19}	Inverse difference normalization	$\sum_{x=0}^{G^L-1}\sum_{y=0}^{G^L-1} \frac{\mid p(x,y)\mid}{1+(x-y)}$
FD^{20}	Inverse difference moment normalization	$\sum_{x=0}^{G^L-1}\sum_{y=0}^{G^L-1} \frac{p(x,y)}{1+(x-y)}$
FD^{21}	Information measure of correlation 1	$\frac{HXY-HXY1}{\max HX,HY}$
FD^{21}	Information measure of correlation 2	$\sqrt{(1-\exp(HXY2-HXY)}$

$$A = \frac{TN+TP}{TN+TP+FN+FP} \tag{2}$$

where *TN* is True Negative, *FN* is False Negative, *TP* is True Positive and *FP* is False Positive

3.3.3 Update Using CSO Algorithm

The next is updation of the solution using the CSO algorithm. In this updating process, two cases are considered.
Case 1: The food owning crow n of food source B_n^t did not realize the cheat crow m it follows so the cheat crow attains to the food location of the owner crow. The updated thief crow position is provided in (3).

$$B_n^t = S_m^t + R_m * FL_m^t * (B_n^t - S_m^t) \tag{3}$$

where, R_m is a random number with the range [0,1], FL_m^t is the flight length of the crow m at t iteration and S_m^t is the present position of the thief crow.
Case 2: The owner crow n realize that the cheat crow m follows him, therefore, the owner crow will mislead crow m by going to any other random location of search space. In this case, the position of crow m is updated by (4).

$$S_m^t = \begin{cases} \text{update position using (3)}, & R_m \geq P_m^k \\ \text{update to random position}, & \text{else} \end{cases} \tag{4}$$

where P_m^k is the probability of awareness of the thief crow m at iteration k.

3.3.4 Termination Criteria

The algorithm stops it's implementation only if it reached a maximum number of iteration. The solution containing the best fitness value taken using CSO and it is given as the best solution for classification. Also, the best solution is utilized as an input of a ANFIS classifier.

3.4 Tumour Classification Using ANFIS Classifier

The selected features are given to classify an image as normal or abnormal. For classification, Adaptive Neuro-Fuzzy Inference System (ANFIS) classifier is used [11]. ANFIS is the combine the finest sorts of fuzzy inference system (FIS) and artificial neural network (ANN). It joints the express data illustration of a FIS and the learning power of the artificial neural network. The important purpose of ANFIS is to combine the optimum features of the neural networks and fuzzy system. ANFIS aid as a principle for building an agreement of fuzzy if-then guidelines with appropriate enrollment capacities to give the stipulated input-output pairs. This is a fuzzy inference scheme with a back engendering that tries to diminish the error and augments the execution. The block diagram of ANFIS building design is assumed in Fig. 2.

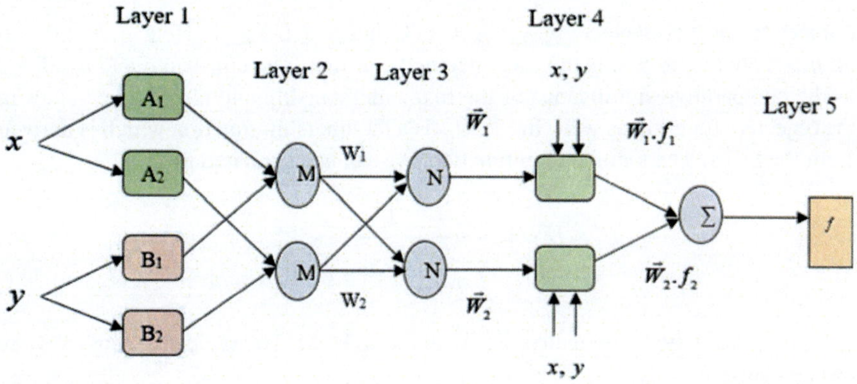

Fig. 2 ANFIS architecture

In Fig. 2, x and y are the crisp inputs. A_i and B_i are the representative labels (low, medium, high, etc.) with i is 1 or 2. w is the weight of ANN procedure. The organization of ANFIS is encompassed of five-layered feed-forward neural network as in Fig. 2. ANFIS architecture contains five layers of nodes. Out of the five layers, the first and the fourth layer obtains adaptive nodes, and the second, third and fifth layer obtains the fixed nodes. This ANFIS process classifies an image as normal and abnormal.

3.5 Segmentation Using PFCM Classifier

The abnormal images are passed to the segmentation stage. In this stage, the stone region (tumour region) is accurately segmented from the abnormal image. For segmentation PFCM algorithm is utilized. PFCM produces possibilities and memberships at the same time, along with the common point prototypes or cluster centres for each cluster. PFCM is a Probabilistic hybridization of C-means (PCM) and through the Fuzzy C-means (FCM), it can often avoid various problems of FPCM, FCM and PCM. PFCM solves the noise sensitivity to detect FCM, overcomes the coincident clusters problem of PCM and eliminate the row sum constraints of FPCM.

Let's consider the input image which consists of stone region and normal region. These two regions are clustered from the input image. So, that the size of the cluster is 2. The proposed PFCM optimize the following objective function which is given in (5).

$$O_{PFCM}(U,T,V;Y)=\sum_{k=1}^{n}\sum_{i=1}^{c}(aU_{ik}^{m}+bT_{ik}^{\eta})*(\|y_k-v_i\|)_A^2\sum_{i=1}^{c}\gamma_i\sum_{k=1}^{n}(1-T_{ik})_\eta \quad (5)$$

Subject to the parameters $\sum_{i=1}^{c} U_{ik} = 1 \forall k$ and $0 \leq U_{ik} \leq 1$. Here $a, b > 0$ and $m, \eta > 1$. In (1), $\gamma_i > 0$ is the user specified constant. The constant a and b defined as the comparative significance of the fuzzy membership and also it give values in the objective function as well. In (5), U_{ik} is a membership function which is derived from the FCM. The membership function U_{ik} can be calculated as

$$U_{ik} = \frac{1}{\sum_{j=1}^{c} \left(\frac{\|y_k - v_i\|}{\|y_k - v_j\|} \right)^{\frac{2}{m-1}}} \tag{6}$$

Similarly, in (7), typically matrix T_{ik} is similar to PCM. The typically matrix T_{ik} can be calculated as

$$T_{ik} = \frac{1}{1 + \left[\frac{D^2(x_k, v_i)}{\eta_i} \right]^{\frac{1}{m-1}}} \tag{7}$$

The cluster center of V_i of PFCM is can be calculated as

$$V_i = \frac{\sum_{k=1}^{n} (au_{ik}^m + bu_{ik}^m) X_k}{\sum_{k=1}^{n} (au_{ik}^m + bu_{ik}^m)}, 1 \leq i \leq c. \tag{8}$$

The clustering process is continued on k-number of iterations. After the clustering process, the image is grouped into 2 numbers of clusters namely, normal region and stone region. Then performance of the approach is analyzed in terms of different metrics.

4 Results and Discussion

The sample image used for experiments is given in Fig. 3, which is based on the Whole Brain Atlas (WBA) open access T1 dataset and size of the image is "512×512" for the performance evaluation [13].

4.1 Evaluation Metrics

The performance of the proposed methodology is analyzed in terms of evaluation metrics: specificity, accuracy and sensitivity.

Sensitivity: It is the actual positives which are properly recognized. It gives the capacity of test to recognize positive results.

$$\text{Sensitivity} = \frac{TP}{TP + FN} * 100 \tag{9}$$

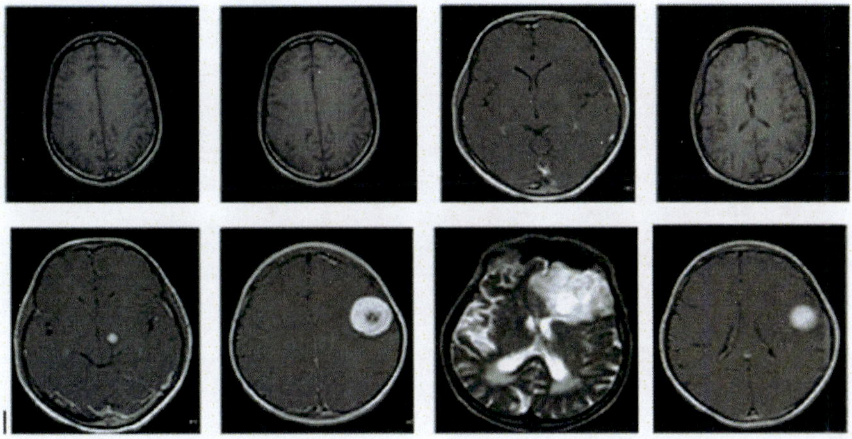

Fig. 3 Sample images from WBA dataset

where *TP* is True Positives and *FN* is False Negatives.
Specificity: It is the proportion of negatives which are properly recognized. It gives the capacity of test to recognize negative results.

$$\text{Specificity} = \frac{TN}{TN + FP} * 100 \tag{10}$$

where *FP* is False Positives and *TN* is True Negatives.
Accuracy: The percentage of tumour and non-tumour parts in an image is correctly segmented by the measurement accuracy.

4.2 Performance Analysis Using Evaluation Metrics

In this approach for classification, optimal ANFIS and segmentation PFCM is utilized. The performance of both the methods is analyzed in terms of accuracy, sensitivity and specificity.

In Fig. 4, the proposed methodology is analyzed in terms of accuracy, sensitivity and specificity. From Fig. 4, the proposed methodology attains the better accuracy of 93.4% which is 90% for using ANFIS based MRI tumour classification, 89% for using neural network and 88% for using FLS based MRI tumour classification. In addition to these parameters, sensitivity performance is also analyzed. The proposed method is attaining the maximum sensitivity of 95% which is 93% for using ANFIS classification, 92% for using NN based MRI image classification and 91% for using FLS based classification. Similarly, the specificity is also compared with other methods. Figure 5 shows the performance of the fitness function. Maximum accuracy is

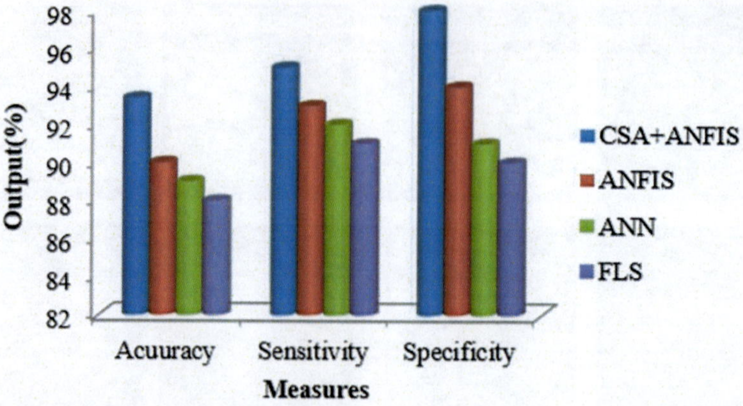

Fig. 4 Performance of classification stage

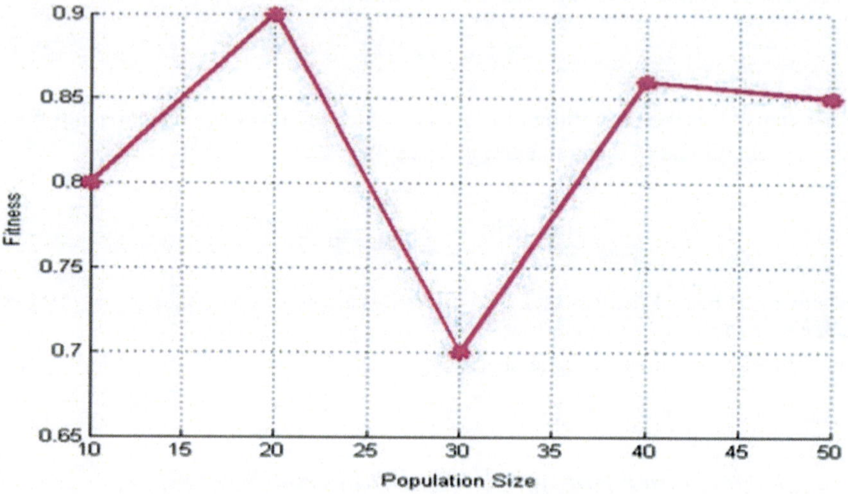

Fig. 5 Population size versus fitness

considered as the fitness function. From the results, it is clearly understood that the proposed methodology attains better results compared to other approaches.

In Fig. 6, segmentation stage performance is analyzed. The segmentation performance of proposed methodology is analyzed in terms of accuracy, sensitivity and specificity. When analyzing Fig. 6, the proposed PFCM based segmentation attain the maximum accuracy of 94.23% which is 91.52% for using FPCM based clustering, 90.32% for using PCM based clustering and 89.37% for using FCM based clustering. This is because, the proposed PFCM overcomes the difficulties of PCM, FCM and FPCM.

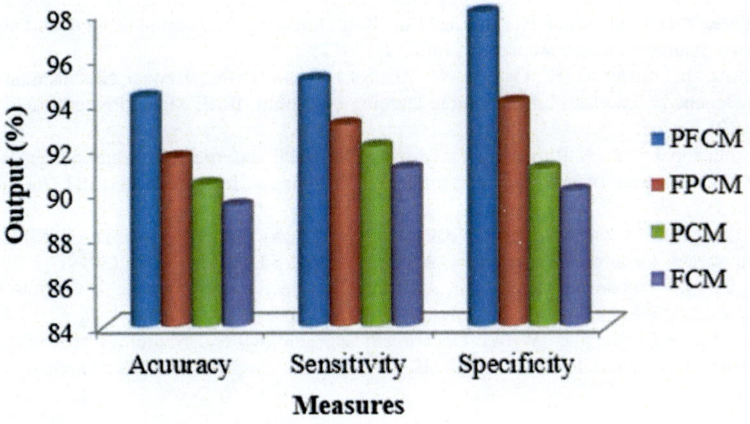

Fig. 6 Performance of segmentation stage

5 Conclusion

Efficient MRI image tumour classification using optimal FIS system has been proposed. Here, for feature selection crow search algorithm is utilized, classification process FIS is utilized and PFCM has been used for the segmentation process. The performance of proposed methodology is compared in terms of accuracy, sensitivity and specificity. Segmentation and classification performance are compared with another algorithm. It is observed that the proposed method outperforms the existing models.

References

1. Bankman, I.N.: Hand Book of Medical Image processing and Analysis. Academic Press in Biomedical Engineering (2009)
2. Sridhar, D., Krishna, I. M.: Brain tumour classification using discrete cosine transform and probabilistic neural network. In: International Conference on Signal Processing, Image Processing and Pattern Recognition (2013)
3. Kabade, Rohit S. & Gaikwad M.: Segmentation of brain tumour and its area calculation in brain MRI images using K-mean clustering and Fuzzy C-mean algorithm. International Journal Computer Science Engineering Technology **4**(5), 524–31 (2013)
4. Borden, Neil M., Forseen, Scott E.: Pattern Recognition Neuro radiology. Cambridge University Press, New York (2011)
5. Osareh, A., Shadgar, B.: A computer aided diagnosis system for breast cancer. Int. J. Comput. Sci. Issues (IJCSI) **8**(2) (2011)
6. Sridhar, D., Muralik, K.: Brain tumour classification using discrete cosine transform and probabilistic neural network. In: International Conference on Signal Processing, Image Processing and Pattern Recognition (ICSIPR) (2013)
7. Huang, M., Yang, W., Wu, Y., Jiang, J., Chen, W., Feng, Q.: Brain tumour segmentation based on local independent projection-based classification. IEEE Trans. Biomed. Eng. **61**(10) (2014)

8. Kharat, K.D., Kulkarni, P.P., Nagori, M.B.: Brain tumour classification using neural network based methods. Int. J. Comput. Sci. Inf. **1**(4) (2012)
9. Sun, Z.-L., Zheng, C.-H., Gao, Q.-W., Zhang, J., Zhan, D.-X.: Tumour classification using Eigengene-based classifier committee learning algorithm. IEEE Signal Process. Lett. **19**(8) (2012)
10. Nanthagopal, A.P., & Rajamony, R.: Wavelet statistical texture features-based segmentation and classification of brain computed tomography images. Image Process. IET **7**(1), 25–32 (2013)
11. Karaboga, D., & Kaya, E.: Adaptive network based fuzzy inference system (ANFIS) training approaches: a comprehensive survey. Artif. Intell. Rev. **52**(4), 2263–2293 (2019)
12. Sayed, G.I., Hassanien, A., & Azar, A.T.: Feature selection via a novel chaotic crow search algorithm. Neural Comput. Appl. **31**(1), 171–188 (2019)
13. The Whole Brain Atlas (WBA): Department of Radiology and Neurology at Brigham and Women's Hospital. Harward MedicalSchool, Boston, http://www.med.harvard.edu/aanlib/navt.html

Follicle Segmentation Using K-Means Clustering from Ultrasound Image of Ovary

Ardhendu Mandal, Debasmita Saha, and Manas Sarkar

Abstract Detection of number, shape and size of follicles in the ovary can play an important role in the diagnosis and monitoring of different diseases like infertility, PCOS (Polycystic Ovarian Syndrome), ovarian cancer, etc. Nowadays the identification of these characteristics of follicles is done manually by radiologists and doctors from the Ultrasound Images of ovaries. Sometimes manual analysis can be tedious and thus may lead to erroneous results. In this paper, a method is proposed for automatic segmentation of follicles from Ultrasound Images using the K-means clustering technique.

Keywords Follicle detection · Ultrasound images · Ovary · Segmentation · K-means clustering

1 Introduction

For diagnosis of different diseases like PCOS (Polycystic Ovarian Syndrome), ovarian cancer, infertility, etc., in woman's body, it is important to determine the ovarian status. Ovary is the reproductive organ of female body. Ova or eggs are produced in the ovary. Inside the ovary, there are some spherical fluid-filled structures which are called follicles. The number of follicles present in the ovary plays an important role in these diagnosis processes [1]. Thus, ovary is frequently scanned by ultrasound

A. Mandal · D. Saha (✉) · M. Sarkar
Department of Computer Science and Application, University of North Bengal, Siliguri 734013, West Bengal, India
e-mail: debasmita.saha@hotmail.com

A. Mandal
e-mail: am.csa.nbu@nbu.ac.in; am.csa.nbu@gmail.com

M. Sarkar
e-mail: sarkarmanas@nbu.ac.in

D. Saha
Department of Computer Science, University of Gour Banga, Malda 732103, West Bengal, India

D. Bhattacharjee et al. (eds.), *Proceedings of International Conference on Frontiers in Computing and Systems*, Advances in Intelligent Systems and Computing 1255,
https://doi.org/10.1007/978-981-15-7834-2_51

imaging. Nowadays in most of the cases the analysis of these Ultrasound Images are done by medical experts manually, which is sometimes a tedious and error-prone job. Thus, there is a need of making the process of detecting the number of follicles from ultrasound images automatically, so that the monotonous work of medical experts can be reduced, as well as the detection accuracy can be improved.

Many researchers have tried different approaches to automatically segment follicles from ultrasound images such as active contour method [2], edge-based method [3], object growing method [4], morphology [5], etc. Still, there is a lot of scope to explore in this field of detection of follicles automatically from ultrasound images of ovaries. Thus, in this paper, another method is proposed in this direction mainly based on the clustering technique called K-means clustering.

Image segmentation can be described as the method of dividing a digital image into different regions such that each region contains pixels which has similar attributes. The resulting images obtained from image segmentation are more meaningful and can be analyzed easily to reach to some decisions. Thus, it can be said that the efficiency of image analysis mostly depends on the solidity of image segmentation.

Clustering is the technique where the same type of objects is grouped together. These groups are called clusters. Objects of one cluster must have different characteristics than the objects of another cluster. There are several algorithms to accomplish the task of clustering. One of these algorithms is the K-means clustering algorithm. Now, these clustering techniques can be used for image segmentation [6], as different pixels of an image has different characteristics and based on those characteristics the pixels can be grouped into several clusters. These clusters of pixels can be viewed as different segments of the image.

2 Data Collection

We have collected 19 ultrasound images of ovaries from two radiology centre, i.e., Gama Imaging and Diagnostic Center, Singatala, Malda, West Bengal, India, and Swagat Diagnostic Centre, Dhubri, Assam, India, respectively. We asked a radiologist to manually detect the number of follicles in each of these sample ultrasound images. These manually segmented images are considered as ground truth for our experiment. We have tried to propose a method which can automatically predict the number of follicles as segmented by the radiologist manually.

3 Proposed Method

Medical images such as ultrasound image can suffer from low contrast issue [7]. So, for enhancing the contrast of the images, histogram equalization is used so that the image quality gets improved.

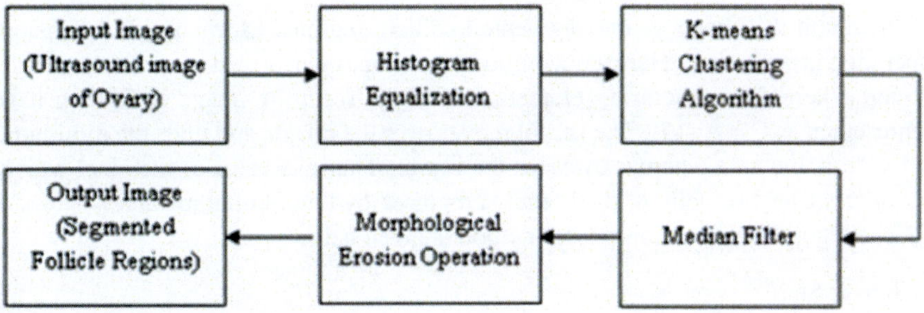

Fig. 1 Flowchart of the method

K-means clustering algorithm is implemented on the equalized image. The basic working mechanism of k-means clustering is as follows:

Step 1: At first K (number of clusters) points are initialized randomly or based on some heuristics, which are the centers of the clusters and are called as centroids [8]. Let us assume there are X = x_1,x_2,x_3…,x_n data points and Y = y_1,y_2…,y_k are the set of centroids.

Step 2: The Euclidean distance of each data points from K centroids are calculated.

Step 3: Each data point is placed in the cluster from whose centroid it has the minimum distance (Fig. 1).

Step 4: After assigning all points, the centroids are updated for each cluster. The new centroids will be the mean of all the points belonging to one cluster. If the ith cluster has q no of data points then the new centroid y_i will be

$$y_i = (1/q)\sum_{j=1}^{q} x_i \tag{1}$$

Step 5: Data points are reassigned based on the distances calculated from the newly obtained centroids.

Step 6: Repeat from step4 until convergence is achieved, i.e., no data points move between clusters and the centroids get stable.

In the proposed method, k-means clustering works on the basis of intensity values of pixels. Here, initially, four cluster centers were chosen, i.e., value of k is 4 and by using the above mentioned method, the centroids were updated until all the data points got stable under one of the four clusters. Thus, the total number of pixels were labeled under one of the four clusters (0–3). Now, the clustered data points were reshaped into the image and we get the segmented image.

On the segmented image, the median filter is applied as this filter removes any unwanted noise, as well preserves the edges of segmented regions (Fig. 1). Thus, we get a segmented image with smoother edges.

To retain the shape of the segmented follicle regions, Morphological Erosion operation is performed. Here, convolution operation is performed between the image A and a kernel or structuring element (B). If the resulting image is C, then it is represented as $C = A \ominus B$. The kernel moves over the image, and finds the minimum value from the area where it overlaps the image. Then the value of the pixel which is under the anchor point of the kernel is replaced by that minimum value.

The algorithm for the proposed method is given below:

Step 1: Start.
Step 2: Load the ultrasound image to be segmented.
Step 3: Calculate the histogram of the image.
Step 4: The histogram is equalized by removing the pixels having highest and lowest intensity values with cutoff percent of 15 and the image is remapped.
Step 5: Reshape the equalized image into 2-dimensional array.
Step 6: K-means clustering algorithm is performed on the array, where number of cluster k is considered as 4.
Step 7: The pixel positions which are labeled under the cluster which specify the follicle regions are extracted and replaced by the intensity value (0), i.e., black. Rest of the pixel is replaced by value (255), i.e., white.
Step 8: The labeled array is reshaped into image.
Step 9: Median filter is applied.
Step 10: Morphological erosion operation performed with a 5×5 structuring element to get the final segmented image.
Step11: Stop.

4 Experimental Results

The proposed procedure is applied to 19 sample ultrasound images. Few of the results are shown below.

4.1 Case I: Correct Segmentation

A sample image for which the medical expert manually segmented 3 follicles and our proposed method also segmented 3 follicles (see Figs. 2 and 3).

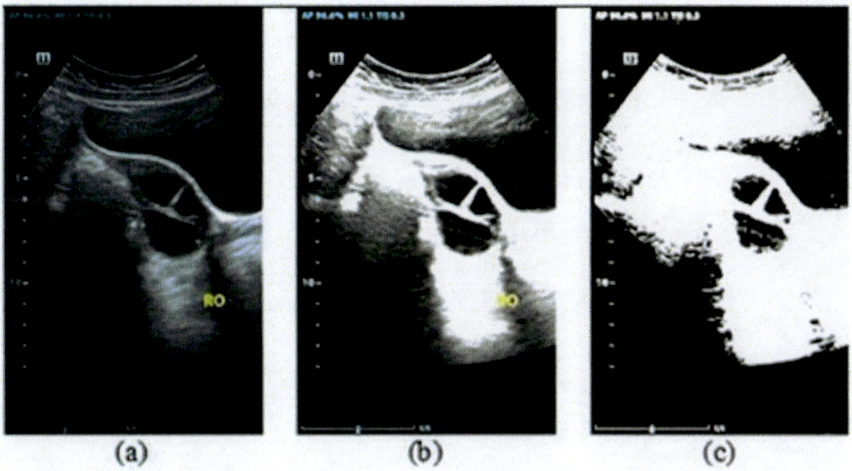

Fig. 2 **a** Sample image; **b** after performing histogram equalization; **c** after performing K-means clustering;

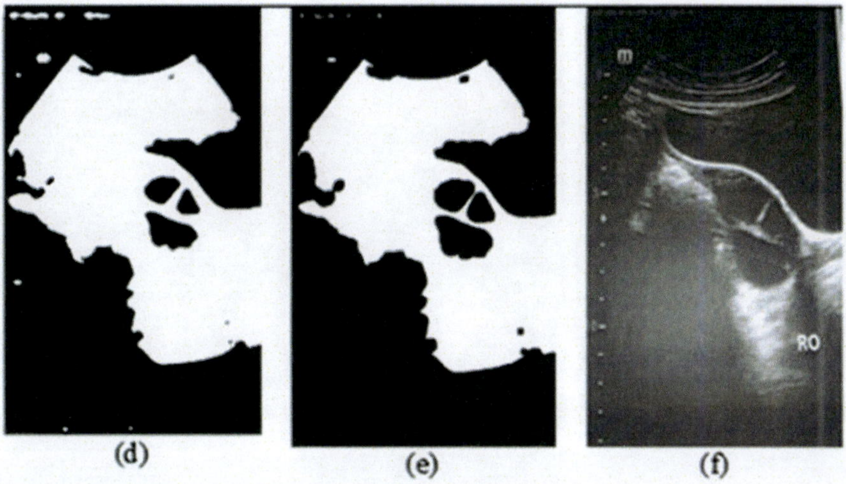

Fig. 3 **d** After applying median filter; **e** after performing morphological erosion; **f** manually segmented image;

4.2 Case II: Segmentation with False Rejection

A sample image for which the medical expert manually segmented 5 follicles, but our proposed method is able to segment only 2 follicles. Here, 3 regions were not segmented in spite of being follicles (see Figs. 4 and 5).

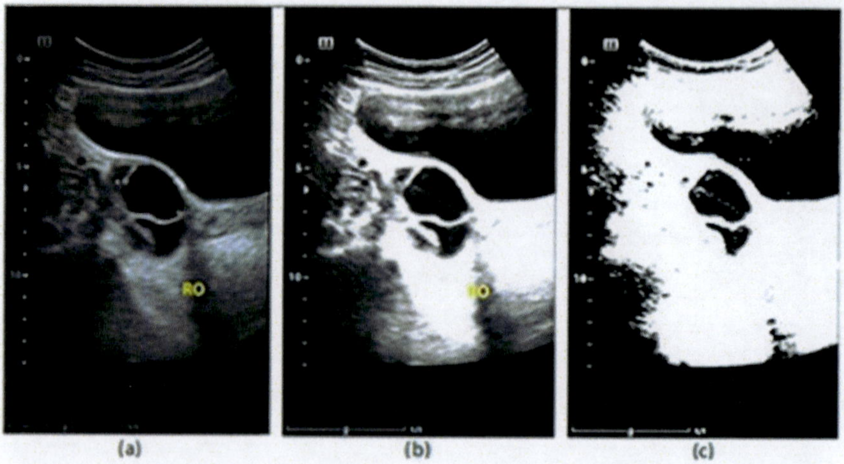

Fig. 4 **a** Sample image; **b** after performing histogram equalization; **c** after performing K-means clustering;

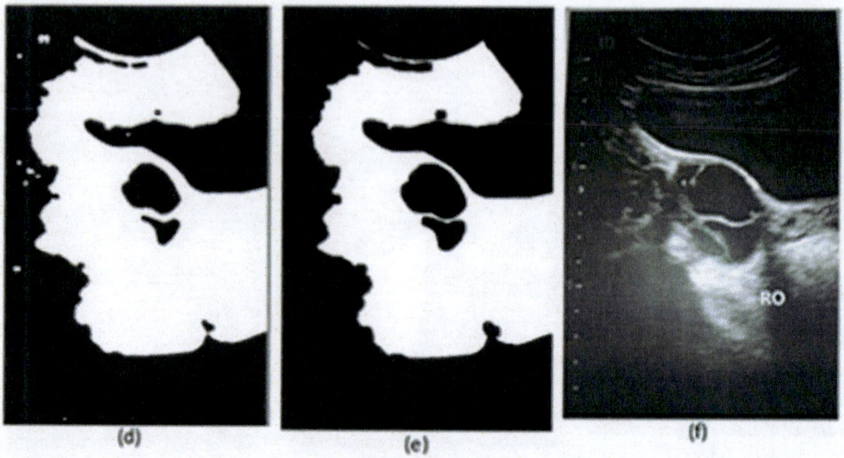

Fig. 5 **d** After applying median filter; **e** after performing morphological erosion; **f** manually segmented image;

4.3 Case III: Segmentation with False Acceptance

A sample image for which the medical expert manually segmented 1 follicle, but our proposed method segmented 2 follicles. Here, 1 extra region was segmented which was not follicle originally (see Figs. 6 and 7).

Other experimental results are not shown in the pictorial form, but the complete analytical result is depicted in the following section.

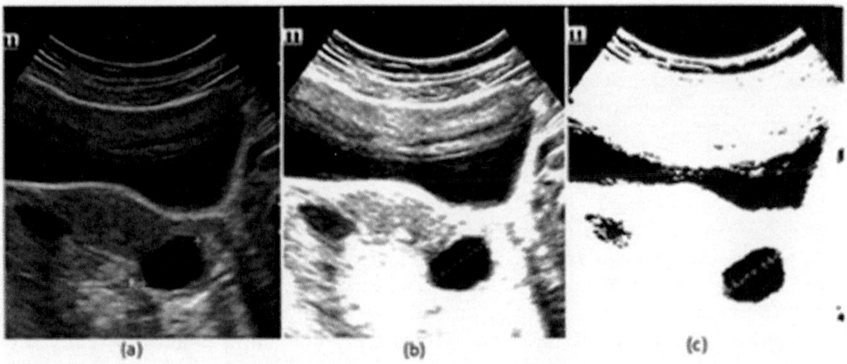

Fig. 6 **a** Sample image; **b** after performing histogram equalization; **c** after performing K-means clustering;

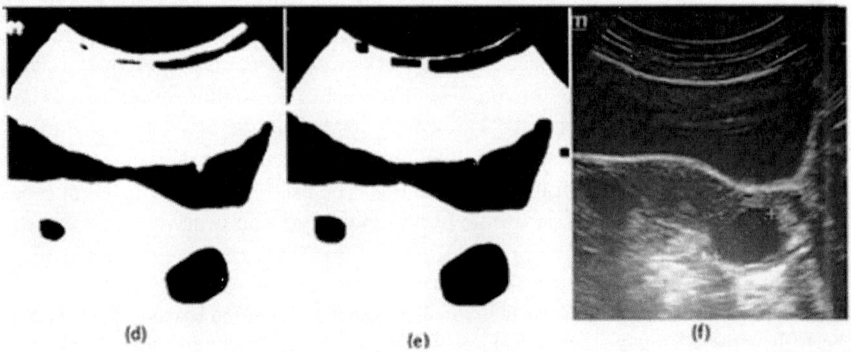

Fig. 7 **d** After applying median filter; **e** after performing morphological erosion; **f** manually segmented image;

5 Performance Evaluation

The proposed method is evaluated on the basis of Type I and Type II error rate. Type I (α) error occurs when the region is not follicle but it is detected as follicle. Type II (β) error occurs when the region is follicle but it is not detected as follicle. For the proposed method, the classification rate is 84.61% and the precision is 90.90%. The False Acceptance Rate or Type I Error is 7.69%. The False Rejection Rate or Type II Error is 23.07%. Table 1 shows the performance evaluation of the method.

Table 1 Performance evaluation of proposed method

No. of input images	No. of follicles detected by proposed method	No. of follicles detected by medical expert	Type I error (α)	Type II error (β)
19	55	65	5	15

6 Conclusion

The proposed method uses the k-means clustering algorithm to automatically segment the follicle regions from the ultrasound images of ovaries. Histogram equalization has been used for Contrast enhancement of the images before applying the clustering algorithm. Afterwards, morphological operation was applied to make the clustered segments more prominent. Finally, Type I and Type II error was calculated to evaluate the performance of the proposed method. From the experimental results, we can conclude that the proposed method works well. It seems that the method can contribute to the process of automatic follicle detection from ultrasound images and reduce the burden of manual detection of follicles by the medical experts. Further, the accurate size and shape of the segmented regions can be calculated to extract more information about the follicles so that it can be used to classify different types of ovaries such as normal ovary, cystic ovary, and polycystic ovary.

Acknowledgements The authors would like to thank Gama Imaging and Diagnostic Center, Singatala, Malda, for providing the ovarian ultrasound images. The authors are also grateful to Dr. Md. Laskar Ali, MBBS, DMRD, Swagat Diagnostic Centre, Dhubri, Assam for helping by manually segmenting the follicle regions and for his guidance and suggestions at different phases of the work.

References

1. Rabiu, O., Usman, A.D., Tekanyi, A.M.S.: A review on computer assisted follicle detection techniques and polycystic ovarian syndrome (PCOS) diagnostic systems. Int. J. Comput. Trends Technol. (IJCTT) **28**(1), 41–45 (2015)
2. Hiremath, P.S., Tegnoor, J.R.: Automatic detection of follicles in ultrasound images of ovaries using active contours method. In: The proceedings of IEEE International Conference on Computational Intelligence and Computing Research (ICCIC), pp. 28–29 (2010)
3. Hiremath, P.S., Tegnoor, J.R.: Automatic detection of follicles in ultrasound images of ovaries using edge based method. Int. J. Comput. Appl. Special Issue Recent Trends Image Process. Pattern Recogn.**1**, 5–16 (2010)
4. Deng, Y., Wang, Y., Shen, Y.: An automated diagnostic system of polycystic ovary syndrome based on object growing. J. Artif. Intell. Med. **51**(3), 199–209 (2011)
5. Lawrence, M.J., Eramian, M.G., Pierson, R.A., Neufeld, E.: Computer assisted detection of polycystic ovary morphology in ultrasound images. In: Fourth IEEE Canadian Conference on Computer and Robot Vision (CRV), vol. 7, pp. 105–112 (2007)

6. Parekh, A.M., Shah, N.B.: Comparative study on ovarian follicle detection using segmentation techniques. Int. J. Innovative Res. Comput. Commun. Eng. **4**(9), 16683–16689 (2016)
7. Dhanachandra, N., Manglem, K., Chanu, Y.J.: Image segmentation using K-means clustering algorithm and subtractive clustering algorithm. In: The Proceedings of Eleventh International Multi-Conference on Information Processing-2015 (IMCIP-2015), Elsevier, Procedia Computer Science, vol. 54, pp. 764–771 (2015)
8. Shehroz, S.: Khan and Amir Ahmad. Cluster centre initialization algorithm for K-means cluster. Pattern Recogn. Lett. **25**(11), 1293–1302 (2004)

Follicle Segmentation from Ovarian USG Image Using Horizontal Window Filtering and Filled Convex Hull Technique

Ardhendu Mandal, Manas Sarkar, and Debosmita Saha

Abstract Ultrasound imaging is the best medical imaging technology to observe and monitor the growth and physiological status of the follicles, most importantly the paramount or dominant follicle in the female's ovary. But ultrasound images are always heavily poisoned by speckle noises although it is extensively used in infertility treatment. In this paper, a segmentation technique has been developed and discussed to completely remove the speckle noises and segment different follicles from ultrasound images. The proposed segmentation technique used a 20 pixel long window and standard deviation of the USG image for smoothing and despeckling the image. Further, morphological opening followed by morphological closing operations have been applied to the image for removing the paper and salt noise. Next, segmentation of the follicles is done by finding the active contours and filled convex hull from the intermediate USG image that contains only the follicles those are bright i.e. white in color with a black background. Follicles are properly classified and detected by applying a set of relevant parameters. Finally, a comparative study has been presented between the experimental results and inferences made by the experts to validate the result towards determining the degree of accuracy of the proposed technique.

Keywords Ultrasound image · Image segmentation · Active contour · Convex hull · Salt and paper noise · Image despeckling · Paramount or dominant follicle · Ovary · Morphological opening and closing

---

A. Mandal · M. Sarkar (✉)
Department of Computer Science and Application, University of North Bengal, Siliguri 734013, West Bengal, India
e-mail: sarkarmanas@nbu.ac.in; sarkarmanas22@gmail.com

A. Mandal
e-mail: am.csa.nbu@nbu.ac.in; am.csa.nbu@gmail.com

D. Saha
Department of Computer Science, University of Gour Banga, Malda 732103, West Bengal, India
e-mail: debasmita.saha@hotmail.com

D. Bhattacharjee et al. (eds.), *Proceedings of International Conference on Frontiers in Computing and Systems*, Advances in Intelligent Systems and Computing 1255,
https://doi.org/10.1007/978-981-15-7834-2_52

1 Introduction

Proper Analysis of the health and condition of developing follicles in female's ovary is the driving force for diagnosing the female's reproductive system. In general, this is performed by scrutinizing the Ultrasound images of ovary by the human experts. But this procedure has two major pitfalls. Firstly, it is time consuming and secondly, the accuracy level totally depends on the experts that may lead to error due to misjudgments by the experts.

Today we live in the modern era of computer applications which is becoming an inseparable part of whole sphere of our life. It plays a vital role in modern medical diagnosis too, especially in medical imaging and analysis techniques. Positron Emission Tomography (PET), Ultrasound Imaging (USG), Computerized Tomography (CT) Magnetic Resonance Imaging (MRI) etc. are the widely used medical imaging techniques used today. Because of its low cost, portability, noninvasiveness nature, Ultrasound medical imaging is highly preferred for diagnosis purpose. But Ultrasound images are always contaminated with speckle noises which are produced due to the reflection of high frequency sound wave from tissues under the epidermis.

To make the Ultrasound Images more useful for diagnosis, we always need a proper despeckling and segmentation techniques. Very limited work has been done on these aspects. Potocnik and Zazula [1] segmented follicular ultrasound images based on an optimal thresholding applied to coarsely estimated ovary. However, this method does not produce optimal segmented results. Cigale and Zazula [2] utilized cellular automata and cellular neural networks for the follicle segmentation. Although the result obtained from this method is very promising, but need further improvement for perfect segmentation of the follicles. Sarty [3] used a semi-automated system for follicle outer wall segmentation which employs watershed segmentation technique [4] using binary mathematical morphology. Hiremath and Tegnoor [5] used edge based segmentation method with Gaussian filter for segmenting follicles from ultrasound images. But due to poor speckle reduction, the performance of these above said techniques have poor while for clear follicle segmentation. Recently, some further works have been done on automatic follicle segmentation. Li et al. [6] have done automatic segmentation using CR-Unet which incorporates the spatial recurrent neural network (RNN) into a plain U-Net. Eliyani, Hartati and Musdholifah [7] recently used an active contour based method to segment object based on the similarity of follicle shape feature. Gopalakrishnan and Iyapparaja [8] designed an algorithm which automatically discovers follicles from the ultrasound images efficiently using active contours with modified Otsu threshold value. Zeng and Jun [9] automatically segment follicles based on faster R-CNN method which uses Ross Girshick's end-to-end neural network that combines object detection and classification. Wanderley et al. [10] has developed the first fully supervised fully convolutional neural network (fCNN) for ultrasound image segmentation which is quite impressive.

This paper proposed a novel technique which will not only remove the speckle noise significantly but also perform a perfect segmentation of follicles.

2 Data Collection

The research study has been done on the data provided by Swagat Diagnostic Centre, Dhubri, Assam, India, under the supervision of Dr. Mohammad Laskar Ali, Radiologist, Dhubri Govt. Hospital, Assam, India. The patients were in the age group of 23–35 years. They were well informed about the study and a consent form was signed by the willing patients. As most of patients were pregnant or had ovaries not containing follicles, only 15 numbers of ovarian ultrasound images containing follicles have been collected within the duration of two months. These collected digital USG images has been used for the study purpose. We also printed these images and got the follicles marked by Dr. Mohammad Laskar Ali that we use as the ground truth for the performance evaluation of the proposed technique.

3 Proposed Method

The ovarian ultrasound images, containing number of follicles of different size and shape, endometrium, blood vessels, are always contaminated with speckle noises. Ovarian follicles are spherical fluid-filled structures which may have diameter within the range of 2–25mm. As follicles are sacs, filled up with liquefied, they appear as gloomy circular or elliptical object because they passes most of the ultrasound waves which makes them darker compared to their neighbor. Proposed method aims to eliminate the speckle noise contamination and properly segment the follicles present in the USG image. An original sample of ovarian ultrasound image is shown in Fig. 1.

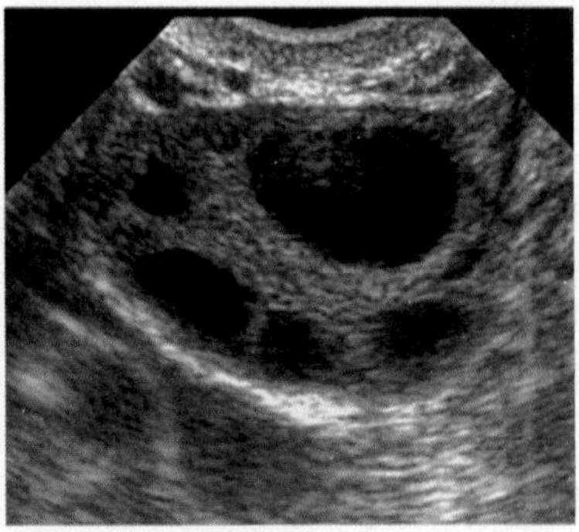

Fig. 1 Ultrasound image of ovary with multiple follicles

4 USG Image Preprocessing

The blessing of ultrasound image comes with the course of speckle noise contamination. De-speckling the USG image is considered to be the first move at the preprocessing stage in the algorithm proposed. To serve this purpose, mean (1) and standard deviation (2) of the noisy image is computed first using the following formula:

$$\mu = \frac{1}{N}\sum_{i=0}^{N} X_i \tag{1}$$

$$\sigma = \sqrt{\frac{\sum_{i=0}^{N}(X_i - \mu)^2}{N}} \tag{2}$$

Here, μ and σ are the mean and standard deviation of the USG image pixels under consideration.

Gray scale ultrasound images are represented by two dimensional arrays of pixel values in computer memory. Initially, this technique flattens the image array in row major order. Then, set a 20 pixel long horizontal window to find the mean of continuous 20 pixels starting from the pixel under consideration in the flatten array. Intensity value of the pixel (P_i) under consideration is decided on the basis of calculated local mean (μ_L) (3) and standard deviation (σ) of the whole image.

$$L = \frac{1}{N}\sum_{i=L}^{L+20} P_i \tag{3}$$

If the local mean is greater than double of the square of image standard deviation σ, then replace the original pixel value by (mean(μ) + 2∗standard deviation(σ)). Otherwise replace the pixel value by the minimum value within the 20 pixel long window. This is done for every pixel in the flatten image array under consideration.

Now, reshape the flatten array to its original shape. Then, if the partially filtered USG image is displayed, it will contain number of white dots (salt noise) in the image (Fig. 2a). Next we apply morphological erosion operation (4) with a disk shape structuring element of size 4 to dispel the salt noises still present in the partially filtered USG image (Fig. 2b).

$$G_f = G_{pf} \Theta B \tag{4}$$

Here, G_{pf}, G_{f} and B are the partially filtered image, filtered image and disk shape structuring element of size 4 respectively. Figure 2b shows the filtered image after removing the high intensity speckle noise by applying the above said techniques.

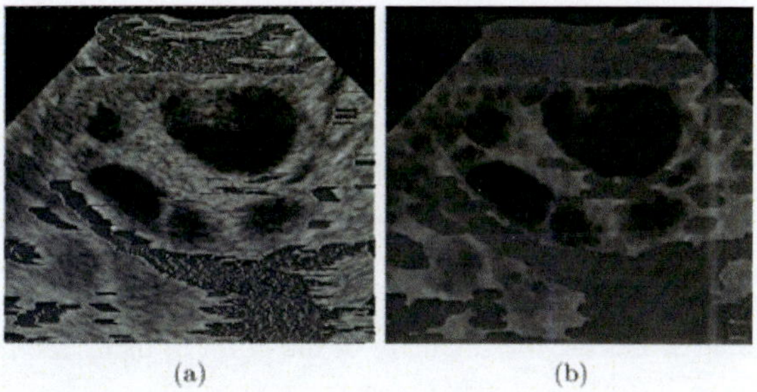

Fig. 2 **a** After performing windowing operation. **b** Final filtered image

4.1 Proposed Follicle Segmentation Technique

In the post processing approach the filtered USG image obtained using Horizontal Window Filter (HWF) is used to segment the follicles present in the image by finding convex hull using the active contours [11] of the homogeneous region. At the beginning, the filtered image is binary thresholded by using 2 * standard deviation (σ) as the threshold value. Then, active contours of the thresholded image are computed. Active contours are the set of points describing the boundaries of the homogeneous regions in an image. Further, the convex hull of each homogeneous region is computed using the related contour points. This will produce the number of bounded areas those might not be completely filled. In the next step, these partially filled hull areas are filled. Further, morphological opening by closing operation (5) using a disk shape structuring element of size 4 is applied on the hull filled image to remove the bright points (i.e. salt noises) and to smooth the boundaries of the segmented follicles.

$$G_S = (G_H \circ B) \bullet B \tag{5}$$

where, G_S, G_H and B are the finally segmented image, filled hulled image and structuring elements respectively.

Finally, Otsu's automatic binary thresholding [12] is applied to make the follicle bright and the background region dark. Any segmented object that touched image border is discarded from the processed image.

Follicles are distinguished based on three geometric parameters which are area (R), ratio of major axis Versus minor axis (ρ) and circularity (C_r) of the object. Here R means total number of pixels inside an object, ρ is the ratio breadth and height of the rectangle which exactly fit the object and (C_r) is the ratio of area of an object to the area of a circle with the same convex perimeter. The parametric values are used for identification of follicles as follows:

An object is follicle if:

1. $100 < R < (\frac{W}{2} \times \frac{H}{2})$, Here W and H are width and height of image
2. $\rho > 0.45$
3. $C_r > 0.237$

4.2 *The Proposed Algorithm:*

Step 1: Start
Step 2: Find the Standard deviation of the original ovarian USG image.
Step 3: Flatten the two dimensional image arrays in row major order.
Step 4: For each point (P_i) in the flatten array find local mean and local minima using 20 pixels long window starting from P_i.
Step 5: Decide the intensity of the point Pi using the following rule:

- If (local mean value(μ_L) is $> 2\times$ (Standard Deviation)^2 then:
  - Intensity value of P_i = (mean+2×Standard Deviation)
- Else
  - Intensity value ofP_i =Minimum (Pixel values among the 20 pixels long window)

Step 6: Reshape the flatten array to its original shape.
Step 7: Apply morphological erosion operation using a disk shape structuring element of size 4.
Step 8: Apply Binary Thresholding on the filtered image using 2*standard deviation (σ) as the threshold value.
Step 9: Find the convex hull of the filtered image by finding the active contours.
Step 10: Fill the bounded hull areas to complete the follicle segmentation process.
Step 11: Apply Otsu's automatic binary thresholding [12] to make the follicles bright and the background.
Step 12: Identify the follicles based on the aforesaid parameters.
Step 13: Stop.

5 Experimental Results

The sequel of images, the original image, intermediate filtered image, filtered image, convex hulled intermediate image, filled hull image and finally segmented images (Fig. 3) are presented in this section. The final segmented image clearly shows the follicles present in the USG image under consideration.

Figure 3 shows how the follicles are beautifully segmented.

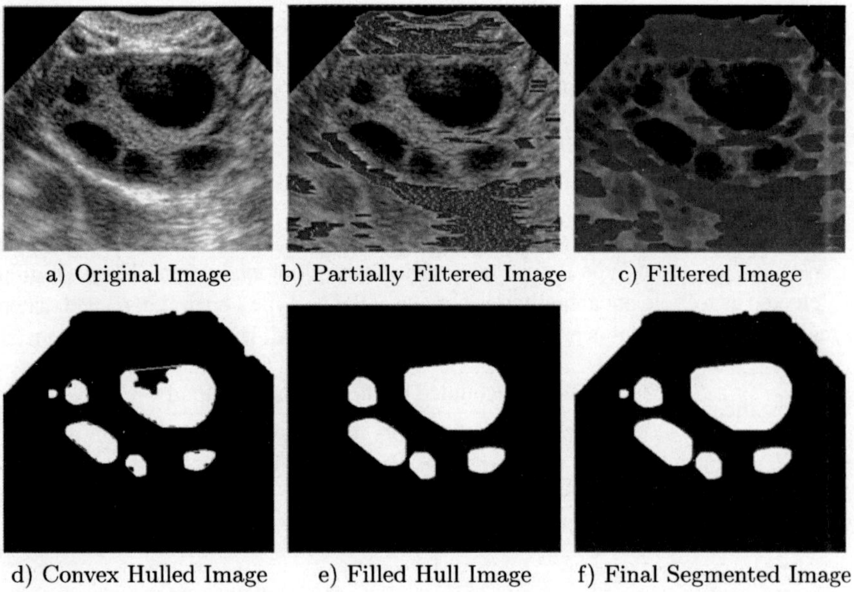

Fig. 3 Sequel of images

6 Performance Evaluation

Comparison between the follicle count segmented by the algorithm designed above and follicle count identified by the Radiologist for the same image provides the Performance Evaluation. Figure 4 shows two images, first one is an USG image on which radiologist (Dr. Mohammad Laskar Ali) marked the follicles and other one contains the segmented follicles of the same USG image.

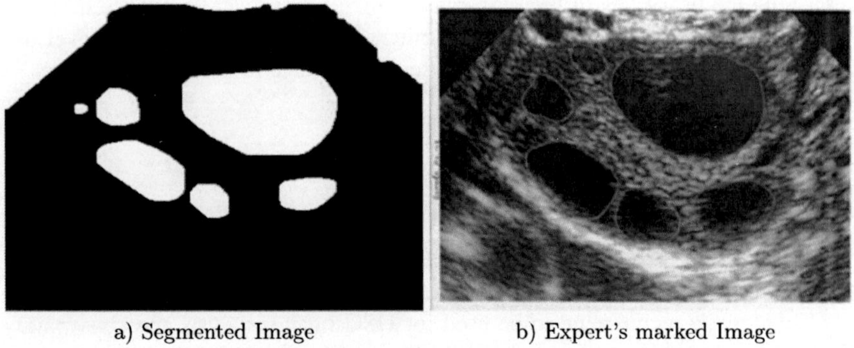

Fig. 4 **a** Segmented image, **b** Expert's marked USG image

Table 1 Performance evaluation of the proposed method

No. of image considered	Follicles tracked out by proposed method	Follicles tracked out by medical expert	Type I error (α)	Type II error (β)
15	47	58	6	17

We use Type I and Type II error rate as the basis of our evaluation. When a region is detected as follicle but actually it is not, we call it as Type I error (i.e. α) and, when a region is follicle but it is not detected as follicle, we call it as Type II (i.e. β) error.

$$\text{Classification rate} = \frac{\text{Follicles counted by the proposed method}}{\text{Follicles counted by the medical expert}} \times 100\% \tag{6}$$

$$\text{Precision rate} = \frac{\text{Follicles counted by the proposed method}}{\text{Follicles counted by the proposed method} + \text{TypeI error}} \times 100\% \tag{7}$$

Applying the proposed method it has been found that the classification rate (6) is 81.03% and the precision rate (7) is 88.68%. The False Acceptance Rate or α error or Type I error is 10.34%. The False Rejection Rate or β error or Type II error is 29.31%. Table 1 shows the Performance evaluation results:

7 Future Scope

This paper discussed a technique that concentrated only on segmenting follicles present in an ovarian USG image. But, it does not describe any metadata or further analysis about the segmented follicles by itself. Further development should be carried out in this direction by setting different suitable parameters for analyzing and describing the properties of the segmented follicles those can be used to diagnose related physical problems and diseases. Further, the precision and accuracy level also need to be improvised towards achieving cent percentage performance measure.

8 Conclusion

As the proposed technique attained success rate of 81.03% in classification and 87.23% in precision to successfully segment the follicles from USG images, hence this technique is recommend to be used for USG image analysis and abnormality diagnosis in the concerned domain. It further can be concluded that, the proposed technique shall guide the future research and development in the concerned field.

Acknowledgements The authors are grateful to the Swagat Diagnostic Centre for providing the ovarian USG images. We are very much thankful to Radiologist Dr. Mohammad Laskar Ali (MBBS, DMRD) for giving his diagnosis report and providing useful medical information about infertility. The authors express their sincere gratitude for consistent support and coordination to the Physician and the Diagnostic Centre mentioned above.

References

1. Potocnik, B., Zazula, D.: Automated ovarian follicle segmentation using region growing. In: International Conference on Information Technology Interfaces, pp. 157–162. IEEE (2000)
2. Cigale, B., Zazula, D.: Segmentation of Ovarian Ultrasound Images Using Cellular Neural Networks, pp. 563–581. World Scientific (2004)
3. Sarty, G.E., Liang, W., Sonka, M., Pierson, R.A.: Semiautomated segmentation of ovarian follicular ultrasound images using a knowledge-based algorithm. Ultrasound Med. Biol. **24**, 27–42 (1998) (Elsevier)
4. Shafarenko, L., Petrou, M., Kittler, J.: Automatic watershed segmentation of randomly textured color images. IEEE Trans. Image Process. **6**, 1530–1544 (1997)
5. Hiremath, P.S., Tegnoor, J.R.: Automatic Detection of Follicles in Ultrasound Images of Ovaries Using Edge Based Method. Special Issue on RTIPPR, pp. 120–125. IJCA (2010)
6. Li, H., Fang, J., Liu, S., Liang, X., Yang, Xin and Mai, Zixin and Van, Manh The and Wang, Tianfu and Chen, Zhiyi and Ni, Dong, CR-Unet: A Composite Network for Ovary and Follicle Segmentation in Ultrasound Images,journal of biomedical and health informatics, IEEE,2019
7. Hartati, S., Musdholifah, A., et al.: Machine learning assisted medical diagnosis for segmentation of follicle in ovary ultrasound. In: International Conference on Soft Computing in Data Science, pp. 71–80. Springer (2019)
8. Gopalakrishnan, C., Iyapparaja, M.: Active contour with modified Otsu method for automatic detection of polycystic ovary syndrome from ultrasound image of ovary. In: Multimedia Tools and Applications. Springer, Berlin (2019)
9. Zeng, T., Liu, J.: Automatic detection of follicle ultrasound images based on improved faster R-CNN. J. Phys. Conf. Ser. **1187**(10), 042112. IOP Publishing (2019)
10. Wanderley, D.S., Carvalho, C.B., Domingues, A., Peixoto, C., Pignatelli, D., Beires, J., Silva, J., Campilho, A.: End-to-end ovarian structures segmentation. In: Iberoamerican Congress on Pattern Recognition, pp. 681–689. Springer, Berlin (2018)
11. Hemalatha, R.J., Thamizhvani, T.R., Josephin Arockia Dhivya, A., Josline Elsa Joseph, Babu, B., Chandrasekaran, R.: Active contour based segmentation techniques for medical image analysis. In: Medical and Biological Image Analysis. BoD–Books on Demand (2018)
12. Kurita, T., Otsu, N., Abdelmalek, N.: Maximum likelihood thresholding based on population mixture models. Pattern Recogn. **25**, 1231–1240 (1992) (Elsevier)

Evolution of E-Sensing Technology

Aramita De Das and Ankita Pramanik

Abstract This is the era of automation, in which the real and the virtual world are fast converging. The established notions are being rapidly changed by the use of robotics, machine learning (ML) and artificial intelligence (AI). Automation can be found in normal day-to-day life applications, industries, space and health care everywhere. In the zeal to achieve better sensing techniques and improved output results from machines, electronic or robotic forms of human body parts and organs are being developed. Robotic arm, electronic nose, electronic tongue, robotic moving fingers, etc. are few such examples. E-sensing is garnering a huge interest due to its ability to mimic human behaviour. Thus, a detailed study into e-sensing technologies is the need of the hour. In this paper, a brief study on the recent works on e-sensing technology is presented. The current study sheds light upon the definition, classification, practical application of e-nose and e-tongue for different type of measurements.

Keywords E-nose · E-tongue · Artificial neural networks (ANNs) · Principal component analysis (PCA)

1 Introduction

Over the last 30 years, electronic sensing (e-sensing) methods have been playing a significant role in different fields [1]. All the e-sensing technologies, electronic nose (e-nose) and electronic tongue (e-tongue) are gaining huge popularity in medical sciences and also in industrial applications. E-nose and e-tongue are also known as artificial nose and tongue. These devices have undergone rapid development to suit industrial applications. E-nose and e-tongue are applied to detect and recognize odours and flavours. Application of e-sensing technology can be found in the

A. De Das (✉)
Institute of Engineering and Management, Kolkata, India
e-mail: aramita.de@gmail.com

A. Pramanik
Indian Institute of Engineering Science and Technology, Shibpur, Shibpur, India
e-mail: ankita@telecom.iiests.ac.in

D. Bhattacharjee et al. (eds.), *Proceedings of International Conference on Frontiers in Computing and Systems*, Advances in Intelligent Systems and Computing 1255,
https://doi.org/10.1007/978-981-15-7834-2_53

field of medical science [2], food industries, food control monitoring, mining and many more. It has been developed as a system of automatic detection, reorganization and classification of odours, taste, etc. [3]. From the imprecise interactions, the majority of odorants and tastes are recognized through a fusion of the global chemical compounds.

The design of e-nose is heavily inspired from the biological nose. By olfactory cortex neural network, the electro-neurographic signals are being interpreted. In the basic design of e-nose system, the components replicate some of the important biological parts of human body. Lung is implemented as a pump. The vibrissae and mucous of human nose are implemented as an inlet sampling system. An array of sensors has been introduced as the olfactory receptors. Neural network of olfactory cortex have been implemented as a signal processing unit [4]. Odour plays a major role in e-nose system. It is defined as a property of physical sensation or substance, occurring from stimulation of human organs. The conventional method to measure odour is olfactometry with the help of test analysers [5].

E-nose technology has an enormous application in the medical sector. It has been implemented in diagnostics, pathology, real time monitoring of patient, development of medicines, treatment of physiology, telemetry healthcare system, healing in wound and grafting, etc. In biomedicine e-nose technology is helping solve numerous, unique and complex problems. The e-sensing technology is serving medical science since 1980, and also helping mankind to discover something new in future. The collective uses of e-nose with other medical instrument will help to improve results in medical sector. It will help in remote health monitoring, helpful diagnosis, for decision-making by the doctors and in telemedicine, etc. [2]. E-sensing technology, especially e-tongue and e-nose have created a bench mark in present century. As the sensors are non-contact type, they are used for real time monitoring. E-nose, e-tongue or a combination of both are deployed in various sectors to detect and solve variety of problems, namely identification of chronic cancers, food testing and tea tasting.

Apart from medical science e-sensing is also used in other industries, namely tea industry and food packaging industry. In tea manufacturing industry, the quality of the tea is traditionally judged manually by a "tea tester". However, this procedure is costly and not accurate for every batch of tea packaging. Testing by e-nose is rapidly replacing the manual testing method. E-nose is also finding use in pollution control, chemical industries, cosmetic industries, etc.

A discussion on the evolution of e-sensing technology to endow with better technologies in near future is the need of the hour. To address the discussed requirements, the present work discusses the progress of e-sensing technology and the various application areas of this technology. This work also suggests the various future works that can be undertaken in the future.

The rest of the paper is presented as follows: A brief write-up on the existing e-sensing technology along with its classification and applications is presented in Sect. 2. Section 3 presents the state of the art of e-sensing technology, e-tongue and e-nose. Finally, the conclusions along with the scope for future work are described in Sect. 4.

2 E-Sensing Technology

Since 1982, researchers have been developing the electronic sensing technology to determine the human senses using different types of sensor arrays. E-nose and e-tongue are most popular e-sensing technologies. E-nose and e-tongue have been deployed in various research and industrial applications. The various typical application areas are given in Fig. 1.

2.1 Traditional E-Nose and E-Tongue Systems

To recognize simple, as well as complex odorants, e-nose technology is being used since the last 40 years. It can be used to identify various gases [4]. E-tongue is an instrument with a set of sensor arrays connected to a data analyzing system. The data analysis part includes artificial neural networks (ANNs), principal component analysis (PCA) [6]. It can detect the quality of food or complex sample for monitoring their quality.

2.2 Components of E-Nose

Sensing system and pattern recognition systems are the basic two components of e-nose. Sensing system measures the sample using different types of sensors. The metal oxide semiconductor (MOS), conducting polymer (CP), piezoelectric sensors and optical sensors are used frequently. The sensing system consists of sensors array. From this array, qualitative and quantitative measurements are collected [4]. After sensing the sample, the data needs to be analyzed. This process of analyzing data is

Fig. 1 Classification of e-sensing technologies

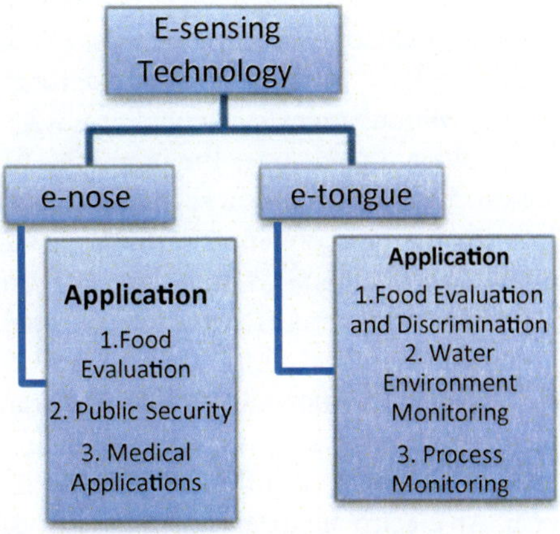

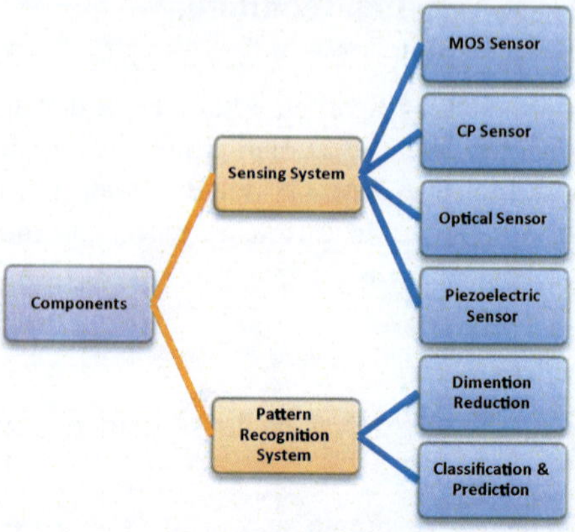

Fig. 2 Methodology of e-nose

known as pattern recognition system. After collecting the data from the sensor array a machine learning algorithm of the pattern recognition system has been processed. PCA, linear discriminate analysis (LDA), partial least squares (PLS) regression and ANN are the most commonly used pattern recognition system in e-sensing technology [4]. The various divisions of the methodology of e-nose are presented in Fig. 2.

3 E-Nose and E-Tongue: State of the Art

To detect gaseous element, MOS sensors are used frequently in e-sensor system. To measure organic compound CP sensor is used widely in the field of e-sensor technology.

E-nose have described by the authors as virtual sensor in paper [7]. It is different from the traditionally used chemical sensors arrays. In this work, the authors have demonstrated two different types of virtual sensors: VS1 and VS2. VS1 is signal based and VS2 is sensor based. Nano-structured ZnO-based film has been used in VS2. As a sensitive layer for quartz crystal microbalance (QCM) sensor, the ZnO-based film is used. Various types of alcohol are used as test samples. Under the illumination, the sensitive film changes the properties of absorption towards alcohols. To get the classification of alcohols a scanning electron microscope (SEM) image of sensitive ZnO film has been obtained. It shows the efficiency of application of virtual sensors. Figure 3 shows the SEM image of ZnO sensor film.

Alcohol, drugs, cocaine and tobacco consumption are increasing in a large amount throughout the world. Drug control board needs to measure more accurately to control and eliminate the rampant storage and transportation of the banned substances. E-tongue and e-nose is the instant solution to measure these kinds of unwanted element. In Fig. 4, analysis of different types of alcohol are being presented in a graphical form. An electrochemical sensor based system has been developed to address these

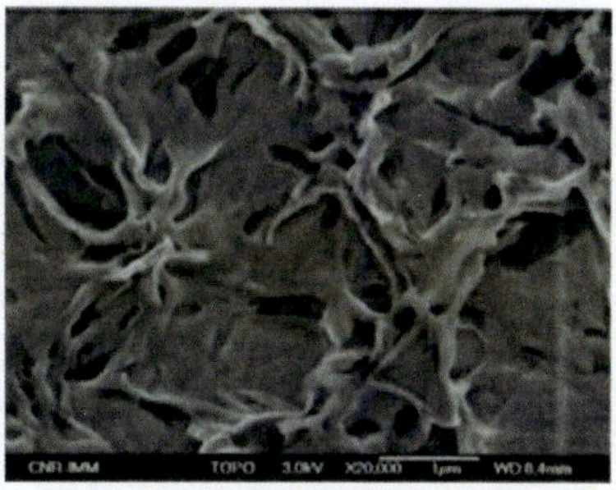

Fig. 3 SEM image of ZnO sensitive film

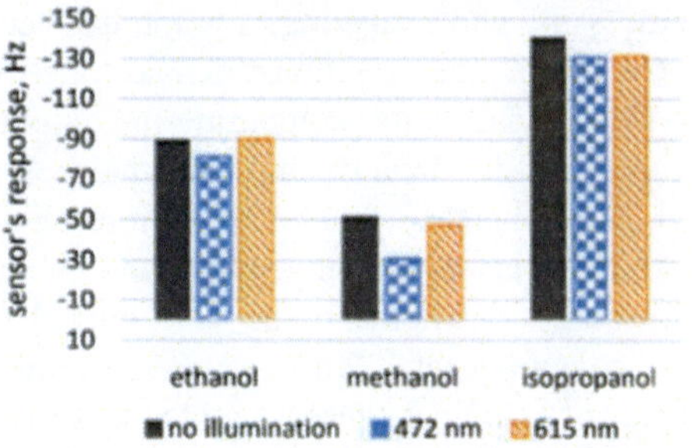

Fig. 4 Responses of sensors for alcohol

problems in work [8]. In their process, the sensor is developed from two different electrodes, glass carbon (GC) and platinum (Pt). The developed sensor was capable of discriminating and classifying five-cutting agents in cocaine. They have faced an unsatisfactory plot without a normalization process. By this means the concentration information stand in the way of the collected data. But finally they have developed a process to remove the concentration information [8]. E-nose has also been used to identify the different types of tobacco [6]. The authors have used three types of sensor array to determine the three different types of tobacco. They have collected the samples of tobacco from three different countries in Brazil, Flue Cured, Burley and Oriental. To calculate the relative response (Ra) the output data has been processed mathematically for each analyzed tobacco.

These days, pollution control board is using e-nose in many of its applications. An e-nose can be used in industrial odour and gaseous emission measurement and monitoring systems [5]. Numerous inorganic pollutants and organic and volatile organic compounds are emitted from industries. The inorganic pollutant consists of chlorine (Cl), ammonia (NH_3), sulphur dioxide (SO_2), carbon dioxide (CO_2), carbon monoxide (CO), nitric oxide (NO), nitrogen dioxide (NO_2) and hydrogen sulphide (H_2S). Common types of organic and volatile organic compounds are propane-2-ol, methanol, ethanol, acetone, etc. The authors have measured the odours concentration and intensity of these compounds by e-sensing technology. The work is also presents the results of field work. Due to the interference caused by environmental parameters such, as humidity and temperature, the measurement and controlling the concentration of gases can be taken up as future work.

A recent study has revealed that junk food is the main reason of obesity in young people. Main reason for this is the fried oil, which is used repeatedly. Monitoring of the fried oil has been done in the work in [9]. The work has developed a data combination of e-tongue and e-nose system with mixed edible-oil as test sample. Figure 5 shows a basic diagram of the combination system.

Authors have developed 2 air purification system for e-nose, one is used for gas injection another one is used for gas purification. Air purification system is connected through a pump. The samples are transported and analyzed by the array of sensors. The sensors are placed in the chamber. The data from e-nose is combined with the e-tongue system. A voltammetric e-tongue is used in this work. This system is consisting of sensor array, electrochemical workstation and computer. After data processing and analyzing, a graph has been obtained. The results show that when PCA is used, the detection efficiency of e-tongue system is improved than the e-nose system. Finally after combining the output data of 2 different systems (e-nose and e-tongue), significantly improves the acceptance rate.

The demand for bio-chemical information is rapidly growing in various sectors like financial sector, atmosphere and the public concern on health and quality, etc. To address this demand, the work in [10], explored a new concept on sensor arrays in e-tongue systems. The e-tongue systems ware manufactured for application in the fields of chemical sensors and biosensors. In Fig. 6, image of their work has been shown.

Tea is an expansively addicted beverage. Tea industries have a growing global market. The important qualities for tea industry are colour, aroma, flavour and strength. Human experts, known as "tea tester" analyze the flavour, aroma and colour of tea manually. In paper [11], the authors have developed an e-nose system to be used

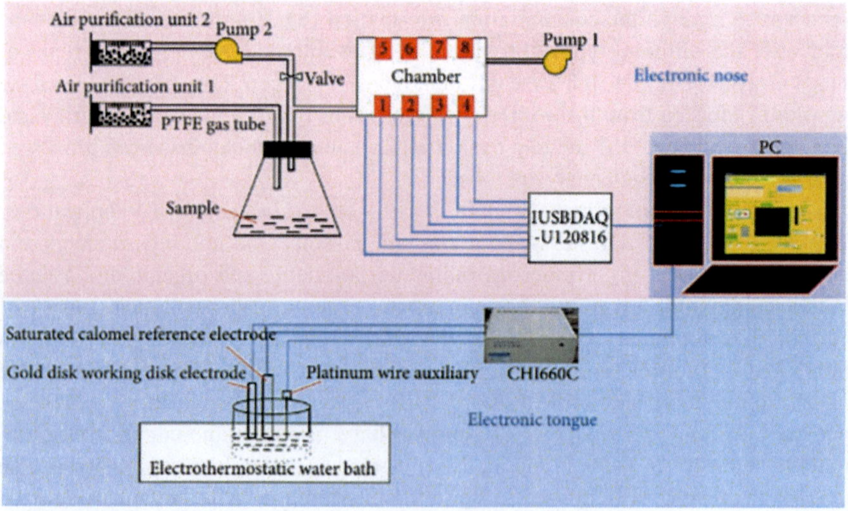

Fig. 5 Basic diagram of the combination system of e- nose and e-tongue

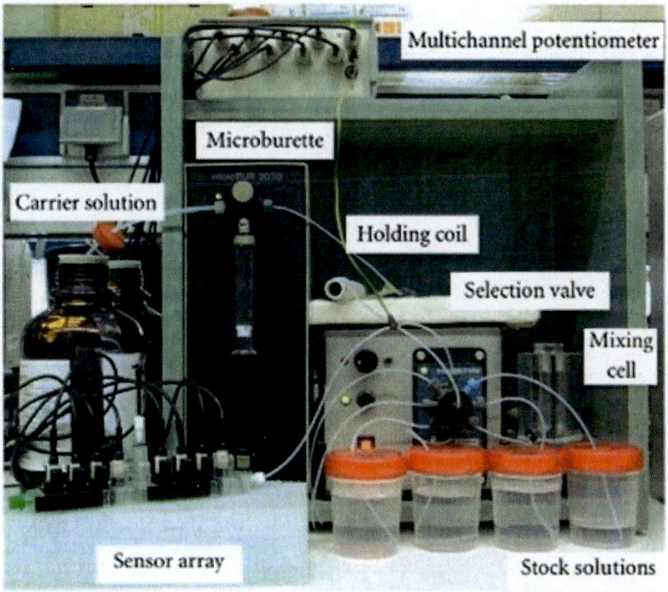

Fig. 6 Diagram of injection analysis system

in tea industry. In Fig. 7, the customized diagram of e-nose has been developed. Few samples of various kinds of black tea are processed by the e-tongue and the samples are accordingly classified. The results are presented in Figs. 8, 9 and 10. Then the data is analyzed through a three-layer multilayer perceptron (MLP) network. An instrument to predict or measure the quality of tea by measuring different kinds of chemical components, presented in tea was also developed. In addition to these, a computational model to obtain a better result from the available information was also developed.

A network system, similar to electrical circuit of e-tongue, for the categorization of different types of beverages like juice, black tea, beer has been developed in [12]. The samples are used as electrolyte in the system. A large amplitude pulse voltammetry (LAPV) voltage signal is used between two electrodes. The output of data samples of current (μA) were collected in a personal computer with developed software. Then it is interfaced in Lab-VIEW and a data acquisition card (DAS) to get the final result [12].

In paper [13], e-nose is used to detect cancer of human body. Samples from human body were processed through sensors, amplified and then data analysis was done through multi-component analysis, dimension reduction and pattern recognition. Though they have concluded that electronic nose is not a perfect solution for cancer diagnostics, but it would be implemented as a valid support in a non-invasive cancer detection strategy. In Fig. 11, the basic structure has been developed of electronic nose.

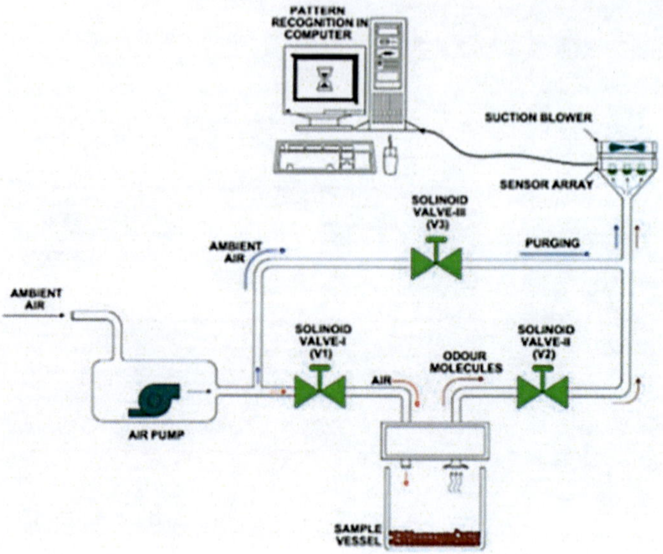

Fig. 7 Customized e-nose setup

Fig. 8 Compounds versus colour of tea

Compounds	Colour
Theaflavins	Yellowish Brown
Thearubigins	Reddish brown
Flavonol glycosides	Light yellow
Pheophorbide	Brownish
Pheophytin	Blackish
Carotene	Yellow

Fig. 9 Compounds versus taste in tea

Compounds	Taste
Polyphenol	Astringent
Amino acids	Brothy
Caffeine	Bitter
Theaflavins	Astringent
Thearubigin	Ashy and slight astringent

Fusion of e-nose and e-tongue has been used for animal source food authentication and quality assessment [14]. Animal foods like pork, cheese, egg, honey, butter, ham, etc. were used as test sample. They have made a chart to check the combination level of e-nose and e-tongue.

The work in [15], has proposed a low cost e-tongue to detect explosive material. The e-tongue consists of (i) electrodes, (ii) electronic equipment and (iii) a personal computer (PC) with a software application. Through different types of electrodes,

Compounds	Flavour
Linalool, Linalool oxide	Sweet
Geraniol, Phenylacetaldehyde	Floral
Nerolidol, Benzaldehyde, Phenyl ethanol	Fruity
Trans-2-Hexenal,n-Hexanal, Cis-3-Hexenol, Grassy, b-Ionone	Fresh Flavour

Fig. 10 Compounds versus flavour in tea

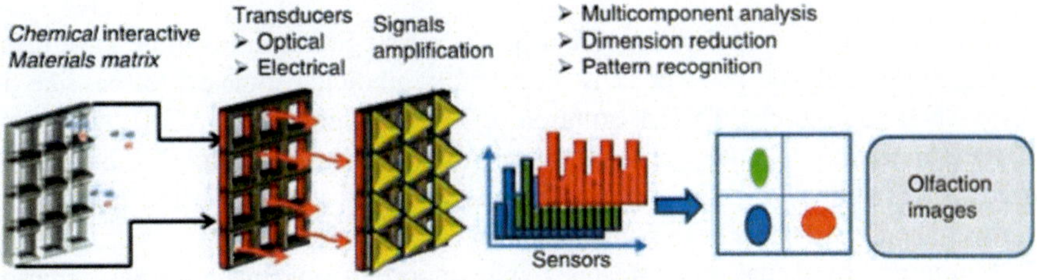

Fig. 11 Overview of the electronic nose measurement

electronic equipment have collected data and processed it in PC using software application. The electronic equipment has a potentiostat, a digital processing unit and an analogue unit. A potentiostat is placed in the electronic equipment, which is configured with three different modes. Counter electrode (CE), working electrode (WE) and reference electrode (RE) are the tree modes of potentiostat. The e-tongue system is planned to perform with various types of electrodes configuration. The designed e-tongue system is depicted in Fig. 12.

A USB to universal asynchronous receiver transmitter (UART) converter with a microcontroller (16 bit) has been connected in digital processing unit. Visual basic6

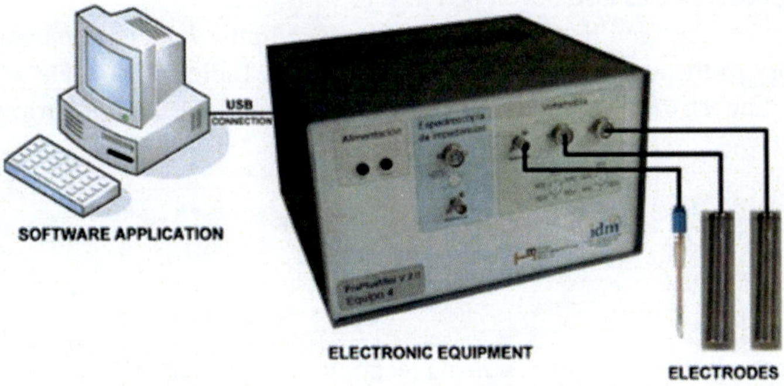

Fig. 12 Electronic tongue systems

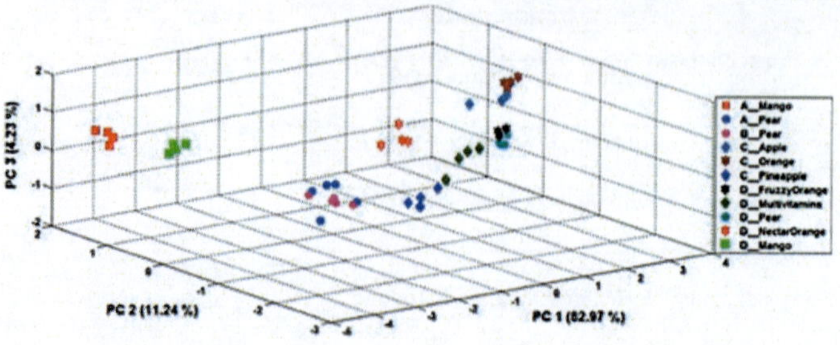

Fig. 13 A three dimension PCA plot analyzed on 11 fruit juice using e-nose

is used as a software application. Data pre-processing and multivariate have been used for data analysis. The work has successfully predicted some explosives like (i) 2,4,6-TNT: commercial TNT dissolution 0.13 M in acetonitrile (ACN); (ii) ACN: CH_3CN minimum 99.84%; (iii) potassium nitrate: KNO_3. PCA and PLS have been used in this study. To calibrate, analyze and authenticate a predicted model, the software application has been combined with electronic equipment.

Monitoring food quality and freshness is an important topic in the current scenario. In [16], authors have combined e-tongue and e-nose to identify the different fruit juices. They have collected total 46 fruit juices from 4 different manufacturers. In their e-nose system, total 5 numbers of gas sensors were used and they had designed their e-tongue system using six types of electrodes. The electrodes are potentiometric type electrode. (i) Sensor array, (ii) sampling vessel and (iii) data acquisition system (DAS) are the three parts of e-nose system. The e-tongue system has been designed with 1 sensor array, which consists of 6 chemical sensors (potentiometric type). From e-nose and e-tongue systems the data have been collected. Then the raw data have been analyzed by data processing techniques like PCA, CA orfuzzy ARTMAP ANN.

In aromatic or medicinal industries e-nose is playing an important role these days. To determine or to monitor different kinds of aromatic plants like saffron, flowers, medicinal plant, e-nose can be used [17]. E-nose was used to measure and monitor the quality of saffron in [17]. A statistical analysis using PCA and ANN based data analyzing process was also developed in Fig. 13.

Thus, it can be seen that, e-tongue and e-nose find a host of application areas. A summary of the above discussions is presented in Table 1. The table will help in understanding the e-sensing technology and its applications easily and in a systematic fashion.

4 Conclusions

In this paper a brief survey on e-sensing technology is presented. It can be concluded that the number of works on e-nose are much more compared to that of e-tongue. Several works combining e-tongue and e-nose to obtain improved results have also

Table 1 Summary table of above mentioned papers

Reference No.	Year of the paper	Types of E-sensor	Sample used	Area of research
[7]	2019	E-nose	Alcohols	Alcohol industry
[5]	2015	E-nose	Industrial odours and gas	Pollution control
[9]	2014	E-nose and E-tongue	Mixed edible-oil	Food oil industry
[10]	2012	E-tongue	Different types of chemical	Electrochemical industry
[11]	2008	E-nose	Black tea	Agro industry
[13]	2012	E-nose	Cancer cell	Medical industry
[14]	2017	E-nose and E-tongue	Food of animals	Animal food-processing manufacturer
[15]	2012	E-tongue	Trinitrotoluene (TNT)	Bomb squad
[16]	2014	E-nose and E-tongue	Fruit juice	Nutrition production company
[12]	2017	E-tongue	Black tea	Agro industry
[18]	2016	E-nose	Medicinal and aromatic plant	Agro industry
[19]	2015	E-nose	Human urine	Cancer treatment
[8]	2017	E-tongue	Cocaine samples	Anti narcotic squad
[17]	2015	E-nose	Saffron	Cosmetic industry
[6]	2018	E-nose	Tobacco	Tobacco industry

been undertaken. The application areas of e-sensing technology might be diverse but the basic parts of it are the same. The e-sensing technology consists of a sensor, signal amplification or any connection with the software module and data analysis. Various works have concentrated on the sensor part, whereas others have concentrated on the data analysis part. Various types of samples have also been used. A table has been created to visualize and understand e-sensing technology vividly.

Using e-nose and e-tongue, different types of parameters can be measured. A wireless system to transmit and receive data can be added with the current e-sensing system. Thus, IOT can be incorporated with e-sensing technology to offer improved solutions.

References

1. Wilson, Alphus, Baietto, Manuela: Applications and advances in electronic-nose technologies. Sensors **9**(7), 5099–5148 (2009)
2. Wilson, A.D.: Future applications of electronic-nose technologies in healthcare and biomedicine. In: Akyar, Isin, ed *Wide Spectra of Quality Control. InTech Publishing, Rijeka, Croatia. 267–290.* (2011): 267-290
3. Keller, P.E., Kangas, L.J., Liden, L.H., Hashem, S., Kouzes, R.T.: Electronic noses and their applications. In: World Congress on Neural Networks (WCNN), pp. 928–931 (1995)
4. Zou, Y., Hao, W., Xi, Z., D, H., Ping, W.: Electronic nose and electronic tongue. In: Bioinspired Smell and Taste Sensors, pp. 19–44. Springer, Dordrecht (2015)
5. Deshmukh, S., Bandyopadhyay, R., Bhattacharyya, N., Pandey, R.A., Jana, A.: Application of electronic nose for industrial odors and gaseous emissions measurement and monitoring–an overview. Talanta **144**, 329–340 (2015)
6. Esteves, C., Henrique, A., Iglesias, B.A., Ogawa, T., Araki, K., Hoehne, L., Gruber, J.: Identification of tobacco types and cigarette brands using an electronic nose based on conductive polymer/porphyrin composite sensors. ACS Omega **3**(6), 6476–6482 (2018)
7. Burlachenko, J., Kruglenko, I., Manoylov, E., Kravchenko, S., Krishchenko, I., Snopok, B.: Virtual sensors for electronic nose devises. In: 2019 IEEE International Symposium on Olfaction and Electronic Nose (ISOEN), pp. 1–3. IEEE (2019)
8. Silva, T.G., Paixão, T.R.L.C.: Development of an electronic tongue to distinguish cutting agents in cocaine samples to understand drug trafficking. In: 2017 ISOCS/IEEE International Symposium on Olfaction and Electronic Nose (ISOEN), pp. 1–3. IEEE (2017)
9. Men, H., Chen, D., Zhang, X., Liu, J., Ning, K.: Data fusion of electronic nose and electronic tongue for detection of mixed edible-oil. J. Sens. **2014** (2014)
10. Del Valle, M.: Sensor arrays and electronic tongue systems. Int. J. Electrochem. **2012** (2012)
11. Bhattacharyya, Nabarun., Bandyopadhyay, Rajib., Bhuyan, Manabendra., Tudu, Bipan., Ghosh, Devdulal, Jana, Arun: Electronic nose for black tea classification and correlation of measurements with "Tea Taster" marks. IEEE Trans. Instrum. Meas. **57**(7), 1313–1321 (2008)
12. Kumar, S., Ghosh, A., Tudu, B., Bandyopadhyay, R.: An equivalent electrical network of an electronic tongue: a case study with tea samples. In: 2017 ISOCS/IEEE International Symposium on Olfaction and Electronic Nose (ISOEN), pp. 1–3. IEEE (2017)
13. D'Amico, A., Di Natale, C., Falconi, C., Martinelli, E., Paolesse, R., Pennazza, G., Santonico, M., Jason Sterk, P.: Detection and identification of cancers by the electronic nose. Expert Opin. Med. Diagn. **6**(3), 175–185 (2012)
14. Rosa, Di., Rita, Ambra., Leone, Francesco., Cheli, Federica, Chiofalo, Vincenzo: Fusion of electronic nose, electronic tongue and computer vision for animal source food authentication and quality assessment—a review. J. Food Eng. **210**, 62–75 (2017)
15. Garcia-Breijo, E., Masot Peris, R., Olguín Pinatti, C., Alcañiz Fillol, M., Ibáñez Civera, J., Bataller Prats, R.: Low-cost electronic tongue system and its application to explosive detection. IEEE Trans. Instrum. Meas. **62**(2), 424–431 (2012)
16. Haddi, Z., Mabrouk, S., Bougrini, M., Tahri, K., Sghaier, K., Barhoumi, H., El Bari, N., Maaref, A., Jaffrezic-Renault, N., Bouchikhi, B.: E-Nose and e-Tongue combination for improved recognition of fruit juice samples. Food Chem. **150**, 246–253 (2014)

17. Heidarbeigi, K., Saeid Mohtasebi, S., Foroughirad, A., Mahdi Ghasemi-Varnamkhasti, Rafiee, S., Rezaei, K.: Detection of adulteration in saffron samples using electronic nose. Int. J. Food Prop. **18**(7), 1391–1401 (2015)
18. Kiani, S., Minaei, S., Ghasemi-Varnamkhasti, M.: Application of electronic nose systems for assessing quality of medicinal and aromatic plant products: a review. J. Appl. Res. Med. Aromatic Plants **3**(1), 1–9 (2016)
19. Westenbrink, E., Arasaradnam, R.P., O'Connell, N., Bailey, C., Nwokolo, C., Dev Bardhan, K., Covington, J.A.: Development and application of a new electronic nose instrument for the detection of colorectal cancer. Biosens. Bioelectron. **67**, 733–738 (2015)

Extraction of Leaf-Vein Parameters and Classification of Plants Using Machine Learning

Guruprasad Samanta, Amlan Chakrabarti, and Bhargab B. Bhattacharya

Abstract The venation network present in plant leaves carries a fingerprint of the species and their analysis is likely to provide deep insights about the identity of the plants and the surrounding climate. Since photosynthesis has a direct correlation with the microfluidic vein networks, such information will be immensely useful to plant biologists from agricultural and ecological perspectives. Albeit research in leaf-venation patterns has recently received attention in computational botany, very little work is known that focuses on the extraction of suitable features of underlying vein network. In this paper, for the first time, we define certain graph-theoretic features considering it as a planar graph, which can be extracted from the skeletons of the leaf-images of a leaf-vein network. We study venation patterns in several leaf-images for three different trees (Jackfruit, Mango and Peepal) that are abundant in the Indian subcontinent. Our analysis on the extracted vein parameters based on K-means clustering and KNN-classification yields encouraging initial results.

Keywords Computational botany · Graph theory · Pattern recognition · Machine learning

1 Introduction

Leaf-Veins are composed of Xylem and Phloem. Xylem consists of dead cells which carry water and other nutrients required for the growth of the plants. Phloem carries prepared food from the leaves and distributes the same throughout the plant. The

G. Samanta (✉) · A. Chakrabarti
A.K. Choudhury School of Information Technology, University of Calcutta Technology Campus, Salt Lake, Sector-III, Kolkata 700106, India
e-mail: samantagp@rediffmail.com

A. Chakrabarti
e-mail: acakcs@caluniv.ac.in

B. B. Bhattacharya
Indian Institute of Technology Kharagpur, Kharagpur 721302, India
e-mail: bhargab.bhatta@gmail.com

D. Bhattacharjee et al. (eds.), *Proceedings of International Conference on Frontiers in Computing and Systems*, Advances in Intelligent Systems and Computing 1255,
https://doi.org/10.1007/978-981-15-7834-2_54

underlying microfluidic network of the vein patterns can be envisaged as an embedded planar graph and ideally, it should satisfy Euler's Formula on a connected planar graph. If $G(V, E)$ is a connected planar graph, the number of vertices (V), the number of edges (E) and the number of faces (F) are related by the Formula ($V - E + F = 2$) [1]. We collect a number of leaves from three different trees and capture their digital images. The skeleton of the venation structure present in a leaf-image is highlighted using an edge-thinning technique following binarisation. Several stamp-sized images are then cut out from a leaf-image to create a sample database. Vein parameters are extracted from each sample and compared to check the extent of relational match with Euler's Formula. The classification of three classes of Plants has been made based on the data set shown in Table 1.

The document is organised in several sections. Section 2 contains related work in this field. Section 3 contains pre-processing of three classes of tree-leaves (Jackfruit, Mango and Peepal) and sample images from each of the classes under study; this section also contains a set of thirty images, ten stamp-size image segment from each class. Section 4 describes the methodology used for processing the images created in Section 3. Section 5 reports the results of the processing in the previous section. Section 6 contains the results on classification. Section 7 describes the method of identification of an unknown leaf. Concluding remarks appear in Section 8.

2 Related Work

The plant is one of the most important forms of life on the planet. Plant recognition and identification are important for agriculture and ecology. Leaf-vein analysis is still an unexplored area which has a link with the genetics of plants [2, 3]. Leaf features play an important role in plant taxonomy [3, 4]. The type of leaf-vein structure is an important morphological feature in botanical science [5–7]. Plants need the right proportion of nutrients for their growth and reproduction. The visual inspections are not enough to identify such symptoms. These can be detected by studying deformation in vein structure and the edges of the leaves [8, 9].

3 Pre-Processing

3.1 Extraction of Leaf-Vein Skeleton

Fresh leaves of three classes of trees (Jackfruit, Mango and Peepal) were collected from the garden and were kept under water for six weeks to get the vein structure after removal of the green pigment and other materials. Figure 1 shows the representative samples of the skeletons of three leaves from three types of trees. The images were captured by the Digital Camera of iPhone 6S Plus (12 Mega pixels).

Fig. 1 **a** Peepal, **b** Jackfruit, **c** Mango

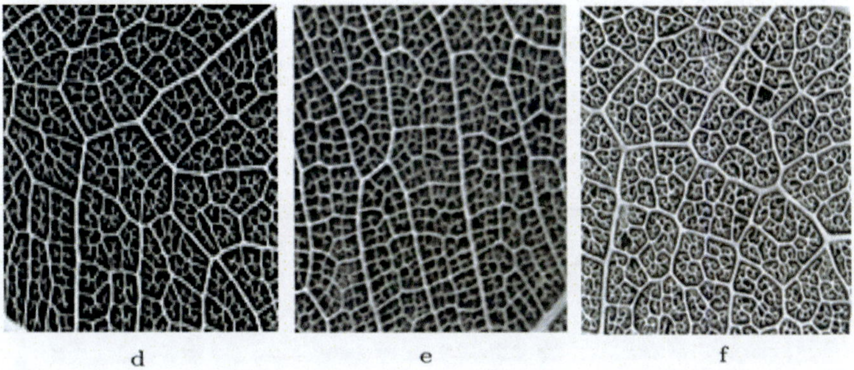

Fig. 2 Sections of 400 × 400 images of **d** Peepal, **e** Jackfruit and **f** Mango

3.2 Selection of Image Segments

The sections of the images where veins were prominently visible were selected as shown in Fig. 2. Three such representative samples were selected, one from each class. Each of these images is of 400 × 400 pixels (not drawn to the scale).

3.3 Stamp-Size Images for Analysis

Ten regions were selected from each of these samples to get a set of thirty samples and the measurements were done on vein parameters as per the method described in Sect. 4. The results of the computations are tabulated in Sect. 5. Figure 3 shows the samples of 100 × 100 pixels (not drawn to the scale). The intermediate images that were obtained during processing are shown in Fig. 4.

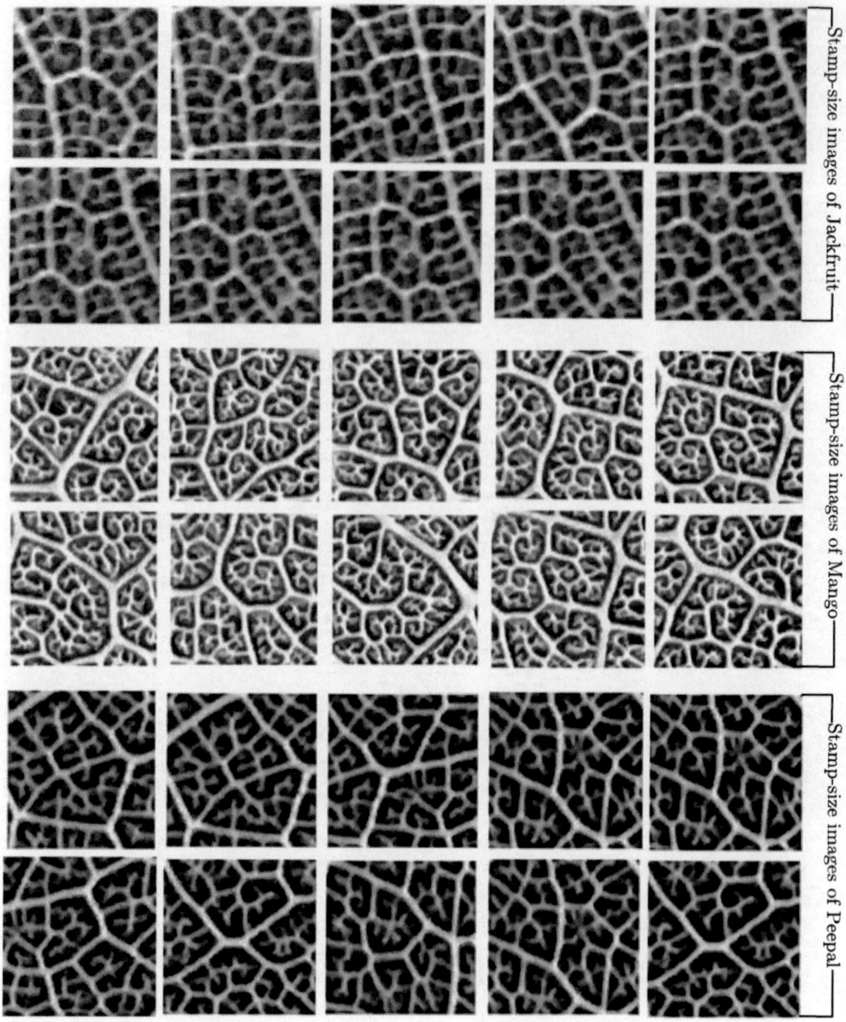

Fig. 3 Thirty samples for analysis—10 from each class

4 Proposed Method

The sample images are processed using a desktop computer with the following configuration: Lenovo Work Station, $Intel@i54460T$, $CPU@1.9GHz$, $4GB$ RAM, 64 bit Windows 10 Operating System with $MATLAB2017b$. The computational steps required to extract vein parameters from a leaf sample are given below:

(a) An image is selected from a set of 30 coloured samples as shown in Fig. 3, and it is converted to a grayscale image for simplicity of processing.

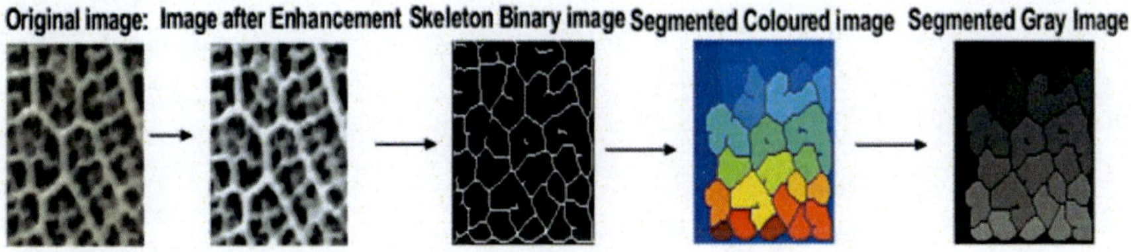

Fig. 4 Output images from $MATLAB2017b$ at various stages of processing

(b) The brightness of the grayscale image is adjusted and then the contrast is enhanced using histogram equalization.
(c) The grayscale image is then converted to a binary image and skeletonized using the morphological tool provided in MATLAB2017b. Our objective is to remove isolated pixels, bridge unconnected neighbouring components and clean the image. The images obtained at various stages of processing are shown in Fig. 4.
(d) The regions of this binary image is coloured using 'Flood-Fill Algorithm' so that each region is differently coloured. The number of colours used gives the number of Faces in the graph.
(e) We also compute the number of Edges and Vertices from the binary image. The number of Edges is calculated by counting the transitions across two adjacent Faces. The number of Vertices is calculated by counting the pixels whose neighbourhood comprises three or more distinct colours.
(f) The Area of each Face is calculated by counting the number of pixels present in it.
(g) The entire process is repeated for all samples.

We observe that the classification of leaves from three trees can be done based on two dominant features: the number of Vertices (V) and the number of Edges (E). Using Euler's formula, the number of Faces can be derived. We select these two features (V, E) based on maximum inter-class separation and minimum intra-class separation. Other features such as the area and shape of faces have not been considered for classification.

5 Experimental Evaluation

Experimental results of 30 samples are given in Table 1, where we have calculated the values of $V - E + F$ for each sample. The graph-theoretic parameters (Vertices, Edges and Faces) have been computed by the algorithm described in the previous section. However, in many cases, there is a deviation from Euler's Formula. This mismatch occurs because of the noise caused by imperfect imaging and processing of samples. Ideally, the value of $(V - E + F)$ should be equal to 2 for a planar map. We have observed that it perfectly matches with very good quality images, i.e., sample #12, sample #25, sample #28, and sample #30. The output of K-means

Table 1 Computed graph-theoretic vein parameters

Serial No.	# Vertices	# Edges	# Faces	V − E + F	Label
01	37	59	22	0	Jackfruit
02	42	68	25	−1	Jackfruit
03	37	61	23	−1	Jackfruit
04	38	62	24	0	Jackfruit
05	44	69	26	+1	Jackfruit
06	44	69	26	+1	Jackfruit
07	42	69	26	−1	Jackfruit
08	42	66	25	+1	Jackfruit
09	42	69	26	−1	Jackfruit
10	39	68	25	−4	Jackfruit
11	26	40	17	+3	Mango
12	34	55	23	+2	Mango
13	27	41	19	+5	Mango
14	32	49	22	+5	Mango
15	24	37	16	+3	Mango
16	27	46	18	−1	Mango
17	30	47	21	+4	Mango
18	29	45	20	+4	Mango
19	34	52	21	+3	Mango
20	23	40	17	0	Mango
21	31	53	21	−1	Peepal
22	34	55	20	−1	Peepal
23	31	54	21	−2	Peepal
24	37	59	23	+1	Peepal
25	35	61	24	+2	Peepal
26	30	52	21	+1	Peepal
27	38	62	24	0	Peepal
28	32	56	22	+2	Peepal
29	34	61	24	−3	Peepal
30	37	57	22	+2	Peepal

clustering is shown in Fig. 5, and the analysis of the result after classification is shown in Sect. 6. We have tabulated the results of 30 samples for classifying the trees of three classes and used another set of 20 samples for identification. We tested our algorithms with more than 100 samples taking more than 40 samples from each class. We have also tested the proposed algorithm directly on the images of green (fresh) leaves (without doing any pre-processing to extract the skeletons). However, the results were not observed to be accurate.

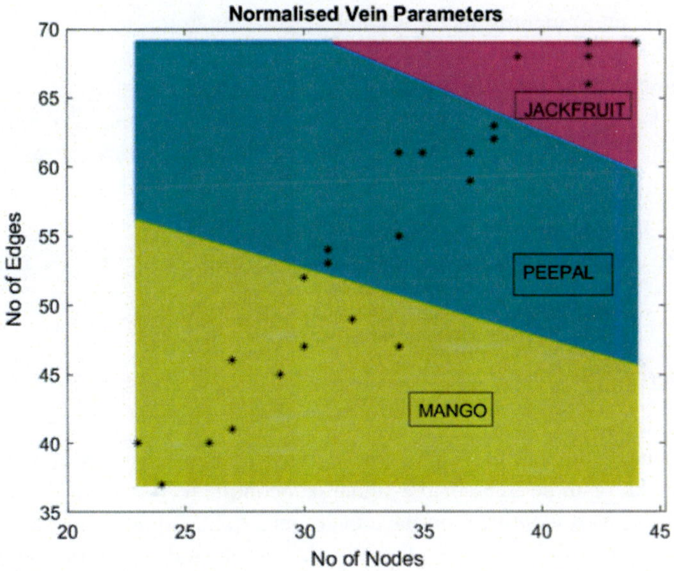

Fig. 5 Output of K-means Clustering

Table 2 True cass versus predicted class

True class		Predicted Class		
		Peepal	Mango	Jackfruit
	Jackfruit	2	0	8
	Mango	1	9	0
	Peepal	9	1	0

6 Analysis of Results for Classification

The consolidated data for calculating accuracy is given in Table-2

1. Overall Accuracy = $\frac{Correctly-Predicted-Samples}{Total-No-of-Samples} * 100 = \frac{26}{30} * 100 = 86\%$
2. Accuracy for Peepal = $\frac{9}{10} * 100 = 90\%$
3. Accuracy for Mango = $\frac{9}{10} * 100 = 90\%$
4. Accuracy for Jackfruit = $\frac{8}{10} * 100 = 80\%$

7 Identification of an Unknown Sample

Given a sample leaf-segment, its vein pattern is captured by a digital camera of a mobile phone and its parameters are extracted by applying the method described in Sect. 4. The data is stored as a text file (USample.txt), and KNN-classification algorithm is run to identify its class label. If the sample lies too far from the class identified by the algorithm, an error message is displayed, otherwise, it is mapped to the nearest class label. We tested it with 20 samples out of which 15 could be properly classified; three samples were classified wrongly and two could not be identified because they were too far from any of the three classes that were studied.

1. Accuracy = $\frac{15}{20} * 100 = 75\%$
2. Error in identification = $\frac{3}{20} * 100 = 15\%$
3. Identification could not be done for $\frac{2}{20} * 100 = 10\%$ cases.

8 Conclusion

We have studied for the first time, some graph-theoretic features present in leaf-vein networks of three trees of the Indian subcontinent. Based on these features, data clustering and identification have been performed using K-means and KNN algorithms, respectively. Our study reveals that among a database of 30 images, 26 samples could be properly classified with an accuracy of 86%. Some misclassification was noted between Jackfruit and Peepal samples whose vein structures are close to each other with respect to the above feature-set. On a positive side, only two instances of misclassification were observed between Peepal of plant species and various ecological parameters that determine the leaf-venation patterns.

References

1. Harary, F.: Graph Theory. Addison-Wesley (1969)
2. Ji, Z., Christopher, A., AlborDobon, A., Daniel, R., Simon, O., Michel, M., Simon, G., Steven, P., Nick, P.: Leaf-GP: an open and automated software application for measuring growth phenotypes for arabidopsis and wheat. Plant Methods **13**(117) (2017)
3. Jonas, B., Louai, R., Pugfelder, D., Huber, G., Scharr, H., Hulskamp, M., Koorn neef, M., Schurr, U., Schurr, S.J.: Pheno vein : a tool for leaf vein segmentation and analysis. Plant Physiol. **169**(4), 2359–2370 (2015)
4. Sack, L., Scooni, C.: Leaf venation: structure, function, development, evolution, ecology and applications in the past, present and future. New Phytol. **198**(4), 983–1000 (2013)
5. Price, C.A., Symonova, O., Yurity, M., Hilley, T., Weitz, J.S.: Leaf extraction and analysis framework graphical user interface, segmentation and analyzing the structure of leaf veins and aureoles. Methods Mol. Biol. **918**, 41-49. ISBN:978-1-61779-995-2 (2012) (online)
6. Zheng, X., Wang, X.: Leaf vein extraction based on gray scale morphology. Int. J. Image Graphics Signal Process. **2**. https://doi.org/10.5815/ijigsp.2010.02.04
7. Li, Y., Chi, Z., Fing, D.D.: Leaf vein extraction using independent component analysis. In: IEEE Conference on Systems, Man, and Cybernetics, October 8–11 (2006)
8. Radha, R., Jeyalakshmi, S.: An effective algorithm for edge and vein detection in leaf images. In: World Congress on Computing and Communication Technologies, pp. 128–131 (2014)
9. Valliammal, N., Gitalakshmi, S.N.: Plant leaf segmentation using non-linear K-means clustering. Int. J. Comput. Sci. (9), 212–218 (2012)

Individual Pig Recognition Based on Ear Images

Sanket Dan, Kaushik Mukherjee, Subhojit Roy, Satyendra Nath Mandal, Dilip Kumar Hajra, and Santanu Banik

Abstract In this paper, individual light coloured pig (Yorkshire) has been recognized based on their ear images captured using mobile phones. The ears have been kept parallel to the light source and the images have been captured from the opposite side of the light source. The auricular venation pattern from each captured ear image has been extracted, the template has been generated and stored in a database. The templates of recaptured ear images have been matched with the stored templates of the same pig using average Euclidean distance. The pig has been verified if average Euclidean distances of matching are <40. The unknown pig ear images have also been captured and the corresponding templates have been matched with all templates stored in the database. The pig has been identified if Euclidean distance between any pair of the template is <40.

Keywords Light coloured pig capturing · Image processing · Matching · Euclidean distance · Auricular venation pattern

S. Dan · K. Mukherjee · S. Nath Mandal (✉)
Department of Information Technology, Kalyani Government Engineering College, Kalyani, Nadia 741235, India
e-mail: satyen_kgec@rediffmail.com

S. Dan
e-mail: sanketdan@gmail.com

K. Mukherjee
e-mail: kaushik8.m@gmail.com

S. Roy
Gangarampur Government Polytechnic College, Kadighat, Gangarampur 733124, West Bengal, India
e-mail: subhojitroykgec@yahoo.com

D. Kumar Hajra
Uttar Banga Krishi Viswavidyalaya, Pundibari, Cooch Behar 736165, India
e-mail: dhajra@gmail.com

S. Banik
NRC on PIG, Rani, Guwahati 781131, Assam, India
e-mail: sbanik2000@gmail.com

D. Bhattacharjee et al. (eds.), *Proceedings of International Conference on Frontiers in Computing and Systems*, Advances in Intelligent Systems and Computing 1255,
https://doi.org/10.1007/978-981-15-7834-2_55

1 Introduction

Animal identification is an important task for disease control, livestock management, production management and also for maintaining the breeding record of the animal. Some of the pig farmers uses a traditional method like tattooing, clipping, branding to keep track of the individual pig. Traditional animal identification methods such as ear notching or methods like radiofrequency IDs are not able to provide accurate and secure animal identification due to possibilities of theft, falsification and duplication. These methods which were used to apply to mark the animal creates the degree of stress on the animal. In addition, many common marking procedures such as branding (heat, cold or chemicals), tattooing, toe clipping, ear notching and tagging are also involved in tissue damage, and therefore, cause pain. There is a requirement of an ideal method which will identify the individual animal reliably and permanently with no adverse effects. In this context, biometric method are non-invasive, painless, economical and not alter during the lifetime. There are several traits which have been used for biometric marker such as iris image, ear venation pattern, retinal image; muzzle/nose print etc. Some of the unique trait(s) are not changed throughout the lifetime of the animal. Iris image is an efficient marker for the identification of the human eye. But in case of pig eyes this method can't be used as pig eyes are covered with a thick mesh of fat, hence the complete pig iris is not visible. Eyelid and Eyelash are covered with a fatty layer. Another method for animal identification is muzzle prints. In this method Scale Invariant Feature Transform (SIFT) [1], has used for detecting the interesting points for muzzle image matching. Saliva secretion is one of the major issues with this method. That is why muzzle print is not clearly visible due to this problem. On the pig ear, venation pattern is easily visible if the light source is being placed on the front of the ear and picture has been taken from the opposite direction. This venation pattern is captured by the mobile camera. This venation pattern is fixed throughout the lifetime of the pig.

Aim of this paper is to recognize pig (Yorkshire) using auricular venation pattern. The ear of the pig has been kept in parallel to the light source. The light source has been placed to the inner side of the ear and ear images have been captured from the outer side of the ear using mobile phones placed parallel with it. The light has been projected at the middle of the ear to visualize ear properly. The pictures were resized into 700×550 pixels to make all pictures uniform in size. The mask has been created for cropping the region of interest (ROI). The green channel has been selected and some morphological operations have been applied to normalize the image. The branching points have been detected and coordinators of these points have been used as a template. In verification, the matching has been performed between each two pair of templates, one from query template and other from the database of the same pig ear and average Euclidean distance has been computed. In identification, the matching has been done between query templates and all stored templates in database and average Euclidean distance has been calculated for each matching pair. It has been observed that the average Euclidean distance of template matching is less than 40 in verification, as well as in identification.

2 Related Work

Iris image is an efficient method for the identification of an individual. Unique identification of humans using iris pattern analysis has been studied extensively over the years. For iris pattern identification purpose such as arching ligaments, furrows, ridges, crypts, rings, etc. has been used to recognize the individual. Since the iris pattern may vary person to person [2]. De and Ghoshal [3] have proposed a method for identification of goat using iris. Iris is non-circular. For identification purpose, they use Structural, Statistical and Fourier Descriptors. The success rate of this proposed method is 97.85% and 4.5% of false acceptance rate and false rejection rate is 2.2%. Lu et al. [4] have proposed a useful method for cow identification system based on iris analysis and recognition. Ali et al. [5] have introduced a method of Fingerprint Recognition System. For recognition purpose, firstly use the sensor to capture the fingerprint data, pre-process the data and extract the feature from the image and finally store this data into a database for matching. Verma et al. [6] have investigated the detection and Classification of Diabetic Retinopathy using Retinal Images. For recognition of retinopathy, the retinal images are being captured from the fundus camera and analyze the images, detect blood vessel, identify heamorrhages and classify different stages of diabetic retinopathy into normal, moderate and non-proliferative diabetic retinopathy (NPDR). The accuracy rate is 90%. Chaudhuri et al. [7] have introduced a detection of blood vessel of the retinal image using an operator for feature extraction based on the optical and spatial properties of objects. This automatic detection of retinal blood vessels is very useful for doctors for early diagnose of retinal disease. Zhang et al. [8] have proposed a method for individual authentication system by extracting the venation structure by using multi-scale filtering from infrared palm vein images.

3 Existing Method of Identification

Traditional animal identification systems like Ear Notching, Ear tag, Branding and Tattooing are used to identify individual pigs which are shown in Fig. 1. The methods are protecting from theft, fraudulent and duplication. The external ear is cut into "V" shape in different locations and the individual number is determined by the position of the cut. Ear tags are the most common method of identification of most of the livestock species. The branding involves putting permanent identification mark on the visible body part of the animal by searing with the help of hot or cold branding irons or chemicals. The tattooing instrument uses alpha-numeric identification code impregnated over hairless part of the animal with the use of sharp, needle like projections that are secured in tattooing forceps. After piercing the tattoo mark, tattooing ink is usually applied over the site. Paint branding is a temporary form of identification of animals mostly used for flock identification. These are of importance during

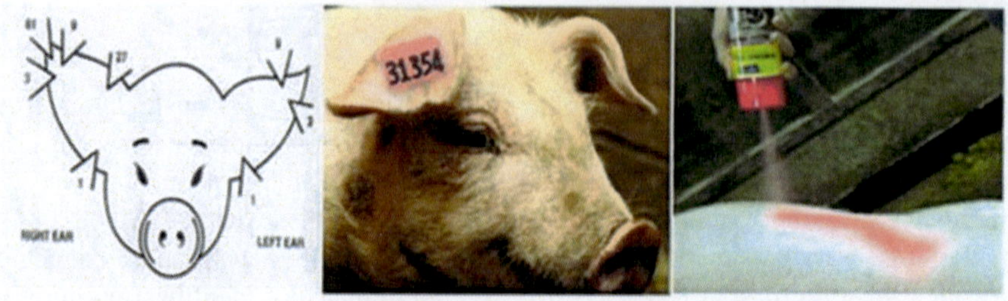

Ear Notching in Pig Ear Tagging in Pig Temporary Paint Branding

Fig. 1 Traditional identification system

animal show/exhibition or after the purchase of animals from a flock until it reaches its final destination.

Most of the traditional methods have a problem of readability as the identification marks remain invisible due to distance from the animal, growth of hair on the marking area, soiling, wear and tear, breakage of the tag and pigmentation. Tags may be lost from the animal's body by tearing away from the ear flap. Due to marking on the animal, it is responsible to create some degree of stress related to capture, handling and restraint. Besides, there are many common methods of marking procedures like as toe clipping, ear notching or branding also responsible for damaging the tissues, and therefore, cause pain.

4 Pig Recognition System

Pig recognition system has been constructed same as human recognition system using fingerprint or using other traits. The system consists of four fundamental stages: (1) Ear image acquisition; (2) Segmentation and auricular tree generation; (3) Features extraction and (4) matching. The comparison among the recaptured query ear template and stored templates previously obtained from the same pig yields matching scores. If scores are lower than the threshold, the query template is originated from the same pig. In identification, matching is performed between query ear template with each stored template in the database. The pig is identified if the matching score below the threshold point. This process is shown in Fig. 2.

5 Pig Identification Phrases

The ears have been captured during ear image acquisition, and finally, templates have been created from each ear image. The phases are involved is shown in Fig. 3.

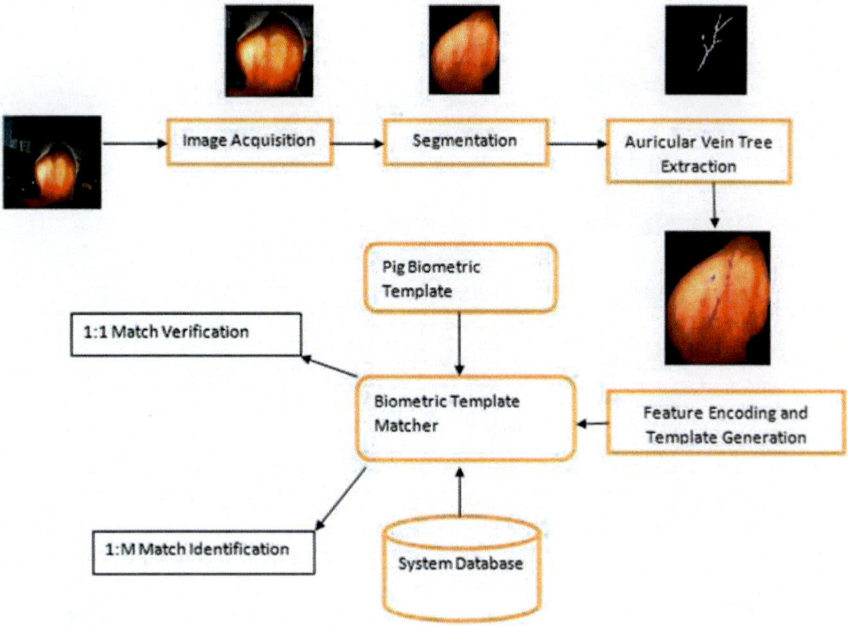

Fig. 2 Pig recognition system

5.1 Image Acquisition

In this experiment, several ear pictures from Yorkshire breed have been captured from the organized farm, NRC on Pig, Rani, Guwahati, Assam. Out of these images, ear images captured from five different individual pigs have been used in this experiment. Ear images have been captured using a mobile camera. The ears had kept parallel to the light source and the images were captured from the opposite side of the light source. The high illumination light source like as torchlight has been projected at the center of the ear so that the light will be equally distributed on the ear's inner surface of the ear. A mobile phone with 13-megapixel camera has been used in this experiment. To reduce the effect of external light, a black cloth has been placed around the experimental setup.

5.2 Preprocessing

The numbers of pictures have been captured during image acquisition. The images while are parallel to the mobile lens and perpendicular to the x-axis have been selected for further processing as shown in Figs. 4 and 5. The selected images have been resized into 700 × 550 pixels. The intensity level of the images is enhanced

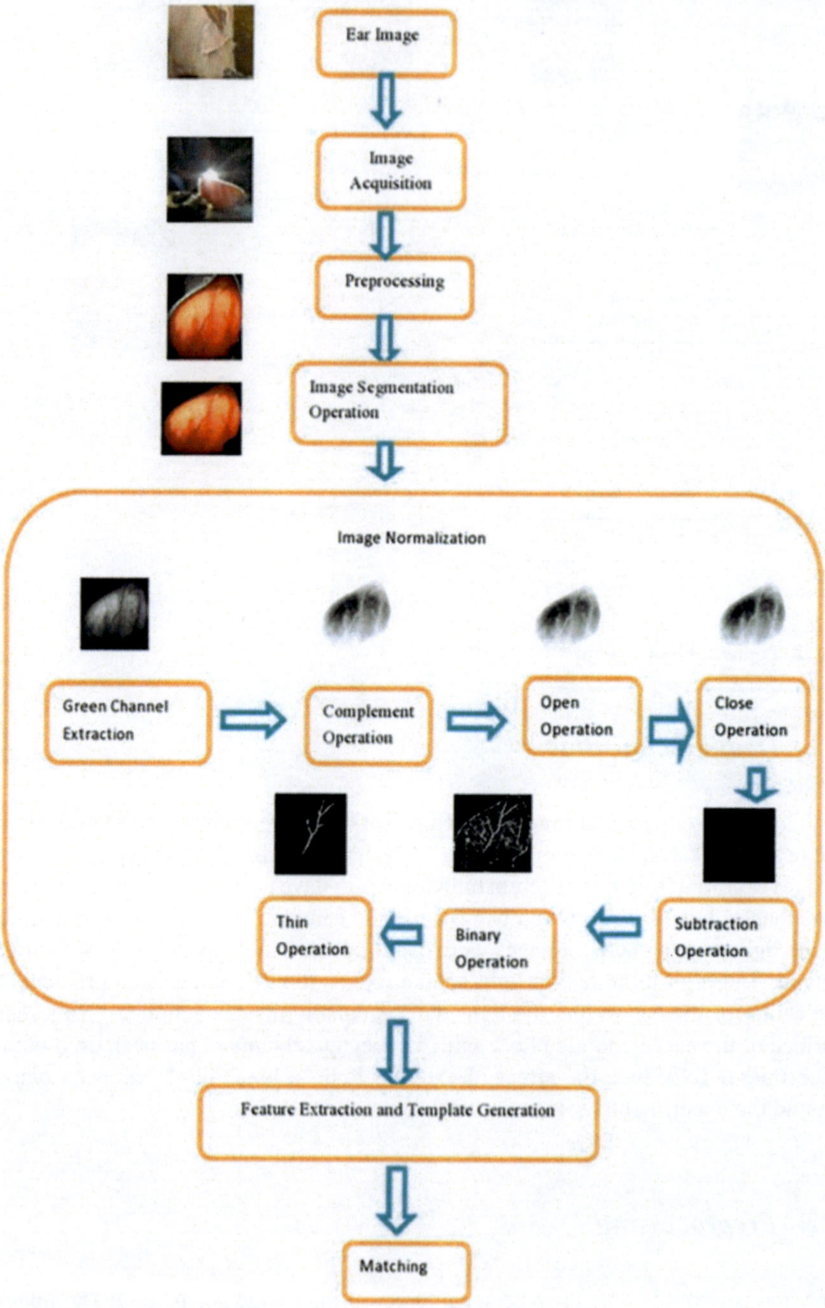

Fig. 3 Phrases of pig recognition

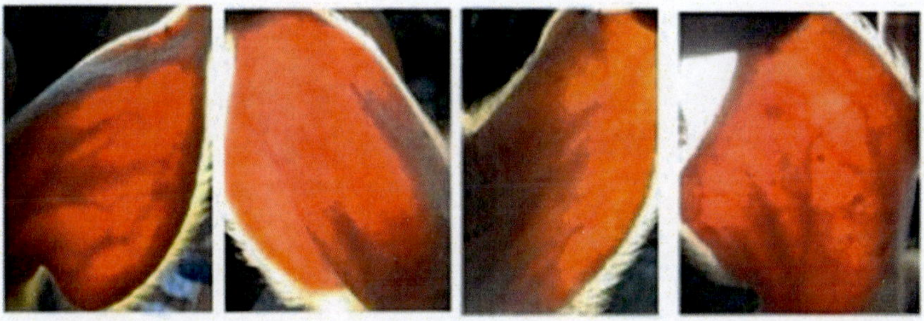

Fig. 4 Different angle, all are discarded

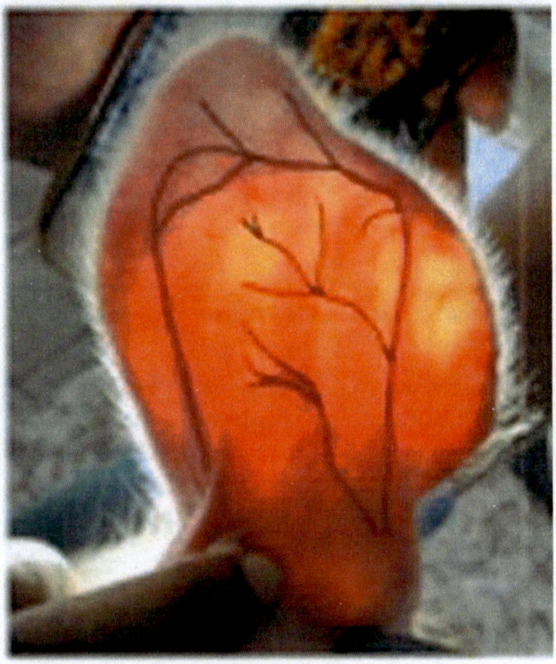

Fig. 5 This is the right way to capture images

by the Contrast-limited adaptive histogram equalization (CLAHE) [9] approach. This method is applied per pixel basis in the particular region using the histogram equalization method based on the region surrounding that pixel. As a result each pixel have arranged by the intensity proportional mapping according to its neighbor pixel. CLAHE approach have been applied to the small region of the image rather than considering the whole image. Noise of the image is reduced for each small region, and the boundary and edges are more prominent than input image (Figs. 4, 5 and 6).

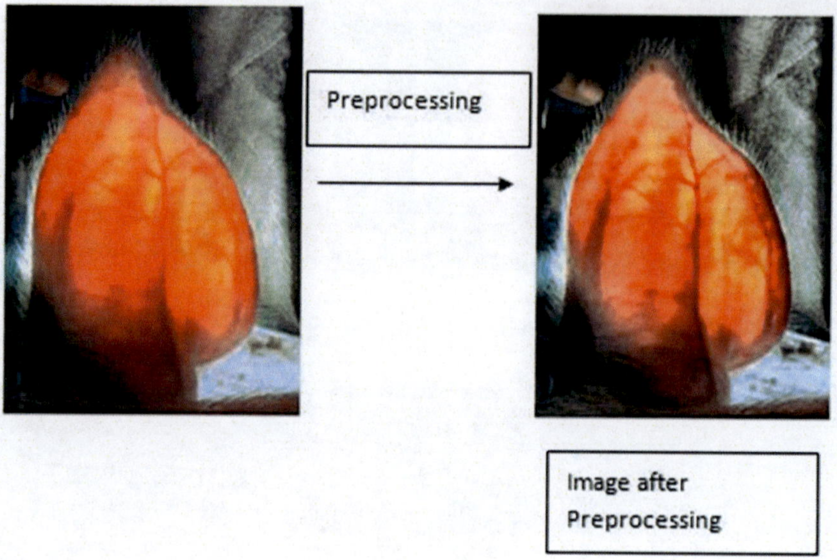

Fig. 6 Image preprocessing

5.3 Ear Vein Segmentation

The snake active contour method [10], has been used for segmentation. Some points have been selected on the border of ear picture and a spline curve has been generated based on these points. One binary mask image has been created by making white inside the spline curve and black outside of this curve. Finally, the segmented ear image has been obtained by overlapping the original and mask image are shown from Figs. 7 to 8.

Fig. 7 Mask generation

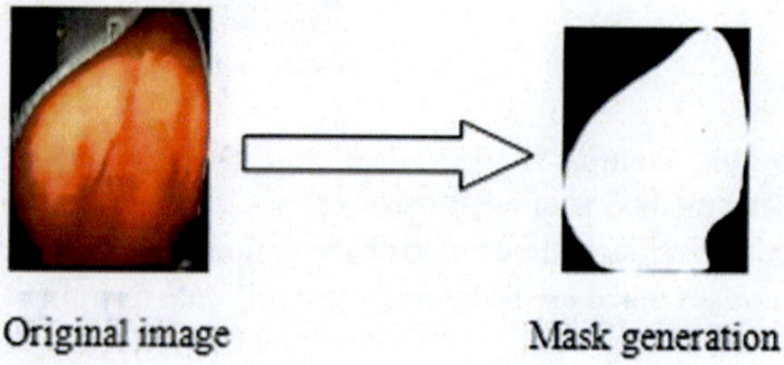

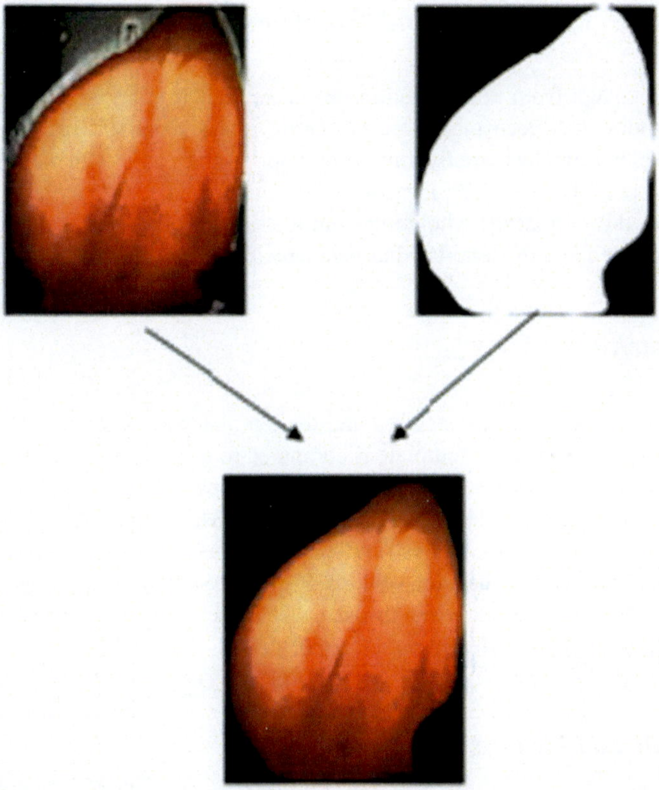

Fig. 8 Segmented image

5.4 *Normalization*

It has been observed that the red and blue channels of the segmented ear images contained high amount of noise from which extracting the exact venation structure is extremely difficult. Here, the green channel has been used for normalization. The green channel image has been complemented to make darker of veins from rest. The erosion of open operation and dilation of closed operation have been applied on complemented image and resultant images of open and close operations have been subtracted to view vein in more precise structure. The resultant image has been converted into a binary image for thinning. To make thin of venation tree, searching operation has been applied on the image. The row wise searching has been applied. The first white pixel has been searched in each row and it will continue until the first black has been encountered. All white pixels except middle have been converted into black. Searching has been performed with 5° intervals. The process of thinning has been repeated until all pixels have been encountered.

5.5 Features Extraction and Template Creation

The output image from normalization has been placed on coordinate. One 3 × 3 sliding window has been moved on the entire image from (0, 0) position. If the central pixel is 1 and has exactly three neighbour's values as 1, then the central pixel is a branch point (Fig. 9). The location of branch point has been stored and moved the window throughout the whole image in the same fashion. All the branch points have been sorted and the template has been created.

5.6 Matching

The pig ear image has been captured and the templates are stored in the database. The unknown pig ear image template is compared to every pig ear images template. Initially, we are taking 5 same and individual pig ear image. Each of the image templates is matched with every individual image. Matching is done based on the Average Euclidean distance analysis. After continuous experiment, it has been seen the maximum Euclidean distance value is 40 for the same pig ear image. So pig is identified if the Average Euclidean distance (AED) between any pair is less than 40. Otherwise, it belong to different pig image (Fig. 10).

6 Result and Discussion

Auricular veneration pattern has been extracted from five images captured from left ear of each pig and template has been created based on branching point from each ear. Generally, the right ear of pig is tagged. The matching has been per-

Fig. 9 Branch point

0	1	0
0	1	0
1	0	1

Fig. 10 Average Euclidean distance

$$\text{Average Euclidean distance} = \frac{\sum\sum \sqrt{(x-a)^2 + (y-b)^2}}{\text{No of element on template}}$$

Table 1 Euclidean distances for verification based on branching points

Sample 1	Sample 2	Euclidean distance
6436_104_ID1	6436_104_ID2	15
6436_104_ID1	6436_104_ID3	30
6436_104_ID1	6436_104_ID4	34
6436_104_ID1	6436_104_ID5	29

formed between the query template and stored template. The average Euclidean distance based on branching points has been calculated in each matching. In this experiment, the matching has been performed between one fixed template obtained from each pig with the rest four templates obtained from the same pig ear stored in the database and their matching scores in verification. The best picture has been selected from five different pictures from each pig. The matching has been performed between one best picture and the rest four pictures from four other pigs. The average Euclidean distances have been calculated. One ear picture with id, 6436_104_ID1, have been compared with four ear pictures with id 6436_104_ID2, 6436_104_ID3, 6436_104_ID4 and 6436_104_ID5 and average Euclidean distances have been given in Table 1. The best ear pictures 6436_104_10, 6434_104_454, 6432_104_211, 6430_104_228 and 6437_104_334 have been selected and matching have been performed between 6436_104_10 and 6434_104_454, 6432_104_211, 6430_104_228 and 6437_104_334. Here the picture id 6436_104_10, 6434_104_454, 6432_104_211, 6430_104_228 and 6437_104_334 have defined as 6436_104_ID1, 6434_104_ID1, 6432_104_ID1, 6430_104_ID1 and 6437_104_ID1 respectively.The average Euclidean distance of each matching is given in Table 2. The verification and identifications among five ear images from five pigs are furnished in Tables 3 and 4. The graphical representation of matching among the five best images from five individuals is shown in Fig. 11.

Table 2 Euclidean distances for identification based on branching points

Sample 1	Sample 2	Euclidean distance
6436_104_ID1	6434_104_ID1	86
6436_104_ID1	6432_104_ID1	52
6436_104_ID1	6430_104_ID1	56
6436_104_ID1	6437_104_ID1	66

Table 3 Euclidean distance comparison of five different images of each individuals

ID	ID1	ID2	ID3	ID4	ID5	Best Picture
6436_104	0	15	30	34	29	6436_104_10
6434_104	0	23	29	35	20	6434_104_454
6432_104	0	17	32	22	25	6432_104_211
6430_104	0	33	30	28	27	6430_104_228
6437_104	0	27	22	37	35	6437_104_334

Table 4 Euclidean distance comparison of five best images from five individuals

ID	6436_104_ID1	6434_104_ID1	6432_104_ID1	6430_104_ID1	6437_104_ID1
6436_104_ID1	0	86	52	56	66
6434_104_ID1	86	0	72	60	60
6432_104_ID1	52	72	0	82	65
6430_104_ID1	56	60	82	0	65
6437_104_ID1	66	60	65	65	0

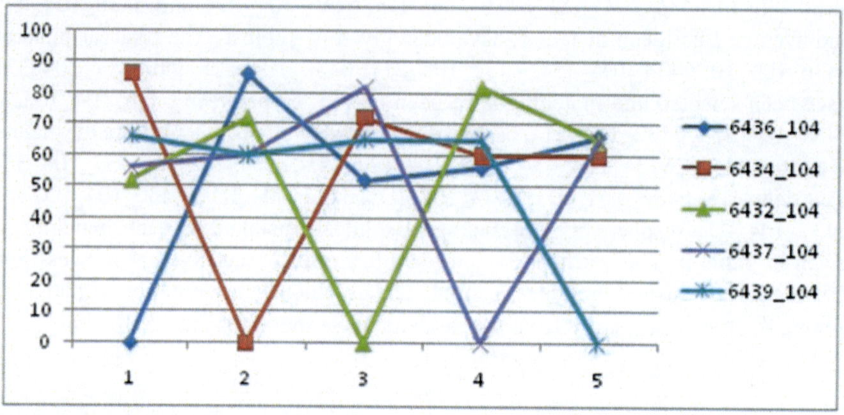

Fig. 11 Graphical representation of matching among five best images from five individuals

7 Conclusion and Future Work

In this paper, individual Yorkshire pig has been recognized using the auricular venation pattern. In this experiment, five ear pictures from five pigs have used for identification and verification. The pig has been recognized if average Euclidean distance is less than 40 based on branching point only. The end points, branch angles, distances among branches will be taken for matching in future. The model will be tested using ear pictures from more pigs. The dark coloured pigs like duroc, ghungroo and Niang megha will also be tested in future.

Acknowledgements The authors would like to thank ITRA-Digital India Corporation (formerly known as Media Lab Asia), Ref. No.: ITRA/15(188)/Ag&Food/ImageIDGP/01 dated 09/11/2016 for funding this research work. The authors would also like to thank Dr. A. Bandopadhyay, Senior consultant, ITRA Ag&Food, Dr. Arnab Sen, Head, Animal Health, ICAR research complex for NEH, Barapani, Dr. Binay Singh, Scientist, ICAR-RC for NEH region, Tripura Center, Agartala, Pritam Ghosh, Subhranil Mustafi and Kunal Roy Kalyani Government Engineering College for helping us to implement this research work.

References

1. Awad, A.I., Zawbaa, H.M., Mahmoud, H.A., Nabi, E.H.H.A., Fayed, R.H., Hassanien, A.E.: A robust cattle identification scheme using muzzle print images. In: 2013 Federated Conference on Computer Science and Information Systems, pp. 529–534. IEEE (2013)
2. Daugman, J.: How iris recognition works. In: The Essential Guide to Image Processing, pp. 715–739. Elsevier (2009)
3. De, P., Ghoshal, D.: Recognition of non circular iris pattern of the goat by structural, statistical and fourier descriptors. Procedia Comput. Sci. **89**, 845–849 (2016)
4. Lu, Y., He, X., Wen, Y., Wang, P.S.: A new cow identification system based on iris analysis and recognition. Int. J. Biometrics **6**(1), 18–32 (2014)
5. Ali, M.M., Mahale, V.H., Yannawar, P., Gaikwad, A.: Overview of fingerprint recognition system. In: 2016 International Conference on Electrical, Electronics, and Optimization Techniques (ICEEOT), pp. 1334–1338. IEEE (2016)
6. Verma, K., Deep, P., Ramakrishnan, A.: Detection and classification of diabetic retinopathy using retinal images. In: 2011 Annual IEEE India Conference, pp. 1–6. IEEE (2011)
7. Chaudhuri, S., Chatterjee, S., Katz, N., Nelson, M., Goldbaum, M.: Detection of blood vessels in retinal images using two-dimensional matched filters. IEEE Trans. Med. Imaging **8**(3), 263–269 (1989)
8. Zhang, Y.-B., Li, Q., You, J., Bhattacharya, P.: Palm vein extraction and matching for personal authentication. In: International Conference on Advances in Visual Information Systems, pp. 154–164. Springer (2007)
9. Zuiderveld, K.: Contrast limited adaptive histogram equalization. In: Graphics Gems IV, pp. 474–485. Academic Press Professional, Inc. (1994)
10. Kass, M., Witkin, A., Terzopoulos, D.: Snakes: active contour models. Int. J. Comput. Vis. **1**(4), 321–331 (1988)

Analysis of Large-Scale Human Protein Sequences Using an Efficient Spark-Based DBSCAN Algorithm

Soumyendu Sekhar Bandyopadhyay, Anup Kumar Halder, Piyali Chatterjee, Jacek Sroka, Mita Nasipuri, and Subhadip Basu

Abstract The development of modern high throughput sequencing techniques has resulted in an exponential growth in meta-genomic sequence accumulation that could greatly enhance large-scale functional annotation. Processing of these large and redundant sequences has become a major challenge for researchers. Clustering by similarity is one of the major steps to reduce the redundancy of these enormous resources and analysis of such large biological sequences. The n–gram feature representation, generally used in sequence clustering and classification, results in high dimensional input spaces, for larger values of n. However, it becomes intractable to cluster such large-scale sequences by current algorithms due to a large number of dimensions. Efficiently designed, clustering approach can easily scale to handle large-scale sequences by utilizing the power of parallel computing with high-performance computing systems. In this work, we propose a spark-based DBSCAN algorithm to cluster large-scale human protein sequence data by exploring n-gram ($n <= 3$) features. Experimental results in human protein sequence data shows efficient speedup in clustering and effective in reducing the redundant sequence (ranges from 31.9 to 48.7%) and more than 84% of non-singleton clusters are of high quality.

S. Sekhar Bandyopadhyay · A. Kumar Halder (✉) · M. Nasipuri · S. Basu
Department of Computer Science and Engineering, Jadavpur University, Kolkata, India
e-mail: anupkhalder.cse.rs@jadavpuruniversity.in

S. Sekhar Bandyopadhyay
e-mail: soumyendusb.cse.rs@jadavpuruniversity.in

M. Nasipuri
e-mail: mita.nasipuri@jadavpuruniversity.in

S. Basu
e-mail: subhadip.basu@jadavpuruniversity.in

P. Chatterjee
Department of Computer Science and Engineering, Netaji Subhash Engineering College, Kolkata, India
e-mail: piyali.gini@gmail.com

J. Sroka
Institute of Informatics, University of Warsaw, Warsaw, Poland
e-mail: j.sroka@mimuw.edu.pl

D. Bhattacharjee et al. (eds.), *Proceedings of International Conference on Frontiers in Computing and Systems*, Advances in Intelligent Systems and Computing 1255,
https://doi.org/10.1007/978-981-15-7834-2_56

Keywords Sequence clustering · n-gram feature · Parallel computing · Apache Spark · DBSCAN

1 Introduction

In computational biology, many problems such as protein homology detection, functional annotation, can be formulated as a sequence clustering task where amino acid sequence compositions of proteins are considered as the key information. Protein sequences are any combination of 20 constituent elements (amino acid) by maintaining intrinsic dependencies between these amino acids. In a protein sequence, the dependencies between neighboring entities can be explored by generating all possible overlapping sub-patterns of a certain length n, called n-grams [4]. N-gram based dependencies in data are explored to improve the richness of the representation. However, in n-gram representation with higher values of n, the dimension of the input space increase exponentially (20^n). It became intractable to apply data clustering algorithm with higher dimensional input space while the dataset itself is large. So, an efficient algorithm needs to be designed for seamless processing of higher dimensional large-scale sequence data. A plenty of methods have been developed in the last decades for sequence clustering [3, 4, 7, 11]. However, the sequence clustering from large sequence data remains a serious challenge for those existing approaches. Based on the structural organization of clustering, these approaches can be categorized into two groups: hierarchical clustering (HC) and greedy heuristic flat clustering (GHFC). In HC based approaches [2, 10], sequences are organized in a hierarchical tree and produce different level of operational taxonomic units (OTUs) at various similarity levels that can explore more biological significance. However, HC based methods have major drawbacks that it requires pairwise distance computation and requires to store the distance matrix. Cd-hit [7] and UCLUST [3] are widely used GHFC approach that employ greedy flat clustering to reduce the computational complexity. Though GHFC are faster than HC based approaches, GHFC produces lower cluster quality than HC based approaches. It is highly desired that clustering methods can be easily scaled to handle massive sequence data using parallel computation with high-performing computing architecture. However, it is inherently difficult to redesign the existing approaches in the parallel mode as in each iteration distance computation is totally depends on its earlier step or each division/merging rely on the previous operations. In this scenario, parallel density-based clustering could be an efficient alternative in large-scale sequence data clustering with higher dimensional input space.

In this paper, we proposed a spark-based DBSCAN algorithm, that efficiently employ on the higher dimensional input space and large-scale human sequence data using parallel computing resources. The proposed method was implemented on a Spark cluster that eventually helps to parallelize the resource and computation efficiently. Experimental result shows efficient speedup in computation compared with different experimental architectural setups. In addition, quality of the clusters is assessed with biological significance such as domain correspondence score (DCS).

2 Methodology

We developed a new computational method for parallel clustering analysis of large-scale human sequence data with higher dimensional input space. High dimensional input features are generated from each protein sequences by exploring the n-grams feature (at $n = 3$, total number of features is $20^3 = 8000$) [4]. The basic idea is to first partition the data into equal-sized small partitions and process DBSCAN on each partition in parallel mode and at final step individual results are merged for the final clustering solution. The DBSCAN algorithm within each partition is further paralleled over the higher dimensional input spaces where each data is processed as fixed length multiple subdimensions. The clustering results from each subdimension-based processing are merged using a consensus mechanism. Finally, merged results from each partition are considered as the final cluster. The flowchart of the proposed method presents in Fig. 1. All parallel computations are processed on Apache Spark architecture.

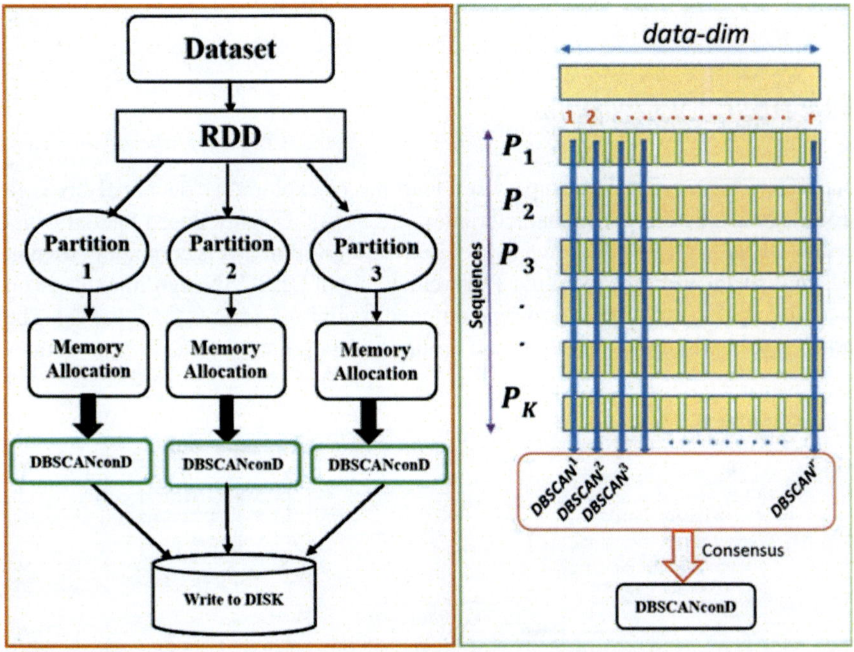

Fig. 1 Overview of the proposed parallel scheme for large-scale sequence clustering using DBSCAN (left). DBSCANconD: consensus-based parallel DBSCAN clustering on fixed length sub dimensional partitioned data (right)

2.1 Apache Spark Architecture

Storing and processing a large amount of data requires splitting the problem into some dependent and independent set for parallel execution. Hadoop with a cluster of commodity hardware can store and process huge dataset in a reasonable time [1]. However, Hadoop lacks some computational issues such as: in-memory computation, computing in real time data efficiently, etc. The intermediate data write-back operations to HDFS results in low latency for Map-Reduce scheme in Hadoop. In the other hand, Spark by Apache software foundation makes the computing faster by using in-memory computation [9] with a new memory abstraction called Resilient Distributed Dataset (RDD). RDDs are partitioned and distributed across multiple nodes and can rebuild if a partition is lost. Spark architecture is consisting of master and worker nodes. The main module of the master node is the driver program that controls the application by creating a spark context. Spark context works with the cluster manager to manage multiple jobs distributed over the worker nodes which execute the tasks on the partitioned RDD and return the result back to the spark context. The block diagram of spark architecture is given in Fig. 2.

2.2 DBSCAN Algorithm

One of the main focus of unsupervised learning or exploratory data analysis is to group the data based on the characteristics of the data. Density Based Spatial Clustering (DBSCAN), a well-known data clustering algorithm that is commonly used in data mining and machine learning. The architecture of DBSCAN algorithm based on two basic parameters, radius (*eps*) and number of minimum points (*MinPts*) [8]. The radius (*eps*) defines the neighborhood around a data point whereas *MinPts* defines

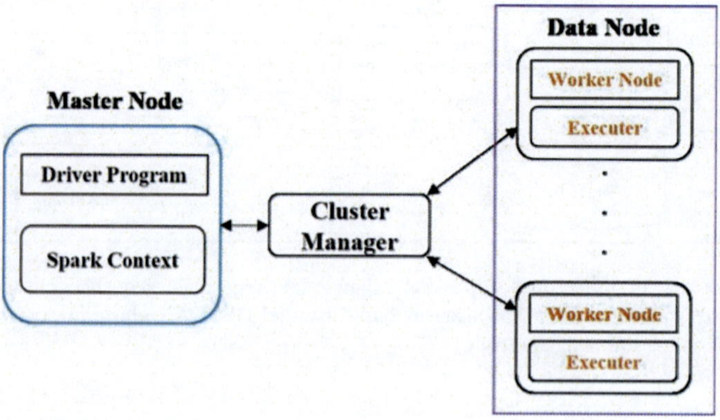

Fig. 2 Apache Spark master node data node architecture

minimum number of neighbors within *eps* radius. The neighbor points $N_\epsilon(p)$ for any point $(p \in D)$ is defines as $N_{eps}(p) = \{n \in D | dist(n, p) \leq eps\}$. Here, Euclidean distance is used for $dist(n, p)$ computation.

Algorithm 1 *DataPartition*

Input: data D, $block_{size}$
Output: < key, list(value)>
Process:

- 1. Transform the data D into RDD.
- 2. Convert the RDD an array.
- 3. noPart ← size(D)/ $block_{size}$
- 4. Parallelize RDD array over *noPart* partition containing the array and map each partition with index as a new_{RDD}.
- 5.**return** new_{RDD} as < key, value > pair

Algorithm 2 *Local_DBSCAN*

Input:*D<key, list(value)>*, ϵ, *MinPts*
Output: < key, *cluster_number*>
Process:

- 1. The partition generates list (values).
- 2. An arbitrary point *p* is selected from the partitioned data and for searching the neighborhood points.If $N_\epsilon \leq MinPts$ then we assign point *p* as the core point else the point is marked as noise.
- 3. If *p* is marked as core point, then make cluster *c* with point *p* and all the other points belonging to cluster *c*. Each cluster is being assigned with a cluster number.
- 4. **Repeat** steps 2 and 3 until all the points of the datasets are assigned to a cluster or are marked as noise
- 5. **return** the data points with their assigned cluster number.

Based on these parameters, data points are classified into 3 categories: core points, border points and noise/outliers. A point is classified as core *point* if it has more than *MinPts* points within *eps* whereas for border point, it has fewer than *MinPts* within *eps*, but is in the neighborhood of a core point. An *outlier/noise* point is neither a *core* point nor a *border* point. For any two points, $\boldsymbol{x}$ and $\boldsymbol{y}$ are said to be connected if *x* is dense and the distance between *x* and *y* is less than *eps*.

To implement the DBSCAN algorithm in parallel, we proposed that distributing the dataset efficiently can minimize the computation time for large-scale datasets. In this work, we have proposed two-level parallel processing for large-scale data clustering. The first one is to split the dataset over the cluster by transforming it into an RDD and then executing the DBSCAN algorithm to each splits in parallel (see Algorithm 1). Within each partition, DBSCAN is further paralleled over the higher

Algorithm 3 *Clustering*

Input: q,ϵ,*MinPts*
Output: *Clusters*
Process:

- 1. D $\leftarrow$ Load data of dimension *(k,n)*
- 2. PartitionRDD $\leftarrow$ DataPartition(D, $block_{size}$)
- 3. For each partRDD in PartitionRDD
  - 1. $r \leftarrow n/q$
  - 2. split data dimension over $\boldsymbol{r}$ number of subdimensions of length $\boldsymbol{q}$ as $partRDD_i$
  - 3. process $Local_{DBSCAN}$ in parallel over all subdimension,
    set, $CL_i \leftarrow Local_{DBSCAN}(partRDD_i, \epsilon,$ MinPts)
  - 4. CL $\leftarrow$Consensus result over all CL_i
- 4. Collect result for all partition of PartitionRDD and merge as final result
- 5. Write to Disk

dimensional input spaces where each data is processed as fixed length multiple subdimensions (Algorithm 2). Finally, the resultant clusters are collected from different partitions and are written into disks as described in Algorithm 3.

2.3 Cluster Validation

The clustering results are analyzed and validated in the contest of biological significance. We have incorporated domain correspondence score (DCS) to quantify the clustering results as described in [4]. DCS score is the measure of purity for any cluster in terms of functional domain annotation for a group of sequences. DCS is a more effective metric for both single and multiple domain proteins that belong to a particular cluster. For detailed DCS computation please see [4]. Higher DCS indicates highly conserved domain annotation within a cluster and lower represent the reverse or even corrupted. A cluster with identical domain annotation (single or multiple) for every point (proteins) ensures DCS as 1 that indicates a high quality pure cluster.

3 Results and Discussion

In this work, human protein sequence data have been chosen for the end to end experiment. Total 20,431 reviewed human protein sequences are collected from UniProtKB/Swiss-Prot (https://www.uniprot.org) as fasta formats and then converted into the desired higher n-gram (here $\boldsymbol{n} \leq 3$) feature representations. The experimental set up is carried out mostly on bigram and trigram features where the input spaces

Table 1 Performance speedup of different execution setup

Data size	Data points	Data Dim.	EE1 (min)	EE2 (min)	EE3 (min)	SpeedUp (EE2/EE1)	SpeedUp (EE3/EE1)
16	20,413	400	0.3	0.5	0.4	1.67	1.33
326	20,413	8000	6.9	32	16	4.63	2.32
653	20,413 × 2	8000	20	71	48	3.55	2.4
981	20,413 × 3	8000	35	$	$	-	-

range from 400 (i.e, 20^2) to 8000 (i.e, 20^3). The experiment has been devised with two objectives, (1) speedup efficiency and (2) quality of cluster. First, we concentrate on the performance benefit of the clustering on a large biological data by leveraging the power of parallel computation on Apache Spark framework. Then, the quality of the clustering result is analyzed and validated with respect to biological relevance. To compare the performance of the proposed method on different parameters such as data size, data points and data dimension, three different experimental environments (**EE**) has been set up. They are 1-Master and 3-Slave multi node-based physical spark cluster referred as ***EE1***, single standalone node as ***EE2***, and spark community edition distribution over cloud, hosted by databricks as ***EE3***. The performance speedup has been reported in Table 1. The parallel execution of DBSCAN clustering algorithm on higher dimension (bigram or trigram) protein sequence data with ***EE1*** setup are faster than the other two (***EE2, EE3***). In all three ***EE***-setup, the algorithm is experimented with different data size and dimensions. The results depict that increasing the size of the dataset, the speedup ratio of ***EE1*** compared to ***EE2*** and ***EE3*** also increases. As the size of the datasets exceed the block size of Spark framework (i.e, 128 MB), the parallel execution becomes faster compare to sequential and pseudo cluster execution. To analyze the dataset, we first obtain the cluster consensus by partitioning the data into equal length subdimensions. From the consensus result, singleton clusters are extracted as noise in all clustering and the remaining are referred to as non-singleton clusters. We compare the results obtained from complete dimension-based approach and subdimension-based consensus approach (see Table 1) in both bigram and trigram features. In complete dimension-based approach, in both features, produces a maximum number of clusters 20,369 and 20,126 which suggest that the redundancy removal power is very low 0.21 and 0.14%. In contrast, Consensus-based approach results with higher redundancy reduction rate 48.7 and 31.9% for bigram and trigram respectively although bigram-based approach produces a maximum number of corrupted clusters. Interestingly, all the clusters resulted from complete dimension are single-domain protein suggests the approach is not suitable for multi-domain proteins. Above results suggest that consensus-based sequence clustering with higher dimensional representation is sensitive to both single-domain and multi-domain protein and highly powerful in removing redundant sequences.

In the second phase, domain-based analysis is incorporated to quantify the cluster quality in the context of biological importance. Domain annotation data for protein

Table 2 Quality assessment of clustering results and comparison with existing methods

Clustering methods	Total clusters	Non-singltn	% non-singltn.	Corrupt	Seq/clust	%Redn. Reduct.
BiGram_con	**10452**	421	83.1	12	**1.954**	**48.7**
BiGram_all	20,369	49	100	0	1.003	0.21
TriGram_con	**14,355**	362	**84.2**	2	**1.423**	**31.9**
TriGram_all	20,126	54	100	0	1.006	0.14
Uniref90 [11]	19,537	293	81.0	2	1.046	0.42
Cd-hit90 [7]	19,544	108	84.1	1	1.045	0.425

sequences are collected from Pfam database (https://pfam.xfam.org/). The quality of the clusters is analyzed through DCS for each non-singleton cluster. If all the proteins in a cluster agree with the same Pfam domain, the cluster is considered as pure otherwise corrupted (see Table 2). Among the non-singleton clusters, TriGram_con surpass other *state-of-the-art* methods in creating high quality compact clusters as 84.% having DCS 1 whereas in Cd-hit is 84.1% but the number of non-singleton clusters is more than 1/3 of TriGram_con.

4 Conclusion

In this paper, we present a two-level parallel DBSCAN clustering for human protein sequences that address the computational issues of large-scale biological data processing and analysis. The experimental result showed efficient speedup in the proposed method and effectively reduces the redundant from sequences as our method have achieved higher sequence/cluster score (1.423) considering only two corrupted clusters. In the proposed method with trigram feature (consensus), the percentage of non-singleton clusters having $DCS = 1$ is higher than other *state-of-the-art* approaches. The quantitative evaluation shows that the clustering results improved with higher values of n and speedup ratio improves with increasing data size which motivated us to extend the values of n for clustering with higher dimension and operates on even large datasets such as cross organism level, considering both reviewed and unreviewed sequences as future work. Our approach has wide range of applicability in different level bioinformatical analyses such as redundancy removal for bias-free classifier design in PPI prediction [5], functional annotation [4], disease detection analysis [6], protein modeling, etc.

Acknowledgements This work is partially supported by the CMATER research laboratory of the Computer Science and Engineering Department, Jadavpur University, India, Department of Biotechnology grant (BT/PR16356/BID/7/596/2016), Govt. of India. For research support, AKH acknowledges the Visvesvaraya Ph.D. scheme for Electronics & IT research fellowship award, under MeitY, Govt. of India.

References

1. Bandyopadhyay, S.S., Halder, A.K., Chatterjee, P., Nasipuri, M., Basu, S.: Hdk-means: Hadoop based parallel k-means clustering for big data. In: 2017 IEEE Calcutta Conference (CALCON), pp. 452–456. IEEE (2017)
2. Cai, Y., Sun, Y.: ESPRIT-tree: hierarchical clustering analysis of millions of 16S rRNA pyrosequences in quasilinear computational time. Nucleic Acids Res. **39**(14), e95–e95 (2011)
3. Edgar, R.C.: Search and clustering orders of magnitude faster than blast. Bioinformatics **26**(19), 2460–2461 (2010)
4. Halder, A.K., Chatterjee, P., Nasipuri, M., Plewczynski, D., Basu, S.: 3gclust: human protein cluster analysis. IEEE/ACM Trans. Comput. Biol. Bioinform. **16** (2018)
5. Halder, A.K., Dutta, P., Kundu, M., Basu, S., Nasipuri, M.: Review of computational methods for virus-host protein interaction prediction: a case study on novel ebola-human interactions. Briefings Funct Genomics **17**(6), 381–391 (2017)
6. Halder, A.K., Dutta, P., Kundu, M., Nasipuri, M., Basu, S.: Prediction of thyroid cancer genes using an ensemble of post translational modification, semantic and structural similarity based clustering results. In: International Conference on Pattern Recognition and Machine Intelligence, pp. 418–423. Springer (2017)
7. Li, W., Godzik, A.: Cd-hit: a fast program for clustering and comparing large sets of protein or nucleotide sequences. Bioinformatics **22**(13), 1658–1659 (2006)
8. Luo, G., Luo, X., Gooch, T.F., Tian, L., Qin, K.: A parallel DBSCAN algorithm based on spark. In: 2016 IEEE International Conferences on Big Data and Cloud Computing (BDCloud), Social Computing and Networking (SocialCom), Sustainable Computing and Communications (SustainCom) (BDCloud-SocialCom-SustainCom), pp. 548–553. IEEE (2016)
9. Qi, Y., Jie, L.: Research of cloud storage security technology based on HDFS. Comput. Eng. Design **8** (2013)
10. Rodrigues, J.F.M., von Mering, C.: HPC-CLUST: Distributed hierarchical clustering for very large sets of nucleotide sequences. Bioinformatics (Oxford, England) **10** (2013)
11. Suzek, B.E., Huang, H., McGarvey, P., Mazumder, R., Wu, C.H.: UniRef: comprehensive and non-redundant uniprot reference clusters. Bioinformatics **23**(10), 1282–1288 (2007)

Biomolecular Clusters Identification in Linear Time Complexity for Biological Networks

Soumyadeep Debnath, Somnath Rakshit, Kaustav Sengupta, and Dariusz Plewczynski

Abstract Identification of biomolecular clusters from biological networks based on structures is a critical task because the existing algorithmic approaches require high computation and also not feasible for complex, large networks. Majority of these clustering techniques are real-time such as Louvain model which is considered as the fastest algorithm, utilized modularity maximization process for clusters or communities identification. Here we explained a faster, accurate and efficient algorithmic approach for biomolecular clusters identification considering the low running time, as well as better cluster quality using network (graph) based traversal techniques. We also justified that this algorithm works on linear time complexity in order to generate firstly the initial cover and final cover after modularity maximization.

Keywords Biomolecular clusters · Biological networks · Linear running time complexity · Network traversal · Threshold · Modularity maximization

---

S. Debnath
Tata Consultancy Services Limited, Kolkata, India
e-mail: soumyadebnath13@gmail.com

S. Rakshit · K. Sengupta (✉) · D. Plewczynski (✉)
Laboratory of Functional and Structural Genomics, Centre of New Technologies, Warsaw, Poland
e-mail: k.sengupta@cent.uw.edu.pl

D. Plewczynski
e-mail: d.plewczynski@cent.uw.edu.pl

S. Rakshit
e-mail: s.rakshit@cent.uw.edu.pl

S. Rakshit
School of Information, The University of Texas at Austin, Austin, TX, USA

D. Plewczynski
Faculty of Mathematics and Information Science, Warsaw University of Technology, Warsaw, Poland

K. Sengupta
Faculty of Mathematics, Informatics and Mechanics, University of Warsaw, Warsaw, Poland

D. Bhattacharjee et al. (eds.), *Proceedings of International Conference on Frontiers in Computing and Systems*, Advances in Intelligent Systems and Computing 1255,
https://doi.org/10.1007/978-981-15-7834-2_57

1 Introduction

Biological Networks can be visualized as three dimensional graphical structure and defined as $N = (M, C)$, where M represents the set of all biomolecules and C represents the list of connections among them. Here, $m = |M|$ and $c = |C|$ are the number of molecules and connections, respectively. These networks have some implicit features such as molecular clusters which are groups of molecules with dense intraconnections and sparse interconnections. These clusters are connected via some common molecules, termed as *Outline-Cluster Molecules* which mark the boundaries among clusters. So finding such outline molecules help to reach new clusters with identifying the molecules inside those clusters, termed as *Cluster Molecules.*

In this algorithm, firstly during the network traversal, all the molecules are classified on the basis of locating inside the clusters or outline of the clusters and then the outline molecules are assigned to appropriate clusters based on the location of most of their neighbors measuring the probability of belonging. In this process, the outline molecules of same clusters having higher *Belonging Probability* are also assigned to appropriate clusters. After that, based on this initial split, the network size is reduced merging every cluster in one *Giant Molecule* having all intra-cluster edges connected by a weighted self-loop. Finally, the final cover is generated using modularity maximization stage. Though this is an iterative process, the traversal technique has been used in sequential manner to identify molecular clusters within a linear running time complexity.

Most of the existing cluster identification algorithms are computationally expensive with high time consumption. Though generally overlapping clusters also exist, but in reality for all networks, majority of them are suitable for only disjoint clusters. Few algorithmic models take input the required number of clusters which never generate the set of best clusters considering efficiency because these estimations never work properly without prior network analysis. Here, our algorithmic model performs comparatively faster and better than existing approaches considering both overlapping and disjoint clusters with better quality or goodness measurement utilizing the combination of two well-known network (graph) traversal techniques such as, depth first and breadth first traversal.

2 Related Works

From the abstract view of biological networks, it can be considered that all biomolecular clusters are nothing but a meaningful formation of a group of community molecules with relevant similarity and dependency on each other. And here we primarily focus at the cluster identification techniques based on community extraction concepts which are popular for lesser running time. This similar algorithmic model was already proposed previously and applied on different social network datasets by

Basuchowdhuri et al. [1], as one of the best solutions for the community detection problem of graph.

A clustering technique base on density was already proposed in OPTICS by Ankerst et al. [2], with different benchmark datasets and also similar density-based clustering was proposed by Li et al. [3], using nearest neighbor graph very recently. Another clustering algorithm for overlapping clusters was proposed in ONDOCS by Chen et al. [4], which took help of visualization like OPTICS with worst case time complexity of $O(n\log n)$. Our technique is also similar of both which utilized network structure but has achieved linear worst case time complexity. An updated approach for overlapping clusters based on betweenness was discovered by Gregory [5], which produces clusters with small-diameter and takes $O(n\log n)$ time for large and sparse networks, respectively. Previously a structural analysis model with centrality and similarity evaluation on social networks was proposed by Sarkar et al. [6]. In a different side, an approach of modularity of a cluster/cover was proposed using intercluster edges minimization with intracluster edges maximization by Newman [7], which is famous for modularity maximization problem. Further for extension, a disjoint cluster identification model was discovered by Clauset et al. [8], which is popular as ‘CNM’ method.

A fast cluster detection algorithm with $O(m+n)$ time complexity was proposed by Raghavan et al. [9], famous as ‘Label Propagation Algorithm’ (LPA). Another clustering model using greedy modularity maximization technique, famous as ‘Louvain’ method was proposed by Blondel et al. [10], which focuses on finding neighbors for modularity increment is highly appreciated for retrieving disjoint clusters from any network. Other cluster detection algorithms, popular for fast performance in reality are ‘Infomap’ by Rosvall et al. [11] and ‘LexDFS’ by Creusefond et al. [12]. Here LexDFS takes $O(n\log n)$ worst case running time using depth first traversal. Also for dynamic and static PPI networks, clustering was shown by Wang et al. in [13].

3 Methodological Process

In this section, a model for identification of molecular clusters from biological networks in linear time complexity is proposed using network traversal. The sequential detailed process is briefly described below with diagram (shown in Fig. 1).

3.1 Stage 1: Types of Input and Output Data Familiarization

In this initial stage for input data, at first different types of biological networks can be selected. The molecule and network relations for popular biological networks are mentioned below (shown in Table 1).

These networks can be directed, undirected, weighted and also mixed, so in order to make them effective for our algorithm as input, all networks are considered as

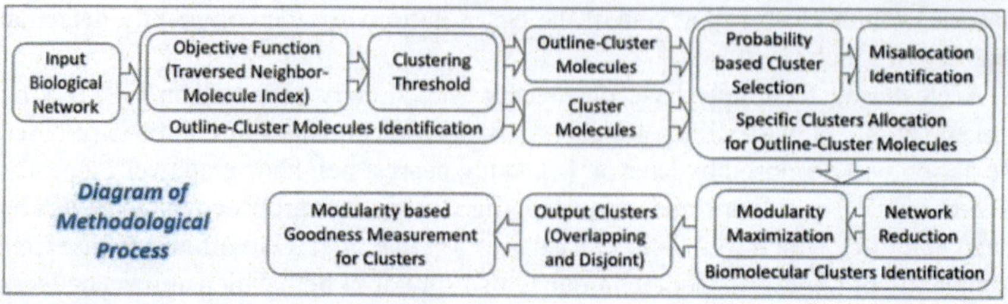

Fig. 1 Biological cluster identification model

Table 1 Popular biological networks with their molecules

Biological networks	Protein–protein interaction networks	Metabolic networks	Genetic interaction networks	Gene/transcriptional regulatory networks	Cell signaling networks
Bio-molecules	Proteins	Metabolites and enzymes	Genes	Genes and transcription factors	Proteins, nucleic acids, metabolites

undirected and unweighted. Now regarding the output clusters, our algorithm can generate both disjoint and overlapping clusters, their mathematical representation is mentioned below;

In any Biological Network $N(m, c)$, let a cover $C = \{C_1, C_2, \ldots, C_k\}$ such that $\cup_i C_i = m$. For *Disjoint Clusters*, $\forall_{ij}$ here $C_i \cap C_j = \emptyset$ and for *Overlapping Clusters*, $\exists_{ij}$ where $C_i \cap C_j \neq \emptyset$.

Here, m and c represents the number of molecules and connections between them respectively.

3.2 *Stage 2: Outline-Cluster Molecules Identification*

Clusters are nothing but the sub-networks with dense intraconnections and sparse interconnections. Molecules lying in the bordering areas of the clusters, termed as *Outline-Cluster Molecules* are the connecting points among two or more clusters. In this stage, all those border molecules are identified through network traversal via each of them based on an *Objective Function* and all other molecules are tried to assign in appropriate molecular clusters by following three sub-stages.

3.2.1 Objective Function Selection

Here the *Objective Function* of this model, termed as *Traversed Neighbor-Molecule Index* (TN_mI) is calculated for each molecule as mentioned below

$$\text{Traversed Neighbor} - \text{Molecule Index}(TN_mI) \text{ of a molecule}$$
$$= \left(\frac{\text{Number of Neighbor} - \text{Molecules which were already traversed}}{\text{Degree of that molecule in that network}}\right)$$

Initially, these values for all molecules are unassigned and each of them is calculated one by one during the network traversal for only once when that molecule is discovered (traversed from the neighbor molecule for the first time). So, this calculation is fully dependent on both the *Traversal Order* and *Starting Molecule* of the traversal process but it will not impact the outputs (justified in the Result Analysis Sect. 5.5). This value for each molecule always lies from 0 to 1. During this traversal process, it is 0 only for the starting molecule and 1 for any other molecule (if all it's neighbors are already traversed).

3.2.2 Clustering Threshold Selection

In this model, to identify the best available molecular clusters from the network, a particular threshold value, termed as *Clustering Threshold* (Th_c) is selected concentrating on the cluster size.

During the network traversal (especially at initial stage), for low value of Th_c by gathering many molecules large size clusters can be generated and for high value of Th_c by fragmenting many small clusters can be generated. The values of this parameter may differ for different networks; such as, for any Dense Network this value is greater than any Sparse Network. Experimentally observed (details shown in the Result Analysis Sect. 5.4) that for the better quality of molecular clusters this threshold value should be from 0.6 to 0.8. Here, this value is considered as 0.7 so that clusters become accurately shaped with appropriate significance and relevant on ground truth.

3.2.3 Cluster and Outline-Cluster Molecules Categorization

During the network traversal all molecules are categorized into two labels based on the *Objective Function, known as Traversed Neighbor-Molecule Index.* Those are *Cluster Molecules* (Ic_m) and *Outline-Cluster Molecules* (Oc_m).

For any molecule at the time of traversal, if its TN_mI value is greater than the Th_c value of the network, then the molecule is identified as *Cluster Molecule* which is already located inside any molecular cluster. Otherwise, the molecule is identified as *Outline-Cluster Molecule* which is lying in the outline of different molecular clusters.

3.3 *Stage 3: Specific Clusters Allocation for Outline-Cluster Molecules*

This stage is responsible for allocation of the *Outline-Cluster Molecules* (Oc_{m}) to the appropriate molecular clusters on which most of their neighbors are belonging.

3.3.1 Probability Based Cluster Selection

Here for each *Outline-Cluster Molecules*, this is decided by comparing a probability ration with all clusters, termed as *Belonging Probability* (*Bp*) as mentioned below

$$\text{Belonging Probability}(Bp)\text{of an } Oc_{\mathrm{m}} \text{ with respect to any Cluster}$$
$$= \left(\frac{\text{Number of Connections from the } Oc_{\mathrm{m}} \text{ to all the molecules of that Cluster}}{\text{Maximum number of Connections possible between the } Oc_{\mathrm{m}} \text{ and that Cluster}}\right)$$

Here increase of cluster size also increases the denominator of this probability ratio, so the allocated clusters for *Outline-Cluster Molecules* are independent of the cluster size with the highest probability value for belonging.

3.3.2 Misallocations Identification

Now, due to the sequential network traversal process, there can be different outcome situations of previous stage for which few *Outline-Cluster Molecules* are unassigned to particular molecular clusters like mentioned below

1. If they have no neighbors in any cluster (for networks with small disconnected components).
2. If they have same *Belonging Probability* value for multiple clusters.

Also, few *Outline-Cluster Molecules* may still be allocated to such molecular clusters which are not expected to be allocated on them as per ground truth. Such problems have been resolved with proper evaluation by the next stage of our algorithm to identify appropriate structured molecular clusters for the whole

3.4 *Stage 4: Biomolecular Clusters Identification*

3.4.1 Reduction of the Network

In this sub-stage, the number of molecular clusters is reduced by conjugating, therefore, the whole network is transformed into a network of *Giant Molecules*. In all clusters, every *Giant Molecule* has a self-loop with a weight which is equal to twice of the number of intracluster connections. Between two clusters, the connections among the couples of *Giant Molecules* have weights which are equal to the number of intercluster connections.

3.4.2 Modularity Maximization

Due to potential merges, the modularity value is changed after the network reduction and increase of it is maximized greedily. To generate the final cover of the clusters, iteratively the *Giant Molecules* are merged and initially the clusters are fragmented by our algorithm with high precision value. So these cluster fragments are merged and achieved the actual cluster structure with high precision and recall value both by the *Modularity Maximization* process. Based on the ground truth these final clusters are more significant than the initial cluster fragments. Primarily this maximized modularity is considered for the accuracy measure of the clusters in our algorithm.

3.5 Stage 5: Goodness Measurement of the Clusters

Theoretically for a cover C_i in the network N, if the value of the *Goodness Function* is given by $Q(N, C_i)$, then the objective is to evaluate c from the below mentioned equation

$$\text{argmax}_c \in C\,Q(N, c) = \{Q(N, C_i)\,|\forall C_i, C_j \in C, Q(N, C_i) \geq Q(N, C_j)\}$$

Here the only condition is that from hypothesis space C, $Q(N, c)$ should be maximum among all covers which are possible. During this, for cover c, the intersection/common part (molecular set) of two clusters is null for considering disjoint, if not, then the cover is overlapping with some common molecular set.

Therefore, for the output molecular clusters from our algorithm, *Goodness* is measured by *Modularity* as *Goodness Function* considering '*modularity*' defined by Newman [7] and '*overlapping modularity*' defined by Nepusz et al. [14], Shen et al. [15], Nicosia et al. [16], for evaluating disjoint and overlapping clusters, respectively.

3.6 Time Complexity Analysis

For a bionetwork $N(M, C)$ where $m = |M|$ is number of molecules and $c = |C|$ is number of connections among them, the *Worst Case Theoretical Time Complexity* has been analyzed based on the *Big-O Asymptotic Notation* for all the stages of our algorithm.

1. For network traversal, both the '*Depth First and Breadth First Graph Traversal*' techniques take with $O(|M| + |C|)$ time.
2. For molecule categorization in Stage 2, traversal through all the connections takes $O(|C|)$ time.
3. For assigning the outline molecules into clusters, the time depends on the number of outline molecules and their respective degrees. And the number of outline molecules can't exceeds total number of molecules because the summation of degrees for all molecules is maximum $2|C|$, so this step takes $O(|C|)$ time.
4. For network reduction procedure, a molecule is appointed as a member of a cluster in a constant execution time in which all connections have been traversed once. So, this takes $O(|C|)$ time.
5. For modularity maximization part, time complexity is unknown [10], but it does not take time more than the order of the network which is already reduced just because each molecule only traverses its neighbors in order to increase modularity. So, this also takes $O(|C|)$ time.

Therefore, here the overall running time complexity for worst case scenario to create the initial bias for the molecular clusters identification is $O(|M| + |C|)$. The performance and quality of our algorithm has also been verified in the result section with some experimental outputs performing on standard biological network dataset.

4 Dataset Details

In order to start the implementation, any type of biological network data has been required with significant properties as the algorithm is applicable for all directed, undirected, weighted and also mixed networks. Therefore, for testing the efficiency of our algorithm, the dataset of the physical and functional protein-protein association network for human has been selected from '*BioSNAP Datasets: Stanford Biomedical Network Dataset Collection*'[1] [17], a repository of biomedical network datasets.

This dataset (*PP-Decagon_ppi.csv.gz*) has 19,081 protein molecules and 715,612 connections among them, known as '*Human Protein-Protein Association Network*'[2] [18] contains all the edge-lists of both directed physical interactions and undirected functional associations between human proteins. Here nodes are represented as protein molecules and edges as associations between them.

5 Result Analysis

In this section, the several modules of our detailed result analysis process has been elaborated below on the previous mentioned protein network dataset.

---

[1] http://snap.stanford.edu/biodata/.

[2] http://snap.stanford.edu/biodata/datasets/10008/10008-PP-Decagon.html.

Table 2 Comparison of modularity with other standard cluster identification algorithms

Our algorithm (disjoint)	Our algorithm (overlapping)	Louvain (disjoint)	CNM (disjoint)	Infomap (disjoint)	Label Prop (disjoint)	LexDFS (disjoint)
0.978641	*0.939827*	0.932874	0.885763	0.82645	0.798457	0.543718

Table 3 Comparison of cover size with other standard cluster identification algorithms

Our algorithm	Louvain	CNM	Infomap	Label Prop	LexDFS
106	211	978	1658	4587	14,396

Here in order to test these experiments, a standard machine with Intel Xeon 2.4 GHz quad-core processor, 32 GB RAM, 1 TB hard drive and Fedora LINUX operating system of version 3.3.4 has been used for these large processing tasks. And '*CoDACom*', a community detection tool[3] has been used for execution of all state-of-the-art algorithms on network clustering. All the source codes of this algorithm are available in a public *GitHub* repository; *BioClusters*.[4]

5.1 *Quality of the Molecular Clusters*

As per the mentioned *Goodness Measurement* (Stage 5) section of our methodological process, the modularity for both the disjoint and overlapping molecular clusters of our dataset have been analyzed here by comparing with other existing standard cluster detection algorithms. The results of this experiment is mentioned below (Table 2).

So, it can be summarized from this analysis module that the goodness or quality of our generated molecular clusters is better than others.

5.2 *Cover Size of the Molecular Clusters*

Here, the quantity of clusters for our dataset has been analyzed based on the cover size of molecular clusters by comparing with other existing standard cluster detection algorithms. The results of this experiment is mentioned below (Table 3).

If the average size of molecular clusters are larger, then only the cover size becomes smaller. A cover with clusters having large size increases the modularity if only it has more significant and meaningful structure than a cover with clusters having large size.

[3] https://codacom.greyc.fr/index.php.

[4] https://github.com/ResearchProjects-codes/BioClusters.

Table 4 Comparison of running time with other standard cluster identification algorithms

Our algorithm (s)	Louvain (s)	LexDFS (s)	Label Prop (s)	Infomap (s)	CNM (s)
4.127368	5.974852	22.335614	46.192718	743.826425	3558.045837

Table 5 Result on molecular clusters for different clustering threshold values

Clustering threshold (Th_c) values, Modularity values of final cover						
0.55, 0.964854	0.6, 0.972163	0.65, 0.973863	*0.7,* *0.978641*	0.75, 0.974875	0.8, 0.973216	0.85, 0.968544

So this proves that our algorithm produces efficient molecular clusters with smaller cover size (marked as italic) which ensures larger average size for our clusters.

5.3 *Efficiency of Our Algorithm*

Here, the efficiency of our algorithm has been analyzed based on the running time (in seconds) by comparing with other existing standard cluster detection algorithms. The results of this experiment is mentioned below (Table 4).

It has been observed that the practical running time of our algorithm is the lowest (marked as italic). So, it can be summarized from this analysis module that our algorithm is faster than others in real life.

5.4 *Variation of Clustering Threshold* (Th_c) *Values*

Here, the quality of the final covers of the molecular clusters for our dataset have been analyzed using modularity values with respect to different values of our *Clustering Threshold* (Th_c) parameter. It has been observed that the clusters from our dataset have their best cover (marked as italic) when the threshold value is 0.7 (Table 5).

So, it can be summarized from this analysis module that the estimated value of our selected *Clustering Threshold* (Th_c) parameter for our algorithm is justified.

5.5 *Variation of Starting Molecule or Order of Traversal*

In our network traversal process, different starting molecules or traversal order may generate different covers. So in order to analyze statistical similarity, a random sample of 500 starting molecules have been selected from our dataset. A graph has

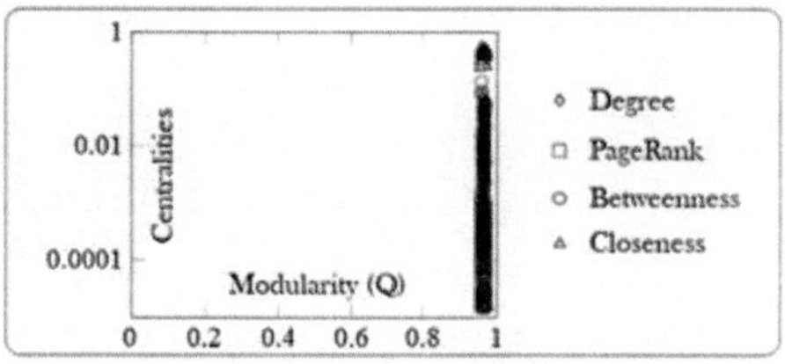

Fig. 2 Centrality measures versus modularity of final cover

been plotted on *Centrality Measures* (*Degree, PageRank, Betweenness and Closeness centralities*) with respect to the *Modularity* values of final cover (disjoint or overlapping) generated for each of those molecules (shown in Fig. 2).

It has been observed that in spite of changing the starting molecule the modularity scores have not been impacted significantly. So, it can be summarized here that the cluster quality of our algorithm is not dependent on selection of the starting molecule or order of the network traversal.

6 Conclusion and Future Scope

At the end, we can conclude that an algorithmic approach has been proposed here for identification of efficient biomolecular clusters from different types of biological networks with high quality of goodness measure in linear running time complexity based on the network size using the combination of two well-known traversal methods like depth first and breadth first traversals. This method works faster than any other state-of-the-art methods and the goodness/quality of these molecular clusters is like either better than or as good as the existing state of-the-art algorithms.

For future scope, this work can be extended with dynamic cluster detection feature based on time. As the major molecular interactions are not static or one-time event, so sometimes it is already observed that the connection between two molecules exists for a certain time duration for many biological networks. In this case, our algorithm needs a time parameter to identify the live molecular clusters specific to that particular time. This work can also be extended by comparisons with different biological complex detection techniques in order to create strong benchmark with the improved result.

Acknowledgements This work has been supported by Polish National Science Centre (2019/35/O/ST6/02484, 2014/15/B/ST6/05082), Foundation for Polish Science (TEAM to DP) and the grant from Department of Science and Technology, Govt. of India and Polish Government under Indo-Polish/Polish-Indo project No.: DST/INT/POL/P-36/2016. The work was co-supported by European Commission Horizon 2020 Marie Skłodowska-Curie ITN Enhpathy grant 'Molecular Basis of Human enhanceropathies'; and by grant 1U54DK107967-01 "Nucleome Positioning System for Spatiotemporal Genome Organization and Regulation" within 4DNucleome NIH program, European Commission as European Cooperation in Science and Technology COST actions: CA18127 "International Nucleome Consortium" (INC), and CA16212 "Impact of Nuclear Domains On Gene Expression and Plant Traits".

References

1. Basuchowdhuri, P., Sikdar, S., Nagarajan, V., Mishra, K., Gupta, S., Majumder, S.: Fast detection of community structures using graph traversal in social networks. Knowl. Inf. Syst. **59**(1), 1–31 (2019)
2. Ankerst, M., Breunig, M.M., Kriegel, H.P., Sander, J.: OPTICS: ordering points to identify the clustering structure. ACM Sigmod Rec. **28**(2), 49–60 (1999)
3. Li, H., Liu, X., Li, T., Gan, R.: A novel density-based clustering algorithm using nearest neighbor graph. Pattern Recogn. **102**, 107206 (2020)
4. Chen, J., Zaïane, O., Goebel, R.: A visual data mining approach to find overlapping communities in networks. In: 2009 International Conference on Advances in Social Network Analysis and Mining, pp. 338–343. IEEE (2009)
5. Gregory, S.: A fast algorithm to find overlapping communities in networks. In: Joint European Conference on Machine Learning and Knowledge Discovery in Databases, pp. 408–423. Springer, Berlin (2008)
6. Sarkar, D., Debnath, S., Kole, D.K., Jana, P.: Influential nodes identification based on activity behaviors and network structure with personality analysis in egocentric online social networks. Int. J. Ambient Comput. Intell. (IJACI) **10**(4), 1–24 (2019)
7. Newman, M.E.: Modularity and community structure in networks. Proc. Natl. Acad. Sci. **103**(23), 8577–8582 (2006)
8. Clauset, A., Newman, M.E., Moore, C.: Finding community structure in very large networks. Phys. Rev. E **70**(6), 066111 (2004)
9. Raghavan, U.N., Albert, R., Kumara, S.: Near linear time algorithm to detect community structures in large-scale networks. Phys. Rev. E **76**(3), 036106 (2007)
10. Blondel, V.D., Guillaume, J.L., Lambiotte, R., Lefebvre, E.: Fast unfolding of communities in large networks. J. Stat. Mech. Theory Exp. **2008**(10), P10008 (2008)
11. Rosvall, M., Bergstrom, C.T.: Maps of random walks on complex networks reveal community structure. Proc. Natl. Acad. Sci. **105**(4), 1118–1123 (2008)
12. Creusefond, J., Largillier, T., Peyronnet, S.: A lexdfs-based approach on finding compact communities. In: From Social Data Mining and Analysis to Prediction and Community Detection, pp. 141–177. Springer, Cham (2017)
13. Wang, R., Wang, C., Liu, G.: A novel graph clustering method with a greedy heuristic search algorithm for mining protein complexes from dynamic and static PPI networks. Inf. Sci. **522**, 275–298 (2020)
14. Nepusz, T., Petróczi, A., Négyessy, L., Bazsó, F.: Fuzzy communities and the concept of bridgeness in complex networks. Phys. Rev. E **77**(1), 016107 (2008)
15. Shen, H.W., Cheng, X.Q., Guo, J.F.: Quantifying and identifying the overlapping community structure in networks. J. Stat. Mech. Theory Exp. **2009**(07), P07042 (2009)
16. Nicosia, V., Mangioni, G., Carchiolo, V., Malgeri, M.: Extending the definition of modularity to directed graphs with overlapping communities. J. Stat. Mech. Theory Exp. **2009**(03), P03024 (2009)
17. Zitnik, M., Rok Sosic, S.M., Leskovec, J.: BioSNAP Datasets: Stanford Biomedical Network Dataset Collection (2018)
18. Zitnik, M., Agrawal, M., Leskovec, J.: Modeling polypharmacy side effects with graph convolutional networks. Bioinformatics **34**(13), i457–i466 (2018)

Security Track

Prevention of the Man-in-the-Middle Attack on Diffie–Hellman Key Exchange Algorithm: A Review

Samrat Mitra, Samanwita Das, and Malay Kule

Abstract In this report, we demonstrate the pros and cons of Diffie–Hellman key exchange algorithm that is commonly used for the exchange of keys in private-key cryptosystem. A very frequently occurring problem of Diffie–Hellman key exchange algorithm during transmission of the keys through some channel is the Man-In-The-Middle Attack. Security of the message gets affected due to this attack because the attacker who is in the middle of the sender and receiver of the message tampers the message and modifies it according to his/her need. There are some well-known solutions available for the prevention of this attack. Our aim is to review those available techniques.

Keywords Cryptography · Private-key cryptosystems · Diffie–Hellman key exchange · Man-in-the-middle attack · Geffe generator · Digital signature

1 Introduction

Cryptography is the science of information security. Cryptography conventionally is referred to the process of converting the plain text (The information which is to be transferred) to the cipher text (Unintelligible gibberish). A special set of secret characters or numbers used to convert the plain text message to cipher text are popularly known as the keys. The method of changing the simple text into cipher text is called Encryption. The inverse process or the alteration of that cipher text to the equivalent or corresponding plain text is known as Decryption [1, 2]. A cryptosystem

S. Mitra (✉) · S. Das
Jalpaiguri Government Engineering College, Jalpaiguri, West Bengal, India
e-mail: samratmitra338@gmail.com

S. Das
e-mail: samanwita.das99@gmail.com

M. Kule
Indian Institute of Engineering Science and Technology, Shibpur, India
e-mail: malay.kule@gmail.com

D. Bhattacharjee et al. (eds.), *Proceedings of International Conference on Frontiers in Computing and Systems*, Advances in Intelligent Systems and Computing 1255,
https://doi.org/10.1007/978-981-15-7834-2_58

comprises of a predetermined set of conceivable plain texts, a finite set of conceivable cipher texts, a finite set of likely keys, an encryption rule for encoding the plaintext into the ciphertext and a decryption rule for decoding the generated ciphertext back to the plaintext.

Private-key Cryptosystems are such that there is a way for sender and receiver to clandestinely share a key k previous to the transmission of plaintext and they can use all the available encryption and decryption techniques described by their undisclosed value of k. One style of sharing those keys is the key agreement procedure whereby sender and receiver together establish the secret key by using those values, they have sent to each other over a public transmission medium. On the other hand, Public-key Cryptosystems are such that receiver preserves his key (and his decoding rule) to himself, whereas the equivalent encoding rule is known to the audience. Therefore, a sender can send encoded messages without any prior sharing of keys, and receiver will be the solitary person, who will be able to decrypt the messages, sent to him.

The Diffie–Hellman Key Exchange Protocol [1, 3], used in private-key cryptosystem, was developed in 1976. Diffie–Hellman Key Exchange is not a mechanism to encrypt data but a method of exchanging the keys securely. Diffie–Hellman key exchange attains this secure exchange of the keys by generating a "shared secret" (sometimes called a "key encryption key") between two devices. The shared secret then encrypts the symmetric key (or "data encryption key" used in DES, Triple DES, CAST, IDEA, Blowfish, etc.) for safe transmission. The safety of the Diffie–Hellman Key Exchange is broadly dependent on the trouble of solving the discrete logarithm problem. Calculating the discrete logarithm of a number modulo is problematic since it takes unevenly the same time which is equivalent to the time needed for factoring the product of two prime numbers, which is the foundation for RSA (Rivest–Shamir–Adleman) algorithm. Therefore, the Diffie–Hellman Algorithm is roughly as protected as RSA (Rivest–Shamir–Adleman) algorithm.

The rest of the paper is organized as the following. The Diffie–Hellman Key Exchange Algorithm is elaborated in Sect. 2. Section 3 presents the Man-In-The-Middle Attack on the Diffie-Hellman Key Exchange technique. Section 4 presents the ways of preventing the Man-In-The-Middle attack, followed by the conclusion and future work in Sect. 5.

2 Diffie–Hellman Key Exchange Procedure

Diffie–Hellman key exchange algorithm, which is also called exponential key exchange algorithm, is a technique for digital encoding which uses the principle of raising some numbers to powers of specific values for calculating the keys. The Diffie–Hellman Key Exchange Algorithm is based on the discrete logarithm problem (DLP). The procedure necessitates two fairly large numbers, supposing p and g, and in that case, p and g are both overtly obtainable numbers. Here, parameter p is taken as a prime number with at least 512-bits and the other parameter g (also known as a generator) is an integer fewer than p, which has the following property: for each of the numbers, say *n*, between 1 to $p - 1$ comprehensive, there is a power *k* over *g* such that

$$n = g^k \bmod p \quad (1)$$

Let Alice and Bob want to settle on a communal undisclosed key using the Diffie–Hellman key exchange Algorithm, they will engender shared key as follows: initially, Alice produces an arbitrary sequestered value *a* and Bob produces a random sequestered value *b*. Both *a* and *b* are drawn from the same collection of integers. Then they originate their public values using the taken parameters *p* and *g* and their sequestered values. The public value for Alice is

$$x = g^{\mathrm{a}} \bmod p \quad (2)$$

The public key value for Bob is the following:

$$y = g^{\mathrm{b}} \bmod p \quad (3)$$

Then they interchange their public values (*x* and *y*) and Alice figures out:

$$ka = y^{\mathrm{a}} \bmod p \quad (4)$$

Then Bob also figures out:

$$kb = x^{\mathrm{b}} \bmod p \quad (5)$$

Subsequently; both of Alice and Bob is computing the same key $ka = kb = g^{ab}$ $mod\, p = k$. Alice and Bob will now be having a communal secret key *k*.

If we take an arbitrary prime number *p* as a very large number of approximately 40 characters, it will take more than 10^{21} stages and each stage encompasses several steps as well. Even using Google's computers which are assessed to perform 300 trillion computations per second, it would take coarsely 5 years to resolve. Alice or Bob, then orates and perhaps alters them before re-encrypting with the suitable key and conveying them to the other party. This susceptibility is present because Diffie–Hellman key exchange does not substantiate the contributors.

If an observer, say Mallory, studies integers *p*, *g*, *x*, *y* but not the discrete logarithm *a* of *x* and *b* of *y* with respect to the base *g*. She wishes to govern the clandestine key which is g^{ab} mod *p* from *p*, *g*, *x*, *y*. She can compute discrete logarithms mod *p*, and try to solve the Diffie–Hellman problem. She governs the discrete logarithm *b* of *y* to the base *g* and computes the key $k = x^b$. This is the only known method for flouting the Diffie–Hellman protocol. Until now, no one has prospered in flouting Diffie–Hellman problem. It is a significant exposed problem of public-key cryptography to find such a proof. As long as the Diffie–Hellman problem is problematic to unravel, no observer can govern the secret key from publicly known data [3].

3 Man-in-the-Middle Attack on Diffie–Hellman Algorithm

The Diffie–Hellman key exchange is susceptible to a Man-In-The-Middle attack. In this attack, an opponent Mallory can not only seize messages form Alice and Bob, but also can send different message and stop sending the original message. The attack is as follows: an opponent Mallory seizes the public value of Alice and directs her own public value to Bob. When Bob conveys his public value, Mallory alternates it with key calculated by herself and redirects it to Alice. Mallory and Alice then settle on one shared key and Mallory and Bob settle on another shared key, without knowing that these are coming from Mallory. Advanced this interchange, Mallory merely decodes any messages which are sent out by either Alice or Bob, and then she reads and possibly alters them before re-encoding with the suitable key and conveying them to the other party. This susceptibility is present because Diffie–Hellman key exchange does not substantiate the participants.

Then Mallory can perform the mentioned actions: pick her own random $e \in (Z\ p)^*$; and then she calculates g^z mod p. Then capture g^a that Alice is sending to Bob and substitute g^z in its place. Note that Bob does not notice any alteration (Because both g^a and g^z are the members of Zp^*) and obediently replies with gb. Mallory seizes g^b and sends g^z instead to Alice. This way Alice ends up discerning that she is sharing $ka = (g^z)^a$ mod p with Bob while Bob thinks that he has end up sharing the $kb = (g^z)^b$ mod p with Alice. Note that, in fact, both Alice and Bob are sharing a key with the intruder Mallory, who is able to calculate both the ka and kb. Now each and every time, Bob is still trying to send some data to Alice, he will apparently encode (and/or authenticate) it using *kb*. Mallory can intercept it, decode with *kb*, re-encode using *ka*, and send it to Alice obediently. Bob and Alice will never be able to comprehend that they are not sharing the key within themselves and that key is being intruded by some third party and the authenticity of their message is being hampered. This major problem while transferring the key using the Diffie–Hellman Key Exchange Algorithm is known as "Man-In-The-Middle" attack, and is just one of the reasons why key transfer between two parties is really a major concern.

4 Prevention of the Man-in-the-Middle Attack

Emerging Cloud computing [4], technology is considered to be a massive preparation of resources, which are presented for the customers by using the internet by some cloud service providers as per their own needs. Also, storage space, called the cloud storage is also given to the users to upload and share the own repository of them [4]. In the field of cloud computing, different types of cryptographic techniques are used to encrypt the data of the users, which are really sensitive and helps their data from getting intercepted or tempered. For safe exchange of keys in case of the symmetric Key cryptosystems, the Diffie–Hellman Key Exchange Algorithm may be used by preventing the Man-In-The-Middle Attack. The aim is to prevent this attack and how to securely transfer data within two parties. The base of the algorithm lies on the theory of arbitrary numbers and related logarithms [4].

4.1 The Existing Diffie–Hellman Key Exchange Algorithm and First Proposed Algorithm

Whitefield Diffie and Martin Hellman has given another secured way of exchanging the keys utilizing asymmetric key cryptosystems. These keys are used to encrypt ensuing communications. Their competence depends on the difficulty of estimating the discrete logarithms. In the proposed solution, two publicly known numbers are used—a prime number 'p' and a primitive root 'w' such that $p < w$. As 'w' is a primitive root of 'p' then the numbers $w \bmod p$, $w^2 \bmod p$, …., $w^x \bmod p$ will be producing all numbers from 1 to $p - 1$ when we take modulus of them with respect to p.

Let Alice and Bob are the two users who are transferring messages between them:

1. Both Alice and Bob agrees on two of the publicly known components, mentioned above.
2. Alice is selecting the private key of her as 'x' considering that $x < p$ and also computing the public key as $A = w^x \bmod p$.
3. Bob is again selecting the private key of him as 'y' considering that $y < p$ and is computing the public key as $B = w^y \bmod p$.
4. Each of the two parties are sharing the same set of public keys A and B with each other.
5. Alice is computing the private key as $Ka = B^x \bmod p$ and Bob is doing the same calculation as $Kb = A^x \bmod p$.
6. Now both the parties have the same private key which will be used in further communication between them.

Both Alice and Bob will be able to compute the same private key as they are only ones who are able to know the key. As there is no such way of authenticating the users, this method is susceptible to the Man-In-The Middle Attack. Also, as the keys remains same for the whole time, it will result in the generation of same sort of decoded text for the corresponding plain text. By using these data, the eavesdropper can easily identify the interdependence of the plain text and its corresponding cipher text. The algorithm reported below is used to protect the keys from the attack

1. Here we are keeping the same Diffie–Hellman algorithm and adding the arbitrary numbers for more security.
2. First five steps in this algorithm are remained intact as the actual Diffie– Hellman Key exchange algorithm.
3. Alice and Bob both are having the same private key. The arbitrary numbers 't' and 's' are selected by them in such a way that $0 < t$ and consequently $s < q$.

4. Now we are calculating another public key by applying logarithm on the keys and having one secret number '*m*', where the secret number '*m*' is taken as the base for calculating the logarithm. Both Alice and Bob know the value of '*m*', such that they can use them further.
5. Now Alice calculates the other public key as: $C_{\text{Alice}} = \log_{\text{m}} (t, K1)$.
6. Similarly, Bob calculates his public key as: $C_{\text{Bob}} = \log_{\text{m}} (t, K2)$.
7. Both Alice and Bob interchange the keys, which will be used in further processes of encoding of the messages.

The private number '*m*' is used as the base of the logarithm by both the two parties, Alice and Bob. The eavesdropper, Mallory, if tries to get the key, it is difficult for her to get the same base. Both Alice and Bob are authenticated in the whole procedure by using this secret number, as it is known to both of them.

Here an arbitrary number, considered in the exchange of keys, is producing the different cipher text for the same piece of plain text. The same piece of plain text is then encoded using different set of keys every time. As a result, the same piece of plain text is generating different cipher text each time. Thus, if the eavesdropper can intercept the message, it will be impossible for him/her to get the original plain text.

4.2 Distribution of Keys Without Being Compromised by the Middle Person

In this technique a password is assumed as a string of eight characters, which is chosen by Alice, let it be 'abcdefgh', having a length of total 64-bits. Each character of this password is converted to the corresponding ASCII characters [1]. To make the password more secure, a pseudo random number generator is used which is popularly known as Geffe Generator, which can generate longer sequences. The Geffe generator is widely used in the stream cipher-based cryptography to generate new sequences. Due to amazing features the Geffe generator is used in the work like, the sequence that is generated by the Geffe generator is well balanced in the number of zeroes (0's) and ones (1's). The whole device could be playing the character of LFSR1 (linear feedback shift registers) in the same arrangement with like generators, and this intricacy would escalate consequently.

The Geffe generator needs three number of linear feedback shift registers, where the length of those must be prime comparatively [5]. That is the greatest common divisor (GCD) of the lengths of the initial values of the inputs into the corresponding registers is one. Here, the initial lengths of inputs to the corresponding registers are taken arbitrarily as 20-bits, 21-bits, and 23-bits, so that the GCD of the corresponding length comes out to be one, and gives us a sequence of 64- bits. The connection geometry is as shown in Fig. 1.

Three sequence will be generated from the corresponding registers and the Geffe Generator will then merge all of them to create a complete sequence. Now, if a

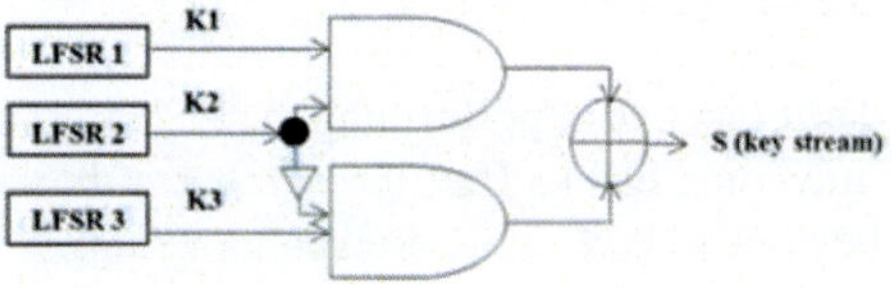

Fig. 1 Geffe generator

sequence generated by the Geffe generator is having a very large length, then the system will lag if not halt for an instance.

The sequences of 1024-bits, 512-bits, 256-bits or 128-bits are generated by the Geffe generator. Then some important statistical tests are performed to analyze the randomness of the generated sequence. If these tests results in a failure then we are to start from the very beginning and the user is asked to give a new password as input to the system. Otherwise on completion of the tests, we will be analyzing the sequences and will take the sequence which is giving us better results from either of the four lengthed sequences. The tests performed are as follows:

Frequency Test

For every generated random binary sequence, it is expected from our side that almost 50% of the sequence is made of 0's and rest of the 50% contains 1's. Let us denote the number of zeroes (0's) by n_0 and number of ones (1's) by n_1. The determination of the test actually depends on the number of zeroes and ones. The statistical formulae used here is

$$X1 = \frac{(n0 - n1)^2}{n} \tag{6}$$

The threshold value for the sequences to pass the test for one degree of freedom is ($X1 < 3.8415$) [6].

Serial Test

The serial test depends on the reappearance of the subsequence like 00, 01, 10, 11 in the sequence, generated by the Geffe Generator correspondingly. And let us assume that each of the subsequence is representing almost nearly a quarter to n. The statistical formulae which is used to perform the Serial Test is

$$X2 = \frac{4}{n-1}\sum_{i=0}^{1}\sum_{j=0}^{1}(n_{ij})^2 - \frac{2}{n}\sum_{i=0}^{1}(n_i)^2 + 1 \tag{7}$$

The threshold value for the sequence to pass the test for two degrees of freedom is ($X2 < 5.9915$) [6].

Poker Test

The objective of performing the Poker test is to break the generated random sequence into some blocks (Let the number of blocks in which the generated sequence is to be divided is K) and corresponding lengths be M. We have to check the repetition of the blocks and has to compare with it, the expected value for any arbitrary sequence, the number of blocks denoted by K is represented by the following equation, $K = n/M$. The statistical formulae being used here is

$$X3 = \frac{2^M}{K}\left(\sum_{i=1}^{2^M}(n_i^2\right)) - K \tag{8}$$

The threshold value for the test for degrees of freedom of $2^m - 1$ is ($X3 < 14.0671$) [6].

Three tests were repeatedly done to check the selected sequence S with 1024-bits, 512-bits, 256-bits, and 128-bits long (with the assumed password) and the outcomes of the tests given in Table 1. From the table it is evident that the sequence having a length of 128-bits is unable to pass the frequency test and thus this sequence is rejected.

Next sequence of 256-bits has successfully passed all the three tests and thus it is selected as the sequence with minimum length and also this sequence is having a good amount of randomness. All the test data are provided in the Table 1.

The 256-bit sequence is selected and each eight bits is to be converted into some decimal numbers using their ASCII values. As a result, we will be getting a total of 32 numbers and then we have to take the modulus of each number by 10 and

Table 1 Results of the statistical tests

Length of sequence	Tests	Values found from test	Degrees of freedom	Threshold value
1024-bits	Frequency test	0.0351625	1	3.8415
	Serial test	2.0401125672	2	5.9915
	Poker test	9.1935483871	7	14.0671
512-bits	Frequency test	0.0703125	1	3.8415
	Serial test	0.4521923924	2	5.9915
	Poker test	7.2235294117	7	14.0671
256-bits	Frequency test	1.00000	1	3.8415
	Serial test	1.4274509804	2	5.9915
	Poker test	7.5176470588	7	14.0671
128-bits	Frequency test	5.281525	1	3.8415
	Serial test	4.5691437008	2	5.9915
	Poker test	7.90476190	7	14.0671

we will get numbers only from 0 to 9. Then all the results, we are getting are to be accumulated and combined to get a large number of 32 digits.

Then assuming 8-digit private key of the user (Which is assumed to be within 2 to $p - 2$ [2], where p is to be provided by the server itself), the task is to break and divide the number of 32 digits (as obtained earlier) in 8-blocks. Finally, we have to take one digit from each block arbitrarily along with a private key (PrK) with eight digits. Here all the users will have their own private key.

Both the end users (i.e., Alice and Bob) are getting two numbers, let them be g and p. Now p is a prime number and g is the primitive root, modulo p. p and g will come from the server itself with the public key. Alice will be getting p, g, the public key from Bob and the username. Then the encryption is done using the private key of the receiver. As cited, the private key should be lesser than p, the prime number.

The algorithm that can solve the problem is given here:

At the time of registration:

1. The password from the user, having a length of eight characters is converted to a binary sequence of 64-bits using the corresponding ASCII codes of the characters.
2. The binary sequence is further divided into three registers (LFSR's) with 20-bits, 21-bits, and 23-bits, respectively, as demonstrated in the use of Geffe generator in Sect. 4. The Geffe generator is used to create a sequence with 256-bits, as it was the minimum length that has passed all the three statistical tests.
3. The generated sequence is checked and tested with frequency test, serial test, and poker test.

   a. If the selected sequence is passing all the tests we have to proceed further.
   b. Else move to step-1 and again ask the user to enter a new password.

4. Convert the sequence, which has passed all the three tests into eight equal size blocks and take one number from each block and construct the private key.
5. The same method is used to find both the private keys called PrK1 and PrK2.
6. The admin will send an activation message, asking Alice and Bob to activate their passwords.

   a. If they receive the message then they will continue.
   b. Else Pw1 and Pw2 is changed by the intruder. Move to Step 15.

7. The server hashes the passwords and encrypts the private keys (i.e., PrK1 and PrK2) using the admin key and then the random number is added and saved into the user information table.

At the time of Login:

1. The server then readdresses Alice and Bob to the next page and delivers them the necessary information such as p, g, Public keys, and usernames. The received private keys such as the PrK1, PrK2 are used in encryption of the same. Alice and

Bob have the passwords to access the information. The public keys are calculated using the following equations [7]:

$$\text{PuK1} = g^{\text{PrK1}} \bmod p \tag{9}$$

$$\text{PuK2} = g^{\text{Pr K2}} \bmod p \tag{10}$$

2. The server redirects them to the next page and provides them with all necessary information like the Public Keys, p, g, and usernames correspondingly. Then they use their passwords to access the data.
3. Then they calculate their shared keys and the server also calculates the same.

$$\text{SK}_{\text{Alice}} = {\text{PuK}_2}^{\text{PrK1}} \bmod p \tag{11}$$

$$\text{SK}_{\text{Bob}} = {\text{PuK}_1}^{\text{PrK2}} \bmod p \tag{12}$$

4. The server then hashes the shared key and keeps in the user information table.
5. Alice and Bob then hash their own shared keys to the server.
6. The server checks the hashed shared keys and with the hash table:
   a. If the hashed shared key matches with any of the entries of the table then continue.
   b. Else shared key is manipulated by the intruder.
7. Now everyone (i.e., Alice and Bob) are ready to share their messages, which is encrypted using the shared key. Move to the end of the process.
8. Pop Up shows the warning that the data is compromised.
9. End.

So, the intruder Mallory can only obtain the hashed data, which is also encrypted, that is not to be used by her to retrieve the messages and she will be unable to interrupt the communication process.

5 Conclusion

We have reviewed some methods that are available to avoid the Man-In-The-Middle attack and also has discussed and analyzed an algorithm from the view of its efficiency to avoid the Man-In-The Middle attack in Diffie–Hellman Key Exchange Algorithm. An effectual technique to strengthen the Diffie–Hellman Key Exchange against Man-In-The-Middle attack was demonstrated in the paper. The Geffe generator is used to create a binary sequence with high degree of arbitrariness and three tests were done to check the generated sequences and also to select the required minimum length for which all the tests are satisfied, before calculating the secret and shared key. The procedure, proposed here guarantees us that no such keys will be transmitted through the channel and will be saved as hashes within the server. The sender and receiver are also identified from the user information table. This method prevents the Man-In-The-Middle attack and gives better performance than the existing algorithms in terms of transferring confidential messages between two parties.

References

1. Khader, A.S., Lai, D.: Preventing man-in-the-middle-attack in Diffie-Hellman Key Exchange Protocol. ICT (2015)
2. Preneel, B., Paar, C., Pelzl, J.: Understanding Cryptography: A Textbook for Students and Practitioners. Springer, London (2009)
3. Ahmed, M., Sanjabi, B., Aldiaz, D., Rezaei, A., Omotunde, H.: Diffie-Hellman and its application in security protocols. Int. J. Eng. Sci. Innovative Technol. (IJESIT) **1**, 69–73 (2012)
4. Gadhavi, L., Bhavsar, M., Bhatnagar, M., Vasoya, S.: Design of Efficient Algorithm for Secured Key Exchange Over Cloud Computing, pp. 180–187 (2016)
5. Wei, S.: On generalization of Geffe's generator. IJCSNS **6**, 161–165 (2006)
6. Hosseini, S.M., Karimi, H., Jahan, M.V.: Generating pseudorandom numbers by combining two systems with complex behaviours. J. Inf. Secur. Appl. 149–62 (2014)
7. Cristian, T.: Security issues of the digital certificates within public key infrastructures. Informatica Economica **13**, 16–28 (2009)

Blind Source Camera Identification of Online Social Network Images Using Adaptive Thresholding Technique

Bhola Nath Sarkar, Sayantani Barman, and Ruchira Naskar

Abstract The ubiquitous use of digital images has led to a new provocation in digital image forensics. Source camera identification authorizes forensic investigators to discover the probable source model that are appointed to acquire the image under investigation in criminal cases like terrorism, child pornography, etc. The techniques proposed till date, to identify the source camera, are inappropriate when applied to images captured with smartphones, subsequently uploaded and downloaded from these online social networks (OSN) such as *Facebook*, *WhatsApp,* etc. Such images lose their original quality, resolution, and statistical properties as online social platforms compress the images substantially, for efficient storage and transmission. Here, we present an efficient forensic image-source identification procedure, which follows an adaptive thresholding mechanism to differentiate between high and low image-camera correlation. The results of our experiments demonstrate that the proposed method drastically improves the performance of blind source camera identification of compressed OSN images as compared to traditional global threshold approach.

Keywords Adaptive threshold · Digital forensics · Peak–to–correlation–energy ratio · Sensor pattern noise · Source camera identification

B. N. Sarkar (✉) · S. Barman
Jalpaiguri Government Engineering College, Jalpaiguri 735102, India
e-mail: bns2151@cse.jgec.ac.in; sarkaraditya817@gmail.com

S. Barman
e-mail: sb2149@cse.jgec.ac.in

R. Naskar
Indian Institute of Engineering Science and Technology, Shibpur 711103, India
e-mail: ruchira@it.iiests.ac.in

D. Bhattacharjee et al. (eds.), *Proceedings of International Conference on Frontiers in Computing and Systems*, Advances in Intelligent Systems and Computing 1255,
https://doi.org/10.1007/978-981-15-7834-2_59

1 Introduction

Digital forensics can be described as a subdomain of forensic field, engirdling the revival and inquisition of materials initiated in digital gadgets, related to cybercrime. Sensor fingerprint-based source identification procedures manipulate by extricating camera/sensor precise fingerprints, and subsequently correlating those test images noise residuals [1].

In most of the existing researches, for blind source camera identification of digital images and related experiments, the images are captured with high quality cameras (such as Digital Single Lens Reflex (DSLR) cameras), which produce very high-resolution images [2]. Such images can be mapped back to their sources considerably efficiently, due to the strong sensor artifacts contained within such images.

However, the scenario is not quite similar when we deal with low quality or compressed images. This happens mainly due to the wide usage of OSN today, (such as Facebook, WhatsApp, YouTube, etc.), all of which compress multimedia data (images or videos) at a high compression rate, for easy storage and efficient transmissions.

Such digital data are many times associated with sensitive personal information, which become all the more relevant when involved in a crime scene. Every OSN possess their own propriety compression features, which when claimed on images uploaded, are further stored or transmitted. Source identification accuracy of such procedures drop abruptly, when provided with OSN images. Here, we deal with the problem of source identification with OSN images.

To solve this problem, we compute the sensor fingerprint or Sensor Pattern Noise (SPN) of each source, and the Photo Response Non-Uniformity (PRNU) noise residual of each image under observation. Therefore, the degree of similarity between each pair (test image (I), source (C)), is evaluated as Peak–Correlation–to–Energy ratio (PCE). Then, we apply a threshold to detect whether the similarity is HIGH or LOW. In this paper, we propose a novel adaptive threshold technique for efficient source identification of compressed OSN images. As proved from our experimental outcomes, the accuracy in source detection has improved by adoption of our proposed adaptive threshold.

The rest of the paper is organized as follows. In Sect. 2, we present the associated researches in the domain of forensic identification of source camera. In Sect. 3, we present the proposed source camera identification for OSN images with an adaptive thresholding approach. In Sect. 4, we present our experimental outcomes and associated discussions. Finally, we conclude the paper in Sect. 5, with some directions for future research in this area.

2 Background

In the recent decades, there have been several researches in the domain of identification of source camera in digital forensics. Most of these research works are based on images captured with high-end DSLR cameras as presented in [3], which produce high image resolution. In this segment, we have presented a recapitulation of state-of-the-art researches as presented in [4], which solved the problem of mapping images back to their source camera models.

Sameer et al. [1], have shown that the conventional source detection techniques are unsuccessful in discerning the correct source in case of presence of inaccessible camera models, and falsely maps images to one of the approachable camera models. The authors have proved that their proposed strategy segregates known and unknown models, helping it to attain high source identification authenticity in comparison to traditional methods.

Gelogo and Kim [5], propose to compress the images before dissemination, in order to diminish the transmission bandwidths usage. In [5], the authors have proved the transmission issues of compressed images, and introduced transmission procedures to address these issues, and have also shown that compressed image transmission procedures are productive due to their diminished transmission duration as those parts of an image are conveyed, which represent the whole image by progressively transmitting the image data.

Shullani et al. [6] have proposed the VISION dataset comprising images captured from smartphones, and uploaded to online social networks including Facebook, WhatsApp, and YouTube, which is a major contribution toward the development of reL-LIFE datasets for standardizing present-day forensic researches.

3 Proposed Source Camera Identification for Online Social Network Images with an Adaptive Thresholding Approach

In this module, we present a camera sensor fingerprint-based image-source identification technique. First, we present the traditionally adopted global thresholding approach, and prove how the implementation of such approach gets unfavorably influenced due to OSN compression of digital images. Next, we propose an adaptive thresholding mechanism to optimize the performance of forensic source camera identification for OSN images, by reducing the false matches.

The lossy image compression, authorizes refurbish only of an estimation of the primeval data, which is usually allowed for upgraded compression rates and smaller sized files. Such lossy compression techniques are used for depletion of the amount of data which would otherwise be required to store, handle, and/or impart the represented content. Using well-designed lossy compression technology, there is often a possibility of reduction of a substantial amount of data before the result is moderately degraded to be perceived by the user.

In order to overcome this problem of JPEG compression induced loss of sensor specific artifacts of camera models, thereby, we propose a method of adaptive thresholding technique for forensic image-source detection, which has been elaborated in Sect. 3.3, later.

3.1 SPN and PCE Computation for Source Camera Identification of OSN Images

The first module computes the Sensor Pattern Noise (SPN) of each camera model and the Photo Response Non-Uniformity (PRNU) noise of each image under observation. Consequently, the degree of similarity between each pair (test image (I), source (C)) is evaluated as Peak–Correlation–to–Energy ratio (PCE). A high PCE rate for (I, C) designates that image (I) is inaugurated from source (C), whereas a low PCE denotes that (C) is not originated from (I).

The PRNU component of a single (ith) image I_i (Noise Residual) is computed as

$$\mathrm{PRNU}_{Ii} = {P_x}^{(i)} - DF({P_x}^{(i)}) \tag{1}$$

where, ${P_x}^{(i)}$ represents the initial image, which passes through a Denoising Filter (DF), in order to induce a denoised image. The Sensor Pattern Noise (SPN) of a source camera C_j is computed as

$$\mathrm{SPN}_{cj} = \frac{\sum_{i=1}^{n} \mathrm{PRNU}_{I_i}.P_x^{(i)}}{\sum_{i=1}^{n} \left(P_x^{(i)}\right)^2} \tag{2}$$

wherein n is the total number of images used to compute the camera fingerprint.

The degree of similarity between PRNU of an image and SPN of camera is generated in form of Peak–to–Correlation–Energy ratio (PCE) as

$$\mathrm{PCE}(I_i, cj) = \frac{\rho_{\mathrm{peak}}^2}{\frac{1}{|r|-|\varepsilon|} \sum_{r \notin \varepsilon} \rho 2} \tag{3}$$

wherein, ρ denotes the normalized cross correlation between PRNU_{Ii} and SPN_{Cj}, ρ peak represents the largest cross correlation value precise to (I_i, Cj).

In Table 1, we present a sample image-source correlation (PCE) matrix. As evident from Table 1, three smartphone camera models are taken in this experiment, viz., Samsung_GalaxyS3Mini, Apple_iPhone4s, and Huawei_P9. The Sensor Pattern Noise (SPN) are extracted from the camera models and correlated with the PRNU of

Table 1 Sample PCE correlation matrix for native images collected from the VISION dataset [6] (results shown for three images per model)

	Sensor pattern noise (SPN) of camera model			
		Samsung_GalaxyS3Mini	Apple_iPhone4s	Huawei_P9
PRNU of images from different source model	Samsung_GalaxyS3Mini	**2828.010697**	−1.556146845	0.521388
	Samsung_GalaxyS3Mini	**1784.629741**	−0.050870101	1.179899
	Samsung_GalaxyS3Mini	**331.986507**	0.54182531	−0.74639
	Apple_iPhone4s	−0.373167916	**1442.119088**	−1.9307
	Apple_iPhone4s	0.80842314	**447.1836881**	−0.50162
	Apple_iPhone4s	−0.333314715	**1434.398127**	−0.7924
	Huawei_P9	0.177209703	0.02183604	**1183.302**
	Huawei_P9	−0.480834577	−0.105823016	**193.2016**
	Huawei_P9	0.033500415	−0.250315108	**178.7577**

the test images from each model, to produce the Peak–Correlation–to–Energy (PCE) matrix.

We can observe from Table 1, that the PRNU of images from Samsung_GalaxyS3Mini has high PCE value when correlated with the SPN of Samsung_GalaxyS3Mini, whereas, has low PCE value when correlated with other camera models. So it can be said that the images here are correctly mapped back to their source models without any false camera detection.

However, the scenario drastically changes when we consider compressed or OSN images. For example, in Table 2, we present the PCE correlation results for Facebook uploaded images, from the same set of three cameras. It is evident from Table 2, that the PCE values are much lower, as compared to those in Table 1. This happens due to high degrees of image compression brought by the social networks as Facebook.

Next in Sect. 3.2, we present the traditional approach of global thresholding, used till date for distinguishing between HIGH and LOW PCE.

Table 2 Sample PCE correlation matrix for Facebook high quality images collected from the VISION dataset [6] (results shown for three images per model)

	Sensor pattern noise (SPN) of camera model			
		Samsung_GalaxyS3Mini	Apple_iPhone4s	Huawei_P9
PRNU of images from different source model	Samsung_GalaxyS3Mini	**6.781891**	0.000652	−0.00465
	Samsung_GalaxyS3Mini	**610.2031**	0.17032	−0.69915
	Samsung_GalaxyS3Mini	**14.737**	−0.3692	0.007427
	Apple_iPhone4s	0.3743	**58.79633**	−6.18059
	Apple_iPhone4s	0.00724	**2.534997**	−0.2025
	Apple_iPhone4s	−0.70379	**763.3681**	2.948701
	Huawei_P9	0.157596	−0.5684	**117.606**
	Huawei_P9	−0.47584	0.160426	**1.21647**
	Huawei_P9	−2.1E−05	0.000127	**8.006714**

3.2 Global Thresholding Approach

In traditional forensic source identification, the HIGH and LOW PCE values are differentiated based on a threshold which is uniform for all camera models that the system needs to deal with, hence termed as a global threshold.

The global thresholding is based on a 2–means clustering approach which is enforced on the image-source correlation or PCE values, in order to differentiate between HIGH(ClusterHj) and LOW PCE(ClusterLj) values as presented in [1]. The threshold is selected precise to each camera model as

$$\text{Global threshold} = \text{mean}\left(\min(\text{ClusterHj}), \max(\text{ClusterLj})\right) \tag{4}$$

The PCE values, which are greater than the threshold evaluated above, are considered HIGH, i.e., True Positives (marked with blue (TP)), and the remaining values LOW, i.e., False Positives (marked with orange (FP)). It can be observed from Figs. 2, 3 to 4 that a huge number of false positives are obtained in case of OSN images, which implies that a vast number of images are mapped to wrong camera models (which are not its true source). Especially, in Figs. 2 and 3, the number of false positives are even greater than that of true positives. In the next section, we present an adaptive thresholding technique to reduce the false matches found in (Figs. 1, 2, 3 and 4).

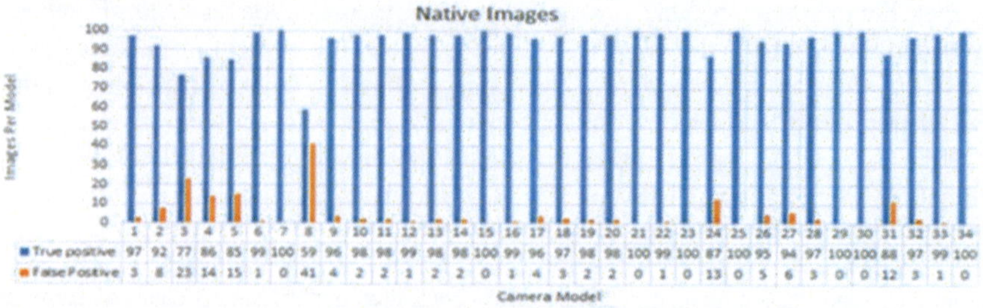

Fig. 1 True positives and false positives for native images using global thresholding

Fig. 2 True positives and false positives for Facebook high quality images using global thresholding

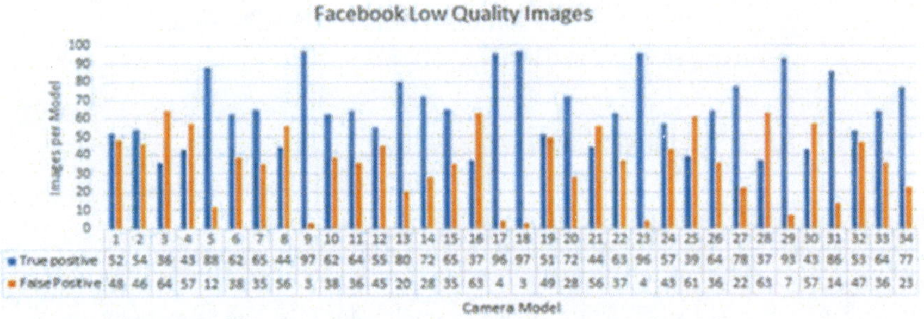

Fig. 3 True and false positives for Facebook low quality images using global thresholding

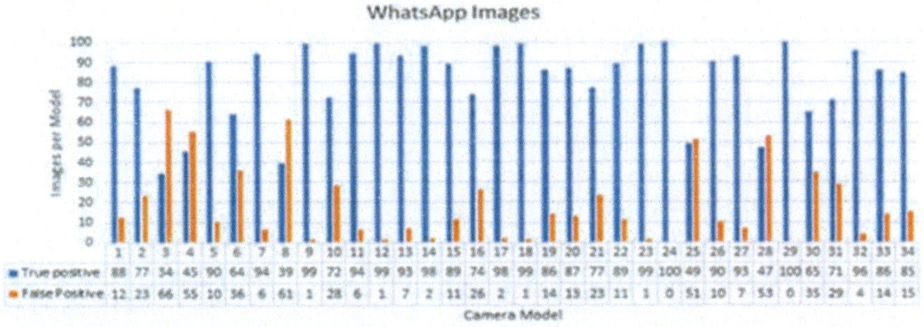

Fig. 4 True positives and false positives for WhatsApp images using global thresholding

3.3 Proposed Adaptive Thresholding Approach

Here, we propose an adaptive thresholding approach, whereby, we select different thresholds for different categories of images, depending on the social networking site where the image has been uploaded. Adaptive thresholding is the approach where the threshold value is framed for limited regions, there will be distinct threshold values for distinct regions.

In this module, we have proposed the selection of adaptive thresholds for four different categories of images, viz., Native images, Facebook high quality images, Facebook low quality images and WhatsApp images, from the 34 camera models from VISION image dataset [6].

We divide the PCE values obtained from the PCE similarity matrix in module 1, into two clusters: a maximum cluster (Cluster_{max}) and a minimum cluster (Cluster_{min}). Here, Cluster_{max} consists of all PCE values of the form PCE (PRNUi, $\text{SPN}_{j)}$ where I_i originated from camera C_j. And Cluster_{min} consists of all PCE values of the form PCE (PRNUi, $\text{SPN}_{j)}$ where I_i did not originate from camera C_j.

Next, we compute the maximum of Cluster_{min} for 34 camera models, including all four categories of images. The adaptive threshold is then computed as follows:

$$\text{mean}\,(\max(\max\ \text{of}\,\text{Cluster}_{\text{min}}),\ \min(\max\ \text{of}\,\text{Cluster}_{\text{min}}))$$
$$+\ ((\text{mean}\,(\max\,\text{of}\,\text{Cluster}_{\text{min}}))/2) \quad (5)$$

PROCEDURE 1: Adaptive Threshold Selection

Input: Minimum Cluster of PCE matrix Cluster_{min}, Test Image Set (I_{test}) from given camera models C_1, C_2, …, PCE Similarity Matrix between (I_{test}) and (C_1, C_2, ….)
Output: Adaptive_threshold

1. Initialize MAX_{TEMP} to a sufficiently small value (-10^9) and sum to 0;
2. For each camera model C_j do
3. For each image $I_i \in I_{test}$ do
4. MAX_{TEMP} = max (MAXTEMP, PCE (I_i, C_j))
5. End for
6. MAX(j) = MAX_{TEMP}; //Storing maximum PCE for each camera model
7. End for
8. For each camera model C_j do
9. sum = sum + MAX(j);
10. End for
11. Average = sum/C_j;
12. Adaptive_threshold = ((max (MAX) + min (MAX))/2) + (Average/2);
13. Return Adaptive_threshold;

In Figs. 5, 6, 7, and 8, we present bar graphs showing the true and false positives statistics for 34 camera models, using adaptive threshold for native images, high and low quality images downloaded from Facebook, images downloaded from WhatsApp, respectively. The true and false positives are shown in blue and orange, respectively. We have plotted the number of images along *Y*-axis and 34 camera models along *X*-axis. From the graphs, we can observe that, huge number of false positives observed in the graphs shown in Figs. 1, 2, 3, and 4, have been drastically reduced, improving the source detection accuracy for OSN images. Our experimental results and related discussions have been presented next.

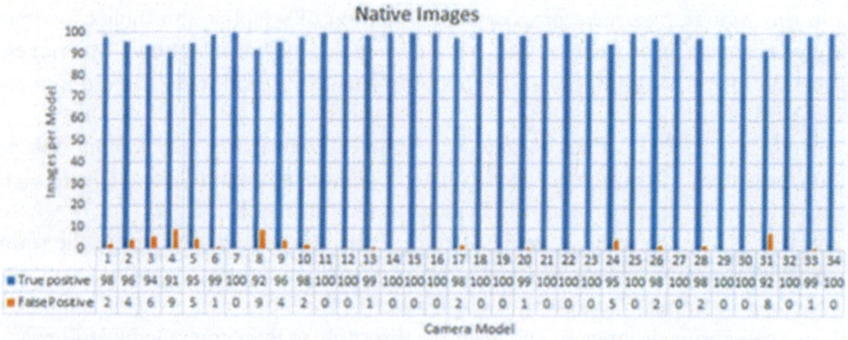

Fig. 5 True positives and false positives for native images using adaptive thresholding

Fig. 6 True and false positives for Facebook high quality images using adaptive thresholding

Fig. 7 True and false positives for Facebook low quality images using adaptive thresholding

Fig. 8 True and false positives for WhatsApp images using adaptive thresholding

4 Experiments, Results, and Discussion

In this section, we put forward our explorations and outcomes for performance estimation of our propounded approach. We have used the Vision image dataset, a benchmark datum for camera–based forensic analysis providing 16,958 images from 74 cameras having 26 distinct makes. Here, we present the outcomes for smartphone images obtained from 26 camera models of 34 distinct makes from Vision Image database.

All our experiments were conducted in MATLAB, on a I5 processor and 8 GB RAM and are organized in three phases. First, to find the SPN and PCE matrix of the images downloaded from online social platforms in source camera. Second, using global threshold, find the number of true positives and false positives and represent it by a bar graph. Third, using *adaptive threshold*, find the number of true and false positives and represent it by a bar graph.

Module 1: Here we find the SPN and PCE matrix of the images downloaded from online social platforms. From each model of 34 camera models, we have considered four different image categories: Native images, Facebook high quality images, Facebook low quality images, and WhatsApp images, with 100 images each. Therefore, summarily from each camera model we consider $4 \times 100 = 400$ images in total.

We first calculate the Sensor Pattern Noise (SPN) of the camera models and store it. Then we compute the PRNU noise residual of each category of test images. Finally, the PCE matrix is obtained by correlating each image (category-wise) with each camera model. Hence, we obtain four different 3400×34 PCE matrices.

Module 2: Here we find the number of true and false positives using global threshold. Our experimental results consistent to the above, are presented in Figs. 1, 2, 3, and 4 and summarized in Table 3. It can be clearly observed that a huge rate of false positives is obtained for each of the 34 camera models.

Table 3 Comparison between global and adaptive thresholding

Image category	With global threshold			With adaptive threshold		
	True positives	False positives	Accuracy (%)	True positives	False positives	Accuracy (%)
Native images	3226	174	94.88	3337	64	98.14
Facebook H.Q	2309	1091	67.91	3097	303	91.08
Facebook L.Q	2186	1214	64.29	3171	229	93.26
WhatsApp images	2766	634	81.35	3320	80	97.64
Average accuracy (%)			77.1			95.03

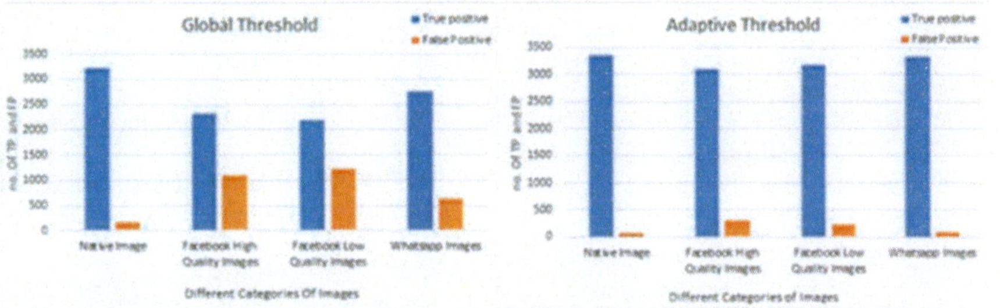

Fig. 9 Number of true positives (TP) and false positives (FP) obtained with global thresholding versus adaptive thresholding

Module 3: In this experiment, we aim to reduce the number of false positives observed in module 2, so that we can maximize the accuracy of the source camera model detection. To solve this, we adopt the adaptive thresholding approach, which is computed, as illustrated in Sect. 3.3. The results are presented in Fig. 9.

The improved source detection accuracy results along with True and False Positives, are presented in Table 3.

5 Conclusion

In this work, we have addressed the problem to identify the source camera for highly compressed online social network images. Here we have proposed an adaptive thresholding approach to distinguish between HIGH and LOW image-source PCE correlations. The proposed adaptive threshold is selected specific to a particular online social networking site, to which an image is uploaded. The results obtained prove that the proposed technique achieves an overall source detection accuracy of over 95%, in contrast to 77% accuracy obtained by traditional forensic approaches.

The future work in this direction involves dealing with large-scale test scenario, where the system comprises of millions of image samples per camera model.

References

1. Sameer, V.U., Sugumaran, S., Naskar, R.: K-unknown models detection through clustering in blind source camera identification. IET Image Proc. **12**(7), 1204–1213 (2018)
2. Sameer, V.U., Sarkar, A., Naskar, R.: Source camera identification model: classifier learning, role of learning curves and their interpretation. In: 2017 International Conference on Wireless Communications, Signal Processing and Networking (WiSPNET). IEEE (2017)
3. Gloe, T., Böhme, R.: The dresden image database for benchmarking digital image forensics. In: Proceedings of the 2010 ACM Symposium on Applied Computing. ACM (2010)

4. Sameer, V.U., Dali, I., Naskar, R.: A deep learning based digital forensic solution to blind source identification of Facebook images. In: International Conference on Information Systems Security. Springer, Cham (2018)
5. Gelogo, Y.E., Kim, T.H.: Compressed images transmission issues and solutions. Int. J. Comput. Graph. **5**(1), 1–8 (2013)
6. Shullani, D., Fontani, M., Iuliani, M., Shaya, O.A., Piva, A.: Vision: a video and image dataset for source identification. EURASIP J. Inf. Secur. **2017**(1), 15 (2017)

Event-B Based Formal Modeling of a Controller: A Case Study

Rahul Karmakar, Bidyut Biman Sarkar, and Nabendu Chaki

Abstract Event-B is an event-driven approach for system development. It has the flexibility to develop different discrete control systems. Event-B is a refinement-based step-by-step modeling methodology. There is a well-tested open-source tool available for Event-B model checking, formalization of mathematical proofs and system validation is done in RODIN. This paper presents a short survey on the usage of an Event-B-based model to locate the research gaps followed by a case study to build a model using the 2-stage refinement strategy to stop the precious groundwater wastage and conserve it. We try to model the behavior required for the environment of the system. The proposed controller then controls the environment. The controller acts accordingly and achieves the goal of groundwater conservation.

Keywords Formal modeling · Event-B · RODIN tool · Industry automation · Eclipse IDE · Water pump controller · Groundwater conservation

1 Introduction

Event-B [1] is a modeling language and its application range is versatile. Not only a sequential program to distributed systems but it also has the privilege to model different control systems. Event-B models the environment which is the necessity to assure the correctness of the proposed systems [2]. The Event-B-based formal modeling proposed by Abrial [1] is a top-down engineering approach consisting of the step-by-step refinement strategy. The designers, therefore, design the refinement

R. Karmakar (✉)
The University of Burdwan, Burdwan, India
e-mail: rkarmakar@cs.buruniv.ac.in; rahulkarmakar6@gmail.com

B. B. Sarkar
Techno International, Rajarhat, Kolkata, India
e-mail: bidyutbiman@gmail.com

N. Chaki
University of Calcutta, Kolkata, India
e-mail: nabendu@ieee.org

D. Bhattacharjee et al. (eds.), *Proceedings of International Conference on Frontiers in Computing and Systems*, Advances in Intelligent Systems and Computing 1255,
https://doi.org/10.1007/978-981-15-7834-2_60

strategies to meet the system requirements and specifications. A model in Event-B consists of two components—context and machine. Context is the static part of a model. It contains all the constants, axioms, sets, and theorems that remain unaffected at any state of a machine. The machine describes a whole model. It sees the context and also contains the dynamic part consisting of variables, invariants, theorems, and events. The dynamic part is affected when an event has occurred (i.e. when the state of a machine is changed). An event consists of a set of actions, guards, and parameters. The axioms are the predicates made-of constants, and the invariants are the predicates made-of variables (and constants). The refinement can be done by imposing features one-by-one to an abstract model. Thus, we can find a set of refined models. At the final refinement, the model has all the desired features. The context of one model can be extended to the context of another model. When a machine (concrete machine) refines another machine (abstract machine), it refines the abstract events of the previous one; it can also extend the concrete events as well [1, 3]. Formal modeling allows to do a more rigorous review of the system and as a result, improve the quality of the system. The Rigorous Open Development Environment for Complex Systems (RODIN) [4, 5] tool supports refinement based rules and generates proof obligations. It is an eclipse-based open-source IDE and can be extended with plugins. The notations of Event-B either automatically or interactively generate the proof obligation. The main advantage of RODIN is its flexibility for proof obligation and model checking [4]. Starting from the initial model, each of the refinements is proved so that if any error is found at any stage of refinement, the error must not be carried forward to the next refinements. The RODIN platform indicates which rules are successfully proved and which contradict. Event-B has drawn reasonable attention both for industrial and academic communities toward modeling problem specifications and iterative refinements. The structure of the paper is given below.

Section 1 gives a glimpse of the Event-B-based formal modeling approaches particularly in industry and safety-critical domains with a short survey. Section 3 represents a case study on a water pump controller design. The control strategy has been designed using Event-B followed by the model validation in Sect. 4 and conclusion in Sect. 5.

2 System Review

This section represents a brief review of different control strategies developed using Event-B. We consider mainly industrial automation and safety-critical systems. A few works are considered for the review though a handsome number of works are there. We start with an article by Michel Butler where he proposed a control strategy of a water pump. He tries to achieve the control goal and control strategy using the Event-B modeling language. The system has a monitor variable for checking the water level of the tank, and a control variable that controls the environment and switches on and off the pump. The Input and Output variables tell the water level value stored in sensors and registers. The event UpdateWaterLevel is used to update the current water level of the tank. DecreseWaterLevel and IncreseWaterLevel events refine the previous event UpdateWaterLevel. Guard variables are used to enforce time-bound in the environment events. The control event is used to switch on and off the pump by checking the threshold value of the tank. The maximum threshold value signifies that the tank is full and it will stop the pump automatically and vice versa [6].

Jean-Raymond Abrial in his book Modeling in Event-B system and software engineering discussed different applications developed using Event-B modeling. In second chapter of the book, "Controlling cars on a bridge" he discussed controlling cars on a bridge. Here, he develops a control system for scheduling the cars from the mainland to Iceland via a bridge. The system development starts with the initial model where only the mainland and Iceland are considered. In the first refinement, the bridge is introduced and assumes that no car is on the bridge. Then traffic lights are introduced. The Green lights allow a car to go and Red light stops a car. Then sensors are introduced in the system for identifying a car whether it is entering the bridge or leaving the bridge. The one-way and both way nature of the bridge are also considered. The controller has different events and functions to refine the environment. In chapter sixteen "A location access controller", we find the design of the location access controller where people can find their locations. The design controls the sensors and other devices to get accurate location information. Finally, in chapter seventeen, "Train system" the design of a train system controller is proposed. It will act as a software agent to control trains. The goal of the controller is to provide the safety and reliability of a train network. All the examples are well formalized with Event-B notation and they are deadlock-free [1].

The authors show in paper [2] how the platforms and trains are managed by the controller. They identified and designed the network topology using the event-driven approach. They consider sensors, switches, and actuators for the purpose. Different events and actions are developed. Safety rules are also imposed on those actions. The model is developed in the RODIN tool, and all the consistencies are checked and are available online.

A Programmable Logic Controller (PLC) is modeled in the paper [7]. The PLC is meant for a huge factory setup where radiopharmaceuticals are produced. The initial model says that the Automatic production line only starts and there is no production. Then the progress event of work is done step by step with the help of the refinements. Safety is the main concern. The starting refinement deals with the safety rule where the action of the cylinders is considered. The action of the cylinder tells the progress of the event to the next step. The next refinements deal with the safety rule for the people. A safety alarm is used for safety. The position of the delivery is very important and to get the correct information about the delivery system, and this rule is used in the succeeding refinements. The final rule tells the successful accomplishment of delivery and for that, a confirmation message from the device is introduced as an end rule. If all the steps of the system are completed and safety rules are followed, then the risk must be minimized that we discussed above. We find an article about an aircraft landing system [8]. The failure of a critical system like aircraft causes a fatal end. Different landing gear arrangement is used in aviation and it has sequential operations like opening the doors of gearboxes, extending and closing the doors. The landing gear system has three main components like the mechanical system, the digital system, and the pilot interface. Event-B is used not only to model the system behavior of the landing gear system but also to prove the safety requirements. The first refinement is about the press up and press down events of the system. The next refinement is about opening and closing the doors of the aircraft. Observation of the gear is done in the next refinement. Next, sensors and actuators are introduced in the system. Failure detection and adding lights are also done in the system. These refinements are modeled by the Event-B notations. The refinements are verified by the RODIN tool. The integration of domain-based features using Event-B modeling is done in an article. A case study of noise gear velocity is shown in this paper [9]. Designers find Event-B-based modeling and RODIN-based proof obligation of the requirements of the system in safety and embedded systems [10].

The findings of the survey suggest that the Event-B-based formal modeling can be very useful for control strategy design of a controller.

3 A Case Study on a Controller Design

The Event-B Modeling Language has been used to design a controller which addresses the aspect of consumable groundwater conservation. The consumable freshwater is essential for the sustainability of mankind and day by day, the water demand exceeds its availability. The careless wastage of water is one of the major reasons behind it. In countries like India, the main source of drinking water is groundwater. Most households have a pump to withdraw groundwater. The main goal of the system is to control the pump connected to a water tank automatically to save the wastage of water. We start with an abstract model of the system. Then in each refinement, we consider the requirements and refine the model. The controller communicates with the environment with the help of sensors and the communication is bidirectional. The controller receives input from the environment via the sensor and produces outputs accordingly to change the environment [6]. Three major steps are followed: Firstly, the environment is modeled to behave accordingly. This is an abstract model and omits the controller completely. Here, only the pump and water tank have been considered. Secondly, the capacitive liquid level sensor and relay switch are introduced to change the environment: Here, these components ensure that they interact with each other correctly and change the environment accordingly. Finally, the controller has been modeled to get statistical analysis: The strategy has been proposed which will enable us to perform some statistical analysis generated by the controller. The main advantage of this development strategy is that one can start from a basic model and using step-by-step refinements reach the final model.

Table 1 The environments and functions of the system

ID	Description
ENV-1	The system consists of a pump, a water tank, a relay switch, and a capacitive liquid level sensor
ENV-2	When the relay is on, the pump is started to withdraw groundwater; when the relay is off, the pump stops
ENV-3	The capacitance level liquid sensor gives the minimum and maximum water level of the water tank
ENV-4	There is a microcontroller that controls the whole system
FUNC-1	The controller will control the groundwater withdrawal by the pump
FUNC-2	When the capacitive liquid level sensor gives minimum value, the controller switches on the relay; hence, the pump is started
FUNC-3	When the sensor gives maximum value, the system switches off the relay; hence, the pump is stopped

3.1 Requirement Document

The initial model has the minimum functionalities like start filling the tank or stop filling. Then new refinements are added and detailed requirements of the system are considered. A capacitive liquid level sensor, relay switch are considered in the system for checking the water level and to start and stop the pump accordingly. The requirement document is given in Table 1.

3.2 Initial Model

The initial model gives the highest level of abstraction of the system. The water tank is only visible in the model. The tank is started to fill or stopped to stop filling based on the water level in the water tank. A microcontroller is there in our system which will sense the current water level of the tank. The block diagram of the system is shown in Fig. 1.

If it finds the water level to be lesser than the minimum threshold, then it will initiate the pump to start filling the tank. When the water level crosses the maximum threshold, then it immediately stops the pump. The static part

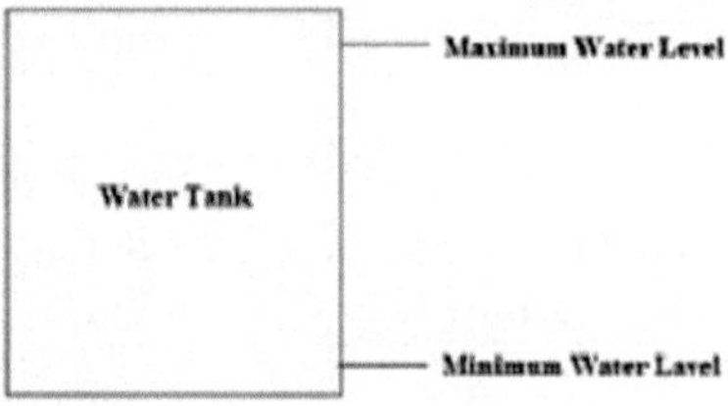

Fig. 1 Block diagram of the initial model

of the initial model has 2 sets, TANK_STATUS and FILLING_STATUS. The TANK_STATUS is either empty or full and FILLING_STATUS is either started or stopped. These set values are constants and their properties are represented as axioms like empty is not equal to full and started is not equal to stopped. The threshold value for minimum and maximum water levels of the tank is represented by two constants, WATER_LEVEL_MINIMUM and WATER_LEVEL_MAXIMUMT. The corresponding axioms of the constants ensure a value which belongs to a natural number.

The dynamic part of the initial model sees the context discussed above. The machine has three variables, PresentWaterLevel represents the current water level in the tank. The tank status variable denotes whether the Water Tank is full or empty. The fillingstatus variable represents whether the tank has started filling or stopped filling the Water Tank.

The requirement properties are established with the invariant rules of the machine. There are 7 invariant properties.

inv1:	presentWaterLevel ∈ ℕ
inv2:	tankStatus ∈ TANK_STATUS
inv3:	fillingStatus ∈ FILLING_STATUS
inv4:	presentWaterLevel ≤ WATER_LEVEL_MINIMUM ⇒ tankStatus = EMPTY
inv5:	tankStatus = EMPTY ⇒ fillingStatus = STARTED
inv6:	presentWaterLevel ≥ WATER_LEVEL_MAXIMUM ⇒ tankStatus = FULL
inv7:	tankStatus = FULL ⇒ fillingStatus = STOPPED

An event is a set of actions that may occur or is to be performed during the execution of the system. The occurrences of an event change the state of the machine. Therefore, one or more variables are modified during the execution of an event. We introduce a guard for the events. The basic events of the initial model START_FILL and STOP_FILL are formalized using Event-B notations and are as follows [1, 4, 6].

START_FILL function	STOP_FILL function
START_FILL WHERE grd1:presentwaterlevel = WATER_LEVEL_MINIMUM THEN act1: tankstatus = EMPTY act2: fillingstatus = STARTED END	STOP_FILL WHERE grd1:presentwaterlevel = WATER_LEVEL_MAXIMUM THEN act1: tankstatus = FULL act2: fillingstatus = STOPPED END

To begin execution, the first event which occurs in every system is the INITIALIZATION. In this event, the initial values are assigned to the variables. Therefore, every machine must have an INITIALIZATION event. It must not have a guard and a parameter. It is assumed that the water tank is initially empty. There are two main events: START_FILL occurs when the water level in the tank reaches its minimum level. STOP_FILL occurs when the water level in the tank reaches its maximum level. When an event makes changes to the variables, it must preserve the hypothesis associated with those variables and the system must be deadlock-free. The proof-obligation rules are provided to confirm these preservations. The RODIN tool automatically generates proof obligations and tries to prove the model. It shows errors when there is any contradiction.

3.3 Refinement 1

This model extends the context and refines the machine of the initial model. The Capacitive Liquid Level Sensor, Relay Switch, and Pump are considered in this refinement. The abstraction is removed in this refinement and the capacitive liquid level sensor, relay switch, and the pump can be seen. Since the system has a sensor, the water level checking is not an issue. The whole setup is shown in Fig. 2.

The sensor's output is taken to decide whether to on or off the relay to start or stop the pump, respectively. The pump will start and stop automatically by the capacitance sensor and relay switch of the pump. When the water tank becomes empty, and the capacitance value becomes lowest, then the microcontroller switch on the relay switch to start the pump. On the other hand, when the water level reaches its maximum level, then the microcontroller swithes off the relay switch to stop the pump. Therefore, in this refinement, we need two sets of statuses, RELAY = {ON, OFF} and CAPACITANCE = {MINIMUM, MAXIMUM}. The variables named capacitance and relay are used to track the current statuses of the sensor and the relay, respectively, in the machine part. The invariant rules used in this refinement are given below.

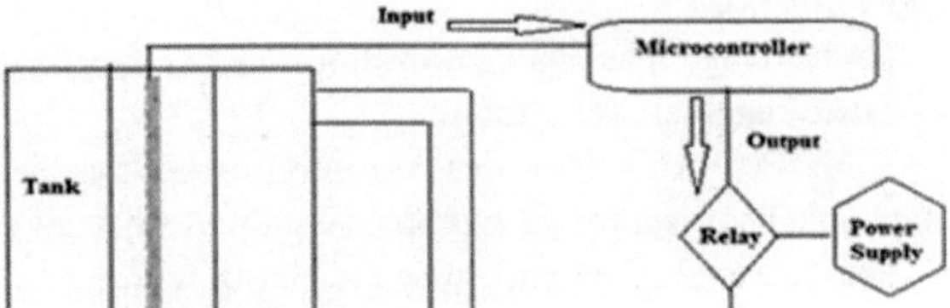

Fig. 2 Overall block diagram of the refined system

Inv8:	capacitance ∈ CAPACITANCE
Inv9:	relay ∈ RELAY
Inv10:	tankStatus ∈ TANK_STATUS
Inv11:	capacitance = MINIMUM ⇒ presentWaterLevel = WATER_LEVEL_MINIMUM
Inv12:	capacitance = MAXIMUM ⇒ presentWaterLevel = WATER_LEVEL_MAXIMUM

The refinement also has three events. The INITIALIZATION event sets the initial value for the variables. Initially, capacitance equals MINIMUM and relay is ON. START_PUMP refines START_FILL to ON the relay when the capacitance value is MINIMUM, and STOP_PUMP refines STOP_FILL to OFF the relay when the capacitance value is MAXIMUM. The events in this model are starting and stopping the pump, and are formalized using Event-B notations as given as follows [1, 4, 6].

START_FILL function	STOP_FILL function
START_PUMP REFINES START_FILL WHERE grd1: capacitance = MINIMUM THEN act1: relay = ON END	STOP_PUMP REFINES STOP_FILL WHERE grd1: capacitance = MAXIMUM THEN act1: relay = OFF END

3.4 *Refinement 2*

Here, some events are proposed for statistical analysis of water usage like EndOfDay, EndOfMonth, and EndOfYear. Based on the statistical data of the previous month, the maximum water consumption limit per day is set for the next month. The model generates a statistical report on a daily, monthly ,and yearly basis for usage, monitoring, and validation.

EndOfDay: The water consumption per day is added to a set to generate a statistical analysis for a month.

EndOfMonth: The water consumption per month is added to a set to generate a statistical analysis for 12 months. A set will store all the data.

EndOfYear: A set will store the yearly report.

Daily water usage can be limited from statistical analysis. One can get the limit from the last few years' daily usages of water. There will be the actions which will withdraw water according to the limit received by the statistical analysis. The pump can be started and stopped according to the water consumption limit also. One household can get a previous month or year data such that they can get the average use of water and use the pump accordingly.

4 Model Analysis and Validation

This analysis helps to maintain water consumption by a household in a great way. We are still working on the refinements and add more features like getting statistical data when draining and filling are performed at the same time. This model may be applied to a municipality pumping station in the future for an optimized water supply. The Rodin Platform is an Eclipse-based open-source integrated development environment (IDE) for Event-B. It provides operative provision for refinement and mathematical proof. We prove all our assumptions and model consistency in the RODIN platform. We generate all the proof obligations in the RODIN tool and validate the water conservation model using the context, sets, constraints, events, and axioms of START_FILL function, STOP_FILL function in refinement 1, and additional events like EndOfDay, EndOfMonth, and EndOfYear. The proposed controller not only controls the whole system but also manages the redundant withdrawal of water. The controller addresses the excessive overflow of water from the water tank and helps to control the state. We are capable to successfully distinguish all the environments and functionalities. We can formalize the environments and functionalities in the final model and validate using the RODIN tool. We have 14 automatic proofs for the initial model and 5 automatic proofs for the 1st refinements. The proof statistics from RODIN toolset are shown in Fig. 3a, b.

These are the simple rules so the proofs are less, and we did not need any iterative proofs so far. The full proof design makes the system implementation and testing smoother, and we can firmly deal with the real-time system design using Event-B.

a

Element Name	Total	Auto	Man.	Rev.	Und.
Machine_0	**13**	**13**	**0**	**0**	**0**
INITIALISATIO...	5	5	0	0	0
inv1	1	1	0	0	0
START_FILL	4	4	0	0	0
inv2	0	0	0	0	0
STOP_FILL	4	4	0	0	0
inv3	0	0	0	0	0
inv4	3	3	0	0	0
inv5	3	3	0	0	0
inv6	3	3	0	0	0
inv7	3	3	0	0	0

b

Element Name	Total	Auto	Man.	Rev.	Und.
Machine_1	**4**	**4**	**0**	**0**	**0**
INITIALISATIO...	2	2	0	0	0
inv1	0	0	0	0	0
inv2	0	0	0	0	0
inv3	0	0	0	0	0
START_PUMP	1	1	0	0	0
STOP_PUMP	1	1	0	0	0
inv4	1	1	0	0	0
inv5	1	1	0	0	0

Fig. 3 **a** Proof statistics of machine 0. **b** Proof statistics of machine 1

5 Conclusion

This paper summarizes the work of Event-B, especially for controller design. A System model has been developed using Event-B and validated using the RODIN tool. The case study addresses the burning issue of the groundwater conservation. We start from an abstract model and design the model incrementally to meet the required goals. A system is modeled into Event-B notations. The proposed system has a machine and context that deals with the environment. For detailing the system, different events are introduced. Event-B based modeling has the facility to consider the environment with the software part. All the variables, invariants, and guards help to meet all constraints of the system. The RODIN platform plays a great role to prove the models. The application domain could be anything but all we need is the ability to formalize it. All the applications we discussed in Sect. 2 are safety-critical and failures can be instantly traced back. Researchers can design plug-in for the RODIN framework for automatic code generation in many languages like Java, C, and C++.

References

1. Abrial, J.-R.: Modeling in Event-B System and Software Engineering. Cambridge University Press, UK (2010)
2. Hudon, S., Hong, T.S.: Development of Control systems Guided by Models of Their Environment. Electronic Notes in Theoretical Computer Science, vol. 280, pp. 57–68. Elsevier (2011)
3. Cansell, D., M´ery, D.: The Event-B modelling method: concepts and case studies. In: Brauer, W., Hromkovic, J., Rozenberg, G., Salomaa, A. (eds.) EATCS Series, Monographs, Theoretical Computer Science (Logics of Specification Languages), vol. 57, pp. 47–152. Springer, Heidelberg (2008)
4. Jastram, M., Butler, M.: RODIN User's Handbook. DEPLOY project (2010)
5. Rezazadeh, A., Evans, N., Butler, M.: Redevelopment of an Industrial Case Study Using Event-B and RODIN. BCS-FACS Christmas 2007, Meeting Formal Methods in Industry. The British Computer Society (2007)
6. Butler, M.: Using Event-B Refinement to Verify a Control Strategy (Unpublished)
7. Fu, K., Fang, B., Li, Y., Li, H.: Research on Event-B based formal modeling and verification of automatic production line. In: 28th Chinese Control and Design Conference (CCDC), IEEE, China 2016, pp. 3690–369 (2016)
8. Méry, D., Singh, N.K.: Modelling an aircraft landing system in Event-B. In: Communication and Computer Science, vol. 433, pp. 154–159. Springer, Berlin (2014)
9. Méry, D., Sawant, R., Tarasyuk, A.: Integrating "domain-based features into Event-B: noise gear velocity a case study. In: Margaria, T., Steffen, B. (eds.) MEDI 2015, LNCS, vol. 9344, pp. 89–102. Springer, Heidelberg (2015)
10. Event-B Homepage: http:// http://www.Event-B.org. Accessed 5 Sept 2017

High Payload RDH Through Directional PVO Exploiting Center-Folding Strategy

Meikap Sudipta, Jana Biswapati, Bera Prasenjit, and Singh Prabhash Kumar

Abstract In this paper, a high payload reversible data hiding scheme (HPRDH) has been designed using Directional Pixel Value Ordering (DPVO) Exploiting Center-Folding Strategy. At first, Center-Folding Strategy has been applied on a cover image to embed secret information which generates Dual Marked Images (DMI). After that, Directional Pixel Value Ordering (DPVO) has been utilized to embed more secret data on interpolated Dual Marked Images (DMI). Here, the advantage of the center-folding method has been employed, which compresses the valuable hidden information through averaging. On the other hand, DPVO provides repeated embedding on overlapped pixels of an image block in different directions like horizontal, vertical, and diagonal one after another. The proposed method is highly adaptive in nature because of its variable-size secret data along with different image block sizes for successful embedding. The experimental outcomes show that the proposed method is very suitable to embed more hidden data without compromising the visual quality of the image compared to other state-of-the-art methods. The intended outcome brought into the limelight some remarkable sublime characteristics in the field of hidden data communication, tamper detection, and digital forgery detection without which the technological life is stunted. Innumerable government and private sector facets including health care, commercial security, defense, and intellectual property rights get immensely benefited from this scheme.

Keywords Reversible data hiding · Center-folding · Directional pixel-value-ordering · Embedding capacity · Psnr · Steganography

M. Sudipta · J. Biswapati (✉) · B. Prasenjit · S. P. Kumar
Department of Computer Science, Vidyasagar University, Midnapore, India 721102
e-mail: biswapatijana@gmail.com

M. Sudipta
e-mail: sudiptameikap@gmail.com

B. Prasenjit
e-mail: munnabera14@gmail.com

S. P. Kumar
e-mail: singhg11@gmail.com

D. Bhattacharjee et al. (eds.), *Proceedings of International Conference on Frontiers in Computing and Systems*, Advances in Intelligent Systems and Computing 1255,
https://doi.org/10.1007/978-981-15-7834-2_61

1 Introduction

The data concealing technique uses cover images, text documents, or any varieties of multimedia as a data bearer where the secrets are embedded within it in a method of the way to retrieve the message successfully by the receiver. Information concealing procedure can be divided into reversible and irreversible which are based upon the reconstruction of the original image. The secret data as well as the cover image can be rebuilt after successfully completing the extracting process. This paper utilizes the RDH method as a core to fix secrets into a pixel of images. After data extraction, irreversible methods are inadequate to reconstruct the real cover image whereas reversible methods can reconstruct the real image. This paper works on the reversible data hiding method to embed secret messages into cover images.

In 2003, Tian [12] described an information concealing scheme through difference expansion (DE) to conceal secrets inside a pixel pair. It describes the redundancy in the image for achieving high embedding data and the distortion was low. In 2005, Lee et al. [2] presented a lossless information hiding scheme where distortion can be fully removed from the marked image. Also, the secret data has been taken out from a marked image. In 2013, pixel value ordering-based information hiding approaches have been described by Li et al. [3] where minimal and maximal pixels are moderated due to message embedding. This shows a high-fidelity RDH scheme for images on a new strategy which embedded the data through prediction error expansion (PEE). Peng et al. [10] upgraded PVO presented by Li et al. where new pixel differences were calculated and new histogram modification was used. High payload-based data hiding method through subsample has been described by Biswapati Jana [1]. Lu et al. [4] approached reversible data hiding using the center-folding technique. To achieve high embedding, it uses two copies of the real image. This scheme folds the secret information and then embeds it into dual stego images. Meikap et al. [6, 7] discussed a PVO-based RDH schemes which increased the high capacity of data bits. The challenge is to increase the payload by considering embedding again and again into overlapped pixels in several directions within a block after folding the secrets and stuffing it into pixels. To overcome this challenge, we introduced an improved center-folding-based information hiding method through directional PVO with different block sizes. Our suggested method increases the strength of security. The secrets are transformed into decimal values on the basis of p data set during the first embedding. Second, the embed process is dependent on the number of image blocks and the size of the image block during second embedding. Third, the information is distributed over two images. For retrieval of the message, two stego images as well as the value of p, the number of image blocks and the size of the image block are needed. If an unauthorized person wants to access the message, the said parameters are needed. It may be hard for an unauthorized person.

The rest of the paper is as follows: in Sect. 2, the proposed method of embedding and extracting is explored. The observational results accompanied by comparisons are explored in Sect. 3. In Sects. 4 and 5, steganographic attack and conclusions of the proposed method are explore, respectively.

2 Proposed Method

In this section, we proposed the activity of secrets embedding as well as extraction of Improved Center-Folding-based Directional Pixel Value Ordering. We attempt to increase the capacity of the embedding by preserving the image quality along our suggested Improved Center-Folding-based Directional Pixel Value Ordering scheme. The total process is grouped by the following steps.

2.1 Data Embedding Phase

Step 1: The real cover image represents $CI = \{pixel_{1,1}, \ldots, pixel_{w,h}\}$, where w and h denoted image width and height, respectively. The hidden information takes every p bits as a set and modifies them to a decimal of hidden symbol d_s. Then hidden symbol d_s is compressed and adjusted to the symbol range from $P = \{0, 1 \ldots, 2^p - 1\}$ to $Q' = \{-2^{p-1}, -2^{p-1} + 1, \ldots, -1, 0, 1, \ldots, 2^{p-1} - 2, 2^{p-1} - 1\}$. The folded hidden symbol d_s' can be calculated as follows:

$$d_s' = d_s - 2^{p-1} \tag{1}$$

where 2^{p-1} represents intermediary values. The secrets embedding process can be accomplished by

$$\begin{cases} d_{s1}' = \left\lfloor \frac{d_s'}{2} \right\rfloor, \\ d_{s2}' = \left\lceil \frac{d_s'}{2} \right\rceil \end{cases} \tag{2}$$

where d_{s1}' and d_{s2}' are derived from d_s'. These values are inserted into pixel $pixel_{i,j}$ to construct dual marked pixels $pixel_{i,j}'$ and $pixel_{i,j}''$ by the following equations:

$$\begin{cases} pixel_{i,j}' = pixel_{i,j} + d_{s1}', \\ pixel_{i,j}'' = pixel_{i,j} - d_{s2}'. \end{cases} \tag{3}$$

The information hides into pixels in between 2^{p-1} and $256 - 2^{p-1}$.

Step 2: Now, we expand these images through the interpolation method shown in Fig. 1 by averaging the neighbor pixels. If the image size is $(v \times v)$, then the interpolated image will be $(v + (v - 1)) \times (v + (v - 1))$, where interpolated row/column is $(v - 1)$. One is added to both row and column to form an even number of rows and columns and copying the last row and column value.

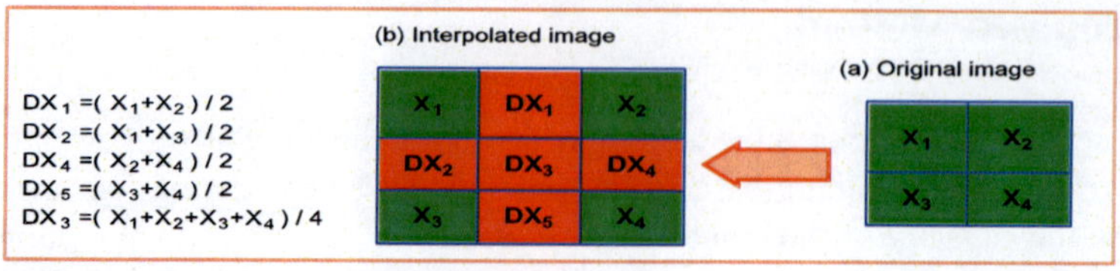

Fig. 1 Interpolated image construction. Red color represents interpolated row and column

Step 3: In this step, the dual interpolated stego images are split into nonoverlapping blocks. In each block, all the pixels are sorted in the ascending order. Here is a chance to change pixel values more than two which may affect the sorting order. To keep the order same, we subtract α from the smallest pixel onwards then adding the same from the largest pixel which is dependent on Lemma 1. The α is dependent on the block size. After each inclusion or exclusion, the α decreases by one.

Lemma 1 *If the size of the block is* $(2v-1)$ *and the sorted pixels in increasing manner are* $(pixel_1, pixel_2, pixel_3, \ldots, pixel_{2v-3}, pixel_{2v-2}, pixel_{2v-1})$*, then the changed pixel value is* $(pixel_1 - \alpha_{(v-2)}, pixel_2 - \alpha_{((v-2)-1)}, pixel_3 - \alpha_{((v-2)-2)}, \ldots, pixel_{v-1} - \alpha_{((v-2)-(v-2))}, pixel_v, pixel_{v+1} + \alpha_{((v-2)-(v-2))}, \ldots, pixel_{2v-3} + \alpha_{((v-2)-2)}, pixel_{2v-2} + \alpha_{((v-2)-1)}, pixel_{2v-1} + \alpha_{(v-2)})$*, where, the value of* $\alpha_{((v-2)-(v-2))} = ((v-2)-(v-2))$*. The* α*'s maximal and minimal values are* $(v-2)$ *and 0, respectively.*

2.1.1 Embedding in Minimum-Modification

In every row of a block X, assume that n pixels $(pixel_1, \ldots, pixel_n)$ are ordered in an arising manner to receive $(pixel_{\sigma(1)}, \ldots, pixel_{\sigma(n)})$. We compute

$$d_{\min_f} = pixel_b - pixel_c, \quad \text{where} \quad \begin{cases} c = \max(\sigma((2)+f), \sigma((1)+f)), \\ b = \min(\sigma((2)+f), \sigma((1)+f)), \\ f = (0, 1, 2, 3, \ldots, \text{fix}(n/2)-1). \end{cases} \tag{4}$$

To round the data toward zero, we use fix(). Now, the minimum pixels are changed into $pixel'$. In each operation, α is changed. The modified minimum pixels are derived by

$$pixel' = \begin{cases} (pixel_{\sigma((1)+f)} - \alpha) - 1, & \text{if } d_{\min_f} < 0 \\ (pixel_{\sigma((1)+f)} - \alpha) - D, & \text{if } d_{\min_f} = 1 \\ (pixel_{\sigma((1)+f)} - \alpha) - D, & \text{if } d_{\min_f} = 0 \\ (pixel_{\sigma((1)+f)} - \alpha) - 1, & \text{if } d_{\min_f} > 1 \end{cases} \tag{5}$$

where $D \in \{0, 1\}$ are inserted into the pixel. Assume the changed value of X=$(cpixel_1, cpixel_2, \ldots, cpixel_n)$ in each row, where $cpixel_{\sigma((1)+f)} = pixel'$ and

$cpixel_i = pixel_i$ for all $i \neq \sigma((1) + f)$. The changed value of $X = (CR1, CR2, ..., CRn)$ are presented in each column, where $CR\alpha((1) + f) = cr$ and $CRi = cri$ for all $i = \alpha((1) + f)$.

2.1.2 Embedding in Maximum-Modification

The data inserting into pixels for Maximum-Modification is discussed below. Calculate

$$d\max_f = pixel_d - pixel_e \text{ where } \begin{cases} d = \min(\sigma((n) - f), \sigma((n-1) - f)), \\ e = \max(\sigma((n) - f), \sigma((n-1) - f)), \\ f = (0, 1, 2, 3, \ldots, \text{fix}(n/2) - 1). \end{cases} \quad (6)$$

The maximum pixels are changed into $pixel'$. In each operation, α is changed. The modified maximum pixels are derived by

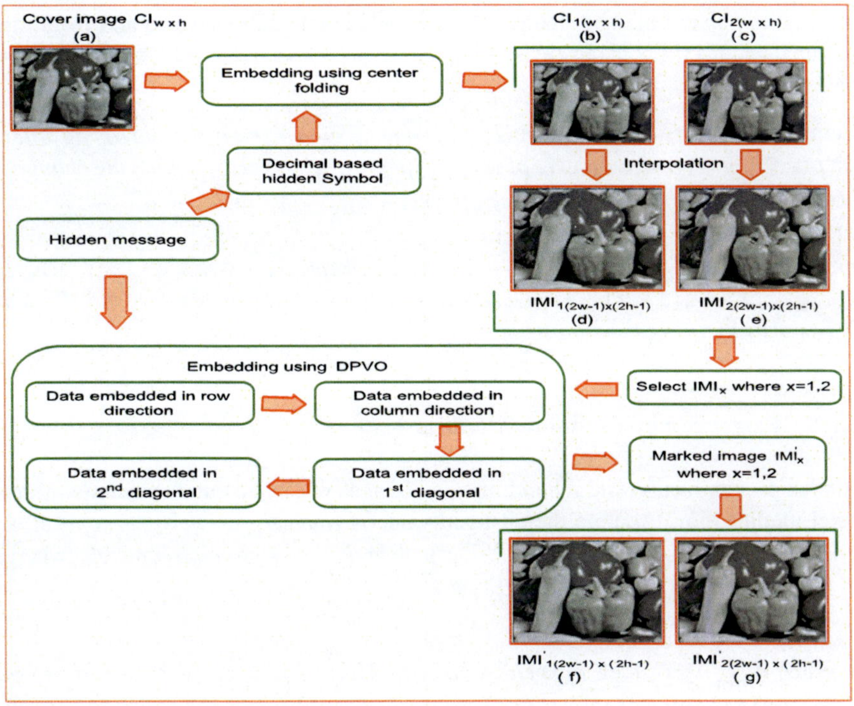

Fig. 2 Overall data embedding process in proposed improved center-folding-based DPVO scheme

$$pixel' = \begin{cases} (pixel_{\sigma((n)-f)} + \alpha) + 1, & \text{if } d_{\max_f} < 0 \\ (pixel_{\sigma((n)-f)} + \alpha) + D, & \text{if } d_{\max_f} = 1 \\ (pixel_{\sigma((n)-f)} + \alpha) + D, & \text{if } d_{\max_f} = 0 \\ (pixel_{\sigma((n)-f)} + \alpha) + 1, & \text{if } d_{\max_f} > 1 \end{cases} \tag{7}$$

where $D \in \{0, 1\}$ are inserting bits. Assume the changed value of X=($cpixel_1$, $cpixel_2, \ldots, cpixel_n$) in each row, where $cpixel_{\sigma((n)-f)} = pixel'$ and $cpixel_i = pixel_i$ for all $i \neq \sigma((n) - f)$. The changed value of $X = (CR1, CR2, ..., CRn)$ are presented in each column, where $CR\alpha((1) + f) = cr$ and $CRi = cri$ for all $i = s((1) + f)$. In the second phase of secrets embedding, the maximum and minimum pixels have been changed in three distinct directions one after the other. The overall embedding process is shown in Fig. 2.

2.2 Extraction Phase

Step 1: The process of secrets extraction is established in the diagonal direction of the interpolated marked image. The α is added and subtracted to and from the minimal and maximal pixels of 1st, 2nd, and so on, respectively, which is dependent on Lemma 2.2.

Lemma 2 *If the size of the block*= $(2v - 1)$ *and ordered pixels in the increasing order is* $(pixel'_1, pixel'_2, pixel_3, \ldots, pixel'_{2v-3}, pixel'_{2v-2}, pixel'_{2v-1})$, *then the changed pixel value is* $(pixel'_1 + \alpha_{(v-2)}, pixel'_2 + \alpha_{((v-2)-1)}, pixel'_3 + \alpha_{((v-2)-2)}, \ldots, pixel'_{v-1} + \alpha_{((v-2)-(v-2))}, pixel'_v, pixel'_{v+1} - \alpha_{((v-2)-(v-2))}, \ldots, pixel'_{2v-3} - \alpha_{((v-2)-2)}, pixel'_{2v-2} - \alpha_{((v-2)-1)}, pixel'_{2v-1} - \alpha_{(v-2)})$, *where* $\alpha_{((v-2)-(v-2))} = ((v - 2) - (v - 2))$. *The* α*'s maximal and minimal are* $(m - 2)$ *and 0, respectively.*

2.2.1 Extraction in Minimum-Modification

Hidden data extraction and rebuilding of an image are carried out through minimum pixel modification. Assume the changed value is $(cpixel_1, cpixel_2, \ldots, cpixel_n)$. The mapping σ lasts unchangedly. We calculate $d'_{\min_f} = cpxel_b - cpixel_c$, where (b, c, f) are discussed in equation (4).

- When $d'_{\min_f} \leq 0$, then $cpixel_b \leq cpixel_c$. Here, $c = \sigma((2) + f)$,$b = \sigma((1) + f)$, and also $\sigma((1) + f) < \sigma((2) + f)$:
  - When $d'_{\min_f} \in \{-1, 0, \}$, secrets are found and they are $D = -d'_{\min_f}$. The restored pixel is $pixel_{\sigma((1)+f)} = (cpixel_b + \alpha) + D$;

- When $d'_{\min_f} < -1$, then, no secrets are found. The restored pixel is $pixel_{\sigma((1)+f)} = (cpixel_b + \alpha) + 1$.

- When $d'_{\min_f} > 0$, then $cpixel_b > cpixel_c$. Here, $c = \sigma((1) + f)$,$b = \sigma((2) + f)$, and also $\sigma((1) + f) > \sigma((2) + f)$:
  - When $d'_{\min_f} \in \{1, 2\}$, then, secrets are found and they are is $D = d'_{\min_f} - 1$. The recovered pixel is $pixel_{\sigma((1)+f)} = (cpixel_c + \alpha) + D$;
  - if $d'_{\min_f} > 2$, then, no secrets are found. The restored pixel is $pixel_{\sigma((1)+f)} = (cpixel_c + \alpha) + 1$.

2.2.2 Extraction in Maximum-Modification

Assume the changed values are $(cpixel_1, cpixel_2, \ldots, cpixel_n)$. The value of mapping σ is not changed. We evaluate $d'_{\max_f} = cpixel_d - cpixel_e$ where in equation (6), (d, e, f) is discussed.

- When $d'_{\max_f} \le 0$, then $cpixel_d \le cpixel_e$. Here, $d = \sigma((n-1) - f)$, $e = \sigma((n) - f)$, and also $\sigma((n-1) - f) < \sigma((n) - f)$:
  - When $d'_{\max_f} \in \{-1, 0\}$, then, secrets are found and $D = -d'_{\max_f}$. The recovered maximum pixel is $piixel_{\sigma((n)-f)} = (cpixel_e - \alpha) - D$;
  - When $d'_{\max_f} < -1$, then, no secrets are found. The restored maximum pixel is $pixel_{\sigma((n)-f)} = (cpixel_e - \alpha) - 1$.

- When $d'_{\max_f} > 0$, then, $cpixel_d > y_e$. Now, $d = \sigma((n) - f)$, $e = \sigma((n-1) - f)$, and also $\sigma((n-1) - f) > \sigma((n) - f)$:
  - When $d'_{\max_f} \in \{1, 2\}$, here, secrets are found and they are $D = d'_{\max_f} - 1$. The pixel is $pixel_{\sigma((n)-f)} = (cpixel_d - \alpha) - D$;
  - When $d'_{\max_f} > 2$, then, no secrets are found. The recovered maximum pixel is $pixel_{\sigma((n)-f)} = (cpixel_d - \alpha) - 1$.

This extraction procedure begins from the 2nd diagonal of the marked image. After that, the 1st diagonal, then vertical, and lastly horizontal directions are utilized to release the secrets from both interpolated marked images. Now, the $IMI1_{(2w-1)\times(2h-1)}$ and $IMI2_{(2w-1)\times(2h-1)}$ interpolated images are recovered.

Step 2: We can restore the marked dual images $CI1_{w\times h}$ and $CI2_{w\times h}$ from $IMI1_{(2w-1)\times(2h-1)}$ and $IMI2_{(2w-1)\times(2h-1)}$ interpolated images, respectively, by eliminating all interpolated rows and columns.

Step 3: In this step, we remove the hidden data from both stego pixel $pixel'_{i,j}$ and $pixel''_{i,j}$ of $CI1_{w\times h}$ and $CI2_{w\times h}$ marked images, respectively, by the following formulas:

$$d_s^{'} = pixel_{i,j}^{'} - pixel_{i,j}^{''} \tag{8}$$

$$d_s = d_s^{'} + 2^{p-1} \tag{9}$$

where p is the number of hidden bits that makes a group. The real pixel $pixel_{i,j}$ is rebuilt as follows:

$$pixel_{i,j} = \lceil \frac{pixel_{i,j}^{'} + pixel_{i,j}^{''}}{2} \rceil. \tag{10}$$

Now, reconstruct the real cover image $CI_{w \times h}$.

3 Experimental Results and Comparisons

In this section, the execution of the proposed scheme is contrasted among other existing dual image-based secret hiding schemes by Qin et al.'s [11], Lu et al.'s [8], and Lu et al.'s [4]. We have taken (512 × 512) grayscale images together with Lena, Fishing boat, Peppers, Airplane F-16, Baboon, and Barbara from USC-SIPI [13] and CXR1000_IM-0003-1001 and CXR1025_IM-0020-1001 from National Library of Medicine [5] database as a test input for analysis from which only USC-SIPI images are shown in Fig. 3. The secrets embedding is dependent on p and block size. It is examined that the image quality is better at a small value of p with a big size of the block than a large value of p with the small size of the block but the message embedding rate is low. For example, when $p = 2$, then, 6,02,118 bits are inserted with an average Peak signal-to-noise ratio (PSNR) 51.81 dB. When $p = 3$, then, 8,78,185 bits are inserted with average PSNR 48.75 whereas, 11,10,259 bits are inserted with average PSNR 46.55 when $p = 4$ in the image Lena as shown in Table 1.

The comparisons in PSNR(dB) among existing PVO-based schemes with the proposed scheme is depicted in Table 2. The introduced method revamps secrets capacity (EC) estimated with other PVO-based schemes. The hidden message capacity is 321133, 353897, and 91753 bits greater than Qin et al.'s [11], Lu et al.'s [8], and Lu et al.'s [4], respectively, for image Lena while the value of p is 3 as shown in Table 2. The introduced method increases the capacity of the message hiding in input images.

Fig. 3 The six standard images are taken as an input for our experiments

Table 1 The secrets embedding (EC) in bits with distinctinction of p of 2 different database images with average PSNR(dB)

Database	Cover image(CI)	p	$PSNR_1$ (Avg.)	$PSNR_2$ (Avg.)	Average PSNR	EC_1	EC_2 $(IMI_1^{'})$	EC = $EC_1 + EC_2$
USC-SIPI	Lena	2	51.06	52.56	51.81	3,30,135	2,71,983	6,02,118
		3	47.93	49.58	48.75	4,72,654	4,05,531	8,78,185
		4	46.14	46.96	46.55	5,12,637	5,97,622	11,10,259
	Airplane F-16	2	50.87	49.50	50.18	2,88,634	3,47,486	6,36,120
		3	49.69	48.53	49.11	4,27,012	5,11,008	9,38,020
		4	46.82	45.37	46.09	5,19,643	6,16,586	11,36,229
	Fishing boat	2	49.20	48.36	48.78	2,83,142	3,26,292	6,09,434
		3	46.31	45.38	45.84	4,11,586	4,61,431	8,73,017
		4	42.58	41.90	42.24	5,80,127	5,42,983	11,23,110
National library of medicine	CXR1025_	2	50.76	50.64	50.70	3,01,062	3,18,359	6,19,421
	IM-0020-	3	47.53	46.91	47.22	4,43,718	3,98,830	8,42,548
	1001	4	39.98	39.06	39.52	5,21,282	6,25,531	11,46,813
	CXR1000_	2	51.27	50.83	51.05	2,77,306	3,10,537	5,87,843
	IM-0003-	3	46.18	47.95	47.06	4,65,677	3,98,866	8,64,543
	1001	4	43.17	40.74	41.95	5,10,320	6,24,437	11,34,757

Table 2 Comparison among the other schemes and the proposed scheme with image quality in PSNR(dB) as well as embedding capacity(EC) in bits

Schemes	Measure	Lena	Baboon	Peppers	Barbara	Fishing boat
Qin et al. [11]	$PSNR_1$	52.11	52.04	51.25	52.12	52.11
	$PSNR_2$	41.58	41.56	41.52	41.58	41.57
	Avg. PSNR	46.85	46.80	46.39	46.85	46.84
	EC	5,57,052	5,57,096	5,57,245	5,57,339	5,57,194
Lu et al. [8]	$PSNR_1$	49.20	49.21	49.19	49.22	49.20
	$PSNR_2$	49.21	49.20	49.21	49.20	49.21
	Avg. PSNR	49.21	49.21	49.20	49.21	49.21
	EC	5,24,288	5,24,204	5,24,192	5,24,288	5,24,284
Lu et al. ($k = 2$) [4]	$PSNR_1$	49.89	49.89	49.89	49.89	49.89
	$PSNR_2$	52.90	52.87	52.92	52.90	52.90
	Avg. PSNR	51.40	51.38	51.41	51.40	51.40
	EC	5,24,288	5,24,172	5,23,780	5,24,288	5,24,286
Proposed method ($p = 2$)	$PSNR_1$	51.06	52.35	51.40	52.28	49.20
	$PSNR_2$	52.56	51.16	51.84	51.43	48.36
	Avg. PSNR	51.81	51.75	51.62	51.85	48.78
	EC	6,02,118	6,87,544	6,00,026	6,13,519	6,09,434

It is evaluated that our method has surpassed compared to other PVO methods in terms of payload while the quality of the image is unchanged. It is also cleared that the image quality is greater than other existing schemes shown in Fig. 4.

4 Steganographic Attack

4.1 Detection of Tamper and Recovery of Image

We have tested the originality of the proposed work by the computation of some statistical parameters such as standard deviation (SD), PSNR (dB), and correlation coefficient (CC) to measure the steganographic attack which is displayed in Fig. 5 This finding explains the effectiveness of the suggested work.

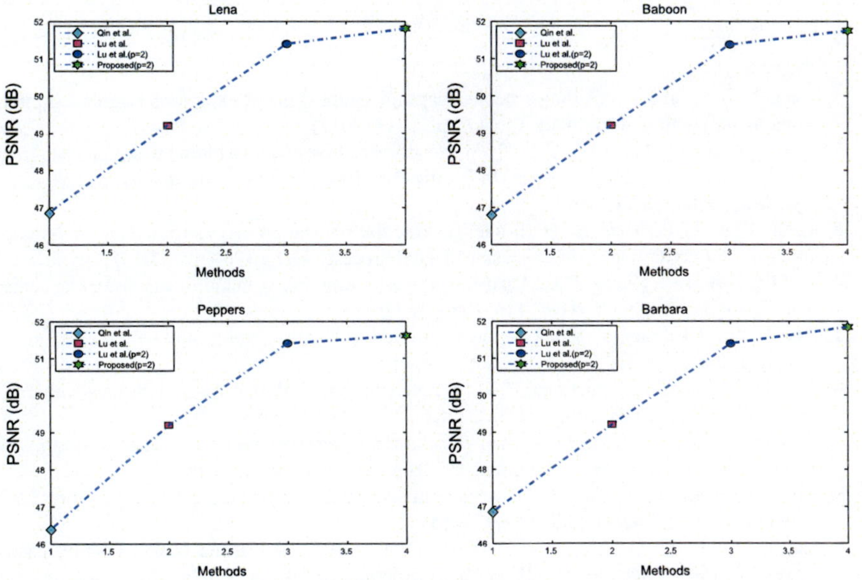

Fig. 4 Comparisons of image quality in PSNR(dB) among the methods of Qin et al. [11], Lu et al. [8], Lu et al. [4], and proposed method with $p = 2$

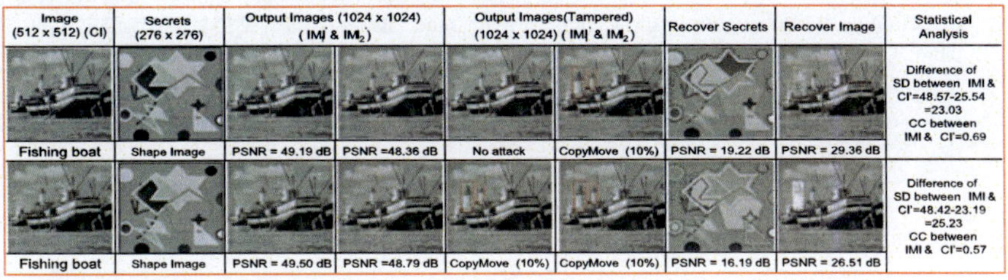

Fig. 5 Secrets recovered using copy and move counterfeit on Sailboat_on_lake image

5 Conclusion

This paper proposed an improved center-folding technique through Directional PVO for reversible data hiding with different sizes of p(set of data) and image block. The algorithms for data inserting and removing are developed in a manner so that huge messages are inserted and remove to and from image pixels, respectively. The suggested method fulfills secure message communication through the insertion of secrets among dual images. Using our method, we get a PSNR value above 48.50 dB and embedding capacity above 5,85,000 bits when p is 2. The proposed method provides good results. It is discovered that the suggested scheme appears to be of superior performance than other PVO works.

References

1. Jana, B.: Reversible data hiding scheme using sub-sampled image exploiting Lagrange's interpolating polynomial. Multimed. Tools Appl. 1–17 (2017)
2. Lee, S. K., Suh, Y. H., & Ho, Y. S. (2004, November). Lossless data hiding based on histogram modification of difference images. In Pacific-Rim Conference on Multimedia, pp. 340-347. Springer, Berlin, Heidelberg
3. Li, X., Li, J., Li, B., Yang, B.: High-fidelity reversible data hiding scheme based on pixel-value-ordering and prediction-error expansion. Signal Process. **93**(1), 198–205 (2013)
4. Lu, T.C., Wu, J.H., Huang, C.C.: Dual-image-based reversible data hiding method using center folding strategy. Signal Process. **115**, 195–213 (2015)
5. The National Library of Medicine presents MedPix®. https://openi.nlm.nih.gov/gridquery.php?q=&it=x
6. Meikap, S., Jana, B.: Directional PVO for reversible data hiding scheme with image interpolation. Multimed. Tools Appl. **77**(23), 31281–31311 (2018)
7. Meikap, S., Jana, B.: Directional pixel value ordering based secret sharing using sub-sampled image exploiting Lagrange polynomial. SN Appl. Sci. **1**(6), 645 (2019)
8. Lu, T.C., Tseng, C.Y., Wu, J.H.: Dual imaging-based reversible hiding technique using LSB matching. Signal Process. **108**, 77–89 (2015)
9. Lu, T. C., Lin, C. Q., Liu, J. W., & Chen, Y. C. (2017). Advanced Center-Folding based Reversible Hiding Scheme with Pixel Value Ordering. In Advances in Intelligent Information Hiding and Multimedia Signal Processing, pp. 83-90. Springer, Cham
10. Peng, F., Li, X., Yang, B.: Improved PVO-based reversible data hiding. Digit. Signal Process. **25**, 255–265 (2014)
11. Qin, C., Chang, C.C., Hsu, T.J.: Reversible data hiding scheme based on exploiting modification direction with two steganographic images. Multimed. Tools Appl. **74**(15), 5861–5872 (2015)
12. Tian, J.: Reversible data embedding using a difference expansion. IEEE Trans. Circuits Syst. Video Technol. **13**(8), 890–896 (2003)
13. University of Southern California, "TheUSC-SIPIImageDatabase". http://sipi.usc.edu/database/database.php?volume=misc

A Robust Audio Authentication Scheme Using (11,7) Hamming Error Correcting Code

Datta Kankana and Jana Biswapati

Abstract In the advancement of high-speed internet technology and vast use of social media, people are sending audio and video message for their daily communication in case of textual information. Due to unprotected nature of communication channel and increasing unauthorized users or adversary, it is essential to measure some protection of valuable multimedia messages which is one of a popular research issue for various human centric applications, especially in law enforcement and military applications. In this context, an effective audio authentication scheme has been proposed using (11,7) Hamming codes which are employed on an audio file. Now, for the first round, any arbitrary position (say nth position) bit will be considered as a secret key, and for rest of the cases data embedded position will be treated as a secret key position for the immediate round. For the first round, secret message bit will be embedded in the LSB position (if possible), otherwise next bit will be checked and for the remaining cases message bit will be embedded in a position where (LSB position bit ! = secret message bit) and (secret key position bit ! = secret message bit). Continue this process to embed all secret data bits within the cover audio file and produce stego audio. Various private, public and government sector will be benefited from this scheme.

Keywords Audio steganography · Hamming error correcting code · Echo hiding [1], Shared secret key

D. Kankana
Department of Computer Applications, Haldia Institute of Technology, HIT Campus, Hatiberia, Haldia, Purba Medinipur 721657, West Bengal, India
e-mail: dattakankana18@gmail.com

J. Biswapati (✉)
Department of Computer Science, Vidyasagar University, Midnapore 721102, India
e-mail: biswapatijana@gmail.com

D. Bhattacharjee et al. (eds.), *Proceedings of International Conference on Frontiers in Computing and Systems*, Advances in Intelligent Systems and Computing 1255,
https://doi.org/10.1007/978-981-15-7834-2_62

1 Introduction

The Steganography [2] is a combination of Latin words.'STEGANOS' implies covered and 'GRAPTER' implies writing, that means the secret message is concealed within the host medium, treated as a covered medium. Steganography [3] is a process of inserting the secret message/data/information within the entertaining medium (textual, sound file, picture, and video), etc. Recently, there are lots of effective, secure information hiding and steganographic algorithms have been developed [4]. On the other hand, there are also many powerful steganography analysis tools/software have been formulated to extract the secret from the cover media files. These steganographic algorithms can be detected by steganalytical detectors due to the lack of efficiency in the point of security and imperceptibility. Bicubic Interpolation Based Audio Authentication (BIAA) has been proposed by [5], for secured communication and authentication which has been suggested by [6]. Some effective authenticating standard is demanded to determine the copyright infringement from altered versions [5]. Therefore, some checking is required to overcome the piracy of music from any type of modification. The establishment of the song audio signal has to be dealt in such a way that any limited alters may identify the general perceptible quality [6]. Here, a secure and efficient audio steganographic algorithm is proposed where the valuable secret information is encoded by employing shared secret key and Hamming error correcting code to make the information more secure. Steganography algorithm should be defined in such a way that the difference between the host and embedded audio signals are negligible. On the other hand, Hamming Code [7], is a technique developed by R.W Hamming which can determine up to two simultaneous bit errors and is able to correct single-bit errors. This scheme can be applied to data bits of any length. According to the Hamming technique the data bits(d) and redundant bits(r) are preserved the following relationship: $2^r >= d + r + 1$. In this paper, we have applied the Hamming error correcting code before embedding the valuable secret information within the host audio signal using the secret key for authentication and tampered detection.

The rest of the article is arranged as mentioned below: the contribution of the paper is enlisted in Sect. 1.1, and the motivation and objective are drafted in Sect. 1.2. In Sect. 2, the proposed method of embedding and extraction are talked about. The experimental results with comparisons are shown in Sect. 3. Finally, in Sect. 4, conclusions are drawn.

1.1 Contribution of the Paper

The contributions of this paper are described below:

(i) The proposed method provides audio authentication through data hiding using Hamming error correction code.
(ii) The suggested method is reversible. We can retrieve both the hidden message and the cover audio media without distortion.

(iii) The advised method uses a shared secret key to protect unauthorized manipulation from illegal users.
(iv) During extraction, the mentioned technique correct one bit error which helps for error recovery at the receiver end.

1.2 *Motivation and Objective*

In social media, users are very much familiar to communicate by sending audio or video message which are easy to create than a text message. But there are a few investigations that has been exercised to protect audio signal from unauthorized manipulation and/or unintentional tampering during communication through an untrusted channel like the internet. Therefore, the motivations and objectives of our studies have been listed below:

(i) **Creation**: There are few authentication schemes that are available and can embed secret data during the creation or recording the audio message, especially for the users in social media. Our motivation and objective is to design an innovative low-cost authentication scheme for an audio signal protection which perform through watermarking scheme by embedding authentication code within an audio signal during recording or creation of audio signal.
(ii) **Authentication and Protection**: Earlier the Hamming code was used for error identification and recovery. Here, our motivation is to use the Hamming error correcting code for audio authentication and protection from unauthorized manipulation of the audio signal, especially audio message in social media.
(iii) **Robustness**: The error creation for protection and authentication in LSB minimize bit error rate (BER) which improve imperceptibility and innocuousness of the communication as well as message protection.
(iv) **Security**: Our objective is to develop reversible data hiding scheme using Hamming error correcting code and secret key to enhance robustness and security which is better than existing works.
(v) **Imperceptibility**: Imperceptibility is the main demand of any steganographic method. So maintaining good imperceptibility is the first and foremost objective of the proposed scheme. Hence we have used data modification in the LSB of the digitized audio signal (Fig. 1).

2 The Proposed Scheme

The performance of the existing steganographic scheme is below satisfactory level and fail to provide user's demand. Here, we have proposed a new data hiding scheme using (11,7) Hamming error correcting code. During the embedding time, we create

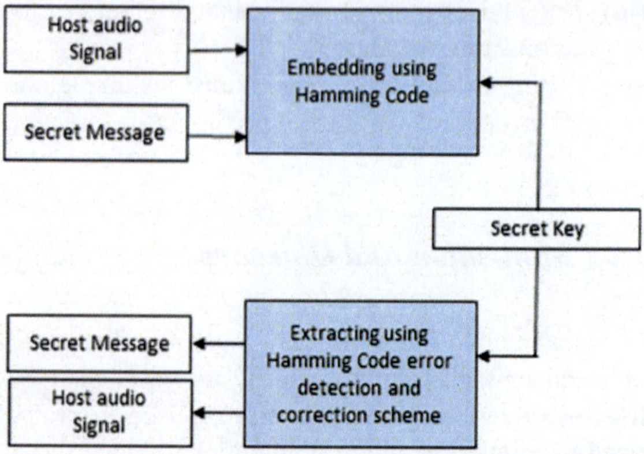

Fig. 1 Pictorial representation of the overall suggested Scheme

an error forcefully. To make this scheme more secure, we considered a secret key which will be required to extract the secret message during decoding time. Generally, LSB position is treated as key position, but if (key bit = LSB) or (key position bit = secret message bit) then the position of the key will be incremented by 1 for a particular iteration.
Figure 2 shows the basic structure of the proposed authentication scheme through watermarking.

2.1 Embedding and Extraction Algorithm

The step by step procedure to embed the secret message bits to produce the stego audio has been given in Algorithm-1. In the embedding phase, use of secret key makes this scheme more secure and strong compared to other existing schemes.

The step by step procedure to retrieve the hidden message string from the stego audio with the help of Hamming code has been given in the Algorithm-2. Secret key helps to extract the secret message from the stego audio signal during the extracting phase.

2.2 Example of Proposed Scheme

According to Fig. 2, a host signal matrix and a secret message bits are considered then convert host signal into 11 bits binary format. After that, choose four redundant bits and perform Hamming adjustment through even parity. For the first round, any arbitrary position (say 3rd position) will be treated as a secret key position, but for the rest of the rounds, embedded data bit position will be treated as a secret key position.

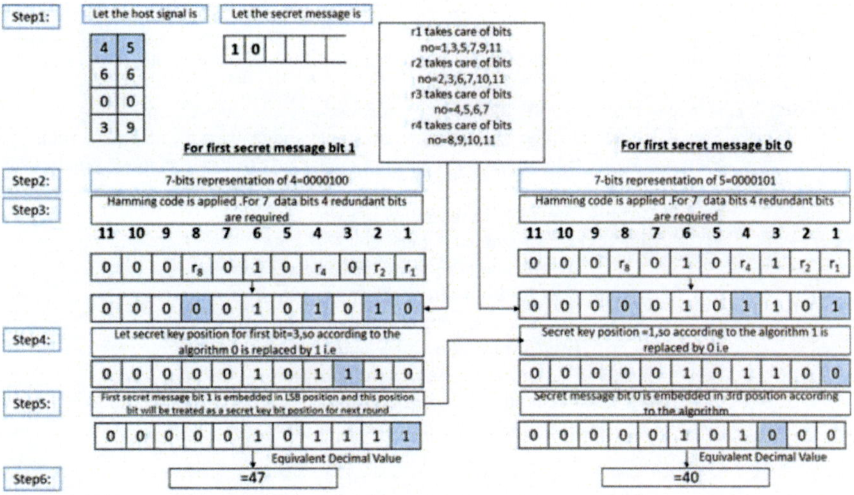

Fig. 2 Numerical representation of the suggested scheme for data embedding

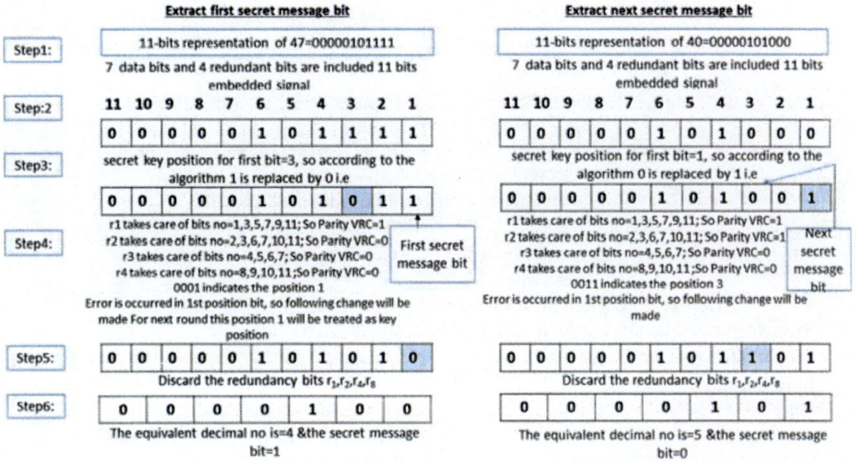

Fig. 3 Diagram of data extraction of the proposed scheme

In step-5, a bit will be embedded in the LSB position if it satisfies the following conditions (LSB position bit ! = secret message bit) and (secret key position bit ! = secret message bit) Then, we convert the binary numbers into equivalent decimal values. The above procedure will be continued until or unless all secret bits will be covered.

Begin
Step 1:Calculate the length of secret message and convert the secret message into the secret binary digits
Step 2:This step is repeated until all data are embedded
  Step 2.1:read the value of the host audio signal and convert it into binary value
  Step 2.2:Apply Hamming code concept to make an error on the above binary value using redundant bits
  Step 2.3:Apply secret key in the following way
    Step 2.3.i:if (int count=0) then
      nth position bit is treated as a secret key
      Step 2.3.i.a:if nth position bit=1 then
        set 0 in nth position
      Step 2.3.i.b:else
        set 1 in nth position
    Step 2.3.ii:else(other than nth position)
      Step 2.3.ii.a:if secret key bit=1 then
        set 0
      Step 2.3.ii.b:else
        set 1
  Step 2.4:Embed a bit from the secret message in the following way:
    Step 2.4.i:if(all bits are 0 and embedding bit=0)then
      embbed 0 in the first right hand side position except LSB and secret key position
    Step 2.4.ii:else if(all bits are 1 and embedding bit=1)then
      embbed 1 in the first right hand side position except LSB and secret key position
    Step 2.4.iii:else if (LSB position bit!=secret message bit) and (secret key position bit!=secret message bit)
      secret message bit will be embedded in LSB
    Step 2.4.iv:else
      next bit will be checked on the basis of above condition
  Step 2.5:Embeded position bit will be treated as a new secret key bit
Step 3:Convert the above binary values into decimal value
Step 4:Send the above stego audio signal to the receiver
Step 5:End(Algorithm)
End

Algorithm 1: Data Embedding Algorithm in Binary Audio Signal

According to Fig. 3, receiver takes the first decimal value of the embedded matrix and converts into 11 bits binary numbers which holds 7 data bits and 4 redundant bits. After that secret key position will be determined and change the bit. For the first round, any arbitrary position (same as encoded part) will be the secret key position. For the rest of the rounds, error bit position will be treated as a secret key position for the rest round. In step-4, identify the error bit position using the Hamming code and change it from either 0 to 1 or 1 to 0. This change bit will be the message bit and finally remove the redundancy bits r1, r2, r4 and r8 and get the 7 bits binary numbers. In step-6, converts it into the decimal value which is the original value of the host signal. These above procedures will be continued until or

Begin
Step 1:This step is repeated until all data are extracted
    Step 1.1:Read the value of embedded signal and convert it into binary value
    Step 1.2:Identify the secret key in the following way
        Step 1.2.i:if (int count=0) then
            nth position bit is treated as a secret key
            Step 1.2.i.a:if nth position bit=1 then
                set 0 in nth position
            Step 1.2.i.b:else
                set 1 in nth position
        Step 1.2.ii:else(other than nth position)
            Step 1.2.ii.a:if secret key bit=1 then
                set 0
            Step 1.2.ii.b:else
                set 1
    Step 1.3.i:Detect and correct the single bit error using Hamming Code and recover the part of the secret message to consider this error message bit as the part of the secret message bit.
    Step 1.3.ii:if no error will be found then check all bits are 0s or1s
        Step 1.3.i.a:if(all bits are 0)
            First right hand side bit will be treated as message bit except LSB and secret key position bit
        Step 1.3.i.b:if(all bits are 1)
            First right hand side bit will be treated as message bit except LSB and secret key position bit
    Step 1.4:Remove the redundant bits.
    Step 1.5:Convert the binary values of original audio signal and secret message
Step 2:End(Algorithm)
End

Algorithm 2: Data Extracting Algorithm from binary audio signal

unless all message bits will be retrieved. This extracting procedure is a reversible process of the embedding procedure described in the proposed Algorithm-2 above. After extraction, the scheme checks the presence of the extracted secret information treated as an authentication code. If some tampering has been taken place in the stego audio file, then the embedded message will not be matched with the original one. Hence, tampered audio file has been identified.

3 Experimental Result

This section represents the experimental results of the proposed scheme. We have observed a strip of 10 parts audio song "100 Miles From Memphis", sung by Sheryl Crow and its related experimental results. The sampled measures of the audio is presented in Table 1. Mean opinion score (MOS) is a standard used in the domain of

Quality of Experience. The amplitude vs time graph of the considered audio signal is constituted in Fig. 4. After embedding the secret code within the audio file the stego file is generated and the amplitude vs time graph is presented in Fig. 5. The difference is minimum and shown in Fig. 6.

Figure 7 shows the histogram of the original audio signal and Fig. 8, shows the histogram of the embedded signal. From the histograms of the host and embedded signal, we can conclude that the the distortion of the stego signal is very negligible. The use of secret key and error detection mechanism by Hamming Code makes the secret message more secure than others scheme. Figure 9 shows a comparison between the host signal and stego signal.

Table 1 The Experimental results of PSNR, BER and MOS of five selected audio files

Audio (10 s)	PSNR	BER	MOS
Audio1	54.54	0.002	5
Audio2	39.24	0.003	5
Audio3	52.12	0.005	5
Audio4	45.32	0.001	5
Audio5	53.21	0.002	5

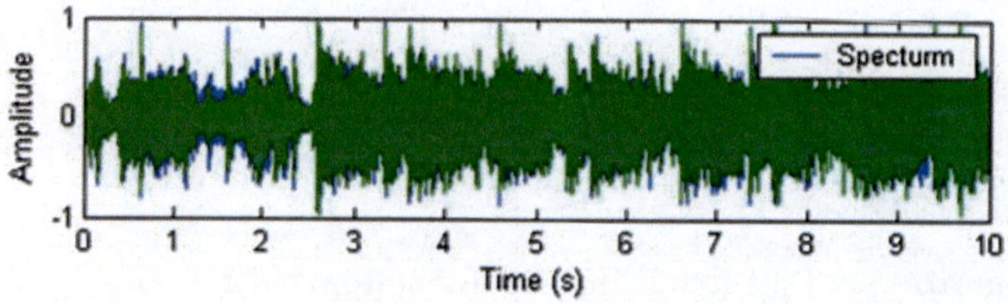

Fig. 4 The original spectrum of audio song ("100 Miles from Memphis" sang by Sheryl Crow)

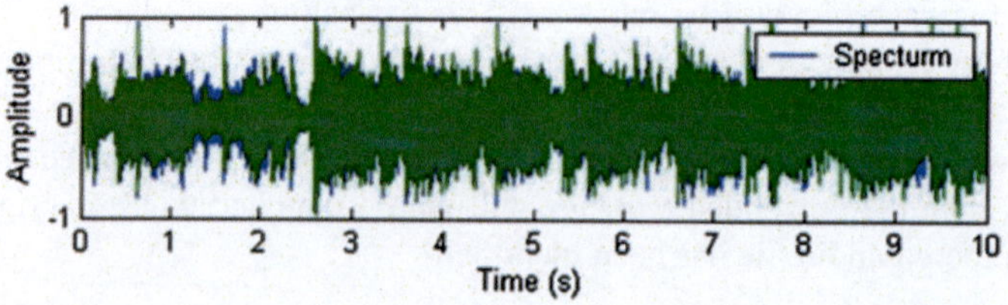

Fig. 5 The stego spectrum of audio song ("100 Miles from Memphis" sang by Sheryl Crow)

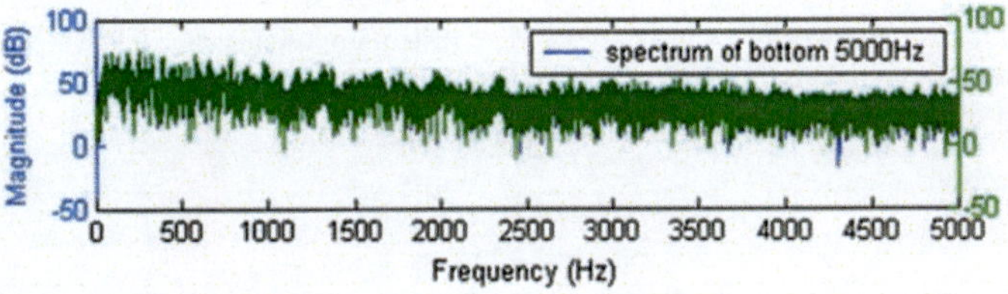

Fig. 6 The difference between original and stego spectrum of audio song ("100 Miles from Memphis" sang by Sheryl Crow)

Fig. 7 Graphical representation of host audio signal

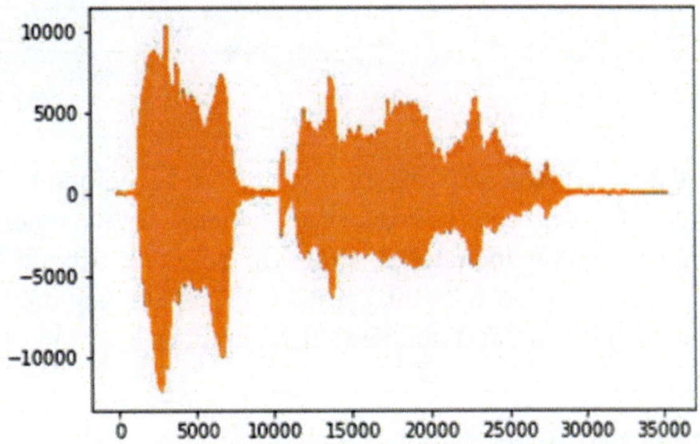

Fig. 8 Graphical representation of stego audio signal

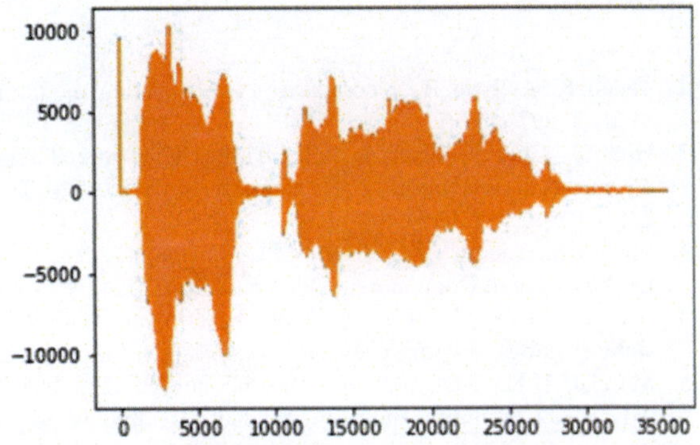

4 Conclusions

In this paper, we have proposed an Audio Steganography method to improve the drawbacks of different conventional audio steganography techniques like LSB Coding, Parity Coding, Echo Hiding etc.

The proposed method is to make an error forcefully during the embedding time using the concept of (11,7) Hamming Code. The use of Secret Key concept makes this method more secure. Hamming code can detect the maximum 2-bits error and correct the 1-bit error. So, using the Hamming code mechanism, we correct error easily during the extraction time. Moreover, our reversible audio steganography scheme

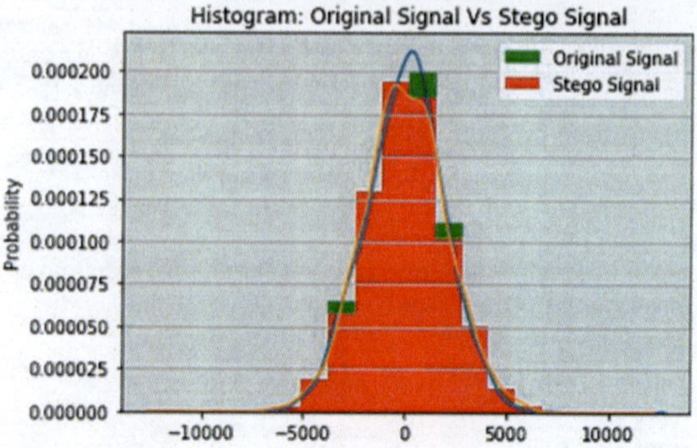

Fig. 9 Comparison between the original audio signal and stego audio signal

with secret key provides a more robust technique to hide the secret message compared to the other conventional schemes. In our scheme stego audio signal remains more or less unchanged, so it will be more difficult by the hackers to retrieve the secret message without knowing the secret key and error bit. On the other hand, authentication and tampered detection has been checked and verified.

References

1. Tekeli, K., Asliyan, R.: A comparison of echo hiding methods. Eurasia Proce. Sci. Technol. Eng. Math. **1**, 397–403 (2017)
2. Hua, G., Goh, J., Vrizlynn, L.L.: Thing. Time-spread echo-based audio watermarking with optimized imperceptibility and robustness. IEEE/ACM Trans. Audio Speech Lang. Process. **23**(2), 227–239 (2015)
3. Mishra, S., Yadav, V.K., Trivedi, M.C, Shrimali, T.: Audio steganography techniques: a survey. In: Advances in Computer and Computational Sciences, pp. 581–589. Springer (2018)
4. Hua, G., Huang, J., Shi, Y.Q., Goh, J., Vrizlynn, L.L.: Thing. Twenty years of digital audio watermarking—a comprehensive review. Signal Process. 128, 222–242 (2016)
5. Mondal, U.K., Mandal, J.K.: Bicubic interpolation based audio authentication (biaa). In: Advanced Computing and Communication Technologies, pp. 163–174. Springer (2018)
6. Zhang, J.: Audio dual watermarking scheme for copyright protection and content authentication. Int. J. Speech Technol. **18**(3), 443–448 (2015)
7. Jana, B., Giri, D., Mondal, S.K.: Partial reversible data hiding scheme using (7, 4) hamming code. Multimed. Tools Appl. **76**(20), 21691–21706 (2017)

Authentication on Interpolated Subsampled Based Image Steganography Exploiting Secret Sharing

Jana Manasi and Jana Biswapati

Abstract In this paper, we developed a secure secret sharing technique with authentication through steganographic scheme using the interpolated sub-sampling methodology. Based on a (k, n)-threshold method, a secret image is shared by n shadow images which are embedded within n subsampled interpolated cover images. On the other hand, k shadow images ($k \leq n$) can be used to recover the secret image. The proposed scheme uses a secure hash algorithm (SHA 12) to enhance authentication that prevents dishonest participants from cheating.

Keywords Secret sharing · Image interpolation · Authentication · Sub-sampling · SHA 512

1 Introduction

Secret data transmission through an untrusted public channel like the internet has become an important way of communication due to low-cost and easy to available media. Cryptography and steganography are two popular schemes used to achieve the goal of secure transmission of valuable information over public channel. In many applications, it will be a high risk to keep the secret information by only one person without keeping the duplicate data because the secret information may be destroyed accidentally. To solve this problem, Shamir [1] proposed the nobel concept of (k, n)-threshold secret sharing scheme. The method is developed to split a hidden message S into n, shares and circulate them to n members. Any k number of shares or more than k number of shares can be used to retrieve the hidden message, where k is used

J. Manasi
Department of Computer Applications, Haldia Institute of Technology, HIT Campus, Hatiberia, Haldia, Purba Medinipur 721657, West Bengal, India
e-mail: manasi.das30@gmail.com

J. Biswapati (✉)
Department of Computer Science, Vidyasagar University, Midnapore 721102, India
e-mail: biswapatijana@gmail.com

D. Bhattacharjee et al. (eds.), *Proceedings of International Conference on Frontiers in Computing and Systems*, Advances in Intelligent Systems and Computing 1255,
https://doi.org/10.1007/978-981-15-7834-2_63

as a threshold and $k = n$ *but no one can retrieve the secret message by using less than k shares*. Thien and Lin [2] designed a (k, n)-threshold secret image sharing method. A noise-like shadow image of size $1/k$ of the secret image has been produced by this scheme for the benefits of storage, transmission, and image hiding. Based on Shamir's secret sharing scheme, some polynomial based image sharing schemes [3–7], are briefly reviewed to improve quality of stego image and authentication ability. Image interpolation is widely used in medical field where a small digital image is amplified into larger one. Some commonly used interpolation [8–11], methods are proposed to implement reversible data hiding.
In light of the above discussion, a secret image sharing using sampling and interpolation method is proposed. This paper is organized as follows. Section 2, presents the proposed secret sharing scheme. Section 3, shows the experimental results and analysis. Conclusions are finally drowned in Sect. 4.

2 Proposed Scheme

For information sharing through image, Shamir's [1] (r, n) threshold method and Thien et al. [2] technique, are used to generate n shadow images. First, split the hidden information D into n shares $(D_1, D_2, ...D_n)$ and at receiver end knowing only r shadow images ($r \leq n$) it is possible to retrieve the hidden information D. The n shares can be generated by using following equation.

$$q(x) = (a_0 + a_1x + a_2x + ... + a_{r-1}x) \quad mod \quad p \tag{1}$$

where $a_0 = D$ and p is a prime number. Then evaluate $D_1 = q(1), ..D_i = q(i), ..., D_n = q(n)$ where each D_i is a shadow image. In Shamir's scheme, the coefficients $a_1, a_2, ..a_{r-1}$ are randomly taken from a range of integers within 0 to (p−1), whereas in Thien et al.'s scheme, the coefficients are taken from the r pixels of the secret image.
Since the gray value of a pixel lies between 0 and 255, we have taken a prime number 251 as p and have truncated all pixel elements 251 to 255 of the secret image to 250 so that all pixel elements are in the range 0–250. But it would be a lossy secret image sharing method.

Sampling [12] is the method of selecting values from an image. Using the following equation we can split a cover image into some subsampled images. Suppose p(i, j) is a pixel of an image I of size $M \times N$ where i = 0,..M−1 and j = 0,..N−1. Two sampling factors, Δ u and Δ v set the required sub-sampling intervals in a row and column direction, respectively, to produce a subsampled image S_k of size $M/\Delta u \times N/\Delta$ v, where k = 1,2,.....$\Delta u \times \Delta v$.

$$S_k(i, j) = I(i.\Delta v + floor(\frac{k-1}{\Delta u}), j.\Delta u + ((k-1) mod \Delta u)) \tag{2}$$

Jung and Yoo [8] proposed an interpolation method (NMI) to implement reversible data hiding using steganography. The pixels of cover image are calculated using Eq. 3, where $0 \leq j \leq i$ and $m, n = 0, 1, 2, ...127$.

$$C(i,j) = \begin{cases} I(i,j) & \text{if i = 2m, j = 2n} \\ I(i,j-1) + I(i,j+1))/2 & \text{if i = 2m, j = 2n+ 1} \\ I(i-1,j) + I(i+1,j))/2 & \text{if i = 2m+1, j = 2n} \\ I(i-1,j-1) + C(i-1,j) + C(i,j-1))/3 & \text{otherwise} \end{cases} \quad (3)$$

2.1 Embedding Algorithm

Figure 1 shows the block diagram of secret sharing and data embedding procedure. The following algorithm describes the proposed secret sharing and embedding procedure.

Input: secret image(S) and an input image I of size $M \times N$.
Output: k stego images.
Step 1: r pixels are taken from secret image and using Eq. 1 n shadow images are produced.
Step 2: A cover image of size M x N is taken and it is subsampled into k images using Eq. 2
Step 3: Each subsampled image is interpolated using Eq. 3, generating cover images.

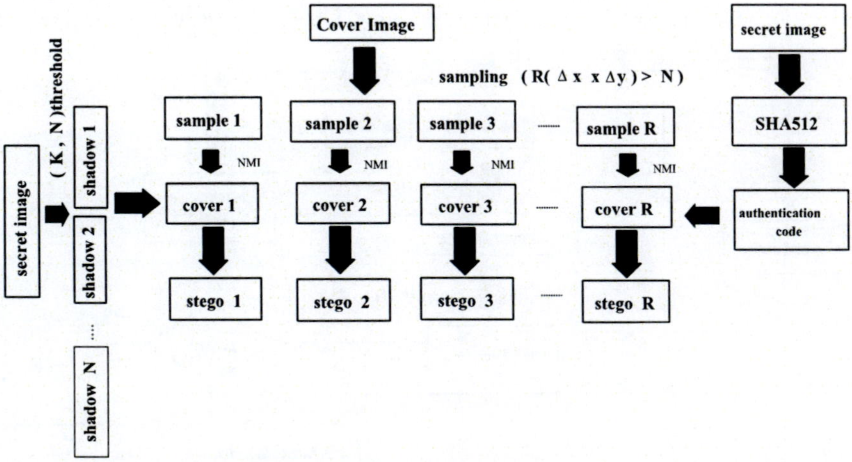

Fig. 1 A schematic representation of proposed embedding method

Step 4: n shadow images are embedded into interpolated pixels of n cover images.
Step 5: For authentication, SHA 512 is applied on secret image to generate 512 bits authentication code and embedded into kth cover image.

2.2 *Extraction Algorithm*

When the recipient has the k stego images, the secret image can be retrieved from k shadow images embedded into k stego images and the input image would be restored to its original form. Figure 2 shows the schematic diagram of information retrieval and restoration procedure. The detailed methodology is given below:
Input: Δ u × Δ v stego images of size $M \times N$.
Output: Input image I of size $M \times N$ and secret data S.
Step 1: Retrieve k shadow images from any k stego images.
Step 2: Take k pixels from each of the k shadow images.
Step 3: The coefficients $a_0 - a_{r-1}$ in Eq. 1 are solved using these k pixels and the Lagrange's interpolation method.
Step 4: Step 2 and step 3 are repeated until all pixels of the k shadow images are processed to construct a secret data S.
Step 5: Retrieve authentication code A_1 from Δu×Δv th stego image .
Step 6: Apply SHA 512 on retrieved secret data S to produce a secure authentication code A_2. Compare A_1 and A_2 to verify whether image is tampered or not.
Step 7: Restore Input image I from all Δu × Δv stego images.

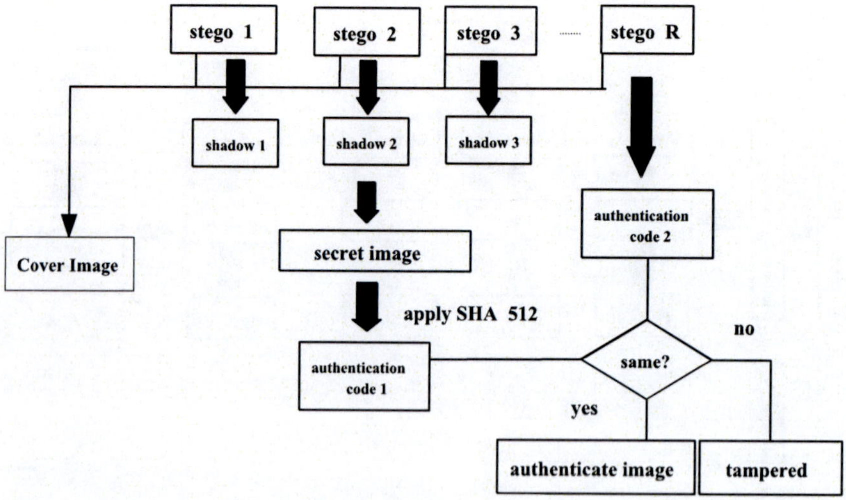

Fig. 2 A block diagram of extracting secret image

2.3 Numerical Example

A numerical example of secret data and its three shadow images for (2, 3)-threshold scheme is depicted in Fig. 3. A numerical example is given to Fig. 4 to embed three shadow images and an authentication code into four subsampled interpolated cover images. Figure 5 shows a numerical example to extract a secret image from any two shadow images.

2.4 Merits of Proposed Secret Image Sharing Approach:

The above algorithm describes the basic idea of the proposed scheme which shows some benefits described in the following.

1. The benefits of sub-sampling

In data hiding using image interpolation method, before interpolation an input image is scaled down to original image results in some data modification. Instead of scaled down, the input image is sampled into some sub-sampling images before interpolation without any data modification. At the receiver end, we can retrieve the Input image without losing any data.

2. The benefits of interpolation

Here a popular image interpolation method (NMI) is applied to implement reversible data hiding, as well as to embed shadow images of the secret image into interpolated subsampled cover images.

3. The benefits of embedding shadow images to different interpolated subsampled images

Using the (r,n)-threshold scheme to the secret image, we generate n shadow images of size 1/r of the secret image (r = 2,3,...n). Instead of transmitting directly, these shadow images are embedded to different interpolated subsampled host images to hide the existence of the shadow images from hackers. Many data hiding methods

Fig. 3 An example of generating shadow images

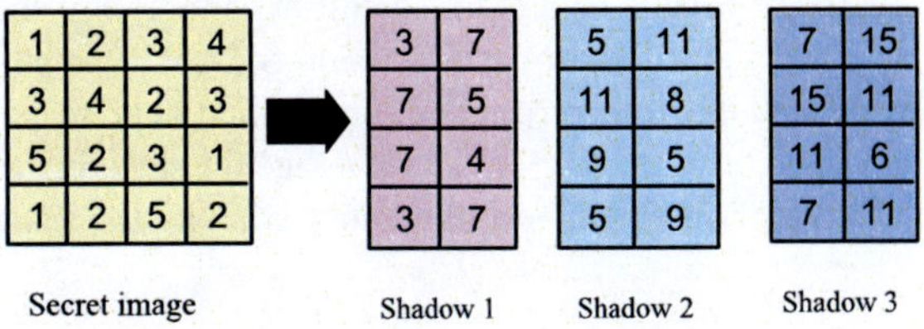

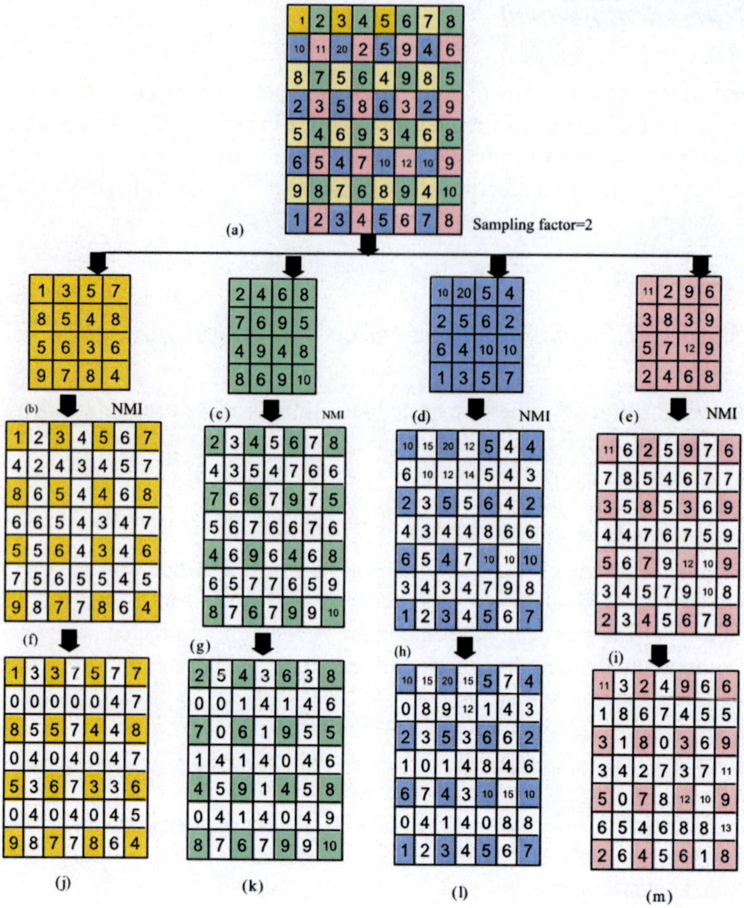

Fig. 4 An example of embedding process

require that the size of the secret image must be at least $\frac{1}{2}$ (or even $\frac{1}{4}$) smaller than the host image. The reduction size of the shadow image is convenient for this purpose.

4. Security analysis

The proposed algorithm improved the security like the (k,n)-threshold secret sharing, the reversible data hiding using steganography and the authentication checking. Here we use a secure hash function SHA 512 to generate an authentication code embedded into an interpolated subsampled host image. At the receiver side, it is not needed to keep the original secret image for checking authentication. Comparing two authentication codes we can identify whether the image is tampered or not.

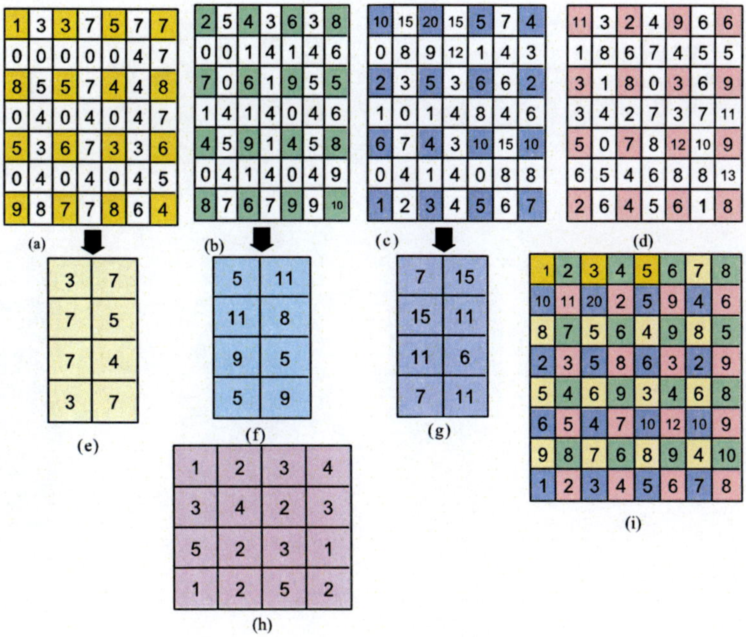

Fig. 5 An example of extracting process

3 Experimental Results:

Here some experimental results are presented to show the feasibility of this scheme. We use a (2, 3)-threshold technique to clarify the effects of our work. A 128 × 128 grayscale image is taken which is shown in Fig. 6a, and three shadow images of size 128 × 64 are produced shown in Fig. 6b through(d). In Fig. 7a, a 512 × 512 gray level input image is taken which is sampled (sampling factor $\Delta v = \Delta u = 2$) into four subsampled images of size 128 × 128 as shown in Fig. 7b through (e). After that, these four subsampled images are interpolated using the NMI method to produce four interpolated cover images of size 512 × 512 as shown in Fig. 7f through i. Figure 8a–c show the three stego images generated after embedding three shadow images of Fig. 6b–d. Figure 8d shows a stego image generated after embedding an authentication code produced by the secure hash algorithm SHA 512. The three shadow images retrieved from stego images of Fig. 8a–c, are shown in Fig. 8e–g, respectively, and a secret image which can be retrieved from any two of these three shadow images is shown in Fig. 8h.

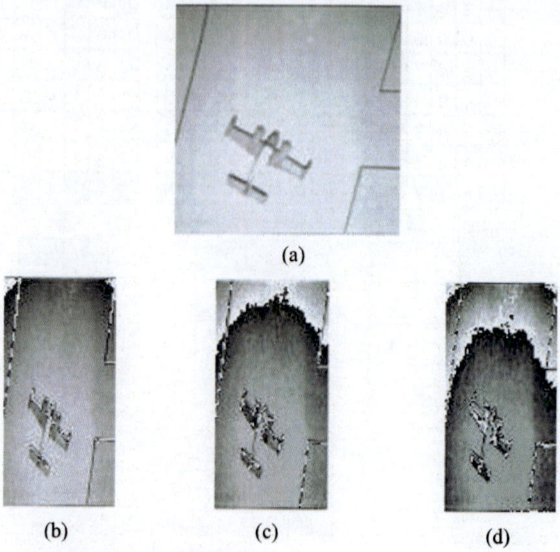

Fig. 6 Experimental results: **a** a 128 x 128 secret image **b–d** three shadow images whose size are all 128 x 64

Fig. 7 **a** a 512 x 512 cover image **b–e** the four subsampled images whose size are all 128 x 128 **f–i** the four subsampled images (512 x 512) after interpolation

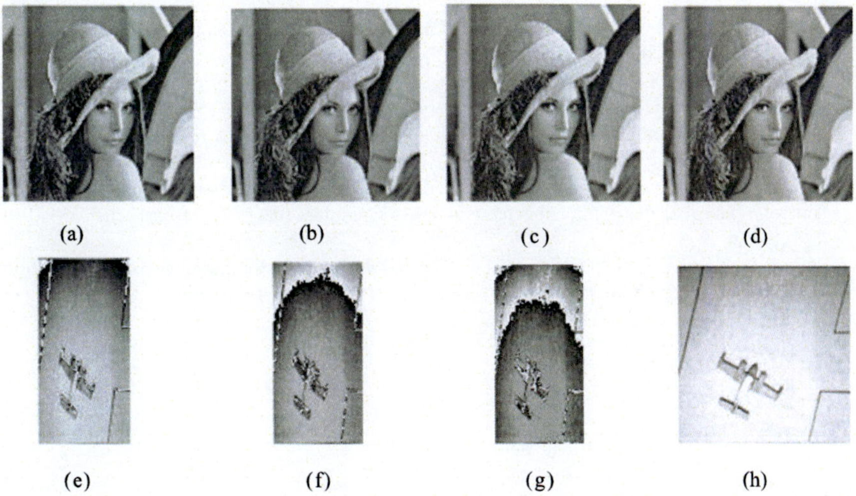

Fig. 8 **a–c** stego images after embedding shadow images **d** stego image after embedding SHA 512 **e–g** shadow images retrived from stego images (h) secret image retrived from any two shadow images

4 Conclusions

A secret image sharing with reversible data hiding and authentication has been proposed. Instead of transmitting shadow images directly, these are embedded into subsampled cover images to hide the existence of the secret image. In most of the reversible data hiding using the interpolation method, the original images are restored but we can't restore the input images. But due to sub-sampling here we can restore the input image without any loss of data. Here the authentication ability is greatly improved by using a secure hash algorithm SHA 512.

References

1. Shamir, A.: How to share a secret. Commun. ACM **22**(11), 612–613 (1979)
2. Thien, C.C., Lin, J.C.: Secret image sharing. Comput. Gr. **26**(5), 765–770 (2002)
3. Lin, C.C., Tsai, W.H.: Secret image sharing with steganography and authentication. J. Syst. Softw. **73**(3), 405–414 (2004)
4. Yang, C.N., Chen, T.S., Yu, K.H., Wang, C.C., : Improvements of image sharing with steganography and authentication. J. Syst. Softw. **80**(7), 1070–1076 (2007)
5. Chang, C.C., Hsieh, Y.P., Lin, C.H.: Sharing secrets in stego images with authentication. Pattern Recognit. **41**(10), 3130–3137 (2008)
6. Yang, C.N., Ciou, C.B.: A comment on sharing secrets in stegoimages with authentication. Pattern Recognit. **42**(7), 1615–1619 (2009)
7. Wu, C.C., Kao, S.J., Hwang, M.S.: A high quality image sharing with steganography and adaptive authentication scheme. J. Syst. Softw. **84**(12), 2196–2207 (2011)

8. Jung, K.H., Yoo, K.Y.: Data hiding method using image interpolation. Comput. Stand. Interfaces **31**(2), 465–470 (2009)
9. Lee, C.F., Huang, Y.L.: An efficient image interpolation increasing payload in reversible data hiding. Expert Syst. Appl. **39**(8), 6712–6719 (2012)
10. Tang, M., Hu, J., Song, W.: A high capacity image steganography using multi-layer embedding. Optik-Int. J. Light Electron Opt. **125**(15), 3972–3976 (2014)
11. Hong, W., Chen, T.S.: Reversible data embedding for high quality images using interpolation and reference pixel distribution mechanism. J. Vis. Commun. Image Represent. **22**(2), 131–140 (2011)
12. Kim, K.S., Lee, M.J., Lee, H.Y., Lee, H.K.: Reversible data hiding exploiting spatial correlation between sub-sampled images. Pattern Recognit. **42**(11), 3083–3096 (2009)

Evolving Secret Sharing with Essential Participants

Jyotirmoy Pramanik and Avishek Adhikari

Abstract Komargodski et al. introduced *Evolving Secret Sharing* which allows an impartial participant, called *dealer*, to share a secret among unbounded number of participants over any given access structure. In their construction for evolving secret sharing over general access structure, the size of share of the ith participant happens to be exponential ($\mathcal{O}(2^{i-1})$). They also provided constructions for (k, ∞) threshold secret sharing. We consider the problem of evolving secret sharing with t essential participants, namely, over t-(k, ∞) access structure, a generalization of (k, ∞) secret sharing ($t = 0$). We further generalize this access structure to a possible case of unbounded number of essential participants and provide a construction for secret sharing on it. Both the constructions are information theoretically secure and reduce the share size of the construction due to Komargodski et al. over general access structure, exponentially. Moreover, the essential participants receive ideal (and hence, optimal) shares in the first construction.

Keywords Evolving access structure · Secret sharing · Essential participants · Information theoretic

1 Introduction

In *secret sharing*, one can so share an information (usually a field element) among n (fixed and pre-decided) participants that certain subsets are able to reconstruct it back, while others are not [18]. Given any access structure on a set of participants,

J. Pramanik (✉)
Department of Pure Mathematics, University of Calcutta, 35, Ballygunge Circular Road, Kolkata 700019, India
e-mail: jyotirmoy.pramanik2@gmail.com

A. Adhikari
Department of Mathematics, Presidency University, 86/1, College Street Rd, Kolkata 700073, India
e-mail: avishek.adh@gmail.com

D. Bhattacharjee et al. (eds.), *Proceedings of International Conference on Frontiers in Computing and Systems*, Advances in Intelligent Systems and Computing 1255,
https://doi.org/10.1007/978-981-15-7834-2_64

there exists a secret sharing scheme realizing it. *Evolving secret sharing* generalizes the notion of usual secret sharing where the participants' set was to be known beforehand. It allows participants to join one by one, and the dealer hands them their shares without refreshing shares already distributed. Komargodski et al. introduced evolving secret sharing in [8]. We discuss few of these notions in detail in Sect. 2. In Sect. 3, we introduce t-(k, ∞) and (t, ∞, k, ∞) secret sharing and provide two constructions. In Sect. 4, we summarize our results and suggest further research directions.

Our Contribution: In this paper, we provide a construction for secret sharing realizing t-(k, ∞) access structure where fixed t participants are essential. Essential participants in this scheme receive a share of size $\mathcal{O}(1)$, whereas ith of the other participants receives a share of the size $(k-1)\cdot\log i + poly(k, \ell)\cdot\mathcal{O}(\log i)$ for an ℓ-bit secret being shared. We further generalize this access structure to (t, ∞, k, ∞) access structure and provide a construction for secret sharing realizing it. In the latter construction, the ith participant receives a share of size $\mathcal{O}((k-1)\cdot\log i + poly(k, \ell)\cdot\mathcal{O}(\log i))$. Share sizes in both the schemes are a huge (exponential) improvement compared to the scheme for general access structure having share size $\mathcal{O}(2^{i-1})$ in [8]. We compare our results with [8] for a single-bit secret in Table 1.

2 Preliminaries

For a given access structure $\Gamma \subset 2^{\mathcal{P}}$ on a participants' set $\mathcal{P}$, a subset $\mathcal{A}$ of participants is called *qualified* if and only if $\mathcal{A} \in \Gamma$; otherwise $\mathcal{A}$ is *forbidden*. A (t, n) threshold access structure on n participants consists of qualified sets which are precisely of size t or more. For secret sharing on any given access structure, an impartial participant $\mathcal{D} \notin \mathcal{P}$ (called the *Dealer*) invokes the *share generation* protocol ShareGen and generates n shares, one for each participant. In the hour

Table 1 Comparison of size of shares for a single-bit secret

Construction	Share size of the ith party
[8] General access structure	2^{i-1}
[8] (k, ∞)	$(k-1)\cdot\log i + poly(k)\cdot\mathcal{O}(\log i)$
1. This paper t-(k, ∞)	
(i) Essential	$\mathcal{O}(1)$
(i) Other	$(k-1)\cdot\log i + poly(k)\cdot\mathcal{O}(\log i)$
2. This paper (t, ∞, k, ∞)	
(i) Essential	$\mathcal{O}((k-1)\cdot\log i + poly(k)\cdot\mathcal{O}(\log i))$
(i) Other	$(k-1)\cdot\log i + poly(k)\cdot\mathcal{O}(\log i)$

of need for reconstruction of the secret, certain participants pool their shares in the *reconstruction* protocol Reconst. The secret sharing scheme is denoted by $\Pi = (\mathsf{ShareGen}, \mathsf{Reconst})$. The *correctness* property in a secret sharing scheme ensures that any qualified set of participants is able to reconstruct the secrets with certainty, i.e., $Pr[s' = s | s' \leftarrow \mathsf{Reconst}(\mathscr{A})\ and\ \mathscr{A} \in \Gamma] = 1$. On the other hand, due to *perfect secrecy*, Reconst outputs the correct secret from share of a forbidden set with probability no more than that derived from the probability distribution of the secret space $\mathscr{S}$, i.e., $Pr[s' = s | s' \leftarrow \mathsf{Reconst}(\mathscr{A})\ and\ \mathscr{A} \in 2^{\mathscr{P}} \setminus \Gamma] = Pr[s \leftarrow \mathscr{S}]$. *Share size* of a participant P_i is the size of collection of all possible shares for him; this collection (called the *share space* V_i of P_i) is generated due to different values of randomness of the share generation algorithm. In an *ideal* secret sharing scheme, the share size and secret size are same.

Secret sharing with essential participants was initiated in a work by Arumugam et al. in [2]. They denoted this type of access structure as $(k, n)^*$ *access structure* where a secret image was shared into n *shadow* images where the presence of shadow corresponding to *one* particular participant was *essential*. Later, this notion was generalized to access structures containing t essential participants as t-(k, n) secret sharing in [5, 7, 14]. A further generalization (t, s, k, n) secret sharing was considered in [10] by Li et al. where at least t essential shadows (among s of those) were necessary to reconstruct the secret, along with the threshold condition being satisfied.

Evolving secret sharing was introduced by Komargodski et al. in [8]. As opposed to usual secret sharing with n participants, they considered a far more practical variant where there is no upper bound on number of participants. Participants join one by one, and they are handed over a share based on shares distributed to previous participants but without interacting with the previous participants. In other words, shares are not refreshed. Most of the secret sharing schemes are linear [18] in nature and require the underlying field of size at least log(field size), where (field size) > #(participants). This creates a problem for the *evolving* setup where the number of participants is not known beforehand. Komargodski et al. provided a beautiful solution for this problem in [8] on general access structure where the ith participant receives a share of size $\ell \cdot 2^{i-1}$ for an ℓ bit secret. They also provided a (k, ∞) secret sharing scheme sharing an ℓ bit string with share size of the ith participant being $(k-1) \cdot \log i + poly(k, \ell) \cdot \mathcal{O}(\log i)$. A few more follow-up works in evolving setup can be found in [3, 4, 6, 9, 11].

3 Evolving Secret Sharing with Essential Participants

Secret sharing with essential participants is a generalized case of usual threshold secret sharing. Though being well studied in traditional secret sharing, this notion is yet unexplored in evolving setup except for a work by Dutta et al. [6]. In the following sections, we introduce secret sharing on t-(k, ∞) and (t, ∞, k, ∞) access structures.

3.1 A Construction for t-(k, ∞) Secret Sharing Scheme

In a t-(k, ∞) secret sharing, qualified subsets are those which are of at least size k and contain t special participants, called the *essential participants*. The essential participants are predefined and fixed and are free to join as and when they wish to, just like other non-essential participants. Of course, until the last essential participant has joined, no subset of participants is qualified. We define an attribute function $f : \mathscr{P} \rightarrow \{0, 1\}$ for each participant P_i as $f(P_i) = 1$ if and only if P_i is an essential participant. The function f can also be interpreted as the characteristic function of the subset of essential participants. Let us demonstrate the simple case of $1 - (2, \infty)$ secret sharing: To share a secret $s \in \{0, 1\}^\ell = \mathscr{S}$, give the essential participant P_α a random number $r \leftarrow \mathscr{S}$ and every other participant $r \oplus s$. Reconstruction is done by XORing two shares. Every participant receives a share of constant size, and this scheme is ideal. This example portrays a somewhat extremal case of evolving secret sharing with essential participants. Another such extremal case of consideration would be $k - (k, \infty)$ secret sharing. In this case, all but the essential participants would receive *dummy shares* which might play no role whatsoever in secret reconstruction. For the rest of this paper, we shall assume that $t < k$. Now that we are warmed up with how two simplest instances of t-(k, ∞) secret sharing schemes work, let us move on to a more general construction. We assume the availability of (k, ∞)—secret sharing schemes Π_k due to Komargodski et al. [8] for every $k \geq 2$. We shall use this scheme as a black box to generically produce a t-(k, ∞) secret sharing scheme.

Theorem 1 *For positive integers t, $(<)k$ and ℓ, there exists a t-(k, ∞) secret sharing scheme sharing an ℓ bit secret, meeting the correctness and perfect secrecy conditions. Moreover, the scheme is ideal for essential participants and for the ith non-essential participant, share size is given by $(k - 1) \cdot \log i + poly(k, \ell) \cdot \mathscr{O}(\log i)$.*

Proof For $k > t$, we demonstrate the following secret sharing scheme (ShareGen, Reconst) attaining the said conditions.

ShareGen : For a secret $s \in \{0, 1\}^\ell = \mathscr{S}$, we describe the share generation protocol below:

1. Generate $t + 1$ random numbers $r_1, r_2, \ldots, r_t, r_{t+1} \xleftarrow{\$} \{0, 1\}^\ell$ such that $s = \bigoplus_{i=1}^{t+1} r_i$.
2. Initialize $c = 0$.
3. On arrival of the ith participant P_i $(i = 1, 2, 3, \ldots)$, if P_i is an essential participant, i.e., if $f(P_i) = 1$, then update c by adding 1 to it and give r_c to P_i as his share; else run the share generation algorithm of Π_{k-t} to generate a share w_i of r_{t+1} and give it to P_i. If at any point of share generation $c > t$, then ShareGen aborts.

Reconst : k participants including the t essential participants pool their shares; the $k-t$ non-essential participants reconstruct r_{t+1} using reconstruction algorithm of Π_{k-t}. Further, they find s by bitwise XORing $r_1, r_2, \ldots, r_t, r_{t+1}$. If a forbidden set submits shares for reconstruction, **FAIL** is output.

Proof of Correctness: Every qualified set of participants $\mathscr{A} \in \Gamma$ contains the t essential participants and at least $k-t$ other participants. Due to correctness property of reconstruction algorithm of Π_{k-t}, these $k-t$ or more participants can uniquely reconstruct r_{t+1}. The secret s is found by XORing r_i's for $i \in [t+1]$.

Proof of Perfect Secrecy: In t-(k, ∞) access structure, there are two kinds of forbidden sets possible, namely, (i) *Type 1* forbidden sets which contain k or more participants but do not contain at least one essential participant; (ii) *Type 2* forbidden sets which contain at most $k-1$ participants in total. For a Type 1 forbidden set $\mathscr{A}$, members of $\mathscr{A}$ possess the following set of information $info^{(1)} = \{r_i$: for $\leq t-1$ values of i from $[t]\} \sqcup \{$shares of $r_{t+1}\}$, where $\sqcup$ denotes disjoint union. Using $info^{(1)}$ $\mathscr{A}$ can reconstruct r_{t+1}, since there are at least $k-(t-1) = k-t+1$ shares of r_{t+1} present. Without loss of generality, let us assume that the first essential participant is not present in $\mathscr{A}$, then participants of $\mathscr{A}$ can reconstruct with a probability $Pr[\text{Finding } s = r_1 \oplus r_2 \oplus \cdots \oplus r_{t+1} | r_2, r_3, \ldots, r_t, r_{t+1}] = Pr[r_1 \xleftarrow{\$} \{0,1\}^\ell] = Pr[s \xleftarrow{\$} \mathscr{S}]$, i.e., the best that a Type 1 forbidden set can do with their shares is guess the secret s (without looking at any share, like any person not present in $\mathscr{P}$). A Type 2 forbidden set either consists of all the essential participants but $k-t-1$ non-essential participants; or, $t-1$ or lesser essential participants. The proof of perfect secrecy for the latter of these two cases can be done in a manner similar to Type 1. We only prove perfect secrecy for the former case now. $\mathscr{A}$ possesses the following set of information: $info^{(2)} = \{r_1, r_2, \ldots, r_t\} \sqcup \{k-t-1$ shares of $r_{t+1}\}$. Due to perfect secrecy of Π_{k-t} used, it follows that $Pr[\text{Finding } s = r_1 \oplus r_2 \oplus \cdots \oplus r_{t+1} | info^{(2)}] = Pr[r_{t+1} \xleftarrow{\$} \{0,1\}^\ell] = Pr[s \xleftarrow{\$} \mathscr{S}]$.

Share Size Analysis: The scheme described above is ideal for essential participants. For the ith non-essential participant, share size is given by $(k-1) \cdot \log i + poly(k, \ell) \cdot \mathscr{O}(\log i)$. It is convenient to assume $k \geq 3$ as for $k = 2$, the access structure reduces to two trivial sub-cases of $1-(2,\infty)$ and $2-(2,\infty)$ access structures, where secret sharing can be done trivially, as shown at the beginning of this section. Due to our construction, share size of the ith non-essential participant preserves the share size of the ith participant in (k, ∞) secret sharing scheme of [8] by Komargodski et al. sharing ℓ bit strings. □

We further generalize t-(k, ∞) secret sharing in the following section. Specifically, we give rise to a new access structure called (t, ∞, k, ∞) access structure in Sect. 3.2 in which qualified subsets are those which contain any t of the possibly infinite collection of *pseudo-essential* participants and also k participants in total. We call these participants pseudo-essential because essentiality of these participants doesn't depend on their individuality but on their grouping with other similar participants in sufficient number. It can be noted that, unlike t-(k, ∞) access structure, in this access structure one may find qualified subsets consisting of only pseudo-essential participants. As a particular case, if no new pseudo-essential participant arrives after the t-th one, it is nothing but a t-(k, ∞) access structure, establishing the fact that (t, ∞, k, ∞) access structure is indeed a generalization of t-(k, ∞) access structure. Moreover, (t, ∞, k, ∞) access structure can be seen as a generalization of another access structure, namely, (t, s, k, n) access structure. Secret sharing was done on the latter access structure by Li et al. in [10].

3.2 A Construction for (t, ∞, k, ∞) Secret Sharing Scheme

We define a new access structure called (t, ∞, k, ∞) access structure in this section where a qualified subset of participants contains at least k participants in total including at least t participants from a subset $\mathscr{P}_{ps}$ of special participants called *pseudo-essential participants*. The subset may not be known at the beginning but this subset can be characterized by defining an attribute function as in Sect. 3.1. To summarize, $f : \mathscr{P} \rightarrow \{0, 1\}$ is a function defined on the collection of participants $\mathscr{P}$ [which is also unknown at the beginning but f can be identified with a function with similar properties being defined on the set $\mathbb{N}$ of natural numbers and, hence, is convenient] as $f(P_i) = 1$ if and only if P_i is a pseudo-essential participant. At the beginning of the scheme, we set $\mathscr{P}_{ps} = \emptyset$ and whenever a new pseudo-essential party joins, we add him to the set $\mathscr{P}_{ps}$. We assume availability of (k, ∞)—secret sharing schemes Π_k [8] for every $k \geq 2$. In this construction, every pseudo-essential participant receives share which is *heavier* than the size of every other participant, the convenience of which we describe in proof of Theorem 2.

Theorem 2 *For positive integers t, $(<)k$ and ℓ, there exists a (t, ∞, k, ∞) secret sharing scheme sharing an ℓ bit secret, satisfying correctness and perfect secrecy conditions. Moreover, the share of size of the ith participant is $\mathscr{O}((k-1) \cdot \log i + poly(k, \ell) \cdot \mathscr{O}(\log i))$ if he is pseudo-essential; otherwise, share size is $(k-1) \cdot \log i + poly(k, \ell) \cdot \mathscr{O}(\log i)$.*

Proof For $k > t$, we demonstrate the following secret sharing scheme (ShareGen, Reconst) attaining the said conditions.

ShareGen : For a secret $s \in \{0, 1\}^{\ell} = \mathscr{S}$, we describe the share generation protocol below:

1. Generate a random number $r \xleftarrow{\$} \mathscr{S}$.
2. On arrival of the ith participant P_i, if $f(P_i) = 1$ then run the share generation algorithms of Π_t and Π_{k-t} to generate a new shares $w_{1,i}$ and $w_{2,i}$ of r and $r \oplus s$, respectively, and give $(w_{1,i}, w_{2,i})$ to P_i as his share; else run the share generation algorithm of Π_{k-t} to generate a new share $w_{2,i}$ of $r \oplus s$ and give it to P_i.

Reconst : Suppose, k parties $P_{i_1}, P_{i_2}, \ldots, P_{i_k}$ pool their shares.

1. Set $\mathscr{P}_{ps,t} = \emptyset$ and $L = \{i_1, i_2, \ldots, i_k\}$.
2. Adjoin the first t pseudo-essential participants present for reconstruction to $\mathscr{P}_{ps,t}$ and delete their corresponding indices from L. In other words,
   $c = 0$
   for $(i \text{ in } L)$:
       if $(f(P_i) = 1)$:
           $\mathscr{P}_{ps,t} = \mathscr{P}_{ps,t} \cup \{P_i\}$
           $L = L \setminus \{i\}$.
           $c \mathrel{+}= 1$
           if $(c = t)$:
               break.
3. Run the reconstruction algorithm of Π_t on $\{w_{1,i} : P_i \in \mathscr{P}_{ps,t}\}$ to reconstruct r. Run the reconstruction algorithm of Π_{k-t} on $\{w_{2,i} : i \in L\}$ to reconstruct $r \oplus s$. XOR r and $r \oplus s$ to reconstruct s. If a forbidden set submits shares, **Reconst** outputs **FAIL**.

Proof of Correctness: Every qualified set $\mathscr{A}$ in this access structure is of size $\geq k$ and contains t pseudo-essential participants. If $\mathscr{A}$ contains more than t pseudo-essential participants, we treat the first t of them as pseudo-essential and the others ordinarily. The (first) t pseudo-essential participants reconstruct r, and the remaining participants reconstruct $r \oplus s$ using respective reconstruction algorithms of Π_t and Π_{k-t}. Since both the algorithms possess correctness, the property is preserved for our construction as well.

Proof of Perfect Secrecy: The proof for perfect secrecy is similar to Theorem 1.

Share Size Analysis: The ith participant receives a share of size of size $\mathscr{O}((k-1) \cdot \log i + poly(k, \ell) \cdot \mathscr{O}(\log i))$ if he is pseudo-essential; otherwise, share size is $(k-1) \cdot \log i + poly(k, \ell) \cdot \mathscr{O}(\log i)$. It can be noted that pseudo-essential participants receive shares which are *heavier* compared to other participants. This is convenient as there are qualified sets consisting of only pseudo-essential participants, and hence, they should possess shares corresponding to both r and $r \oplus s$. □

4 Conclusion and Future Research

To sum up, we provide a secret sharing scheme realizing t-(k, ∞) access structure where t (fixed) participants are essential. Essential participants in this construction receive a share of size $\mathscr{O}(1)$, whereas ith of the other participants receives a share of the size $(k-1) \cdot \log i + poly(k, \ell) \cdot \mathscr{O}(\log i)$ for an ℓ-bit secret being shared. We further generalize this access structure to a new access structure called (t, ∞, k, ∞) access structure and provide a secret sharing scheme realizing it. In the latter construction, the ith participant receives a share of size $\mathcal{O}((k-1)\cdot\log i + poly(k,\ell)\cdot\mathcal{O}(\log i))$. Share sizes in both the schemes are a huge (exponential) improvement compared to the scheme for general access structure having share size $\mathcal{O}(2^{i-1})$ in [8].

A further research direction would be considered dynamic thresholds (both in t and k) like [9] in both the access structures demonstrated. Another interesting follow-up work would be to introduce secret sharing with cheaters [1, 12, 13, 15–17, 19] in evolving setup.

References

1. Adhikari, A., Morozov, K., Obana, S., Roy, P.S., Sakurai, K., Xu, R.: Efficient threshold secret sharing schemes secure against rushing cheaters. In: ICITS 2016, Revised Selected Papers, pp. 3–23 (2016)
2. Arumugam, S., Lakshmanan, R., Nagar, A.K.: On (k, n)*-visual cryptography scheme. Des. Codes Cryptogr. **71**(1), 153–162 (2014)
3. Beimel, A., Othman, H.: Evolving ramp secret-sharing schemes. In: SCN 2018, Proceedings, pp. 313–332 (2018)
4. D'Arco, P., Prisco, R.D., Santis, A.D., del Pozo, A.L.P., Vaccaro, U.: Probabilistic secret sharing. In: MFCS 2018, pp. 64:1–64:16 (2018)
5. Dutta, S., Adhikari, A.: XOR based non-monotone t- $(k, n)^*$ visual cryptographic schemes using linear algebra. ICICS **2014**, 230–242 (2014)
6. Dutta, S., Roy, P.S., Fukushima, K., Kiyomoto, S., Sakurai, K.: Secret sharing on evolving multi-level access structure. In: Information Security Applications - 20th International Conference, WISA 2019, Jeju Island, South Korea, August 21–24, 2019, Revised Selected Papers, pp. 180–191 (2019). https://doi.org/10.1007/978-3-030-39303-8_14, https://dblp.org/rec/conf/wisa/Dutta0FKS19.bib
7. Guo, T., Liu, F., Wu, C.K., Ren, Y., Wang, W.: On (k, n) visual cryptography scheme with t essential parties. In: ICITS 2013, Proceedings, pp. 56–68 (2013)
8. Komargodski, I., Naor, M., Yogev, E.: How to share a secret, infinitely. In: TCC 2016-B, Proceedings, Part II, pp. 485–514 (2016)
9. Komargodski, I., Paskin-Cherniavsky, A.: Evolving secret sharing: Dynamic thresholds and robustness. In: TCC 2017, Proceedings, Part II, pp. 379–393 (2017)
10. Li, P., Yang, C., Wu, C., Kong, Q., Ma, Y.: Essential secret image sharing scheme with different importance of shadows. J. Vis. Comm. Im. Rep. **24**(7), 1106–1114 (2013)
11. Paskin-Cherniavsky, A.: How to infinitely share a secret more efficiently. IACR Cryptol. ePrint Arch. **2016**, 1088 (2016)
12. Pramanik, J., Adhikari, A.: Ramp secret sharing with cheater identification in presence of rushing cheaters. Gr. Complex. Cryptol. **11**(2), 103–113 (2019)
13. Pramanik, J., Roy, P.S., Dutta, S., Adhikari, A., Sakurai, K.: Secret sharing schemes on compartmental access structure in presence of cheaters. In: ICISS 2018, Proceedings, pp. 171–188 (2018)
14. Praveen, K., Rajeev, K., Sethumadhavan, M.: On the extensions of (k, n)*-visual cryptographic schemes. In: SNDS 2014, Proceedings, pp. 231–238 (2014)
15. Roy, P.S., Adhikari, A., Xu, R., Morozov, K., Sakurai, K.: An efficient robust secret sharing scheme with optimal cheater resiliency. SPACE **2014**, 47–58 (2014)

16. Roy, P.S., Adhikari, A., Xu, R., Morozov, K., Sakurai, K.: An efficient t-cheater identifiable secret sharing scheme with optimal cheater resiliency. IACR Cryptol. ePrint Arch. **2014**, 628 (2014)
17. Roy, P.S., Dutta, S., Morozov, K., Adhikari, A., Fukushima, K., Kiyomoto, S., Sakurai, K.: Hierarchical secret sharing schemes secure against rushing adversary: Cheater identification and robustness. In: ISPEC 2018, Proceedings, pp. 578–594 (2018)
18. Shamir, A.: How to share a secret. Commun. ACM **22**(11), 612–613 (1979)
19. Tompa, M., Woll, H.: How to share a secret with cheaters. J. Cryptol. **1**(2), 133–138 (1988)

A New Lossless Secret Image Sharing Scheme for Grayscale Images with Small Shadow Size

Md. K. Sardar and Avishek Adhikari

Abstract The current paper offers a lossless (k, n)-threshold scheme with reduced shadow size using the algebraic properties of the polynomial ring $\mathbb{Z}_{251}[x]$ over the field $\mathbb{Z}_{251}$. Unlike most of the existing secret image sharing schemes, our scheme does not require any preprocessing steps in order to transform the image into a random image to avoid data leakage from the shares of the secret image. Moreover, the efficiency of our proposed scheme is explained through security analysis and demonstrated through simulation results.

Keywords Polynomial-based secret sharing · Lossless recovery · Finite field

1 Introduction

Security is one of the most important issues when storage or transmission of secret image is considered. Secret sharing (SS) can resolve this issue. A (k, n)-SS scheme encodes the secret data into shares and distributes such shares to n participants in such a manner that in the recovery phase, when k or more participants pool their shares, the secret is recovered. On the other hand, fewer than k participants, even if collect their shares, no extra information is revealed. The notion of secret sharing was further extended to visual cryptography [1, 3–5] by Noar and Shamir [6]. In 2002, Thien and Lin [8] used the polynomial secret image sharing scheme of Shamir to share secret images in $\mathbb{Z}_{251}$. Instead of using the first coefficient, they [8] used all the coefficients of the $k-1$ degree polynomial in Shamir (k, n)-scheme. As a result, the size of shadow images reduced to $\frac{1}{k}$ times the secret image. However, in

Md. K. Sardar (✉)
Department of Pure Mathematics, University of Calcutta, Kolkata, India
e-mail: mks.pubm@gmail.com

A. Adhikari
Presidency University, Kolkata, West Bengal, India
e-mail: avishek.adh@gmail.com

D. Bhattacharjee et al. (eds.), *Proceedings of International Conference on Frontiers in Computing and Systems*, Advances in Intelligent Systems and Computing 1255,
https://doi.org/10.1007/978-981-15-7834-2_65

[8] all gray values greater than 250 were truncated. As a result, the scheme becomes lossy. However, to achieve the lossless recovery, Thien and Lin [8] split those pixels whose values are larger than 250 into two pixels, resulting in the increase of the share size. In 2013, Yang et al. [9] developed Lin et al.'s [8] model to achieve a lossless scheme by using Galois Field $GF(2^8)$. Ding et al. [2] successfully proposed a lossless (k, n)-scheme in $\mathbb{Z}_{251}$ without pixel expansion. But, in this scheme, the share size remains same with the original secret image.

We have successfully overcome these issues. Using the algebraic properties of the polynomial ring $\mathbb{Z}_{251}[x]$ over the field $\mathbb{Z}_{251}$, in the present paper, we propose a lossless (k, n)-scheme with diminished shadow size. The main advantage of our scheme is that, unlike most of the existing scheme, it requires no preprocessing to transform the secret image into a random-looking image, resulting in our scheme an efficient one.

The arrangement of the rest of the paper is as follows. Section 2 offers certain preliminary methods that will be required throughout the paper. In Sect. 3, the proposed scheme is discussed. Security of our scheme is analyzed in Sect. 4. In Sect. 5, experimental effects and analyses are offered. Comparisons are presented in Sect. 6. Finally, Sect. 7 draws the conclusion of our work that leads to the future research direction of our work.

2 Preliminaries

For better understanding, we will first present Shamir's secret sharing scheme in this segment as the idea for the proposed scheme. The SS scheme, proposed by Shamir in 1979 [7], is based on the truth that there is a uniquely $k-1$ degree polynomial in a plane across k points. Using this fact, Shamir divided the secret $\mathbb{S}$, from a suitably chosen field F_p, p a prime greater than n, into n shares $\mathbb{S}_1, \mathbb{S}_2, \ldots, \mathbb{S}_n$ and distributed them to n participants $\mathbb{P}_1, \mathbb{P}_2, \ldots, \mathbb{P}_n$ in such a way that any k $(1 < k \leq n)$ or more participants are required to reconstruct the original secret data $\mathbb{S}$. It was proved that if fewer than k shares are collected by the enemy, then they will not get any extra information about the secret data $\mathbb{S}$. Shamir defined $k-1$ degree polynomial as

$$h(x) = c_0 + c_1 x + c_2 x^2 + \cdots + c_{k-1} x^{k-1} \pmod{p}, \tag{1}$$

where the coefficient c_0 is the secret ($c_0 = \mathbb{S}$) and all the remaining coefficients $c_1, c_2, \ldots, c_{k-1}$ are randomly chosen elements from the field F_p, having p elements. Finally, to each participant $\mathbb{P}_\alpha$, the point $(x_\alpha, h(x_\alpha))$ on the polynomial is given as share, where $x_\alpha \in F_p$ and $x_\alpha \neq x_\beta$ for all $\alpha \neq \beta$; $\alpha, \beta = 1, 2, \ldots, n$. During the reconstruction phase, when any k or more participants come together, they will be able to reconstruct the polynomial as described in Eq. (1) by using Lagrange's interpolation formula as follows:

$$\phi(x) = \sum_{\alpha=1}^{k} h(x_\alpha) \prod_{\substack{\beta=1 \\ \beta \neq \alpha}}^{k} \frac{x - x_\beta}{x_\alpha - x_\beta} \pmod{p}. \tag{2}$$

Then, they will calculate the value of $\phi(x)$ at $x = 0$ to get back the secret $c_0 = \mathbb{S}$ as follows: $\mathbb{S} = c_0 = \phi(0) = \sum_{\alpha=1}^{k} h(x_\alpha) \prod_{\substack{\beta=1 \\ \beta \neq \alpha}}^{k} \frac{-x_\beta}{x_\alpha - x_\beta} \pmod{p}$.

3 The Proposed Scheme

In this segment, we are going to propose a completely lossless polynomial-based secret image sharing (PSIS) scheme with small shadow size using modulo a prime 251 within the pixel range. So as to acquire lossless recovery, the proposed scheme manipulates the grayscale pixels values which are larger than 250 when sharing the secret image. We are assuming that the dealer and all the participant involving with this sharing scheme are honest, i.e., every one of them follows the protocol and never goes astray from the protocol.

3.1 The Sharing Phase

For a (k, n)-PSIS scheme, the sharing polynomial of degree $(k - 1)$ is defined as follows:

$$f(y) = a_0 + a_1 y + a_2 y^2 + \cdots + a_{k-2} y^{k-2} + a_{k-1} y^{k-1} \pmod{251}, \tag{3}$$

where the coefficients are chosen as follows : split the secret image $\mathbb{S}$ into $H \times \frac{W}{d}$ non-overlapping segments, $\Delta\mathbb{S}[i, j] = \{\mathbb{S}[i, d * j + l] : 0 \leq l < d\}, 0 \leq i < H, 0 \leq j < \frac{W}{d}$, where $d = \frac{k-2}{2}$ or $\frac{k-1}{2}$ according to k is even or odd. For the consecutive d pixels of each (i, j)-th segment $\Delta\mathbb{S}[i, j]$, we set $a_{2r} = \Delta\mathbb{S}[i, j][r]$ in consideration of the coefficient pairs $(a_{2r}, a_{2r+1}), r = 0, 1, 2, \ldots, d - 1$.

If k is even and $a_{2r} < 250$, then a_{2r+1} is chosen randomly from $[0, 250]$ and if $a_{2r} \geq 250$, then $a_{2r+1} = (255 - a_{2r}) + 6t, r = 0, 1, ..., d - 1$, where t is chosen randomly from $[0, 40]$, then set $a_{2r} = 250$ and the last two coefficients a_{k-2} and a_{k-1} are chosen randomly from $[0, 250]$.

If k is odd and $a_{2r} < 250$, then a_{2r+1} is chosen randomly from $[0, 250]$ and if $a_{2r} \geq 250$, then $a_{2r+1} = (255 - a_{2r}) + 6t, r = 0, 1, ..., d - 1$, where t is chosen randomly from $[0, 40]$, then set $a_{2r} = 250$ and the last coefficient a_{k-1} is chosen randomly from $[0, 250]$. Then compute $\mathbb{S}_q[i, j] = f(q), q = 1, 2, \ldots, n$. Repeat this process until all the pixels are exhausted. Finally, $(q, \mathbb{S}_q)$ is given to the q-th participant $\mathbb{P}_q$ as share for $q = 1, 2, \ldots, n$. There are in total $H \times \frac{W}{d}$ sharing phases in the whole program, and size of each shadow image becomes $H \times \frac{W}{d}$. The complete process is described in Algorithm 1.

Algorithm 1: Share generating algorithm of our (k,n)-PSIS scheme.

Input: A secret grayscale image $\mathbb{S}$ of size $H \times W$ with (i,j)-th pixel $\mathbb{S}[i,j], 0 \le i \le H-1, 0 \le j \le W-1$, and two positive integers k and n, such that $2 \le k \le n$.

Output: The n shadows $\mathbb{S}_1, \mathbb{S}_2, \ldots, \mathbb{S}_n$.

Split $\mathbb{S}$ into $H \times \frac{W}{d}$ non-overlapping segments, $\Delta\mathbb{S}[i,j] = \{\mathbb{S}[i, d*j+l] : 0 \le l < d\}, 0 \le i < H, 0 \le j < \frac{W}{d}$, where $d = \frac{k-2}{2}$ or $\frac{k-1}{2}$ according as k is even or odd.
for $(i = 0; i < H; i{+}{+})$ **do**
  for $(j = 0; j < \frac{W}{d}; j{+}{+})$ **do**
    //Assign the coefficients $a_0, a_1, \ldots, a_{k-1}$ as follows:
    if $(k\%2 == 0)$ **then**
      for $(r = 0; r < d; r{+}{+})$ **do**
        $a_{2r} = \Delta\mathbb{S}[i,j][r]$
        if $(a_{2r} < 250)$ **then**
          a_{2r+1} is chosen randomly from $[0, 250]$
        else
          $a_{2r+1} = (255 - a_{2r}) + 6t$, //where t is chosen randomly from $[0, 40]$
          $a_{2r} = 250$
        end
      end
      Randomly choose the last two integer coefficients a_{k-2} and a_{k-1} from $[0, 250]$
    else
      for $(r = 0; r < d; r{+}{+})$ **do**
        $a_{2r} = \Delta\mathbb{S}[i,j][r]$
        if $(a_{2r} < 250)$ **then**
          a_{2r+1} is chosen randomly from $[0, 250]$
        else
          $a_{2r+1} = (255 - a_{2r}) + 6t$, // where t is chosen randomly from $[0, 40]$
          $a_{2r} = 250$
        end
      end
      Randomly choose the last integer coefficient a_{k-1} from $[0, 250]$
    end
    for $(q = 1; q \le n; q{+}{+})$ **do**
      $f(q) = a_0 + a_1 q + a_2 q^2 + \cdots + a_{k-1} q^{k-1} \pmod{251}$
      $\mathbb{S}_q[i,j] = f(q)$
    end
  end
end

3.2 Lossless Recovery Phase

When at least some k participants meet, the secret is revealed during the reconstruction phase. Accordingly, without loss of generality, we may assume that the k shadows $\mathbb{S}_1, \mathbb{S}_2, \ldots, \mathbb{S}_k$ are submitted by means of the individuals $\mathbb{P}_1, \mathbb{P}_2, \ldots, \mathbb{P}_k$ to reconstruct $\mathbb{S}$ successfully. Using Lagrange's interpolation, the shareholders will reconstruct a sequence of polynomials from these share images. They will then reconstruct each section of the image. By correctly placing the image sections next to each other, the original image will be obtained. The steps used in the reconstruction phase are given in Algorithm 2:

Algorithm 2: Recovery phase of our (k, n)-PSIS scheme
Input: At least k shadows are randomly selected from n secret shadows $\mathbb{S}_1, \mathbb{S}_2, \ldots, \mathbb{S}_n$
Output: $\mathbb{S}$, original secret

Step1: For each $(i, j), 0 \le i < H, 0 \le j < \frac{W}{d}$, where $d = \frac{k-2}{2}$ or $\frac{k-1}{2}$ according as k is even or odd, repeat the steps 2-4.
Step2: The k participants will construct the following system of linear equations (4) over the finite field $\mathbb{Z}_{251}$.
Step3: As V is a Vandermonde matrix of order k, it is invertible and so (5) has always a unique solution say $(\bar{a_0}, \bar{a_1}, \bar{a_2}, \ldots, \bar{a}_{k-1})^T$.
Step4: From the above system of solutions, if $\bar{a}_{2r} < 250$ do $\mathbb{S}[i, d * j + r] = \bar{a}_{2r}$, else $\mathbb{S}[i, d * j + r] = \bar{a}_{2r} + (\bar{a}_{2r+1} \mod 6)$ for $0 \le r < d$.
Step5: This way, all the $H \times W$ pixels of the secret image are retrieved and the same $\mathbb{S}$ secret image is restored, resulting in the scheme's lossless recovery.

$$\sum_{q=0}^{k-1} a_q * r \equiv_{251} \mathbb{S}_r[i, j], \text{ for all } r = 1, 2, \ldots, k. \tag{4}$$

The above system of equations can be written as

$$VX \equiv_{251} B, \tag{5}$$

where $V = V(1, 2, \ldots, k-1)$, $X = (a_0, a_1, \ldots, a_{k-1})^T$, $B = (\mathbb{S}_1[i, j], \mathbb{S}_2[i, j], \ldots, \mathbb{S}_k[i, j])^T$.

4 Security Analysis

In this section, we will show that although any $k - 1$ or fewer participants collude, not enough information can be obtained about the secret image. To measure the probability of getting information regarding the secret image, without loss of generality, we may assume that the first $k - 1$ participants with shares $\mathbb{S}_1 = (1, \mathbb{S}_1[i, j]), \mathbb{S}_2 = (2, \mathbb{S}_2[i, j]), \ldots, \mathbb{S}_{k-1} = (k - 1, \mathbb{S}_{k-1}[i, j])$ collude with each other to guess about the secret image. Then from Eq. (3), they can construct only $k-1$ Eq. (6) with k unknowns $a_i, i = 0, 1, \ldots, k-1$.

$$\sum_{q=0}^{k-1} a_q * r \equiv_{251} \mathbb{S}_r[i, j], \text{ for all } r = 1, 2, \ldots, k-1 \tag{6}$$

Since there are $k-1$ equations with k unknown, they must guess correctly the share of the k-th participant in order to obtain the correct solution of the system of equations (6). For each (i, j)-th pixel $0 \leq \mathbb{S}[i, j] < 256$ and $0 \leq \mathbb{S}_r[i, j] < 251$ for $r = 1, 2, \ldots, k-1$, assume that the image contains m pixels whose value is ≥ 250. If the pixels of the image are < 250, the coefficients $0 \leq a_{2r} < 250, r = 1, 2, \ldots, d-1$. The $k-1$ unknown can be uniquely solved by the $k-1$ equations for a fixed value of each $a_{2r}, r = 1, 2, \ldots, d-1$. Therefore, in this case the probability of guessing the exact solution is $\frac{1}{250}$. Thus, the probability of guessing correctly $H \times W - m$ pixels of the secret image is $\left(\frac{1}{250}\right)^{\frac{H \times W}{d} - m}$. Also, the probability of guessing rightly the remaining m pixels with pixel value ≥ 250 is $\left(\frac{1}{41}\right)^{m}$. Hence, the possibility of guessing the proper image correctly is $\left(\frac{1}{250}\right)^{\frac{H \times W}{d} - m} + \left(\frac{1}{41}\right)^{m}$.

5 Experimental Results and Analyses

This section contains various statistical analyses such as histogram, correlation coefficient, MSE, PSNR, etc. for the effectiveness of the proposed scheme.

The proposed (5, 7)-PSIS scheme generates seven random shadows as shown in Fig. 1. From these individual shadows, no information regarding the secret will reveal. The secret can be retrieved without losing any pixel by only eligible participants. On the other hand, forbidden set of participants get random-like image.

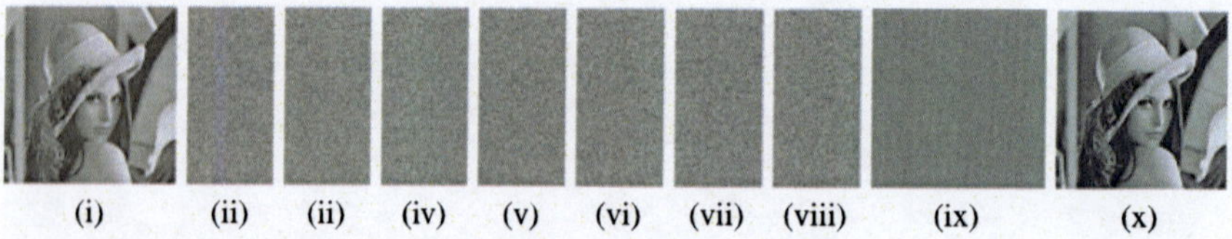

Fig. 1 Experimental results of our (5, 7)-PSIS scheme. (i) Original image (510 × 510) of Lena, (ii)–(viii) shadow images (510 × 255) of Lena, (ii) recovered resulting from shadows (ii)–(v), and (iii) reconstruction resulting from shadows (ii)–(vi)

Table 1 The correlation coefficient of the secret image Lena and its corresponding shadow images

Image	Horizontal (H)	Vertical (V)	Diagonal (D)	Average (H,V,D)
Lena	0.96607670	0.97504514	0.95318401	0.96476868
Shadow1	0.01600685	0.00645891	−0.00692287	0.00518097
Shadow2	0.00731571	−0.00315786	−0.00893320	−0.00159178
Shadow3	0.00030135	0.01579561	0.00560986	0.00723561
Shadow4	−0.00900259	0.02003986	0.01592961	0.00898896
Shadow5	0.00296330	−0.01203020	0.00936662	0.00009991
Shadow6	0.00676792	−0.00646033	0.01464378	0.00498379
Shadow7	−0.01668509	0.00662455	−0.00019139	−0.00341731

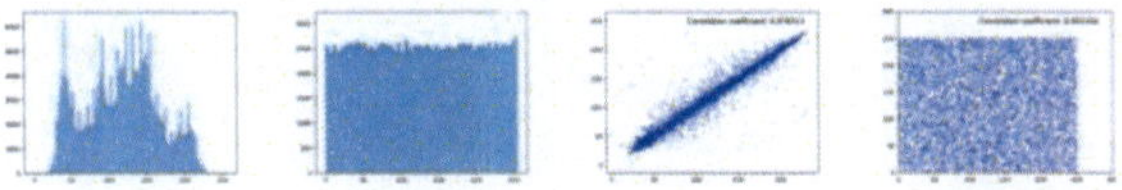

(i) (ii) (iii) (iv)

Fig. 2 Simulation results for our (k, n)-PSIS scheme: (i)–(ii) Histograms of original secret image and corresponding shadow image of Lena, (iii)–(iv) Scatter plot of correlation coefficients of the original secret and the corresponding shadows of Lena

From the tabular data as shown in Table 1, it is evident that the correlation between the adjacent pixels in the shadow images is very less. Visually, from the scatter plots as shown in Fig. 2(iii)–(iv), it is observed that pair of pixels in the meaningful secret image is highly correlated. However, the scatter plot of the shadow images shows that these images provide no meaningful information, and it appears random-like images.

Figure 2(i)–(ii) shows the histogram of secret and the corresponding shadows.

6 Comparison

When compared with the work of Thien and Lin [8], we can see that before the secret sharing, the proposed system does not need to encode the secret image. In addition, theoretical and numerical evidence in Table 2 show that our scheme is totally lossless. Ding et al. [2] successfully proposed a lossless (k, n)-scheme in $\mathbb{Z}_{251}$ without pixel expansion, i.e., the share size remains the same as the original secret image, whereas our proposed scheme generates reduced shadow size with lossless recovery (Table 3).

Table 2 MSE, PSNR, correlation values of secret image, and recovered images of the proposed (k, n)-PSIS scheme

Schemes	MSE	PSNR	Correlation
Lena	0	∞	1
Baboon	0	∞	1
Fruit	0	∞	1
Bird	0	∞	1

Table 3 Comparisons of notable properties for (k, n)-PSIS scheme

Schemes	Pre-computation	Share size	Lossless recovery
Shamir [7]	No	1	No
Thien and Lin [8]	Yes	$\frac{1}{k}$	No
Ding et al. [2]	No	1	Yes
Our	No	$\frac{2}{k-1}$ or $\frac{2}{k-2}$	Yes

7 Conclusion

In this paper, we propose a totally lossless polynomial-based secret image sharing scheme with small shadow size. The main benefit of our proposed scheme is that reduced shadow images are generated using the prime 251 which is smaller than 256. Security analysis and simulation results show the efficiency of our (k, n)-PSIS scheme. Construction of more efficient algorithms in terms of share size could be an interesting further work.

Acknowledgements The author would like to express his special thanks to the Council of Scientific and Industrial Research (CSIR), Government of India for providing financial support (Award No.09/028(0975)/2016-EMR-1).

References

1. Adhikari, M.R.A. (2014). Basic Modern Algebra with Applications. Springer, New Delhi. 978-81-322-1598-1
2. Ding, W., Liu, K., Yan, X., Liu, L.: Polynomial-based secret image sharing scheme with fully lossless recovery. Int. J. Digit. Crime For. **10**(2), 120–136 (2018)
3. Dutta, S. and Adhikari, A. (2017). Contrast Optimal XOR Based Visual Cryptographic Schemes. In *Information Theoretic Security - 10th International Conference, ICITS 2017, Hong Kong, China, November 29 - December 2, 2017, Proceedings*, pp. 58–72
4. Dutta, S., Adhikari, A., Ruj, S.: Maximal Contrast Color Visual Secret Sharing Schemes. Designs, Codes and Cryptography (2018)
5. Dutta, S., Rohit, R.S., Adhikari, A.: Constructions and Analysis of Some Efficient t- $(k, n)^*$-visual Cryptographic Schemes using Linear Algebraic Techniques. Des. Codes Cryptogr. **80**, 165–196 (2016)

6. Naor, M., Shamir, A.: Visual Cryptography. In: De Santis, A. (ed.) Advances in Cryptology—EUROCRYPT'94, pp. 1–12, Berlin, Heidelberg. Springer Berlin Heidelberg (1995)
7. Shamir, A.: How to share a secret. Commun. ACM **22**(11), 612–613 (1979)
8. Thien, C.-C., Lin, J.-C.: Secret image sharing. Comput. Gr. **26**(5), 765–770 (2002)
9. Yang, C.-N., Ouyang, J.-F., Harn, L.: Steganography and authentication in image sharing Without Parity Bits. Opt.Commun. **285**(7), 1725–1735 (2012)

Multi-factor Authentication-Based E-Exam Management System (EEMS)

Sharthak Mallik, Shovan Halder, Pranay Saha, and Saswati Mukherjee

Abstract With the explosion of Internet-based technologies like online examination system, a lot of manual tasks have been automated, thereby saving time and effort. But deployment of such automated system has made the process more vulnerable to various risks, such as confidentiality loss, intrusion of critical data, and impersonation. In this paper, we proposed an online exam system using multi-factor authentication that allows instructors to set exams for different subjects via an integrated high text editor. Students can take exams after code-based verification and biometric authentication. It automates the process of exam scheduling, grading, and reporting, thereby reducing the workload of the instructors. A survey has been conducted to prove the efficacy of the system in terms of reliability, scalability, and availability.

Keywords Authentication · Face recognition · Grading · Code generation · Encryption

1 Introduction

In recent years, online system has gained a lot of popularity as they facilitate distance-mode learning and online examination among a large number of students across any demographic location. The procedures of paper-based examination are time-consuming and error-prone. To overcome such constraints, online examination has

S. Mallik · S. Halder · P. Saha
Jalpaiguri Government Engineering College, Jalpaiguri, West Bengal, India
e-mail: sharthakmallik@gmail.com

S. Halder
e-mail: shovanhalder9@gmail.com

P. Saha
e-mail: sahapranay077@gmail.com

S. Mukherjee (✉)
School of Education Technology, Jadavpur University, Kolkata, India
e-mail: swati.mukherjee@jadavpuruniversity.in; saswatimuk@gmail.com

D. Bhattacharjee et al. (eds.), *Proceedings of International Conference on Frontiers in Computing and Systems*, Advances in Intelligent Systems and Computing 1255,
https://doi.org/10.1007/978-981-15-7834-2_66

become imperative due to automated checking, grading, result publication, etc. However, developing a suitable system that fulfills the goals of a particular organization can be challenging due to security, customization, ease of use, and cost. The objective of this paper is to propose a web-based secure examination system using multi-factor authentication like verifying student's identity and authorship. Multi-factor authentication scheme is used as an alternative to single-factor authentication with the goal of better security [1]. Following are the security measures used in conventional online tests such as, knowledge-based authentication, username and password verification, and a webcam-based face recognition. But the majority of an online assessment system is based on a single-factor authentication that monitors the fraudulent activities and unauthorized access.

In this work, a secure online exam assessment system is proposed that includes multiple ways of verifying the students' authenticity appearing for exam to prevent academic malpractices and quality assurance in education. Following are the key features of the proposed system.

- It provides an exam administration portal with three major functional modules, admin module, student module, and instructor module.
- Admin module is responsible for defining course with subject specification, exam management, result generation, management of student's login status, and access permission.
- Admin module also manages user-specific roles of both students and instructors.
- Instructor module is responsible for defining the exam types, specifying the exam topics, creating question bank for different disciplines, and evaluating the answers.
- Exam dashboard enables the students to take designated exams.
- Security measures are implemented to prevent unauthorized access and impersonation; answers are encrypted to prevent impersonation during submission.
- Student's authenticity is verified using secret code, and biometric authentication is done by applying existing face recognition API.
- A rich text editor is integrated to create various types of questions and relevant answers. Questions are randomized to eliminate unfair activities while giving exam.
- The proposed system provides an easy-to-use environment with prompt response, reliability, and flexible accessibility.

The rest of this paper is organized as follows. Section 2 presents the related works. In Sect. 3, the methodology of the proposed system is described. Experimental analysis is presented in Sect. 4. Finally, we conclude the paper in Sect. 5.

2 Related Works

An automated web-based [1] exam system is developed that includes multi-factor authentication measures but without face recognition features. Authors [2] use pining and unpinning function to restrict students from browsing other services and random capturing and exchanging of pictures during exam.. Three different sub-sections [3], user information, audit logs, and user management and system security are used. User management creates and manages student and teacher's account and grant access rights. Audit section traces student's activities. Similarly, authors investigated the use of face and iris method in an intervening manner. Several video analytic applications have emerged especially in surveillance and security [4]. TeSLA is an e-assessment system [5] that comes with authentication and authorship verification. Teacher re-designs or creates new assessments with two security measures: enrollment and real. In enrollment, teacher creates a different learning model, and students perform their intended tasks which are authenticated and validated. Integrated palm printing [6] is used along with traditional name–password combo for authentication. The system demands webcam for constant monitoring of examiners. Keystroke pattern [7] matches the username and password for authentication purposes, and cost is low since no special hardware is needed. In [8], IP addresses and time stamp are used to monitor distance learning tests given by students. IP addresses although unique can easily be manipulated using VPNA profile-based authentication framework combined with user-id and password. Authentication [9] follows a simple implementation and requires no special hardware. Possibility of cheating increases if answers are shared. Yao angle variation [10] uses a webcam to capture audio and video. From video input, feature points are extracted, Yao angle variations are calculated, and audio input is compared to a marginal value. Symmetric and asymmetric crypto-system [11] with digital signature schemes are applied for preventing fraudulent activities in an online exam. Examiner uses authentication components registered with the trusted article authority.

3 Proposed Work

The proposed system supports three functional modules: admin (administrator), instructor, and student as shown in Figs. 1 and 2. Admin module creates course, and examination, assigns instructor and student, notify exam and publish result as given in Fig. 3. Admin can also grant or revoke rights to/from both instructor and student. Instructor module creates questions according to question type, set marks, and monitors the assigned exams through the instructor dashboard. A rich text editor is integrated to define questions. The only exception is that the instructor has to evaluate manually the subjective-type answers, whereas objective-type answers are automatically evaluated and generated. Before taking the designated exam, a student needs to register via the exam dashboard by giving personal, academic, and professional details and uploading its own facial image. All these details are stored in the database.

When a student is assigned to an exam as shown in Fig. 4, admin sends an exam notification to the student's email mentioning the date and time along with the exam URL. Students are allowed to give exam after proper authentication which is accomplished in two steps as follows.

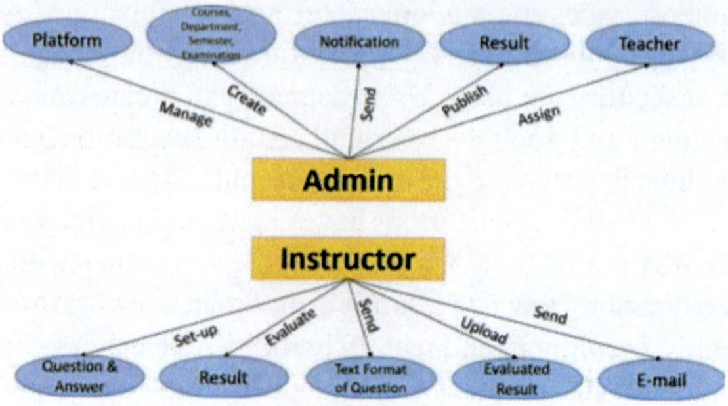

Fig. 1 Admin and instructor modules

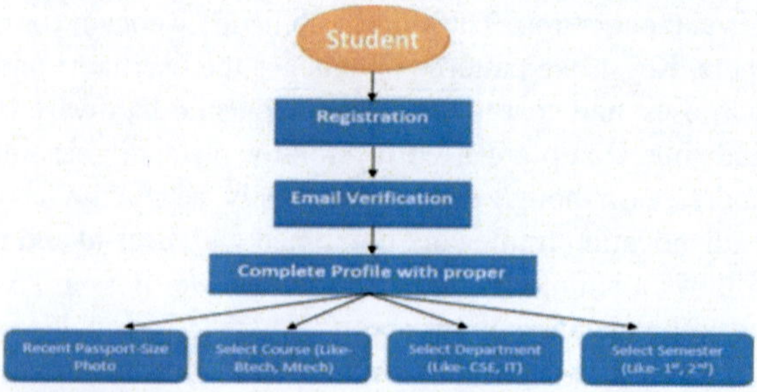

Fig. 2 Student module

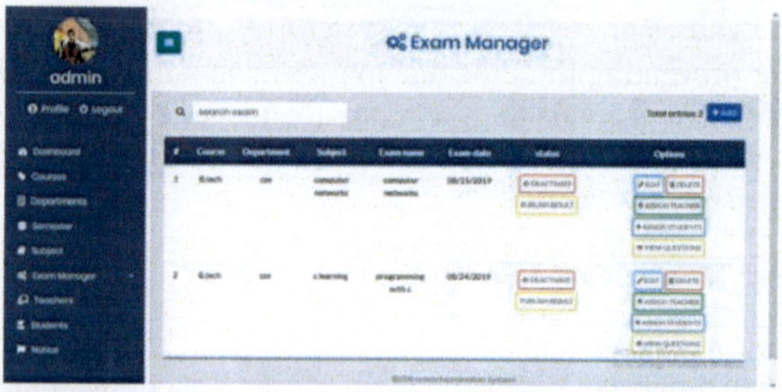

Fig. 3 Admin manages exam manager

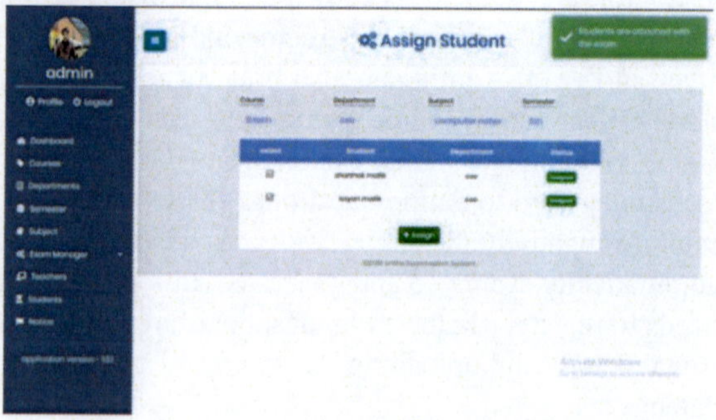

Fig. 4 Admin assigns students to exams

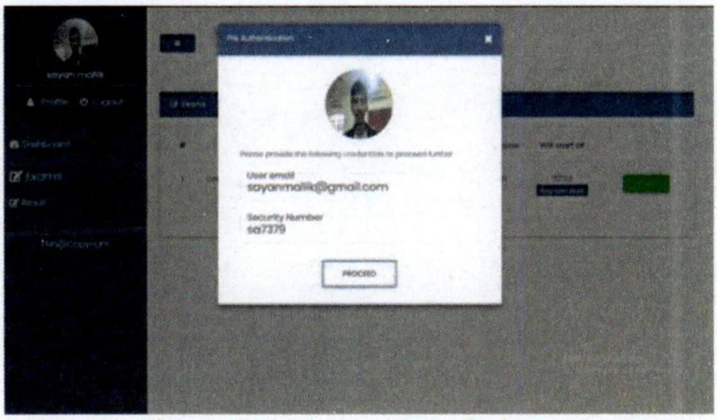

Fig. 5 Verification of security number

- Code verification module and
- Biometric-based face recognition module.

3.1 Code Verification Module

When a student clicks the exam URL in the notification mail, a unique secret number is sent to his email. Student enters this secret number for login as shown in Fig. 5. Once the login is successful, student is directed to the next interface for biometric authentication using face recognition module.

3.2 Face Recognition Module

In face recognition module, candidate's image is captured by webcam and stored in the database as a reference image along with his name. The input image is compared with the stored reference image to find the most similar reference image. If both images are similar, the candidate's name appears in the exam panel automatically. In, during face-based authentication, it is very likely that images of more than one candidate may be captured via webcam. In biometric-based face recognition module, built-in face detection method tries to locate faces which are present more than once in the database. If the detection count is more than one, access is denied with a message that there exists more than one candidate. The face recognition module has two stages: face detection and face comparison as shown in Figs. 6 and 7.

For face detection, we have used an existing Tiny Face Detector method [12], a tinier version of Tiny Yolo v2 that utilizes depthwise separable convolutions instead of regular convolutions. This face detector is much faster but slightly less accurate than SSD MobileNet [13] face detector method.

In face comparison stage, the system compares the captured image with the reference image using a deep convolutional neural network (DCNN)-based API [14].

Fig. 6 Face detection (Stage-I)

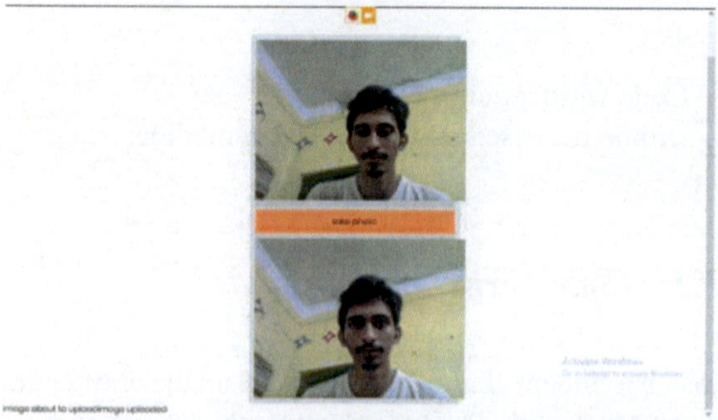

Fig. 7 Face comparison (Stage-II)

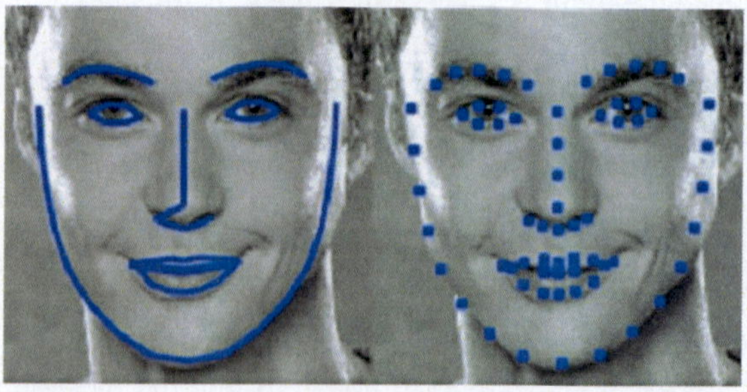

Fig. 8 Face point markers

This API can localize a large number of facial landmarks of 68 points considering the geometric orientation of the face as depicted in Fig. 8.

The extracted image is fed into the deep network model based on ResNet-34 architecture. The network trains 128 feature vectors of human face descriptors to map the face. The face descriptor of the extracted image is compared with the face descriptors of the reference image. The Euclidean distance is used to compare the face descriptors of two images. If the summation of distance (considering 128 feature vectors) between the two face descriptors is too high, then face recognition module gives access to a large number of unauthorized users than desired. On the other hand, if it is too low, then the system is preventing the access even to the authorized user. The reason is, the matching between the two images will not be accurate if the distance threshold is too high or too low. On repeated execution of the face

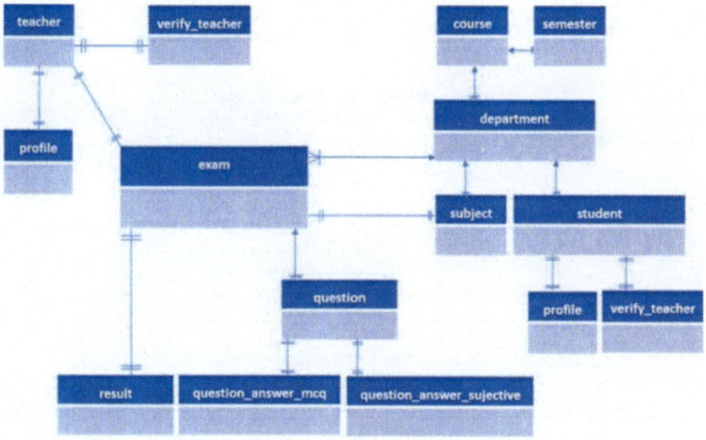

Fig. 9 Number of tables designed for the proposed system

recognition module, we noticed that for an image of dimension 150x150, 0.6 can be considered as the optimum distance threshold for accurate recognition. It is to be noted that, for different size images, the optimum distance threshold may vary.

3.3 Encryption Module

After the exam completion, submitted answers are encrypted using AES algorithm [15] to prevent impersonation during transmission. The stored answers are decrypted at the time of evaluation. Encryption can be used in both hardware and software, and it comes with three variations based on the key sizes (128, 192, and 256 bits). Encryption consists of 10 rounds of processing for 128-bit keys, 12 rounds for 192-bit keys, and 14 rounds for 256-bit keys.

3.4 Database Design

Figure 9 represents the database design of the proposed exam system. The teachers' table contains the records of the administrator and the teachers. Each teacher has a profile information, which is stored in the profile table and associated with the teachers' table.

A verify_teacher table associated with the teachers' table contains the verification token, which is passed as a parameter to the email verification URL. The crowfoot notations are used to represent the database design. The student table stores the student-related information, such as first name, last name, email, pass-

Fig. 10 Instructor's response

SL. No	Parameters	Questions	Answers					Total Answer	Average Score
			5	4	3	2	1		
1.	Performance	How is the overall performance of the system?	4	3	1	0	0	8	4.375
2.	Readability	How easy the exam system to use/operate?	5	2	0	1	0	8	4.375
3.	Usability	How responsive, fast and easy to handle the system?	3	4	0	1	0	8	4.125
4.	Reliability	How reliable is the system?	5	3	0	0	0	8	4.625

word, course, department, semester, etc. The exam table consists information of different exams for different subjects. The students who are assigned to an exam can take the exam. Each question may be of two types: multiple-choice and subjective. Based on the question types, two tables are created, question_answer_mcq and question_answer_subjective. For each student, a result is generated for each exam, and each result instance is associated with one exam at a time which is shown by one-to-one relationship between the exam and the result table in Fig. 9.

4 Experimental Analysis

A comparative analysis of CNN-based face recognition method is not performed in our work because we have used an existing optimized Tiny Face Detector method [12] and DCNN-based API [14] for face recognition of the students during examination. A survey is conducted for measuring the efficacy of the proposed system considering certain parameters. The parameters are system's performance, readability, smoothness, usability, and reliability. Students are required to respond to three parameters, performance, readability, and smoothness. Instructors are required to respond to four parameters, i.e., system's performance, readability, usability, and reliability. There were a total of 97 students and 8 instructors participated in this survey. Both students and instructors are required to rate the parameters on a scale of 1–5 as given here. The ratings of instructors and students are shown in Figs. 10 and 11.

5—Strongly Agree
4—Agree
3—Neutral
2—Disagree
1—Strongly Disagree

Students' rating shows that 80% has agreed that the interface is well readable, 84% has found that the system is easy to use, and 82% has rated the overall system's performance as good. On the other hand, 87% instructors agreed that the overall

Fig. 11 Student's response

SL. No	Parameters	Questions	Answers					Total Answer	Average Score
			5	4	3	2	1		
1.	Performance	How is the overall performance of the system?	47	24	12	11	8	97	4.097
2.	Readability	How easy the exam system to use/operate?	43	25	18	7	4	97	4.030
3.	Smoothness	Did you encounter any technical issues?	46	27	14	6	4	97	4.144

system performance is favorable, 100% agreed that it is well designed, 100% assured that it is faster to use, and 87% agreed that it is reliable.

5 Conclusion

The need for a secure online examination system has become extremely important in today's world. In this paper, web-based examination system has been developed that employs several authentication measures to verify the integrity and authorship of students who are giving exams. An exam dashboard is designed and integrated with the three functional modules that helps to manage students and instructor and generate results after evaluation. The functionalities of face recognition module, code verification module, and data encryption enhance the security of the proposed system. As a future work, design of improved database design can reduce the processing complexity, and scanning of QR codes instead of security codes can improve the system usability.

References

1. Al-Hawari, F., Alshawabkeh, M., Althawbih, H., Abu Nawas, O.: Integrated and secure web-based examination Management system. Comput. Appl. Eng. Educa. **27**(4), 994–1014 (2019)
2. Nagal, R., Nemkul, P., Kumar, D., Kumar, N., Joseph, A.: Android based secure exam management system to prevent impersonation. Int. J. Latest Technol. Eng. Manage. Appl. Sci. **VI**, 46–49 (2017)
3. MisikirTashu, T., P Esclamado, J., Horvath, T.: Intelligent on-line exam management and evaluation system (2019)
4. Garibotto, G., Murrieri, P., Capra, A., De Muro, S., Petillo, U., Flammini, F., Esposito, M., Pragliola, C., DI Leo, G., Lengu, R., Mazzino, N., Paolillo, A., D'Urso, M., Vertucci, R., Narducci, F., Ricciardi, S., Casanova, A., Fenu, G., De Mizio, M., Ferone, A.: White paper on industrial applications of computer vision and pattern recognition, vol. 8157(2013)

5. Okada, A., Noguera, I., Aleksieva, L., Rozeva, A., Kocdar, S., Brouns, F., Ladonlahti, T., Whitelock, D., Guerrero- Roldán, A.: Pedagogical approaches for e-assessment with authentication and authorship verification in higher education. Br. J. Educ. Technol. (2019)
6. Al-Saleem, S., Ullah, H.: Security considerations and recommendations in computer-based testing. Sci. World J. **2014**, 562787 (2014)
7. Ramu, T.S., Arivoli, T.: A framework of secure biometric based online exam authentication: An alternative to traditional exam (2013)
8. Gao, Q.: Using ip addresses as assisting tools to identify collusions. Int. J. Bus. Hum. Technol. **2** (2012)
9. Ullah, A., Xiao, H., Lilley, M., Barker, T.: Using challenge questions for student authentication in online examination. Int. J. Infonomics **5**, 631–639 (2012)
10. Asha, S., Chellappan, C.: Biometrics: An overview of the technology, issues and applications. Int. J. Comput. Appl. **39**, 35–52 (2012)
11. C K, S., Narasimhan, H., Padmanabhan, T.: Cryptography and Security (2011)
12. Liu, W., Anguelov, D., Erhan, D., Szegedy, C., Reed, S., Fu, C.Y., Berg, A.C.: Ssd: Single shot multibox detector. In: European conference on computer vision, pp. 21–37. Springer (2016)
13. Hu, P., Ramanan, D.: Finding tiny faces. 2017 IEEE Conference on Computer Vision and Pattern Recognition (CVPR), pp. 1522–1530 (2016)
14. Zhou, E., Fan, H., Cao, Z., Jiang, Y., Yin, Q.: Extensive facial landmark localization with coarse-to-fine convolutional network cascade. In: The IEEE International Conference on Computer Vision (ICCV) Workshops (2013)
15. Mukhopadhyay, D.: Cryptography: advanced encryption standard (aes). Encycl. Comput. Sci. Technol. **279** (2017)

A Novel High-Density Multilayered Audio Steganography Technique in Hybrid Domain

Dipankar Pal, Anirban Goswami, Soumit Chowdhury, and Nabin Ghoshal

Abstract The work in this paper presents a novel data hiding technique with high density of payload data for transmission of sensitive message in a concealed form within a digital audio signal used as cover. To enhance security, the payload message is embedded into discrete Fourier transformed coefficients of time-domain audio samples, to make any potential hacking attempt harder. The proposed technique increases the density of the transmitting message by inserting multiple bits of message data at multiple LSB layers, selected pseudo-randomly within the binary counterpart of each transformed frequency coefficient. Furthermore, a 160-bit message digest is inserted along with the message data to validate the integrity of the extracted message by the receiver. Thorough experimentation has been carried out to evaluate the effectiveness of this method taking different types of cover audio and message as input to various standard performance metrics.

Keywords Information security · Digital audio steganography · Time domain · Frequency domain · DFT · LSB · SNR · PSNR · MOS

D. Pal (✉)
Department of Computer Science and Engineering, Techno India (Main), Kolkata 700091, India
e-mail: mail2dpal@yahoo.com

A. Goswami
Department of Information Technology, Techno India (Main), Kolkata 700091, India
e-mail: angos.kol@gmail.com

S. Chowdhury
Government College of Engineering & Ceramic Technology, Kolkata 700010, India
e-mail: soumitchowdhury2019@gmail.com

N. Ghoshal
Department of Engineering and Technological Studies, University of Kalyani, Kalyani 741235, West Bengal, India
e-mail: nabin_ghoshal@yahoo.co.in

D. Bhattacharjee et al. (eds.), *Proceedings of International Conference on Frontiers in Computing and Systems*, Advances in Intelligent Systems and Computing 1255,
https://doi.org/10.1007/978-981-15-7834-2_67

1 Introduction

The area of communication technologies has much evolved in the recent times and consequently, transmitting sensitive data over a public network has become one of the most important concerns for every organization. Till date secured data transmission mainly involves application of cryptographic techniques, which represent the content of a message in a distorted form to unauthorized people. But these techniques are susceptible to attract the attention of any unintended party about the content of the message being conveyed. This shortcoming led the researchers to find new techniques and experimentation has begun with techniques based on steganography, where sensitive data is kept in disguise into another insusceptible cover medium (e.g. digital image or audio signal) to avoid any suspicion about any hidden data being transmitted. Identity of the sending and receiving parties also become hard to reveal as well.

Over time, digital steganography has become more popular and an important research topic for the present research community to explore more novel and efficient techniques of hiding data for different purposes and areas of communication technologies.

1.1 Literature Review

One attribute of Human Auditory System (HAS) is that, it has a wider dynamic range than Human Visual System (HVS) [1, 2]. This makes data hiding in audio signals more challenging compared to images or video sequences. Also, the quality of audio steganographic methods are influenced by various factors. The most influencing factors are security, imperceptibility, hiding capacity, compression and robustness to noise [3–5]. In general, most audio steganography techniques are based on either time domain [6, 7] or, frequency-transform domain [8–10]. Some recent techniques have exploited the features of both [11].

Time-domain techniques commonly use the Least Significant Bit (LSB) substitution method, echo hiding method, etc., whereas frequency domain techniques involve the modification of the magnitude or phase of the transformed coefficients to insert the transmitting message or payload. To retrieve the payload at the receiving end, inverse frequency transform is used.

The main advantage of transform domain techniques is that they can achieve more robustness, and the fidelity of the original audio signal is seriously distorted in case any unauthorized attempt is made to remove the embedded payload. Discrete Cosine Transform (DCT) [12], Discrete Fourier Transform (DFT) [13] and Discrete Wavelet Transform (DWT) [14] are the most widely used transform domain techniques used for this purpose.

The proposed work in this paper employs DFT along with LSB substitution on DFT coefficients and is capable of hiding a large volume of digital data of any type into an uncompressed audio signal and produces the corresponding stego signal, without affecting its property and audio quality.

2 The Technique

There are two major process components in the proposed algorithm, insertion of the message at the source and extraction of the message at the destination. In this context, the message consists of three types of data, which are (1) length of the message data, in bytes (size: 32 bit), (2) a message digest, generated by SHA-1 hashing (size: 160 bit) of the message data and (3) the message data itself (size: depends on the number of audio samples in the carrier). Collectively, we call them payload, which is going to be inserted in the cover medium for secret transmission.

The entire payload data is considered as a stream of bits during the process of insertion, and the original cover audio signal is considered as a collection of nonoverlapping frames, each having two consecutive samples of cover audio. In order to apply signal-processing rules effectively, audio signals are analyzed in terms of short time intervals (frames) due to their nonstationary behavior [15]. To convert each of these frames or blocks into transformed domain, one-dimensional discrete Fourier transform (1D-DFT) is applied, producing two frequency components corresponding to two sample values. This (1D-DFT) can be denoted in series form as

$$F(u) = \frac{1}{N} \sum_{x=0}^{N-1} f(x) e^{-2\pi i x u / N} \tag{1}$$

For each frequency coefficient, leaving the imaginary component untouched, the integer part of only the real component is considered for inserting multiple (2 or 3) payload data bits at multiple LSB layers, between the 1st and 4th LSB layers. Also, Insertion positions are not static and are decided pseudo-randomly in the range of 0–3 using a function. To get the time-domain samples back, inverse DFT is applied on each frame containing modified frequency components, which can be denoted in a series form as

$$f(x) = \sum_{u=0}^{N-1} F(u) e^{2\pi i x u / N} \tag{2}$$

The extraction process involves DFT transformation of the stego audio samples to the frequency domain. Extraction of payload bits is performed considering only the real component (excluding the floating point part) of each transformed frequency coefficient. In this process, the extraction positions of bits are also determined pseudo-randomly using the same function as was used at the time of insertion. The extracted bits, in sequence, form the payload consisting of three components, i.e. message length in bytes, message digest (SHA-1 hash) and actual message data. At the

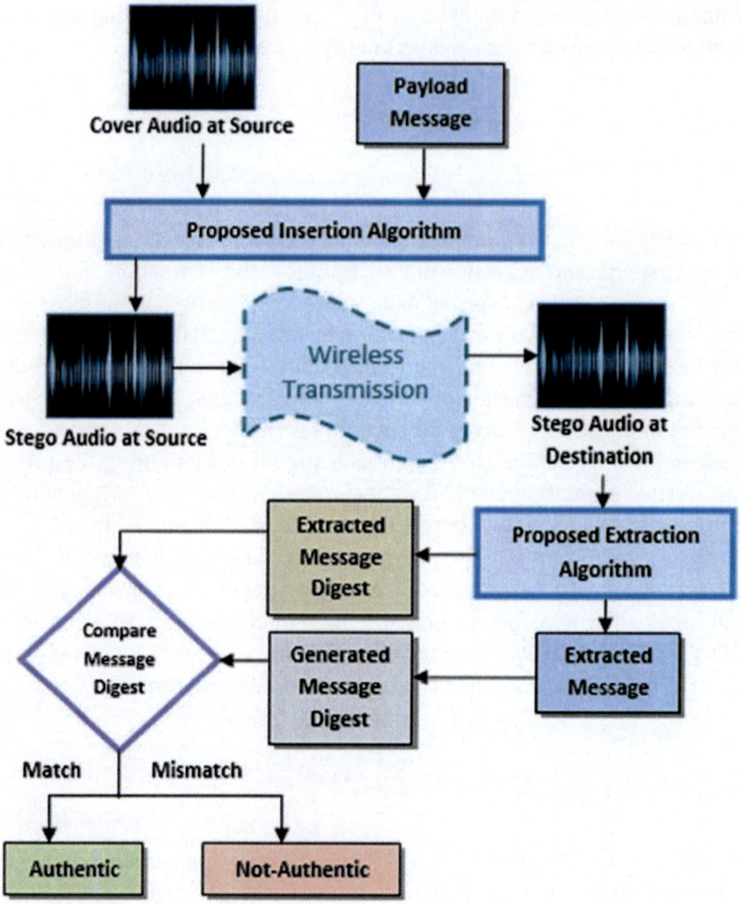

Fig. 1 The overall workflow of the proposed technique

receiving end, another message digest is computed on the received message and then matched against the extracted message digest, to verify the integrity and authenticity of the transmitted message.

Also, the extraction process is entirely blind, where the original cover audio is not necessary for the purpose of extraction. Overall process workflow of this technique is depicted in Fig. 1.

2.1 Algorithm for Insertion

Input: A digital audio (PCM WAVE) and any message data.

Output: A digital carrier audio (PCM WAVE) with the same property as the original.

Steps:

1. Store the entire header part from the source audio into the output audio, as it is.
2. Compute a 160-bit digest (SHA-1) of the input message.
3. Repeat steps from 3.1 to 3.7 until all the payload bits are inserted into the cover audio samples.

    3.1. Store two consecutive audio samples from the source audio into a two-element block.
    3.2. Transform the samples into their corresponding frequency components by applying one-dimensional DFT on the block.
    3.3. Using a pseudorandom function, produce 2 or 3 integers, between 0 and 3.
    3.4. For each frequency coefficient, consider only the real component and convert the integer part of it into a binary equivalent.
    3.5. Overwrite the original bits (at the positions obtained in step 3.3) with payload bits using binary masking.
    3.6. Get the corresponding time-domain samples back by applying one-dimensional inverse DFT on each modified block.
    3.7. Store the resultant audio samples into the target file preserving their sequence.

4. Transfer the remaining source audio samples (if any) to the target file.
5. Stop.

2.2 Algorithm for Extraction

Input: A digital carrier audio (PCM WAVE) produced by the Insertion algorithm.

Output: The extracted components of payload data.

Steps:

1. Repeat steps from 1.1 to 1.6 until the entire payload is extracted.

    1.1. Store two consecutive audio samples from the input file into a two-element block.
    1.2. Transform the samples into their corresponding frequency components by applying one-dimensional DFT on the block.
    1.3. Produce 2 or 3 integers (between 0 and 3) pseudo-randomly, utilizing the same function used at the time of insertion.
    1.4. For each frequency coefficient, consider only the real component and convert the integer part of it into a binary equivalent.
    1.5. Read the payload bits from the bit positions obtained in step 1.3 and store them in memory.

1.6. Combine eight consecutive extracted bits and form a byte of payload data to produce the entire payload incrementally.

2. Compute a SHA-1 digest of the message received as part of the extracted payload.
3. Validate the integrity and authenticity of the received message by comparing the message digest within the payload and the one computed in the previous step.
4. Stop.

3 Performance Test

To evaluate the outcome of the proposed algorithm in different scenarios, we have thoroughly experimented with various combinations of carrier audio and message format. After applying the insertion process on each combination of carrier audio and message format, various types of objective and subjective metrics [16, 17] were used to evaluate the quality of each stego audio.

Two objective metrics were used in the result analysis, which are (1) *Signal-to-Noise Ratio (SNR):* defined as the ratio of signal power to the noise power, and a ratio higher than 1:1 indicates more signal than noise; (2) *Peak Signal-to-Noise Ratio (PSNR):* defined as the ratio between the maximum possible power of a signal and the power of corrupting noise that affects the fidelity of its representation.

One limitation of objective metrics is that they are unable to take into consideration some particular properties of Human Auditory System. In view of that, for the sake of accuracy a different metric is used to measure the subjective quality of audio signals, which is called Mean Opinion Score (MOS).

After carefully listening to the source and the message embedded audio, a number of listeners rate them in the scale of 1 (very annoying/distorted) to 5 (absolutely imperceptible/same as original). The final MOS is then determined by computing the arithmetic mean of all the individual scores, which may range from 1 (worst) to 5 (best).

In the test phase, three types of audio signals (each of 16 bits/sample resolution and with stereo channel) have been used as carrier and also three types of messages were inserted distinctly in all three types of carrier audio signals. Two sets of experimentation has been done, where in the first phase 2 bits of payload are inserted in two different LSB layers of each carrier frequency coefficient and in the second phase 3 bits of payload are inserted in three different LSB layers of each carrier frequency coefficient.

3.1 *Results from Objective Tests*

Tables 1 and 2 show different test parameters, i.e. the maximum number of bytes that each category of carrier audio can hide and the number of bytes actually inserted for each message type. For the sake of comparison, results are shown as SNR and

Table 1 Results obtained by 2-layer embedding

Cover audio	Capacity (Bytes)	Message type, Size (Bytes)	Proposed algorithm		Standard LSB method	
			SNR (dB)	PSNR (dB)	SNR (dB)	PSNR (dB)
Instrumental	665890	Text, 638936	64.2929	84.1545	57.1671	77.0287
Pop	681968		64.8577	84.3530	57.6424	77.1377
Speech	640468		64.4471	84.0971	57.1953	76.8453
Instrumental	665890	Image, 633435	64.2771	84.1387	57.1118	76.9733
Pop	681968		64.8234	84.3187	57.5825	77.0778
Speech	640468		64.4035	84.0535	57.1334	76.7834
Instrumental	665890	Audio, 638223	64.2395	84.1010	57.1637	77.0253
Pop	681968		64.7943	84.2896	57.6422	77.1376
Speech	640468		64.3860	84.0360	57.2077	76.8577
Average		636865	64.5023	84.1713	57.3162	76.9852

Table 2 Results obtained by 3-layer embedding

Cover audio	Capacity (Bytes)	Message type, Size (Bytes)	Proposed algorithm		Standard LSB method	
			SNR (dB)	PSNR (dB)	SNR (dB)	PSNR (dB)
Instrumental	998835	Text, 958416	60.3566	80.2182	55.1370	74.9985
Pop	1022952		60.8867	80.3821	55.6254	75.1207
Speech	960702		60.4560	80.1060	55.1671	74.8172
Instrumental	998835	Image, 954300	60.2318	80.0933	54.8859	74.7475
Pop	1022952		60.7653	80.2606	55.3679	74.8633
Speech	960702		60.3568	80.0068	54.8978	74.5478
Instrumental	998835	Audio, 959216	60.2503	80.1119	54.9068	74.7684
Pop	1022952		60.7861	80.2814	55.3903	74.8857
Speech	960702		60.3829	80.0330	54.9497	74.5998
Average		957310	60.4969	80.1659	55.1475	74.8165

PSNR values for two different test scenarios, one using the proposed algorithm and the other using the standard LSB substitution method without any transformation.

It can be observed from the above results that despite a significant increase (320445 bytes) in the amount of embedded message data, there is a marginal decrease in the average SNR and PSNR values (4.0054 dB) in case of 3-layer embedding compared with 2-layer embedding of the proposed algorithm. Also, in comparison with the standard LSB substitution method, the proposed algorithm shows better SNR and PSNR values in case of both 2-layer and 3-layer embedding based on other common test parameters. Table 3 presents a comparative analysis on average SNR and PSNR

Table 3 Comparison of results obtained

Method	2-layer embedding		3-layer embedding	
	Average SNR (dB)	Average PSNR (dB)	Average SNR (dB)	Average PSNR (dB)
Proposed algorithm	64.5023	84.1713	60.4969	80.1659
Standard LSB method	57.3162	76.9852	55.1475	74.8165
Increase over standard LSB method	7.1861	7.1861	5.3494	5.3494

values obtained from the test results, which shows the efficiency of the proposed algorithm.

3.2 *Results from Subjective Test*

Opinions are collected from more than 20 listeners about their listening experience for 24 pairs of original and stego audio, experimented with the proposed algorithm. The final results of MOS, computed for each carrier audio and message type pair, are shown in Table 4.

The values in Table 4 exhibit the subjective quality of the audio signals where there is hardly any perceptible distortion experienced in the resultant stego audio signal in case of 2 layers and very little (for some combinations) or no perceptible distortion in case of 3 layers.

These results, achieved from various perspectives, prove the efficiency of the proposed algorithm for secure transmission of a sensitive message of any kind within

Table 4 Final MOS obtained using Subjective Test

Carrier audio	Message type	MOS Grade (2 Layers)	MOS Grade (3 Layers)
Instrumental	Text	5	5
Pop		5	5
Speech		5	5
Instrumental	Image	5	5
Pop		5	5
Speech		5	4
Instrumental	Audio	5	5
Pop		5	5
Speech		5	4

an audio signal, and this technique can be further extended for more than 3 layers of embedding as well.

4 Conclusion

The proposed algorithm is appropriate for transmitting high volume of sensitive message in any form (text, image or audio) using an audio signal as carrier. Instead of hiding message bits directly in the time domain, it uses frequency domain based on DFT to strengthen the security of the hidden message.

Also, multiple LSB layers within the transformed frequency coefficients are selected pseudo-randomly for embedding payload bits, making any unauthorized retrieval of the hidden message extremely difficult. This approach increases the hiding capacity of the cover audio signal as well. The extraction procedure is also very straightforward and blind. After thorough experimentation, the algorithm has been evaluated by different quality metrics and results reveal its effectiveness for covert transmission of sensitive data with a high bandwidth.

References

1. Bender, W., Gruhl, D., Morimoto, N., Lu, A.: Techniques for data hiding. IBM Syst. J. **35**(3 and 4), 313–336 (1996)
2. Zwicker, E., Fastl, H.: Psychoacoustics. Springer Verlag, Berlin (1990)
3. Cvejic, N.: Algorithms for audio watermarking and steganography. Oulu 2004, ISBN: 9514273842
4. Djebbar, F., Ayad, B., Meraim, K.A., Hamam, H.: Comparative study of digital audio steganography techniques. EURASIP J. Audio Speech Music Process. **2012**, 25 (2012)
5. Cvejic, N., Seppnen, T.: Increasing the Capacity of LSB-Based Audio Steganography, FIN-90014. University of Oulu, Finland (2002)
6. Pal, D., Goswami, A., Ghoshal, N.: Lossless audio steganography in spatial domain. In: International Conference on Frontiers of Intelligent Computing: Theory and Applications, AISC 199, pp. 575–582. Springer (2012). https://doi.org/10.1007/978-3-642-35314-765
7. Acevedo, A.: Digital watermarking for audio data in techniques and applications of digital watermarking and content protection. Artech House, USA (2003)
8. Arnold, M.: Audio watermarking: features, applications and algorithms. Proceedings of the IEEE International Conference on Multimedia and Expo, pp. 1013–1016, 2000
9. Ghoshal, N., Mandal, J.K.: Discrete Fourier Transform based multimedia color image authentication for wireless communication (DFTMCIAWC). In: 2nd International Conference on Wireless Communications, Vehicular Technology, Information Theory and Aerospace & Electronic Systems Technology, Wireless Vitae 2011, ISBN: 978-1-4577-0787-2/11, Le Royal Meridian Chennai, India, 28th Feb–3rd March, 2011
10. Zhang, Y., Lu, Z.M., Zhao, D.N.: Quantization based semi-fragile watermarking scheme for H.264 video. Inf. Technol. J. **9**(7), 1476–1482 (2010)
11. Foo, S.W., Ho, S.M., Ng, L.M.: Audio watermarking using time-frequency compression expansion. In: IEEE International Symposium on Circuits and Systems, pp. 201–204 (2004)
12. Dhar, P.K., Khan, M.I., Ahmad, S.: New DCT based watermarking method for copyright protection of digital audio. Int. J. Comput. Sci. Inf. Technol. (IJCSIT) **2**(5), 91–101 (2010)

13. Yang, S., Tan, W., Chen, Y., Ma, W.: Quantization-Based digital audio watermarking in discrete Fourier transform domain. J. Multimed **5**(2) (2010)
14. Al-Haj, A., Mohammad, A.A., Bata, L.: DWT–Based audio watermarking. Int. Arab J. Inf. Technol. **8**(3), 326–333 (2011)
15. Cox, I.J., Kilian, J., Leighton, F.T., Shamoon, T.: Secure spread spectrum watermarking for multimedia. IEEE Trans. Image Process. **6**(12), 1673–1687, 1997, 1997
16. Cvejic, N., Seppnen, T.: Digital audio watermarking techniques and technologies: applications and benchmarks, InformatIon Science Reference. Hershey, New York, 2008, p. 79, 239. ISBN 978-1-59904-515-3 (ebook)
17. Campbell, D., Jones, E., Glavin, M.: Audio quality assessment techniques-A review and recent developments. J. Signal Process. Elsevier, 0165-1684/2009 Elsevier B.V. https://doi.org/10.1016/j.sigpro.2009.02.015

Multi Data Driven Validation of E-Document Using Concern Authentic Multi-signature Combinations

Soumit Chowdhury, Sontu Mistry, Anirban Goswami, Dipankar Pal, and Nabin Ghoshal

Abstract The idea raises a novel data security protocol for ensuring reliable online affirmation of e-documents with stronger compliance of authentication, integrity confidentiality, and non-repudiation scenarios. Initially the parent authority shares four secret signatures to the authorized client and stores them on server database. The protocol is initiated by server with hiding of four parent body signatures on the e-document image with each one hosted on a separate region. Critically, this hiding is dictated by self-derived hash functions on client id-number for precisely locating the concern region hosting the particular applicable signature. Importantly, this idea is implemented through hash values tracing the start indexes for circular sequencing of regions as well as signatures both. Next, this certified e-document is endorsed at the client end with segmented concealment of each shared signature on each separate region. Interestingly, here also similar region based circular sequencing of signatures are adopted with self-designed hash operations on client name, client date-of-birth and server issued session random challenge. Finally, this authenticated e-document is ratified by server with validation of all embedded signatures through those same hash functions to establish data authenticity cum legality of both parties. Additional enhancement is also addressed by fabricating signature bits on distinctly transformed pixel byte elements of an image block and adopting standard deviation

S. Chowdhury (✉) · S. Mistry
Government College of Engineering & Ceramic Technology, Kolkata 700010, India
e-mail: sc_study123@yahoo.in

S. Mistry
e-mail: mistrysontu@gmail.com

A. Goswami · D. Pal
Techno India, Salt Lake City, Kolkata-700091, India
e-mail: angos.kol@gmail.com

D. Pal
e-mail: mail2dpal@yahoo.com

N. Ghoshal
Department of Engineering & Technological Studies, University of Kalyani, Kalyani, West Bengal 741235, India
e-mail: nabin_ghoshal@yahoo.co.in

D. Bhattacharjee et al. (eds.), *Proceedings of International Conference on Frontiers in Computing and Systems*, Advances in Intelligent Systems and Computing 1255,
https://doi.org/10.1007/978-981-15-7834-2_68

based block transformations to promote excellent robustness. Overall, this protocol clearly confirms superiority over other works both in terms of security principals and standardized evaluation of exhaustive simulation results from different angles.

Keywords E-Document validation · Dynamic authentication · Standard-Deviation based encoding · Variable encoding · Multi-Signature hiding

1 Introduction

Tremendous growth in digital data transmission has really stressed the importance of validating the transmitted digital documents to actually judge their credibility from different angles. In the current scenario this is achieved with secret embedding of copyright signatures on the e-document clearly conferring its ownership claims. However, such signatures has to be securely encoded so that their existence will not be revealed to the unauthorized parties and also they are not destroyed by external attacks. Importantly, this kind of approach is widely adopted for authenticating the e-documents related to e-governance systems along with bio-metric features of the client incumbent. In this aspect Hasan [1] has shown an idea where multi-signatures were securely dispersed on different color intensity components of the e-certificate in non-overlapping manners. Then Chowdhury [11] has recently tried a concept of validating the e-document online form both the issuer and incumbent perspectives. This work [11] has clearly complied all the major data security issues with hiding of signatures controlled by hash information computed on crucial sensitive data. Importantly most works in this domain have also tried such signature embedding concepts as Behnia [2] used a chaotic map idea for hiding different signatures on color intensity planes with encryption of both embedding positions and data bits. More secure bit encoding ideas are then used with transform domain concepts as Bhatnagar [3] coded multi-signature bits on segmented Discrete Cosine Transform (DCT) blocks by utilizing the threshold value of block energy. Further elevations continued in this area with Discrete Wavelet Transformed (DWT) based works as Babaei [5] shown segmented fabrication of multiple signatures on uniform blocks of wavelet. Then Natarajan [5] used LL2 sub-bands of DWT for multi-signature concealment while Thanki [6] targeted the HH3 and HH4 coefficients of wavelets for multi-signature castings. Further, Mohananthini [8] focused better performance for the segmented multi-signature dispersing with LL2 sub-bands of DWT. Next, sadh [9] has used LH1, LL2, LL3 sub-bands of DWT on different color intensity planes for multi-signature hiding while Hassan [1] has also adopted such similar technique in recent times. So, this whole survey clearly reflects that the transform domain data hiding concepts are widely used for better robustness. However, these works have rarely focused on all the critical data security issues related to both client-server data authentications and hence this proposed idea enhances them as-

1. Confirming legality of the e-document from both parent and client perspectives to confirm total authentications. This is done with region wise hiding of client

and server copyright signatures both on the e-document dispersed separately at both end.
2. Complying crucial data security issues as authenticity, integrity, confidentiality, and non-repudiations with a unique protocol for online validation of e-documents.
3. Employing a novel and improved data hiding technique with standard-deviation based pixel byte transforms for the sub image blocks. Also by variably concealing the copyright data on separate transformed pixel-byte elements of the image block with different threshold ranges and pixel-byte transforms vastly lifts the robustness.

Now to discuss these issues in detail, the paper is organized as follows—Sect. 2 presents the client-server authentication protocol, with data bit hiding and sensing algorithms. Next Sect. 3 shows experimental observations and their comparisons while conclusion is stated in Sect. 4, and the related references given at the end.

2 Proposed Methodology for Validation of the E-Document

At first the parent body shares four mutually secret signatures to the concern client and also stores them in its server database. To start of the server hides four of its copyright signatures on the e-document image with each one hosted on the first partition of each equal separate regions (Fig. 1). Vitally these signatures are hidden on the basis of hash values computed on the client identity number and this stamped e-document is then transmitted to the authorized client. Next, the client also hides four shared signatures on this certified e-document with each signature resides on the second partition of each equal separate regions (Fig. 1). Critically, this signature casting is also based on hash data computed on client-name, client date-of-birth and server given session random challenge. Ultimately, this authenticated e-document is ratified by server with detection of all concealed signatures by using those same hash functions and then validating them from the signatures stored in the database.

A. Proposed Data Authentication Protocol

Step1: Server hides four copyright signature images (S1,S2,S3,S4) on e-document (as Fig. 1) by utilizing dual hash data computed on client Identity Number (ID_NO). Here the first hash value RN$\epsilon\{1,2,3,4\}$ decides the starting region number while the second hash value SN$\epsilon\{1,2,3,4\}$ gives the starting signature index for region based circular embedding of signature sequences. This idea is reflected in Fig. 2 while the hash functions are given in Eqs. 1, 2 with r_i denoting the ith digit (i = 1 to n) of ID_NO.

$$RN = (ID\_NO + \sum_{i=1}^{n} r_i) Mod 4 + 1 \quad (1)$$

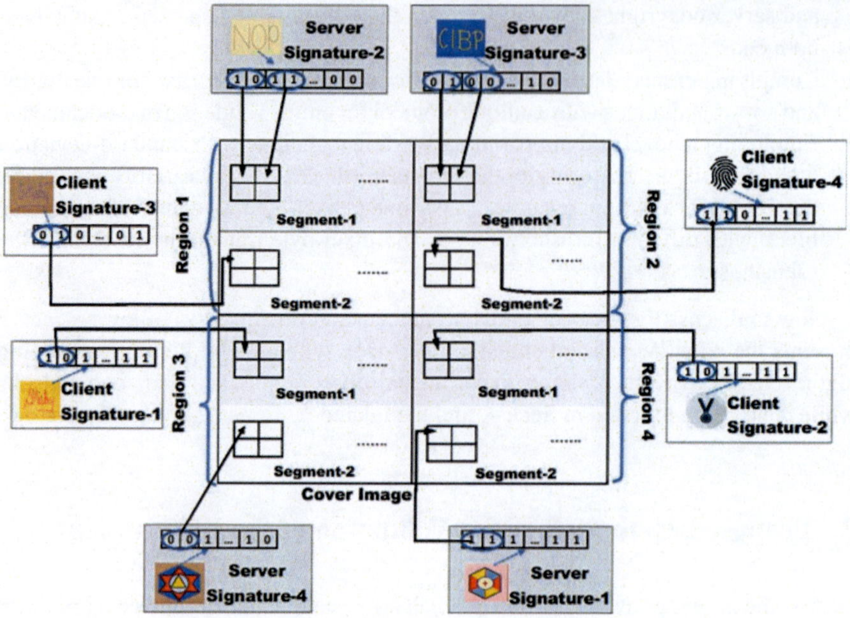

Fig. 1 Region wise embedding of server and client side signatures

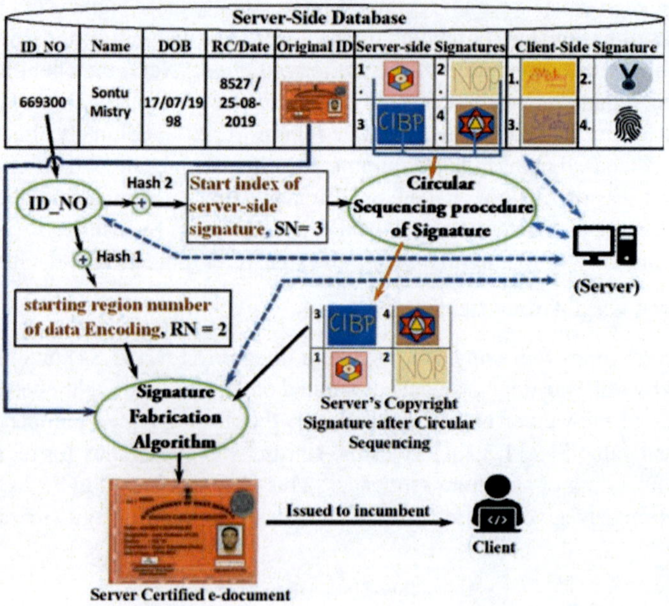

Fig. 2 Initial embedding of signatures at the server-side

$$SN = (\text{Reverse of } \text{ID}_{\text{NO}} + \sum_{i=1}^{n} r_i) Mod4 + 1 \quad (2)$$

Step2: This stamped document and session random challenge (RC) is sent to client.

Step3: Client now secretly disperses four shared signatures (C1,C2,C3,C4) on this e-document (as Fig. 1) and this is also dictated by dual hash data (H1,H2)ϵ{1,2,3,4} derived from client name and client date-of-birth. Here H1 tracks the starting region number found from hash operations on client name (taking ASCII of ith character of client name as a_i) and RC. Further, H2 identifies the starting signature index found with hash operations on client date of birth and RC. Hence, just like server similar region based circular embedding of signature sequences are adopted. This is shown in Fig. 3 while hash operations are given in Eqs. 3, 4 where d_i is the ith digit of client date of birth (as Int_Date) and its decimal value is (ddmmyyyy), with iϵ{1,2,..,n}.

$$\text{H1} = \left[\left(\sum_{i=1}^{n} a_i * 2^i\right) + \text{RC}\right] \text{Mod4} + 1 \quad (3)$$

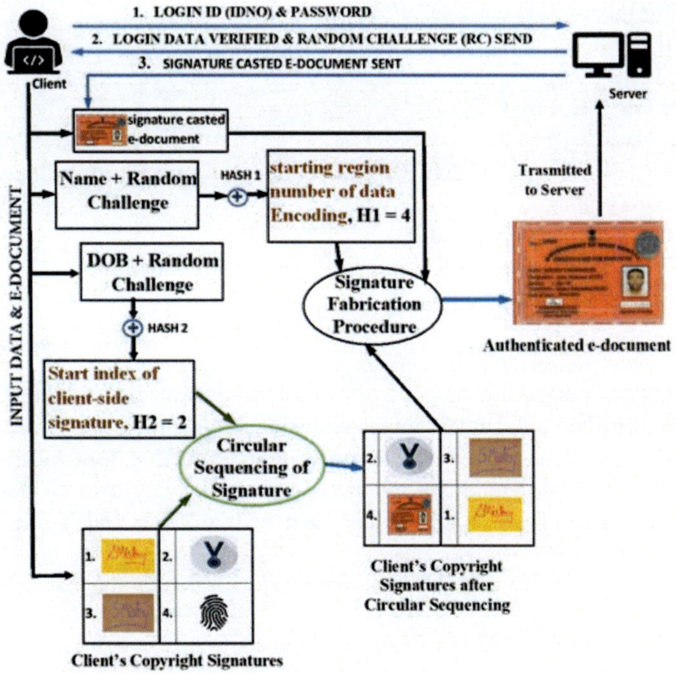

Fig. 3 Client-side signature fabrication process

$$H2 = (\text{Int\_Date} + RC + \sum_{i=1}^{n} d_i)\text{Mod}4 + 1 \tag{4}$$

Step 4: This authenticated e-document (IMG) is now sent to server for affirmation.

Step5: Server finds same hash values (RN,SN,H1,H2) by opting same Eqs. 1–4.

Step6: Server senses all signatures from each region of the received e-document by the following function F^{-1} and sorts them in valid orders based on the concern hash.

Step7: These sorted signatures are individually matched with their stored forms for ensuring the validation and is shown in Fig. 4 while this algorithm is stated below.

Let array ARR[P][Q] stores all the sensed signatures with Pϵ{1,2,3,4} and Qϵ{1,2} indicating the region number and segment number respectively. Next, the signatures in ARR[P][Q] are sorted according to the storage sequences based upon the concern hash value and is kept in array SARR[P][Q]. Now each signature of SARR[P][Q] is matched with its stored form in array STORE[P][Q] and if every match complies a certain threshold then only the e-document is validated. This idea is shown below with 'CNT' is the count of signature match and 'Sort' function sorts the signatures.

Start: CNT = 0;

$STORE[P][Q]_{P = 1 \text{ to } 4,\ Q = 1 \text{ to } 2} := [[S1, S2, S3, S4]\ [C1, C2, C3, C4]]$;

$ARR[P][Q]_{P= 1 \text{ to } 4,\ Q = 1 \text{ to } 2} = F^{-1}$ (IMG);

$SARR[P][Q]_{P = 1 \text{ to } 4,\ Q =1 \text{ to } 2} = Sort(ARR[P][Q]_{P=1 \text{ to } 4,\ Q = 1 \text{ to } 2}$, RN, SN, H1, H2);

Loop1: For P = 1 to 4

Loop2: For Q = 1 to 2

If (SARR [P][Q] = = STORE[P][Q]) CNT = CNT + 1; Else Continue;

End Loop1

End Loop2

If (CNT == 8) "Validation is Confirmed"; Else "Validation Failed";

End;

B. Signature Data Fabrication Algorithm

Input: Eight color signature images and one color e-document image of valid sizes.

Output: Certified e-document image with eight signature images hidden on it.

Method: e-document is splitted into consecutive sets of 2 × 2 sub block matrices of pixel bytes and two signature bits are encoded per pixel byte element of the matrix. **Step1:** Read 8 signature bits and next matrix M1 = $[X_1,X_2,X_3,X_4]$ with $X_i \epsilon \{0,1..,255\}$

Step2: Compute the Standard Deviation (SD) [10] value for the matrix block M1 as

$$SD = \sqrt{\frac{1}{N}\sum_{i=1}^{N}(X_i - X)^2} \tag{5}$$

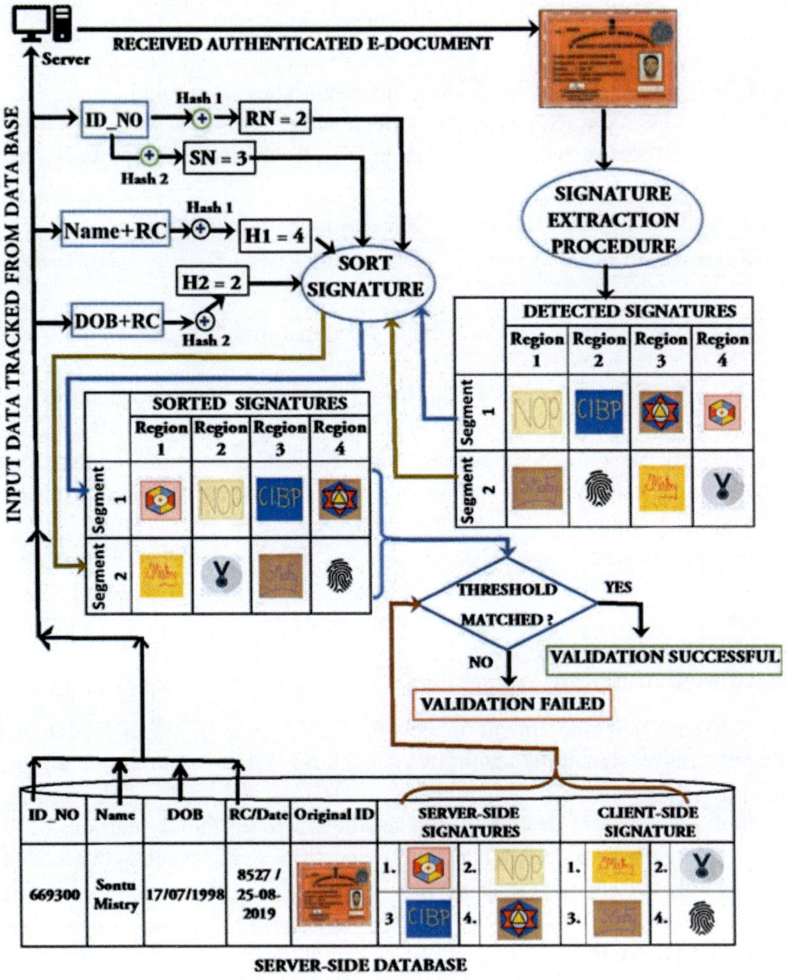

Fig. 4 Server-side signature validation process

where N = 4 denotes the total elements in M_1, and X represents their average value.

Step3: If last four decimal digits of M_1 is $D_4D_3D_2D_1$, then M_1 elements altered as-

$X_{C1} = X_1\text{-}(SD + D_1)$	$X_{C2} = X_2\text{-}(SD + D_1 + D_2)Mod4$
$X_{C3} = X_3\text{-}(SD + D_1 + D_2 + D_3)Mod6$	$X_{C4} = X_4\text{-}(SD + D_1 + D_2 + D_3 + D_4)Mod8$

Step4: Replace each X_{Ci} with the concern threshold range value X_{Bi} as tracked from Eq. 6 based on the concern input bit pairs $b_jb_{j+1}\epsilon(00,01,10,11)$, with $i\epsilon\{1,2,3,4\}$ as-

$$X_{Bi} = [UB(X_{Ci}) + LB(X_{Ci}) + 1]/2 \quad (6)$$

where $UB(X_{Ci}) = (X_{Ci} - X_{Ci} \text{ Mod } 16) + [\text{decimal}(b_j b_{j+1}) + 1]*4 - 1$
and $LB(X_{Ci}) = (X_{Ci} - X_{Ci} \text{ Mod } 16) + \text{decimal}(b_j b_{j+1})*4$

Step5: Transform each bit coded element X_{Bi}, as obtained in step5 to X_{Ti} as-

$X_{T1} = X_{B1} + D_1$	$X_{T2} = X_{B2} + (D_1 + D_2)\text{Mod}4$
$X_{T3} = X_{B3} + (D_1 + D_2 + D_3)\text{Mod}6$	$X_{T4} = X_{B4} + (D_1 + D_2 + D_3 + D_4)\text{Mod}8$

Step6: Compute SD_{new} by Eq. 5 for these transformed values X_{Ti} as found in step5.

Step7: Each X_{Ti} is converted to X_{Wi} for $i \in \{1,2,3,4\}$ to find the coded matrix M2 as $[X_{W1} = (X_{T1} + SD_{new}), X_{W2} = (X_{T2} + SD_{new}), X_{W3} = (X_{T3} + SD_{new}), X_{W4} = (X_{T4} + SD_{new})]$. Now if needed perform some delicate adjustments on transformed pixel bytes considering their bit coding ranges to keep this final bit coded pixel byte values in spatial ranges.

Step9: Repeat step1 to step7 for all the 2 × 2 sub-block matrices for a specific region.

Step10: Stop.

C. Signature Data Sensing Algorithm

Input: Authenticated e-document image with eight signature images hidden on it.

Output: Eight signatures, with two signatures sensed per region of the e-document.

Method: E-document is splitted into sets of 2 × 2 sub block matrices of pixel bytes and two embedded bits $b_i b_{i+1}$ are sensed from every pixel byte element of the matrix. **Step1:** Read next matrix $M2 = [X_{W1}, X_{W2}, X_{W3}, X_{W4}]$ and find its SD as SD_{new} by Eq. 5.

Step2: Transform the elements of M2 based on the digits of the matrix number as-

$X_{B1} = X_{W1} - (SD_{new} + D_1)$	$X_{B2} = X_{W2} - [SD_{new} + (D_1 + D_2)\text{Mod}4]$
$X_{B3} = X_{W3} - [SD_{new} + (D_1 + D_2 + D_3)\text{Mod}6]$	$X_{B4} = X_{W4} - [SD_{new} + (D_1 + D_2 + D_3 + D_4)\text{Mod}8]$

Step3: Find the value Y_i as $Y_i = (X_{Bi} \text{ Mod } 16)$ for all matrix elements $i \in \{1,2,3,4\}$.

Step4: If(Y_i >=0 AND Y_i < 4) Then $b_i b_{i+1}$=00; If(Y_i >=4 AND Y_i < 8) Then $b_i b_{i+1}$=01; If (Y_i >=8 AND Y_i < 12) Then $b_i b_{i+1}$=10; If (Y_i >=12 AND Y_i < 16) Then $b_i b_{i+1}$ = 11;

Step5: Repeat step1 to step4 for all the 2 × 2 sub-block matrices for a specific region.

Step6: Stop

3 Simulation Results & Comparisons

This scheme is tested with color e-document and signature images of size 512 × 512 and 24 × 24 respectively taken in ppm format where Table 1 reflects identical visual qualities for the signature coded e-documents. The data hiding potency is evaluated on Gimp (2.10), Irfanview (4.52) and MATLAB (R2018a) with Table 1 justifying superb hiding invisibility through Pick Signal to Noise Ratio (PSNR) over **38 dB** while Correlation Coefficient (CC) and Structural Similarity Index Measurement (SSIM) [6–8] values close to 1. This fact is also supported through various histogram analysis in Table 2 mostly focusing similar histograms for the signature dispersed e-document preventing statistical and visual attacks. Further, Table 3 also confirms

Table 1 Visual representation of signature hidden e-documents in contrast to the original form

e-document used	Original form	Modified form	SSIM	CC	PSNR (dB)
CARD-1	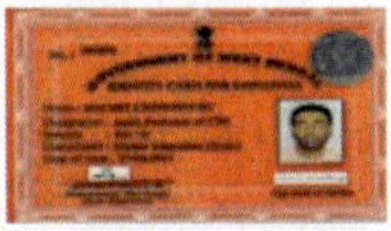		0.9897	0.9957	38.7810
CARD-2	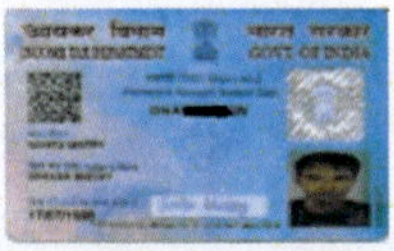	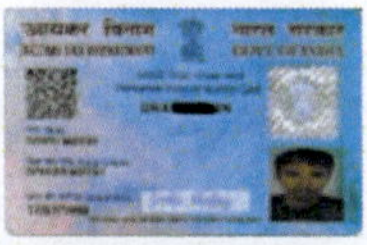	0.9612	0.9927	38.7346
CARD-3	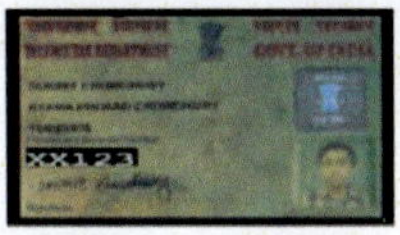	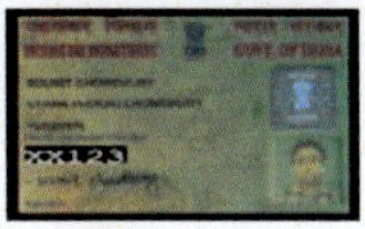	0.9714	0.9934	38.6537

Server side signatures		Client side signatures			

Table 2 General and RGB histogram analysis for signature embedded ID-CARD-1 image

Histogram type	Original histogram	Modified histogram
General	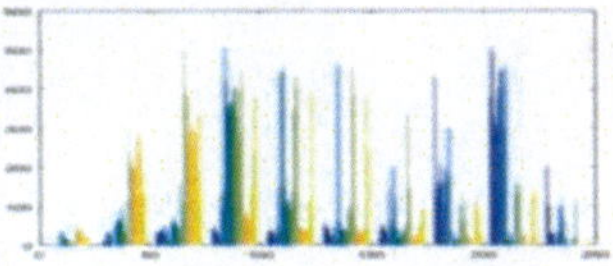	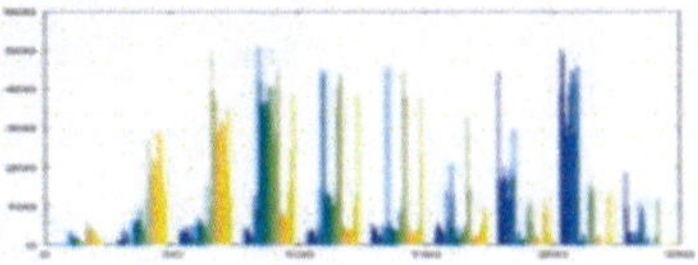
R-G-B domain		

Table 3 Comparison of data hiding invisibility scenario with other current schemes

Works/Schemes	Signature type	Data hidden	Hiding type	PSNR(dB)
Benia [2]	Grey scale image	6144 bytes	Segmented	30.11
Bhatnagar [3]	Grey scale image	5,120 bytes	Segmented	33.85 (max.)
Babai [4]	Binary image	4,096 bits	Segmented	28.44 (max.)
Thanki [6]	Grey scale image	320 bytes	Successive	30.79
Mohantini [8]	Color image	13,824 bytes	Segmented	38.06 (max.)
Sadh [9]	Binary image	8,192 bits	Segmented	38.9060
Chowdhury [12]	Color image	12,288 byte	Segmented	39.5428
This work (on ten test images)	**Color image**	**13,824 bytes**	**Segmented**	**38.75 (avg.)**

this better hiding invisibility with PSNR rise (mostly > 1%) over existing ideas under higher payload (mostly > 10%). Here number and size of signatures are optimized as per the protocol need to keep the hiding invisibility within a satisfactory level. Next, Table 4 proves enhanced robustness with CC value match for the four lesser quality signatures sensed from the attacked CARD-1 mostly better than other works. Then Fig. 5 proves at least one quality signature recovery against attacks of Table 4 and some other cases with highest CC value among all sensed signatures for any specific attack mostly >0.7. This resistance against several noise and filtering attacks is due to SD and threshold based data coding, but various file format alteration attacks like .ppm to .jpg, .ppm to .png, .ppm to .tif and finally back to.ppm can also be tackled.

Table 4 Signature recoveries against attack in comparison to the cCurrent works

Attack used	Tested in Works	Attack strength	CC of Sign-1	Four low Sign-2	Quality Sign-3	Signatures Sign-4
Salt & peeper noise	[5]	Density of 5%	0.76	0.42	–	–
	[8]	Density of 3%	0.8856	0.9978	–	–
	[12]	Density of 5%	0.9187	0.8495	0.7676	0.4953
	This work	**density of 5%**	**0.9475**	**0.9798**	**0.9769**	**0.9478**
Median filtering	[5]	Block of 3 × 3	0.97	0.93	–	–
	[7]	Block of 3 × 3	0.9011	0.9240	–	–
	This work	**Block of 13 × 13**	**1.0**	**0.9997**	**0.9998**	**0.9996**
Winner filtering	[8]	Block of 3 × 3	0.7679	0.8142	–	–
	This idea	**Block of 3 x 3**	**1.0**	**0.9996**	**0.9998**	**0.9996**
Crop	[5]	No data found	0.85	0.65	–	–
	[7]	No data found	0.6562	0.6851	–	–
	[8]	No data found	0.36	0.3374	–	–
	This work	**Removed 75%**	**1.0**	**1.0**	**1.0**	**0.9999**
Row(R), Column (C) blanking	[3]	20 Col, 20 Row.	0.9898	0.9876	0.7332	–
	[7]	No data found	0.6279	0.6866	–	–
	[8]	No data found	0.6705	0.6686	–	–
	This idea	**60 Col, 60 Row**	**0.9121**	**0.9999**	**0.9999**	**0.9999**
Translation	[5]	No data found	0.99	0.35	–	–
	[7]	No data found	0.8141	0.7055	–	–
	[8]	No data found	0.9359	0.9586	–	–
	This work	**[-0.4 and -0.4]**	**1.0**	**0.9999**	**0.9999**	**0.9999**
Smooth	[5]	No data found	1.0	0.98	–	–
	This work	**Density of 30%**	**1.0**	**0.9897**	**0.9288**	**0.8941**

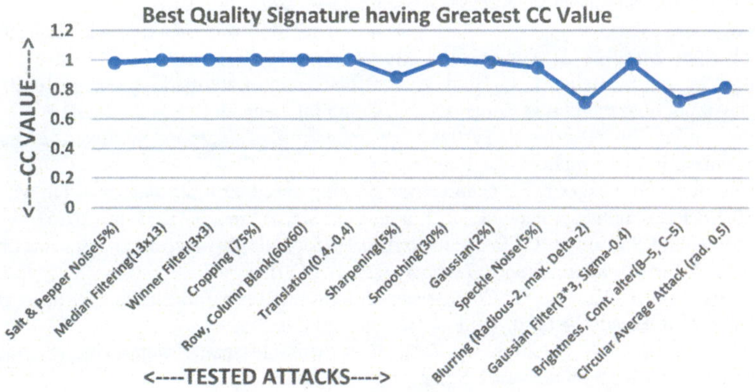

Fig. 5 Best quality of the sensed signature under individual attacks with highest CC value

4 Conclusion

This proposed idea raises a novel data authentication protocol for online validation of e-documents from different perspectives and the major research contributions are

1. Firmly meeting all the crucial issues of confidentiality, integrity, non-repudiation, and authentications in a better and stronger manner in contrast to the existing works.
2. Novel signature hiding scheme with cover image region wise circular ordering of signatures securely decided through the self-defined hash functions on critical data.
3. Imparting superior data hiding concept with standard deviation based pixel byte transforms for the sub image blocks and variable encoding of signature bits on them.

Finally, the implementation confirms better hiding imperceptibility with minimum 1% rise of PSNR over most of the other works with high data payload embedding. Further, enhanced robustness is materialized with improved recovery of signatures against wider ranges of attacks in contrast to the existing works with best recovered signature CC value mostly staying over 0.7. Hence, this scheme can suit e-document validations in e-governance systems especially under the wireless domain which clearly needs greater security and robustness. However, this raised idea can be more improved to resist various forms of image compression and geometrical attacks.

References

1. Hasan, H.R.: Copyright protection for digital certificate using blind watermarking technique. Kurd. J. Appl. Res. **3**(1), 75–79 (2018)
2. Behnia S., Teshnehlab M., Ayubi P.: Multiple watermarking scheme based on improved chaotic maps. Comm. Nonlinear Sc. Numerical Simul. **15**(9), 2469–78 (2010)
3. Bhatnagar G., Wu Q.M.J.: A new robust and efficient multiple watermarking scheme. Multimed. Tools Appl. **74**(19), 8421–8444 (2013). https://doi.org/10.1007/s11042-013-1681-8
4. Babaei M., Ng K., Babei H., G. Niknajeh H.: Robust multi watermarking scheme for multiple digital input images in DWT domain. Int. J. Comp Inf. Tech. **3**(4), 834–840 (2014)
5. Natarajan M., Govindarajan Y.: Performance comparison of single and multiple watermarking techniques. Int. J. Comput. Netw. Inf. Secur. **6**(7), 28–34 (2014)
6. Thanki R.M., Borisagar K.R.: Compressive sensing based multiple watermarking technique for biometric template protection. Int. J. Image Gr. Signal Proc. **7**(1), 53–60 (2015)
7. Mohananthini N., Yamuna G.: Image fusion process for multiple watermarking schemes against attacks. J. Net. Commun. Emerg. Technol. **1**(2), 1–8 (2015)
8. Mohananthini, N., Yamuna, G.: Comparison of multiple watermarking techniques using genetic algorithms. J. Electr. Syst. Inf. Technol. **3**(1), 68–80 (2016)
9. Sadh, R., Mishra, N., Sharma, S.: Dual plane Multiple spatial watermarking with self-encryption. Sadhana Indian Acad. Sci. **4**(1), 1–14 (2016)
10. Calculate Standard Deviation. (n.d.). Explorable, https://explorable.com/calculate-standard-deviation. Last accessed 30 Aug 2019

11. Chowdhury, S., Mistry, S., Ghoshal, N.: Multi-Phase digital authentication of e-certificate with secure concealment of multiple secret copyright signatures. Int. J. Innov. Technol. Explor. Eng. **8**(10), 3365–3380 (2019)
12. Chowdhury, S., Mukherjee, R., Ghoshal, N.: Dynamic authentication protocol using multiple signatures. Wirel. Personal Comm. **93**(3), 1–32 (2017). https://doi.org/10.1007/s11277-017-4066-x

A Survey Report on Underwater Acoustic Channel Estimation of MIMO-OFDM System

Avik Kumar Das and Ankita Pramanik

Abstract Underwater acoustic (UWA) channels are highly challenging in nature and from the past few decades, many researchers are investigating the efficient way of UWA channel estimation. Different approaches have been taken for UWA channel estimation like OFDM, Compressed Sensing, etc. In this paper, a survey is presented on various estimation techniques involved in the UWA channel. Work is presented in three different parts. The first part presents different challenges in the UWA channel. In the second part, works on MIMO-OFDM-based UWA channels are presented. In the last part, a survey on compressed sensing based channel estimation of the UWA-MIMO-OFDM system is presented. A performance comparison is given on the different channel estimation algorithms.

Keywords Underwater Acoustic (UWA) Communication · Channel Estimation (CE) · OFDM · MIMO · Compressed Sensing (CS)

1 Introduction

Recurrence-specific blurring in channel estimation caused by Extended multi-path, refractive properties of medium, low proliferation, speed of sound in water, unnatural transmission capacity, and frequent time-fluctuation, which makes Under Water Acoustic (UWA) channel very challenging for establishing communication [1].

In recent years, many researchers are finding different ways to estimate channel utilizing air as a medium in MIMO-OFDM system (Multiple Input Multiple Output-Orthogonal Frequency Division Multiplexing) [6]. Of late, OFDM [2] has developed as a promising option for UWA interchanges on account of its strength to channels

A. K. Das (✉) · A. Pramanik
Indian Institute of Engineering Science and Technology, Shibpur, Howrah, India
e-mail: avik.rs2018@telecom.iiests.ac.in

A. Pramanik
e-mail: ankita@telecom.iiests.ac.in

D. Bhattacharjee et al. (eds.), *Proceedings of International Conference on Frontiers in Computing and Systems*, Advances in Intelligent Systems and Computing 1255,
https://doi.org/10.1007/978-981-15-7834-2_69

that show long postpone spreads and recurrence selectivity. However, the application of OFDM to UWA channels [11] does not help much in solving the channel estimation problem.

Over the years, the field of communication is evolving faster and in a better way. Starting from point-to-point communication nowadays communication is being included in an underwater scenario. Communication scenario in underwater is different from conventional established communication system in terms of having time-varying multi-path propagation and scarcity of bandwidth that creates hindrance in achieving desirable throughput.

From decades many research is being carried out and a brief history gathered from different papers is presented here. In the survey report, first, there is a survey on the initial challenges and the initial researches on UWA channels [1, 3]. Due to the demand for higher rate and bandwidth, MIMO was introduced in UWA. MIMO is tagged with different techniques from the years for UWA communication applications. But the performance of the MIMO-OFDM is one of the best as compared with the previous works [14]. Later the massive MIMO is introduced as the number of users and the demand for speed increased [8]. Then the compressed sensing techniques are merged with the UWA-MIMO-OFDM as most of the UWA channel contains sparse [5]. Different compressed sensing algorithms are applied in many MIMO systems for channel estimations. Here few papers on few efficient compressed algorithms are discussed.

The objective of this survey papers are as follows:

1. Finding a better algorithm, UWA CE algorithm, for massive MIMO-based system.
2. Exploiting the optimum channel model and denoising technique by which better channel response in the UWA channel in a massive MIMO system can be obtained.
3. Sort out the best-compressed sensing algorithm for the better sparse recovery.

In Sect. 2, a literature survey on the existing UWA channel estimation algorithms is presented. The survey is divided into three main sections. The first part deals with the historical perspective of UWA channel estimation. System performance is surveyed in the second part of the survey. The third section presented the advantages of introducing compressed sensing in MIMO-OFDM system. The compressed sensing algorithm is very helpful because the UWA signals contain many sparse signals. A system model is presented of an MIMO-OFDM system using space-time block-code (STBC) code as its performance is improved in the channel estimation. Then in Sect. 3, a comparative table is presented where different techniques and features are compared from the year 2011 to 2019 in the UWA-MIMO-OFDM system. Then a comparative graph is shown for different algorithms of compressed sensing in the UWA channel estimation problem. Finally, the conclusions are provided in Sect. 4.

2 Literature Reviews

The following sections the few existing and efficient technique are discussed and compared to find out one of the best ways to estimate the UWA-MIMO-OFDM channel.

2.1 The Preliminary UWA Communication Channel Estimation and Introduction to MIMO-OFDM

Bandwidth limitation, extensive multi-path fading, massive Doppler shift, frequent time variation, refractive nature of the medium makes UWA channel very challenging for high-speed communication. The paper authored by Milica et al. [1] has explained the improved performance and the robust nature and compared it with the previous existing communication systems. Many UWA challenges, as mentioned earlier, are explained in this paper in a simple way. A theoretical flavor of spatial modulation is also explained here. According to [3], MAP and Maximum Likelihood Sequence Estimation (MLSE) are optimal detection techniques that are performed in the early stages and the result grows exponentially with the complexity associated with a number of antennas involved.

In the initial stages of UWA development, many experiments were done to observe the characteristics of MIMO channel. One was carried out in the Mediterranean Sea with 2 and 4 transmit projectors as mentioned in [4] and the other one involved 6 transmitting projectors to observe the performance of MIMO channel. The experimental results verified the advantages of MIMO over SISO. A complex homotopy method [24] has also been proposed for efficient channel estimation(CE) of the UWA channel.

2.2 UWA Channel Estimation of MIMO System Using OFDM Techniques

An adaptive system performance for the UWA-OFDM system is shown in [9] by changing the parameters. A wavelet denoising technique is used in the UWA-OFDM system for removing the noise [10]. To reduce the noise of the channel and also for efficiency, interpolation and diversity methods are used. To improve the overall performance of MIMO, OFDM system feedback system is introduced. Feedback blocks are also responsible to make certain decisions for the betterment of the performance [16]. UWA channels are very adverse in terms of their selectivity; they can be selective separately for time and frequency [15]. For this dual-selective nature, cyclic prefix (CP) is used to solve inter-block-interference. For the sparsity in the UWA channel, an OMP algorithm is also applied here in the channel estimation.

In this time-domain oversampled OFDM system citech6915, the two most applied time-varying channel models are described here [18]. One model is DPS-BEM and another one is DRM.

2.3 UWA Channel Estimation of MIMO-OFDM System Using Compressed Sensing Techniques

In recent times, compressed sensing is rapidly used for the high-speed UWA communication system design. By combining the MIMO system, fading problem was solved in a huge number of percentages. The system model of the MIMO-OFDM system presented here [7] has given a better channel performance graph. In the MSE performance comparison graph, we can see the OMP and CS-MP compressed sensing algorithm in the MIMO-OFDM system of 2 transmitters and 2 receiver systems have a very good performance. A comparative performance analysis is done is this paper [5] in LMSE MMSE and LS estimation in the MIMO-OFDM system. Rayleigh channel with 2 T_x and 2 R_x is taken here and the CoSaMP technique is applied.

Paper [18] presented the effect of Doppler frequency. For joint CE, the procedure can be improved by using an application of structured compressed sensing in the MIMO-OFDM system. Data block in OFDM system causes multi-path effects, which creates inter-block-interference (IBI) in conventional time-domain synchronous OFDM (TDS-OFDM) [8]. As a solution, [8] proposed CE based on time-frequency training OFDM (TFT-OFDM) for transmission in the OFDM system. Also, [23] says that wireless communication, LTE (Long-Term Evolution) can also be useful for UWA channel estimation and MIMO channel shall be estimated by compressed sensing method. Paper [22] introduced Distributed Compressed Sensing (DCS) for channel estimation. MIMO-OFDM system in association with space-time block coding technique yields the maximum performance [13]. DCS gave the best performance with interval 6. Orthogonal STBC or the Alamouti coding can be used for efficient CE in MIMO-OFDM [19].

2.4 An Optimum Model for the UWA-MIMO-OFDM System

The UWA channel is estimated in the receiver end from the data from the receiving antennas [12]. Transmitting and receiving antennas are used in association with transducer and hydrophone in the following block diagram (Fig. 1).

In the system model of the transmitter part, the message is first QPSK modulated. Then the output is fed to the STBC section. The STBC module then divides the QPSK measurement matrix into vectors for transmission. According to the MIMO system the STBC encoded data are divided to the number of Transmitting antennas.

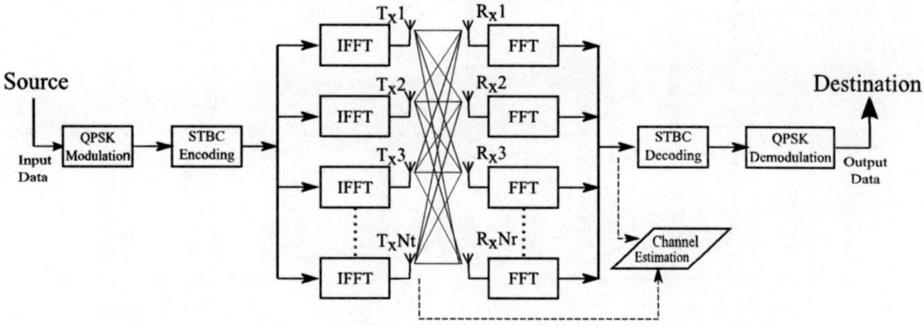

Fig. 1 STBC-OFDM system model for the UWA-MIMO-OFDM system [12]

If it was a 2x2 MIMO system then the two input sequence from STBC blocks will be following ST1 and ST2.

$$ST1 = [(t_1), (-t_2^*), (t_3), (-t_4^*), \ldots\ldots\ldots, (t_{N-1}), (t_N^*)]$$

$$ST2 = [(t_2), (t_1^*), (t_4), (-t_3^*), \ldots\ldots\ldots, (t_N), (t_{N-1}^*)]$$

The vectors are then transmitted by the transmitting antennas. The number of transmitter antennas T_x and receiver antennas R_x depend on the MIMO system. After the receiving is done by the R_x antennas, the channel estimation operation is performed. The received vectors with added noise from the channel are then estimated. The output is then fed to the STBC section and the QPSK demodulation for the recovery of the signal.

2.5 *Table of Comparison*

A table of comparison is represented in the following Table 1 for the different UWA-OFDM system. From the comparison of Table 1, it is clear that OFDM when applied to UWA-MIMO system, a better channel estimation is obtained. It also helps in noise reduction. The adaptive power allocation technique in the OFDM is also useful for the MIMO channel estimation in UWA communication channel.

3 Comparison of Optimum Existing Methods in MIMO-OFDM System

In the previous section, a brief literature survey of few selective papers were presented. In this section, the comparison of the performances of some existing meth-

Table 1 Table of comparison of different UWA-OFDM system

Sl no	Paper topic	Author and Reference	Publishing year	Technique used	Characteristics and features (channel, no of antennas)	Special features
1	OFDM with pilot aided CE	Huang and Lawrence [26]	2010	Channel impulse response, Bellhop Algorithm	Time-varying shallow water acoustic channels	Linear interpolation
2	A sparsity-aware approach for NBI estimation	Gomaa and Al-Dhahir [25]	2011	OFDM, CS, Sparsity, MIMO	T_x = 1,2 R_x= 2	Narrow-Band Interference (NBI)
3	CE in distributed CS using Kalman filter	Yun et al. [21]	2011	(LS) method, pure DCS (SOMP algorithm is used)	T_x = 2 R_x= 2	DCS-KF method to estimate the channels
4	UWA CE via complex homotopy	Qi et al. [24]	2012	Sparse recovery, homotopy	Gaussian	Complex homotopy
5	Channel equalization algorithm based on wavelet denoising and decision feedback in UWA-OFDM system	Zheng et al. [10]	2012	Channel equalization	Wavelet denoising and decision feedback	Interpolation and diversity method, decision feedback wavelet denoising
6	MIMO channel estimation based on distributed compressed sensing for LTE advanced	Xu et al. [23]	2013	Distributed compressed sensing	Interval= 6	Spatial channel model extension DCS estimation method in LTE-A MIMO-OFDM system
7	Time-domain oversampled OFDM communication in doubly selective UWA channels	Bo Peng et al. [11]	2015	OFDM, 16-QAM	Rayleigh Distribution Fs=48kHz, Fc= 12kHz, CP 50ms	Oversampling
8	Adaptive power allocation for noncooperative OFDM systems in UWA interference channels	Pottier et al. [22]	2016	Noncooperative OFDM Systems	T_x = 3 R_x = 3	Adaptive power allocation technique
9	Structured CS based time-frequency joint CE for MIMO-OFDM systems	Fan et al. [8]	2018	Structured compressed sensing, time-frequency, joint channel estimation	ITU-VB T_x = 4 R_x = 4	Time-frequency training OFDM (TFT-OFDM)
10	Massive MIMO-OFDM CE via distributed CS	Akbarpour-Kasgari and Ardebilipour [20]	2019	Compressed sensing, massive MIMO, OFDM	StFBP-based channel estimation algorithm	Stage-Wise forward–backward pursuit (StFBP)

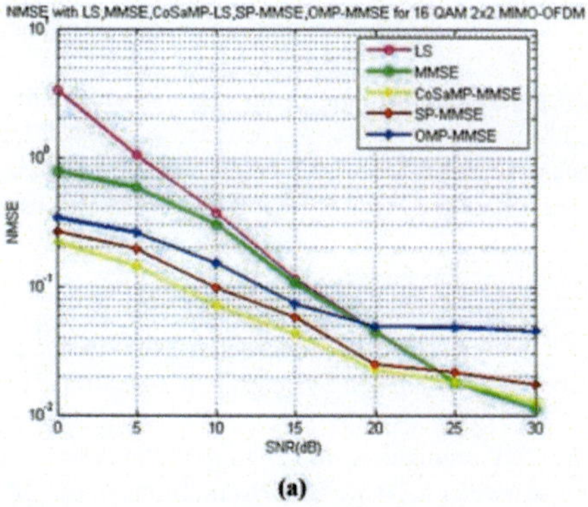

Fig. 2 Comparison between the different techniques of compressed sensing **a** NMSE versus SNR for CE in 2 × 2 MIMO-OFDM system using SP-LS, SP-MMSE [5], and **b** NMSE versus SNR CE in 2 × 2 MIMO-OFDM system using CoSaMP-LS and SP-LS [5]

ods in UWA-MIMO system is presented [25]. Following figures are the comparison between the different techniques of compressed sensing as shown in Fig. 2.

From the above comparative channel estimation graph, some ideas about the application of different compressed sensing algorithms like OMP, CoSaMP, StOMP in UWA channel estimation is obtained. The STBC codes using full MIMO system for sparse recovery and gives improved MMSE versus SNR performance.

4 Conclusion

From the given survey report, it can be concluded that the MIMO-OFDM system along with the application of compressed sensing algorithms yields better performance in the channel characteristics graphs in highly challenging UWA channel. With the advent of technology when the rapid data rate is required, channel estimation should be improved and faster. The STBC codes are going to help in the channel estimation field.

References

1. Mandar, R.C., Shiraz, S., Lee, A., Milica, S.: Recent Advances in underwater acoustic communications & networking, 978-1-4244-2620-1/08/IEEE
2. Li, B., Zhou, S., Stojanovic, M., Freitag, L., Willett, P.: Multicarrier communication over underwater acoustic channels with nonuniform Doppler shifts. IEEE J. Ocean. Eng. **33**(2), 198–209 (2008)
3. Catipovic, J.A.: Performance limitations in underwater acoustic telemetry. Oceanic Eng. IEEE J. **15**, 205–216 (1990)

4. Stojanovic, M.: Recent advances in high-speed underwater acoustic communications. Oceanic Eng. IEEE J. **21**, 125–136 (1996)
5. Jayanthi, P.N., Ravishankar, S.: Sparse channel estimation for mimo-ofdm systems using compressed sensing, 978-1-5090-0774-5/16/2016 IEEE
6. Li, G., Li, T., Xu, M., Zha, X., Xie, Y.: Sparse massive MIMO-OFDM channel estimation based on compressed sensing over frequency offset environment. EURASIP J. Adv. Signal Process (2019). Springer
7. Jianfeng, S., Xiaomin, Z., Yihai, L.: Channel estimation based on compressed sensing for high-speed underwater acoustic communication, 978-1-4799-5835-1/14/ 2014 IEEE
8. Yujie, F., Hui, L., Shuangshuang, S., Weisi, K., Wenjie, Z.: Structured compressed sensing-based time-frequency joint channel estimation For MIMO-OFDM Systems, 978-1-5386-3758-6/18/IEEE
9. Chude, O., Uche, A.K., Razali, N., Nunoo, S., Al-Samman, A.M., Tharek, A.R.: Adaptive transmission technique for short range mobile underwater acoustic OFDM communication. In: Globecom Workshops (GC Wkshps), 2013 IEEE, pp. 1361–1366. IEEE (2013)
10. Zheng, C., Tao, J., Gang, Q., Xuefei, M: Channel equalization algorithm based on wavelet de-noising and decision feedback in under water acoustic OFDM system. In: Wireless Communications, Networking and Mobile Computing (WiCOM), 2012 8th International Conference on, pp. 1–4. IEEE (2012)
11. Peng, B., Rossi, P.S., Dong, H., Kansanen, K.: Time-domain oversampled OFDM communication in doubly-selective underwater acoustic channels. IEEE Commun. Lett. **19**(6), 1081–1084
12. Xinmin, R., Min, Z., Jing, H., Xuguang, L., Zongwei, Y.: Estimation of Underwater Acoustic MIMO-OFDM Channel Based on Compressed Sensing, pp. 1–6. IEEE Underwater Technology (UT) (2019)
13. Berger, C., Zhou, S., Preisig, J., Willett, P.: Sparse channel estimation for multicarrier underwater acoustic communication: from subspace methods to compressed sensing. IEEE Trans. Signal Process. **58**(3), 1708–1721 (2010)
14. Chen, Z., Wang, J., Zhang, C., Song, J.: Time-domain oversampled receiver for OFDM in underwater acoustic communication. In: Proceedings IEEE VTC Fall, Sep. 2014, pp. 1–5
15. Wang, Z., Zhou, S., Giannakis, G., Berger, C., Huang, J.: Frequencydomain oversampling for zero-paddedOFDMin underwater acoustic communications. IEEE J. Ocean. Eng. **37**(1), 14–24 (2012)
16. Tao, J., Wu, J., Zheng, Y., Xiao, C.: Oversampled OFDM detector for MIMO underwater acoustic communications. In: Proceedings OCEANS, Sep. 2010, pp. 1–5
17. Wu, J., Zheng, Y.: Oversampled orthogonal frequency division multiplexing in doubly selective fading channels. IEEE Trans. Commun. **59**(3), 815–822 (2011)
18. Pottier, A., Socheleau, F.-X., Laot, C.: Adaptive power allocation for noncooperative OFDM systems in UWA interference channels. In: Underwater Communications and Networking Conference (UComms), 2016 IEEE Third, pp. 1-5. IEEE (2016)
19. Imad, S.A., Abdelhalim, Z., Fatma, N., Reem, I.: A proposed new schemes to reduce PAPR for STBC MIMO FBMC systems,Communications on Applied Electronics (CAE), ISSN : 2394-4714, **6**(9), April 2017
20. Akbarpour-Kasgari, A., Ardebilipour, M: Massive MIMO-OFDM channel estimation via distributed compressed sensing. Iranian J. Sci. Technol. Trans. Electr. Eng. **43**(1), 159–170 (2019). Springer
21. Yun, T., Wenbo, X., Zhiqiang, H., Baoyu, T., Donghao, W.: Channel estimation; distributed compressed sensing; Kalman filter. In: 2011 IEEE International Conference on Signal Processing, Communications and Computing (ICSPCC), 1–4. IEEE (2011)
22. Pottier, A., Socheleau, F.-X., Laot, C.: Adaptive Power Allocation for Noncooperative OFDM Systems in UWA Interference Channels. 978-1-5090-2696-8/16/2016 IEEE
23. Xu, L., Niu, K., He, Z., Xu, W., Zheng, Z.: MIMO channel estimation based on distributed compressed sensing For LTE-advanced, 978-1-4799-0434-1/13/2013 IEEE
24. Qi, C., Wu, L., Wang, X.: Underwater acoustic channel estimation via complex homotopy, 978-1-4577-2053-6/12/2012 IEEE

25. Gomaa, A., Al-Dhahir, N.: A sparsity-aware approach for NBI estimation in MIMO-OFDM, 1536-1276/11, 2011 IEEE. Trans. Wirel. Commun. **10**(6), June 2011
26. OFDM with Pilot Aided Channel Estimation for Time-Varying Shallow Water Acoustic Channels, "OFDM with Pilot Aided Channel Estimation for Time-Varying Shallow Water Acoustic Channels", 978-0-7695-3989-8/10, 2010 IEEE

Vulnerability of Cloud: Analysis of XML Signature Wrapping Attack and Countermeasures

Subrata Modak, Koushik Majumder, and Debashis De

Abstract Simple Object Access Protocol (SOAP) uses text-supported eXtensible Markup Language (XML) messaging configuration to interchange web encrypted information over a multitude of internet protocols such as SMTP, FTP, HTTP, etc. SOAP is one of the backbone solutions to exchange data between cloud and cloud service users via the Internet. The wide use of decentralized distributed cloud computing systems and rapidly increasing demand for cloud services result in a substantial increase in security vulnerabilities in the cloud. A comprehensive analysis of signature wrapping attack of XML Data and it's countermeasures to detect and prevent this web security threats have been analyzed in this paper. In the context of security vulnerabilities, we have discussed the use of Xpath expression, ID referencing, relative XPath referencing, absolute XPath referencing, fastXPath Structure-based referencing, and their weaknesses.

Keywords XML wrapping attack · SOAP · Cloud · XPath expression · ID referencing · Relative XPath referencing · Absolute XPath referencing · FastXPath

1 Introduction

With the increased adoption of IoT, security vulnerabilities in the cloud environment are the biggest challenges in cloud computing. Applications such as Software-as-a-Service are impossible without Web Services. In distributed and grid-based environment, the most common security that is implemented is Secure Socket Layer (SSL) or Transport Secure Layer (TSL). It is used for point to point security in data communication. Security does not only rely on transport between each hop—it also needs to secure the message at the final destination point. Since SSL and TSL are used as

S. Modak (✉) · K. Majumder · D. De
Department of Computer Science and Engineering, Maulana Abul Kalam Azad University of Technology, Kolkata, West Bengal, India
e-mail: subratamodak.cs@gmail.com

K. Majumder
e-mail: koushik@ieee.org

D. Bhattacharjee et al. (eds.), *Proceedings of International Conference on Frontiers in Computing and Systems*, Advances in Intelligent Systems and Computing 1255,
https://doi.org/10.1007/978-981-15-7834-2_70

lower level security protocols, it does not fulfill the security requirements of Web Services in distributed and grid-based environment. For implementing data communication among cloud users, SOAP-based web service is a very adaptable software system setup, compatible in diverse environments, and is a propelling technology for commercial cloud applications. One of the better examples is Amazon's Elastic Compute Cloud (EC2) that uses web services to build infrastructure and storehouse as services for cloud service users. Web Services intrinsically have their own security issues. One of the most critical security issues is XML signature element wrapping attack [1]. XML signature ensures reliability, integrity, and message authentication [2].

2 Foundation of XML Signature Wrapping Attack

2.1 *SOAP—Simple Object Access Protocol*

SOAP is intended to use structured XML-based messaging framework across the Internet. The protocol is used to exchange encoded data using a variety of protocols such as HTTP, SMTP, etc., in a decentralized, distributed environment.

2.2 *XML Signature Wrapping Attacks*

The aim of the attacker is to inject malicious or fake data into signed XML documents in a way that it cannot be detected. It also ensures that authenticity and integrity are still verified by application processing logic. As a result, the attacker injects malicious code or fake data that can execute arbitrary code in web service applications [3].

Figure 1. shows the structure of the SOAP message that is created in a right way and signed by the sender. WS-Security is applied to authenticated SOAP message Departure time and manifest identifier "msf1". XML signature is used to protect all <manifest> and <Departure> elements by hashing and digitally signing. The <manifest> element is to be protected by attribute reference Id="msf1" and <Departure> element by the XPath Expression. In wrapping attacks, the <manifest> element tag is changed from its original position in the SOAP Header that is unknown to the web application logic [3]. As a result of this, web service application never processes the wrapped element. An example is shown in Fig. 2. SOAP Header has added <wrapper> element tag and its child <Manifest Id="msf1"> without invalidating the signature digest value. Original body position new <Manifest Id="newmsf1"> has been added logically having the right structure but with completely different content.

Colored structure which is processed and validated by the signature verification technique. Application ideology for "Sid = S1" is to contain `/Envelope/Body/Manifest [@Sid="s1"]` and other `/Envelope`

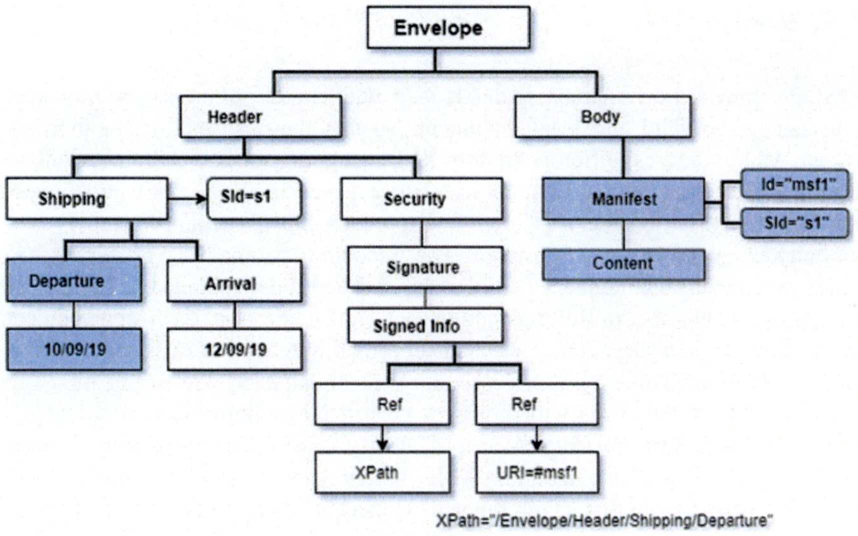

Fig. 1 SOAP element with XML digital signature

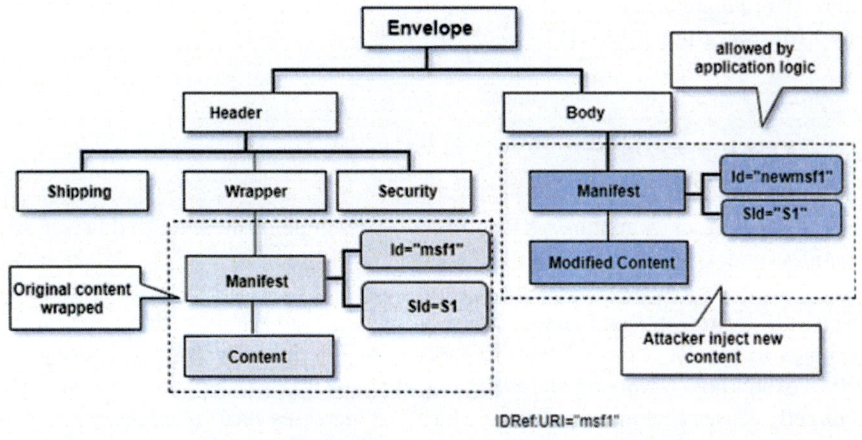

Fig. 2 Valid signature IDRef but modified SOAP element

`/Header/Shipping [@Sid="s1"] /Departure`. It ignores namespace and prefixes in the text. In signature verification steps, it identifies with the Id="msf1". As the sender's content of the message is not altered, therefore, it can be concluded that the signature is still valid [3]. Moreover, it processes the `<manifest Id="newmsf1">`, which is badly injected by an attacker and it can execute arbitrarily web request with arbitrary parameters without the user's permission.

2.3 Related Work

XML documents are vulnerable to detect fake modifications of messages. Attackers can change the valid and legal documents so that they can attain unauthorized access. Authors address different kinds of XML wrapping attacks—"Simple Context Optional Element, Optional Element in security header (sibling value) and Namespace injection (Sibling order)" [4]. In Inline approach, SOAP account concept is used for quick diagnosis of wrapping attack. The Account has some basic properties—a. Envelope contains the number of child nodes, b. The header contains the number of child nodes, c. Number of Reference Id in each signed object, d. Each signed object has predecessor and successor. If the attacker provides his own SOAP account to the victim then immediately the SOAP account is invalidated [5]. The verification key of the signature should be provided by a trust certificate authority such as X.509 [6]. Streaming-based Web Services Security Gateway (WeSSeGa) mechanism is used to validate SOAP elements including XML Signatures with XPath transformations. XPath Transformation uses XPath Filter. This mechanism also helps to enhance the performance of XPath function, and signed subtree [7]. A solution XSpRES that makes sure that the XML signatures are created and validated by using the signature library. The authors have also introduced an architecture, a hardened XML Signature Library for obtaining evaluation results that provide the targeted robustness against any XML Signature attacks [8].

2.4 XPath Filtering

Transformation of Signature, XPath filtering is implemented to find a set of complex nodes. The expression is Boolean in nature. In Fig. 1, the expression: "`/Envelope/Header/Shipping/Departure`" produces output with the whole document. Due to the complex issues of XPath Filtering, firstly, it is very complex to specify a right XPath Expression in filtering. Secondly, according to XPath expression, traversing and evaluating of all the nodes is very time-consuming. It directly causes performance issues. Thirdly, it increases security issues.

2.5 XPath Filter 2

Standard XPath Filter2 transformation is applied to overcome XPath Filtering issues and also used to reduce performance issues. This expression is used to define and sign XML documents. XPath Filtering produces Boolean results for all nodes based on evaluation. As XPath Filter 2 uses standard XPath expression, it permits sequences of expressions to find the set of nodes. And then set of nodes are combined by applying set operations—intersection, union, and subtraction. Since this Filter permits XPath

Expression arbitrarily, therefore, it becomes difficult to specify a correct XPath Filter Expression.

2.6 XML Canonicalization

In Web Service, XML canonicalization is used to transform XML documents into canonical form by applying a set of rules. One rule is syntax-based rule and another is context-based rule. Normalization of attribute value, line breaking, empty element—mainly this type of conversion is belonging in syntax-based Rule.

3 XML Signature Wrapping Scenario

In signature wrapping attacks, information of location is a very important part of XML Signature's Semantics. Id attribute is used to check signed content to detect whether the signature is valid or not. If the hash value of data which is referred by id is the same as the <Digest value> element, it is a valid signature otherwise the documents are located in the base documents. XPath is emphasized on the location of signed content. It is fully dependent on XPath Expression. After analyzing different XPath cases, every signed data is referenced by a particular Id.

3.1 Use of Identifier Referencing

Identifier-related referencing is mostly used for its simplicity. Every ID reference always has its alternative XPath Filtering 2 and XPath Filter expression. For example, the Id="msf1" it is found either using ID via URI="msf1" and with the XPath filter ancestor-or-self:: node()[@Id="msf1"] or with XPath Filter 2 (Fig. 3).

Signed element location is not a factor, it can present any place in the search domain. To perform XML wrapping attack, the position of the signed data is changed in the SOAP document in order to make sure that it can be easily validated by

Fig. 3 Id referencing protected subtree [3]

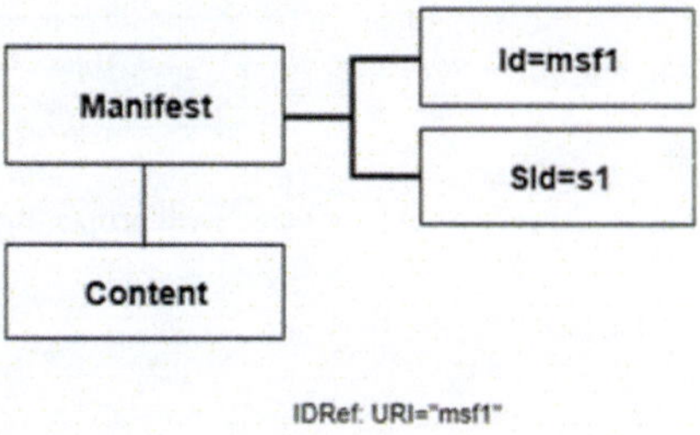

the verification logic. Only one protected subtree and one hash value are present which are identical. In signature wrapping attacks, ID reference URI="msf1". The semantics of XML signature are not violated because the same protected subtree and hash value are present.

3.2 XPath Referencing with Self–or- Descendant

In XPath, "/descendant-or-self: node ()" selects all the descendants and nodes. The expression: "/Envelope/Header/Shipping/Departure" is used to set the signed data in the XML message in Fig. 1.

In the XPath Filter2, expression is as follows: Ancestor_or_self:: Departure parent :: node() =/Envelope/Header/Shipping /Departure. The expression finds the <Departure> tag whose root element is <Envelope> and parent node is <Shipping>. The tag <Shipping> is under <Header> tag. Wrapping attacks are successfully performed if an attacker changes the actual element with the false one, and also shifts the original <Shipping> tag to a different location. That modification is under the root <Envelope>. Therefore, the modification is still unknown and it is undetectable because the logic of verification only verifies the hash value of the original element. In Fig. 4, message is altered but signature is not changed by the attacker (Fig. 5).

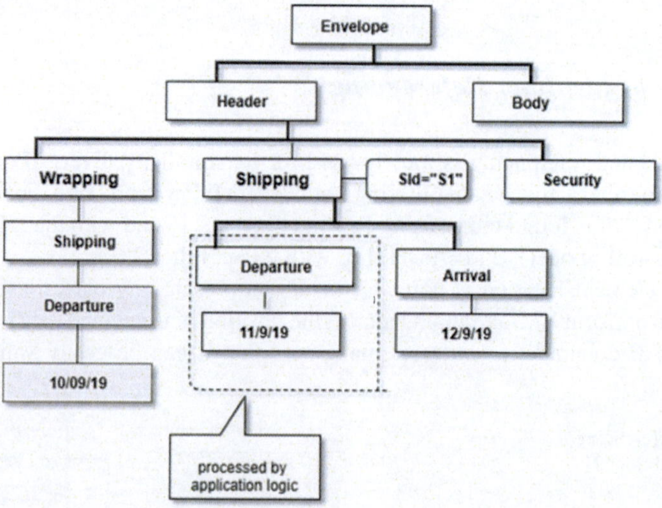

Fig. 4 Altered SOAP message with proper signature

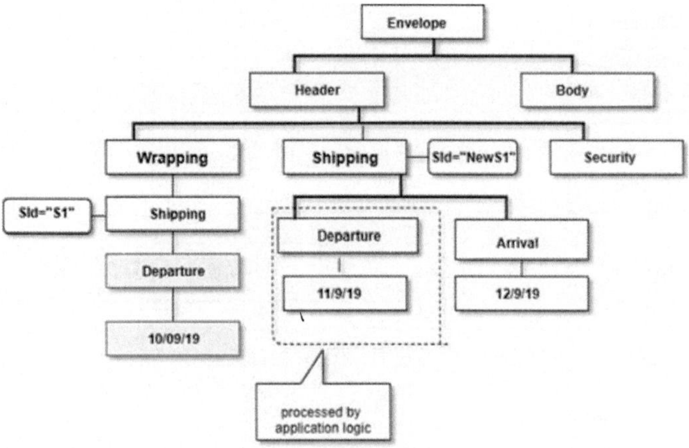

Fig. 5 Changed SOAP Element with proper Digital Signature, XPath: As attribute predicate [3]

3.3 Use of Relative XPath as a Reference but with XPath Filtering

Sign elements along with its ancestors are required in some applications. In such a case, XPath Filtering may be used. If someone wants to sign the <Departure> element in XML documents and whose parent node is <Shipping Sid = " S1"> the XPath Filter Expression: `Ancestor-or self:: Departure /parent :: Shipping [@Sid="S1"] /parent :: Header/parent::Envelope`. But the filter expression is difficult to define this type of relative reference in XPath filtering 2 because this filter is used to filter all reference documents. In such case leading is inserted implicitly to the relative XPath expression. In XML signature wrapping attack, an attacker can change the actual <Shipping> tag by injecting the malicious new element without SId="S1", and shift the original element to a different unknown place. In the previous example, modified message and original message both have the same protected subtree as well as hash subtree. Therefore, semantic of Signature is not different.

3.4 Use of Absolute XPath as a Referencing but Without Descendant-or-Self

Absolute XPath referencing selects the root node of XML documents. In XPath Filter 2, the root node is known as a context node. XPath Expression is as follows: i. "`/Envelope/Header/Shipping/Departure`" or its equivalent: "`ancestor-or-self:: Departure[parent::node() = /Envelope /Header/Shipping]`", ii."`/Envelope/Header/Shipping[@Sid="S1"]`

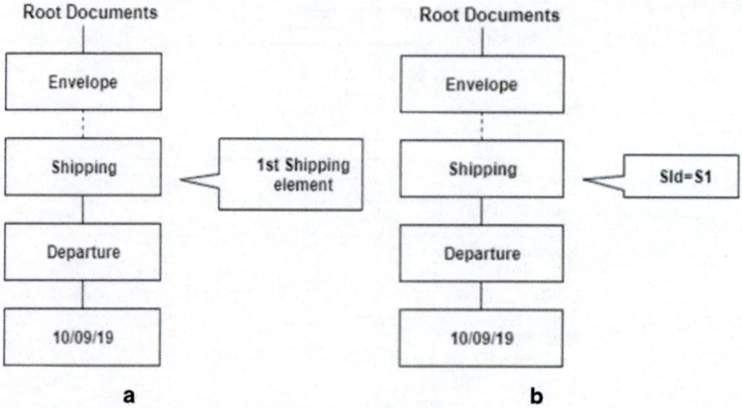

Fig. 6 **a** Absolute XPath with location predicate. **b** Attribute Predicate Protected Subtree

or its equivalent: "`ancestor-or self:: node() = /Envelope/Header /Shipping [@SID ="s1" ]"` (Fig. 6).

Every step of the vertical position of the XML document is fixed. Here the XML document is digitally signed. SOAP-based message has some limitations for unique header element. So the position of horizontal <Shipping> element is fixed. The results of XPath expression 2 are the same excluding the position of horizontal element <Shipping> which is fixed by SID="s1". Application logic executes one of the rules: i. <Departure> whose parent node is <Shipping> element, and is under /Envelope/Header/. ii. <Departure> whose parent node is <Shipping> element (reference ID attribute is Sid="s1"), under tag/Envelope/Header. If rule no i and rule ii both are used, then XML signature is not vulnerable to XML wrapping attacks.

3.5 *FastXPath*

FastXPath approach is used to fix the position of the vertical signed element by applying absolute XPath without beginning from the root node. For any relocation of a signed element, it invalided the signature. It is difficult to execute a signature wrapping attack. In the earlier, the approach of XPath expression is very low performance due to process of XPath Transform. It is very important to parse the XML Document elements into DOM Tree. This representation increases the overhead in evaluating the XPath expression of each element of DOM tree representation. Application of this kind expression reduces the performance for using XPath Transform in WS (Fig. 7).

Fig. 7 Structure of FastXPath [3]

```
"{  FastXPath ::='/' RelativeFastXPath
    RealtiveFastXPath::=Step
                            | RealtiveFastXPath '/' Step
    Step::= QName PredicatePosition?
    PradicatePosition::= Position Predicate?
    Position::=`[`n[1-9][0-9]*`]`
    Predicate::=`[` PredicateExpr`]`
    PredicateExpr ::=PredicateStep
          | PredicateExpr `and` PredicateStep
    PredicateStep::='@' QName '=' Lieral
    Literal ::="`"[^`]*"`"
              | "'"[^`]*"'"    }"
```

4 Comparative Study of Different Countermeasures Weakness of Wrapping Attack

To deal with Web Security vulnerability of XML signature wrapping attack, server-side security policies are required to prevent the attack [9]. Policies are—i. The security header must contain Signature. ii. The element is defined by SOAP Envelope and SOAP Body have to be referenced from a signature. iii. A Key must be issued by a trusted CA like X.509 certificate. That key is used to verify Signature, iv. Should not contain this type of tag -"/soap: Envelope/soap: Header/wsa:#ReplyTo", v. Proper use of absolute XPath expression [10]. But Policies are not enough and it is still unable to detect unwanted modification of XML data by the Attacks. Timestamp must be referenced by extra XPath Expression that creates additional issues. The use of XPath reference is not recommended by basic security protocols [6, 11]. To detect XML signature wrapping attacks, policy verification and policy advisor are used to test and generate security policies. The advisor policy has introduced the following security assertions—i. compulsory element: "wsa: To, wsa: Action, Soap: Body". ii. Signed element:"wsa: MessageID, wsu: Timestamp". iii. Authentication purpose recommendation: X.509 certificate issued by a trusted CA [12]. Inline Mechanism [4] to find the wrapping attack, SOAP element is inserted into a well-structured SOAP account. This SOAP account is signed by the sender and then it is also inserted into SOAP Header. The Account has some basic properties—i. Envelope contains the number of child elements, ii. Header contains the number of child elements, iii. Number of Reference Id, iv. Each signed object has predecessor and successor. If an attacker wants to inject some message, he or she has to change the structure of the message. After that number of a child node of SOAP has to be modified from one to many. This Inline mechanism has the ability to detect unwanted modification by checking the number of child elements. It has some disadvantages. Verification of child element of well-known SOAP account is not standard according to WS policy. In Inline approach, Solutions are—a. insert an element that has a detailed information of signed elements. Find out the parent node and insert "id "attribute into it; c. Other information is kept under the parent node (Table 1).

Still, it is vulnerable because it is possible to move the signed into other places but the depth of the structure is not changed. Another disadvantage is SOAP Account information protects the bonding among parent node and child node. It is possible to

Table 1 Comparative analysis of prevention methods of wrapping attack

Methods	Weakness and solution
WS Policy	**Weakness**: Weak policies do not fulfill the requirement to detect security attacks. Complex security policy was not in XML syntax. **Solution:** Effective security policies
Inline Approach	**Weakness**: **1. The** relation between the parent node and child node is preserved by using SOAP account. But it cannot fix the location of SOAP element information. **2.** One can easily move the signed message to another place so that the depth of the structure is not fixed. **3**. The verification of SOAP account element is not standard. **Solution**: Absolute XPath that ensures the position of the signed documents is fixed
XPath Referencing with successor-or itself	**Weakness**: Wrapping attacks are successfully performed if an attacker shifts the original message to a different location. That modification happens under the root <Envelope>, therefore the modification is still unknown to us and it cannot be detected because the logic of verification only verifies the hash value of the original data **Solution**: Use of Absolute XPath, Relative XPath, and also Fast XPath

change the location of the element to a new position it cannot be fixed the location change of the SOAP element. So wrapping attack is still possible. [4], their strong recommendation is to use Absolute XPath which is vertical fixing and horizontal fixing. It ensures that moving of the signed element of the SOAP message to a new place is not feasible with a valid signature.

5 Conclusions and Future Scope

SOAP-based XML is playing a very important role to handle sensitive information in cloud computing applications. Many security-related applications are already being used in the web service technology to handle XML Signature wrapping attack and XML Rewrite attack. To handle the man-in-the-middle attack is really tough. Mostly security is needed to process and store sensitive data in the cloud. Recent research shows that XML security specifications including message flow do not fulfill the requirement. Research activities are focused on XML schema handling process. In a real-world scenario, thousands of XML-based SOA messages are exchanged on a daily basis in the cloud. Verifying and processing every SOAP account header is a costly and time-consuming process. In future work, emphasis should be on time- and cost-based models. Many possible solutions have been proposed by researchers. After the analysis of mechanisms such as XPath expression, FastXPath, Inline approach, it is clear that the signed element is still under the wrapping attack. So, before accepting XML data, the need for proper verification at server side as well as client side is very important.

References

1. Gruschka, N., Lo Iacono, L.: Vulnerable cloud: SOAP message security validation revisited. In: 2009 IEEE International Conference on Web Services, Los Angeles (2009)
2. Gruschka, N., Jensen, M., Lo Iacono, L., Luttenberger, N.: Server-side streaming processing of ws-security. IEEE T. Serv. Comput. **4**, 272–285 (2011)
3. Gajek, S., Jensen, M., Liao, L., Schwenk, J.: Analysis of signature wrapping attacks and countermeasures. In: ICWS 2009. IEEE International Conference on Web Services, Bochum (2009)
4. McIntosh, M., Austel, P.: XML Signature Element Wrapping Attacks and Countermeasures. IBM Research Report, NewYork (2005)
5. Rahaman, M.A., Schaad, A., Rits, M.: Towards secure SOAP message exchange in a SOA. In: 3rd ACM workshop on Secure web services, New York (2006)
6. Kouchaksaraei, H.R., Chefranov, A.G.: Countering wrapping attack on XML signature in SOAP message for cloud computing. arXiv preprint arXiv:1310.0441 (2013)
7. Somorovsky, J., Jenson, M., Schwenk, J.: Streaming based verification of XML signature in SOAP messages. In: Proceedings of the 2010 6th World Congress on Services, SERVICES 2010, pp. 637–644. IEEE Computer Society, Washington, DC
8. Mainka, C., Jensen, M., Lo Iacono, L., Schwenk, J., Ivanov, I., Sinderen, M., Leymann, F., Shan, T.: Making XML signatures immune to XML signature wrapping attacks. In: Cloud Computing and Services Science, Springer International Publishing, vol. 367, pp. 151–167 (2013)
9. Gruschka, N., Lo Iacono, L.: Vulnerable: SOAP message security validation revised. In: ICWS 2009, Processing of the IEEE International Conference on Web Services. IEEE, Los Angeles (2009)
10. Gajek, S., Liao, L., Schwenk, J.: Breaking and fixing the inline approach. In: SWS '07: Proceedings of the 2007 ACM workshop on secure web services, pp. 37–43. ACM, New York, NY, USA (2007)
11. Bhargavan, K., Fournet, C., Gordon, A.D.: A semantic for Web Services authentication. Theor. Comput. Sci. **340**(1), 102–153 (2005)
12. Jensen, M., Schwenk, J.O., Gruschka, N., Iacono, L.L.: On technical security issues in cloud computing. In: IEEE International Conference on Cloud Computing (CLOUD-II 2009), pp. 109–116 (2009)

VLSI Track

Verification of Truth Table Minimization Using Min Term Generation Algorithm Weighted Sum Method

Rohit Kumar Baranwal, Debapriyo Saurav Mazumdar, Niladri Pramanik, and Jishan Mehedi

Abstract A conventional approach for preparing truth table(TT) is from Boolean logic function which is then demonstrated in the form of the sum of min terms corresponding to the rows in which it appears. This paper uses the idea of reduction of truth table forming new truth table in which one of the input of expression is reduced and the final expression is described using the exhausted input. The same is continued till all the inputs are used up and the final expression is generated. This paper proposed to verify the reduction of expression obtained from truth table with modified minterm generation algorithm with lesser difficulty A reduced truth table is proposed to avoid bigger size of the truth table for all circuits using multiple inputs, which is proposed in [6] and There after this paper discusses about the verification of the final expression using weighted sum method to check the veracity of truth table minimization, over all the truth table given with same input and outputs described in any form Min term algorithm uses the idea of input system which is could of any number is amounted in a matrix form and is recounted in SOP form. This paper shows the constructive and better logical understanding about truth table minimization and generation of min term which can be used any large number of inputs in an easier way.

Keywords Minimal logic expression · Multi input system · Truth table minimization · Digital system · Boolean function · Weighted sum algorithm

R. K. Baranwal (✉) · D. S. Mazumdar · N. Pramanik · J. Mehedi
Department of Electronics and Communication Engineering, Jalpaiguri Government Engineering College, Jalpaiguri, West Bengal, India
e-mail: rohitbaranwal.810@gmail.com

D. S. Mazumdar
e-mail: debapriyosaurav@gmail.com

N. Pramanik
e-mail: nilpramanik3@gmail.com

J. Mehedi
e-mail: jmehedi2007@yahoo.co.in

D. Bhattacharjee et al. (eds.), *Proceedings of International Conference on Frontiers in Computing and Systems*, Advances in Intelligent Systems and Computing 1255,
https://doi.org/10.1007/978-981-15-7834-2_71

1 Introduction

After the invention of microprocessor, many of the analyst and researchers worked on the field on simplification of inputs of combinational circuits working on minimization in various ways. Major researches where based on increasing the efficiency of system, with reduction in time-consumption, as one of basic motives to simplify the Boolean logic.

One of the popular and common method was developed by Karnaugh [1] about the simplification of the min terms described in SOP form. But K map is difficult to use for minimization of large no of variables. Due to increment in complexity, another method was developed by Quine [2] for minimization of the boolean functions even with large no of variables. But this method was also not so easy as no concrete methods were said for cyclic prime implicant tables, that is, for tables with more than one cross in each column. And so the process was tedious and time taking to generate min terms with more complexity.

So for the minimization of expression a different hybrid approach, which can be used for simplification of any number of variables is used from [6] which is noteworthy and has power to minimize expression easily with lesser difficulty. The method goes for further minimization. But for the truthfulness of this minimization for any number of inputs, we have proposed a Verification of this approach done using truth table which uses the weighted min term algorithm and is used for getting back the similar inputs.

2 Terminologies

TT = truth table BB = Basic Block

3 Literature Review

This section provides an overview regarding all the different techniques of how Input function simplification is done from past and is used in present and is mentioned in the given references. With lot of work done from many researchers, Research done by Quine [2] was more acknowledged and popular with his easier approach on simplification on theoretical and practical purposes. This paper provides a way to reduce the expressions to simpler expressions. But this approach did not worked well with reasonably difficult expressions and it lead to failure of this method. Some more development in this field were accounted in minimization process with a process proposed by author McCluskey [3] which simplifies some more level of difficulty from the expression functions from a cyclic prime implicant table. This process required a list of multiple realizable different min term. K map solution is also another process for smaller no of variables and it gives a different approach of minimization in field of Boolean expression.

In paper [6] the process involved is minimization of truth table with reduction of one variable which reduces the size of truth table by 50% at a time and then using the properties of Boolean algebra for further reduction. So it was difficult for simplification of min terms whereas paper [4] and [5] discusses about an algorithm for min term generation using weighted min term method. We used the method of truth table minimization as minimization technique and thereby checking it using min term generation algorithm of weighted min term method so that it could even used as for calculation for more than 20 variables without any software.

3.1 *Proposed Solution With Example:*

Here the paper proposes the verification of TT output for which some inputs have been considered to system shown in below example

For example:

1. Simplify : f= ∑(0,2,6,8,10,14)

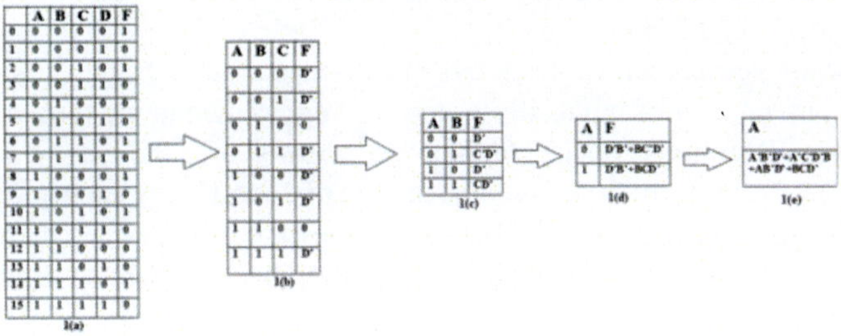

A sequence of min terms is considered for this minimization It uses the way of removing the Nth bit(Least Significant Bit) from N bit Truth table. And so on removing from N to N_1, N_1 to N_2 bit, and so this process is repeated again and again till the final expression is obtained with all bits of a input represented in binary form is exhausted This final expression from truth table minimization is possible following diminution. The last expression represented in above diagram for f, achieved by TT reduction is a more simpler expression from classical method shown below and is best suited for further minimization using weighted min term method. This exercise bring down the bulkiness of TT by 50% with decrease the number of final outputs to half in the succeeding table and it is continued until all inputs are used up

The function is represented by the TT shown in Figure above. The five BBs are identified by rectangular boxes. The following points are worth noting.

From paper [6], steps for reduction from truth table is shown,

(1) Input represented in a series of Truth table in binary form should be sorted in way of lower to higher of input of decimal format
(2) Binary format of Input should go with Maximum Significant Bit to Least Significant Bit followed in way of Left to Right.
(3) The values represented in Binary are always represented in 0 s and 1 s as per Digital Electronics Standard.

Using all 3 rules described above is used for reduction of BB by 50% at one time. Firstly TT represented in 1(a) is decreased to half shown in 1(b) with output in the form of one input. Similar process is applied to get the final expression from TT but it is not fully minimized as shown in 1(e). Some more techniques are added to TT minimization to get full minimized expression which is described in [6].

Expression without reduction is described below

$$f = A'B'C'D' + A'B'CD' + A'BCD' + AB'C'D' + AB'CD' + ABCD'$$

This is the direct Boolean expression from input with no minimization. So 1(e) describes the intermediate minimization using the truth table Some further steps have been In used in [6] to get the complete minimized expression. For checking the validity of final TT output, this paper proposes verification of TT's output to get the same described and used inputs to system. Now weighted min term minimization is further used for verification of any large number of variables without software for the above 1(e) final TT expression.

Now for the verification for truth table result matrix form from [4] is described below, For any number of inputs illustrated and represented in the form of a matrix described in SOP form is considered. Now this method uses the idea of d terms for each row where d term will be represented in form of 0 s and 1 s for each undetermined value to TT final output in 1(e). Below matrix shown describes the total number of available amalgamation i.e., combination for getting back all number of inputs with correct decimal values.

The result of truth table will have 2p–q terms where p denotes number of variables before the truth table operation and q number of variable after truth table operation (p–q) number of variables that can have both 0 s and 1 s. The matrix shown has each row filled with product term of Boolean expressions and so the table obtained from each row after min term generation is hereby

$$P_1 = \begin{bmatrix} B_n & B_{n-1} & B_{n-2} \cdot & B_i \cdot & \cdot & B_3 & B_2 & B_1 \\ x_n & - & - & - - - & - - - & & x_2 & x_1 \end{bmatrix}$$

Matrix P_1 has x_1, x_2 and x_n which gives the fixed value of 0 or 1 depending on expression So it can be considered that inputs B_1, B_2 and B_n are constant inputs while other inputs that is B_{n-1}, B_{n-2},…B_2 don't have any fixed input and can take any composition of values consisting of 0 and 1. For above matrix there are n-3 inputs for the given system. From above truth table minimization the expression obtained,

$$F(A, B, C) = A'B'D' + A'BC'D' + AB'D' + BCD'$$

For SOP form Matrix P for above logic function,

A	B	C	D
0	**0**	–	**0**
0	**1**	**0**	**0**
1	**0**	–	**0**
–	**1**	**1**	**0**

From initial the weighted value of first row is 0. In 1st row there is one inconstant inputs in A,B,C,D. So, in this row the inconsistent input is C for the inconsistent input C we take the value to be 0 and 1. So the table is prepared for the weighted min term generation

Table 1. e:

	Input to System					
C			Binary Number			Min term
0				0		$m_{(1,0)} = 0 + 0 + 0 + 0 = 0$
1				**1**		$m_{(1,1)} = 2 + m_{(1,0)} = 2$

Moving to second row there are no inconsistent input.

So, for the third row, C is again inconsistent input and can now the have the values of 0 and 1 Now considering C as 0 the value of the row is m(3,0) 8 + 0+0 + 0=8 and now considering C equal to 1 m(3,1) 8 + 0+2 + 0=10 here 2 is added because weighted value of 21 = 2. Here it can be seen that two min terms are generated thereby.

Table 2. f:

	Input to System					
C			Binary Number			Min term
0				**0**		$m_{(3,0)} = 8 + 0 + 0 + 0 = 8$
1				**1**		$m_{(3,1)} = 2 + m_{(3,0)} = 10$

Similarly for the 4th row. The table is formed below,

Table 3. g:

	Input to system					
A			Binary Number			Min Term
0				**0**		$m(4,0) = 0 + 4 + 2 + 0 = 6$
1				**1**		$m(4,1) = 8 + m_{(3,0)} = 14$

So total weighted min terms from

Row 1: 0 and 2 while Row 3 has weighted min terms = 8 and 10 and Row 4 has weighted min term 6 and 14.

So, after applying the above algorithm to the truth table reduced expression, It can be concluded that 1st row in contain 1 d-term that is $2^{1} = 2$. While in 2nd row there are no inconsistent input In the 3rd row the total possible values are $2^1 = 2$ and again in 4th row we have 1 d term. The three tables for min term generation is calculated in Table e, f and g.

Therefore total weighted min terms for the matrix are 0,2,6,8,10,14 which is clearly the reversed min term expression taken above in example. So by calculating these d terms the min term expression is verified.

Here we can see that with the use of truth table simplification, the Boolean expression got shorter and less complex and with the use of the weighted min term algorithm we can see initial min terms taken in account which gives a better idea to approach to problem

4 Comparison

In paper [5] the min term algorithm proposed was easy and computable using d term but initially the Boolean expression was considered as any random example. From student point of view the paper becomes difficult to understand as no deduction of the same min term is done to get the minimized Boolean expression. This paper is proposed from student point of view for easy understanding of the verification of truth table reduction using weighted sum min term algorithm which verifies the above taken min term.

5 Conclusion

This paper has proposed a method for verification of truth table minimization using the simplification of Boolean expression from min term generation for SOP form. This paper used the truth table method simplification algorithm which directly reduces truth table by half and then simplification of lesser complex expression with the use of algorithm of modified weighted min term generation. The above verification is done for SOP form of logic expression and can be applied to POS form of logic expression which is not given here and can be worked and studied further.

References

1. McCluskey Jr, E.J.: Minimization of Boolean Functions. Manuscript received June 26, 1956
2. Mehedi, J., Paul, S., Upadhayay, R.: "Modified minterm generation algorithm using weighted sum method" on NOV 2017. In: Third International Conference on Research in Computational Intelligence and Communication Networks (ICRCICN)
3. Karnaugh, M.: Map Method for Synthesis of Logic Circuit
4. Roy, S., Tilak Bhunia, C.: Minterms generations algorithm using weighted sum method. Int. J. Curr. Sci. Technol. **1**(2), l July–December 2013
5. Rathore, T.S.: Minimal realizations of logic functions using truth table method with distributed simplification. IETE J. Educ. **55**(1), 26–32 (2014). https://doi.org/10.1080/09747338.2014.921412
6. Quine, W.V.: The problem of simplifying truth functions. Am. Math. Month. **59**(8), 521–531 (1952)

A Supervised Trajectory Anomaly Detection Using Velocity and Path Deviation

Suman Mondal, Arindam Roy, and Sukumar Mandal

Abstract Nowadays, CCTV Surveillance applications have been significantly grown all over the world. Many methods have been implemented for detecting anomalies of moving objects in video. Implementation of fuzziness on features is one of the most robust detection techniques. The proposed model illustrates a novel detection technique with fuzziness on velocity and path deviation. This paper also covers a small survey on different techniques of detection of abnormalities and also shows the result on Queen Mary University of London junction dataset (QMUL).

Keywords Fuzzy · QMUL · Trajectory · Abnormalities · Anomalies

1 Introduction

Nowadays, our real-world system require automation of surveillance system. It is not possible to follow and review every object in a surveillance system from everywhere in the real-time area. To prevent our society from massive illegal activities, it is necessary to detect and identify the abnormal and unusual events from video surveillance. To reduce human resource and to protect our important organization, we require automated detection of the anomaly on a surveillance system in a multidimensional way. In today's life, the detection of abnormality is a very crucial factor for different research fields of computer vision. This is integrated as well as employed in distinct research sectors, like object trailing, person activity detection, human iden-

S. Mondal (✉) · S. Mandal
Reseache Centre in Natural Science, Raja N L Khan Women's College (Autonomous), Medinipur 721102, West Bengal, India
e-mail: asuman.mondal2014@gmail.com

S. Mandal
e-mail: sm5971@rediffmail.com

A. Roy
Prabhat Kumar College, Purba Medinipur, Contai, West Bengal, India
e-mail: royarindamroy@yahoo.com

D. Bhattacharjee et al. (eds.), *Proceedings of International Conference on Frontiers in Computing and Systems*, Advances in Intelligent Systems and Computing 1255,
https://doi.org/10.1007/978-981-15-7834-2_72

tification, visual surveillance, etc. Implementation of different applications has been done by the researchers in this domain [1]. The observation of various proposition to innovative approaches is furnished by the investigators in the publication [10]. In whatever way, it is not enough to convey essential circumstances and challenges exist in this domain. An extensive view and understanding of the presence of dataset are important and no such endeavors are built in the orientation depending on our expertise. Here, we present a novel approach to detect anomalies in video. This will help the automated visual surveillance search society. This paper also influences the new investigators, scientists, and students to take part in solving problems in this domain.

The details of the paper are arranged as a sequence like Sect. 2 discusses literature survey of the detection of anomaly in the video. We focus to build an expert surveillance system for tracking uncommon behaviors of moving objects in Sect. 3. Origin, destination, path, deviation, and velocity are taken as inspection to analyze the situation. In Sect. 4, we give a description of dataset and experimental result. Conclusion and future scope of the research work are presented in Sect. 5. In the last section, we derive the references.

2 Literature Survey

In the present climate, to suppress unlawful offence or traffic violation, CCTV (Closed-Circuit Television) surveillance system has made more desired in many surroundings such as different stations, airports, critical convergence, etc. Suitable identification and evaluation of unpredictable activities within a domain of interest is a crucial part of the video surveillance system. Timely detection of these relatively infrequent events critical for enabling pro-active measures involves a constant analysis of all the trajectories. These events are typically difficult to human analysts due to the excessive amount of information overload, fatigue, and inattention. For this reason, an automatic or semi-automatic intelligent system is essential to inspect anomalies in a realistic and random environment with the degree of fuzziness. The concept of fuzzy set theory has been introduced by Lotfi Zadeh [15, 16] $[x_0, x_1]$ in order to define the concept of grades in class membership. For this, different processing techniques have been illustrated in [1] of video surveillance system [9]. We can get solutions for the problem using some supervised and unsupervised methods such as unsupervised techniques using low-level optical features like motion and texture [2] with various environments [11], anomaly detection using training data [3], topic-based model [6] reply on unusual situation [14], and tracking moving objects [13]. Simultaneous advancement in multi-object tracking [13] has presented development in moving object tracking in random and difficult environments. For that, investigators have started to concentrate more on detection of abnormality [4] due to adequate tracking results.

3 Proposed Method

Our proposed method has been divided into two stages, training stage and testing stage. These two stages are explained in Fig. 1.

According to the above diagram, the proposed method has started with segregating plotted visual images into a set of segments [12] depending on each trajectory. After dividing the image, we can get some number of regions as illustrated in Fig. 2.

We also have each trajectories size and image size. So, we have calculated a Δ_w and Δ_h in Eq. 1. Δ_w and Δ_h are helping to determine the number of rows and number of columns in a region.

$$\Delta_w = \frac{imagewidth}{trajectorysize}, \Delta_h = \frac{imageheight}{trajectorysize} \tag{1}$$

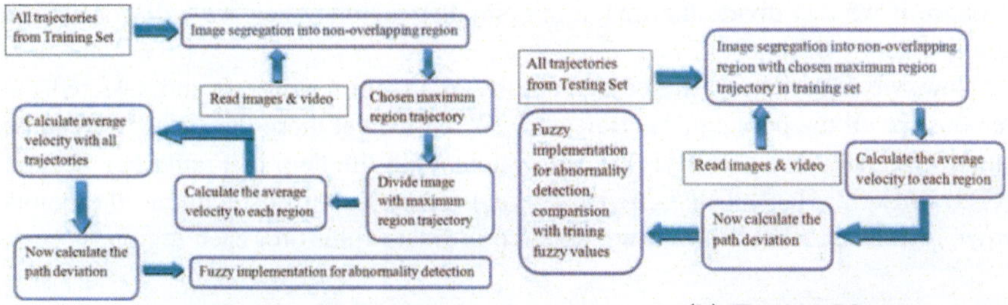

(a) Training Phase

(b) Testing Phase

Fig. 1 **a** Training procedure **b** testing procedure

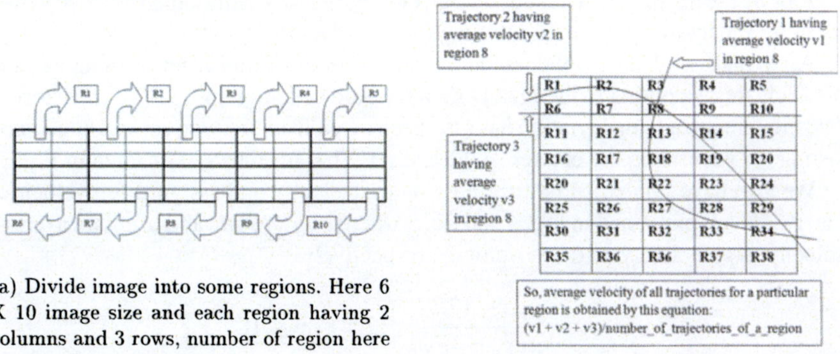

(a) Divide image into some regions. Here 6 X 10 image size and each region having 2 columns and 3 rows, number of region here R1, R2, R3, ... ,R10

(b) Explanation of obtaining the average velocity of all trajectories for a particular region

Fig. 2 Here **a** Region division **b** trajectories understanding

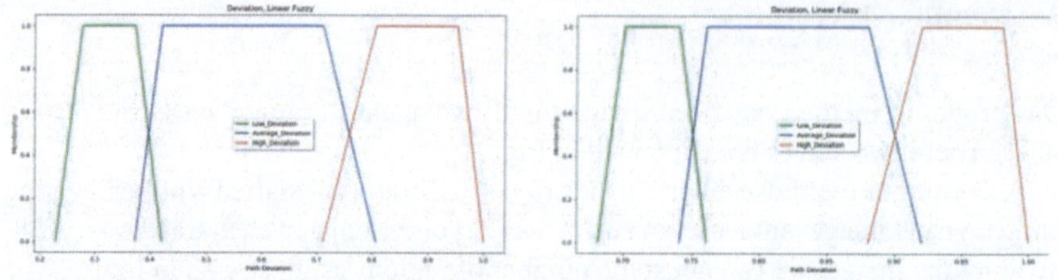

(a) Fuzziness of path deviation for training set

(b) Fuzziness of path deviation for testing set

Fig. 3 After applying trapezoidal member function on **a** training data and **b** testing data for path deviation

Our first objective is to find out the average velocity of each region. To find average velocity, we have chosen the maximum number of regions from all the trajectories because if we can divide the image into smaller regions, then we will not get the actual velocity of object.

Now, we have taken a trajectory one by one from testing set and selected the regions for all the points of the trajectory. When we get these distances, we can be easily computed the velocity(v) of all regions with the help of Frame Per Second (FPS) in Eq. 2. Here, dist(p_1,p_2,p_3, ... ,p_n) means the distance among the points p_1,p_2,p_3, ... ,p_n. These points are pixel co-ordinate values for each region.

$$v = \frac{dist(p_1, p_2, p_3, ..., p_n)}{FPS} \tag{2}$$

So, our first objective has been achieved after getting the velocity of all the selected regions by the trajectory. After obtaining velocity, we have determined average velocity of all trajectories for a particular region described in Fig. 2.

After that, we have implemented average path deviation from training set and made classes according to the fuzziness of the path deviation through all trajectories. The implemented path deviation has also been taken from training set and applied on testing for generating the classes in the test set. That fuzziness has shown in Fig. 3

We also have implemented fuzziness on the velocity and assign those velocities into two levels explained in Fig. 4 and Eq. 3 where avg_v, $high_v$ and low_v are average velocity, high velocity, and low velocity, respectively.

$$\begin{aligned} \mu &= \frac{x - 12.5}{avg_v - low_v}, 12.5 \leq x \leq 18 \\ &= \frac{25 - x}{high_v - avg_v}, 18 \leq x \leq 25 \end{aligned} \tag{3}$$

After implementing Fuzziness, we can be easily detected an abnormality of an object of a particular region with the help of the SVM (Support Vector Machine) [7] classifier.

4 Experimental Results

In this context, we have shown the results of the proposed detection technique of abnormality. Primarily, we have illustrated the data that are used for eventuations. Then, we have shown the outcomes.

4.1 Datasets

In our experiments, we have used Queen Mary University of London junction dataset (QMUL) [8]. This dataset is having 360 × 288 image resolution. It also has 50 min duration of video and 166 trajectories. We have taken 112 trajectories as training data and other is having a testing data. We have also shown the overall movement patterns of the dataset in Fig. 4.

4.2 Results

We have applied a trapezoidal fuzzy membership function on region-wise velocity values to draw a diagram for understanding the low, medium, and high velocity depicted in Fig. 5. The estimated values on QMUL junction dataset for generating the diagram are given below (Table 1).

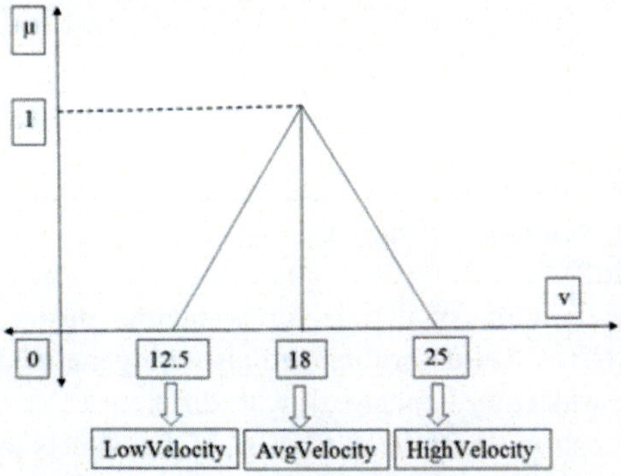

(a) Fuzzy Implementation on velocity. Here we have depicted low velocity, high velocity and average velocity Fuzziness of path deviation for training set

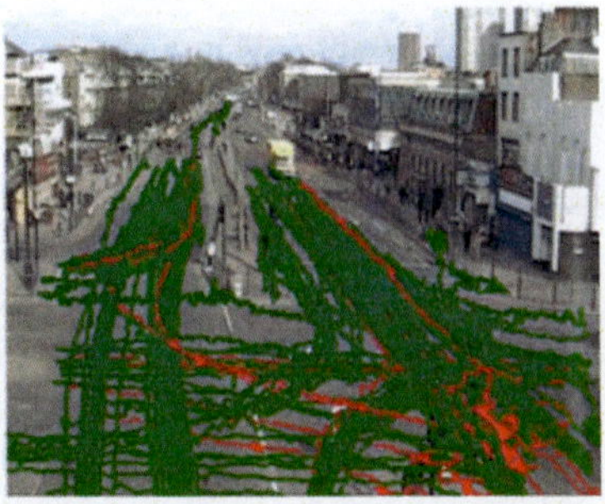

(b) Fifty minutes cumulative sample of motion in QMUL Junction video

Fig. 4 **a** Fuzzy implementation **b** QMUL junction video

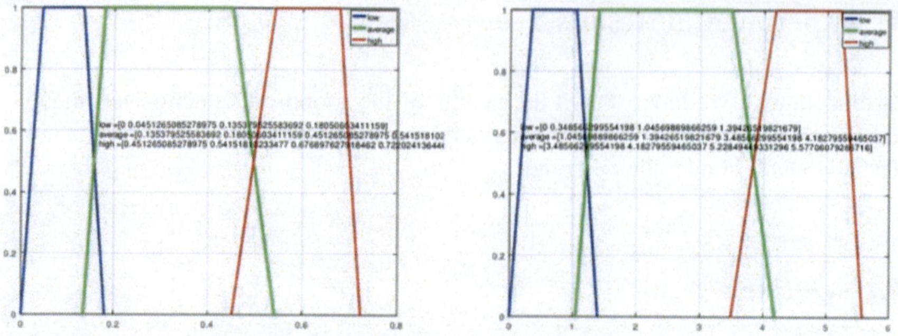

(a) Fuzziness of velocity for training set (b) Fuzziness of velocity for testing set

Fig. 5 After applying trapezoidal member function on **a** training set and **b** testing set for velocity

Table 1 Fuzzy implemented value of regions with respect to the region velocity

Dataset	Low velocity	Medium velocity	Max velocity
Train set	0.0000, 0.04513, 0.13538, 0.18051	0.13538, 0.18051, 0.45127, 0.54152	0.45127, 0.54152, 0.67690, 0.72202
Test set	0.0000, 0.3485, 1.0456, 1.39426	1.0456, 1.39426, 3.4856, 4.18279	3.4856, 4.18279, 5.22849, 5.577060

Table 2 Fuzzy implemented value of path deviation

Dataset	Low velocity	Medium velocity	Max velocity
Train set	0.22777, 0.27604, 0.37256, 0.42083	0.37256, 0.42083, 0.71041, 0.80694	0.71041, 0.80694, 0.95173, 1.00000
Test set	0.68115, 0.70108, 0.74094, 0.76086	0.74094, 0.76086, 0.88043, 0.92028	0.88043, 0.92028, 0.98007, 1.00000

Trapezoidal fuzzy membership function has also been used on path deviation to understand the low, medium, and high velocities of moving objects in Fig. 3 and estimated values are given below (Table 2).

We have extracted seven regions of entry points of all trajectories and four clusters of exit points of all trajectories using DBSCAN algorithm. We also are generated a heat map to analysis the region-wise trajectory benchmark with different colors. With this, we can be shown a clear perception about the movement of the objects in a video in Fig. 6. It has shown some colors like red indicating high-speed velocity, yellow indicating average speed velocity and white indicating low velocity and so on.

Now, we have trained those region-wise trajectory velocities with SVM [5, 7] to get a model through which we have tested other trajectories for detecting the abnormalities. Here, region velocities and path deviation are features of the SVM. We have used radial basis function kernel and soft margin with no cross-validation. After testing, we got 80.6994% abnormality among 473 region velocities. We have also

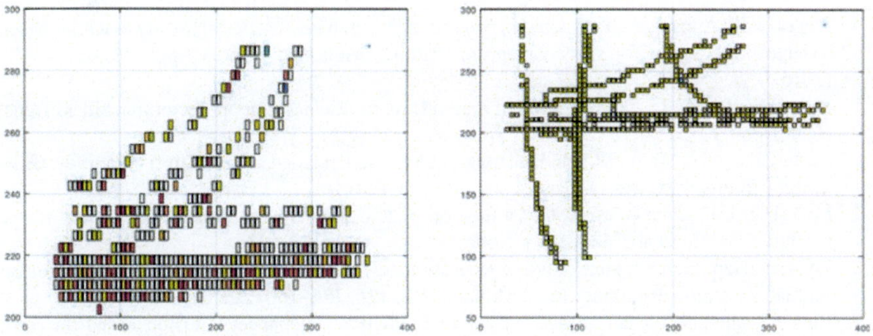

(a) For training set (b) For testing set

Fig. 6 Heat map for **a** training set and **b** testing set

trained with path deviation with SVM getting 70.1493% accuracy on the testing set. We have combined the velocity and path deviation. Then, we got 91.253% accuracy on QMUL junction dataset.

5 Conclusion

Understanding and analyzing of random surveillance environment is an existing problem all over the world. With respect to this, it is very important to construct a successful and smart artificial methodology to integrate with the continuous changeable environment and a combination of several information. In this paper, we have introduced a detection technique of abnormalities from moving video objects. Here, we have only used the basic fuzzy technique to detect the anomalies of moving objects using velocity and path deviation.

Now, our recent aim to extend this method collaborating among the path deviation, velocity, size of the objects, and fuzzy implementation which will show exact alarm activation for all regions of all trajectories.

References

1. Chandola, V., Banerjee, A., Kumar, V.: Anomaly detection: a survey. ACM Comput. Surv. **41**, 15:1–15:58 (2009)
2. Chen, D.Y., Huang, P.C.: Motion-based unusual event detection in human crowds. J. Vis. Comun. Image Represent. **22**, 178–186 (2011)
3. Cong, Y., Yuan, J., Liu, J.: Abnormal event detection in crowded scenes using sparse representation. Pattern Recogn. **46**, 1851–1864 (2013)

4. Dogra, D.P., Ahmed, A., Bhaskar, H.: Smart video summarization using mealy machine-based trajectory modelling for surveillance applications. Multimedia Tools Appl. **75**, 6373–6401 (2016)
5. Hsu, C.W., Lin, C.J.: A comparison of methods for multiclass support vector machines. Trans. Neur. Netw. **13**, 415–425 (2002)
6. Li, T., Chang, H., Wang, M., Ni, B., Hong, R., Yan, S.: Crowded scene analysis: a survey. IEEE Trans. Circuits Syst. Video Technol. **25**, 367–386 (2015)
7. Lin, H.T., Lin, C.J., Weng, R.C.: A note on platt's probabilistic outputs for support vector machines. Mach. Learn. **68**, 71–74 (2007)
8. Long, C., Kapoor, A.: A joint gaussian process model for active visual recognition with expertise estimation in crowdsourcing. Int. J. Comput. Vis. **116**, 136–160 (2015)
9. Rhodes: Anomaly detection and behaviour prediction: Higher-level fusion based on computional neuroscientific principles. In Sensor and Data Fusion In Tech. I-Tech Education and Publishing (2009)
10. Sodemann, A., Ross, M., Borghetti, B.: A review of anomaly detection in automated surveillance. IEEE Trans. Syst. Man Cybern. Part C: Appl. Rev. **42**, 1257–1272 (2012)
11. Song, X., Shao, X., Zhang, Q., Shibasaki, R., Zhao, H., Cui, J., Zha, H.: A fully online and unsupervised system for large and high-density area surveillance: Tracking, semantic scene learning and abnormality detection. ACM Trans. Intell. Syst. Technol. **4**, 35:1–35:21 (2013)
12. Sultani, W., Chen, C., Shah, M.: Real-world anomaly detection in surveillance videos. arXiv:1801.04264, pp. 6479–6488 (2018)
13. Walia, G.S., Kapoor, R.: Recent advances on multicue object tracking: a survey. Artif. Intell. Rev. **46**, 1–39 (2016)
14. Xiao, T., Zhang, C., Zha, H.: Learning to detect anomalies in surveillance video. IEEE Signal Process. Lett. **22**, 1477–1481 (2015)
15. Zadeh, L.A.: From circuit theory to system theory. Proc. Radio Eng. **50**, 856–865 (1962)
16. Zadeh, L.A.: Fuzzy set, information and control **8**, 338–353 (1965)

Modular Secured IoT Using SHAKTI

Soutrick Roy Chowdhury, Aishwarjyamoy Mukherjee, S. Madan Kumar, Kotteeswaran, Anand, N. Sathya Narayanan, and Shankar Raman

Abstract Hardware-planted bugs is a major security concern in today's Internet of Things (IoT) devices. Security testing of a hardware chip that integrates a computation and communication module together is complex, even though the software would only be few lines of code. In such a device, conventional software testing approaches such as code walk-throughs and reviews can be used for detecting security vulnerabilities. However, testing hardware security in such integrated IoT devices is a challenge. In this paper, we use the open-source hardware SHAKTI E-Class as the computation element. Since the hardware design and code are available, security test of computation part can be performed independently of the communication part. For the communication part, we externally integrate the widely used commercial ESP8266 Wireless LAN (WLAN) module through the Universal Asynchronous Receiver Transmitter (UART) with SHAKTI E-class. This ESP8266 WLAN chip can be security tested separately. Finally, security testing of the computation and communication interface can be performed. This leads to a modular security testing approach that could be of immense importance in strategic sectors. This is a first successful attempt to increase the security of IoT devices using the SHAKTI E-class open-source processors.

Keywords Internet of things · SHAKTI · Untrusted computing · Smart cities

S. R. Chowdhury (✉) · A. Mukherjee
Jalpaiguri Government Engineering College, Jalpaiguri, West Bengal, India
e-mail: src2003@cse.jgec.ac.in
URL: https://jgec.ac.in/

S. Madan Kumar · Kotteeswaran · Anand · N. Sathya Narayanan · S. Raman
Indian Institute of Technology, Madras, India
e-mail: snarayan@cse.iitm.ac.in

D. Bhattacharjee et al. (eds.), *Proceedings of International Conference on Frontiers in Computing and Systems*, Advances in Intelligent Systems and Computing 1255,
https://doi.org/10.1007/978-981-15-7834-2_73

1 Introduction

Internet of Things (IoT) is a promising technology which can be used for connecting, controlling, and managing computing devices connected to the Internet using an IP address [1]. IoT deals with many connected devices that not only sense and collect the data but also communicate with humans using various wired and wireless interface technologies. The main objective is to manage and control physical objects around an individual remotely, in an intelligent and meaningful manner. The devices at the edge collect useful contextual data autonomously from IoT-sensor-based devices and send the data to remote servers for offering context-aware services [2]. Because of the work efficiency, convenience, and cost-effectiveness of IoT technologies, private critical networked systems have started adopting IoT in an ad hoc manner [1] (Fig. 1).

However, in the past decade, IoT has developed without appropriate considerations for the profound security challenges. The futuristic appeal IoT presents makes life more comfortable and it is enticing to many. When the infrastructures are operated as separate networks, and if a malicious code is injected in the network. The consequences of these can be catastrophic. Because these networks, it can lead to service paralysis, information spills, and even interruption of connected infrastructures may occur. For example, today, smart cities are promoted as an effective way to counter and manage, uncertainty, and urban risks. Many Governments promote smart cities among the public. The smart cities might offer better and quick services. The smart cities may open an avenue for new vulnerabilities, threats, and attacks making city infrastructure and services insecure [3].

To realize IoT over national infrastructure systems or strategic installations, cybersecurity should be considered from the system design stage itself and must not be

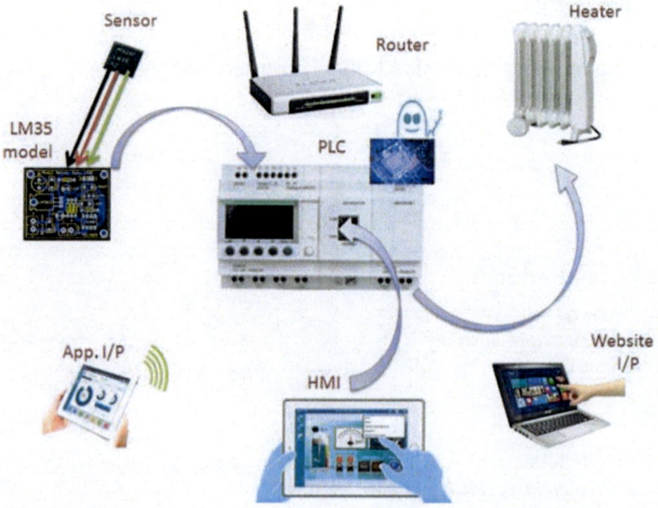

Fig. 1 Internet of things [4]

considered as an add-on technology. Artificial Intelligence (AI) based cyberattacks might be carried out in the future regardless of the techniques that have been used in the past [5]. Having a secure IoT system-on-chip (SoC) in the IoT network and ensuring the devices implement "root of trust" across the IoT network can make users trust the IoT services more [6]. In the past, Untrusted computing platforms have decreased the credibility of IoT in strategic installations. The main hardware security threat is the manufacturing of malicious Integrated Circuits (ICs). This is mostly because of outsourcing the design and migration of fabrication to low-cost areas across the globe [7]. Most of the attacks in the cyber world have not solved the vulnerabilities from an untrusted SoC perspective. This is a serious matter of concern when the smart cities turn into smart maintenance of nuclear or strategic installations. As a first part to mitigate the issue of untrusted SoC, the authors of this paper proposed and implemented a novel idea that shows the working of IoT over an indigenous Open-Source chip. Since the open-source chips can be independently tested for security, a modular approach for security testing can be performed for the whole system.

The rest of this paper is organized as follows: Section 2 presents the related work. Section 3 gives a brief introduction to SHAKTI. Section 4 lists the contributions of the paper. Section 5 presents the IoT implemented over SHAKTI and briefly describes how modular security testing was carried out. Section 6 concludes the paper.

2 Related Works

A lot of IoT devices get deployed in uncontrolled, complex and often hostile environments. Securing these IoT systems presents a number of unique challenges. The report, "State of Internet" (Q3, 2016), observed 138% increase in DDoS attacks (compared to the previous year) with traffic exceeding 100 Gbps. These attacks are attributed to the increase in the IoT device utilization and ease of carrying out DoS attacks. The commercial DoS services are easily available for hire. One can anonymously order a 5–6 Gbps DDoS attack lasting approx. 10 min for as little as $6.

Code injection, Buffer overflow, and Phishing are some of the attacks at the application layer that are done by exploiting the program architecture [5, 9]. These type of attacks can cause the system to lose control and compromise the user's privacy, or even to a complete system shutdown [5]. The attackers have a very good understanding of the underlying system's architecture vulnerabilities and developers of application don't have the insight into the microarchitecture design [8, 9].

Tag cloning, eavesdropping, spoofing, and sleep deprivation attack are some of the common attacks that exploit the privacy. Sleep deprivation can lead to powering off the system completely. This can be catastrophic, in case of automatic vehicles. The attack at the perception layer usually aims at destroying the data collection and the communication [8].

All the papers have talked about the vulnerability assuming the processing units in the IoT devices are secure. There are many attacks that make use of the unused or reserved bits in the packets to launch attack [10]. To overcome these problems, some countries take national-level security measures and make concrete implementation. These steps ensure that critical infrastructure can be operated and maintained safely over Internet [5]. In addition to technical measures, the National Cybersecurity enhancement scheme proposes some security guidelines for individual fields, such as organization, policy, manpower, industry, budget, education, and R & D [5]. All these can ensure the security of IoT networks to a large extent. Still, the persistence of an untrusted computing device is left unsolved.

An interesting attack that has been practically seen is the failure of Syrian Radar due to the presence of Kill Switch in the hardware. An open-source hardware can mitigate or trace such kill switches by auditing the design. Since SHAKTI is a open-source hardware audits can be done on the hardware design itself thereby ensuring safety against such attacks.

3 SHAKTI

SHAKTI is an open-source initiative by the Reconfigurable Intelligent Systems Engineering (RISE) group at IIT-Madras. The aims of SHAKTI initiative include building open-source production-grade processors, complete System on Chips (SoCs), Development boards and SHAKTI-based software platform. The project has currently developed an Embedded class (called E-Class) and edge computing class (called C-Class) of processor based on the RISC-V ISA. RISC-V is a free and open Instruction Set Architecture (ISA) enabling a new era of processor innovation through open standard collaboration [11].

In this paper, we consider the E-Class Base Processor. The E-Class is a 32/64 bit microcontroller supporting a subset of RISC-V ISA with the low area and power consumption having an operational frequency up to 200MHz on silicon. It is positioned against ARM's M-class cores. The major anticipated use of the E-class of processors is in low-power compute environments, automotive and IoT applications such as smart-cards, motor-controls, and home automation. The E-class is also capable of running Real-Time Operating Systems (RTOS) like Zephyr OS [12] and FreeRTOS [13].

3.1 SHAKTI-SDK

Software Development Kits (SDK) are an integral part of any product development. The main objective behind using an SDK is to reduce the development time. The SHAKTI-SDK for the E-class is simple and easily customizable. Some of the essential features like debug codes and board support libraries are provided. The SHAKTI-SDK architecture is shown in Fig. 2.

In this experiment, ESP 8266 module was integrated with SHAKTI E class, and AT commands over UART were used for communication. The UART was configured to send and receive messages, using drivers in the SDK. Tables 1 and 2 describe the message frame format for communication between client and server.

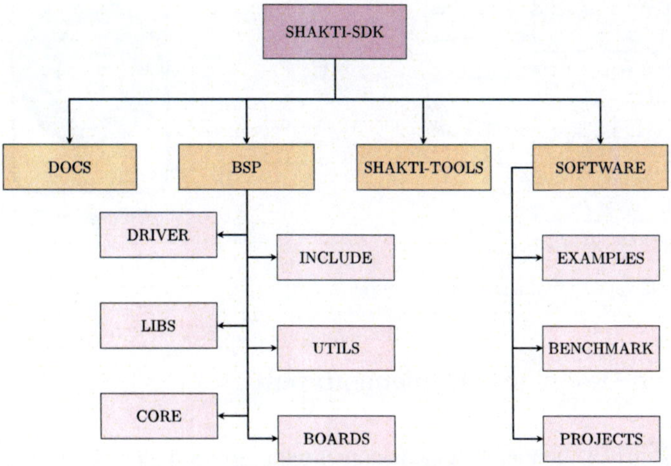

Fig. 2 SHAKTI-SDK architecture

Table 1 Generic packet format for send/receive of messages

⟨ 1 byte ⟩	⟨ 2 bytes ⟩	⟨ 2 bytes ⟩	⟨ 4 bytes⟩	⟨ variable⟩	⟨ 1 byte ⟩	⟨ 1 byte ⟩
Start	Sender	Sensor type	Info type	Payload	End	Checksum

Table 2 Frame format description

S. no	Field name	Field detail
1	Start	$
2	Sender	0 - device, 1 - server, 2 - mobile app
3	Sensor type	0 -LM75, 1 - RTC, 2 - ADC, etc.
4	Info type	LM75 - {0 - periodic update, 1 - threshold update}
5	Payload	sensor data
6	End	*
7	Checksum	XoR of data

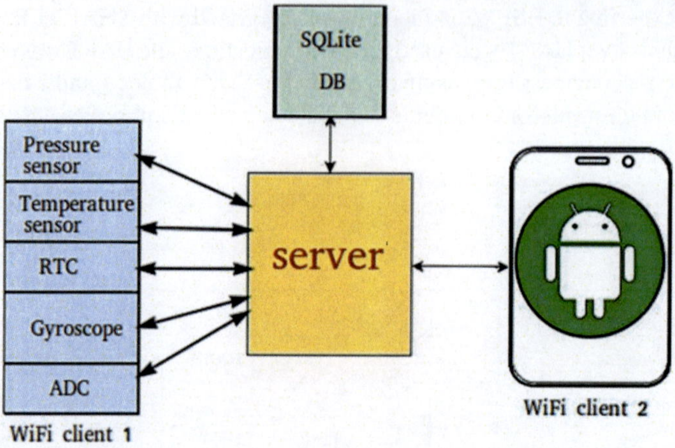

Fig. 3 High-level design diagram of IoT system

4 System Design and Implementation

The setting up of a complete IoT chain comes with certain challenges. Designers need to take into account requirements like prolonged battery life, low-power consumption, microcontroller unit (MCU), secured network and many more. An MCU is the central nervous system of the IoT design. Data collected by sensor nodes is continuously processed by the MCU. In this paper, the SHAKTI E class is chosen, which is designed for embedded and IoT platform. The E class is an in-order 3 stage pipeline with a lean computational unit. The MCU is programmed with Assembly instructions and Embedded C to make it more simple and deterministic. The High-Level Design of IoT setup involves 2 WiFi clients and one server. The MCU with sensors and ESP 8266 is referred to as sensor board. The sensor board and an instance of the Android application are the WiFi clients. The server is a central repository which stores sensor data and configuration messages. The server uses SQLite database for data management. The high-level design of the IoT system is shown in Fig. 3.

4.1 Interfacing E Class to Server

The server acts as a WLAN access point. Both the clients connect to the server using the server's SSID and password. Once connected, the sensors can be configured and information can be exchanged. The sensor information read by the MCU is sent as data packets to the server. The sensor board uses the WiFi module in ESP 8266, to send and receive data. The MCU commands the ESP 8266 module using AT commands [14] over a configurable UART. The sensor board sends data packet periodically to the server. At the server side, the received data is stored and forwarded

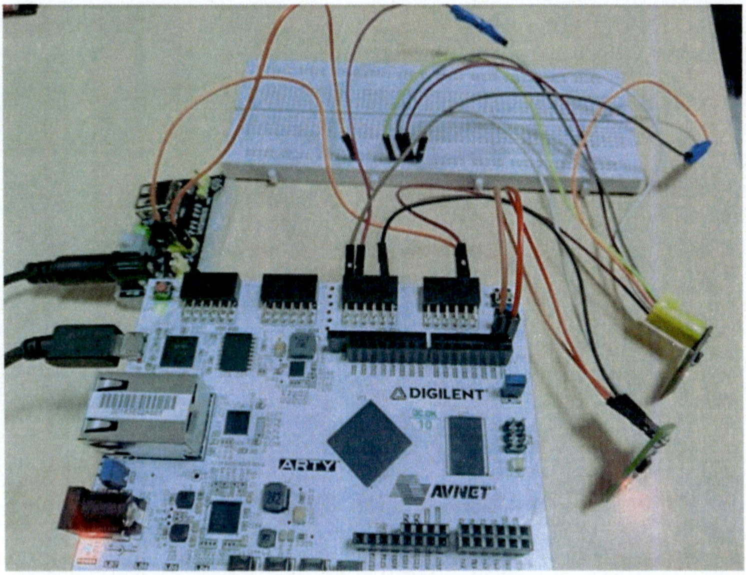

Fig. 4 Interfacing ESP and LM75 with Artix7-35t

to the mobile application on request. The data transfer to/from the client/server is accomplished over TCP/IP protocol. The MCU client side and server code are written in embedded C, which can handle dual client, i.e., one is mobile application and another is sensor board a socket connection. Whenever a mobile application requests for a particular data, server sends the data over TCP/IP. The TCP/IP packet is processed and the appropriate result displayed on the mobile screen. Figure 4 shows the high-level design of the IoT system.

4.2 Sensor Controlled Android Application

An Android-based mobile application based on IntelliJ IDEA is used as a WiFi client. The Android Application is used to request the server for a specified sensor data. The server queries the SQLite database and returns the value. There is also an SQLite database in Android Application, which stores the configuration messages. The Android application developed for controlling the sensor connected in SHAKTI SoC is shown in Fig. 5. The Application uses the IP address and Port id to configure and read the temperature sensor values. The application also has provision for adding and configuring new sensors to make the system more dynamic. The name and image for the new sensor is stored in an SQLite3 database and is loaded every time the application is reloaded. By storing the configuration values in the local database, repeated ping to the server is avoided. The configuration values are fetched from

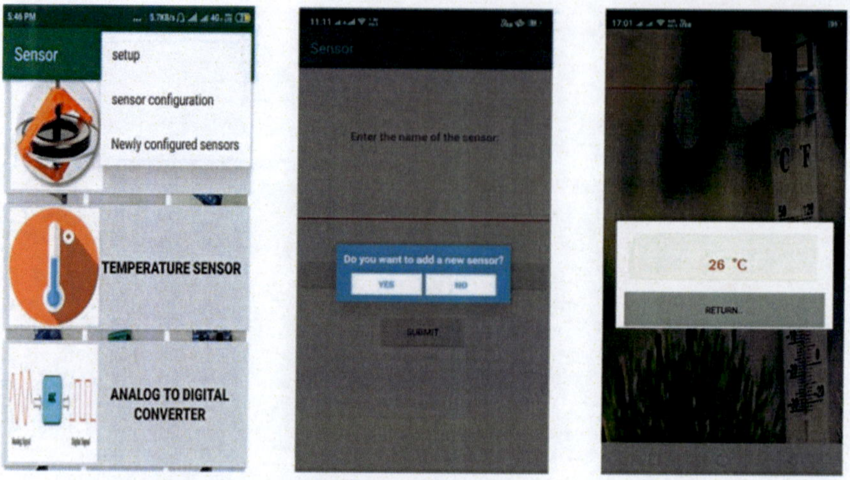

Fig. 5 Snap shot of Android application

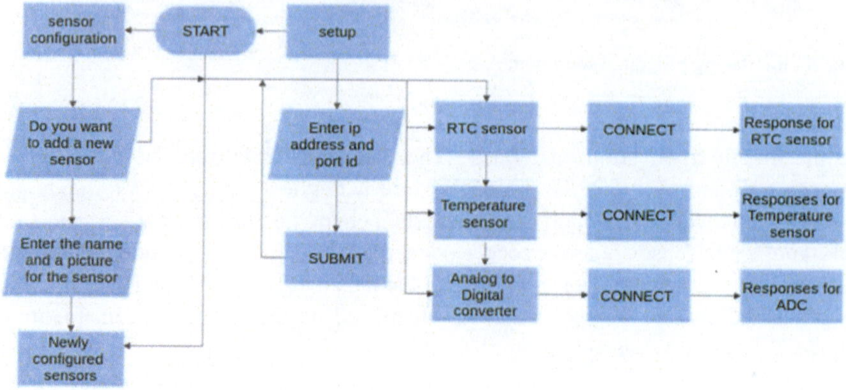

Fig. 6 Flowchart at Android client side

the SQLite database on restart. The server on receiving the ping from the client application responds accordingly and the whole communication can be secured using SSL. The same is explained in the following flowchart Fig. 6.

For example, to read the temperature sensor value, the mobile application configures the server to receive sensor. The specific format of the message that is sent from the application identifies the required sensor that the user wishes to sense. After that, the server sends a message to the sensor board to periodically send sensor data. On receiving the message, the sensor board starts sending sensor data to the server. Based on the periodicity, the server sends data to the mobile application. The micro-architectural design is available Open Source and verified by Code walk (Fig. 7).

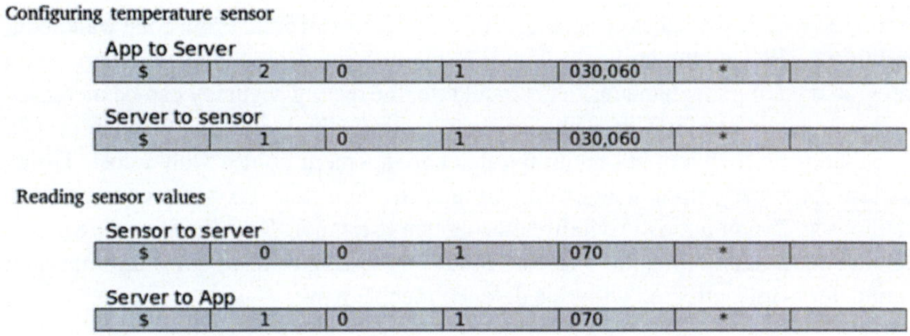

Fig. 7 Messages exchanged for knowing the temperature sensor in app

5 Security Testing

The micro-architectural design is available Open Source and verified by Code walk through [15]. After that, it was compared with other standard designs for any security threats. A malicious change in design at production is none in an Open-Source Indigenous design. The integrity and functionality of the E class were verified using RISC-V tests [16]. RISC-V tests are test suites provided by the RISC-V foundation to evaluate the integrity and quality of the Processor designs. The next testable component is the ESP 8266, which was verified by standard Wireless LAN tests. And Wireshark was used to capture and analyze the WLAN packets [17]. The interface between the ESP 8266 and E class is the UART. The UART was tested with standard UART test cases, probing and signal verification over oscilloscope. Previous test results were performed on only one integrated test case but present technology allows us to modularize the test cases (computing, communications (for the ESP device) and the interface of devices). More such detailed testing is done and the number of test cases in such scenarios can double. The security, however, is on testing and this need to be considered by the designers.

6 Conclusion

In this paper, the authors have proposed a novel idea of implementing a secure IoT network by using an Open Source indigenous SoC design. The usual approach to secure the IoT framework is to fix the weak links in the network architecture or software architecture. As discussed in [7], there is a greater scope for vulnerabilities arising out of Untrusted computing device. The proposal in this paper takes care of the untrusted part of computing. This can be very useful in developing smart cities, connecting and securing strategic installations. The demands for privacy assurances increased with IoT devices collecting all sorts of personal data. From code walkabout and using RISC-V standard compliance test suites, the design is made secure. With integrated testing mechanisms being available, the testing accuracy can be increased up to an accuracy of 100%. Thus, security testing which is a major blueprint can be enhanced effectively but naturally, this enhancement comes with a cost. Trojan attacks are most common with this community and with proper security testing methods as Trojan-agnostic defense mechanism (Channel Noise Profiling) [18]. This defence mechanism uses more testing time. The testing time which is an important parameter is thus affected with this defence mechanism.

References

1. Sarin, G.: Developing smart cities using internet of things: an empirical study. In: 3rd International Conference on Computing for Sustainable Global Development
2. Adjih, C. et al.: In 2015 IEEE 2nd World Forum on Internet of Things
3. Kitchin, R. et al.: The (In)security of smart cities: vulnerabilities, risks, mitigation, and prevention. J. Urban Technol.
4. Abed, A. A.: IoT: Architecture and design. In 2016 AIC-MITCSA (2016)
5. Andrea, I. et al.: Internet of Things: Security vulnerabilities and challenges. In: 2015 IEEE Symposium on Computers and Communication
6. Smith, N.M. et al.: Establishing hardware roots of trust for IoT devices (2015)
7. Jin, Y. et al.: Exposing vulnerabilities of untrusted computing platforms, ICCD (2012)
8. Zhen-hua , D. et al.: A taxonomy model of RFID security threats
9. Chen, K. et al.: Internet-of-things security and vulnerabilities: taxonomy, challenges, and practice. J. Hardware Syst. Secur. (2018)
10. Naik, N. et al.: Discovering hackers by stealth: predicting fingerprinting attacks on honeypot systems. In: 2018 IEEE International Systems Engineering Symposium
11. Waterman, A.S.: Design of the RISC-V instruction set architecture (2016)
12. Zephyr project and Zephyr OS kernel, [online]
13. Barry, Richard and others FreeRTOS, Internet (2008)
14. https://www.espressif.com/en/products/hardware/esp8266ex/overview
15. Shakti, E.: Class https://gitlab.com/shaktiproject/cores/e-class
16. RISC V tool chain
17. Sanders, C.: Practical packet analysis: Using Wireshark to solve real-world network problems
18. Subramani, K.S. et al.: Hardware Trojans in Wireless Networks. http://www.hostsymposium.org/host2018/hwdemo/HOST_2017_hwdemo_22.pdf

Test-Bench Setup for Testing and Calibration of a Newly Developed STS/MUCH-XYTER ASIC for CBM-MUCH Detectors

Jogender Saini, Gitesh Sikder, Amlan Chakrabarti, and Subhasis Chattopadhyay

Abstract Compressed Baryonic Matter (CBM) (Senger, J Phys Conf Ser **50**(1):357 (2006), [1]) is one of the experiments of the upcoming Facility for Antiproton and Ion Research (FAIR) (Senger, J Phys Conf Ser **50**(1):357 (2006), [1]) in Germany. CBM will take data in very high particle interactions rate of 10^7 MHz and thus fast and high granular detectors are required to cope up with the rate and multi-hit probability. With about 10% overall detector hit occupancy, each readout channel may see the hit rate of up to 200 200 KHz. For such a high density and high rate detector readout, a specialized ASIC STS/MUCH-XYTER is designed. This ASIC is made with dual gain so that it can be used for the multiple detectors of the CBM experiment, e.g., Muon Chambers (MUCH) and Silicon Tracking Station (STS). This is a self-triggered hybrid ASIC consisting of 128 analog channels along with a digital back-end for reading out the data. Each channel along with pre-amplifier and shaping circuits also have a 5-bit flash ADC where all the 32 comparators of this ADC can be configured for a particular threshold setting. In this ASIC, low-gain setting is called as MUCH mode. We have developed a test-bench setup to test and calibrate this ASIC in low-gain mode to use it with MUCH detectors. The major task of this test-bench is to optimize all the bias parameters of this ASIC and to calibrate all the ADC channels such that the ASIC performs as per our expectation. The present paper focuses on the details of the test-bench setup and the testing methodologies used to test and calibrate/trim the registers on this ASIC. The paper will be concluded with the basic test results of the ASIC after calibration.

J. Saini (✉) · S. Chattopadhyay
VECC, 1/AF, Bidhan Nagar, Kolkata 700064, India
e-mail: jsaini@vecc.gov.in

S. Chattopadhyay
e-mail: sub@vecc.gov.in

G. Sikder · A. Chakrabarti
University of Calcutta, Kolkata 700106, India
e-mail: gitesh.sikder@gmail.com

A. Chakrabarti
e-mail: achakra12@yahoo.com

D. Bhattacharjee et al. (eds.), *Proceedings of International Conference on Frontiers in Computing and Systems*, Advances in Intelligent Systems and Computing 1255,
https://doi.org/10.1007/978-981-15-7834-2_74

1 Introduction

STS/MUCH-XYTER [2] is a 128-channel dual gain, self-triggered highly configurable hybrid ASIC with about 30,000 settable registers. This ASIC utilizes 180 180 nm CMOS technology in its design and development process. In low-gain setting, this ASIC will be used to readout the CBM [1] MUCH [3] Gas Electron Multiplier (GEM) [4, 5] detectors while high gain setting will be used for STS [6] detectors in the CBM experiment. Outputs of the configurable registers are fed to an associated digital to analog converter (DAC), which controls biasing/trimming voltages of the ASIC at different stages. Without optimizing these bias/trim values, this ASIC cannot be used for any practical application with the detectors. The analog output is fed to a 5-bit flash ADC, where each comparator threshold setting has its own 8-bit trim DAC. The DAC output can drift the threshold voltage of the flash ADC comparator from its nominal value in a limited range. With this flexibility, the ADC in this ASIC can be configured in both linear as well as nonlinear fashion. As there are a huge number of settable registers, hence an automatized test-bench was developed to optimize all the register settings and trim values for the MUCH mode of the ASIC. Figure 1 shows the picture of the STS/MUCH-XYTER board connected with charge injector circuit board at the input connector which can inject charge to all the channels simultaneously. A known charge is injected to the ASIC and the output is readout via the backlink connector as can be seen from the Fig. 1. In this setup, bias/trim parameters are modified automatically in a predefined fashion until an expected response is received at the output for a given input charge. Several software algorithms were applied to the readout data in order to quickly reach the final bias/trim values of an ASIC under test.

2 Architecture of STS/MUCH-XYTER

Figure 2 shows the simplified internal block diagram of the STS/MUCH-XYTER ASIC. As can be seen in Fig. 2, detector input is fed through the point DET_IN. Input stage of each channel of this ASIC consists of a charge sensitive pre-amplifier (CSA). CSA has one more input from the internal pulse generator (AMP_CAL), which can feed to 32 channels at the same time. These 32 channels are controlled by a selection switch with predefined sets of 32 channels. Output of the CSA is fed to two shaping circuits, i.e., a fast shaper and a slow shaper via a polarity selection circuit (PSC). With the help of PSC, this ASIC can be used for both positive and negative input charges. The slow shaper has a bigger time constant which can be set in the range of 80–240 ns and is used for energy measurement while the fast shaper has a smaller time constant of 30 ns and is used to determine the timing information of the incoming pulse. Fast shaper has a high gain and its output is fed to a comparator. The comparator output latches the output of a 12-bit counter driven by internal free running clock, where the output corresponds to the timing

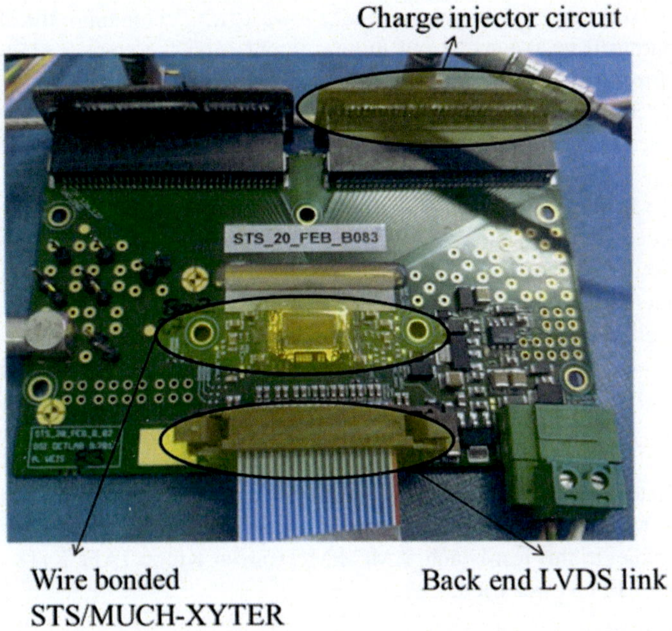

Fig. 1 STS/MUCH-XYTER test front end board

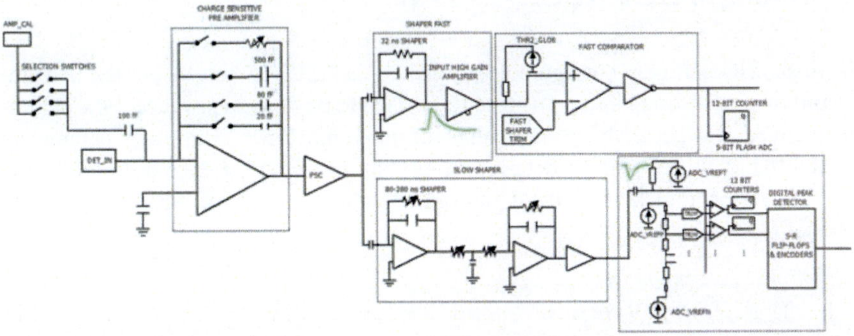

Fig. 2 Simplified internal block diagram of STS/MUCH-XYTER ASIC

information of the incoming pulse. This is a self-triggered ASIC where the threshold of the lowest ADC comparator decides the trigger to readout any channel. As shown in Fig. 4 slow shaper input generates the pulse BLOCK_TS after pulse crosses the threshold in any comparator of the ADC. When fast shaper comparator crosses the threshold, LATCH_TS as shown in Fig. 4 is also asserted and time information from the counters are latched. After the slow shaper pulse crosses back the threshold, DATA_VALID pulse followed by RESET is generated and the channel is ready to take the next signal for processing. Output of the slow shaper is fed to the 5-bit flash

ADC and as soon as any channel crosses the lowest ADC comparator threshold than the data output from the ADC and timing output from the counter is stored in a buffer for further readout from the digital back-end through high-speed LVDS links. Threshold settings of both first comparator as well as the fast shaper of any channel are very crucial as if the threshold is too low then noise might trigger more frequently and we may acquire more noise data as compared to the useful information while on the other hand if we put too high threshold, than we might cut more real signal and may lose overall efficiency of the detector. Hence the first threshold needs to be carefully optimized in order to collect more relevant information without losing much of the efficiency. The digital interface of this ASIC consists of 3 types of differential electrical links (e-links): clock, 1 downlink and up to 5 uplinks. STS/MUCH-XYTER e-links can work in one of the three speed modes. Accepted clock e-link frequency can be 40, 80 or 160 160 MHz, respectively, the downlink speed can be 40, 80, or 160 Mbit/s and uplink speed can be 80, 160, or 320 Mbit/s. Uplink e-links use double data rate (DDR) signaling. E-links are designed for synchronous communication, but clock/data phase aligner circuits are required for proper data transmission. ASIC is designed to be interfaced with rad-hard GBTx [7] chip, but interface with FPGA is also possible. In this test-bench, AMC FMC Carrier Kintex (AFCK) [8, 9] board with Kintex 7 FPGA was used.

3 Testing of STS/MUCH-XYTER

As this ASIC is highly configurable, to make this ASIC operate as per the detector requirements, several bias and trim settings need to be done in advance. Few parameters are easy to get as those are similar for maximum ASIC. But many parameters need to be identified using a detailed calibration procedure. Hence testing of these ASIC is very complex and crucial part of the readout integration with the detector.

3.1 Test-Bench Setup

This ASIC has an internal pulse generator which is designed to generate an internal pulse of range 0–15 fC as shown in Fig. 2 as AMP_CAL. In MUCH mode, the maximum dynamic range supported is 0–100 fC and the internal pulse generator covers about 15% of the required range. Hence an external pulse generator along with a charge injector circuit is required to calibrate all the ADC comparators threshold in the full range of the MUCH mode. As can be seen from Fig. 3, the STS/MUCH-XYTER-based Front End Electronics (FEE) board is connected with a charge injector circuit at the input which is fed by a pulse using Tektronix arbitrary waveform generator. The back-end of the FEE board is connected with a twisted pair flat ribbon cable to AFCK board. FPGA firmware of AFCK board emulates GBTx e-link interface and later performs initial data preprocessing. AFCK can be readout by either

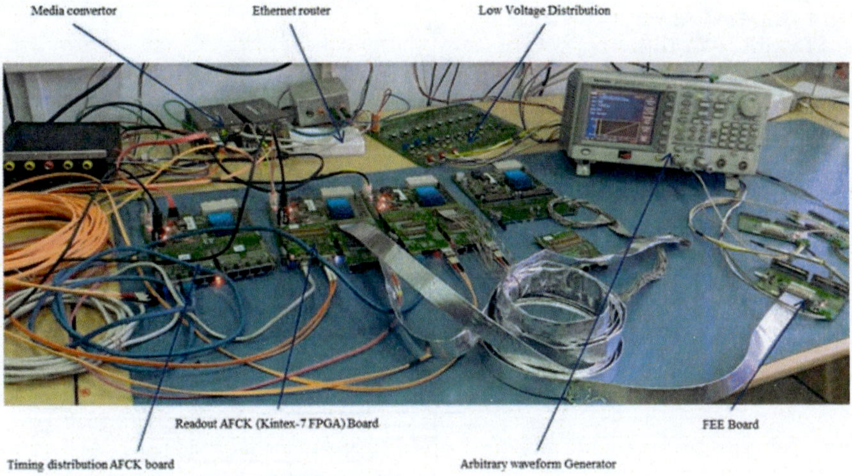

Fig. 3 Test-bench picture of the setup

Internet Protocol Bus (IPBus) or via 10G optical link to a First Level Event Selector (FLES) [10] via a custom FPGA PCIe board Interface Board (FLIB) [10] board of CBM. All the configuration data to the ASIC is sent via IPBus while for readout both IPBus as well as 10G backlink can be used. IPBus is connected to Ethernet switch using the media converter. The data acquisition (DAQ) computer is also required to be in the same network and RARP protocol is running on this computer which distributes the IP address to all the boards connected to this network. Agilent 6700 series power supply system is used to feed the power to all the electronics associated with the test-bench. The waveform generator and the low voltage power supply are kept in the same network as DAQ machine so as to have the full control over these instruments. To vary the input charge to the FEE board, pulse height needs to be varied. Hence, waveform generator setting needs to be changed several times during the calibration procedure. Hence, an automation script in python is made to control the waveform generator pulse height and embedded into the main calibration script to completely automatize the optimization procedure.

3.1.1 Procedure Used to Calibrate the FLASH ADC

As can be seen from Fig. 2, each FLASH ADC has two global settings, namely, ADC_VREFP and ADC_VREFN which is controlled by a 6-bit DAC. These biases set the baseline and the range of the FLASH ADC in analog voltages. There is one more bias setting ADC_VREFT, which shifts the baseline value of the incoming pulse to the ADC as can be seen in Fig. 2. Later in the ADC, each comparator has further trim setting, which is controlled by an 8-bit register connected to the DAC, which can be further used to trim the value of each of the comparators. The

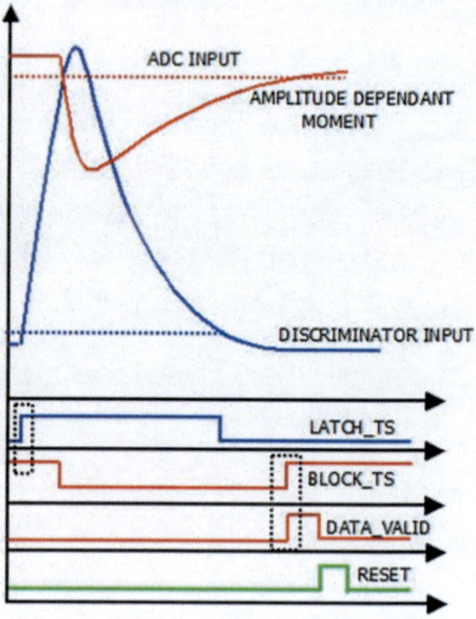

Fig. 4 Trigger waveform of the STS/MUCH-XYTER

output of the detector is readout using a serial differential e-link which operates on a maximum rate of 320 Mbps where each data point in this ASIC corresponds to 32 bits. As this is a self-triggered ASIC, even a single noisy channel may occupy full output bandwidth. Due to this bandwidth limitation, the lowest threshold is decided in MUCH mode such that it doesn't generate too much noise without injecting any external charge. For the present setup, the lowest threshold value is 3 fC and above due to this limitation. To determine the bias settings of the FLASH ADC, all the trim values of the comparators are set to be about 128 which is the middle range of the 8-bit trim DAC. Now data is continuously acquired using the DAQ and the data rate and channel hit frequency are monitored using online monitoring plots. After having a setting for ADC_VREFP, ADC_VREFN, and ADC_VREFT along with all the ADC comparator trim values set to 128, when many channels start firing with reducing ADC_VREFT by one or two values from set value and none of the channels is firing with the set ADC_VREFT, we can assume all the bias are set to the noise limit of the ASIC. Once the bias is set for a reasonable level, the procedure of the calibrating individual comparator of the ADC is initiated (Fig. 5).

3.1.2 Fast Shaper Comparator Calibration

After the ADC calibration is finished, fast shaper value needs to be calibrated matching to the lowest ADC comparator. The logic is that if there is a trigger, we need to know the time information when it occurred. If the fast comparator does not fire while the lowest ADC fired, then we will get the wrong timing information. Hence

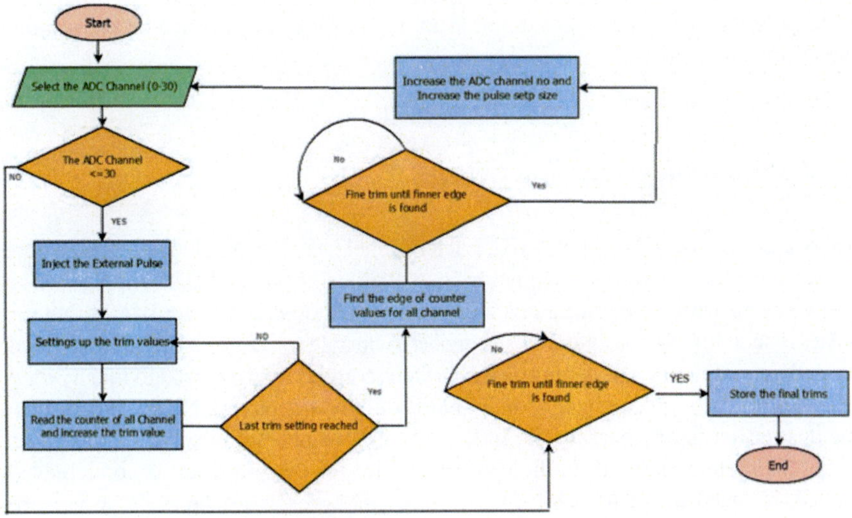

Fig. 5 Linearity plot of ASIC after calibration

waveform generator is again set to the lowest comparator ADC setting. Before setting the trimming values at the input of the comparator, global bias setting for the fast shaper needs to be determined which is indicated as THR2_GLOB in Fig. 2. This bias sets up the baseline of the incoming pulses. This bias is required to be set such that with the injected charge equivalent to the lower comparator threshold of ADC and the fast shaper comparator trimming values are in their mid range, then at least more than 50% fast shaper channel should be firing with respect to the input injected pulse.

After getting these bias parameter, trim values are tuned with the same procedure applied for the slow shaper comparator trimming as explained earlier.

3.1.3 The Calibration Algorithm

In Fig. 6, the calibration technique has been described. First of all, a certain amount of charge has been injected to the ADC channel at which it will be calibrated. Then a sequence of trim values ranging from 30 to 150 are sent to the DAC of a particular comparator with a step value 5. Each and every comparator of the Flash ADC has a 12-bit counter associated with it. It counts how many times a comparator gets fired for a particular trim value. The set of counter values that have been read in the previous step for a comparator are stored in an array. These sets of values are called coarse values. Then we determine the trim value for which the counter value is just greater than 50. This process is called coarse switching. In the end, fine trimming is done to get more closer trim values. All the trim values for all the 31 comparators of all the 128 analog channels are stored in a text file which is uploaded before running the DAQ system.

4 Automation of the Test Setup

There are 128 channels in this ASIC. Each channel has 32 ADC channels which need to be calibrated. Each ADC channel is controlled using an 8-bit DAC input which is known as the trim setting for a given channel. The motivation behind designing this ASIC is such highly configurable settings is to be able to set this ASIC in both linear as well as nonlinear ADC settings as and when required. It is required to find a proper trim setting for a given channel such that it only responds to a charge input above the desired/set input charge only. In this process we need to inject about 50 pulses to get the systematics and set the trim value until the desired result is not achieved. Hence, we need to set 4096 8-bit DAC setting in order to set the ASIC to perform with desired range and linearity.

4.1 Automation of Calibration Procedure

With the manual operation of changing of pulse height, it was taking about 30 min to calibrate the full ASIC for either positive or negative charge input. On various occasions due to some human error as well as setup glitches, these calibrations got stuck in the middle and the whole procedure needs to be repeated to generate a proper calibration file. Later on, the automation of the waveform generator was incorporated in the calibration routine and after that it is only taking 8 min to calibrate 4096 register values one ASIC. The waveform generator used for this purpose was Tektronix AFG 3252 and the standard software library was used namely PyVISA. Python scripts were made to automatically change the input voltage levels when one comparator-type calibration is finished.

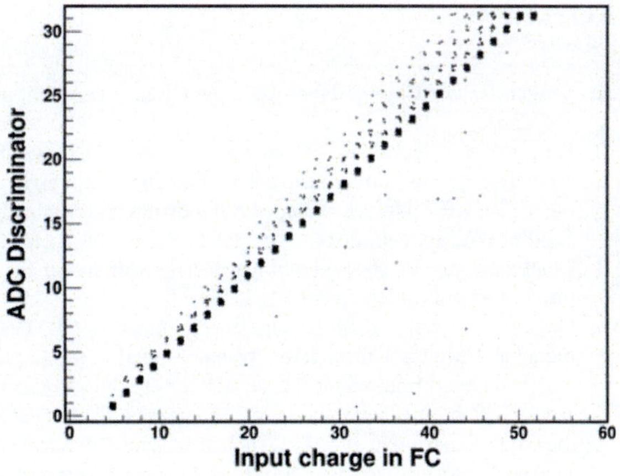

Fig. 6 Flowchart of the full calibration procedure

4.2 *Testing of Noise and Other Parameters*

Apart from these bias parameter, there are about 12 number of global bias parameters which needs to be optimized like shaping time constant, bias current to the input and output stages, etc. To optimize these bias parameters, a known charge is injected to the FEE board with various rise time and fall time with varying frequency. Later on, a continuous DAQ acquires the data and the responses were observed using online monitoring plots. For noise test of this system, we rely on the data record rate of the DAQ system. Further study is going on to fully optimize and understand the behavior of these bias parameters on the behavior of the electronics circuit of the ASIC.

5 Summary

A linearity check after calibration shows that the ADC channels are behaving linearly as can be seen in Fig. 6 which signifies that the calibration procedure developed is working as per expectations. Fast shaper test results show that qualitatively the threshold setting is set close to the lowest slow shaper ADC comparator setting as expected. Other bias parameters and noise test are ongoing. Integration with the MUCH detector will follow after these tests are completed.

Acknowledgements Thanks to Department of Atomic Energy Government of India (DAE), Department of Science and Technology, Government of India (DST) for funding of the CBM-India project.

References

1. Senger, P.: The Cbm collaboration, the CBM experiment at FAIR. J. Phys.: Conf. Ser. **50**(1), 357 (2006)
2. Kasinski, K., Kleczek, R., Otfinowski, P., Szczygiel, R., Grybos, P.: STS-XYTER, a high count-rate self-triggering silicon strip detector readout IC for high resolution time and energy measurements. In: IEEE Nuclear Science Symposium and Medical Imaging Conference (NSS/MIC). Seattle, WA, pp. 1–6 (2014)
3. Chattopadhyay, S. et al.: Technical Design Report for the CBM: Muon Chambers (MuCh). http://repository.gsi.de/record/161297
4. Dubey, A.K. et al.: GEM detector development for CBM experiment at FAIR, Nuclear Instruments and Methods in Physics Research Section A: Accelerators, Spectrometers, Detectors and Associated Equipment, ISSN: 0168-9002, Vol. 718, pp. 418–420 (2013)
5. Adak, R.P. et al.: Performance of a large size triple GEM detector at high particle rate for the CBM Experiment at FAIR, Nuclear Instruments and Methods in Physics Research Section A: Accelerators, Spectrometers, Detectors and Associated Equipment, ISSN: 0168-9002, Vol: 846, pp. 29-35
6. Heuser, J. et al.: Technical Design Report for the CBM Silicon Tracking System (STS). http://repository.gsi.de/record/54798
7. Antonioli, P., Baldanza, C., Falchieri, D., Giorgi, F.M., Mati, A., Tintori, C.: Design and test of a GBTx based board for the upgrade of the ALICE TOF readout electronics. In: IEEE-NPSS Real Time Conference (RT). Padua, pp. 1–3 (2016)
8. Zabołotny, W.M., Kasprowicz, G.: Data processing boards design for CBM experiment. Proc. SPIE (2014)
9. Zabołotny, W.M., Gumiski, M., Pozniak, K.T., Romaniuk, R.S.: Time and clock synchronization with AFCK for CBM, Proceedings of SPIE, **9662**(1), 9662–143 2015
10. Hutter, D., de Cuveland, J., Lindenstruth, V.: CBM first-level event selector input interface demonstrator. J. Phys. Conf. Ser. **898**(3), 032047 (2017)

A New Function Mapping Approach in Defective Nanocrossbar Array Using Unique Number Sequence

Tanmoy Chaku, Mahaswata Kundu, Debanjan Dhara, and Malay Kule

Abstract The designs at nanoscale exhibit much higher defect rates than in conventional lithography-based VLSI designs. It demands new defect-tolerant schemes to achieve high yield at this scale. One of the most promising nanoscale computational architectures is the crossbar-based architecture. In order to realize various logic circuits using nanoscale crossbar arrays; different logic functions need to be mapped within these nanoscale crossbars containing defective crosspoints. In this work, we use a novel technique to find a proper assignment of different logic functions in the nanoscale crossbar arrays having defective crosspoints. Our proposed method is based on the generation and use of unique number sequence during function mapping. The unique sequence accelerates the matching of the functions and nanowires in an efficient way. Experimental results show that our algorithm provides satisfactory results in terms of success percentage of function mapping.

Keywords Nanoscale · Lithography · VLSI · Crossbar · Nanowires

T. Chaku (✉)
Jalpaiguri Government Engineering College, Jalpaiguri, West Bengal, India
e-mail: tanmoy.chaku@gmail.com

M. Kundu (✉)
Central Calcutta Polytechnic, Kolkata, India
e-mail: mahaswatakundu@gmail.com

D. Dhara (✉)
National Institute of Technology, Durgapur, India
e-mail: debanjan.dhara@gmail.com

M. Kule (✉)
Indian Institute of Engineering Science and Technology, Shibpur, Howrah, India
e-mail: malay.kule@gmail.com

D. Bhattacharjee et al. (eds.), *Proceedings of International Conference on Frontiers in Computing and Systems*, Advances in Intelligent Systems and Computing 1255,
https://doi.org/10.1007/978-981-15-7834-2_75

1 Introduction

Nanotechnology-based fabrication is emerging as an alternative to replace the lithography-based silicon technology [1] in the near future. Recently, various nanoscale devices have been developed using nanoscale components such as silicon nanowires (SiNWs) and carbon nanotubes (CNTs) and at the same time new computing architectures are also being proposed using them as basic building blocks [2–4].

Nanoscale fabrication is expected to dominate the future high-speed VLSI technology utilizing the rapid shrinking of the dimension of the traditional lithography-based VLSI devices. These molecular-electronic devices can be built with very high densities (10^{12} devices per cm^2), and operated at very high frequencies of the order of THz [3]. Such nanodevices require modern technology for fabrication and newer methodology for circuit design and analysis.

One of the most promising computational architectures is the crossbar-based architecture [7, 12, 13]. A nanoscale crossbar, as shown in Fig. 1, consists of two parallel planes of nanowires separated by a thin layer of an electrochemical species [9–11]. Each plane consists of a number of parallel nanowires of the same type. The wires in one plane cross the wires in the other plane at right angles. The region where two perpendicular wires cross is called a crosspoint. Few of these crosspoints may become defective during manufacture or at some later stages. If there exist some defective crosspoints in the crossbar, then instead of rejecting the entire circuit, we can try to use the same circuit avoiding the defective crosspoints and making use of the usable crosspoints. Here, we restrict our attention to the defects leading to inoperative connections (i.e., with no ability to activate them), rather than the defects that short out a wire, or prevent routing the output of one gate to the input of another.

Various techniques for mapping different logic functions in the nanoscale crossbar circuit have been reported in the recent past. Integer Linear Programming based method for mapping the functions onto the defective crossbar has been presented in [6]. The greedy algorithmic approach for mapping functions has been reported in [8]. A graph matching based technique has been presented in [9], where the authors

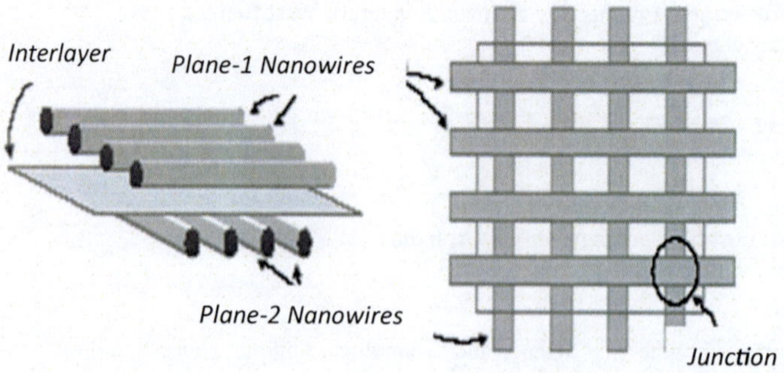

Fig. 1 Nanoscale crossbar

constructed the circuit graph and crossbar graph to find the matching between them. But the time complexity to construct the graphs itself is very large, especially for large circuits.

In paper [5], the authors proposed a new technique for the analysis of manufacturing yield of nanoscale crossbar architectures for different values of defect percentage and crossbar-size. Their technique is based on logical merging of two defective rows (or two columns) that emulate a defect-free row (or column).

In this paper, a new function mapping technique is proposed where all the configurable junctions of a row are identified by elements of a unique number sequence such that the product of the elements of the sequence determines whether a given function can be mapped there or not. The rest of the work is organized in three different sections. Section 2 elaborates the proposed work, Sect. 3 describes the experimental results, and finally Sect. 4 concludes the work.

2 The Proposed Method of Function Mapping

Here, **M** is the crossbar matrix which determines whether the crosspoints are stuck-at-open or not. When the cell value is low (i.e., 0) then the point is considered to be open, otherwise it is workable point.

$$M = \begin{bmatrix} 1\,1\,0\,1 \\ 1\,0\,0\,1 \\ 1\,1\,1\,0 \\ 1\,1\,0\,1 \end{bmatrix} \quad F = \begin{bmatrix} 1\,1\,0\,0 \\ 1\,0\,0\,1 \\ 1\,1\,1\,0 \\ 1\,1\,0\,1 \end{bmatrix}$$

Here, **F** is the matrix which represents the function. When the cell value is low (i.e., 0) then the function does not require the corresponding column in the crossbar matrix otherwise required.

S is the array of a sequence of integer numbers. Its generate by $\boldsymbol{X_{n+1} = 1 + X_n + n/4}$.

$$S = [2\ 3\ 4\ 5\ 6\ 7\ 9\ 11\ 13\ 16\ 19]$$

MD is 2D array which has two columns, index and value. The index value of MD matrix will be the row number of M matrix. Default value of value column of MD matrix is 1. The values of value column will be calculated using M matrix and S array as shown in Fig. 2:

Fig. 2 Flow diagram to generate MD array

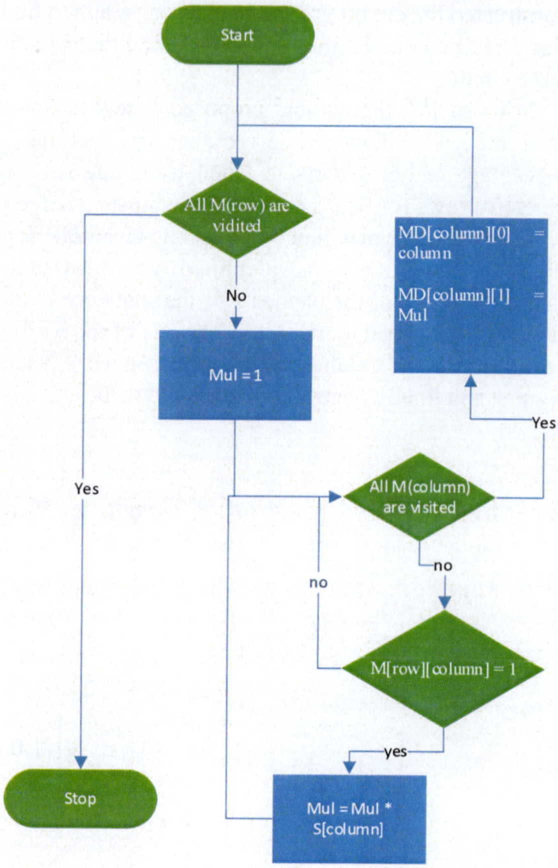

The product of values of the indices of S array where the values of M matrix are 1 for a particular row (Fig. 2).

Algorithm to Generate MD array

```
While(until all M(row) are not visited)
    Mul ← 1
    While(until all M(column) are not visited)
        If M[row][column] = 1 then
            Mul ← Mul * S[column]
        End if
    End while
    MD[column][0] = column
    MD[column][1] = Mul
End while
```

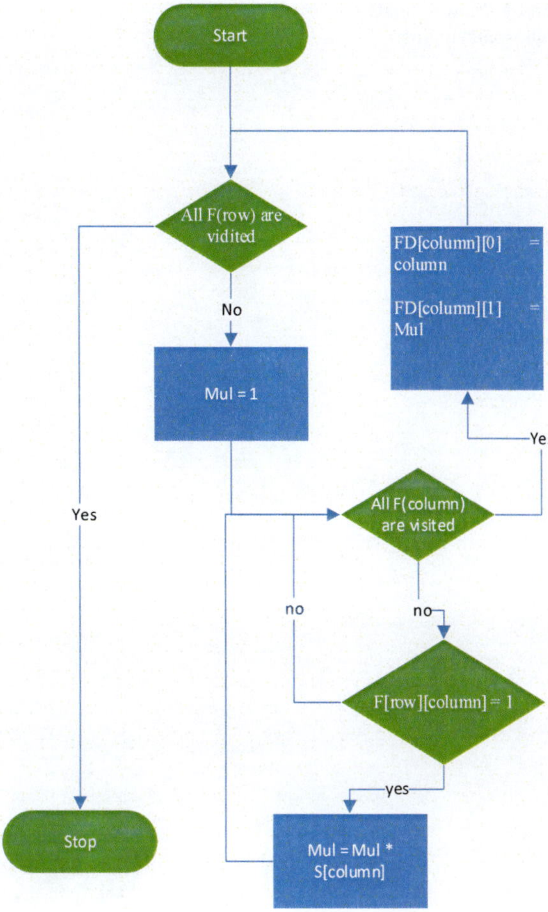

Fig. 3 Flow diagram to generate FD array

Consider the following two as an example of MD and FD (discussed below).

$$\text{MD} = \begin{bmatrix} 0 & 30 \\ 1 & 10 \\ 2 & 60 \\ 3 & 30 \end{bmatrix} \quad \text{FD} = \begin{bmatrix} 0 & 30 \\ 1 & 10 \\ 2 & 60 \\ 3 & 30 \end{bmatrix}$$

FD is 2D array which has two columns, index and value. The index value of FD matrix will be the row number of F matrix. Default value of value column of FD matrix is 1. The values of value column will be calculated using F matrix and S array as shown in Fig. 3:

The product of values of the indices of S array where the values of F matrix are 1 for a particular row (Fig. 3).

Our proposed function-mapping algorithm is depicted in the flow diagram in Fig. 4.

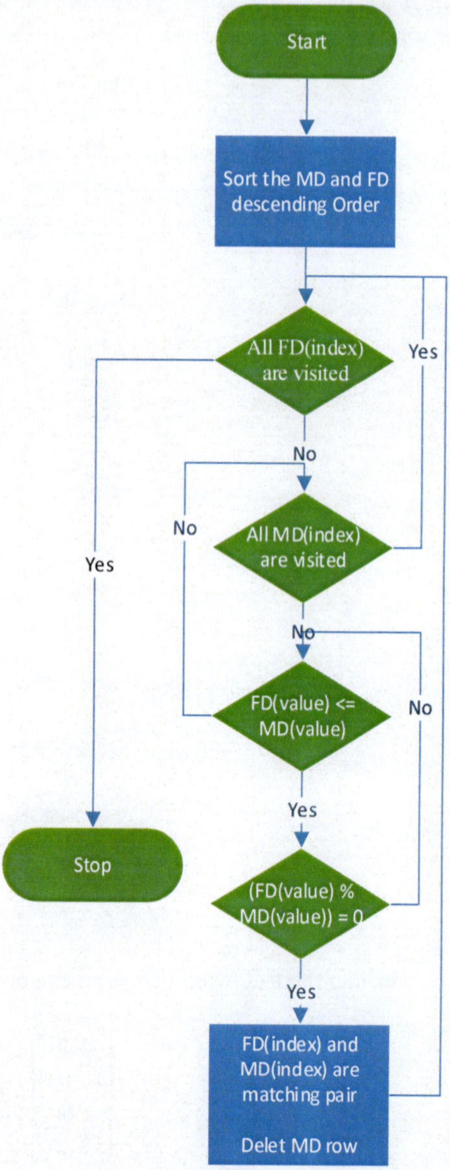

Fig. 4 Flow diagram of function mapping

Algorithm to Generate FD array

```
While(until all F(row) are not visited)
    Mul ← 1
    While(until all F(column) are not visited)
        İf F[row][column] = 1 then
            Mul ← Mul * S[column]
        End if
    End while
    FD[column][0] = column
    FD[column][1] = Mul
End while
```

Algorithm for Function Mapping

```
Sort the MD and  FD descending order
While(until all FD(index) are not visited)
    While(until all MD(index) are not Visited)
        İf FD(value) <= MD(value) then
            If ( MD(value) % FD(value)) = 0 then
                MD(index) and FD(index) are matching pair
                    Delete MD row
                    break while
            End If
        End If
    End while
End While
```

3 Experimental Results

The binary representations of 1–255 decimal numbers are used as a function. There are 255 rows and 8 columns in the function matrix, which stores all functions. Now, crossbar matrix is generated randomly with 255 rows and 8 columns.

Number of defects are calculated using defect percentage (d) as (255 * 8 * d). The experiment is performed 1000 times for every value of defect percentage and among them random 5 results are shown in Table 1. We have calculated the average of 1000 results and then have converted into success percentage (average value * 224/100).

Figure 5 graphically shows that the proposed algorithm provides good function mapping success percentage even with the higher values of defect percentage.

Table 1 Function mapping success percentage

Defect percentage	No of functions mapped					Average	Percentage
0	254	254	254	254	254	254	100
5	254	254	254	254	254	254	100
10	254	254	254	254	254	254	100
15	254	254	254	254	254	254	100
20	254	254	254	254	254	254	100
25	254	254	254	254	254	254	100
30	254	254	254	254	254	254	100
35	254	254	254	254	254	254	100
40	254	254	254	254	254	254	100
45	254	254	254	254	254	254	100
50	254	254	253	254	254	253.8	99.92126
55	247	241	241	245	248	243.6	95.90551
60	217	222	224	221	220	221.8	87.32283
65	200	193	196	195	194	195.3	76.88976
70	161	168	168	168	669	215.1	84.68504
75	128	148	147	141	142	138.2	54.40945
80	110	104	103	102	103	104.6	41.1811
85	76	69	80	78	77	77.5	30.51181
90	47	55	53	53	59	53.6	21.10236
95	30	26	28	34	24	28.8	11.33858
100	0	0	0	0	0	0	0

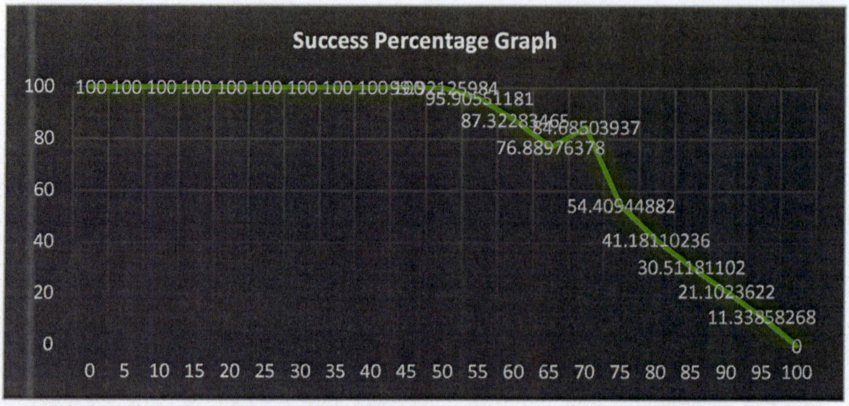

Fig. 5 Success percentage versus defect percentage graph

4 Conclusion

The work presented in this paper reports a new logic function mapping technique in defective nanoscale crossbar arrays using a unique number sequence. It shows good function mapping performance even in the higher range of defect percentages. The work may further be extended with other types of defects that may occur in nanocrossbar circuits.

References

1. Bourianoff, G.: The future of nanocomputing. IEEE Comput. **38**(8), 44–53 (2003)
2. Goldstein, S.C., Budiu, M.: Nanofabrics: spatial computing using molecular nanoelectronics. In: 28th International Symposium Computer Architecture, pp. 178–189 (2001)
3. Goldstein, S.C., Rosewater, D.: Digital logic using molecular electronics. In: International Solid-State Circuits Conference, p. 125 (2002)
4. Mishra, M., Goldstein, S.: Scalable defect tolerance for molecular electronics. Non-Silicon Computation (NSC-1), p. 78 (2002)
5. Kule, M., Rahaman, H., Bhattacharya, B.B.: On finding a defect-free component in nanoscale crossbar circuits. In: 4th ICECCS 2015, Procedia Computer Science, vol. 70, pp. 421–427 (2015)
6. Yang, J.-S., Datta, R.: Efficient function mapping in nanoscale crossbar architecture. In: IEEE International Symposium on Defect and Fault Tolerance in VLSI Systems, pp. 190–196, October 2011
7. Tahoori, M.B.: Low-overhead defect tolerance in crossbar nanoarchitectures. ACM JETC, July 2009
8. Naeimi, H., DeHon, A.: A greedy algorithm for tolerating defective crosspoints in NanoPLA design. In: IEEE International Conference on Field-Programmable Technology (FPT 2004), 6–8 December 2004
9. Hogg, T., Snider, G.: Defect-tolerant logic with nanoscale crossbar circuits. In: HP Labs, Palo Alto, CA, 25 May 2004
10. Butts, M., DeHon, A., Glodstein, S.C.: Molecular electronics: devices, systems and tools for Gigagate, Gigabit chips. In: ICCAD 2002, pp. 440–443
11. DeHon, A., Wilson, M.J.: Nanowire-based sublithographic programmable logic arrays. In: FPGA, pp. 123–132, February 2004
12. Ziegler, M., Stan, M.: Design and analysis of crossbar circuits for molecular nano-electronics. In: IEEE Conference on Nanotechnology (IEEENANO), pp. 323–327 (2002)
13. Su, Y., Rao, W.: Defect-tolerant logic mapping on nanoscale crossbar architectures and yield analysis. In: 24th IEEE International Symposium on Defect and Fault Tolerance in VLSI Systems, DFT'09, pp. 322–330, 7–9 October 2009

Tunneling Barrier Modulation in Negative Capacitance-TFET

Anup Dey and Ruben Ray

Abstract An analytical model for surface potential is developed in this paper for calculating tunneling barrier width modulation of a ferroelectric gate negative capacitance tunneling FET (NC-TFET) device. The influence of using ferroelectric material as gate stack on surface potential is investigated and steep subthreshold slope is indicated. However, it is found that the improvement of drain current is rather limited at high drain bias indicating internal voltage amplification due to ferroelectric negative conductance effect useful for low bias. The underlying physics of the device is discussed and envisaged that coupling ferroelectric material will improve the device performance as a switch.

Keywords Band-to-band (BTBT) tunneling · Ferroelectric material · Negative capacitance · Tunneling FET · Surface potential

1 Introduction

In conventional field-effect transistors (FETs), at best 60 mV/decade subthreshold can be achieved and this causes great restriction on the power dissipation in general MOSFET-based switches [1]. In order to improve subthreshold slope, several efforts are being made in the recent past. The tunneling FET(TFET) comes out as a very remarkable candidate because it can operate in sub-kT/q subthreshold slope regimes, which in turn allows the reduction of supply voltage, an essential requirement for low power applications. Recently, negative capacitance FET (NC FET) attracted great interest due to its high on-off ratio, low subthreshold slope, and improved output conductance performance. The ferroelectric material can exhibit negative capacitance under certain conditions, and this property is exploited for improving the

A. Dey (✉)
Jalpaiguri Government Engineering College, Jalpaiguri 735102, India
e-mail: a_dey2002@yahoo.com

R. Ray
Government College of Engineering and Leather Technology, Salt Lake, Kolkata 700106, India
e-mail: ruben.ray@gmail.com

D. Bhattacharjee et al. (eds.), *Proceedings of International Conference on Frontiers in Computing and Systems*, Advances in Intelligent Systems and Computing 1255,
https://doi.org/10.1007/978-981-15-7834-2_76

subthreshold characteristics of the FET [2–6]. Based on this idea, few efforts are already been made and demonstrated remarkable improvement of drain current and transfer characteristics in conventional MOSFET [6]. Very recently, Lee et al. [7] and few others [8–10] examined the ferroelectric (FE) as gate dielectric material to improve *Ion* of a TFET utilizing the NC effect of ferroelectric (NC-TFET). In [8], the NC effect for improving a long channel TFET is demonstrated while in [9] the analysis has been made for a Gate all around NC-FET. Kobayasi et. al. [11] investigated ferroelectric HfO_2 NC-TFET performance through simulation. However, the critical study on the physical origin of NC TFETs performance enhancement due to ferroelectric material incorporation in gate stack is still scarce in the literature.

In this paper, we, therefore, focus on demonstrating the influence of tunneling barrier modulation of dual gate NC-TFET (DG NC-TFET) by systematically developing a surface potential model along the channel direction and provide the insight of underlying physical reasons for improving the drain current.

2 Model

The structure of the n-channel Dual Gated NC-TFET (DG NC-TFET) is shown in Fig. 1, which consists of metal-FE-insulator-semiconductor (MFIS) structure from the metal-gate to the semiconducting-channel direction. The channel is intrinsic silicon whose thickness (t_{si}) is 8 nm and the oxide thickness (t_{0x}) is 1 nm for the device under consideration. The ferroelectric layer thickness (t_{fe}) is 11 nm with crystalline hafnium silicon oxide as ferroelectric material. The length of the channel region (L_g) is 30 nm and having p-type body doping concentration N_s equal to 1×10^{15} cm^{-3} while doping levels of heavily doped source (p^+) and drain (n^+) are 1×10^{21} cm^{-3}. Crystalline hafnium silicon oxide (HfSiO) is used as the FE material because of its excellent compatibility to silicon and lowest achievable dimensionality essential for the implementation of modern VLSI systems [12]. The structure of a conventional

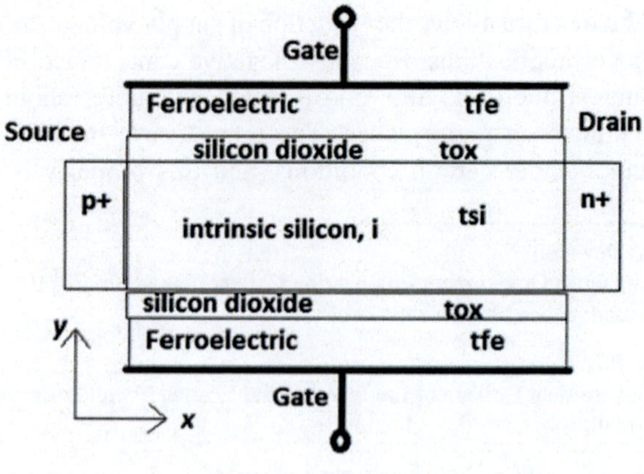

Fig. 1 Cross-sectional schematic view of the DG NC-TFET

DG-TFET is similar to the assumed MFIS device structure when the ferroelectric layer stack is removed. Besides, the insulator layer which acts as a buffer in between the ferroelectric layer stack material and the semiconducting channel layer [13] is likely to suppress interdiffusion caused across them.

The charge–voltage relationship of a ferroelectric can be obtained from the Landau theory [1] which essentially describes the analytical electric field(E) and the polarization (P) relation widely used for ferroelectric thin film, and is given by

$$E = V_{fe}/t_{fe} = 2\alpha P + 4\beta P^3 + 6\gamma P^5 \tag{1}$$

in which V_{fe} is the voltage drop occurring across the ferroelectric layer, and $\alpha(<0),\beta$ and γ are the material-dependent Landau parameters. Assuming that the polarization taking place in a ferroelectric layer is much larger compare to the displacement vector, the charge–voltage relationship across the ferroelectric film is given by

$$V_{fe} = 2\alpha t_{fe} Q + 4\beta t_{fe} Q^3 + 6\gamma t_{fe} Q^5 \tag{2}$$

where the displacement or charge per unit area $Q = P + \varepsilon_0 E \approx P$. When gate voltage is applied we can have

$$V_{gs} - V_{FB} - V_{fe} = \psi_s \tag{3}$$

where ψ_s is the potential at the semiconductor–oxide interface and V_{FB} is the flatband voltage. Referring to Fig. 1, for given gate voltage V_{gs} drop across the ferroelectric layer has only x-dependence due to the influence of band bending $\psi_s(x)$, caused by the drain bias and structural doping asymmetry. Therefore, under fixed gate bias voltage the following second-order differential for surface potential can be obtained [14]

$$\frac{\partial^2 \psi_s}{\partial x^2} - \lambda^2 \psi_s = \beta_1. \tag{4}$$

Here,$\lambda^2 = 2\varepsilon_{ox}/t_{si} t_{ox} \varepsilon_{si}$, ε_{ox} and ε_{si} are the permittivity of the oxide and semiconductor, respectively;$\beta_1 = \left(qN_s/\varepsilon_{si}\right) - qV_{gs1}\lambda^2$, $V_{gs1} - V_{gs} + V_{FB} = -V_{fe}$ and $V_{FB} = \Phi_m - \Phi_{si}$ is the difference between work functions of the gate metal and semiconductor. The general solution of Eq. (4) can be written as

$$\psi_s(x) = A\exp(\lambda x) + B\exp(-\lambda x) - \frac{\beta_1}{\lambda^2} \tag{5}$$

Following boundary conditions may be employed for the considered device

(i) At source channel interface

$$\psi_s(0) = V_{bi} \tag{6}$$

(ii) At drain end

$$\psi_s(L_g) = V_{ds} \tag{7}$$

(iii) Inside the channel region common displacement charge density $Q = \varepsilon_{si} d\psi_s/dx$ and V_{fe} are to be self-consistent where V_{fe} is given by (2).

From (2), (5), (6), and (7), we obtain

$$A = \frac{(V_{ds} + \frac{\beta_1}{\lambda^2})\exp(\lambda L) - V_{bi}}{\exp(2\lambda L) - 1} \tag{8}$$

$$B = \frac{\exp(2\lambda L)V_{bi} - (V_{ds} + \frac{\beta_1}{\lambda^2})\exp(\lambda L)}{\exp(2\lambda L) - 1} \tag{9}$$

2.1 Current Calculation

In order to develop a drain current model, one needs to find the electric field along the channel direction. By taking the derivative of the surface potential ψ_s as defined in (5) with respect to x, the expression for electric field in the channel direction is obtained as

$$\Xi_s(x) = \frac{\lambda(V_{ds} + \frac{\beta_1}{\lambda^2})[\exp(\lambda(L+x)) + \exp(\lambda(L-x))] - V_{bi}\lambda[\exp(-\lambda x) + \exp(\lambda(2L+x))]}{\exp(2\lambda L) - 1} \tag{10}$$

It can be inferred from the band profile that the conduction band (CB) in the channel gradually decreases from source to drain end, and hence the band-to-band tunneling current will start to occur when the conduction band of the channel just align with the quasi-fermi level (QFL) of the hole in the valence band (VB) of the source. The distance from the source end at which the surface potential in the channel drops to the potential corresponding to the source valance band QFL is denoted by $X_{\min}$. The CB of the channel will be even lowered with increase in positive gate voltage and consequent increase in tunneling current till it aligns with the QFL of electron in the drain conduction band. This corresponds to the gate voltage for which maximum tunneling current will occur and any further increase in gate voltage will lead the device in negative conductance region. Mathematically $X_{\min}$ may be expressed as

$$X_{\min} = d(\psi_s(0) - \psi_s((E_G + \varepsilon_f)/q) - d(\psi_s(0)) \tag{11}$$

in which $d(\psi_s(0))$ denotes the distance from the source channel junction where channel surface potential and the source potential($\psi_s(0)$) are equal, and $d(\psi_s(0) -$

$\psi_s((E_G+\varepsilon_f)/q)$ is the distance at which the surface potential drops by $(E_G+\varepsilon_f)/q)$ from $\psi_s(0)$. E_G is the band gap of the semiconductor while the parameter ε_f is used for accounting the energy difference between the source VB and the QFL, if QFL lies inside the VB.

The drain current can be evaluated from the BTBT model suggested by Kane

$$I_d = A_{kane} \times X_{\min} \times F_{avg}^2 \exp(-B_{kane}/F_{avg}) \quad (12)$$

in which F_{avg} is the effective electric field causing tunneling of carriers from source VB to channel CB, and can be expressed as $q/X_{\min}(E_G + \varepsilon_f)$; A_{kane} and B_{kane} are Kane's tunneling parameters, strongly influenced by material's effective mass and band gap. In order to find the value of $X_{\min}$, we note from (5) that $\psi_s(0) = Vbi - \frac{\beta_1}{\lambda^2}$, and with little algebra the distance $L_{\min}$ at which the surface potential drops by $(E_G + \varepsilon_f)/q)$ can be evaluated as $L_{\min} = (1/\lambda)\ln\left[\left\{\left(\frac{E_G+\varepsilon_f}{q} + \frac{\beta_1}{\lambda^2}\right) + \sqrt{\left(\frac{E_G+\varepsilon_f}{q} + \frac{\beta_1}{\lambda^2}\right)^2 - 4AB}\right\}/2B\right]$ where A and B are constants as defined earlier in (8) and (9). Knowing $\psi_s(0)$ and $L_{\min}$, one can easily find the value of $X_{\min}$.

3 Results and Discussion

Using (5), the surface potentials for both DG NC-TFET and DG TFET are plotted along the source to drain direction for various gate voltages in Fig. 2. The gate metal is assumed to have work function as 4.3 eV for analyzing the device performances. It may be observed that NC-FET curves show significant barrier modulation due to ferroelectric NC characteristics. The NC effects are more pronounced at the source side and have a reverse impact at the drain side. The lowering of the barrier is almost similar for variation of gate voltages. This lowering of barrier causes reduced shortest tunneling distance $X_{\min}$ and thereby improved transfer characteristics of the device is achieved. In Fig. 3 we provide the plot of the band diagram and show the reason for lower off-state characteristics of the device and its threshold voltage. The lateral electric field dependences on various gate voltages along the channel direction are plotted in Fig. 4. It is found that the chief NC effect is to increase the field strength at the source end while substantial reductions of field strength at the drain end. This facilitates the tunneling of carriers to take place which eventually reach the drain end in NC-TFET. Using (10) and (12), we extract the transfer output characteristics and found that the drain characteristics are significantly improved when compared NC-FET characteristics with that of the conventional TFET as indicated in Fig. 5.

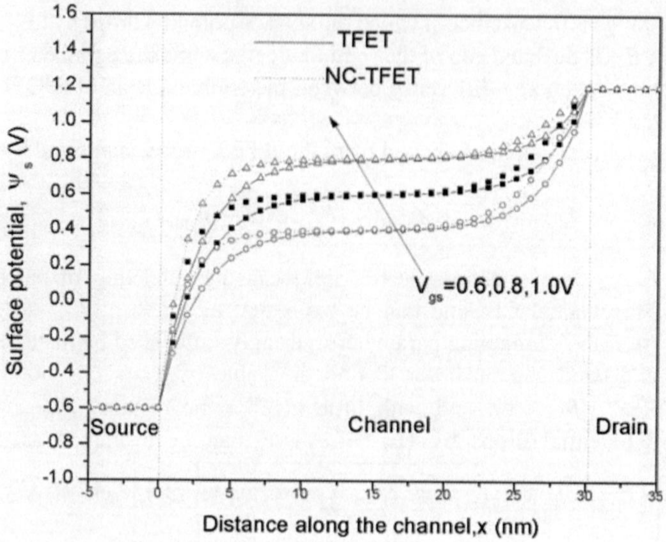

Fig. 2 Comparative surface potential plot of NC-TFET and that of conventional TFET along the channel direction to show the impact of NC

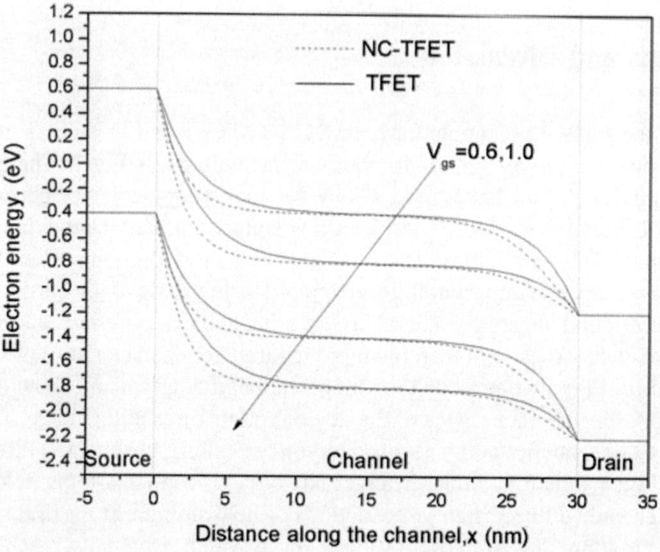

Fig. 3 Band profile along the source to drain direction at the oxide–semiconductor interface for $t_{fe} = 0$ nm and $t_{fe} = 11$ nm when $V_{gs} = 0.6$ and 1.0 V to demonstrate the NC effect of the ferroelectric layer on tunneling barrier modulation

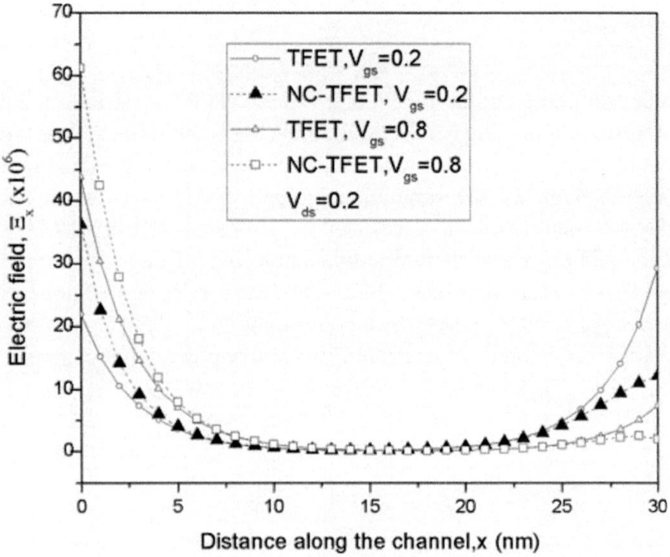

Fig. 4 The plot of electric field in the source to drain direction for NC-TFET and conventional TFET showing impact of NC for improving tunneling probability

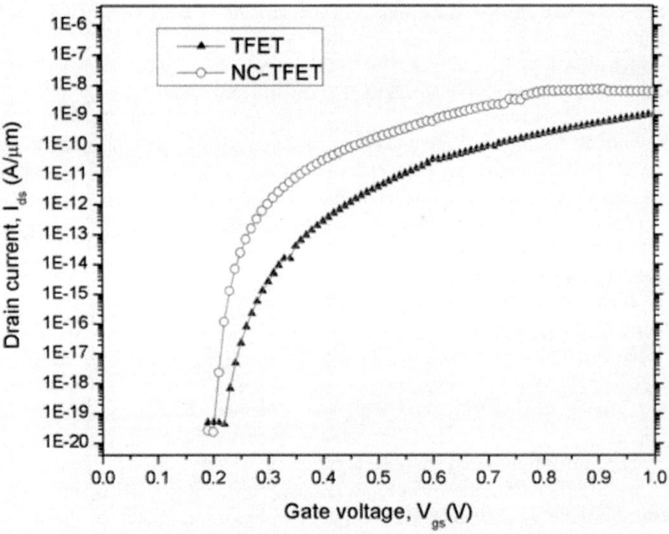

Fig. 5 Comparative plot of drain current as a function of gate voltage (I_D-V_G characteristics) for NC-TFET and conventional TFET for drain bias $V_{ds} = 0.5$ V. The transfer characteristics show the impact of NC effect on device tunneling current

4 Conclusions

In conclusion, surface potential model of a DG NC-TFET is derived and the underlying mechanism of tunneling barrier modulation due to NC effect of the ferroelectric layer is demonstrated. The device characteristics were investigated using this model and validated through TCAD simulation as the limiting cases when ferroelectric thickness layer is made zero and agrees well with our analytical results. It is observed that the DG NC-TFET has substantial subthreshold and off-on drain current improvement in comparison to conventional DG TFET due to barrier modulation by NC effect using ferroelectric stack. Besides, more controllability on device characteristics can be achieved and hence predicts its suitability for low power applications.

References

1. Salahuddin, S., Datta, S.: Use of negative capacitance to provide voltage amplification for low power nanoscale devices. Nano Lett. **8**(2), 405–410 (2008)
2. Lee, M.-H. et al.: Physical thickness 1. x nm ferroelectric HfZrOx negative capacitance FETs. In: 2016 IEEE International Electron Devices Meeting (IEDM) 12–1. IEEE (2016)
3. Zhang, X., Gong, X., Liang, G.: Analysis on performance of ferroelectric NC-FETs based on real-space gibbs-free energy with atomic channel structure. IEEE Trans. Electron Dev. **66**(9), 1100–1106 (2019)
4. Lin, C.-I., Khan, A.I., Salahuddin, S., Hu, C.: Effects of the variation of ferroelectric properties on negative capacitance FET characteristics. IEEE Trans. Electron Dev. **63**(5), 2197–2199 (2016)
5. Dong, Z., Guo, J.: A simple model of negative capacitance FET with electrostatic short channel effects. IEEE Trans. Electron Dev. **64**(7), 2927–2934 (2017)
6. Pahwa, G., et al.: Analysis and compact modeling of negative capacitance transistor with high ON-current and negative output differential resistance—Part II: model validation. IEEE Trans. Electron Dev. **63**(12), 4986–4992 (2016)
7. Lee, M.-H., et al.: Ferroelectric gate tunnel field-effect transistors with low-power steep turn-on. AIP Adv. **4**(10), 107117 (2014)
8. Chowdhury, N., Azad, S.F., Khosru, Q.D.: Negative capacitance tunnel field effect transistor: a novel device with low subthreshold swing and high on current. ECS Trans. **58**(16), 1–8 (2014)
9. Jiang, C., Liang, R., Xu, J.: Investigation of negative capacitance gate-all-around tunnel FETs combining numerical simulation and analytical modeling. IEEE Trans. Nanotechnol. **16**(1), 58–67 (2016)
10. Liu, C. et al.: Simulation-based study of negative-capacitance double-gate tunnel field-effect transistor with ferroelectric gate stack. Jpn. J. Appl. Phys. **55**(4S), 04EB08 (2016)
11. Kobayashi, M., Jang, K., Ueyama, N., Hiramoto, T.: Negative capacitance for boosting tunnel FET performance. IEEE Trans. Nanotechnol. **16**(2), 253–258 (2017)
12. Böscke, T.S., Müller, J., Bräuhaus, D., Schröder, U., Böttger, U.: Ferroelectricity in hafnium oxide: CMOS compatible ferroelectric field effect transistors. In: 2011 International Electron Devices Meeting 24–5. IEEE (2011)
13. Schlom, D.G., Chen, L.-Q., Pan, X., Schmehl, A., Zurbuchen, M.A.: A thin film approach to engineering functionality into oxides. J. Am. Ceram. Soc. **91**(8), 2429–2454 (2008)
14. Reddy, G.V., Kumar, M.J.: A new dual-material double-gate (DMDG) nanoscale SOI MOSFET-two-dimensional analytical modeling and simulation. IEEE Trans. Nanotechnol. **4**(2), 260–268 (2005)

Page Replacement Technique on the Basis of Frequency of Occurrence of Pages

Sayak Das, Nirvik Ranjan Das, Soumik Kr. Basu, Hriddhi Mondal, and Avijit Bose

Abstract The problem with the past generation computers was less primary memory. Present-day computers also face the problem of availability of memory both cache and primary memory in terms of cost and space. Age-old algorithms like LRU, FIFO, and Optimal did exist though it was almost impossible to implement optimal algorithm practically. So, experimentation has been made in this paper where a new algorithm has been proposed and it is noticed to have better hit ratio than the existing ones. For this, a frequency-based page indices mechanism has been proposed and it is seen that in many cases of reference string it gives the better result.

Keywords Memory · Page replacement mechanism · Frequency-based page replacement

1 Introduction

In Computer Architecture, primary memory plays a crucial role. From past to present, the problem is arising between cost and size of memory. Of course, though research has made the secondary storage possible but in case of cache and primary memory still the gap exists. However, a crucial role is being played by operating system which tries to optimize the page fault rate by bringing in the pages from the secondary

S. Das (✉) · N. R. Das · S. Kr. Basu · H. Mondal · A. Bose
Computer Science and Engineering, MCKV Institute of Engineering, Howrah, India
e-mail: sayakdas1999@gmail.com

N. R. Das
e-mail: nirvik.nrd2010@gmail.com

S. Kr. Basu
e-mail: soumikbasu18@gmail.com

H. Mondal
e-mail: hriddhi1990@gmail.com

A. Bose
e-mail: avijit2691@gmail.com

D. Bhattacharjee et al. (eds.), *Proceedings of International Conference on Frontiers in Computing and Systems*, Advances in Intelligent Systems and Computing 1255,
https://doi.org/10.1007/978-981-15-7834-2_77

storage to the physical primary storage. Researchers have always tried to optimize new methodologies and some of the major page replacement strategies like LRU, FIFO already exists. Optimal strategy is probabilistic in nature and hence a stochastic optimization algorithm will be required to solve which is practically infeasible.

1.1 Literature Survey

Lots of research has taken place in this area since the past. LRU (Least recently used), FIFO (First in First Out) have already been implemented and are mostly used. Apart from that theoretically Optimal Page replacement strategy exists. Some of the other algorithms like LDF (Longest Distance First), NRU (Not Recently used) have been used for research. Here we take on some of the other algorithms. Rexha et al. [1] discusses about a synchronous LRU algorithm and at the same time NRU algorithm implementation but the feasibility is still not achieved. Shastri et al. [2] discusses about Belady's anomaly and finds that Optimal Page Replacement strategy is a better method. But as discussed earlier optimal is stochastic in nature and is thus unpredictable. The paper says LRU is better compared to FIFO. Khulbe et al. [3] discuses about a hybrid LRU model and this basically gives a better result compared to the other two papers. Meggido et al. [4] from IBM Research came out with ARC (Adaptive Replacement Cache) algorithm, which gives a better result in terms of hit ratio. Famous texts like Hwang [5] describe the LRU as the most efficient one. Govind Prasad et al. [6] describe "Block retrieval of pages algorithm" which claims a hit ratio of 30%. Hitha Paulson [7] compares several page replacement schemes that are both synchronous and asynchronous. MA et al. [8] describes a new policy aka weighting size and cost replacement policy based on the weighting replacement policy. The algorithm proposes to replace the largest weighting and cost object and thereby achieves optimization to a larger extent.

2 Page Replacement Algorithm Already Present

2.1 First in First Out (FIFO) Page Replacement Algorithm

The FIFO (First In First Out) algorithm is the simplest form of page replacement algorithm. As the name suggests, the page which is inserted into the main memory first goes out first when a page fault occurs. Let us consider a 20-page reference string.

Reference String: 1 2 3 4 5 3 5 4 2 1 2 3 4 5 9 8 7 6 5 1

1	1	1	4	4	4	4	4	4	1	1	1	1	5	5	5	7	7	7	1
	2	2	2	5	5	5	5	5	5	5	3	3	3	9	9	9	6	6	6
		3	3	3	3	3	3	2	2	2	2	4	4	4	8	8	8	5	5
F	F	F	F	F				F	F		F	F	F	F	F	F	F	F	F

Fig. 1 Number of hits = 4

1	1	1	1	5	5	5	5	5	5	5	5	5	5	5	8	8	8	8	1
	2	2	2	2	2	2	2	2	1	1	1	1	1	1	1	7	7	7	7
		3	3	3	3	3	3	3	2	2	3	3	3	3	3	3	6	6	6
			4	4	4	4	4	4	4	4	4	4	4	9	9	9	9	5	5
F	F	F	F	F					F		F			F	F	F	F	F	F

Fig. 2 Number of hits = 7

1	1	1	1	1	1	1	1	1	1	1	1	1	1	9	9	9	9	9	1
	2	2	2	2	2	2	2	2	2	2	2	2	2	2	8	8	8	8	8
		3	3	3	3	3	3	3	3	3	3	3	3	3	3	7	7	7	7
			4	4	4	4	4	4	4	4	4	4	4	4	4	4	6	6	6
				5	5	5	5	5	5	5	5	5	5	5	5	5	5	5	5
F	F	F	F	F										F	F	F	F		F

Fig. 3 Number of hits = 10

3 Page Frame (Fig. 1):

4 Page Frame (Fig. 2):

5 Page Frame (Fig. 3):

2.2 *Least Recently Used (LRU) Page Replacement Algorithm*

The LRU (Least Recently Used) Algorithm is one of the simplest forms of page replacement algorithm. Here, the idea is to replace the least recently used page in the main memory. Let us consider a 20-page reference string.

Reference String: 1 2 3 4 5 3 5 4 2 1 2 3 4 5 9 8 7 6 5 1

1	1	1	4	4	4	4	4	4	4	4	3	3	3	9	9	9	6	6	6
	2	2	2	5	5	5	5	5	1	1	1	4	4	4	8	8	8	5	5
		3	3	3	3	3	3	2	2	2	2	2	5	5	5	7	7	7	1
F	F	F	F	F				F	F		F	F	F	F	F	F	F	F	F

Fig. 4 Number of hits = 4

1	1	1	1	5	5	5	5	5	5	5	3	3	3	3	8	8	8	8	1
	2	2	2	2	2	2	2	2	2	2	2	2	2	9	9	9	9	5	5
		3	3	3	3	3	3	3	1	1	1	1	5	5	5	5	6	6	6
			4	4	4	4	4	4	4	4	4	4	4	4	4	7	7	7	7
F	F	F	F	F					F				F	F	F	F		F	F

Fig. 5 Number of hits = 8

1	1	1	1	1	1	1	1	1	1	1	1	1	1	9	9	9	9	9	1
	2	2	2	2	2	2	2	2	2	2	2	2	2	2	8	8	8	8	8
		3	3	3	3	3	3	3	3	3	3	3	3	3	3	7	7	7	7
			4	4	4	4	4	4	4	4	4	4	4	4	4	4	6	6	6
				5	5	5	5	5	5	5	5	5	5	5	5	5	5	5	5
F	F	F	F	F										F	F	F	F		F

Fig. 6 Number of hits = 10

3 Page Frame (Fig. 4):

4 Page Frame (Fig. 5):

5 Page Frame (Fig. 6):

2.3 *Optimal Page Replacement Algorithm*

In optimal page replacement algorithm, if page fault occurs then that page that won't be used for the maximum time in the future will be removed. It is stochastic nature and a bit hypothetical. Let us consider a 20-page reference string.

Reference String: 1 2 3 4 5 3 5 4 2 1 2 3 4 5 9 8 7 6 5 1

1	1	1	4	4	4	4	4	4	4	4	4	4	4	4	8	8	6	6	6
	2	2	2	5	5	5	5	5	1	1	3	3	3	9	9	7	7	7	1
		3	3	3	3	3	3	2	2	2	2	2	5	5	5	5	5	5	5
F	F	F	F	F				F	F		F		F	F	F	F	F		F

Fig. 7 Number of hits = 6

1	1	1	1	5	5	5	5	5	1	1	1	1	5	5	5	5	5	5	5
	2	2	2	2	2	2	2	2	2	2	2	2	2	9	9	9	6	6	6
		3	3	3	3	3	3	3	3	3	3	3	3	3	8	8	8	8	1
			4	4	4	4	4	4	4	4	4	4	4	4	4	7	7	7	7
F	F	F	F	F					F				F	F	F	F	F		F

Fig. 8 Number of hits = 8

1	1	1	1	1	1	1	1	1	1	1	1	1	1	9	9	9	9	9	1
	2	2	2	2	2	2	2	2	2	2	2	2	2	2	8	8	8	8	8
		3	3	3	3	3	3	3	3	3	3	3	3	3	3	7	7	7	7
			4	4	4	4	4	4	4	4	4	4	4	4	4	3	6	6	6
				5	5	5	5	5	5	5	5	5	5	5	5	4	5	5	5
F	F	F	F	F										F	F	F	F		F

Fig. 9 Number of hits = 10

3 Page Frame (Fig. 7):

4 Page Frame (Fig. 8):

5 Page Frame (Fig. 9):

3 Proposed Algorithm on Page Replacement

3.1 Frequency Page Replacement Algorithm

`f = Frame_size, N = Number of pages INF = infinity. a[] is an array where all pages are stored.` b[] is the table where the first column indicates page number and the second column indicates their frequency and their rows are of size f.

```
Freq_Page_Replacement(f,N,a[],b[][])
  for(i=f upto N){
    for(j=0 upto f)
      if(a[i] != b[j,0])then replace = replace+1
      if(replace=f) then min=INF
        frequency=0
        for(k=0 upto f)
          if(b[k,1] <= min)
            min=b[k,1]
            temp = k   //min frequency index extract
        b[temp,0]=a[i][end of if]
  for(k=i upto N) //calculating frequency of the page inserted after a page fault
    if(a[k]=a[i]) //storing required frequency
      frequency=frequency+1
      b[temp,1]=frequency //inserting frequency in the page frame table
    else then hit_no=hit_no+1
      for(k=0 upto f)
        if(a[i]=b[k,0])
          b[k,1]=b[k,1]-1 [end of else] //as hit occurred so decrement
  return b[][]
```

In this algorithm, it is considered that initially the F number of frames contains F unequal pages. Now if the frequencies of the pages within the frame are same then according to the algorithm if a page fault occurs then the most recently entered page will get removed by the new page in the list. Values in the second column indicate the frequencies, i.e., the unique IDs. The whole mechanism of the algorithm implementation is indicated in Figs. 10 (3 Page Frame), 11, (4 Page Frame) and 12 (5 Page Frame) respectively.

Reference String: 1 2 3 4 5 3 5 4 2 1 2 3 4 5 9 8 7 6 5 1

Time Complexity:

The time complexity of the above algorithm can be calculated as T(N) = (N−F)*(N−i + 2*F), where N is the number of pages and F is the frame size. This is because there

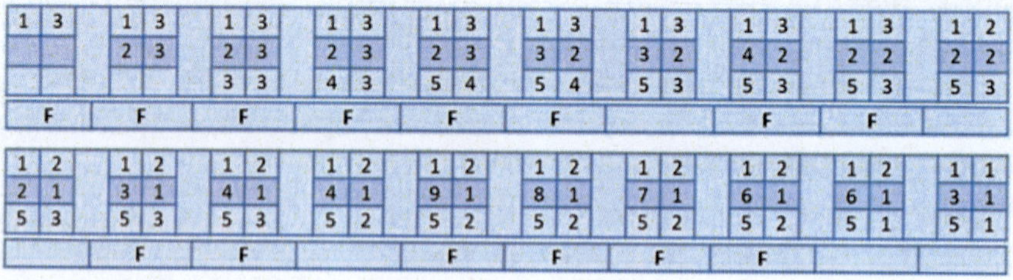

Fig. 10 Number of hits = 6

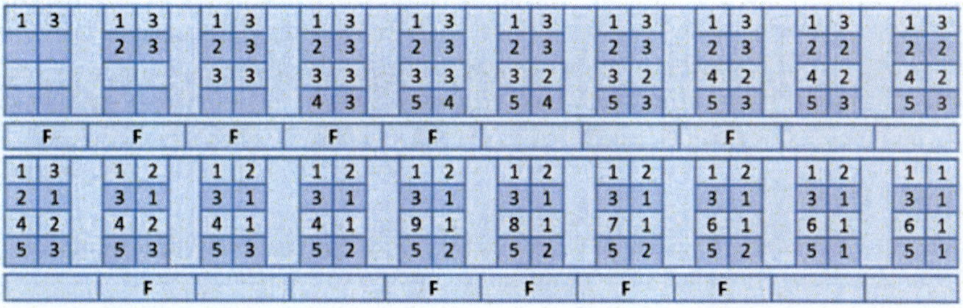

Fig. 11 Number of hits = 9

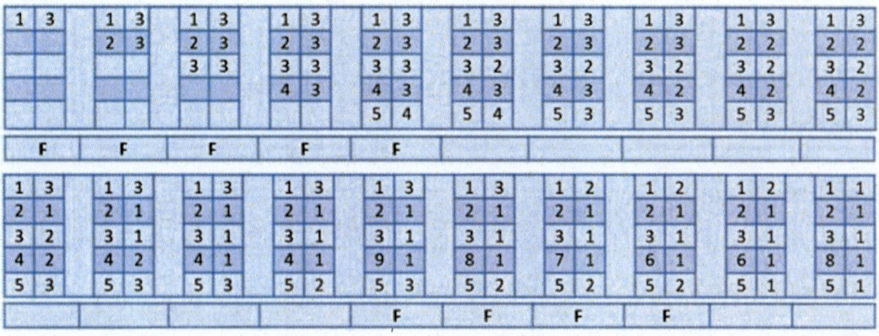

Fig. 12 Number of hits = 1

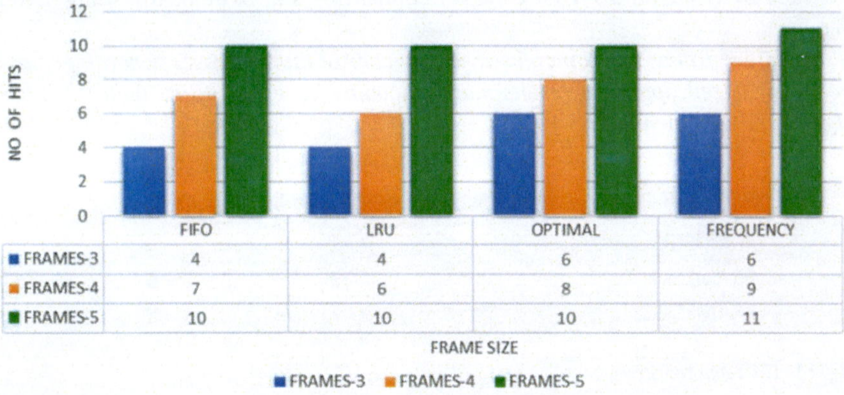

Fig. 13 Frequency base page replacement showing most efficiency taking the reference string provided randomly in the previous sections

are already F elements in the page frames so we can calculate frequencies from the array having N-F elements. So, the approximation of the above equation may lead to O (N^2), as the frame size is very less compared to the number of pages. So we can say the upper bound can be estimated as O (N^2).

Space Complexity:

Here an extra array is taken of size F where F is the frame size for storing the frequency of the ith page in the frame at ith index. So, space complexity is one for storing the pages which takes O(F) and another for frequency which also takes O(F) which leads to a value of O(2F) = O(F).

4 Comparison Between Different Page Replacement Techniques

Few Other Random Strings:

Ref String 1: 1 2 3 4 5 9 8 5 6 3 2 4 1 7 8 0 4 3 2 1 comparison shown in Fig. 14.

Ref String 2: 9 7 8 0 1 2 3 4 5 6 7 9 0 2 7 3 1 6 9 5 comparison shown in Fig. 15

Reference String 3: 4 5 6 6 7 1 0 8 4 0 7 7 3 0 3 2 4 5 2 6 1 2 8 0 5 3 6 9 3 5 4 0 6 4 2 3 7 2 8 6 0 0 3 6 2 5 0 2 0 8 7 4 7 9 9 7 5 9 4 5 1 9 6 1 1 7 1 6 1 2 3 7 0 0 2 5 3 6 2 0 4 0 0 9 6 3 5 1 9 4 8 7 4 1 7 8 9 2 4 7 comparison shown in the following figure Fig. 16.

Now if the frame size is increased, the number of hits increases accordingly and frequency-based approach results more efficiently in terms of hits than the other algorithms as shown in Fig. 16.

	LRU	OPTIMAL	FREQUENCY
FRAMES-3	1	2	**4**
FRAMES-4	1	3	**6**
FRAMES-5	2	7	**8**

Fig. 14 Hit Table for String 1

	LRU	OPTIMAL	FREQUENCY
FRAMES-4	1	2	**5**
FRAMES-6	1	8	**8**
FRAMES-8	6	10	**10**

Fig. 15 Hit Table for String 2

	LRU	OPTIMAL	FREQUENCY
FRAMES-8	70	73	75
FRAMES-16	76	81	82
FRAMES-26	73	73	73

Fig. 16 Hit Table for String 3

5 Conclusion

The purpose of this paper was to design an algorithm with the best possible page replacement scheme. As shown in Fig. 13, we notice the proposed algorithm going at par with the LRU and Optimal Page Replacement scheme. Even in some cases, it is performing better than LRU page replacement scheme. If we take into consideration optimal page replacement scheme it is stochastic in nature and sometimes hypothetical. FIFO already suffers from Belady's anomaly and very often found experimentally to give wrong prediction. As far as the space and time complexity is concerned, the proposed algorithm is having a time complexity of O (N^2) where N is the number of pages. The space complexity is O (F), where F stands for frame size. Our proposed algorithm is for a sequence where the starting of the sequence consists of as many numbers of unequal page indices greater than or equal to the number of frame size. In future more work is required so that the algorithm may run on real time Operating system.

References

1. Rexha, G.(Univeristy of Elabasan), Elmazi, E., Tafa, I. (University of Tirana): A comparison of three page replacement algorithms: FIFO, LRU, optimal. Acad. J. Interdisciplinatory Stud., MCSER Publishing VOL 4 NO 2S2 (2015)
2. Shastri, S., Sharma, A., Mansotra, V. (Univeristy of Jammu & Kashmir): Study of page replacement algorithms and their analysis in C#. Int. J. Eng. Sci. (IJES) **5**(1), 53–57 (2016)
3. Khulbe, P., Pant, S.: Hybrid(LRU) page replacement algorithm. Int. J. Comput. Appl. (0975–8887) (2014)
4. IBM RESEARCH REPORT, One Up On LRU; Nimrod Megiddo, Dharmendra S. Moddha, RJ 10285 (A0302-014) (2003)
5. Hwang, K. (Purdue University), Briggs, F.A. (Rice University): Computer Architecture and Parallel Processing McGraw-Hill International Edition (1985)
6. Arya, G.P. et al.: An improved page replacement algorithm using block retrieval of pages. IJET, 32–35 (2018)
7. Paulson, H. et al.: Page replacement algorithms—challenges and trends. IJCMS, 112–116 (2017)
8. MA et al.: An Improved Web Cache Replacement Algorithm Based on Weighting and Cost https://doi.org/10.1109/access.2018.2829142. IEEE Access 2018

Fault-Tolerant Implementation of Quantum Arithmetic and Logical Unit (QALU) Using Clifford+T-Group

Laxmidhar Biswal, Chandan Bandyopadhyay, Sudip Ghosh, and Hafizur Rahaman

Abstract The quest of efficient quantum circuit is to achieve quantum supremacy in theory as well as in practice. The foremost obstacle is to protect the cohesive time of extremely fragile quantum states from inherent noise. To address this issue, Quantum Error Correction Code (QECC) with fault-tolerant quantum circuit is most desirable. Aiming to contribute toward designing an efficient Quantum Information Processor (QIP); in this work, we have shown the design of an important QIP module, i.e., Arithmetic Logic Unit (ALU). The entire design has been made on top of quantum Clifford+T-group. In the design phase, initially, we formulate a 1-bit design and then to make a generalized representation of the ALU, multiple smaller modules have been integrated. For ensuring improved features in this component, the design has been made fault-tolerant, circuit optimization rules are executed to minimize the design metrics and parallelism in high latency T-gate is ensured. In a way to check the functional correctness of our proposed design, several logical operations have been successfully tested over it.

Keywords Clifford+T · QECC · ALU · T-depth · T-count.

Please note that the LNCS Editorial assumes that all authors have used the western naming convention, with given names preceding surnames. This determines the structure of the names in the running heads and the author index.

L. Biswal (✉) · S. Ghosh · H. Rahaman
School of VLSI Technology, IIEST, Shibpur, India
e-mail: laxmidhar.cvrce@gmail.com

S. Ghosh
e-mail: sudipghosh2005@gmail.com

H. Rahaman
e-mail: hafizur@vlsi.iiests.ac.in

C. Bandyopadhyay
Department of Information Technology, IIEST, Shibpur, India
e-mail: chandanb.iiest@gmail.com

D. Bhattacharjee et al. (eds.), *Proceedings of International Conference on Frontiers in Computing and Systems*, Advances in Intelligent Systems and Computing 1255,
https://doi.org/10.1007/978-981-15-7834-2_78

1 Introduction

Nowadays, quantum computer (QC) has become a hot topic within the academia and industries that could spur the development of new breakthroughs in science, engineering, and technologies. Since finding of quantum algorithms, the QC has outperformed over some of the well-known classically intractable problems with exponential faster speed [1, 2] . Basically, quantum computer works on the principle of quantum mechanics [3] where so-called powerful quantum algorithms are implemented with the help of quantum operators whose close Hamiltonian is reversible in nature for which the quantum computation becomes lossless computation. Furthermore, quantum computer relies on **Qu**antum **Bit** (Qubit) in lieu of classical bit. The formal definition and properties on qubit is presented in Sect. 2. Though the revolutionary QC has many more advantages but has an immense challenge in its realization due to catastrophic obstacles from noisy sources as well as from other constraints in logical and physical levels. For example, direct feedback and fan-out circuit are not allowed in the logical level of quantum circuit due to the limitation posed by the no-cloning theorem [4].

In a way to achieve so-called quantum supremacy, the inherent noisy sources either to be suppressed or to be corrected by using Quantum Error Correction Code (QECC); otherwise, the coherent superposition of basis states will be lost prior the computation ends caused due to decoherence [5]. In this regard, the most promising Surface code has been used systematically to detect and correct the errors but has some threshold limit (approximately 0.75% per gate) [6]. However, fault-tolerant circuit is most desirable so as to contain the error rate per gate of the QECC within the threshold limit.

Currently, the Clifford+T-group has been used as a universal set of primitive transversal operators for the implementation of fault-tolerant quantum algorithms. Thus, Clifford+T-group with surface code becomes a default platform for the realization of fault-tolerant quantum computation [5]. The details about the popular Clifford+T-group is described in Sect. 2. In fact, non-Clifford T-gate has high latency as well as high resource overheads in compared to other Clifford gates [7]. To address this issue, a new design strategy must be followed for efficient realization of a high scalable quantum algorithm. In this conjecture, fault-tolerant quantum circuit for 1-bit Arithmetic and Logical Unit (ALU) has been proposed in the logical level using Clifford+T-group.

This ALU is an integral part of the CPU, embedded into Quantum Information Processor (QIP) and has two unit, *viz.* logical unit and arithmetic unit. In the logical unit, the ALU performs NOT, OR, AND, XOR, and other operations too, whereas addition and subtraction operations are performed in the arithmetic unit.

Here, we are highlighting our key contributions as follows:

- we have designed fault-tolerant 1-bit QALU using Clifford+T-group.
- Further, different strategies are followed to optimize the number of non-Clifford T-gate, to operate T-gates in minimum cycle by means of parallelism, to optimize garbage output *vis-a-vis* make the design more efficient.

The remainder of this paper is organized as follows: a background of quantum gates and circuits is given in Sect. 2. In Sect. 3, our method is proposed and explained briefly. In Sect. 4, experimental result and comparative analysis are described. Finally, the work is concluded in Sect. 5.

2 Background

Here, we introduce the terms, notations related to our work, and also discuss the preliminaries on reversible and quantum logics.

2.1 *Reversible Gate and Reversible Circuit*

Definition 1 (*Reversible Gate*) It is a special kind of logic gate which does not lose any information at the time of operation. Feynman gate, Toffoli gate, Peres gate, and Fredkin gate are well-known reversible gates.

Definition 2 (*Reversible Circuit*) It is widely known as information-lossless, fan-out free acyclic logic circuit in which all gates are reversible. Both reversible gates and circuit have an equal number of input and output lines.

2.2 *Quantum Circuits*

Definition 3 (*Qubit*) It is defined as the coherent superposition of classical basis states with complex probability and used to represent quantum states. The term "Qubit" is derived from the term "**Qu**antum" "**bit**". Mathematically, it is represented by *Ket*-function as $|\psi\rangle$.

$$|\psi\rangle = \alpha|0\rangle + \beta|1\rangle \tag{1}$$

where $\alpha, \beta \in C \ and \ |\alpha|^2 + |\beta|^2 = 1$

Definition 4 (*Quantum circuit*) It is a directed acyclic graph where all vertices are presented by a finite set of primitive quantum operators and take the input as a tensor product of all the input states. Basically, all elementary primitive quantum gates are either 2×2 or 4×4 unitary transfer metrics. For better apprehension, here, a list of the quantum gates and its properties are summarized in Table 1.

Definition 5 (*Clifford+T-group*) It is a finite set of universal transversal primitive gates which contains CNOT-gate with 1-qubit H-gate, S-gate, Pauli(X, Y, Z)-gate,

Table 1 Elementary Quantum Gates and Their Properties

Name of elementary quantum gate	Block diagram	Transformation matrix	Properties	Name of elementary quantum gate	Block diagram	Transformation matrix	Properties
NOT (X)	$\oplus$	$\begin{bmatrix} 0 & 1 \\ 1 & 0 \end{bmatrix}$	$X\|0\rangle = \|1\rangle$ $X\|1\rangle = \|0\rangle$	CNOT(CN)	$\bullet$ / $\oplus$	$\begin{bmatrix} 1&0&0&0 \\ 0&1&0&0 \\ 0&0&0&1 \\ 0&0&1&0 \end{bmatrix}$	$\|00\rangle \rightarrow \|00\rangle$ $\|01\rangle \rightarrow \|01\rangle$ $\|10\rangle \rightarrow \|11\rangle$ $\|11\rangle \rightarrow \|10\rangle$
Z gate	$\boxed{Z}$	$\begin{bmatrix} 0 & 0 \\ 0 & -1 \end{bmatrix}$	$X\|0\rangle = \|0\rangle$ $X\|1\rangle = -\|1\rangle$	S gate	$\boxed{S}$	$\begin{bmatrix} 1 & 0 \\ 0 & i \end{bmatrix}$	$S\|0\rangle = \|0\rangle$ $S\|1\rangle = e^{\frac{i\pi}{2}}\|1\rangle$
$S^{\dagger}$ gate	$\boxed{S^{\dagger}}$	$\begin{bmatrix} 1 & 0 \\ 0 & e^{\frac{-i\pi}{2}} \end{bmatrix}$	$S^{\dagger}\|0\rangle = \|0\rangle$ $S^{\dagger}\|1\rangle = e^{\frac{-i\pi}{2}}\|1\rangle$	T gate	$\boxed{T}$	$\begin{bmatrix} 1 & 0 \\ 0 & e^{\frac{i\pi}{4}} \end{bmatrix}$	$T\|0\rangle = \|0\rangle$ $T\|1\rangle = e^{\frac{i\pi}{4}}\|1\rangle$
$T^{\dagger}$ gate	$\boxed{T^{\dagger}}$	$\begin{bmatrix} 1 & 0 \\ 0 & e^{\frac{-i\pi}{4}} \end{bmatrix}$	$T^{\dagger}\|0\rangle = \|0\rangle$ $T^{\dagger}\|1\rangle = e^{\frac{-i\pi}{4}}\|1\rangle$	Hadamard(H)	$\boxed{H}$	$\frac{1}{\sqrt{2}}\begin{bmatrix} 1 & 1 \\ 1 & -1 \end{bmatrix}$	H $\|0\rangle = \frac{1}{\sqrt{2}}(\|0\rangle + \|1\rangle)$ H $\|1\rangle = \frac{1}{\sqrt{2}}(\|0\rangle - \|1\rangle)$

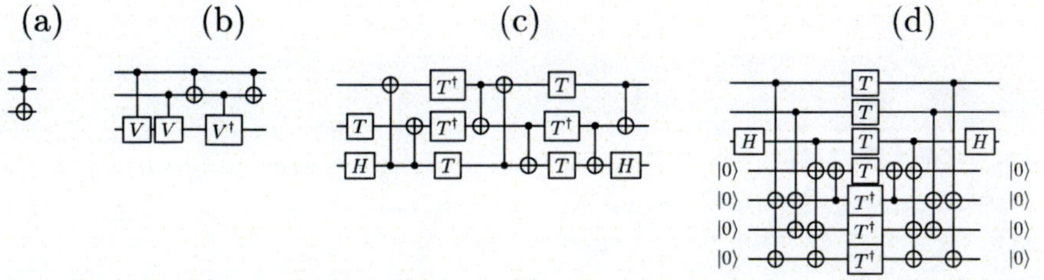

Fig. 1 **a** Reversible Toffoli gate. **b** Equivalent NCV library [9]-based decomposition of Fig. 1a. **c** Equivalent Clifford+T-based realization of Fig. 1a as depicted in [5, Fig. 1c]. **d** Equivalent unit T-depth-based realization of Fig. 1a as depicted in [10, Fig. 1]

and non-Clifford T-gate [8]. Mathematically, T-gate represents fourth root of Pauli-Z-gate and denoted by transfer matrix

$$T = \begin{bmatrix} 1 & 0 \\ 0 & e^{\frac{i\pi}{4}} \end{bmatrix} \tag{2}$$

Here, we present fault-tolerant representation of some well-known reversible gates.

Definition 6 (*Clifford+T-based fault-tolerant representation of Toffoli gate*) Fig. 1a presents universal reversible Toffoli gate in which the state of the target bit toggled when the state of both the control bit are set to 1 and its fault-tolerant implementation using Clifford+T-library is presented in Fig. 1c. Figure 1d presents unit phase depth-based implementation of Fig. 1a where additional ancilla inputs are used to configure the design.

Definition 7 (*Fault-tolerant implementation of reversible Fredkin gate*) Fig. 2a presents a schematic diagram of reversible Fredkin gate which swaps the states of the two input lines when control state is high. In Fig. 2a, line I_0 and I_1 represents input states, whereas C represents control lines. The outputs (O_0 and O_1) for the input lines I_0 and I_1 will be swapped for two input states if control state is high; otherwise it remains as it was. For example, on measurement, the output remains same as input I_0 state for $SC = 0$, and as I_1 for $C = 1$. Figure 2b presents Toffoli-based realization of Fig. 2a.

2.3 Performance Parameter of Fault-Tolerant Circuit

Though there exist a pool of cost matrices to evaluate the efficiency of the designs, here, we discuss only those parameters that are related to our work.

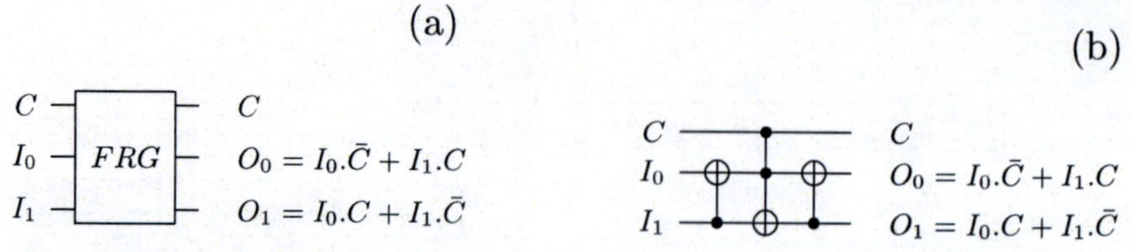

Fig. 2 **a** Schematic representation of reversible Fredkin. **b** Toffoli realization of Fig. 2*a*

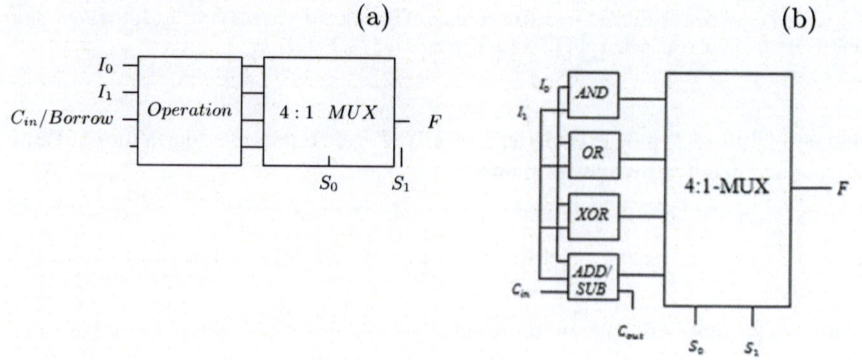

Fig. 3 **a** Schematic representation of ALU. **b** 1-bit conventional ALU

Definition 8 ($T - count$) The minimum number of logical T-gates required to realize a Boolean function into quantum circuital form is termed as $T - count$. Basically, T-gate changes the phase of the quantum state with null effect on changing of and that is why this T-gate sometimes is termed as phase gate.

Definition 9 ($T - depth$)The number of $T - cycle$ requires to execute all the T-gates of quantum circuit is known as $T - depth$.

Definition 10 (*Garbage Output*) It is defined as the unused output in reversible logic gate/circuit.

2.4 Arithmetic and Logical Unit (ALU

ALU is an important part of the information processor which performs various tasks such as arithmetic operations, bitwise logical operation, bit-shifting operation, and also performs the comparative operation through its two units, *viz.* logical unit and arithmetic unit. Here, we concentrate on designing of reversible ALU and its fault-tolerant quantum realization Fig. 3 presents conventional representation of 1-bit ALU.

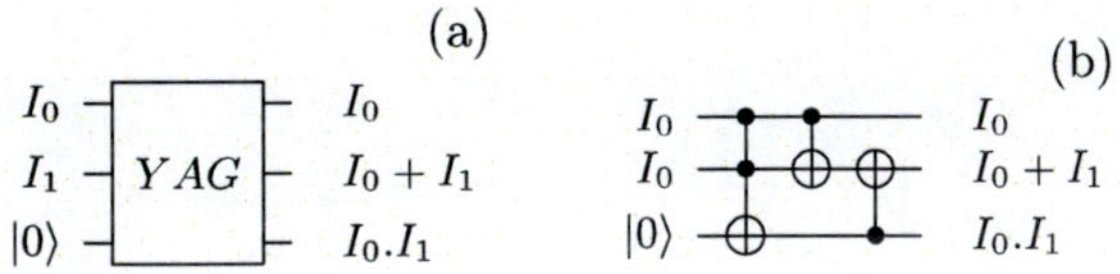

Fig. 4 **a** Reversible YAG-gate as depicted in [11, Fig. 13]. **b** Toffoli-based realization of Fig. 4a

3 Proposed Technique

In this section, we aim to present a fault-tolerant quantum implementation of 1-bit ALU. Here, we use template matching scheme over-optimized reversible ALU which can compute four tasks, *viz.* AND, OR, XOR, and ADD/SUB operations. This section divided into two sub-sections. In Sect. 3.1, a reversible 1-bit ALU is proposed. In Sect. 3.2, we map 1-bit reversible ALU into Clifford+T-based design.

3.1 Design of Reversible **1***-bit ALU*

The schematic diagram of 1-bit ALU is shown in Fig. 3a, where ALU has three sections, *viz.* input section (I_0, I_1, and $C_{in}/Borr$) followed by operation section which performs four tasks and the final output is shown by the multiplexer circuit which depends upon the input select lines. According to the reversible logic paradigm, if each small block of combinational circuits replaced by its equivalent reversible counterpart then the overall circuit will be reversible. Here, we have followed the strategies and focused on efficient implementation for 1-bit R-ALU.

Operation section of ALU: This section performs logical AND, OR, XOR operation and arithmetic operations like ADD/SUB operation. Logical AND, OR operation can be performed using reversible YAG-gate presented in [11, Fig. 13]. Figure 4 presents that reversible YAG-gate. Further, logical XOR and arithmetic ADD operation can be executed using Double Peres Gate (DPG). However, the subtraction operation can be performed with the same DPG-gate with the addition of one CNOT-gate followed by one Fredkin gate. Figures 5 and 6 presents that DPR gate and proposed reversible sub-tractor module. Depending on the state of S_2, *e.g..*, for $S_2 = 0$, the DPR performs ADD operation as well as SUB operation too when $S_2 = 1$. Our proposed reversible full adder or subtractor design are depicted in Fig. 7.
Design of reversible 4:1 reversible Multiplexer (RMUX): 4 : 1 MUX plays pivotal role in the ALU design which displays selective outputs as presented in the operation section. For such selection, 4 : 1 MUX takes two input select lines, here, that input lines are S_0 and S_1. Here, Fredkin gate-based 2 : 1 reversible MUX that reported in [12, Fig. 4], is used to implement 4 : 1 MUX and is shown in Fig. 10. 4 : 1 RMUX can also be implemented directly by using three 2 : 1 MUX. Figure 9 presents flow diagram from 2 : 1 MUX into 4 : 1 MUX and its equivalent 4 : 1 RMUX can be

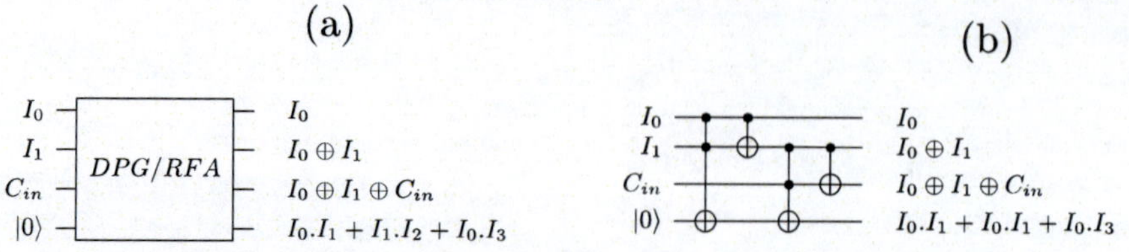

Fig. 5 **a** Reversible DPG-gate (Reversible Full Adder(RFA)) as depicted in [11, Fig. 11]. **b** Toffoli-based realization of Fig. 5a

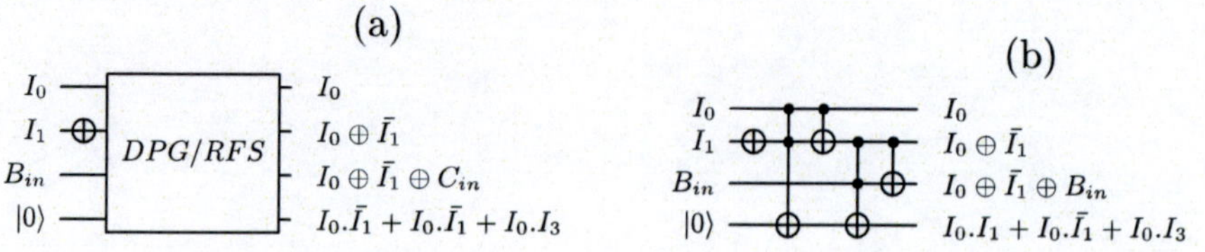

Fig. 6 **a** Reversible DPG-gate-based reversible full subtractor. **b** Toffoli-based realization of Fig. 6a

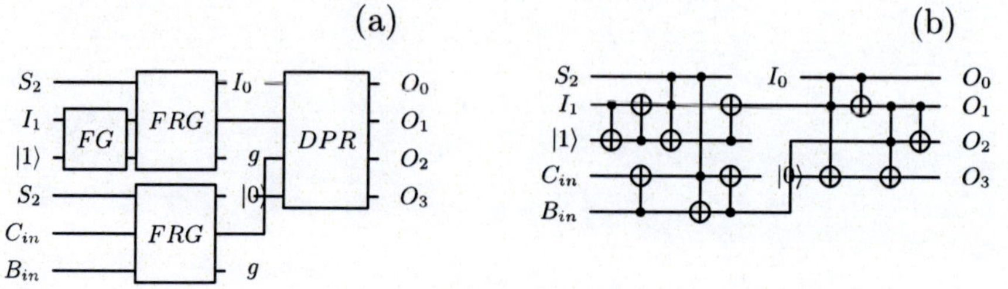

Fig. 7 **a** Reversible DPG-gate-based reversible full adder–subtractor. **b** Toffoli-based realization of Fig. 7a

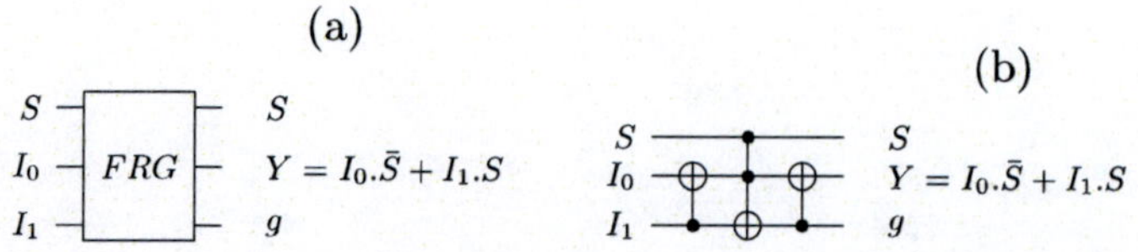

Fig. 8 **a** Fredkin gate-based 2:1 RMUX. **b** Toffoli-based realization of Fig. 10a

directly realized by replacing each 2 : 1 MUX with Fredkin gate. Figure 10 presents that proposed 4 : 1 RMUX. Considering all the above realization of combinational block of Fig. 3 in reversible logic, a Toffoli-based 1-bit R-ALU is proposed in Fig. 11.

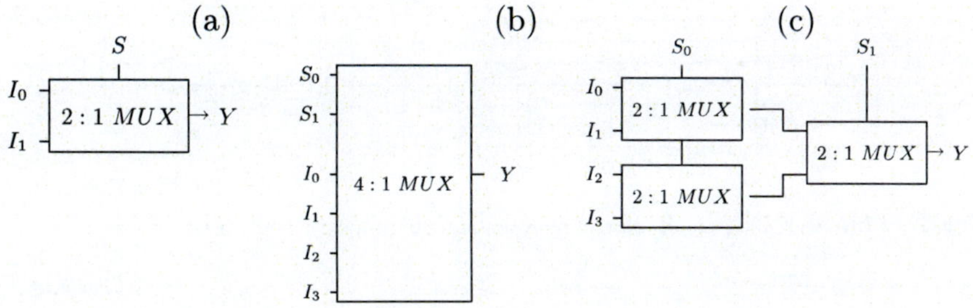

Fig. 9 **a** Schematic diagram of 2:1 MUX. **b** Schematic diagram of 4:1 MUX. **c** Realization of 4:1-MUX using 2:1 MUX

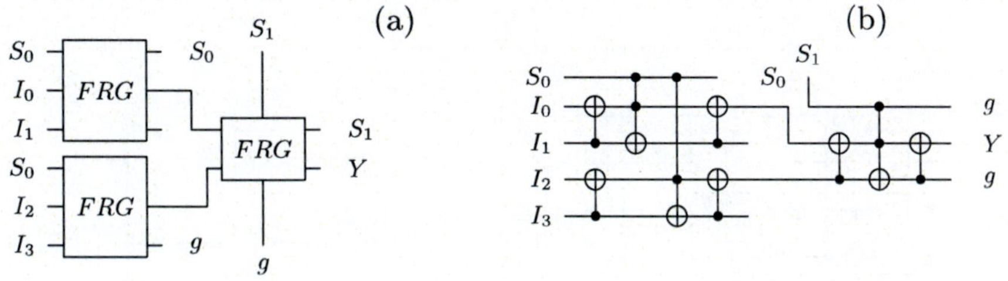

Fig. 10 **a** Fredkin gate-based 2:1 RMUX. **b** Toffoli-based realization of Fig. 10a

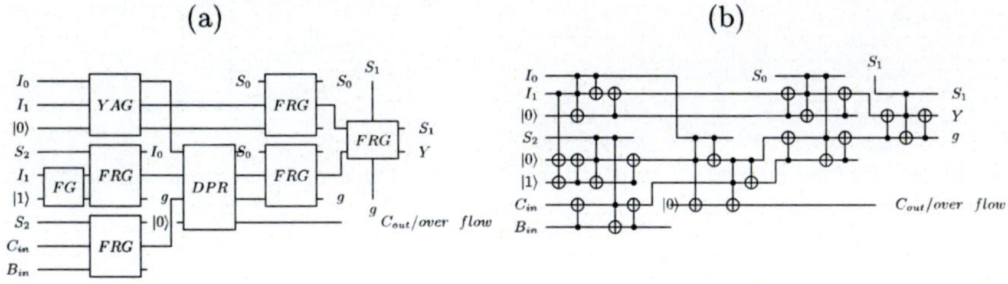

Fig. 11 **a** Proposed R-ALU design **b** Toffoli-based realization of proposed R-ALU (Fig. 11a)

3.2 *Fault-Tolerant Implementation of* 1*-bit ALU*

Here, proposed 1-bit R-ALU is mapped into Clifford+T-group in a way to further convert it into fault-tolerant design. On scanning Fig. 11, it can be found that 1-bit R-ALU consists of reversible Toffoli gate, DPR and CNOT-gate only. So, the fault-tolerant implementation of 1-bit R-ALU can be realized directly by replacing each Toffoli gate and DPR-gate by its equivalent Clifford+T-based architecture. In fact, the Clifford+T-based realization of Toffoli gate, as well as DPR-gate, is already defined in Sect. 2. So, we skip the brief mathematics behind this realization from the reversible gate into fault-tolerant Clifford+T-based architecture as stated in [13]. For better apprehension of this mapping approach, Clifford+T-based DPR-gate is shown in Fig. 14a as shown in [14, Fig. 1] and Toffoli gate is presented in Fig. 1c. Figure 14a can

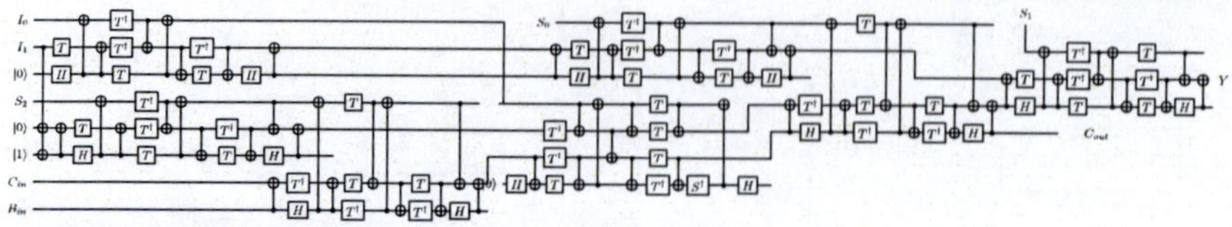

Fig. 12 Clifford+T-based fault-tolerant representation of proposed 1-bit ALU

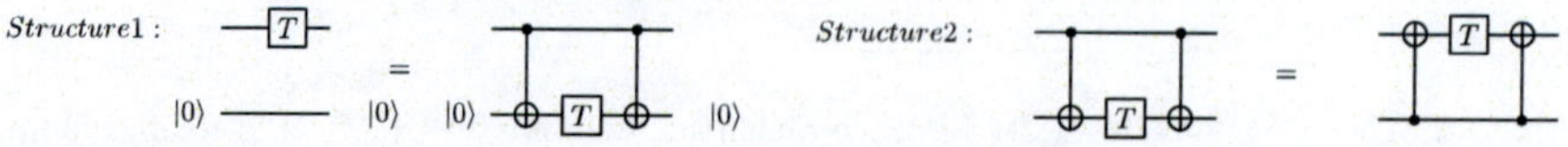

Fig. 13 Equivalent fault-tolerant circuit identities as depicted in [13, Fig. 3f]

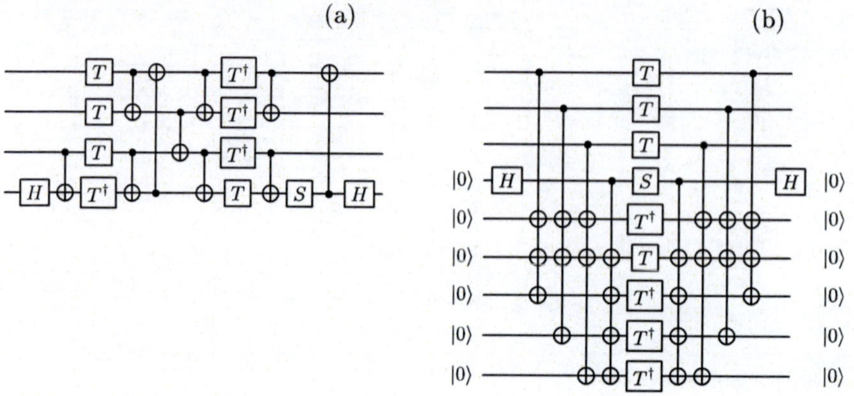

Fig. 14 **a** Clifford+T-*based DPR as depicted in* [14, Fig. 1]. **b** Unit $T-depth$*-based realization of Fig.* 14*a*

further be translated into Fig. 14b with the help of additional ancilla input initiated by $|0\rangle$ where all the T-gates are executed in the unit cycle. Such translation is possible by using mathematical model of [13] to translate all those randomly placed T-gates into one column. In this mapping, two structures depicted in [13, Fig. 3f], is extensively used. Figure 13 illustrates that two structure. Now replace all the Toffoli gate, and DPR-gate by its equivalent Clifford+T-based structure defined in Fig. 1c, and Fig. 14a respectively, and the subsequent resulting circuit is shown in Fig. 12. However, the $T-depth$ of the overall circuit can be reduced by replacing Toffoli gate and DPR-gate by its unit $T-depth$ structure as depicted in Fig. 1d and Fig. 14b, respectively, which eventually reduces the $T-depth$ of the overall circuit into by a factor of 3 with the help of additional cost in terms of ancillary lines (Fig. 8).

Table 2 Opcode for different operation in ALU

Opcode for various task			Operation
S_2	S_1	S_0	
0	0	0	OR
0	0	1	AND
0	1	0	XOR
0	1	1	ADD
1	1	1	SUB

4 Experimental Result and Comparative Analysis

The experimental results are summarized in Table- 2, where the state of selected lines is provided to perform a different task. Currently, there hardly exists any fault-tolerant quantum implementation of ALU circuit. So, we are unable to compare our design methodology with any existing works. However, some works related to ALU are reported over reversible logic and NCV-based quantum circuit in [15–17]. Besides, the proposed designs not only ensures fault-tolerant property but also incurs minimum design overhead as we have used the optimal $T - count$ and $T - depth$-based template matching process. Apart from these, to contain the effect of decoherence, we have used opposite polarity templates which eventually cancels unwanted phase flips in the design.

5 Conclusion

In this work, we have shown Clifford+T-based implementation of 1-bit ALU. In the design process, we have used the functional power of Clifford+T library to transform them into fault-tolerant architectures. Further to make these circuits more efficient, we have used low $T - depth$-based templates to protect the coherence time of extreme fragile quantum state.

References

1. Shor, P.W.: Polynomial-time algorithms for prime factorization and discrete logarithms on a quantum computer. SIAM Rev. **41**(2), 303–332 (1999)
2. Grover, L.K.: Quantum mechanics helps in searching for a needle in a haystack. Phys. Rev. Lett. **79**(2), 325–328 (1997)
3. Feynman, R.P.: Quantum mechanical computers. Found. Phys. **16**(6) (1986)
4. Gottesman, D.: Stabilizer codes and quantum error correction (ph.d. thesis) (1997). Accessed 26-May-2019

5. Biswal, L., Bandyopadhyay, C., Chattopadhyay, A., Wille, R., Drechsler, R., Rahaman, H.: Nearest-neighbor and fault-tolerant quantum circuit implementation. In: 2016 IEEE 46th International Symposium on Multiple-Valued Logic (ISMVL), pp. 156–161. IEEE (2016)
6. Fowler, A.G., Mariantoni, M., Martinis, J.M., Cleland, A.N.: Surface codes: towards practical large-scale quantum computation. Phys. Rev. A **86** (2012)
7. Amy, M., Maslov, D., Mosca, M.: Polynomial-time t-depth optimization of clifford+t circuits via matroid partitioning. IEEE Trans. Comput. Aided Design Integ. Circuit. Syst. **33**(10) (2014)
8. Biswal, L., Bhattacharjee, D., Chattopadhyay, A., Rahaman, H.: Techniques for fault-tolerant decomposition of a multicontrolled toffoli gate. Phys. Rev. A **100**, 062326 (2019)
9. Biswal, L., Bandyopadhyay, C., Wille, R., Drechsler, R., Rahaman, H.: Improving the realization of multiple-control toffoli gates using the ncvw quantum gate library. In: 2016 29th International Conference on VLSI Design and 2016 15th International Conference on Embedded Systems (VLSID), pp. 573–574, Jan 2016
10. Selinger, P.: Quantum circuits of t-depth one. Phys. Rev. A **87** (2013)
11. Syamala, Y., Tilak, A.V.N.: Reversible arithmetic logic unit. In: 2011 3rd International Conference on Electronics Computer Technology, vol. 5, pp. 207–211. IEEE (2011)
12. Kole, D.K., Dutta, J., Kundu, A., Chatterjee, S., Agarwal, S., Kisku, T.: Generalized construction of quantum multiplexers and de-multiplexers using a proposed novel algorithm based on universal fredkin gate. In: 2016 Sixth International Symposium on Embedded Computing and System Design (ISED), pp. 82–86. IEEE (2016)
13. Biswal, L., Das, C.R., Bandyopadhyay, A.C., Rahaman, H.: A template-based technique for efficient clifford+t-based quantum circuit implementation. Microelectron. J. **81** (2018)
14. Soeken, M., Miller, D.M., Drechsler, R.: Quantum circuits employing roots of the pauli matrices. Phys. Rev. A **88** (2013)
15. Gaur, H.M., Singh, A.K., Ghanekar, U.: Design of reversible arithmetic logic unit with built-in testability. IEEE Design Test, 1 (2019)
16. Babaie, S., Sadoghifar, A., Bahar, A.N.: Design of an efficient multilayer arithmetic logic unit in quantum-dot cellular automata (qca). IEEE Trans. Circuits Syst. II: Express Briefs **66**(6), 963–967 (2019)
17. Morrison, M., Ranganathan, N.: Design of a reversible alu based on novel programmable reversible logic gate structures. In: 2011 IEEE Computer Society Annual Symposium on VLSI, pp. 126–131. IEEE, July 2011

Approximation of Fractional-Order Integrator in Delta Domain

Jaydeep Swarnakar

Abstract Fractional-order system (FOS) is a topic of relevance among the researchers working in the field of system and control. This paper presents a methodology for realizing a fractional-order integrator (FOI) in a discrete delta domain. The indirect methodology has been adopted for this work. In the first stage, Modified Oustaloup approximation method has been employed for approximating the FOI in the s domain. The later stage of the work capitalizes the unification property of the delta operator for discretizing the integer order counterpart of the original FOI. Necessary simulation results have been provided to analyze the efficacy of the discrete-time approximation scheme presented over here.

Keywords Fractional-order integrator · Discretization · Delta operator · Modified Oustaloup approximation

1 Introduction

The Fractional-order integrator (FOI) has a strong relevance in the field of control theory and its application [1]. During the research pertaining to the design of a robust feedback amplifier, Bode first proposed the ideal open-loop transfer function as a reference model for control application which was in fact a fractional-order integrator [2]. The transfer function of the FOI is $s^{-\eta}$, where $0 < \eta < 1$. The FOIs are characteristically infinite dimensional. Therefore, finite-dimensional approximation is a must to employ them practically. This calls for the rational approximation or the conversion of an infinite-dimensional system to a finite-dimensional one so that the frequency response of the approximated FOI becomes similar to the ideal FOI. Oustaloup approximation is one of the most prevalent methods for finding the rational approximation of FOI [3]. But, this method suffers from maintaining consistency especially at the boundary region of the frequency range. In [4], an extended

J. Swarnakar (✉)
Department of Electronics and Communication Engineering, School of Technology, North-Eastern Hill University, Shillong 793022, Meghalaya, India
e-mail: jaydeepswarnakar@gmail.com

D. Bhattacharjee et al. (eds.), *Proceedings of International Conference on Frontiers in Computing and Systems*, Advances in Intelligent Systems and Computing 1255,
https://doi.org/10.1007/978-981-15-7834-2_79

form of the Oustaloup approximation method or Modified Oustaloup method was first proposed to get rid of this difficulty. Later on, this revised method has been used for approximating the FOS as seen from many current literature [5–7]. Digital realization of the FOI can be obtained either by employing the indirect discretization method or employing the direct discretization method [8]. In this work, the indirect discretization has been applied through finding an equivalent integer-order transfer function of the FOI in the s domain employing the Modified Oustaloup method first and subsequently discretizing the same. Conventionally, the discretization rules are based on s to z mapping relation in the form of Euler's transform, Tustin transform, Al-Alaoui transform etc. [9]. However, in this work, the delta operator has been used to discretize the FOI in the form of s to γ mapping. In [10], the delta operator was first proposed as an alternate modeling approach for the discrete-time system based on the principle of signal differencing rather than signal shifting. The delta operator facilitates the unified representation of the discrete-time system and the continuous-time system together, which implies that the discrete-time system merges with its continuous-time counterpart at a lower sampling time limit. In contemporary literature, the delta operator has also been familiarized for implementing the FOS digitally [11–15]. Therefore, the same has been involved to discretize the FOI in this work as well. The entire work is divided into four sections. Section 1 gives the introduction. Section 2 describes the indirect realization of the FOI employing the Modified Oustaloup approximation method followed by the discretization in the delta domain. Section 3 analyzes the impact of reducing the sampling time in the discretization process with supporting simulation results. Section 4 draws the final conclusion of the overall work.

2 Discretization of FOI

Let, a fractional-order integrator (FOI) be represented as below

$$L(s) = s^{-\eta} \tag{1}$$

where $0 < \eta < 1$. Now, Eq. (1) is rearranged as below:

$$L(s) = \frac{s^{\mu}}{s} \tag{2}$$

where $\mu = 1 - \eta$ and obviously, $0 < \mu < 1$. So, Eq. (2) represents the reframed continuous-time FOI. The delta-domain realization of the FOI is done through two stages. At first, the Modified Oustaloup approximation is applied to convert s^{μ} into a rational form. The Modified Oustaloup approximation [4] of s^{μ} within a frequency band (ω_l, ω_u) is given as below:

$$s^{\mu} \approx \left(b^{-1} d\omega_u\right)^{\mu} \times \left[\frac{s^2 + bd^{-1}\omega_u s}{(1-\mu)s^2 + bd^{-1}\omega_u s + \mu}\right] \times \prod_{i=-P}^{P} \frac{s + \omega_i^{/}}{s + \omega_i} \quad (3)$$

where ω_l refers to the low frequency, ω_u refers to the high frequency, $\omega_i^{/} = \omega_l \omega_x^{(2i-1-\mu)/P}$, $\omega_i = \omega_l \omega_x^{(2i-1+\mu)/P}$, $\omega_x = \sqrt{\omega_u / \omega_l}$, $b = 10$ and $d = 9$. So, from Eq. (3), it is understandable that the order of rational approximation for s^{μ} will be $(2P + 3)$. Hence, replacing the integer-order equivalent of s^{μ} as obtained from Eq. (3) into Eq. (2), the final rational approximation of the FOI will be of order $(2P + 4)$. After obtaining the rational approximation of the FOI in the s domain, discretization is carried out subsequently. In this work, s to γ mapping has been chosen instead of conventional s to z mapping for the discretization purpose. The delta operator [10, 11] used for discretizing the continuous-time FOI is given as below:

$$s = \frac{1}{\Delta} \ln(\Delta\gamma + 1) \quad (4)$$

From Eq. (4), it results as follows:

$$\gamma = \frac{e^{s\Delta} - 1}{\Delta} \quad (5)$$

Expanding $e^{s\Delta}$ and limiting the sampling time $\Delta \to 0$, the following is obtained:

$$\lim_{\Delta \to 0} \gamma = s \quad (6)$$

Hence, the frequency variable 'γ' of the discrete-delta domain merges with the contintinious-time frequency variable 's'. This is the clear implication that the discrete-time system amalgamates with its continuous-time counterpart at a high sampling frequency as anticipated. Therefore, this fascinating property of the delta operator has been well exploited in this work to approximate the FOI through employing the indirect discretization method.

3 Simulation and Results

The transfer function of the FOI is given as below:

$$L(s) = s^{-0.6} \quad (7)$$

Equation (7) is rearranged as below:

$$L(s) = \frac{s^{0.4}}{s} \tag{8}$$

Using Eq. (3), the term $s^{0.4}$ is approximated by choosing $P = 2$ within the frequency band (ω_l, ω_u), where $\omega_l = 0.01$ rad/sec and $\omega_u = 100$ rad/sec. So, it will result in $(2P+3)^{th}$ or seventh-order approximation of $s^{0.4}$ as shown below:

$$s^{0.4} \approx \frac{\begin{array}{r} 10.08s^7 + 1450s^6 + 38140s^5 + 165100s^4 + 113600s^3 \\ +12140s^2 + 177.5s \end{array}}{\begin{array}{r} s^7 + 253.6s^6 + 13300s^5 + 119300s^4 + 171700s^3 \\ +38860s^2 + 1306s + 4.206 \end{array}} \tag{9}$$

Using Eq. (9), Eq. (8) is represented as follows:

$$L(s) \approx \frac{\begin{array}{r} 10.08s^7 + 1450s^6 + 38140s^5 + 165100s^4 + 113600s^3 \\ +12140s^2 + 177.5s \end{array}}{s\left(\begin{array}{r} s^7 + 253.6s^6 + 13300s^5 + 119300s^4 + 171700s^3 \\ +38860s^2 + 1306s + 4.206 \end{array}\right)} \tag{10}$$

Equation (10) represents the $(2P+4)^{th}$ or eighth-order approximation of FOI in the s domain. Then, the s domain approximant of the FOI is discretized employing the delta operator as specified in Eq. (4). The approximations of FOI have resulted in $L_{\delta 1}(\gamma)$ and $L_{\delta 2}(\gamma)$ as shown in Eq. (11) and Eq. (12) which have been obtained by taking two different sampling instants $\Delta = 0.01$ sec and $\Delta = 0.001$ sec, respectively.

$$L_{\delta 1}(\gamma) = \frac{\begin{array}{c} 6.63\gamma^7 + 627.3\gamma^6 + 13430\gamma^5 + 54970\gamma^4 + \\ 37370\gamma^3 + 3984\gamma^2 + 58.24\gamma + 5.616 \times 10^{-28} \end{array}}{\gamma\left(\begin{array}{c} \gamma^7 + 138.5\gamma^6 + 5038\gamma^5 + 40390\gamma^4 + \\ 56680\gamma^3 + 12760\gamma^2 + 428.5\gamma + 1.38 \end{array}\right)} \tag{11}$$

$$L_{\delta 2}(\gamma) = \frac{\begin{array}{c} 9.558\gamma^7 + 1314\gamma^6 + 33870\gamma^5 + 145800\gamma^4 + \\ 100300\gamma^3 + 10710\gamma^2 + 156.6\gamma + 1.438 \times 10^{-33} \end{array}}{\gamma\left(\begin{array}{c} \gamma^7 + 235.8\gamma^6 + 11900\gamma^5 + 105600\gamma^4 + \\ 151600\gamma^3 + 34290\gamma^2 + 1152\gamma + 3.711 \end{array}\right)} \tag{12}$$

The frequency response graphs for $L_{\delta 1}(\gamma)$ and $L_{\delta 2}(\gamma)$ have been plotted together along with the frequency response of the ideal FOI as shown in Fig. 1. It is obvious that the magnitude responses of both $L_{\delta 1}(\gamma)$ and $L_{\delta 2}(\gamma)$ are close to the magnitude characteristics of the ideal FOI. However, the phase response of $L_{\delta 1}(\gamma)$ deviates larger than the phase response of $L_{\delta 2}(\gamma)$ at the high frequency end as far as the

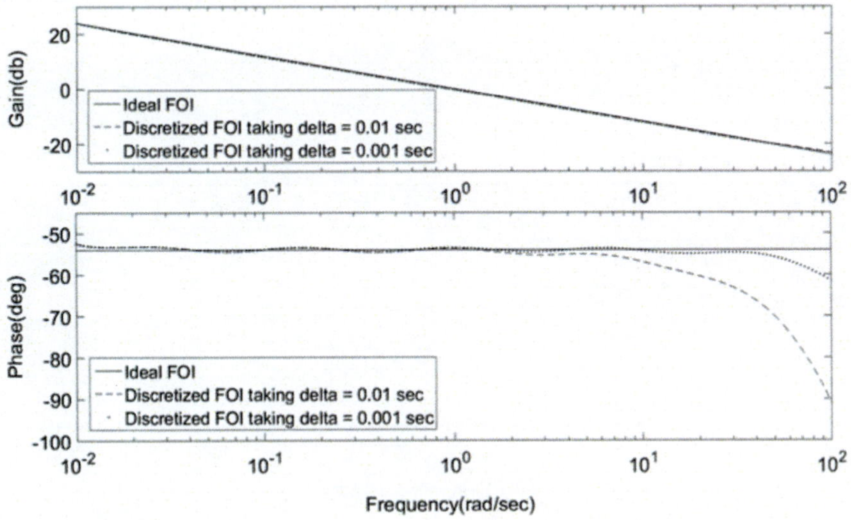

Fig. 1 Frequency responses of $L_{\delta 1}(\gamma)$ and $L_{\delta 2}(\gamma)$ corresponding to two different sampling instants $\Delta = 0.01$ sec and $\Delta = 0.001$ sec along with the frequency response of ideal FOI. $\left(s^{-0.6}\right)$

phase characteristics of the ideal FOI is concerned. This signifies that the overall frequency response of the discretized FOI improves when the sampling time limit reduces from $\Delta = 0.01$ sec to $\Delta = 0.001$ sec. This also reveals the fact that the discretized FOI is getting closer to the ideal continuous-time FOI in terms of frequency response characteristics when the sampling frequency goes high. The error graphs have been plotted in Fig. 2. From the error graphs, it is clear that the frequency response of the discretized FOI delivers smaller errors at a high sampling frequency when $\Delta = 0.001$ sec. The absolute errors for the magnitude approximation and the phase approximation are summarized in Table 1. Undoubtedly, the approximation of the FOI has produced a better result at the lower sampling time as both the magnitude and phase errors have been reduced by taking $\Delta = 0.001$ sec.

4 Conclusions

In this paper, an indirect methodology is implemented to realize an FOI in the delta domain. The first stage of the work involves the Modified Oustaloup method to approximate the FOI in the s domain. The later stage of the work employs the delta operator to discretize the FOI taking two different sampling instants. It has been established that the frequency response of the discretized FOI exhibits almost same characteristics as obtained from the frequency response of the continuous-time FOI especially at the lower sampling time limit. This unification of the discretized FOI and

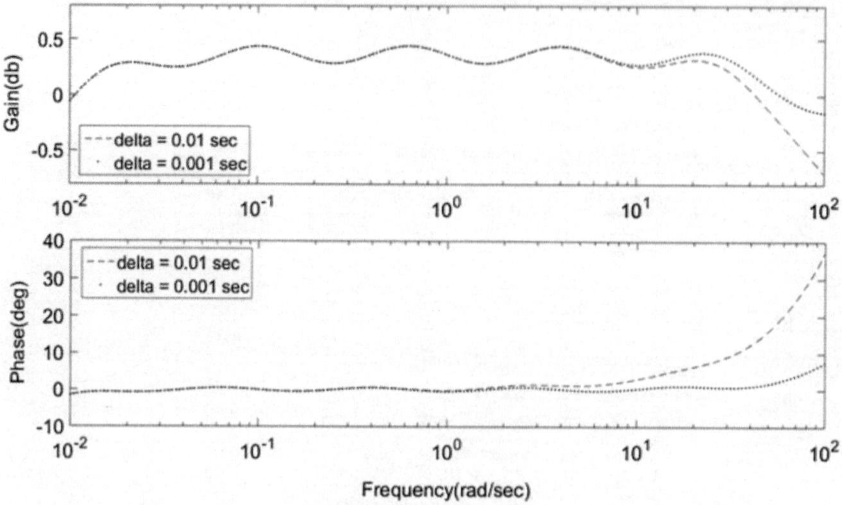

Fig. 2 Comparison of errors resulting after the discretization of ideal FOI $\left(s^{-0.6}\right)$ taking two different sampling instants $\Delta = 0.01$ sec and $\Delta = 0.001$ sec

Table 1 Comparison of errors at two different sampling instants while approximating FOI in the delta domain

Sampling time (sec)	Maximum absolute magnitude error (dB)	Maximum absolute phase error (°)
$\Delta = 0.01$	0.708	37.161
$\Delta = 0.001$	0.152	7.542

its continuous-time counterpart at a high sampling frequency validates the efficacious discrete-time realization of FOI in the delta domain.

References

1. Bode, H.W.: Network Analysis and Feedback Amplifier Design. Princeton, Van Norstand (1945)
2. Monje, C.A., Vinagre, C.A., Feliu, V., Chen, Y.Q.: Tuning and auto-tuning of fractional order controllers for industry applications. Control Eng. Pract. **16**(7), 798–812 (2008)
3. Oustaloup, A.: CRONE Control: Robust Control of Non-integer Order. Hermes, Paris (1991)
4. Xue, D., Zhao, C., Chen, Y. Q.: A modified approximation method of fractional-order system. In: Proceedings of the International Conference on Mechatronics and Automation (ICMA), pp. 1043–1048, IEEE, Luoyang, China (2006)
5. Sheng, H., Chen, Y.Q., Qiu, T.S.: Fractional Processes and Fractional-Order Signal Processing. Springer, London (2011)
6. Nowak, T.K., Duzinkiewicz, K., Piotrowski, R.: Numerical investigation of nuclear reactor kinetic and heat transfer fractional model with temperature feedback. In: Proceedings of the

20th International Conference on Methods and Models in Automation and Robotics (MMAR), pp. 585–590. IEEE, Miedzyzdroje, Poland (2015)
7. Buchade, P.C., Vyawahare, V.A., Bhole, V.V.: Fractional-order control of Z-source inverter using Bode's ideal transfer function. In: Proceedings of the Students' Conference on Electrical, Electronics and Computer Science, pp. 1–6. IEEE, Bhopal, India (2014)
8. Chen, Y.Q., Petras, I., Xue, D.: Fractional-order control—a tutorial. In: Proceeding of the American Control Conference (ACC), pp. 1397–1411. IEEE, St. Louis, USA (2009)
9. Krishna, B.T., Reddy, K.V.V.S.: Design of fractional-order digital differentiators and integrators using indirect discretization. Fract. Calculus Appl. Anal. Int. J. Theory Appl. **25**(33), 143–151 (2008)
10. Middleton, R.H., Goodwin, G.C.: Digital Control and Estimation: A Unified Approach. Prentice Hall, Englewood Cliffs, New Jersey (1990)
11. Maione, G.: High-speed digital realizations of fractional operators in the delta-domain. IEEE Trans. Autom. Control **56**(3), 697–702 (2011)
12. Swarnakar, J., Sarkar, P., Singh, L.J.: Realization of fractional-order operator in complex domains—a comparative study. In: Bera, R., Sarkar, S., Chakraborty, S. (eds.) Lecture Notes in Electrical Engineering, vol. 462, pp. 711–718. Springer, Singapore (2018)
13. Swarnakar, J., Sarkar, P., Singh, L.J.: A unified direct approach for discretizing fractional-order differentiator in delta-domain. Int. J. Model. Simul. Scientific Comput. **9**(6), 185055 (2018)
14. Swarnakar, J., Sarkar, P., Singh, L.J.: Rational approximation methods for a class of fractional-order SISO system in delta domain. In: Bera, R., Sarkar, S.K., Singh, O.P., Saikia, H. (eds.) Lecture Notes in Electrical Engineering, vol. 537, pp. 395–402. Springer, Singapore (2019)
15. Swarnakar, J., Sarkar, P., Singh, L.J.: Direct discretization method for realizing a class of fractional order system in delta domain–a unified approach. Autom. Control Comput. Sci. **53**(2), 127–139 (2019)

Design of Ternary Content-Addressable Memory Using CNTFET

Vikash Prasad and Debaprasad Das

Abstract A ternary content-addressable memory cell using carbon nanotube field-effect transistors (CNTFETs) is presented in this paper. The proposed ternary CAM cell can effectively reduce ∼14% search delay, ∼35% average power consumption with higher noise immunity in comparison with recently presented CNTFET-based ternary CAM cell.

1 Introduction

Content-addressable memory (CAM) can perform search operation in a single clock cycle. This memory can search for data parallelly over all memory cells at very high speed. This feature makes it a very suitable memory element in various applications such as network routers, image coding, pattern recognition, etc. It compares the search data with the stored data. Depending upon the logic levels, CAM can be two types: (i) binary CAM and (ii) ternary CAM. Binary CAM can store two logic levels: 0 and 1, and it provides exact data search, while ternary CAM can store three logic levels, and performs search operation based on partial search data with additional don't care or mask data. The search operation is based on pattern matching. In network router, the routers compare the destination address of all packets to send them through appropriate routes. As the demand of nanoelectronic applications grows, implementation of low operating voltage CAM with high speed, low power dissipation, and large data capacity is a major challenge.

Carbon nanotube field-effect transistor (CNTFET) is one of the rising nanodevices for designing high-performance nanoelectronic circuits because of its near-ballistic transport operation. Like traditional MOSFET, CNTFET has a similar struc-

V. Prasad (✉) · D. Das
Department of Electronics and Communication Engineering, TSSOT, Assam University, Silchar, India
e-mail: vkshprsd@gmail.com

D. Das
e-mail: dasdebaprasad@yahoo.co.in

D. Bhattacharjee et al. (eds.), *Proceedings of International Conference on Frontiers in Computing and Systems*, Advances in Intelligent Systems and Computing 1255,
https://doi.org/10.1007/978-981-15-7834-2_80

ture except it utilizes single-wall semiconducting CNT as the channel. In addition, the threshold voltage of CNTFETs can easily be controlled by controlling the diameter of CNT. This is one of the major characteristics of CNTFET that makes it a suitable device for designing multivalued circuits.

Various ternary memories have been proposed in the literature that utilizes CNTFET. For example, a CNTFET-based ternary static random access memory (SRAM) that utilizes transmission gates and buffer to perform read and write operation has been designed and evaluated by Lin et al. in [1]. The binary SRAM design using CNTFET is presented in [2]. The design of CAM using three-valued circuit structures is presented by Nepal et al. [3]. The three-valued CAM memory cell with low capacitance comparison logic using CNTFET has also been presented by Murotiya et al. [4]. However, these designs use two-power supply voltage that causes higher power consumption and poor noise immunity. The 4-bit binary CAM has been presented by Das et al. [5] that improved the speed of CAM as compared to CMOS-based design. In this paper, the design of lower power, high-speed, and high-noise immunity ternary CAM cell using CNTFET is proposed.

2 Proposed Ternary CAM Cell

The proposed CAM cell is based on ternary logic that it can store and search three logic levels: 0, 1, and 2. Figures 1 and 2 show the schematic of the ternary CAM cell. In this design, two cross-coupled six-transistor-based standard ternary inverter (6T-STI) designs are utilized to store bits [7] because 6T-STI have better noise immunity than three-transistor-STI design [8]. Two n-type CNTFETs, N7 and N8, act as access transistors. These access transistors are used to write and read the data inputs. The access transistors allow writing to the memory (Q and NQ) and reading from the memory when wordline (WL) is asserted. The transistors N9, N10, and N11 constitute the search network. The data to be searched are placed at searchline (SL) and complement of searchline (SLB).

All the transistors are designed using three tubes. The chirality indices of transistors N9 and N10 are kept as (19, 0) for threshold voltage 0.293V, whereas the threshold voltage of N10 is kept as 0.783V using chirality vector of (10, 0) to obtain the match condition whenever logic 1 is compared.

2.1 Write Operation:

The write operation of the proposed ternary CAM is performed as follows. To write data into the CAM cell, the data inputs in its true and complement form are applied to TL and TLB. For example, in order to write data 0 into the memory (i.e., to set node Q to 0 and node NQ to 2) the logic 0 and 2 are applied to TL and TLB, respectively. Now, when the wordline is made high, the pass transistors N7 and N8 allow writing

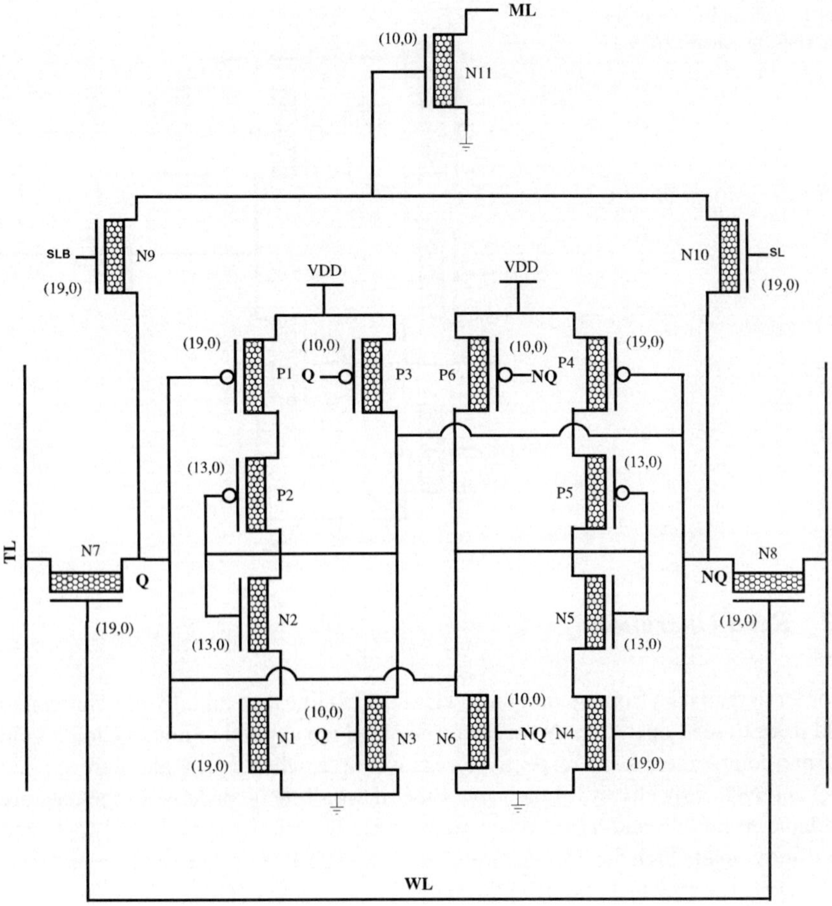

Fig. 1 Proposed ternary CAM cell

data at nodes Q and NQ. If wordline (WL) is made low in the next phase, then the memory cell holds the stored value until the wordline is made high again. When logic 1 is to be written (i.e., both node Q and NQ to be set to 1), logic 1 is placed on both TL and TLB and wordline is made high, which turns ON the transistors N7 and N8 and the node Q is charged to logic 1 and node NQ is discharged to logic 1. In order to write logic 2 in the cell, logic 2 and 0 are applied at TL and TLB and wordline is asserted, the access transistors allow new data to be written in node Q and NQ. The node Q is charged to 2, whereas node NQ is discharged to 0.

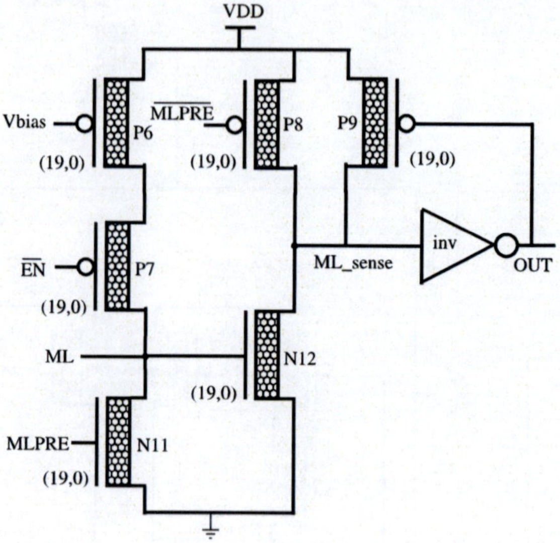

Fig. 2 Schematic of current race sensing scheme [6]

2.2 *Read Operation*

In order to read data from memory, the TL and TLB lines are initially pre-charged to Vdd prior to read operation. Now, data to be read are applied to node Q and NQ in true and complement form, respectively. For example, when logic 0 and 2 are applied to Q and NQ, respectively; at the same time, if wordline is made high, the data are available in the TL and TLB. When both nodes Q and NQ are charged to 1, and wordline is made high the TL is charged to 1 while TLB is discharged to 1. When Q and NQ are charged to 2 and 0, TL is charged to 2, whereas TLB is discharged to 0.

2.3 *Search Operation:*

Ternary data search operation is based on partial data matching. When 0 and 2 are stored and searched, it provides exact data match but to compare data 1 or don't care it gives some pattern matching. The simulated waveform of search operation is shown in Fig. 3. For instance, if logic 1 is stored in the memory and search data 2 is placed in the search lines, the output of the match line sense amplifier (MLSA) indicate logic 2. Again, if 1 is placed in the search lines to compare data against logic 1, the MLSA produces logic 2 output. Hence, the search operation of ternary CAM is based on pattern matching instead of exact data matching.

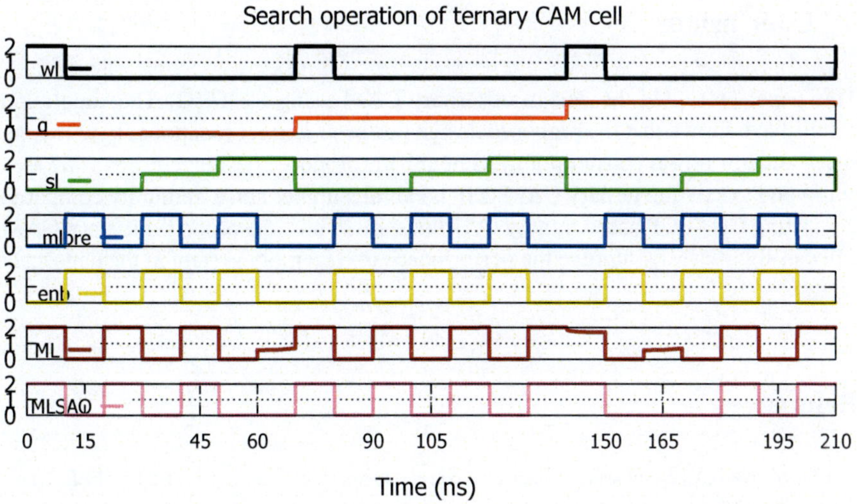

Fig. 3 Simulated waveform of search operation

Table 1 Comparison of CNTFET-based ternary CAM cell

Design	Matchline size	Match delay	Avg. power consumption	#Transistors per cell
Proposed	20 bits	5.97 ps	57.82 μW	17
Nepal and You [3]	20 bits	22.20 ps	43.49 μW	16
Murotiya and Gupta [4]	20 bits	6.90 ps	88.30 μW	11

3 Experimental Results

To make a comparison of CNTFET-based proposed ternary CAM cells with existing CNTFET-based ternary CAM cell, all simulations have been performed on a 1×20 memory array at room temperature with 0.9V power supply voltage using 32-nm CNTFET technology [9]. Comparison results show that the proposed 6T ternary CAM cell consumes less power compared to 3T-STI-based ternary CAM presented in [8] and can perform read, write, and search operation at very high speed. However, the memory cell consists of 17 transistors in 6T-STI-based ternary CAM, which occupies large area compared to 3T-STI-based ternary CAM cell (Table 1).

4 Conclusions

This paper presented the design of ternary CAM using CNTFET. The simulation results have shown that the proposed design performs correct functionality with high speed and low power consumption. Although the proposed design requires more area compared to existing ternary CAM cell, it exhibits higher noise immunity compared to existing CNTFET-based ternary CAM designs. Thus, the proposed ternary CAM cell can significantly improve the performance of a memory system in the nanoscale regime.

References

1. Lin, S., Kim, Y., Lombardi: Design of a ternary memory cell using CNTFETs. IEEE Trans. Nanotechnol. **11**(5), 1019–1025 (2012)
2. Lin, S., Kim, Y., Lombardi: Design of a CNTFET-based SRAM cell by dual-chirality selection. IEEE Trans. Nanotechnol. **9**(1), 30–37 (2010)
3. Nepal, K., You, K.: Carbon nanotube field effect transistor based content addressable memory architectures. IET Micro Nano Lett. **7**(1), 20–23 (2011)
4. Murotiya, S.L., Gupta, A.: Design of content-addressable memory cell using CNTFETs. Int. J. Electron. Lett. **3**(3), 131–138 (2015)
5. Das, D., Roy, A.S., Rahaman, H.: Design of content addressable memory architecture using carbon nanotube field effect transistor progress in VLSI design and test. Lect. Comput. Sci. **7373**, 233–242 (2012)
6. Pagiamtzis, K., Sheikhholeslami, A.: Content addressable memory (CAM) circuits and architectures: a tutorial and survey. IEEE J. Solid-State Circuit. **41**(3), 712–727 (2006)
7. Lin, S., Kim, Y.-B., Lombardi F.: A novel CNTFET-based ternary logic gates design. IEEE NWSCAS, pp. 435–438 (2009)
8. Prasad, V., Banerjee, A., Das, D.: Design of ternary logic circuits using CNTFET. In: International Symposium on Devices, p. 2018. Circuits and Systems (ISDCS), IEEE (2018)
9. Stanford CNFET Model, Stanford Univ., Stanford, CA, USA, 2008. [Online]. https://nano.stanford.edu/stanford-cnfet-model

3-D IC: An Overview of Technologies, Design Methodology, and Test Strategies

Pranab Roy, Arko Dutt, and Hafizur Rahaman

Abstract The emergence of 3-D integration is significant in relation to technology scaling with reference to lower delay, less power consumption, higher bandwidth, and greater performance. In this paper, we have presented a general survey on 3-D technologies, its merits over 2-D integration, demerits, and applications. The major objective of this work is to provide a brief review of preeminent 3-D-IC design techniques and testing mechanism which may enable the development of future effective design automation methods for efficient 3-D-IC integration and its reduced design time.

Keywords 3-D-IC · Technologies · Through-silicon-via · Physical design · Design automation · Testing

1 Introduction

Three dimensional integration plays a pivotal role in order to continue with the Moore's prediction on micro-electronic circuits. A 3-D integrated circuit is basically a singular device that contains a stack of silicon wafers or dies that are vertically interconnected with vias, essentially aimed at improving performance while reducing area when compared to their equivalent 2-D counterparts. The major features that motivate integration in the third dimension against planar 2-D [1] can be stated as follows:

P. Roy (✉) · H. Rahaman
School of VLSI Technology, Indian Institute of Engineering Science and Technology, Shibpur, Howrah, India
e-mail: ronmarine14@yahoo.co.in

H. Rahaman
e-mail: rahaman_h@yahoo.co.in

A. Dutt
Nanyang Technological University, Singapore, Singapore
e-mail: arko.dutt19@gmail.com

D. Bhattacharjee et al. (eds.), *Proceedings of International Conference on Frontiers in Computing and Systems*, Advances in Intelligent Systems and Computing 1255,
https://doi.org/10.1007/978-981-15-7834-2_81

- Lower Form Factor—Form factor is defined as the size of the final product to the functionalities embedded within it. 3-D integration offers lower volume, reduced weight, and fewer footprints.
- Higher performance—3-D integration avails an improved integration density, lower interconnect length and power consumption, and a higher transmission speed. Integration density refers to the number of circuit components (often transistors) contained within a certain area.
- Heterogeneous integration—It is the integration of multiple functionalities in one package. Multiple 3-D device layer structure ensures the potential to integrate varied technologies pertaining to different semiconductor process nodes such as 28 and 45 nm. (in different planes posed as separate layers), as well as disparate components such as digital circuits, RF modules, and memory and analog circuits in a single package.
- Stacking—Stacking helps to overcome the memory to processor performance bottleneck that plagues 2-D ICs.

The rest of the paper is organized as follows. In Sect. 2, predominant 3-D-IC technologies are discussed. Section 3 briefs about the design flow for 3-D-IC design and selective EDA approaches. Section 4 deals with current advances in 3-D-IC testing. Finally, Sect. 5 provides the conclusive remarks.

2 3-D Technologies

The concept of 3-D integration in the semiconductor research domain was developed in the earlier century. Banerjee et al. [2] stated that the concept of three-dimensional integration was first expressed in 1979 [3]. Thereafter, between the 80s and the 90s of the nineteenth century, several works are reported on its device nature and fabrication processes. 3-D technology is basically categorized into 3-D System-in-Package and 3-D Integrated Circuit [4]. The main difference between 3-D SiP and 3-D IC lies in the structure of the vertical interconnect used to connect stacks in the third dimension. 3-D SiP utilizes vertical vias with high aspect ratios, whereas 3-D ICs involve the use of short and narrow vertical vias. Obviously, 3-D IC will ensure greater via integration density and hence allow fine-grain communication between circuit blocks on different stacks in the third dimension.

3-D System-in-Package utilizes bare or packaged die along the vertical direction. It can be classified into IC stacking and packaging technology [5]. IC stacking technologies include SiP involving vertical interconnects as bonded wires, solder balls at the perimeter, array of long, wide, and low-density vertical interconnects, and metal layers on its faces, as shown in Fig. 1. 3-D Packaging has already advanced to a great extent and finds applications in many electronic products such as cellular devices. As far as the 3-D ICs are concerned, they are classically classified into three major categories. The three categories are Monolithic 3-D IC, 3-D ICs with inter-plane vias placed anywhere across the plane (i.e. Through Silicon Vias that are not occupied by

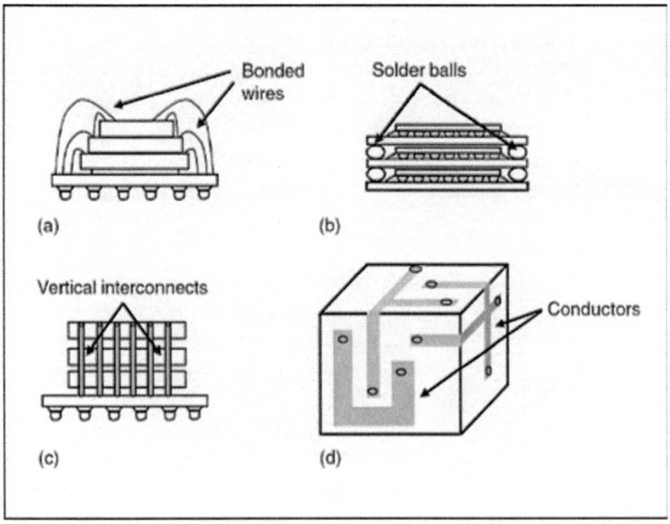

Fig. 1 3-D SiP stacking alternatives—**a** bonded wires, **b** solder balls at the periphery, **c** array of high aspect ratio and low density vertical interconnects, and **d** Metallic interconnections on the 3-D faces. Figure courtesy [4]

transistors on the tier), and contactless 3-D ICs with capacitive or inductive coupling ensuring communication in the third dimension. Manufacture of 3-D ICs may be a sequential as well as a parallel process.

Sequential manufacturing process involves the growth of metal layers and transistors of the upper level stacks on top of the bottommost stack. These methods are used for fabrication of Monolithic 3-D ICs. Monolithic 3-D ICs are of two major types—Stacked 3-D ICs and 3-D Fin-FETs. The growth of active devices on the upper layers need various matured fabrication process techniques such as laser crystallization, seed crystallization, and other techniques involving local clustering approaches. In case of 3-D Fin-FETs, stacking of transistors happens to be on top of one other, the difference being sharing of the same gate for different stacks.

Parallel manufacturing process involves stacking of wafers or dies and bonding the stacks with vias (TSV) in parallel. This process of fabrication is utilized in the manufacture of 3-D ICs with inter-plane vias and contactless 3-D ICs. 3-D ICs with TSVs provide very high wire length reduction with integration along the z-axis. Through Silicon Via is one of the significant features of 3-D IC technology which provides electrical connectivity between tiers in the z-direction. It is named so because it forms a vertical interconnect structure through silicon or oxide layers. Contactless 3-D ICs do not require a conducting path between two stacks in the third dimension, rather they employ the coupling of either electric or magnetic fields for vertical communication. For electric field coupling, the vertical interconnection is assured by small on-chip parallel plate capacitors. For magnetic field coupling, each upper plane contains a spiral inductor located at the same horizontal geometrical position as the

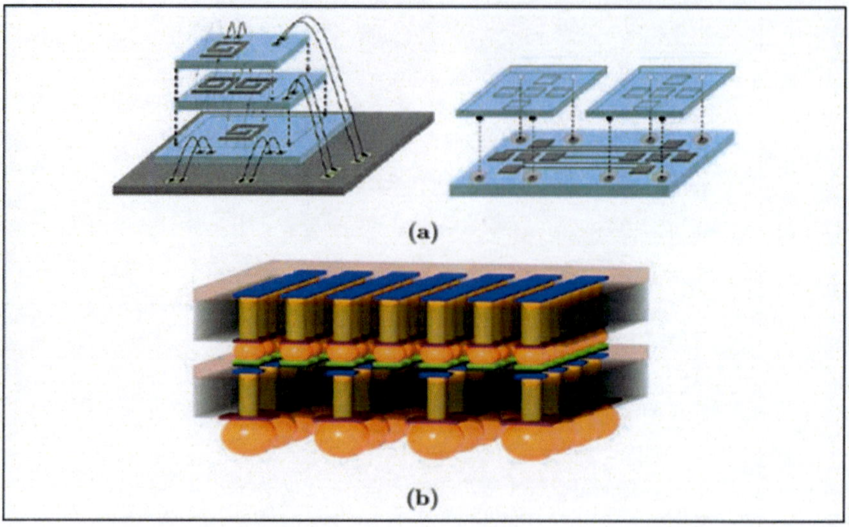

Fig. 2 **a** Coupled 3-D-ICs—Inductive coupling (left) and capacitive coupling(right), **b** A simple 3-D-IC with TSV Figure courtesy [6, 7]

lower plane, thereby enabling vertical communication through magnetic induction. 3-D ICs with TSVs and contactless 3-D ICs are pictorially represented in Fig. 2.

3 3-D IC Design Flow

Design flow in the third dimension mostly resembles 2-D chip design steps. The basic steps in 3-D VLSI design are—3-D system specification, architectural and micro-architectural design, functional design, logic and circuit design (Synthesis), physical design, fabrication, packaging, testing, and debugging. Our study mostly includes 3-D IC design flow (primarily with TSVs) with utmost relevance to physical design because the characteristic trait that distinctly differentiates 3-D ICs from 2-D counterparts is found in the physical design splitting a single stack into two or more stacks and TSV insertion to provide vertical communication among them.

Numerous works presented different 3-D design flow approaches based on the optimization with respect to TSV assignment, thermal, mechanical, power, and timing awareness. The careful study of these approaches helped us to frame a 3-D IC design flow based on which we have continued our survey. The proposed design steps are illustrated in Fig. 3. Existing algorithms, methodologies, and techniques related to the design steps shown in this figure are presented in the following subsections. These methods reveal intended or observed optimization of design parameters as mentioned alongside in Fig. 3. Massive work is still going on to actualize development of advanced algorithms to minimize power requirements and dissipation, heat

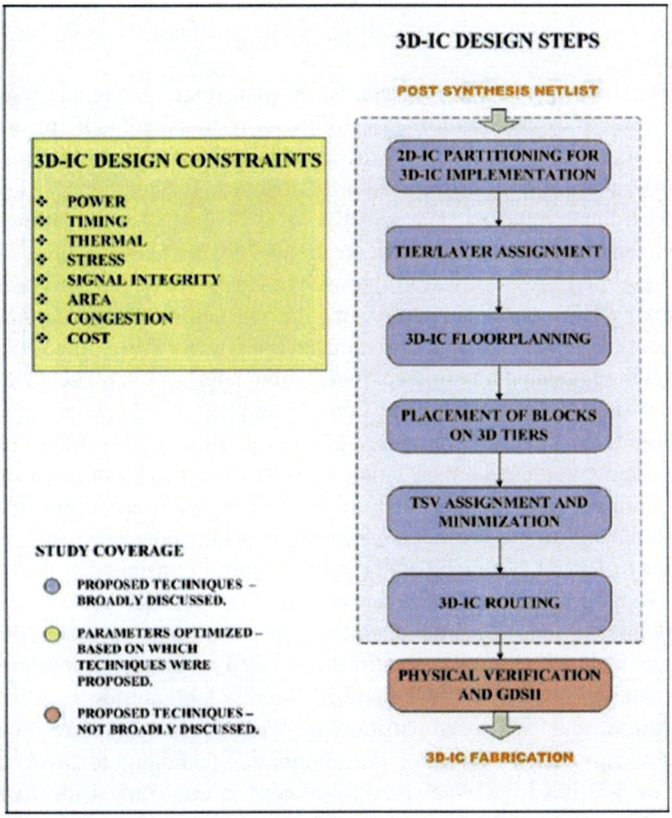

Fig. 3 Proposed 3-D IC design flow with consideration or optimization of specified design constraints

dissipation, area, congestion, TSV overhead, overall cost, and ensure better signal integrity and timing-aware design with a mindset to promote such activities in the near future.

3.1 2-D Partitioning for 3-D Implementation, Clustering, and Layer Assignment

3-D IC physical design starts with the partitioning of 2-D synthesized gate netlist to place circuit blocks on two or more dies in the third dimension. More precisely, we may call it repartitioning of 2-D blocks into 3-D equivalent structure to achieve better footprint, power consumptio,n and delay benefits. The blocks are then clustered, aligned, and assigned to an individual die. Recently proposed methods and algorithms related to blocks' repartitioning, clustering, and tier assignment are discussed in this subsection.

Kumar et al. proposed a hypergraph-based partitioning technique that ensures pre-bond testing of each die in 3-D [8]. Additionally, this method helps in reducing the number of additional flip-flops needed. Their analysis showed that the cut-size remained low, and flip-flops drove a majority of nets on the cut thereby lowering the number of hardware required. The selection of 2-D blocks for the arrangement in the third dimension and tier assignment are major 3-D partitioning goals. Cheng-Chi Chan et al. proposed a cost evaluation approach based upon estimating the tier number (for minimizing TSV count) and partitioning the gate netlist to achieve minimization in the total cost [9]. For a netlist, they evaluated a single die, determined possible tier number, and implemented a multilevel partitioning approach to reduce overall TSV count. They suggested to reformulate Rent's rule [10] in order to efficiently forecast TSV count. Ho-Lin Chang proposed a 2-way partitioning algorithm (being 3-D aware), repeatedly implemented on a multilevel structure that partitions a circuit into k-tiers to minimize area overhead and number of the power and signal TSVs [11]. The major input constraints are the number of tiers (k) and power density constraints. Power density of a tier is defined as the ratio of power consumed at that tier to its total area. Moreover, their analysis showed that this technique proved to be better than the extended hMetis and simultaneous k-way partitioning approaches in respect of runtime as well. The hMetis algorithms are based on efficient multilevel hypergraph partitioning approaches developed in Karypis laboratories, favouring high-quality partitions and high-speed partitioning. Wenchao Gao assumed area density constraints and proposed a layer or tier assignment technique to favor analytical placement of 3-D ICs [12]. Analytical placement in this context means mapping 3-D continuous space to different connected planar IC layer spaces by tier assignment. The technique uses minimum cost flow to optimize the total number of TSVs and total wirelength. A drawback of this method is that—it may prohibit prior 3-D optimized solution. A work by Banerjee et al. in 2014 revealed that almost 45% reduction in TSV count can be achieved as compared to several previous techniques assuming area constraints [13]. They proposed a graph-based partitioning algorithm that initiates with a breadth first search solution and uses a heuristic to improve it. Results show that this algorithm also lowers IR drop (voltage drop at a node) and RC delay associated with TSVs.

3.2 3-D IC Floorplanning

Early planning of overall chip area is very important in any VLSI design flow. In normal 2-D, the power ground network is usually designed after floorplanning—whereas in the 3-D ,co-synthesis of 3-D IC floorplan and power/ground network provides even more benefits.

Falkenstern et al. introduced a tool which develops both the 3-D IC floorplan and power ground network simultaneously [14]. The tool basically uses a 3-D B*-tree floorplan, a resistive power-ground mesh, and a Simulated Annealing engine; it analyzes and insures optimization with regard to the overall chip area, total wirelength, routing space, and IR drops. Their analysis reveals that the total IR drop is minimized at the cost of more power/ground area. Byung-Gyu Ahn et al. proposed an early estimation technique at the floorplanning stage to predict routing congestion at later stages [15]. These types of thoughts and analyses are much needed since failure at the routing stage may lead to redesign and incur additional cost. Synthesized signal nets, clock network, and power-ground network are fed as input to this technique for effective estimation. They took TSV location under consideration as well while implementing the method. They performed experiments comparing their congestion results with the ones found at the routing stage and found their prediction technique to be quite accurate. Since this method may insure higher cost benefits, it may be widely explored by current researchers and checked whether this technique can really offer such accuracy taking considerations of all respects on 3-D IC design implementation stages. Knechtel et al. proposed a design style called Legacy-2-D-integration style [16]. TSVs are grouped in a cluster. Each such cluster is termed as a TSV island. They developed methods to insert TSV islands over the planar chip area of 3-D IC and deadspace1 creation and its redistribution. Their techniques initially suffered from higher interconnect overhead, but are gradually optimized and found to deliver sound results within effective tolerance. This team also discussed the effectiveness of block alignment in 3-D IC interconnection planning [17]. They used expanded corner block list [18] and proposed methods for placing and aligning blocks. The best thing is that they have embedded their techniques in a simulated annealing-based open-source tool which is able to support efficient floorplanning (having fixed outlines) with optimization in wirelength, temperature, and packing density. Robert Fischbach et al. introduced a new layout representation—3-D Moving Block Sequence (3-D MBS), able to handle complex-shaped 3-D blocks (such as orthogonal-polyhedral spread onto several stacks), 3-D floorplanning of rectangular as well as rectilinear 2-D blocks [19]. MBS refers to 2-D layout representation by constructive block insertion. Extending this concept to the 3-D design space gives rise to 3-D MBS. Their experiments showed that this technique offered highly efficient results in terms of packing density, predicted wirelength, and cost. It also supported 2-D legacy IP reuse. According to authors, this proposition was the first attempt to support the floorplanning of complex 3-D blocks.

3.3 Block Placement on 3-D Tiers and TSV Assignment

Implementing the output from the floorplanning step refers to the placement of circuit elements, vias and routing them through wires. Placement incurs placing the circuit blocks on 3-D planes and assigning TSVs between stacks relevant to practical, efficient, and optimized routing in both horizontal and vertical directions. Optimal placement and routing are quite significant in the physical design since these steps are practically applied to chip fabrication. A number of relatively recently proposed methods for 3-D IC placement have been discussed in this subsection. Jason Cong and Guojie Luo were the first to propose a 3-D IC placement procedure based on many-leveled nonlinear coding. This technique considers area density constraints and reduces TSV count and overall wirelength. It is based on concurrently removing overlaps and assigning transistor layers by considering a novel density penalty function which minimizes both the 2-D area and the TSV number. According to authors, they achieved 12% lower wirelength, 29% reduced TSV count with best wirelength, and 20% lower wirelength and 50% reduced TSV count with best TSV count, in comparison to their previous 3-D placement methods. They extended their analyses to encompass reduction of thermal constraints often found in 3-D ICs [20], presenting an efficient technique to lower the temperature by 34% on an average (with a relatively low hike in wirelength considered as overhead). Hsu et al. proposed a cell placement method favoring 3-D IC with TSV, enabling effective TSV size and its physical location while adding TSVs during the placement phase [21]. It follows specifically three stages—1) 3-D IC analytical placement for TSV density and whitespace optimization, 2) adding or assigning TSV and 3) tier-wise detailed placement. In accordance with the authors, it takes a low execution time while ensuring better optimization in terms of TSV numbers, device area, and routing wirelength. Whitespace is similar to dead space, the only difference being that the former is used during placement while the latter is used during floorplanning. Temperature rise in 3-D ICs has been one of the key challenges that need to be taken care of and optimized to a great extent. Prasun et al. proposed an algorithm to place standard cells and gates on active 3-D stacks, insuring uniform thermal distribution in each layer, least maximum temperatures in each layer and the gradient of the maximum temperatures (from bottom to top level dies) [22]. Authors demonstrated that this technique effectively handled the issue of heat dissipation in 3-D ICs. Yi-Rong Chen outlined a 3-D IC-based methodology flow that included cells' partitioning on a single plane post tier assignment, applying Simulated Annealing that insured timing and wirelength optimization. The proposed technique finally implemented a greedy method for overlap (between TSV base and gates on a plane) removal [23]. When compared to 2-D ICs with relatively lower total negative slack, this partitioning-based placement process rendered greater wirelength minimization and consequently higher performance. 3-D IC acceptance and sustenance in the global market requires highly efficient methodology-driven 3-D system designs. In an attempt to do so, researchers of Circuits and Systems group from Delft universitypresented a novel methodology that encircled placement topology analyses considering 2 dies. This procedure accounted for the freedom to choose any CMOS process nodes and was implemented in a SystemC integration platform, whose results showed optimal topologies at 32 nm/45 nm technologies [24]. Keep-out-zones (KOZ) generally secure gates from the adverse impact of TSV-induced stress. It refers to the area around a TSV where no gates can be located to avoid via stress. Athikulwongse and his team provided an analysis in this context with regard to the effects that KOZ have on various reliability parameters and finally presented an innovative 3-D IC placement approach managing TSV stress and following a force-directed method. Experiments performed on this technique were found to ascertain better results compared to wirelength-driven methods in terms of WNS and TNS3. His team also surveyed tier-to-tier temperature-aware coupling and presented two 3-D IC placement techniques. The First method entailed spreading and alignment of TSVs, and the second one involved moving cells considering thermal conduction to heat-sink and TSVs on account of power densities of nearby tiers. Authors advocated that their propositions offered the best 3-D IC-based thermal results over other prevalent placement methods at the time of publishing this issue [25].

3.4 3-D IC Routing

Routing in the third dimension becomes relatively easier once TSVs are assigned or placed between dies. A recent study exploring a 3-D IC physical design put forward global routing and obstacle avoiding Steiner tree algorithms involving multiple dies which can efficiently enable 3-D IC routing [26]. In multilayer global routing flow, initially wires and interconnects are mapped to a planar structure. Then global routing in two-dimension takes place following coherent die assignment. This employs the third-dimension route with relevant consideration of TSV count and positions. However, 3-D Steiner tree algorithms are preferred which attempt to efficiently handle placement hurdles allowing efficient three-dimensional routing. The only major disadvantage of this algorithm is that it requires longer execution time. So researchers may think of optimizing the time complexities of these methods in the future.

Lin et al. proposed an efficient algorithm for the first time which obtains a multilayer obstacle-avoiding rectilinear Steiner minimal tree (also called ML-OARSMT) solution by using multilayer obstacle-avoiding spanning graph technique [27]. In order to ensure a minimal total-cost tree, ML-OARSMT joins the pins on dies and vias connecting two dies by rectilinear edges as well as avoids obstructions for a specified number of pins and obstacles on dies. S.K. Lim proposed an efficient algorithm focused on 3-D IC routing based on Steiner tree construction techniques optimizing temperature, wirelength, and performance [28].

4 3-D IC Testing

Testing of 3-D ICs is necessary for identifying manufacturing defects and reduction in the overall cost. In 3-D IC (preferably with TSVs), defects primarily arise during thinning, alignment, bonding of copper-copper or copper-tin, and fabrication of TSVs. Testing of 3-D ICs requires proper planning for detection and diagnosis of faults. Test challenges have been organized into three categories—test flows, test

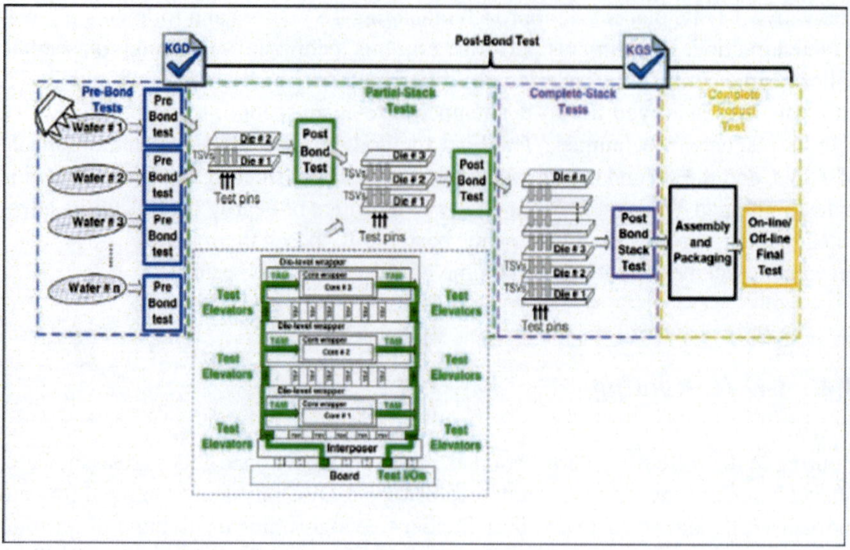

Fig. 4 Detailed 3-D IC test flow. Figure courtesy [29]

contents, and test access [30]. The 3-D IC test flow consists of four phases—pre-bond test, mid-bond test, post-bond test, and final test [31]. In 3-D IC, yield losses are compounded if untested dies are stacked. So the pre-bond test is performed to identify the known good dies (KGDs) prior to stacking. In the second phase, mid-bond test is applied to the tested good dies that might have got corrupted after pre-bond test during the process of die thinning, alignment and bonding. Using mid-bond test, we can avoid the wastage of KGDs. Once the complete stack is formed, post-bond test is applied. In this step, both die and interconnect of the 3-D structure are examined. The final test is carried out after complete 3-D IC packaging and assembling. Figure 4 illustrates that multiple test insertions are used to detect faults in TSVs and circuitry for 3-D IC. It explains post-bond test solution and stacking of KGDs to identify good stack (KGS). Newer methods were proposed to tackle 3-D IC testing through implementation of BIST, test architecture optimization, die-level test interfaces, and test scheduling optimization [32]. Researchers also put forth wireless input-output 3-D IC testing approach using mm-wave to improve the test speed and reduce the test cost [29]. TSV cores are deeply embedded in 3-D stacked ICs. Test access mechanism (TAM) is the process of transporting test stimuli from on-chip hardware to the embedded core. Test access mechanism (TAM) is applied for testing of such cores. TAM uses the concept of modular testing in which each module that constitutes the IC is tested separately. Under the test condition, each module consists of a built-in self-test (BIST) or built-in test (BIT), TAM and test wrapper. Test wrapper translates test stimulus into patterns which can be applied directly to the cores. In 3-D SIC, it is necessary to design test architecture being capable of supporting testing of individual dies in addition to testing of partial and complete stacks. Test architecture optimization needs to reduce the test length (test time) and the number of TSVs used to route the TAM. Based on fixed or changeable wrapper, there are three different test architecture problems—hard dies, soft dies, and firm dies. Those dies with existing test architecture are known as hard dies. In soft dies, the test architecture is not predefined. In firm dies, test architecture already exists but it becomes necessary to add serial or parallel conversion hardware in order to reduce the test pin [33].

Design-for-Testability (DFT) is an IC design procedure which adds definite testability features to a hardware product design. A generic DFT architecture was presented for 3-D stacked IC with multiple towers. A 3-D SIC consisting of a single tower means a 3-D SIC in which each stack contains just a single die. A 3-D SIC with multiple towers means stacking multiple dies on top of a common base die [34]. Each die consists of one or multiple test modules. In 3-D stacked IC, the different DFT techniques that play a major role in fault detection can be classified as follows:

- Reduced Pin-Count Testing (RPCT)—A technique used for reducing the number of IC pins that need to be contacted during testing. It helps in testing multiple chips in parallel.
- Test Data Compression (TDC)—A procedure for reducing the test data volumes of 3-D SICs. It uses don't care bits of Automatic Test Pattern Generator (ATPG) test patterns to compress the off-chip test data.
- Built-In Self-Test (BIST)—In BIST, a test logic circuit is incorporated in the chip. Though this extra circuit increases the chip size, it reduces the test length. It is an effective DFT technique that performs the entire generation of test stimuli and interpretation of test responses on chip. BIST minimizes the test data volume even more than TDC.

5 Conclusion

The success of 3-D IC rests on finding its applications and desired benefits in consumer electronic products. Research in successful development of 3-D IC EDA tools and testing will not only ensure design reliability but also fast adoption in the marketplace. In an attempt to promote rigorous research related to design automation and testing, we have collated a variety of well-recognized algorithmic approaches pertaining to 3-D physical design based on any of area, power, thermal, clock-timing, cost, congestion, stress and/or signal integrity considerations. Existing 3-D IC design, verification, and test tools still suffer from many drawbacks and thus need more research activities for their success.

References

1. Ramm, P.: Shrinking 3-D ICs–Capabilities and frontiers of through Silicon Via Technologies. http://www.researchgate.net/profile/Peter_Ramm/publication/264942174_3-D_ICs_Capabilities_and_Frontiers_of_Through_Silicon_Via_Technologies
2. Banerjee, K., Souri, S.J., Kapur, P., Saraswat, K.C.: 3-D ICs: a novel chip design for improving deep-submicrometer interconnect performance and systems-on-chip integration. Proc. IEEE **89**(5), 602–633 (2001)
3. Geis, M., Flanders, D., Antoniadis, D., Smith, H.I.: Crystalline silicon on insulators by graph oepitaxy. In: Electron Devices Meeting, 1979 International, vol. 25, pp. 210–212. https://doi.org/10.1109/iedm.1979.189581
4. Pavlidis, V.F., Friedman, E.G.: Three-Dimensional Integrated Circuit Design. The Morgan Kaufmann Series in Systems on Silicon. Elsevier and Morgan Kaufmann, Amsterdam (2009)
5. Whipple, T., Kukal, T., Felton, K., Gerousis, V.: IC-package co-design and analysis for 3-D-IC designs. In: IEEE International Conference on 3-D System Integration, 3-DIC 2009, pp. 1–6. IEEE (2009)
6. Bernstein, K.: New dimension in performance: harnessing 3-D integration technology (2007). https://nanohub.org/resources/3596
7. Huang, C.L., Chang, N.S., Chen, C.S., Lin, C.P., Wu, C.M., Huang, C.M.: A novel design methodology for hybrid process 3-D-IC. In: 2012 International Symposium on VLSI Design, Automation, and Test (VLSI-DAT), pp. 1–4 (2012). https://doi.org/10.1109/vlsi-DAT.2012.6212640
8. Kumar, A., Reddy, S.M., Pomeranz, I., Becker, B.: Hyper-graph based partitioning to reduce DFT cost for pre-bond 3-D-IC testing. In: Design, Automation & Test in Europe Conference & Exhibition (DATE), pp. 1–6. IEEE (2011)
9. Chan, C.C., Yu, Y.T., Jiang, I.H.R.: 3-DICE: 3-D IC cost evaluation based on fast tier number estimation. In: 2011 12th International Symposium on Quality Electronic Design (ISQED), pp. 1–6. IEEE (2011)
10. Christie, P., Stroobandt, D.: The interpretation and application of rent's rule. In: IEEE Transactions on Very Large Scale Integration (VLSI) Systems, vol. 8, no. 6, pp. 639–648 (2000). https://doi.org/10.1109/92.902258
11. Chang, H.L., Lai, H.C., Hsueh, T.Y., Cheng, W.K., Chi, M.C.: A 3-DIC designs partitioning algorithm with power consideration. In: 2012 13th International Symposium on Quality Electronic Design (ISQED), pp. 137–142. IEEE (2012)
12. Gao, W., Zhou, Q., Qian, X., Cai, Y., Wang, S.: Chip layer assignment method for analytical placement of 3-D ICs. In: 2013 International Conference on Communications, Circuits and Systems (ICCCAS), vol. 1, pp. 403–407. IEEE (2013)
13. Banerjee, S., Majumder, S., Bhattacharya, B.B.: A Graph-Based 3-D IC Partitioning Technique, pp. 613–618. IEEE (2014). https://doi.org/10.1109/isvlsi.2014.82
14. Falkenstern, P., Xie, Y., Chang, Y.-W., Wang, Y.: Three-dimensional integrated circuits (3-D IC) floorplan and power/ground network cosynthesis. In: Proceedings of the 2010 Asia and South Pacific Design Automation Conference, pp. 169–174. IEEE Press (2010)
15. Ahn, B.G., Kim, J.H., Li, W., Chong, J.W.: Effective estimation method of routing congestion at floorplan stage for 3-D ICs. JSTS: J. Semiconduc. Technol. Sci. 11(4), 344–350 (2011). https://doi.org/10.5573/jsts.2011.11.4.344
16. Knechtel, J., Markov, I., Lienig, J.: Assembling 2-D blocks into 3-D chips. IEEE Trans. Comput.-Aided Des. Integr. Circ. Syst. **31**(2), 228–241 (2012). https://doi.org/10.1109/TCAD.2011.2174640
17. Knechtel, J., Young, E.F., Lienig, J.: Structural planning of 3-D-IC interconnects by block alignment. In: 2014 19th Asia and South Pacific Design Automation Conference (ASP-DAC), pp. 53–60. IEEE (2014)

18. Hong, X., Huang, G., Cai, Y., Gu, J., Dong, S., Cheng, C.-K., Gu, J.: Corner block list: an effective and efficient topological representation of non-slicing floorplan. In: IEEE/ACM International Conference on Computer Aided Design, ICCAD-2000, pp. 8–12. https://doi.org/10.1109/iccad.2000.896442
19. Fischbach, R., Knechtel, J., Lienig, J.: Utilizing 2-D and 3-D rectilinear blocks for efficient IP reuse and floorplanning of 3-D-integrated systems. In: Proceedings of the 2013 ACM International Symposium on International Symposium on Physical Design, pp. 11–16. ACM (2013)
20. Luo, G., Shi, Y., Cong, J.: An analytical placement framework for 3-D ICs and its extension on thermal awareness. IEEE Trans. Comput.-Aided Des. Integr. Circ. Syst. **32**(4), 510–523 (2013). https://doi.org/10.1109/TCAD.2012.2232708
21. Hsu, M.K., Chang, Y.W., Balabanov, V.: TSV-aware analytical placement for 3-D IC designs. In: 48th ACM/EDAC/IEEE Design Automation Conference (DAC), pp. 664–669. IEEE (2011)
22. Ghosal, P., Rahaman, H., Dasgupta, P.: Cell level thermal placement in 3-D ICs. In: 2010 Annual IEEE India Conference (INDICON), pp. 1–4. IEEE (2010)
23. Chen, Y.R., Chen, H.M., Liu, S.Y.: TSV-based 3-D-IC placement for timing optimization. In: 2011 IEEE International SOC Conference (SOCC), pp. 290–295. IEEE (2011)
24. Jagtap, R., Kumar, S.S., van Leuken, R.: A methodology for early exploration of TSV placement topologies in 3-D stacked ICs. IEEE 382–388 (2012). https://doi.org/10.1109/dsd.2012.9
25. Athikulwongse, K., Ekpanyapong, M., Lim, S.K.: Exploiting die-to-die thermal coupling in 3-D IC placement. IEEE Trans. Very Large Scale Integr. (VLSI) Syst. 22(10), 2145–2155 (2014). https://doi.org/10.1109/tvlsi.2013.2285593
26. Kim, D.H., Lim, S.K.: Physical design and CAD tools for 3-D integrated circuits: challenges and opportunities. IEEE Des. Test **32**(4), 8–22 (2015). https://doi.org/10.1109/MDAT.2015.2440317
27. Lin, C.W., Huang, S.L., Hsu, K.C., Li, M.X., Chang, Y.W.: Efficient multi-layer obstacle-avoiding rectilinear Steiner tree construction. In: Proceedings of the 2007 IEEE/ACM International Conference on Computer-Aided Design, pp. 380–385. IEEE Press (2007)
28. Lim, S.K.: Design for high performance, low power, and reliable 3-D integrated circuits. Springer, New York, NY (2013)
29. Lu, J., Loke, W.F., Jung, B.: Millimeter-wave wireless interconnect for 3-D SIC testing. IEEE Des. Test **31**(6), 29–37 (2014). https://doi.org/10.1109/MDAT.2014.2343191
30. Marinissen, E.: Challenges and emerging solutions in testing TSV-based 2 1 over 2-D- and 3-D-stacked ICs. In: Design, Automation Test in Europe Conference Exhibition (DATE), pp. 1277–1282 (2012). https://doi.org/10.1109/date.2012.6176689
31. Chou, C.W., Li, J.F., Yu, Y.C., Lo, C.Y., Kwai, D.M., Chou, Y.F.: Hierarchical test integration methodology for 3-D ICs. IEEE Des. Test **32**(4), 59–70 (2015). https://doi.org/10.1109/MDAT.2015.2427257
32. Noia, B., Chakrabarty, K.: Testing and design-for-testability techniques for 3-D integrated circuits. In: Test Symposium (ATS), pp. 474–479 (2011). https://doi.org/10.1109/ats.2011.67
33. Noia, B., Chakrabarty, K., Goel, S., Marinissen, E., Verbree, J.: Test-architecture optimization and test scheduling for TSV-based 3-D stacked ICs. IEEE Trans. Comput. Aided Des. Integr. Circ. Syst. **30**(11), 1705–1718 (2011). https://doi.org/10.1109/TCAD.2011.2160177
34. Chi, C.C., Marinissen, E., Goel, S., Wu, C.W.: DFT Architecture for 3-D-SICs with multiple towers. In: 2011 16th IEEE European Test Symposium (ETS), pp. 51–56 (2011). https://doi.org/10.1109/ets.2011.52

Oh Dear! It's Just a Tool to Plan the Deployment of a Post-Disaster Network!

Partha Sarathi Paul, Krishnandu Hazra, Sujoy Saha, and Subrata Nandi

Abstract In this paper, we present a software tool for planning the deployment of a rapid-deployable opportunistic network for post-disaster communication establishment in the absence of mobile and other conventional networks. Such a network may be used by Government Agencies and NGOs working for disaster management to coordinate their rescue-relief workers when there is a complete or partial communication blackout. Having a rough knowledge on the affected area (through satellite imagery or otherwise), the location of the control station, and the maximum tolerable delay bound by the opportunistic (DTN) network, the proposed tool could provide with the required communication resources to achieve the desired goal and a deployment plan of the same so that the obtained network has packet delivery delay within the desired delay bound.

Keywords Network planning · Software tool · Delay-Tolerant Networks · Post-disaster Networks

1 Introduction

Natural disasters made human beings vulnerable by damaging normal human activities and disconnecting them from the rest of the world by destroying communication services such as the mobile network and the Internet. As a consequence, the population are forced to be shifted into shelters, their movements are dominated by various

P. S. Paul (✉) · K. Hazra · S. Saha · S. Nandi
National Institute of Technology Durgapur, Mahatma Gandhi Avenue, Durgapur 713209, India
e-mail: mtc0113@gmail.com

K. Hazra
e-mail: krishnandu91@gmail.com

S. Saha
e-mail: sujoy.ju@gmail.com

S. Nandi
e-mail: subrata.nandi@gmail.com

D. Bhattacharjee et al. (eds.), *Proceedings of International Conference on Frontiers in Computing and Systems*, Advances in Intelligent Systems and Computing 1255,
https://doi.org/10.1007/978-981-15-7834-2_82

agents working for the sake of the affected population (e.g., rescue, utility, fire, and police). A well-prepared disaster management authority having a well-managed disaster recovery mechanism could speed up the rescue-relief operation and the further rehabilitation of the victims after the disaster. Japan during the Tsunami in 2011 has shown that effort in disaster preparedness makes rescue-relief prompt, which in turn leads to a reduction in death toll as well as less suffering to the affected people. A similar observation could be made during Cyclone Phailin (2013) and during Cyclone Fani (2019); both made landfalls in the state of Odisha in India, where massive disaster preparedness by the authority has reduced the death toll due to disaster strike into a bare minimum. However, it is also observed that the post-disaster sufferings of the disaster victims could not be avoided after most of the large-scale disasters including the ones mentioned earlier. The situation is worse in the case of developing and underdeveloped countries. Developed countries usually could manage the post-disaster situation by deploying high-calibre machines and equipment that they could afford. On the other hand, developing countries like ours could spend less money and effort and less sophisticated machinery in disaster preparedness and rescue-relief.

Let us consider a large-scale natural disaster (super-cyclone, flood, tsunami, etc.) and the rescue-relief operations just following it. Such operations are often hampered due to complete or partial damage of conventional communication networks and the crisis of power. A well-managed disaster recovery mechanism requires an effective communication infrastructure to build coordination among these working agents. One of the significant aspects of deployment of communication infrastructure is resource (e.g., portable mobile towers, mobile data carriers, and UAVs equipped with mobile communication facilities) allocation and its substantiation. As a solution, Legendre et al. pointed out in [2] that a temporary networking solution using portable wireless communication devices owned by the rescue-relief workers and/or the victims, that can forward data packets within an acceptable delay bound, is necessary till the infrastructure is restored.

These observations motivated us in developing a system titled *CRIMP* [3] that is capable of providing the disaster management personnel with a map-based disaster management service platform on an opportunistic delay-tolerant communication network formed in a seamless manner using the Wi-Fi interface of the user smartphones running the "CRIMP" tool in their smartphones plus some optional special-purpose devices (termed as ISBs and Data Mules in the paper which we in our case built using cheap computing devices like Raspberry Pi and similar boards) depending on their availability in the affected region. Most of the devices being mobile, the connections among them will be random and short-lived, and no end-to-end connection among devices, which leads the network to become a Delay-Tolerant Network (DTN) whose end-to-end packet delay will be in hours or even days instead of seconds or milliseconds. These component devices, in practice, may have a wide variety from user's smartphones and tablet PCs to Laptop computers and special-purpose portable store-forwarding devices (ISBs and Data Mules). The major concern here is to have a framework using which the desired temporary network may be built, and that framework is expected to be a multi-tiered one than single-tier. It is observed that the efficiency and utilization factor of such systems depend highly on the deployment plan of the special-purpose units [5].

In this paper, we present a resource planning tool, that we named *PDCRPT* (Post-Disaster Communication Resource Planning Tool), which is capable of allocating previously mentioned special-purpose resources for efficient communication infrastructure deployment when a precise map of the affected area with strategic (shelter) locations[1] marked on it are given. The proposed tool is capable of providing a concrete deployment plan of the said resources along with associated deployment cost, given the affected area map with marked prospective shelter locations and a given maximum delay bound that the underlying delay-tolerant network may be allowed to tolerate.

2 Related Work

In [5], which was indeed the foundation work of the previously mentioned "CRIMP" system, we noticed that a planned deployment for the same is much superior to an unplanned ad hoc ones, and proposed a four-tier hybrid ad hoc architecture for such purposes which works in the DTN mode. Given an affected area with a number of locations as *Shelter Points (SP)* clearly pointed on that area with a special one marked as *Master Control Station (MCS)*, the paper proposed a temporary network deployment plan only using portable devices like smartphones working in the DTN mode, and some other rapid deployable devices. The rapid deployable devices include (1) Temporary data store-forwarding units with communication facility, called Information Dropbox (IDB) stationed at convenient places, (2) Similar devices attached to rescue-relief vehicles/UAVs, which we called Data Mules (DM), and (3) The long-range WiFi communication (LWC) units placed at a few places of convenience for performance improvement. We proposed in the paper algorithms for suggesting deployment locations for the LWC units, a trajectory for the DMs, etc., using which disaster management personnel can find a stable deployment plan by which nearly cent per cent packet delivery from any *SP* to the *MCS* and vice versa can be ensured within a given delay bound (hours/days). However, our cost-benefit analysis (refer Fig. 1) through the elimination of the devices one by one, and then observing the maximum delivery probability observed by the revised deployment plan with reduced device set, shows a non-linear relationship between the deployment cost and the packet delivery probability. The nature of the diagram indicates the existence of a preferable operating point where the trade-off between cost and performance is balanced. The paper also showed that the packet delivery performance corresponding to a deployment plan is highly dependent on factors like (a) packet load, (b) data rate, (c) Speed of DMs, etc. This information may be obtained from the past

[1]Locations where the affected people take temporary shelter; rescue-relief operations usually coordinated locally from these locations.

disaster records for that area, or from an aerial survey (through Satellite or UAV) of the affected area. However, finding such an operating point that could balance cost and benefit in the current context is quite a non-trivial challenge. Saha et al. [4] introduced an initial layout of such a system, though it lacks a lot of functionalities and could cover the only portion of the problem. The proposed "PDCRPT" tool is supposed to bridge the gap.

3 Design Principle and Software Components

The outline of the tool will be as shown in Fig. 1. Such a tool may be used by big NGOs or Government Agencies for disaster preparedness for the areas like Coastal belt in Eastern India, where natural disasters are quite recurrent. It may also be used for post-disaster deployment planning of temporary ad hoc networks in the affected area.

Figure 2 gives an elaboration of *PDCRPT* black-box shown in Fig. 1. The desired tool requires a convenient and user-friendly interface through which users can interact. An interactive map-based interface would be the best choice here, and Google Maps might be the best and most convenient option for this, as it is richer in path and landmark information than its other counterparts like OpenStreetMap for the zone (Indian Subcontinent, say) we are considering.

Getting a network deployment plan using algorithms in [5], the affected area map needs to be represented as a weighted graph with SPs as the vertices and path segments joining them as the edges, weight of the edge being a representative cost required to move between the pair of edges concerned. For this, the path information from the map is extracted with the help of Google Map APIs (refer Fig. 2a–c).

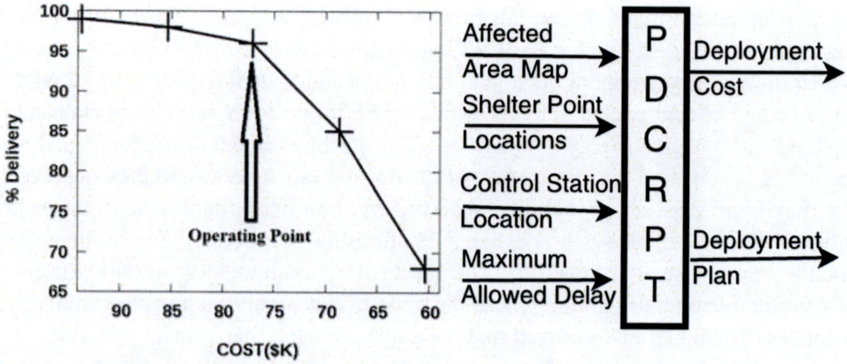

(a) (b)

Fig. 1 **a** Change in Maximum packet delivery probability observed against the reduction of an estimated overall deployment cost (Courtesy [5]), **b** Black-box of the proposed resource planning tool with its input and output sets

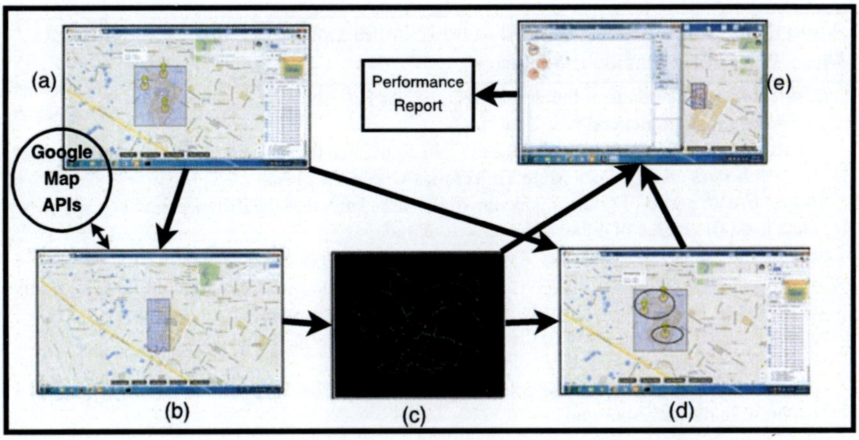

Fig. 2 Process Flow Diagram: **a** Area and Shelter Point Selection, **b** Path Extraction in a distributed manner with the help of Google Map APIs, **c** Final Extracted Path, **d** Extraction of Deployment Plan, **e** Validation of Deployment Plan using ONE Simulator

The challenge here is Google does not provide path information directly for Google Maps. It only allows limited extraction of pathways in *KML* format, download of the above being limited to 10 routes and 23 waypoints per user. So we developed our own distributed algorithm for generating the complete path information in *KML* format exploiting the limited information from Google. The algorithm divides the selected rectangular area into rectangular grids (0.001-degree separation for both latitude and longitude). It takes each corner point on the grid, and snaps that to the nearest road using *Google Maps Road API*. For each pair of snapped points, *Google Maps Direction API* is used to find the path joining the points. By removing possible overlapping segments using *RouteBoxer class Google Map Utility*, full path information for the segment is obtained. If the area is large, then the above algorithm has high running time, and it often fails due to Google's restriction mentioned earlier. To bypass the request limit and also speed up the execution, the area is split into equal-sized rectangular segments by the server machine, and segments are pushed to a pool of client machines. The clients extract the path information as above for that smaller segment it is assigned, and information is pushed back to the server. The server, on receiving the path components stitches them together to form the complete path information for the whole area (Fig. 2c). On obtaining path information, the software generates a deployment strategy as in Fig. 2d following the algorithms proposed in [5]. Algorithm 1 illustrates the procedure mentioned above in a formal manner.

The theoretical study usually fails to capture other external factors for network performance like data rate, packet load, DM speed, etc., due to a lack of closed-form expressions. To verify the effect of these factors on performance, simulation of the scenario under variations of the said parameters is desired. We choose ONE Simulator [1] for this, as it is an accepted simulator for DTN. For interfacing ONE with an interactive map, the path information has to be represented in *WKT* format. For this,

Algorithm 1: The algorithm used to find Shelter Location Clusters and the Data Mule Trajectories inside the Clusters

Input: Rectangle bounds of the drawn Rectangular Box chosen on the map with shelter locations pointed on it

Result: A single WKT file as an input to ONE Simulator that consists of the shelter location clusters and the Data Mule Trajectories within the cluster

Step 1: Create a grid of points inside the drawn area such that the distance between any two points is 0.001 degree of either latitude or longitude.

Step 2: Take each point and snap it to the nearby road to get its coordinate.

Step 3: Create a 2D matrix and store the snapped coordinates of each point into the matrix in the same way as the grid was made. If the point is too far from any road to return a coordinate then add null value to the corresponding 2D value of that point.

Step 4: For all columns in the 2D matrix:

1. Take the first point from the column, find direction from this point to the next non-null point in the same column.
2. Continue for all points.

Step 5: For all rows in the 2D matrix:

1. Take the first point in the row, find direction from this point to the next non-null point in the same row.
2. Continue for all points.

Step 6: Remove all overlapping portions and add the paths to the WKT file.

the software represents the extracted path information in *KML* format, converts *KML* files into *OSM* format, and finally that to *WKT* format using *osm2wkt.jar* tool. The *playfield graphics* of ONE Simulator is scaled properly so that map-based movement models can be used during the simulation.

4 Result and Discussion

Figure 3 shows some screenshots for our software when it is in action. Figure 4 shows the increase in extraction time in hours of the WKT file, that we observed during our experiments for a given map, as the area of the map segment is increasing. However, since the extraction time involves Google API calls, the time required is very much dependent on the speed of the Internet, and hence may vary from network to network. Figure 4 shows the increase in the number of Google API calls by the system as the area of the map segment increases. The later might be a better measure of the system performance as it is independent of the Internet speed.

Fig. 3 Screenshots (clockwise from top-left): **a** Area selection, **b** path extraction started using distributed processing, **c** path extraction continuing, **d** simulation using ONE Simulator

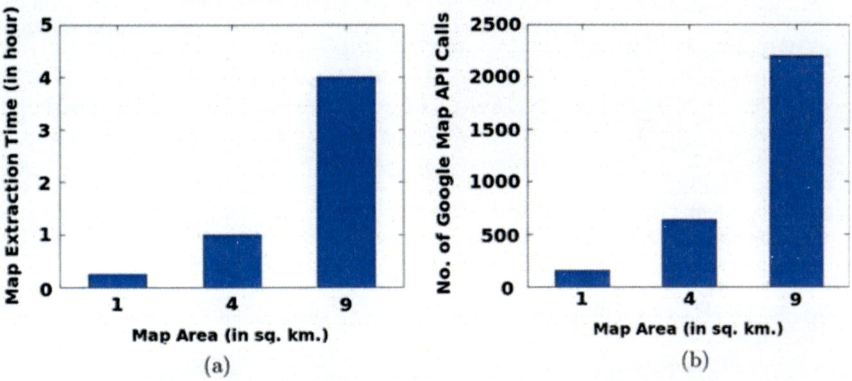

Fig. 4 **a** The time (in hours) required to extract the WKT file for given map segment of a certain size (in sq. km.) **b** The number of Google API calls by the system to extract the WKT file for the given map segment

5 Conclusion and Future Scope

The proposed *PDCRPT* tool has great scope to be used for disaster preparedness as well as post-disaster planning by the Government agencies as well as competent NGOs for better serving the victims after large-scale disasters. The decent and user-friendly interface of the tool would enable the use of the system by computer non-experts. However, when the communication resources at hand are limited, then finding even a near-optimal solution for the deployment plan is non-trivial and beyond the scope of the present implementation. As a future scope, we plan to integrate a module to the software which, given an available resource pool, will provide at least a near-optimal deployment plan and a packet delivery delay bound for such a deployment.

Acknowledgements This Publication is an outcome of the R&D work undertaken partially in the ITRA project of Media Lab Asia entitled "Post-Disaster Situation Analysis and Resource Management Using Delay-Tolerant Peer-to-Peer Wireless Networks (DISARM)", and partially in the project entitled "Developing ICT-based kiosks for post-disaster situational information management using opportunistic networking framework" funded by Department of Higher Education, Science and Technology and Biotechnology, Government of West Bengal, India.

References

1. Keränen, A., Ott, J., Kärkkäinen, T.: The ONE Simulator for DTN protocol evaluation. In: Proceedings of the 2Nd International Conference on Simulation Tools and Techniques, ICST (Institute for Computer Sciences, Social-Informatics and Telecommunications Engineering), ICST, Brussels, Belgium, Belgium, Simutools '09, pp. 55:1–55:10 (2009)
2. Legendre, F.: 30 Years of Ad Hoc networking research: what about humanitarian and disaster relief solutions? What are we still missing? In: Proceedings of the 1st International Conference on Wireless Technologies for Humanitarian Relief, ACM, New York, NY, USA, ACWR '11, p. 217 (2011)
3. Paul, P.S., Ghosh, B.C., Datta, H.S., De, K., Basu, A.P., Pramanik, P., Saha, S., Chakraborty, S., Ganguly, N., Nandi, S.: CRIMP: Here crisis mapping goes offline. J. Network Comput. Appl. (2019). Elsevier
4. Saha, S., Agarwal, N., Dhanuka, P., Nandi, S.: Google map based user interface for network resource planning in post disaster management. In: Proceedings of the 3rd ACM Symposium on Computing for Development, ACM, New York, NY, USA, ACM DEV '13, pp. 22:1–22:2 (2013)
5. Saha, S., Nandi, S., Paul, P.S., Shah, V.K., Roy, A., Das, S.K.: Designing delay constrained hybrid ad hoc network infrastructure for post-disaster communication. Ad Hoc Netw. **25**, 406–429 (2015)

A Reconfigurable Architecture to Implement Linear Transforms of Image Processing Applications

Atri Sanyal and Amitabha Sinha

Abstract A reconfigurable architecture capable of implementing flow graph-based computationally intensive image processing transforms is introduced. The conceptual design of the architecture is deduced using a graph theoretical approach. The graph theoretical model is transformed into a reconfigurable design. The design of the overall architecture as well as the design of the processing element is implemented in VHDL using Xilinx Vivado. The architecture is synthesized. The result is presented in terms of size and timing behaviour. The processing elements inside the architecture are simulated by a test bench program executing two major operations to see the possibility of further improvement of the design. Future direction of the research is discussed.

Keywords Conceptual design · Graph theoretical approach · Actual reconfigurable architecture · Instruction set · Synthesis

1 Introduction

A dedicated processor to implement computationally intensive transformation in image processing application is in the field of study for quite some time. A lot of architecture have been proposed in different papers. Reconfigurability of the FPGA device has made it a popular choice for this kind of architecture. We have seen a number of reconfigurable architectures implementing a single computationally intensive transformation like FFT, FDCT or FDWT or any others in [1–14]. There has also been attempts to generalize some of these transformations and prepare a

A. Sanyal (✉)
Amity Institute of Information Technology, Amity University, Kolkata, Major Arterial Road, Action Area II, Rajarhat, New Town, Kolkata, India
e-mail: atri.sanyal@gmail.com

A. Sinha
Institute of Micro Electronics and VLSI Technology, Maulana Abul Kalam Azad University of Technology, West Bengal, BF Block, Sector 1, Bidhannagar, Kolkata 700064, India
e-mail: becas_amits83@hotmail.com

D. Bhattacharjee et al. (eds.), *Proceedings of International Conference on Frontiers in Computing and Systems*, Advances in Intelligent Systems and Computing 1255,
https://doi.org/10.1007/978-981-15-7834-2_83

co-processor capable of implementing all of them in [15] and a variety of algorithms of single transformation in [16, 17]. Reconfigurable architectures capable of more general operations in image or video processing applications or in general applications are proposed in [18–20]. Here in this paper, we have identified some computationally intensive transformations whose operations can be explained by a flow graph. Using the graph theoretical approach, we have deduced the probable architecture of a processor which is capable of implementing transformations of this type. Next, we have proposed an MIMD architecture along with the instruction set. The architecture is synthesized and implemented using Xilinx Vivado IDE. The result of this synthesis in terms of size and speed is presented. Then primary results of some basic operations are obtained by simulating them using a test bench program. The rest of the paper is organized in this way: Sect. 2 discusses how the Graph Theoretical approach is used to deduce the conceptual architecture of this type of processor. Section 3 discusses the actual reconfigurable implementation and issues related to it. Section 4 describes the instruction set of the processor architecture and the Processing Elements (PE). Section 5 presents simulation results of two major operations of PE and discusses Result and analysis of the synthesized model obtained from Xilinx Vivado in terms of size and speed. Finally, Sect. 6 discusses the conclusion of this study and directs the future improvements of the model.

2 Conceptual Architecture of the Processor Using Graph Theoretical Approach

Different image transformations like DCT, FFT, DWT or Hilbert are explained using a flow graph. A flow graph representing an improved variation of the Loeffler's Fast DCT algorithm implementing 1-dimension 8-point DCT is shown below [21] (Figs. 1 and 2).

From the graph theoretical approach [22], this is an Equi-Vertex K-partite graph of $K = 5$, and the data is travelling from one vertex to another through 5 stages and finally getting accumulated in the vertices of the last stage. In a K-partite graph, the vertex set can be partitioned into K disjoint sets where the vertices of a single set are not connected with each other but vertices of two different sets are connected. If $K = 2$ then we have a bipartite graph. In a Complete bipartite graph, the vertex set can be partitioned into two disjoint subsets, where no of vertices are connected with each other in the same subset and connected with all other vertices of the second one. If all the subsets contain the same number of vertices, then it is called an Equi-Vertex K-Partite or Bipartite graph. An Equi-Vertex K-Partite graph and an Equi-Vertex Complete Bipartite graph is shown below.

Next, we introduce a lemma which will help us to deduce the architecture of our proposed processor.

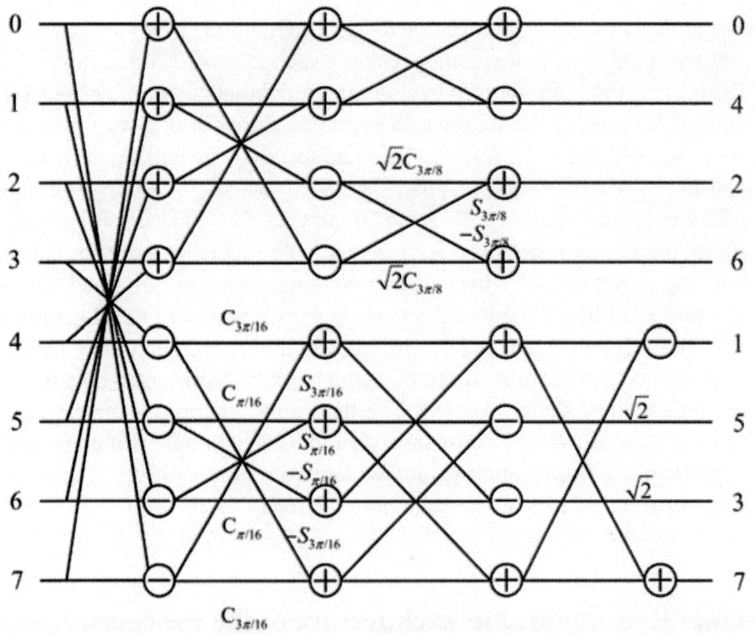

Fig. 1 Flow graph of Loeffler's FDCT algorithm [21]

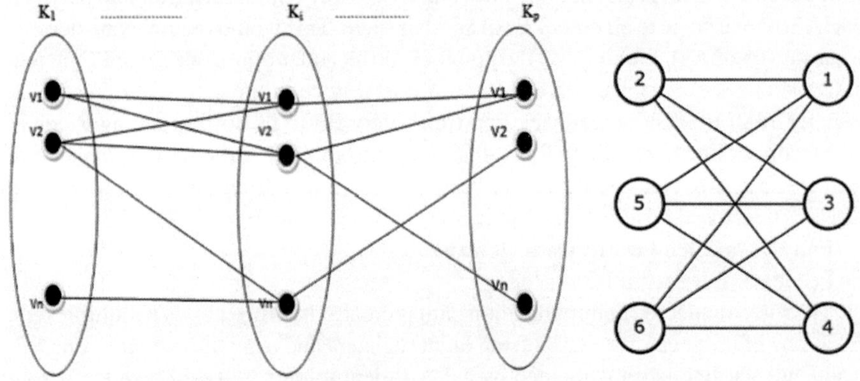

Fig. 2 Equi-Vertex K-partite graph and Equi-Vertex Complete bipartite graph

Lemma 1 All paths constructed from any vertex of one set to any other vertex of another set of an Equi-Vertex K-Partite Graph G1 can be represented by an Equi-Vertex Complete Bipartite Graph G2 in which vertices of two sets have a one-to-one mapping with the vertices of K stages of the K-partite graph.

Explanation: Suppose there is a path from one vertex to another vertex through different sets in G1 denoted as $V_1K1 \rightarrow V_2K2\ldots\ldots \rightarrow V_iKi$ (where Vi is the vertex number and Ki is the stage number of K-Partite Graph G1). We assumed that the vertex set has a one-to-one mapping between every stage of G1 and the two stages of G2. This means that all vertices having the same number in K stages of G1, and two stages of G2 are essentially the same vertices. Suppose one such path in G1 is V1K1→V3K2→V5K3→V2K4...... It can be successfully replicated by the bipartite graph G2 as V1S1→V3S2→V5S1→V2S2...(where Vi is the vertex number and S1,S2 are the two stages of Bipartite Graph G2). Since G2 is a fully connected Bipartite graph, all we have to do is to take an edge which would travel from the corresponding vertex of Set 1 to the corresponding vertex of Set 2 and vice versa. Since the vertex set has a one-to-one mapping with K stages of the K-partite graph and two stages of Bipartite graph, it can be replicated with a K of any size. This lemma helps us to deduce that any transformation which can be represented by a flow graph of arbitrary number of stages, where the vertices represent processing elements and the edges represent a two-way communication line can be implemented by an architecture representing Equi-Vertex Complete Bipartite graph of Fig. 2 provided the number of vertices, i.e. PEs are same in both the cases.

3 Actual Reconfigurable Architecture of the Processor

The conceptual architecture of the processor deduced in Sect. 2 contained two sets of Processing Elements (PE) in both sides and a two-way communication line between each PE of one stage to all other PEs of another stage. This would require considerable hardware overhead. We simplify the graph a little bit and propose one set of PEs at one side and one set of registers in the other. A one-way communication line from each register to all PEs and a feedback line from every PE to its corresponding Register (i.e. feedback line from PE1 to Reg1, PE2 to Reg2, etc.)—this alternative architecture will create a fully connected path between PEs but in two different clock pulses. The Graph Theoretical design of the processor is given below (Figs. 3, 4 and 5).

Fully Connected Feedforward Network

Feedback Edge from PE to Reg

The fully connected communication path is created by 8 nos of 3 × 8 multiplexers. Each one of them can take input from eight registers and send the output to one PE. The feedback line is implemented by a 1 × 2 demultiplexer. It will take the output of PE and can either loop it back to the corresponding register for loading it into the next clock pulse or store the final value to the memory after the entire operation is finished. An 2 × 1 input multiplexer in the same way can either take the external input from memory or can take the output of the previous stage presented in the feedback line as current input. Two registers are used to latch the values while entering or exiting from PE. There are 8 such stages. The figure given below shows one stage of the actual architecture of the reconfigurable architecture. There are 8 such stages.

The proposed architecture contains eight Processing Elements (PE) for calculation of transformations. Since the transformations were primarily of image processing applications, PE actually implements four types of arithmetic operations of floating

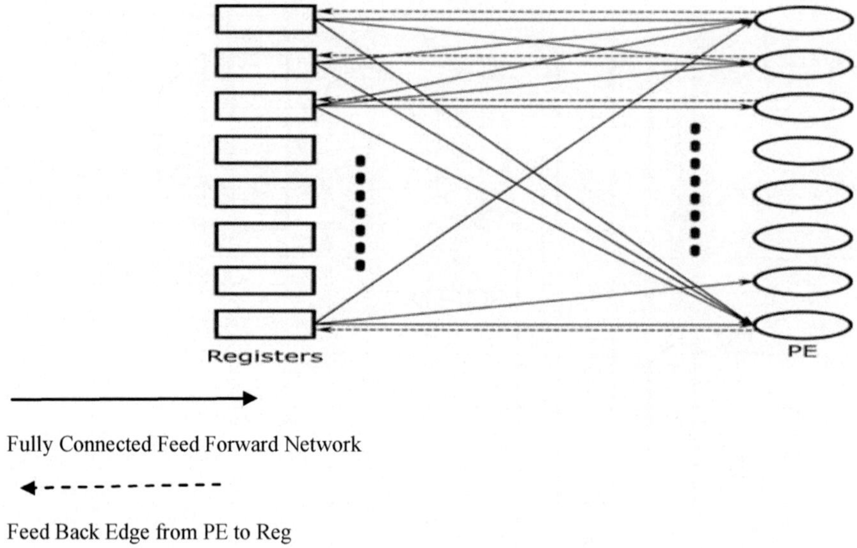

Fig. 3 Directed graph representation of the proposed architecture

point and integer types. These are addition, multiplication, addition after multiplication and multiplication after addition. It is capable of calculating subtraction and division with the same design. In order to execute these operations, PE contains one floating point adder (which can also do 2's complement subtraction), one floating point multiplier, six 32-bit D registers with enabled control, three multiplexers and three demultiplexers to move the data internally within PE or externally to outer registers. The figure below shows the internal structure of the Processing Element (PE).

There is one important specification we should mention here. Though the registers outside the PE had an 8-bit size to contain pixel values of 0–255, the registers inside the PE are of size 32 bit because the IEEE floating point standard 754 is defined on a 32-bit size data. In order to make this design modular, this has been done purposefully. The 8-bit values are padded with leading zeroes upon entering the PE and while exiting from PE, the last 8 bits are extracted.

4 Instruction Set for Processing Element and the Processor

The overall architecture and the PE architecture were given in the previous section. In order to calculate different transformations, all we need is to create the instruction set of the processing element and the overall architecture. The set of instructions in the overall architecture level include instructions to load and store the data to/from memory, load and direct the data from/into the feedback path, and route the data

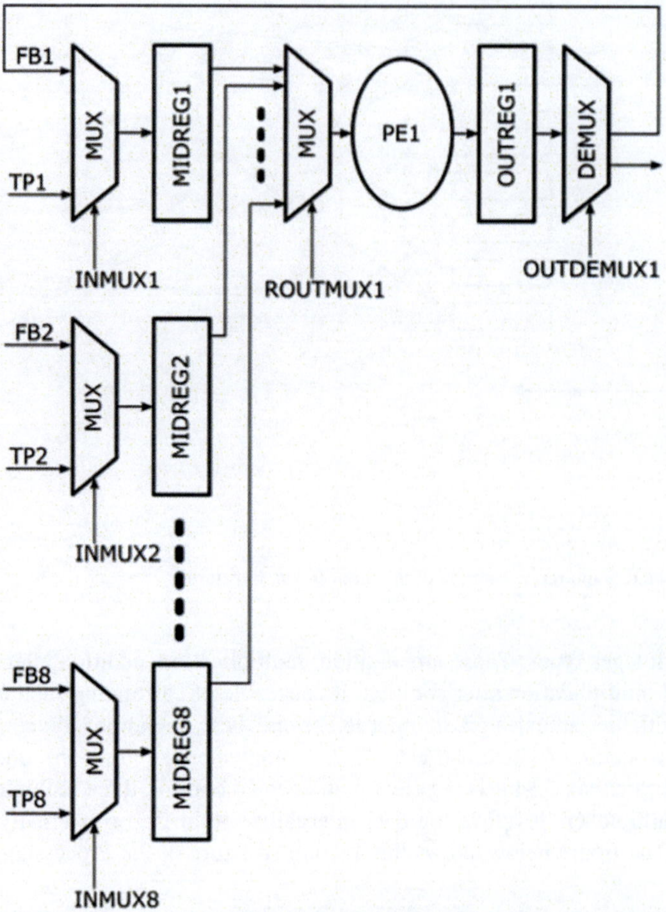

Fig. 4 Design of the first stage of the Proposed architecture along with control lines

from any register to any PE. A list is given below. From the list, we can see that in the overall architecture, the number of bits of the control line will be equal to 40 (Tables 1 and 2).

Inside PE, there are instructions to load and write back the data to/from the overall architecture-level processor, to move data from two internal output registers to any one of the four internal input registers, to add and to multiply. A list is given below. From the list, we can see that the number of bits of the control line in every PE is 12.

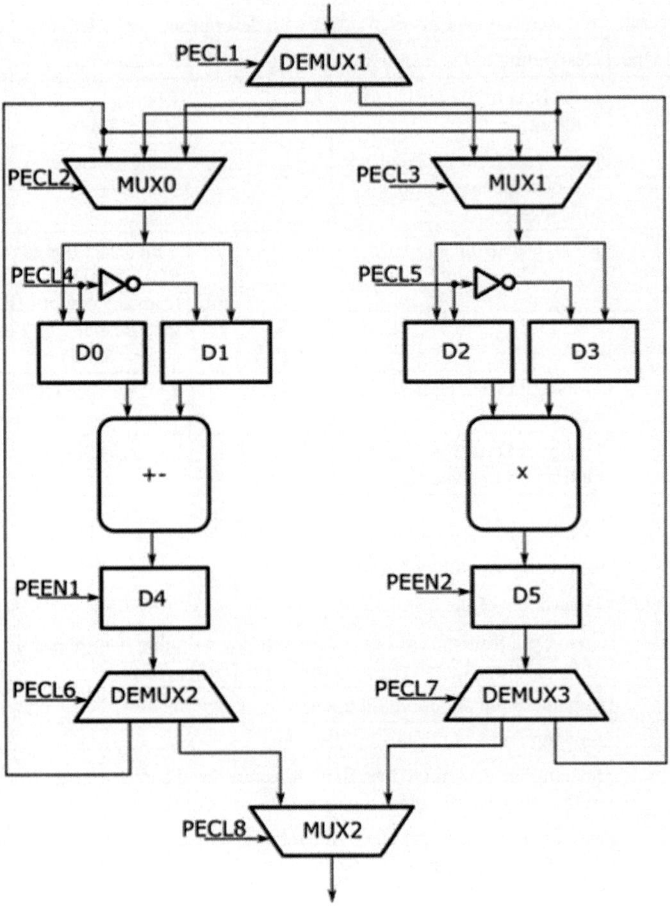

Fig. 5 Design of the Processing Element (PE) along with control lines

5 Primary Results and Analysis

The reconfigurable architecture and its PEs are coded in VHDL and synthesized in Xilinx Vivado. The result of the synthesis is given in the table below.

PE-Level Synthesis Report:

Bandwidth of the PE: 73.12 Mflops.

It is calculated in the following manner. The maximum delay of the setup time found from the timing report in the unconstrained path is 3.419 ns. So the Architecture is capable of executing at a speed of 292.48 MHz. From the Architecture of PE, it is evident that the adder and the multiplier would require 4 clock cycles to execute a floating point addition or multiplication. So it is 292.48/4 = 73.12 Mflops.

Table 1 Overall Architecture-Level Instruction Set with description and its length

Instruction type	Description of the instruction	Length in bits	Explanation
1	Load Data from Outside to MidReg1-8	8	1 bit for each INMUX. Total 1 × 8 = 8 bits
2	Load Data from the feedback line of respective PE to MidReg1-8	8	Same as above
3	Route Data from any MidReg1-8 to any PE 1-8	24	3-bit select line for each ROUTEMUX to route data from any one of MIDREG1-8 to corresponding PE. Total 3 × 8 = 24 bits
4	Store Data from OutReg1-8 to Memory	8	1 bit for each OUTDEMUX. Total 1 × 8 = 8 bits
5	Feedback Data from OutReg1-8 to respective PE	8	Same as above

Table 2 PE-level instruction set with description

Instruction type	Description of the instruction
1–4	Load Data from the overall architecture-level routing multiplexer to any one of the 4 internal input registers of the PE (To D0–D3)
5	Floating Point Addition/Subtraction
6	Floating Point Multiplication/Division
7–8	Sending the data out to overall architecture-level output register from any one of the two internal output registers (From D4 or D5)
9–16	Moving Data from any one of the internal output registers to any four of the internal input registers (From D4 to D0/D1/D2/D3), (From D5 to D0/D1/D2/D3)

Power Utilization Report of PE: Static Power: 0.148 W (84%); Dynamic Power: 0.029 W (16%) (Tables 3, 4, 5 and 6).

Overall Architecture-Level Synthesis Report

Bandwidth: 73.12 Mflops. This is same as PE, showing that the highest delay path is inside the PE, also indicating that any delay in the operation inside PE would decide the timing behaviour of the entire architecture.

Table 3 Utilization report (Summary) of PE

Name	Used
LUT	1240
FF	618
IO	47

Table 4 Utilization report (Primitives) of PE

Name	Used	Functional Category
LUT6	617	LUT
LUT5	181	LUT
LUT4	372	LUT
LUT3	32	LUT
LUT2	69	LUT
LUT1	133	LUT
FDRE	358	Flop & Latch
FDCE	196	Flop & Latch
LDCE	64	Flop & Latch
CARRY4	21	CarryLogic
IBUF	46	IO
OBUF	1	IO
BUFG	3	Clock

Table 5 Utilization report (Summery) of overall architecture

Name	Used
LUT	10160
FF	5192
IO	159

Table 6 Utilization report (Primitives) of overall architecture

Name	Used	Functional category
LUT6	4944	LUT
LUT5	1432	LUT
LUT4	2984	LUT
LUT3	320	LUT
LUT2	552	LUT
LUT1	1272	LUT
FDRE	3056	Flop & Latch
FDCE	1624	Flop & Latch
LDCE	512	Flop & Latch
CARRY4	168	CarryLogic
IBUF	94	IO
OBUF	65	IO
BUFG	12	Clock

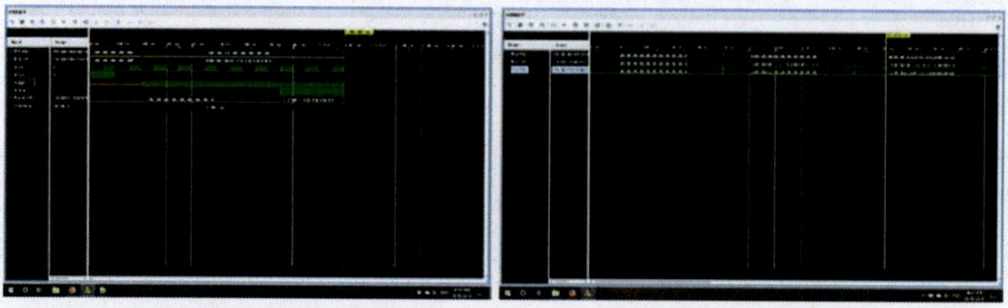

Fig. 6 Simulation Result of the floating point adder and multiplier unit of the PE

Power Utilization Report of Overall Architecture: Static Power: 0.148 W (52%); Dynamic Power: 0.135 W (48%).

A test bench program is written for the simulation of the floating point addition and floating point multiplication and using ISim simulator of Vivado, the result is obtained. The result of the simulation indicates the correctness of the design. The figure of the simulation is given below (Fig. 6).

6 Conclusion and Future Direction of the Study

The conclusion that we can draw from this study is that this architecture is able to perform all the basic arithmetic and data movement operations common to the image transform operations that we considered. The bandwidth and the utilization report suggest a fairly applicable performance. Although the comparison with popular reconfigurable FFT or DCT processor implementations as in [23–25] are avoided as the architecture has not yet been a fully grown processor, the future direction of this study would be to organize the reconfigurable architecture presented here into a fully grown processor. In order to do so, firstly we have to build a control unit for generating and sequencing control signals to implement the instruction set for both the processor level and the PE level. The control unit should be built in such a way that it accommodates the instruction set of PEs to be executed in the MIMD mode. Secondly, we need to build memory buffers and RAM capable of exchanging data to/from the main processor and the transform processor handling all speed mismatch issues by generating appropriate handshaking signals. Thirdly, a number of popular transforms which can be implemented by this processor would be simulated; obtained result with respect to time and size will be used to validate the capability of this processor and would be compared with the previous instances.

References

1. Tseng, P.-C. et al.: Reconfigurable discrete cosine transform processor for object-based video signal processing. In: ISCAS '04 Proceedings of the 2004 International Symposium on Circuits and System (2004)
2. Tseng, P.-C., Huang, C.-T., Chen, L.-G.: Reconfigurable discrete wavelet transform processor for heterogeneous reconfigurable multimedia systems. J. VLSI Signal Process. Syst. Signal Image Video Technol. (2005)
3. Donohoe, G.W.: The fast fourier transform on a reconfigurable processor. In: Proceedings NASA Earth Sciences Technology Conference, Pasadena, CA, June 11–13, 2002
4. Srivatsava, P.S.V., Sarada, V.: Reconfigurable MDC architecture Based FFT processor. Int. J. Eng. Res. Technol. (2014)
5. Joe Hass David, K., Cox, F.: Transform processing on a reconfigurable data path processor. In: 7th NASA Symposium on VLSI Design (1998)
6. Sarada, V., Vigneswaran, T.: Reconfigurable FFT processor—a broader perspective survey. Int. J. Eng. Technol. (IJET) (2013)
7. Shahbahrami, A., Ahmadi, M., Wong, Bertels, K.: A new approach to implement discrete wavelet transform using collaboration of reconfigurable elements. In: Proceedings 2009 International Conference on Reconfigurable Computing and FPGAs
8. Manolopoulos, K.E., Nakos, K.G., Reisis, D.I., Vlassopoulos, N.G.: Reconfigurable fast fourier transform architecture for orthogonal frequency division multiplexing systems (2003). https://pdfs.semanticscholar.org/dd5c/263725af00e5dd4d42d573c269f57d917c8d.pdf?_ga=2.84059166.640751657.1573804365-914446569.1569299704
9. Sinha, A., Sarkar, M., Acharyya, S., Chakraborty, S.: A novel reconfigurable architecture of a DSP processor for efficient mapping of DSP functions using field programmable DSP arrays. ACM SIGARCH Comput. Architect. News **41**(2), May 2013
10. SumitWadekar, Laxman P. Thakare, Dr. A.Y. Deshmukh, " Reconfigurable N-Point FFT Processor Design For OFDM System, International Journal of Engineering Research and General Science Volume 3, Issue 2, March-April, 2015
11. Petrovsky, A., Rodionov, M., Petrovsky, A.: Dynamic reconfigurable on the lifting steps wavelet packet processor with frame-based psychoacoustic optimized time-frequency tiling for real-time audio applications. Design Architect. Digital Signal Process. http://www.intechopen.com/books/design-and-architectures-fordigital-signal-processing2013
12. Thomas, S., Sarada, V.: Design of reconfigurable FFT processor with reduced area and power. ITSI Trans. Electrical Electron. Eng. (ITSI-TEEE) (2013)
13. Rajaram, U.: Design Of Fir Filter For Adaptive Noise Cancellation Using Context Switching Reconfigurable EHW Architecture. Ph.D dissertation, Anna University, Chennai (2009). https://shodhganga.inflibnet.ac.in/handle/10603/27245
14. Reddy, P.S., Mopuri, S., Acharyya, A.: A reconfigurable high speed architecture design for discrete hilbert transform. IEEE Signal Process. Lett. **21**(11), 1413–1417 (2014). https://doi.org/10.1109/LSP.2014.2333745
15. Sanyal, A., Samaddar, S.K., Sinha, A.: A generalized architecture for linear transform. In: Proceedings IEEE International Conference on CNC 2010, Oct 04-05, 2010, Calicut, Kerala, India, IEEE Computer society, pp. 55–60, ISBN: 97-0-7695-4209-6
16. Sanyal, A., Samaddar, S.K.: A combined architecture for FDCT algorithm. In: Proceedings 2012 Third International Conference on Computer and Communication Technology, Allahabad, 2012, pp. 33–37. https://doi.org/10.1109/iccct.2012.16
17. Sanyal, A., Kumari, S., Sinha, A.: An improved combined architecture of the four FDCT algorithms. Int. J. Res. Electron. Comput. Eng., (IJRECE) **6**(4), December 2018, ISSN: 2348–2281
18. Rossi, D., Campi, F., Spolzino, S., Pucillo, S., Guerrieri, R.: A heterogeneous digital signal processor for dynamically reconfigurable computing. IEEE J. Solid-State Circu. **45**(8) (2010)

19. Purohit, S., Rahul Chalamalasetti, S., WimVanderbauwhede, M.M.: Throughput/resource-efficient reconfigurable processor for multimedia applications. IEEE Trans. Very Large Scale Integr. (VLSI) Syst. **21**(7) (2013)
20. Vikram, K.N., Vasudevan, V.: Mapping data-parallel tasks onto partially reconfigurable hybrid processor architectures. IEEE Trans. Very Large Scale Integr. (VLSI) Syst. **14**(9) (2006)
21. Heyne, B., Sun, C.C., Goetze, J., Ruan, S.J.: A computationally efficient high-quality Cordic based DCT. In: 14th European Signal Processing Conference (EUSIPCO 2006), Florence, Italy, September 4–8, 2006
22. Deo, N.: Graph Theory with applications to engineering and computer science. PHI (2007)
23. Wang, M., Wang, F., Wei, S., li, Z.: A pipelined area-efficient and high-speed reconfigurable processor for floating-point FFT/IFFT and DCT/IDCT computations. Microelectron. J. **47** (2016). www.elsevier.com/locate/mejo
24. Han, F., Li, L., Wang, K., Feng, F., Pan, H., Yu, D.: An improved FFT architecture optimized for reconfigurable application specified processor. In: 2015 IEEE 11th International Conference on ASIC (ASICON), Chengdu, 2015, pp. 1–4. https://doi.org/10.1109/asicon.2015.7517201
25. Kirchhoff, M., Kerling, P., Streitferdt, D., Fengler, W.: A real-time capable dynamic partial reconfiguration system for an application-specific soft-core processor. Int. J. Reconfig. Comput. **2019**, Article ID 4723838, 14 p. (2019). https://doi.org/10.1155/2019/4723838

Introduction of Fuzzy Logic Controller in a Modified Phase-Locked Frequency Divider Leading to an Exceptional Noise Rejection

B. Chatterjee and S. Sarkar

Abstract A frequency divider is commonly implemented using a phase-locked loop by placing a frequency multiplier into the feedback path of the loop. In this paper, a novel analog frequency divider is reported where a fuzzy logic controller is introduced in the loop to adjust the control voltage of the voltage controlled oscillator (VCO) using its human knowledge base. In addition, IQ VCO followed by a signal multiplier is used to replace the frequency multiplier. This frequency divider is particularly configured for the divide-by-two operation. In the presence of additive white Gaussian noise (AWGN) in the input signal, the comparisons of simulation results of the proposed loop over the uncontrolled loop show a phenomenal improvement in the divider output on parameters such as total harmonic distortion by nearly 100dBc, spurious-free dynamic range by around 150dBc. This configuration can be further extended for divide-by-odd number as well.

Keywords Phase-locked divider · Fuzzy controlled PLL · Frequency divider · Harmonic distortion · Spurious-free dynamic range · VCO

1 Introduction

There are various applications of the frequency synthesizer where a frequency divider is used as the key component [1, 2]. A frequency divider is useful to improve the phase noise of the carrier signal only on paper. Practically, there is degradation in the signal spectra due to the divider's own noise contribution [1]. Phase-locked loop is very popularly used as a frequency divider where a frequency multiplier is placed into the feedback path of the PLL.

B. Chatterjee (✉)
Academy of Technology, Hooghly, India
e-mail: cbasab@yahoo.co.in

S. Sarkar
Govt. College of Engineering, Kolkata, India

D. Bhattacharjee et al. (eds.), *Proceedings of International Conference on Frontiers in Computing and Systems*, Advances in Intelligent Systems and Computing 1255,
https://doi.org/10.1007/978-981-15-7834-2_84

Here, an analog phase-locked frequency divider is demonstrated for the divide-by-two operation. But this proposed loop is distinctly different from the traditional loop in three aspects. Firstly in this loop, the traditional voltage controlled oscillator (VCO) in the feedback path is replaced by the IQ VCO, and secondly this VCO is connected to a signal multiplier instead of a frequency multiplier. Next, a fuzzy logic controller is introduced after the RF filter to optimize the control voltage of the VCO.

The block diagram of the frequency divider is depicted in Fig. 1. This phase-locked frequency divider loop primarily contains a phase detector (PD), a low-pass filter (LPF), a voltage controlled oscillator and a fuzzy logic controller (FLC). The loop VCO is said to be an IQ VCO, because it has two outputs, in-phase (I) and quadrature phase (Q). The IQ VCO outputs are connected to a signal multiplier. The phase detector along with LPF generates the error signal from the phase difference between the incoming signal and the reference signal coming from the signal multiplier. The fuzzy logic controller drives the VCO in such a way that the phase error is minimized and the signal multiplier output phase gets locked with the incoming signal phase.

In this paper, the proposed fuzzy controlled phase-locked frequency divider (FC-PLFD) is studied theoretically, where the justification of introducing a fuzzy logic controller in the loop is included. This loop along with a uncontrolled phase-locked divider are simulated on the software platform to compare the loop performance in terms of the parameters such as total harmonic distortion (THD) and spurious-free dynamic range (SFDR) when additive white Gaussian noise (AWGN) is present in the input signal.

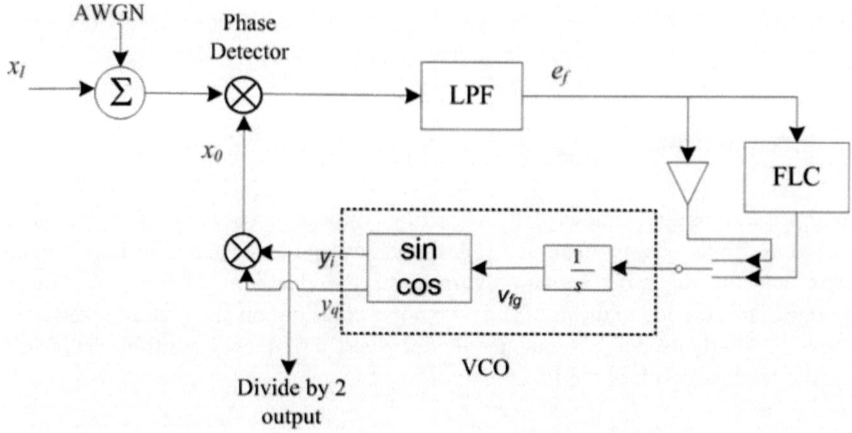

Fig. 1 Block diagram of Fuzzy controlled phase-locked frequency divider

2 Theory

The incoming signal, whose frequency is to be divided, is connected to the phase detector. This is represented by

$$x_I(t) = Acos[2\omega_i t + \psi(t)] \tag{1}$$

Here, $2\omega_i$ and $\psi_i(t)$ are input signal frequency and phase, respectively.
Output of the multiplier connected to the VCO can be written as

$$x_0(t) = 2\sin[2\omega_0 t + 2\theta(t)] \tag{2}$$

Here, ω_0 and $\theta(t)$ are VCO output signal frequency and phase, respectively.
Therefore, output of the phase detector, where $x_I(t)$ and $x_0(t)$ are multiplied, after passing through the RF filter becomes

$$e_f(t) = Asin\phi(t) \tag{3}$$

where the phase error from (1) and (2) is

$$\phi(t) = 2(\omega_0 - \omega_i)t + 2\theta - \psi \tag{4}$$

As $e_f(t)$ is fed to the IQ VCO with the sensitivity μ, then VCO output signal frequency is given by

$$\frac{d\theta}{dt} = \mu Asin\phi(t) \tag{5}$$

The change in phase error can be obtained from (4) and (5),

$$\frac{d\phi}{dt} = 2(\omega_i - \omega_0) + 2\mu Asin\phi(t) - \frac{d\psi}{dt} \tag{6}$$

Considering the input initial phase, $\psi = 0$, it is evident from (6) that the steady state will be attained if

$$\mu A \geq (\omega_0 - \omega_i) \tag{7}$$

and under phase-locked condition $\omega_0 = \omega_i$ and on assumption of small phase excursion (6) reduces to

$$\frac{d\phi}{dt} = 2\mu A\phi(t) \tag{8}$$

Therefore, the VCO output signal is the desired frequency divided output when a phase lock arrives. In the presence of AWGN in the input signal, (8) can be rewritten as

$$\frac{d\phi}{dt} = \mu[2A\phi(t) + N(t)] \tag{9}$$

where $N(t)$, the noise present in the phase detector output, is expressed by

$$N(t) = n_C(t)cos\phi(t) - n_S(t)sin\phi(t) \tag{10}$$

Here, $n_C(t)$ and $n_S(t)$ are quadrature components of the noise mentioned in (10). The phase detector output noise is minimized with the help of LPF, but cannot be eliminated completely. Therefore, the fluctuation or ripples in the LPF output causes the harmonic distortion to be present at the VCO output signal. The noise component at the VCO input also contributes spur at the output signal.

Therefore, to obtain a clean frequency divider output, the input of the VCO must be constant. In the following section, it is going to be shown that an introduction of the fuzzy logic controller can produce a constant output from an input filled with ripples.

3 Fuzzy Logic Controller

Fuzzy logic controller provides an automatic control strategy using a set of linguistic rules [3]. A two-input single-output fuzzy inference (FIS) is considered to avoid computational complexity [4–6]. The two-input crisp variables of FIS are the phase error signal $\phi(t)$ and change in phase error $d\phi/dt$, respectively. The crisp variable output of the fuzzy logic controller is the control voltage of the VCO.

The ranges of input and output variables, i.e., universe of discourses are partitioned into three subsets. They are assigned with linguistic connectives of positive (P), zero (Z) and negative (N). All variable subsets are chosen with triangular shape membership functions (MFs) as depicted in Fig. 2. The following simple set of rules provides the human knowledge base to the controller.

$$R_i : \text{If } x_1 \text{ is } A_1^i \text{ and } x_2 \text{ is } A_2^i \text{ then } y \text{ is } B^i$$

where x_1, x_2 and y are linguistic variables and A_1^i, A_2^i and B^i are fuzzy subsets of MFs, also referred to as linguistic levels of MFs.

The rule table is formed following the above rules as shown in Table 1. This rule table indicates that the absolute value of the fuzzy output rises very fast when the absolute phase error or its change is large. It helps to improve the acquisition performance of the loop. On the contrary, when absolute phase error and its change

Fig. 2 Membership functions of inputs and output

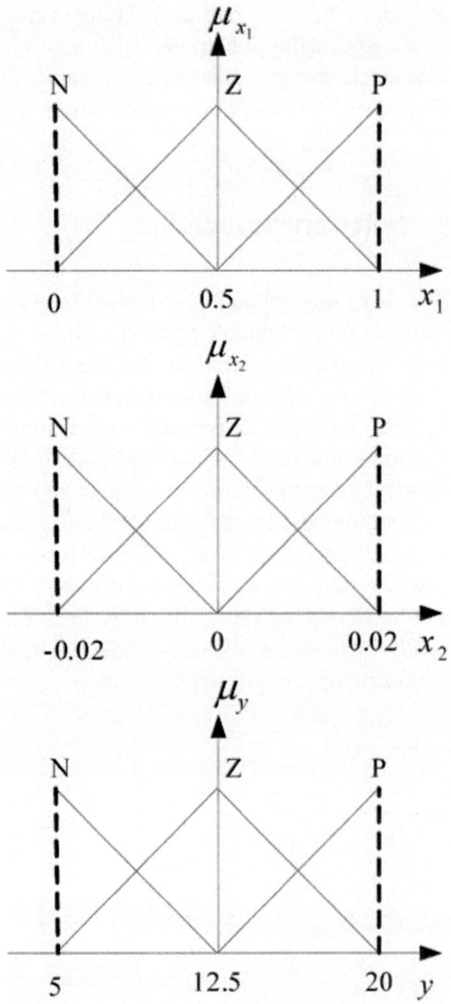

Table 1 Fuzzy Rule Table

$d\phi/dt$	N	Z	P
$\phi(t)$			
N	N	N	Z
Z	N	Z	P
P	Z	P	P

are small, the fuzzy output will remain constant. Therefore, the ripple present in the LPF output will not be reflected at the fuzzy output and the VCO will be driven by a constant voltage. This suggests that an introduction of fuzzy logic controller can produce a VCO output completely free of spur and harmonic distortion.

4 Software Simulation

The fuzzy controlled phase-locked frequency divider (FC-PLFD) is realized using software tools available in MATLAB/Simulink from Mathworks [7, 8]. The design of this loop in Fig. 3 follows the block diagram (Fig. 1) of the earlier section.

In the simulation, the input signal frequency, which is to be divided, is chosen as 160 Hz. The AWGN is added to degrade the input signal quality. The signal, with the difference frequency between the input signal and the output signal, is selected by the RF filter from the phase detector output.

The simulations are carried out both with and without a fuzzy controller in the feedback path after the RF filter. The RF filter output is connected to VCO input through a scalar gain in the uncontrolled loop. In the controlled loop, the RF filter output and its change rate help the fuzzy logic controller to adjust the VCO input control voltage. Therefore, the simulation results are compared to justify the inclusions of the fuzzy logic controller in the loop.

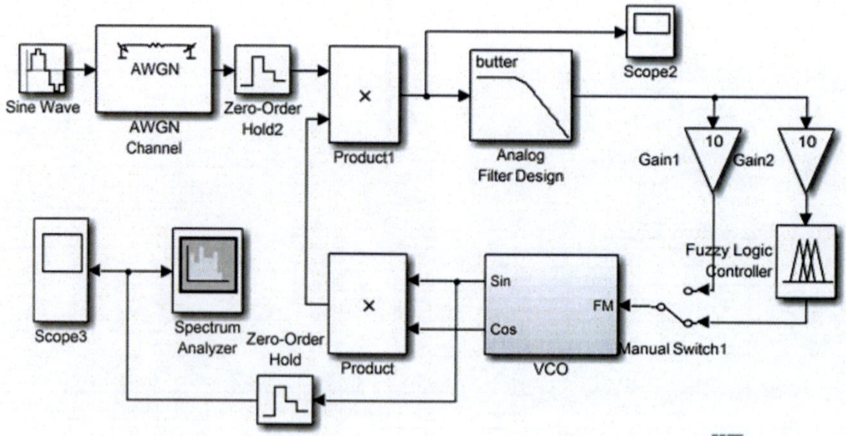

Fig. 3 Simulation model of fuzzy controlled phase-locked frequency divider

5 Simulation Results

The simulation results of the fuzzy controlled phase-locked divider, carried out in a MATLAB/Simulink environment, are compared with those of its uncontrolled counterpart. The simulation outputs are obtained in the presence of AWGN with input SNR 16.5dBc. The noise present in the phase detector output is partially removed by the filter. In the absence of the fuzzy controller in the loop, the filter output is connected to the VCO through a scalar gain. This input control voltage to the VCO is shown in Fig. 4.

When FLC is introduced in the loop, it produces a constant output from the fluctuating input signal coming from the filter to drive the VCO. This constant input signal to the VCO is shown in Fig. 5. The comparison of Figs. 4 and 5 clearly indicates that the VCO input signal fluctuations, which are mainly responsible for distortion in the VCO output signal frequency, approach negligible after addition of FLC in the loop.

The spectrum of the frequency divided output signal with and without FLC in the loop are shown in Figs. 6 and 7, respectively. The presence of harmonic distortion and spur in the output signal spectrum of uncontrolled PLFD is noticeable in Fig. 6. These aberrations are almost eliminated in the spectrum of the output signal of FC-PLFD as shown in Fig. 7. The comparison of the total harmonic distortion (THD) and spurious-free dynamic range (SFDR), which are obtained from the frequency divided output of both versions of phase-locked divider, shows phenomenal improvement in the phase-locked divider output after the addition of FLC. These comparisons of harmonic distortion and SFDR in the FC-PLFD output with uncontrolled PLFD output are shown in Table 2.

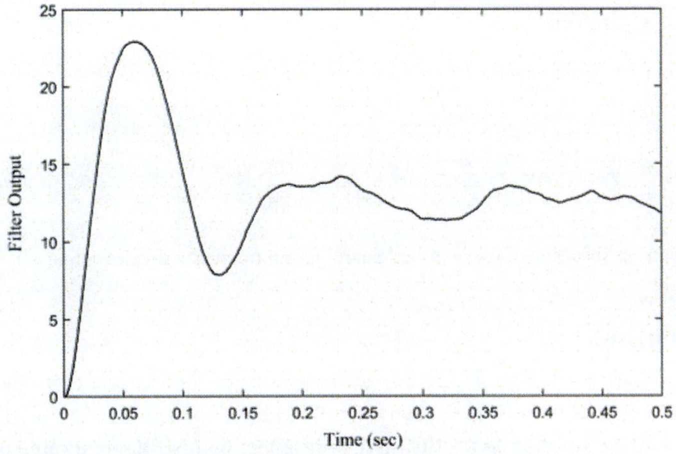

Fig. 4 VCO input control voltage coming from filter through a scalar gain in uncontrolled phase-locked divider in the presence of AWGN in the PLFD incoming signal

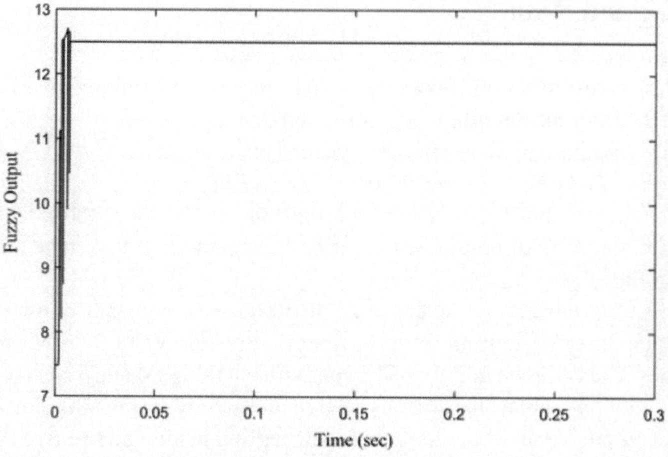

Fig. 5 VCO input control voltage coming from fuzzy logic controller in fuzzy controlled phase-locked divider in the presence of AWGN in the PLFD incoming signal

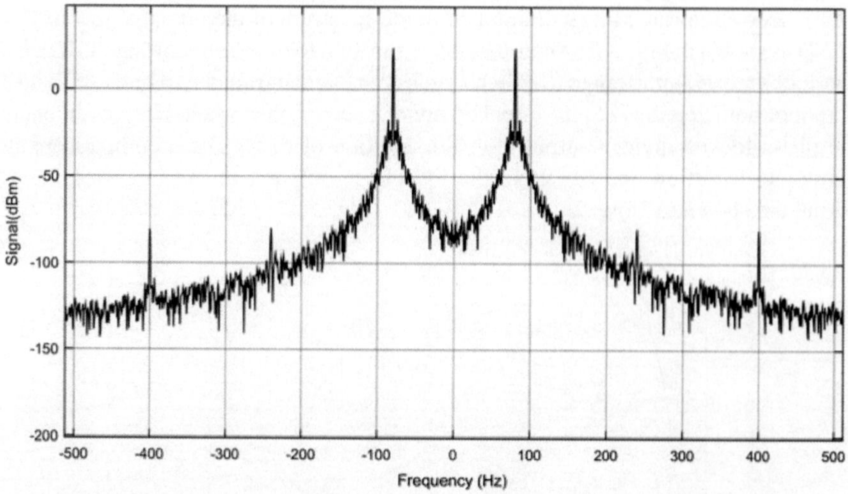

Fig. 6 The power spectrum of uncontrolled phase-locked frequency divider output

6 Conclusion

The theoretical analysis of the fuzzy controlled phase-locked divider has been well established by the simulation results. In the presence of noise, a remarkable improvement is witnessed in the various output parameters of the frequency divider after the introduction of the fuzzy logic controller. This approach is not only limited to

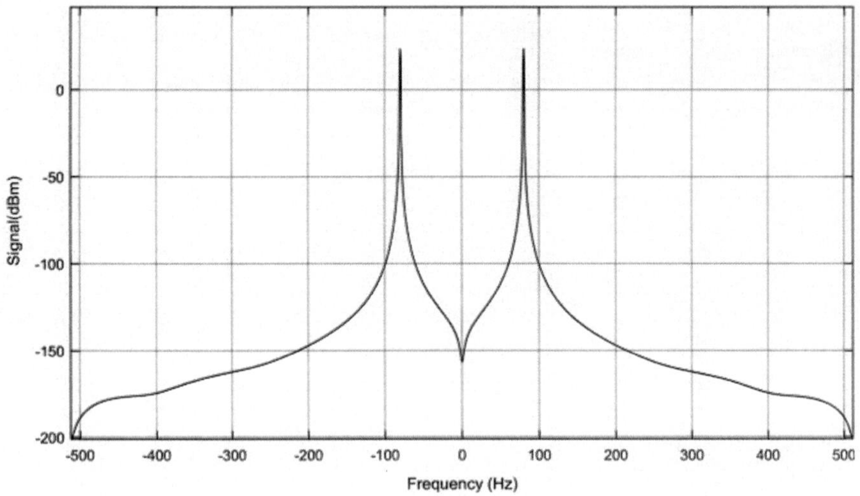

Fig. 7 The power spectrum of fuzzy controlled phase-locked frequency divider output

Table 2 Comparison of output signal from FC-PLFD over Uncontrolled PLFD in presence of AWGN (SNR = 16.5dBc) in the input signal

Signal	Total harmonic distortion	Spurious-free dynamic range
Input	−40.79dBc	35.62 dBc
Uncontrolled PLFD output	−99.81dBc	41.06 dBc
FC-PLFD output	−204.88dBc	199.81dBc

divide-by-two operation, but can also be extended to other frequency dividers with the operation of division by odd numbers.

References

1. Best, R.E.: Phase-Locked Loops: Design, Simulation and Applications, 6th edn. McGraw-Hill, New York (2005)
2. Egan, W.F.: Frequency Synthesis by Phase Lock, 2nd edn. Wiley, New York (2000)
3. Yen, J., Langari, R.: Fuzzy Logic: Intelligence, Control, and Information. Prentice Hall, India (1990)
4. Chatterjee, B., Biswas, B.N., Ray, S.: A novel DSP-based PFC-DPLL with fuzzy controlled acquisition aid to improve acquisition performance and noise immunity. Int. J. Comm. Syst. **28**(15), 2051–2066 (2015)
5. Simon, D., El-Sherief, H.: Fuzzy logic for digital phase-locked loop filter design. IEEE Trans. Fuzzy Syst. **3**(2), 211–218 (1995)
6. Mo, J., Deng, Z., Jia, B., Jiang, H., Biana, X.: Novel FLL-Assisted PLL with fuzzy control for TC-OFDM carrier signal tracking. IEEE Access. **6**, 52447–52459 (2018)

7. Shehova, D.A., Yakimov, P.I.: Teaching PLL fundamentals using MATLAB/simulink. Annu. J. Electron., 139–142 (2014)
8. Godave, A., Choudhari, P., Jadav, A.: Comparison and simulation of analog and digital phase locked loop. In: Proceedings of 9th International Conference on Computing, Communication and Networking Technologies 2018, ICCCNT, pp. 106–109. IEEE Xplore, Bangalore, India (2018)

IoT in Agriculture: Smart Farming Using MQTT Protocol Through Cost-Effective Heterogeneous Sensors

Santanu Mandal, Imran Ali, and Sujoy Saha

Abstract In Agriculture, the demand for better crops is increasing rapidly and seeks proper management in the development of crop production for a better supply. The principle objective here is to implement the usage of technology, i.e. IoT and sensors to enhance the efficiency of the farming systems by providing farmers with the facility to utilize their time and interests in complete potential. This paper presents a custom-built cost-effective device by utilizing the notion of the MQTT IoT protocol that transfers data through the network systems and collects reports about the soil condition. It takes into consideration several factors that affect the health and living of a plant. These factors include the criteria of pH, temperature and moisture for soil, light availability, air temperature and humidity to take into account the present conditions of the plant. Our IoT device for agriculture enables seamless data collection and availability to cloud even for challenged Internet connection.

Keywords Smart farming · AMD · IoT · ESP8266 · Wi-Fi · MQTT

1 Introduction

The world population is expected to reach 9.8 billion by 2050. So, the demand for food will also have increased, and estimations show that the food demand will double itself by 2050. Achieving this increase in the amount of food production becomes tougher due to lowering water levels but International Food Policy Research Institute says that data-driven techniques can boost farm produce by 67% by 2050. So, collecting important data from the field will help in reducing the use of water by 35% and

S. Mandal (✉) · I. Ali · S. Saha
National Institute of Technology Durgapur, Mahatma Gandhi Avenue, Durgapur 713209, India
e-mail: santanucse16@gmail.com

I. Ali
e-mail: imraninsider@gmail.com

S. Saha
e-mail: sujoy.ju@gmail.com

D. Bhattacharjee et al. (eds.), *Proceedings of International Conference on Frontiers in Computing and Systems*, Advances in Intelligent Systems and Computing 1255,
https://doi.org/10.1007/978-981-15-7834-2_85

increase the farm production by 45%; this technique can be applied also in the use of soil nutrients and pesticides.

So, as this technique shows promising results in agriculture, we made our device with 5 sensors and 2 modules: pH sensor for collecting the pH value of soil, Soil moisture Sensor for soil temperature and moisture, grove temperature-humidity sensor for temperature and humidity of the air, and light luminosity sensor for light intensity. Two modules are RTC for timestamp of the data and ESP8266 for sending the collected data to cloud over Wi-Fi. We tried to make our device as cost-effective as possible.

We used the MQTT protocol which is a machine to machine message data transfer protocol originally created by Andy Stanford-Clark of IBM and Arlen Nipper of Cirrus Link in 1999, especially for IoT devices [7].

Our focus is to resist the data loss during data transfer using the MQTT protocol, and our algorithm works even if there is poor/unstable Internet connection, so we will be able to collect all these data to cloud.

2 IoT in Agriculture Field

An IoT-based smart farming is considerably more effective compared to the conventional method. Nowadays, it has an impact on agriculture Industries. Farmers can easily monitor the conditions of the agriculture field from all locations. Against the challenges of erratic weather conditions and elevating climate change, resulting from intensive farming practices, the increased demand has to be supplied with. The objective of this project is implementing a smart device using Internet of Things (IoT) technology and collect sensor data to cloud even if there is an unreliable or weak network connection because the usage of Agriculture Monitoring Device (AMD) for the agriculture field reduces the probability of getting a strong network connection in the agriculture field. Network or power failure should not exhaust the platform. Data collection from sensors should be uninterruptedly supplied to farmers. Diverse sensors like pH sensors which report a few bytes should also be supported. So, sensors should have a wide range for the data. Many farming applications depend on long-term data such as production prediction, crop cycle prediction, water and nutrient content. Also, the farmer may want to know the sensor data when he is away from the field. So, it is mandatory that the device sends the data to cloud. Time lag in the dataset might lead to some fallacy in the application. Thus the device should give as currently updated data as possible.

3 Related Work

A lot of work has been done in the area of the agricultural field. This section discusses a few of the popular works that are related to this domain.

Ganesh et al. [1] have introduced a database that is needed for storing information like maximum and minimum temperature, moisture, pH, soil texture and location for better crop production. In their model, Soil Moisture sensor and DHT11, which collects current temperature and rainfall are connected to Raspberry Pi. The collected data are stored in AWS IoT. A message broker distributes messages to clients based on the topic of the message. They have not mentioned the exact specifications of their sensors and have used a very less number of sensors for testing purpose, also it is not clear whether they are using Internet/Wi-Fi or not.

Syafarinda1 et al. [2] have introduced the Precision agriculture technique which monitors and controls plant growth in greenhouses, suitable for countries with erratic weather conditions. Precision agriculture demands supplying fertilizer needs and smart irrigation. This paper incorporates the usage of Wireless Sensor Network (WSN) with the MQTT protocol. WSN involves IoT devices in monitoring and control. In this paper, the SHT3x and TSL2561 sensors are introduced to record temperature, humidity and light intensity of the greenhouse. Test results of the data transmission speeds prove that the MQTT protocol runs with smaller bandwidth, transmitting data on a 4G network of AXIS provider and with download capacity 0.02kbps and upload capacity 1.65 kbps, the average data transmission speed being 1 1 s. The main limitation of this paper is that this system is not suitable for poor Wi-Fi and Internet connectivity, i.e. in those agricultural areas which are not Internet friendly.

Waleed Al-Gumaei et al. [3] have designed a wireless sensor network, an automated irrigation system that waters street trees, and the performance was enhanced by a graphical user interface. ESP 8277 is induced with a moisture sensor for measuring the water content of soil, rain sensor for rain reports. MQTT Broker acquires data and control from NodeMCU, with which Node-Red is interfaced wirelessly. GUI operates with wireless control and R-Pi allows secured data transmission. An automatic and efficient procedure was carried out without encountering any delay or loss. It is not clear from this paper as to how they make the device more reliable and help the farmers through it.

Srirampaavn et al. [4] have introduced an embedded agriculture systems to improve conservations. Raspberry Pi 3 links sensor networks with cloud. MQTT improves cloud connectivity. NRF24L01 manages communication between sensor networks. They have been used as a plug and play device which is built by a microcontroller with LCD and an interfaced GPS which also enables multicasting between networks and gateway. NRF24L01 sends processed data to the gateway. Data is encapsulated and MQTT stores data on cloud which is accessed through computer or mobile. The author did not show the dataset and did not validate the result. So, it is not clear how the sensors get the accurate result for agriculture fields.

Abhishek et al. [5] studied the various weather factors that are taken into consideration with which the best suitable crop is predicted with the help of advanced knowledge of Decision Tree, Classifier, Regression, etc. Various sensors are introduced in the soil, and atmospheric conditions are checked through the multi-hop system to the server which monitors the crop's growth and health rate for which

TDMA is used. They have not mentioned as to how it is very much useful for farmers and also how the data are stored in a particular storage location.

Maha Kavya Sri et al. [6] have introduced several factors of Agriculture monitoring. IoT has been a boon to precise agriculture. It uses sensors, nodes and cloud, edge and fog computing. IoT architecture uses various nodes, gates and gateways. Each layer operates using a different protocol. IoT should be used properly as it uses low power sources. MQTT and HTTP can be compared in terms of their suitability in IoT application. They have only stated the power, bandwidth rate and comparison of protocols, nowhere mentioning the detailing of sensors and the cloud storage activities.

Problem Statement: *Is it possible to monitor agriculture data using a cost-effective agriculture sensor? How does it store real-time sensor data with/without any Wi-Fi connection?*

Thus, we are concerned about the farmer's activity, so we have designed a small Agriculture Monitoring Device (AMD) incorporating some low-cost sensors, which is easily portable, acquiring some data, storing it into the device, also in the cloud storage if Wi-Fi connection is available, thus analyzing the data and finally plotting it through the system or mobile application. This also makes the device more reliable and helps the farmers with "Smart-Farming", a mobile application that tries to detect the plant condition. Also, we have proposed an algorithm to back up the sensor data in case of poor availability of Wi-Fi connection.

4 Research Methodology

This section consists of feasibility and requirement of sensor specifications, followed by proposed work, data transfer to cloud, implementation of MQTT and data transfer with poor Internet connection.

4.1 Requirement and Feasibility

We have to use some low-cost sensors as shown in the table (Table 1) below.

4.2 Proposed Work

The overall architecture of our proposed system is shown in Fig. 1. In this particular experiment, the execution of the main idea has been performed on a croton plant. The pH value is tested by the sensors and accordingly, the nature and acidity of the soil are determined. The various values of each day are noted. The water content of the soil and availability of daylight is of major significance in determining the

Table 1 Sensor specifications for Agriculture Monitoring Device (AMD)

Sensor name	Type of sensor	Detecting scope	Measuring range	Operating Temp/Humidity	Purpose of use	Image
Analog pH Sensor / Meter Pro Kit for Arduino	Electrolyte	pH of Soil	0–14 pH	0–60°C	Measurement of soil pH for a particular soil	
Soil Moisture-Temperature sensor/ SHT10	Electro-chemical	Soil Moisture and Temperature	–40°C – 120°C & 0 – 100% RH	–40°C – 120°C & 0 - 100% RH	Measurement of soil temperature and moisture	
Grove - Temperature & Humidity Sensor	Semi-conductor	Air Temperature & Air Humidity	Temperature-:0-50°C, Humidity:20%-90%	0-50°C & 20-80%	Measurement of air temperature and moisture	
Adafruit TSL2561 Digital Luminosity/Lux	Semi-conductor	Light Intensity in the Environment	0.1–40,000 Lux	-30–80 °C	Measurement of the luminosity of light in Lux unit	
Grove - Multichannel Gas Sensor	Metal Oxide Semiconductor	No2 in the Environment			To measure presence of nitrogen di-oxide	
I2C RTC DS1307 RTC/Serial Esp-01 Wifi	Real-Time Clock Module for Raspberry Pi - Red	Real-Time Clock			Measures real Date and time. Helps to gain time-series data	
ESP8266 WiFi Module	REES52-ESP8266 Arduino Compatible	WiFi Module			Sends sensor data into cloud	
MicroSD card module	Brand: DFRobot	Interface: SPI, Compatible: MicroSD (TF)	Size: 20x28mm (0.79x1.10")	Working Voltage: 5V	Save and store sensor data	

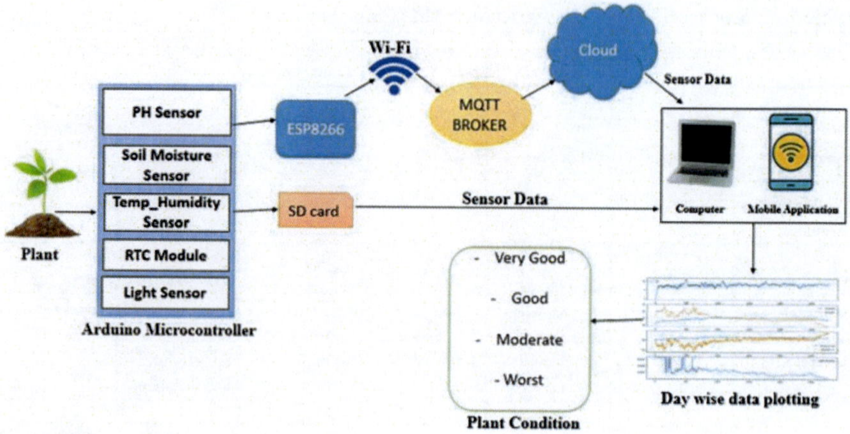

Fig. 1 Architecture of our proposed work

health of the plant. The sensors include pH sensor, soil moisture and temperature sensor, air temperature and air humidity sensor, luminosity/light sensor, working and supply voltage sensors, including a modification using Arduino. The presence of various gases in the atmosphere can be detected as well through sensors. The date and time range along with the various data have been taken into consideration, which forms the basic fieldwork of the experiment. In yet another datasheet, besides the mentioned factors, parameters like availability of water, sunny or rainy conditions, and outdoor or indoor classifications have also been made. This data is received by ES2866 and sent to cloud via MQTT Broker which further forwards it to the mobile or system application. An alternate path is provided by the SD Card. Accordingly, a graph is plotted, giving an immediate report about the day-wise status of the plant determining whether the plant is in very good, good, moderate or worst condition. This will help in determining the optimum conditions for the best desired results from the plant.

4.3 *Data Transfer to Cloud*

We are using ESP8266 to transfer data to cloud over Wi-Fi. ESP8266 is a microcontroller used with Arduino and both the devices communicate on the serial monitor. We collect the data from Arduino and the collected data is transferred to ESP8266 over a serial monitor, collected by the ESP and sent to the Internet over Wi-Fi. ESP8266 Wi-Fi module is configured to a specific network with a specific SSID and password using AT commands. Once the AMD device is on, then it starts sending data to ThinkSpeak cloud; it's the open IoT platform with MATLAB analytics where time-series data is plotted with "Date-Time" in X-axis. A CSV file is automatically

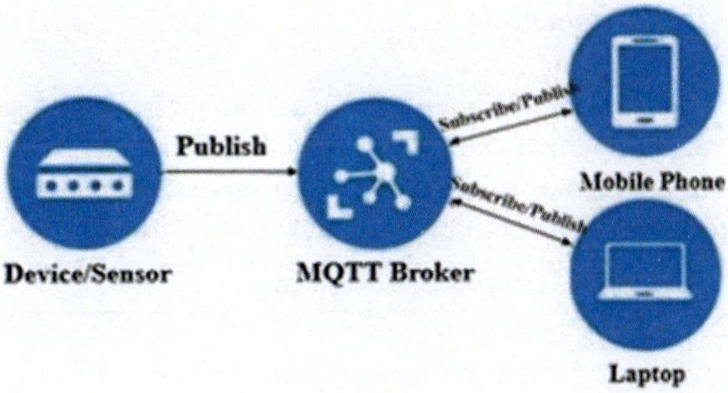

Fig. 2 Working of MQTT protocol

generated, but here we have used the MQTT protocol as it is specially designed for the Internet of Things (IoT). MQTT stands for Message Queuing Telemetry Transport. MQTT is a lightweight messaging protocol for small sensors and mobile devices, optimized for high-latency or unreliable networks created by Andy Stanford-Clark of IBM and Arlen Nipper of Cirrus Link in 1999 [7]. The MQTT protocol uses Publish and Subscribe method. The broad working of the MQTT protocol is shown in the figure (Fig. 2). A client can publish and subscribe both to the broker. A broker is a bridge between many clients, which collect any data over a topic and send it to all the subscribers who subscribed to that topic. Topics are meant to generate a path for incoming messages and decide where to publish it. Each topic level is represented by a forward slash [8].

4.3.1 Implementation of MQTT Protocol

We are using the CSV format and the MQTT protocol to transfer the sensor data into cloud. ESP8266 module works as an MQTT client which publishes the data to the broker using PubSubClient Library and sends the data to our subscribed client into cloud. The MQTT protocol is best in terms of Machine to Machine data transfer. Every Sensor data are sent with a timestamp so that it can be used for any time-series data analysis. Details of the diagram are given in our proposed architecture, Fig. 1.

4.3.2 Data Transfer in Intermittent Network Connection

We want our system works to be good, even if there is a weak or unstable Internet connection. So, we implemented our algorithm in Arduino which ensures about every data received by the broker. We check every time whether the Internet connection is active, if yes then we send the data to cloud, and if "no" we store the collected data in an SD card connected with Arduino. We know that the Internet connection could not be stable all the time in agriculture farms. So we overcome this using the SD card module. Hence, no data miss occurs. We have proposed an algorithm in which we have tried to solve the problem that we have to face.

4.3.3 Algorithm

When we get our Internet connection back, we first send the data from the SD card and delete them one by one until all the collected sensor data are sent to cloud. Then the datagram is sorted in the cloud database with the timestamp of every dataset. The proposed algorithm is as follows.

Algorithm 1: Proposed Algorithm

Step 1: Start
Step 2: Collect the sensor value
Step 3: Check Internet is active or not
if *yes* **then**
  Send the current sensor value in cloud;
  if *SD card has previously saved data* **then**
    Pop the top data, send it to cloud and delete;
  else
    Save the current sensor value in SD card in new line;
  end
end
Step 4: Stop.

5 Results and Discussions

The Complete setup of the prototype with detailed labelling is shown in Fig. 3. Also, the sensor specifications are shown in Table 1, mentioned above.

Fig. 3 Prototype Setup of our work

Fig. 4 Different Test Cases of our work

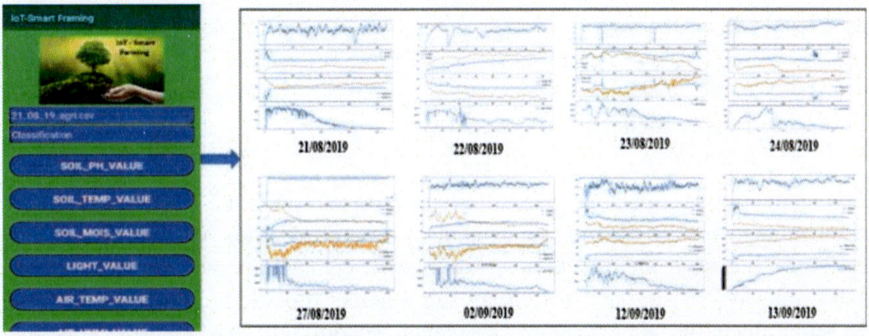

Fig. 5 Smart Farming App and Day-wise Graphs for Different real-time sensor data

In this particular experiment, the execution of the main idea has been performed on a croton plant. A real-time sensor has been used to record the data under various conditions. These data are analysed and graphs are plotted and displayed through an application. We have taken the day-wise sensor data and plant conditions which are shown in Fig. 4 as different test cases.

5.1 *IoT-Smart Farming Application*

We are developing an application titled "IoT-Smart Farming", shown in Fig. 5, through which, we are plotting and viewing the sensor data which will transfer data from cloud/SD card in CSV format. Figures 5 and 6 shows the date-wise graph plotting for the different sensors data.

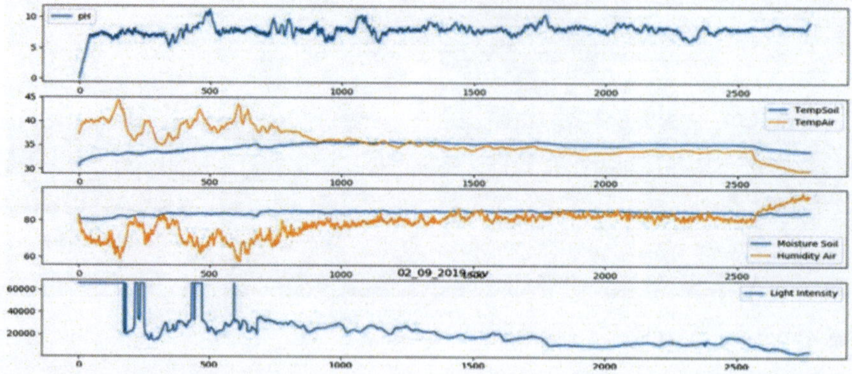

Fig. 6 Broad View of a graph

6 Conclusion and Future Scope

This method of farming is expected to help farmers utilize the various advantages of technology for the all-rounded benefit of crop production. We can collect data for 24*7 7 h even if there is an unstable Internet connection and this IoT device is stable anyway. Sensor data is accurate and reliable. It could help in understanding the requirement of the environment of plants. Accordingly, as soon as any disorder or disruption in the normal requirements of the plant occurs, the sensors detect the abnormality and send an information notice to the farmer in his Android phone via SMS, E-mail, etc. The farmer can immediately take the necessary steps for the development of the plant. Our AMD device can be used for long-term data collection and use those data for forecasting possible humidity or temperature or pH value, so that farmers can use those predictions for farming. We are trying to use the LoRa module for a large farm without Internet coverage. We can use the LoRa module in place of ESP8266. It can transfer low bandwidth data over long distances. All the data will be collected to one device or AMD which has Internet connection and it will send all the data to cloud. Also, we have developed a dashboard to find the previous data of the plant conditions and the exact conditions of our systems. Also, we are trying to increase the power backup in the agriculture fields using solar panels.

References

1. Ganesh, P., Tamilselvi, K., Karthi, P.: Crop prediction by monitoring temperature and rainfall using decision tree with Iot and cloud based system. In: International Conference on Computing Intelligence and Data Science (ICCIDS), pp. 1–9. Coimbatore (2018)
2. Syafarinda, Y., Akhadin, F., Fitri, Z.E., Yogiswara, B.W., Rosdiana, E: The precision agriculture based on wireless sensor network with MQTT protocol. In: 1st International Conference on Food and Agriculture. IOP Publishing (2018)

3. Al-Gumaei, W.: Sathish Kumar Selvaperumal, Raed Abdulla and Chandrasekharan Nataraj: Smart Tree Care System with Internet of Things, Research Journal of Applied Sciences, Engineering and Technology 15(9), 328–336. Maxwell Scientific Publication Corp, Malaysia (2018)
4. Srirampavan, J.: Secured real time agriculture monitoring system. Int. J. Eng. Technol., 281–285 (2018)
5. Abhishek, L., Rishi Barath, B.: Automation in agriculture using IOT and machine learning. Int. J. Innov. Technolo. Exp. Eng. (IJITEE), ISSN: 2278-3075, **8**, BEIESP (2019)
6. Maha Kavya Sri, J., Narendra, V.G.: VidyaPai: implementing and testing of iot technology in agriculture. Int. J. Recent Technol. Engi. (IJRTE), ISSN: 2277-3878, **7**(6S), BEIESP (2019)
7. https://en.wikipedia.org/wiki/Andy_Stanford-Clark
8. http://randomnerdtutorials.com/what-is-mqtt-and-how-it-works/

High-Speed Low-Power CML Technique-Based Frequency Divider in 180 nm Technology

Swarup Dandapat, Shrabanti Das, Moumita Das, and Sayan Chatterjee

Abstract A low-power high-speed CML (common mode logic) frequency divider has been implemented in 180 nm CMOS technology. The PMOS varactors have been set at the source of the transistors, which are cross-coupled, to upgrade the one self-oscillating frequency of dividers. Accompanied by the tuning of PMOS load and the varactors, broad frequency span of 1 GHz–18 GHz has been observed. Accordingly, 3 mW power consumption has been observed.

Keywords Low power · High speed · Frequency divider · PMOS varactor

1 Introduction

Frequency dividers have been considered as the most useful circuit in numerous communication applications such as clock generations, Timing recovery circuits and Frequency synthesizers. Accordingly, a frequency divider has also been used in locking arrangement within a Phase Lock Loop (PLL) system [1]. To realize a low-power high-speed divider, different efforts have been employed on the optimization of the size of transistors [2, 3]. Basic gates and flip-flops have been used to implement basic frequency divider circuits. Common design logic for a divider are TSPC, CML and multi-module divider (MMD). High-speed latches have been proposed using current-mode logic (CML) circuits. In the traditional CML latch topology [4], the

S. Dandapat (✉) · S. Das · M. Das · S. Chatterjee
Electronices & Telecommunication Engineering, Jadavpur University, Kolkata 700032, West Bengal, India
e-mail: swarupdandapat92@gmail.com

S. Das
e-mail: shrabanti.08@gmail.com

M. Das
e-mail: moumitad50524@gmail.com

S. Chatterjee
e-mail: sayan1234@gmail.com

D. Bhattacharjee et al. (eds.), *Proceedings of International Conference on Frontiers in Computing and Systems*, Advances in Intelligent Systems and Computing 1255,
https://doi.org/10.1007/978-981-15-7834-2_86

route and latch method has been implemented with the clock signal for proper operation of the differential pair. Consequently, many techniques have been developed to improve the low power performance of the CML divider. Despite these efforts, the division factor of the frequency divider has been one of the major limitations to achieve better results using CMOS technology. As the frequencies step-up in the direction of the millimeter-wave region, the restriction for divider design increase and huge power is required to achieve high running freq. and multi-modulus freq. division ratios. A millimeter-wave multi-module divider (MMD) has been proposed in [5], in which the circuit has to overcome the problem of higher operating frequency and large power consumption.

Design of the divider has a trade-off among several fundamental parameters such as power consumption, high speed, and chip area [6]. Recently, reduction of power consumption of the frequency divider has tuned into the foremost concern. Subsequently, a lot of dividers have been reported with the current reuse and other techniques (CML, TSPC, Multi-module). A high performance Injection Locked frequency divider has been proposed with the power consumption of 30 mW [7]. Consequently, a Low power e-TSPC 2/3 prescaler technique has been developed to make less the power consumption in [8], and total power dissipation has been reported as 27 mW. In [9,] an Ultra high-speed CML latch has been designed to achieve high speed and low power. Subsequently, a Regenerative frequency divider using Even Harmonic mixer has been proposed in [10] with 10.8 mW power consumption. Eventually, a 19 GHz MMD with programmable dividing up ratio 16–31 has been suggested, which consumes 39.8 mW at .94 V power supply [2]. A 13-49 GHz MMD has been executed in a 130 nm SiGe technology [12] with frequency division ratio 64–127 attaining power consumption of 92 mW. On the other hand, TSPC D-FF circuit has been developed at less than 3.14 GHz clock and 2.34 mW power consumption [6].

Hence, it has been tough to fabricate a high frequency divider with a low-power high-speed and compact circuit structure. To beat this complication, new circuitry has been proposed for high frequency division. This proposed design has been implemented in CML architecture with the tuning of PMOS load and varactor to achieve wide frequency range of 1 GHz–18 GHz. Accordingly, the whole circuit consumes 3 mW power.

2 Basic Operation of CML Frequency Divider

A schematic diagram of a standard high-speed CML-based frequency divider is shown. Flip-flops have been depicted in Fig. 1. The circuit can be noticed as a D flip-flop, perceived using CML logic, with positive reaction connected to its input ports from its dissimilar outputs Q_P/Q_N. Its performance as a divider-by-2 circuit can be demonstrated based on the common study using static operation of the flip-flop.

The factors like the merit of the load resistor and the area of Mc, M_D, and M_L control the working of the divider. For it to work at a high frequency, the length of

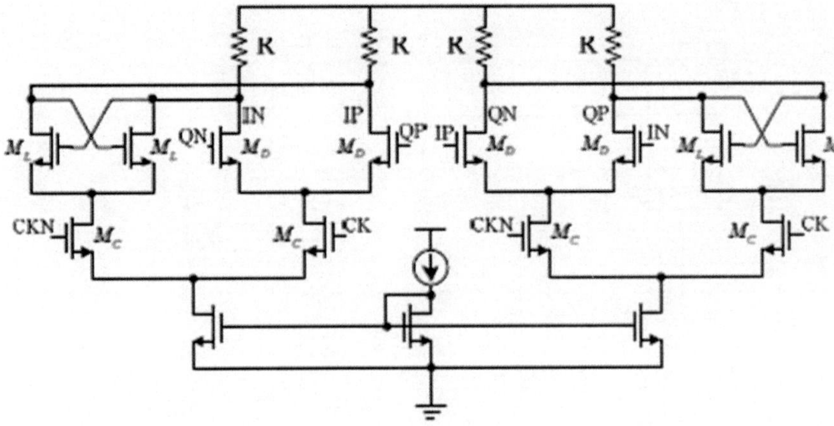

Fig. 1 The conversional CML frequency divider

the channel of those transistors should be the smallest amount of length, allowed by the operating technology, and further, the channel width is also an important specification that is required to be adjudged.

High-speed performance at the rate of concentrated o/p amplitude can be achieved by reducing R, L and W. Also, with the huge area of M_c, additional energy gets attached to familiar origin node for a particular outer introducer signal, which conducts a broad span of working frequency. But it has several disadvantages, for example, the well-built M_c makes the charge of forestage heavier and leads to the power utilization.

An enhanced configuration has been suggested in Fig. 2, where PMOS transistors have been used as linear loads. PMOS loads have been forced to operate in the triode or cut-off region to perform as a resistance. The frequency span can be adjusted

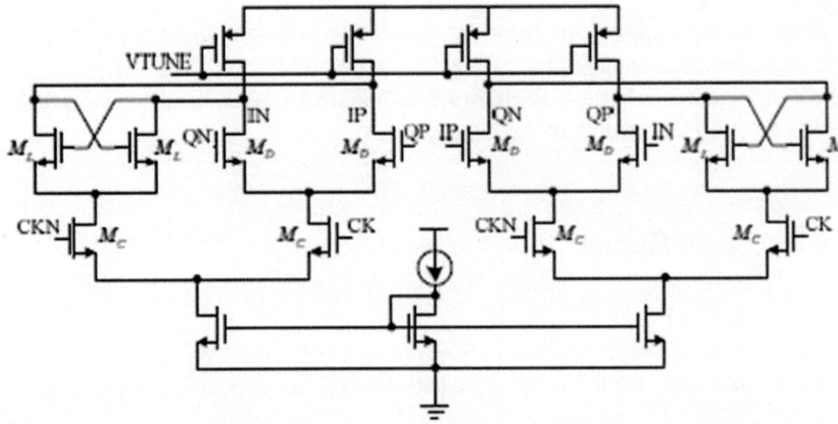

Fig. 2 An improved CML divider

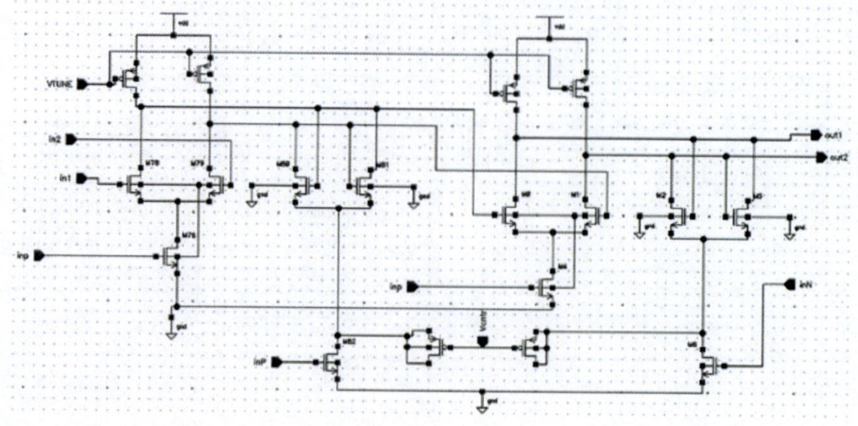

Fig. 3 Schematic of modified divider

by the control voltage V_{TUNE} so that the configuration of this circuit can exhibit a comparatively large operating frequency of 1 GHz–16 GHz

3 Circuit Design

In this paper, a novel configuration CMOS frequency divider using a degeneration MOS varactor capacitor in the origin of the latch transistor Fig. 3 has been introduced and subsequently, the performance has been observed. The circuit guides to low power consumption.

Lessening the value of V_{TUNE} could guide to a compact loading resistance, and growing Vctrl can convey a huge negative capacitor; all the techniques can build the frequency divider working at higher speed.

To attain lower supply voltage, the tail transistor has been omitted in Fig. 3. Vctrl is the voltage which manages the variable capacitor, which has large control on the RC constant particularly the capacitor merit of the load. Consequently, the modified circuit operation has been described in section A.

4 Equivalent Circuit

It is a known fact that the self-oscillation frequency (f_{so}) has an important role in designing dividers. The divider can only run in the region of f_{so}. As the circuit works in a huge signal area, the fabric has been calculated using a small signal time region study where the MOS transconductance (g_m) has been averaged on to mean time interval that is identical to one spell of the self-oscillation frequency.

Fig. 4 Impedance of cross-coupled circuit

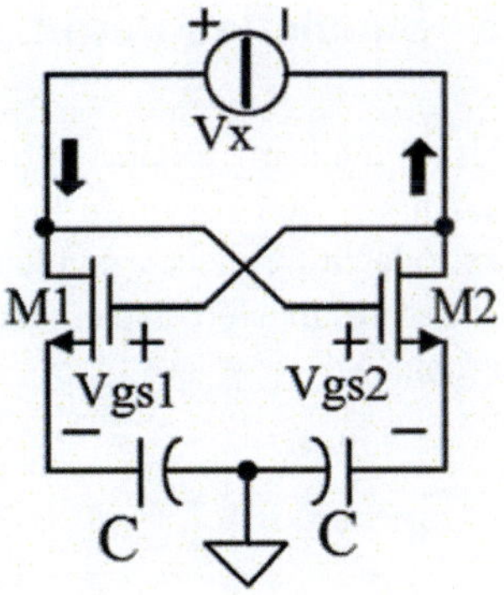

Under these presumptions, the cross-coupled latch transistors with PMOS varactor capacitors have been shown in Fig. 4.

The deduction of the input impedance for the two ports is as pursues:

$$\begin{aligned} I_x &= g_{m1}.V_{gs1} = -g_{m2}.V_{gs2} \\ V_x &= V_{d1} - V_{d2} \\ &= (V_{gs2} - I_x.\frac{1}{sC}) - (V_{gs1} + I_x.\frac{1}{sC}) \\ &= (V_{gs2} - V_{gs1}) - 2I_x.\frac{1}{sC} \\ \therefore Z &= \frac{V_x}{I_x} = -\frac{2}{g_m} - \frac{2}{sC} \end{aligned} \quad (1)$$

Imagine that $g_{m1} = g_{m2} = g_m$ because of the symmetric device. The outcome on the divider frequency tune typically can be calculated by assessing the admittance based on a series-to-parallel change in the circuit of Fig. 4. The admittance can be exhibited as pursues:

$$Y = -\frac{g_m}{2}.\frac{4C^2\omega^2}{g_m^2 + 4C^2\omega^2} - j\omega C.\frac{4C^2\omega^2}{g_m^2 + 4C^2\omega^2} \quad (2)$$

From (2), it has been seen that PMOS varactor capacitive degeneration cross-coupled latch transistor not at most donates a −ve transconductance which maintains the self-oscillation of the divider but as well carries a -ve capacitor. In that method, the capacitive load has been decreased to establish an excessive oscillating frequency. Furthermore, the capacitor can be changed by a varactor also so that the frequency span can mostly increase in size by varying the control voltage of the varactor.

5 Simulation Result

Using the same circuit with input frequency divided by different ratios like 2 and 5/7 have been shown in Figs. 5 and 6, respectively. Moreover, by using only one extra latch frequency divided by 15, the divider's output has been shown in Fig. 7. Different divisions have been found by controlling the VTune and Vcontrl voltage (Table 1).

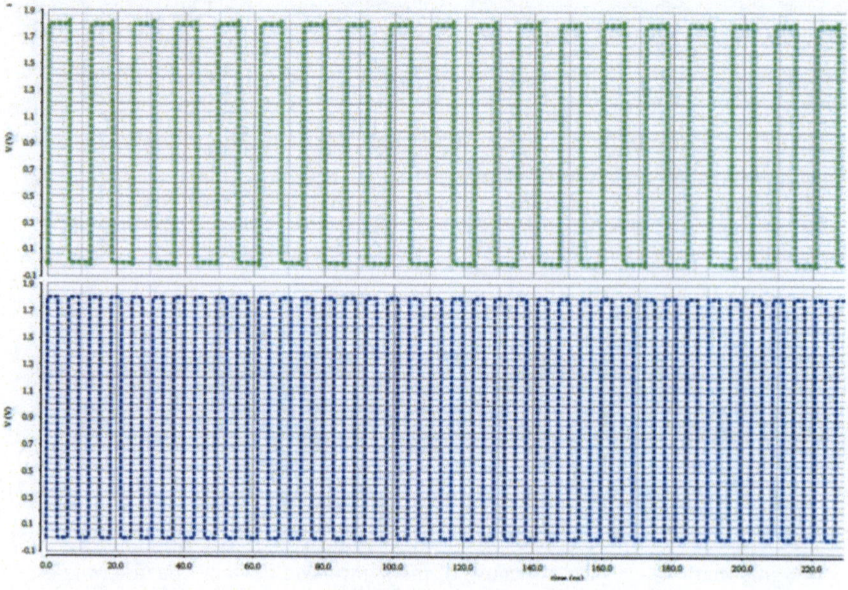

Fig. 5 Simulation result of circuit with input frequency divided by 2

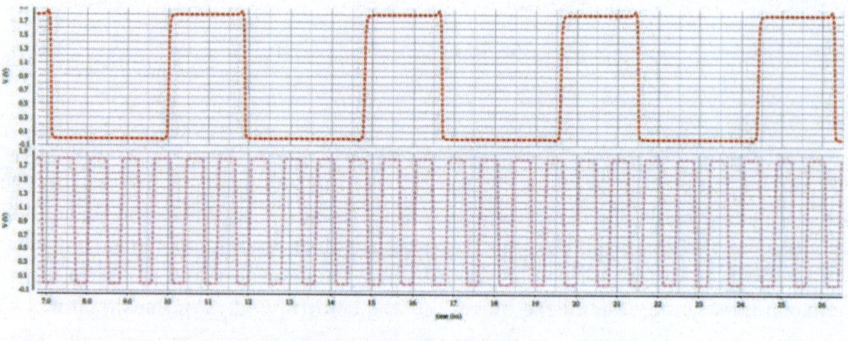

Fig. 6 Simulation result of circuit with input frequency divided by 5/7

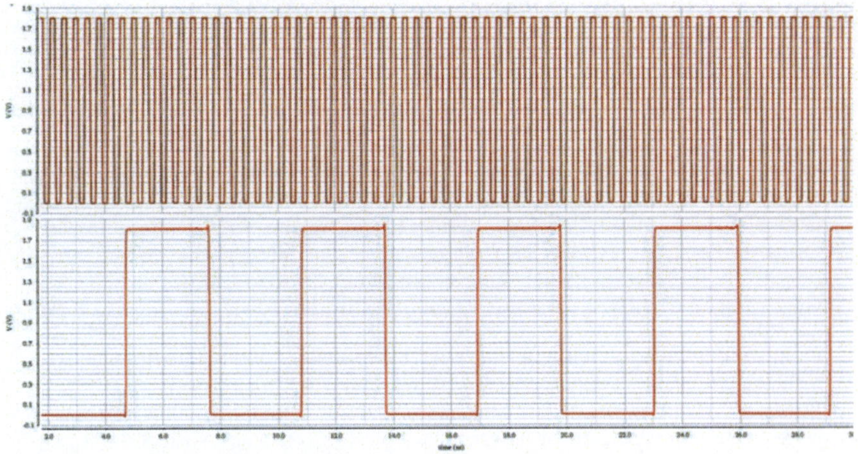

Fig. 7 Simulation result of circuit with input frequency divided by 15

Table 1 Performance comparison with other publications

Reference	Technology	Vdd[V]	P(mW)	Maximum frequency(GHz)	Extra bias
[1]	130 nm	1.2	4.3	17.2	No
[5]	180 nm	2.3	2.34	3.14	No
[6]	180 nm	1.2	30	50	Yes
[7]	180 nm	1.8	28	4.2	No
[9]	90 nm	1.2	10.8	18.4	Yes
This work	180 nm	1.8	3	18	No

6 Conclusion

A modified CML frequency divider using 180 nm CMOS technology has been presented. The divider operated at low power consumption of 3 mW. Injected locked process could facilitate the circuit operate at speed, however, it has bounded locking span and consume space. The modified divider can decide the issue successfully.

Acknowledgements The authors are extremely thankful to SMDP-III(C2SD), for providing them with the required Laboratory facilities to carry out the research works, and they also extend their sincere gratitude towards IC CENTER, Jadavpur University.

References

1. Zhou, H., Li, W., Niu, Y., Zhou, Z., Li, N., Ren, J.: A 0.3–17.2 GHz frequency divider with capacitive degeneration in 130 nm CMOS. In: 2012 IEEE 11th International Conference on Solid-State and Integrated Circuit Technology (2012)
2. Mark, R., Sounder, W., Ratcliff, M., Dai, F., Irwin, J.D.: A 13 GHz lowe power milti-modulus divider implimented. In: 13umSiGe technology IEEE Silocon Monolithic Integrated Cuircuits in RF System, San Diego, CA, USA, January, pp. 1–4 (2009)
3. GU, Q.J., Gao, Z.: A CMOS high speed multi module divider with retiming for jitter suppression. IEEE Microw. Wirel. Compon Lett **23**(10), 554–556 (2013)
4. Gu, Z., Thiede, A.: 18 GHz low-power CMOS static frequency divider. Electron. Lett. **39**(20), Oct. 2003
5. Li, Z., Guo, T., Yang, G., Wang, Z., Liu, F., Li, Q.: Low power 25.4–33.5 GHz programmable multi-module frequency divider. Electron. Lett. **50**(15), 1067–1068 (2014)
6. Ikebe, M., Takada, Y., Ohuchi, M., Motohisa, J., Sano, E.: The design of high frequency true single PhaseClocking divider-by-3 circuit. Int. J. Circuit. Syst. Signal Process. **2**(3) (2008)
7. Ann, S., Park, J., Yu, J., Kim, N.: High Performance Injection-Locked Frequency Divider with 50 GHz LC Cross-Coupled Oscillator in 0.18 μm CMOS Process
8. Nemitha, B., Pradeep Kumar, B.P.: A survey on low power TSPC and E TSPC CMOS 2/3 Prescaler. IJSR **3**(5), May 2014
9. Zhang, X., Wang, Y., Jia, S., Zhang, G., Zhang, X.: "A novel CML Latch for Ultra High Speed Application" Key Laboratory of Microelectronic Devices and Circuits (MoE), Institute of Microelectronics, Peking University, 100871, Beijing, China
10. Lin, Y.-S., Huang, W.-H., Lu, C.-L., Wang, Y.-H.: Wide-locking-range multi-phase-outputs regenerative frequency dividers using even-HarmonicMixersand CML loop dividers. IEEE Trans. Microwavetheory Techn. **62**(12), December 2014